C000128887

Materials Science and Technology

Volume 2 A
Characterization of Materials
Part I

Materials Science and Technology

Materials Science and Technology

A Comprehensive Treatment

Edited by
R.W. Cahn, P. Haasen, E.J. Kramer

Volume 2 A
Characterization of Materials
Part I

Volume Editor: Eric Lifshin

VCH   Weinheim · New York · Basel · Cambridge

Editors-in-Chief:

Professor R.W. Cahn
University of Cambridge
Dept. of Materials Science
and Metallurgy
Pembroke Street
Cambridge CB2 3QZ, UK

Professor P. Haasen
Institut für Metallphysik
der Universität
Hospitalstraße 3/5
D-3400 Göttingen
Germany

Professor E.J. Kramer
Cornell University
Dept. of Materials Science
and Engineering
Bard Hall
Ithaca, NY 14853-1501, USA

Volume Editor:
Dr. E. Lifshin
Building K 1 – Room 2 A 18
General Electric Company
Corporate Research and Development
PO Box 8
Schenectady, NY 12301, USA

Published jointly by
VCH Verlagsgesellschaft mbH, Weinheim (Federal Republic of Germany)
VCH Publishers Inc., New York, NY (USA)

Editorial Directors: Dr. Christina Dyllick-Brenzinger, Karin Sora, Dr. Peter Gregory
Production Manager: Dipl.-Wirt.-Ing. (FH) H.-J. Schmitt
Indexing: Borkowski & Borkowski, Schauernheim
The cover illustration shows a semiconductor chip surface and is taken from the journal "Advanced Materials", published by VCH, Weinheim.

Library of Congress Card No.: 90-21936

British Library Cataloguing-in-Publication Data
A catalogue record for this book is available from the British Library

Die Deutsche Bibliothek – CIP-Einheitsaufnahme
Materials science and technology : a comprehensive treatment /
ed. by R.W. Cahn ... – Weinheim ; New York ; Basel ;
Cambridge : VCH.
   ISBN 3-527-26813-8 (Weinheim ...)
   ISBN 1-56081-190-0 (New York)
NE: Cahn, Robert W. [Hrsg.]
Vol. 2. Characterization of materials.
   A = Pt. 1 (1992)

Characterization of materials / vol. ed.: Eric Lifshin. –
Weinheim ; New York ; Basel ; Cambridge : VCH.
   Materials science and technology ; Vol. 2)
   ISBN 3-527-26815-4 (Weinheim ...)
   ISBN 0-89573-690-X (New York)
NE: Lifshin, Eric [Hrsg.]
   A = Pt. 1 (1992)

© VCH Verlagsgesellschaft mbH, D-6940 Weinheim (Federal Republic of Germany), 1992

Printed on acid-free and low chlorine paper

Composition, Printing and Bookbinding: Konrad Triltsch, Druck- und Verlagsanstalt GmbH,
D-8700 Würzburg
Printed in the Federal Republic of Germany

Preface to the Series

Materials are highly diverse, yet many concepts, phenomena and transformations involved in making and using metals, ceramics, electronic materials, plastics and composites are strikingly similar. Matters such as transformation mechanisms, defect behavior, the thermodynamics of equilibria, diffusion, flow and fracture mechanisms, the fine structure and behavior of interfaces, the structures of crystals and glasses and the relationship between these, the motion or confinement of electrons in diverse types of materials, the statistical mechanics of assemblies of atoms or magnetic spins, have come to illuminate not only the behavior of the individual materials in which they were originally studied, but also the behavior of other materials which at first sight are quite unrelated.

This continual intellectual cross-linkage between materials is what has given birth to *Materials Science,* which has by now become a discipline in its own right as well as being a meeting place of constituent disciplines. The new Series is intended to mark the coming-of-age of that new discipline, define its nature and range and provide a comprehensive overview of its principal constituent themes.

Materials Technology (sometimes called Materials Engineering) is the more practical counterpart of Materials Science, and its central concern is the processing of materials, which has become an immensely complex skill, especially for the newer categories such as semiconductors, polymers and advanced ceramics but indeed also for the older materials: thus, the reader will find that the metallurgy and processing of modern steels has developed a long way beyond old-fashioned empiricism.

There exist, of course, other volumes and other series aimed at surveying these topics. They range from encyclopedias, via annual reviews and progress serials, to individual texts and monographs, quite apart from the flood of individual review articles in scientific periodicals. Many of these are essential reading for specialists (and those who intend to become specialists); our objective is not to belittle other sources in the cooperative enterprise which is modern materials science and technology, but rather to create a self-contained series of books which can be close at hand for frequent reference or systematic study, and to create these books rapidly enough so that the early volumes will not yet be badly out of date when the last ones are published. The individual chapters are more detailed and searching than encyclopedia or concise review articles, but less so than monographs wholly devoted to a single theme.

The Series is directed toward a broad readership, including not only those who define themselves as materials scientists or engineers but also those active in diverse disciplines such as solid-state physics, solid-state chemistry, metallurgy, construction engineering, electrical engineering and electronics, energy technology, polymer science and engineering.

While the Series is primarily classified on the basis of types of materials and their processing modes, some volumes will focus on particular groups of applications (Nuclear Materials, Biomedical Materials), and others on specific categories of properties (Phase Transformations, Characterization, Plastic Deformation and Fracture). Different aspects of the same topic are often treated in two or more volumes, and certain topics are treated in connection with a particular material (e.g., corrosion in one of the chapters on steel, and adhesion in one of the polymer volumes). Special care has been taken by the Editors to ensure extensive cross-references both within and between volumes, insofar as is feasible. A Cumulative Index volume will be published upon completion of the Series to enhance its usefulness as a whole.

We are very much indebted to the editorial and production staff at VCH for their substantial and highly efficient contribution to the heavy task of putting these volumes together and turning them into finished books. Our particular thanks go to Dr. Christina Dyllick on the editorial side and to Wirt. Ing. Hans-Jochen Schmitt on the production side. We are grateful to the management of VCH for their confidence in us and for their steadfast support.

Robert W. Cahn, Cambridge
Peter Haasen, Göttingen
Edward J. Kramer, Ithaca

April 1991

Preface to Volume 2 A

Characterization is essential to the systematic development of new materials and understanding how they behave in practical applications. Throughout all of the volumes of *Materials Science and Technology* the theme of linking properties with microstructure and chemical composition is repeated. Volume 2 focuses on the principal methods required to characterize metal alloys, semiconductors, polymers and ceramics. Modern materials such as high-temperature alloys, engineering thermoplastics and multilayer semiconductor films have many elemental constituents distributed in more than one phase. Details of these phases can be very complex and usually vary with processing. The scale of structural features of importance can differ by orders of magnitude, and sometimes critical features exist only with atomic dimensions. Even after an object is fabricated it can change when exposed to variations in temperature, pressure and different physical environments. Since these physical or chemical changes can affect the performance of a finished component, they must also be understood so that fabrication, processing or compositional modifications can be made to achieve product life extension. The challenge to the materials characterization expert is to understand how specific instruments and analytical techniques can provide detailed information about what makes each material or fabricated product unique. The challenge to the materials scientist, chemist or engineer is to know what information is needed to fully characterize each material or fabricated product and how to use this information to explain its behavior, develop new and improved properties, reduce costs, or ensure compliance with regulatory requirements.

Today there are thousands of techniques used to characterize materials. They often depend on how a given sample or a region of a sample responds to a probe. The probe may be electrons, neutrons, ions, electromagnetic radiation or even a physical object such as a stylus. The response may be an alteration of the probe itself as exemplified by scattering or absorption. However, it can also be the stimulation of a secondary response like the emission of particles or radiation. The range of available characterization techniques is so broad that no individual can master more than a few. Each year new developments are described in hundreds of papers found in dozens of journals and presented at numerous technical conferences. The goal of Volume 2, which will be published in two parts, is to provide an introductory understanding of a number of the most important techniques. Each is described in sufficient detail so that the reader will develop an awareness of the instrumentation used, how it works, what kind of information it provides, and what are its limitations.

Volume 2 A concentrates on microscopy techniques that span a range of magnifications from that of the low magnification available with light optical microscopy, through scanning electron microscopy and finally to the very high resolution capability of transmission electron microscopy. Elemental analysis is described based on both optical and X-ray emission techniques. Questions of crystal and grain structure are addressed in articles on X-ray diffraction, optical microscopy and through the use of electron diffraction techniques achievable with the transmission electron microscope. Other chapters examine the variety of information that can be gained by such different techniques as thermal analysis and the uses of synchrotron radiation. Since many of the techniques used for polymer characterization are quite distinct from those used to study metals and ceramics, a separate article focuses on some of the more frequently used methods for studying this important class of materials. Volume 2 B adds to the methods described in 2 A. It includes articles describing surface analysis techniques such as Auger spectrometry, high-energy ion probes, X-ray photo-electron spectroscopy, as well as the exciting new techniques of scanning tunneling and force microscopy. Other articles cover neutron diffraction and X-ray scattering, stereology, mechanical spectroscopy and electron microprobe analysis. The reader who is interested in obtaining just one type of information, such as elemental analysis, is encouraged to explore more than one article and discover a variety of approaches and then select the one best suited to his or her needs and is compatible with the resources available.

In addition to recognizing the hard work of the authors, I would like to give my special thanks to Professor Robert Cahn for our numerous discussions, his editorial advice, and his help in locating authors.

Eric Lifshin
Schenectady, NY, February 1992

Editorial Advisory Board

Contributors to Volume 2 A

Professor Severin Amelinckx
Universiteit Antwerpen
Elektronenmickroskopie voor
Materiaalonderzoek
Groenenborgerlaan 171
B-2020 Antwerpen
Belgium
Chapter 1

Dr. Patrick K. Gallagher
The Ohio State University
Department of Chemistry
and Materials Science and
Engineering
120 West 18th Avenue
Columbus, OH 43210
USA
Chapter 7

Dr. Andrea R. Gerson
University of London
Department of Chemistry
King's College
The Strand
London WC2R 2LS
UK
Chapter 8

Dr. Peter J. Halfpenny
University of Strathclyde
Department of Pure and Applied
Chemistry
295 Cathedral Street
Glasgow G 1 1 XL
Scotland
Chapter 8

Dr. Ernest L. Hall
General Electric Corporate
Research and Development
Building K-1, Room 1 C 10
PO Box 8
Schenectady, NY 12301
USA
Chapter 2

Dr. Ronald Jenkins
JCPDS
International Centre for Diffraction
Data
1601 Park Lane
Swarthmore, PA 19801-2389
USA
Chapter 9

Professor David C. Joy
University of Tennessee
Department of Zoology
F 239 Walters Life Sciences Building
Room M 313
Knoxville, TN 37996-0810
USA
Chapter 3

Professor Peter N. Keliher †
Villanova University
Department of Chemistry
Villanova, PA 19085-1699
USA
Chapter 6

Professor Günter Petzow
Max Planck Institute for Metals
Research
Institute for Materials Science
Heisenbergstraße 5
D-7000 Stuttgart 80
Germany
Chapter 5

Dr. Stefania Pizzini
Centre Universitaire Paris-Sud
LURE
Bâtiment 209 D
F-91405 Orsay Cedex
France
Chapter 8

Dr. Radoljub Ristic
University of Strathclyde
Department of Pure and Applied
Chemistry
295 Cathedral Street
Glasgow G1 1XL
Scotland
Chapter 8

Dr. Kevin J. Roberts
SERC Daresbury Laboratory
Warrington WA4 4AD
UK
and
University of Strathclyde
Department of Pure
and Applied Chemistry
295 Cathedral Street
Glasgow G1 1XL
Scotland
Chapter 8

Dr. David B. Sheen
University of Strathclyde
Department of Pure and Applied
Chemistry
295 Cathedral Street
Glasgow G1 1XL
Scotland
Chapter 8

Professor John N. Sherwood
University of Strathclyde
Department of Pure and Applied
Chemistry
295 Cathedral Street
Glasgow G1 1XL
Scotland
Chapter 8

Dr. Eileen M. Skelly Frame
GE Corporate Research and
Development
Materials Characterization
Laboratory
Schenectady, NY 12301
USA
Chapter 6

Dr. Robert L. Snyder
New York State College of
Ceramics
Alfred University
Institute for Ceramic
Superconductivity
Alfred, NY 14802
USA
Chapter 4

Dr. Rainer Telle
Max Planck Institute for Metals
Research
Institute for Materials Science
Heisenbergstraße 5
D-7000 Stuttgart 80
Germany
Chapter 5

Dr. Elizabeth A. Williams
General Electric Corporate
Research and Development
Building K-1, Room 2A 22
PO Box 8
Schenectady, NY 12301
USA
Chapter 10

Characterization of Materials

Edited by E. Lifshin

Volume 2 A

Volume 2 B

Contents

1 Electron Diffraction and Transmission Electron Microscopy

Severin Amelinckx

Laboratorium voor Algemene Natuurkunde en Fysica van de Vaste Stof,
Universiteit Antwerpen, Antwerpen, Belgium

List of Symbols and Abbreviations

$A(\boldsymbol{h})$	scattered amplitude for scattering vector $\boldsymbol{h}$
$A(u, v)$	aperture function
$A(x, y)$	amplitude of a beam scattered by a column at (x, y)
$A_{\mathrm{B}}(\boldsymbol{h})$, $A_{\mathrm{D}}(\boldsymbol{h})$	scattering amplitude of a Bragg diffracted, diffuse scattering peak
a_i	base vectors of the crystal lattice
$\boldsymbol{b}$	Burgers vector of dislocation
$\boldsymbol{b}_i$	base vectors of the reciprocal lattice
$B^{(1,\,2)}(\boldsymbol{r})$	Bloch waves excited in a crystal
c	velocity of light in vacuum
C	contrast
C_{C}	chromatic aberration constant
C_{S}	spherical aberration constant
d_{g}	interplanar spacing of the active reflection
d_{H}	interplanar spacing with indices h, k, l
$\boldsymbol{e}$	unit vector normal to a set of planar interfaces
e	electron charge
E	accelerating potential
$\boldsymbol{e}_{\mathrm{n}}$	unit normal
f	focal distance
$f_{\mathrm{e}}(\theta)$	atomic scattering factor for electrons
F_{g}	structure factor of reflection $\boldsymbol{g}$ with indices (h_1, h_2, h_3)
$f_{\mathrm{X}}(\theta)$	atomic scattering factor for X-rays
$F_{u,\,v}$, $F_{x,\,y}$	Fourier transform
Δf	defocus distance
$\boldsymbol{g}$	reciprocal lattice vector
$\Delta \boldsymbol{g}_{\parallel,\,\perp}$	components of $\Delta \boldsymbol{g}$ parallel, perpendicular to $\boldsymbol{g}$
$\boldsymbol{h}$	diffraction vector or position vector in reciprocal space
h	Planck constant
h, k, l	Miller indices
H	indices h, k, l (or h_1, h_2, h_3)
I_{g}	intensity of the diffracted beam
$I_{\mathrm{g,\,max}}$	intensity of I_{g} at $s_{\mathrm{g}} = 0$
$I(\boldsymbol{h})$	scattered intensity for scattering vector $\boldsymbol{h}$
I_{S}	intensity of scattered beam
I_{T}	intensity of transmitted beam
$I_{\mathrm{S,0}}$, $I_{\mathrm{T,0}}$	I_{S}, I_{T} if $s_{\mathrm{g}} = 0$
$\boldsymbol{k}$	wave vector
$\boldsymbol{K}$	wave vector of incident electron beam in the crystal corrected for refraction by the mean inner potential
K	radius of the Ewald's sphere
$\boldsymbol{k}_{\mathrm{g}}$	wave vector of diffracted beam in crystal; $\boldsymbol{k}_{\mathrm{g}} = \boldsymbol{k}_0 + \boldsymbol{g}$
$\boldsymbol{K}_{\mathrm{t}}$	tangential component of $\boldsymbol{K}$
$\boldsymbol{k}_0$	wave vector of incident beam in crystal

K_0	wave vector of incident electron beam in vacuum
$k_{0,t}$, $K_{0,t}$	tangential component of k_0, K_0
m	fraction of long spacings
m	relativistic electron mass
$M(s,z)$	transfer matrix of perfect crystal slab
$m_{A,B}$	atomic fractions of A, B atoms
m_0	rest mass of the electron
n	$\equiv g \cdot b$, image order of a dislocation
N	number of unit cells
$q(x,y)$	2D-transmission function
q	satellite spacing or modulation wave vector
$R(r)$	displacement field of defect
R_g	ratio of the amplitudes ψ_g/ψ_0
r_j	position of scattering unit j
r_n	position of the n-th scattering center
r_L	lattice vector
$R_{\parallel,\perp}$	radial, normal component of $R(r)$
s, u, x, β	alternative deviation parameters
S	amplitude of scattered beam
s_g	deviation vector parameter of reflection g_s
S_g	modulus of the amplitude of φ_g
$s_{g,e}$, s_{eff}	effective local s value in a deformed crystal
S^-	amplitude of scattered beam with $s \Rightarrow -s$
S', S'', T', T''	S, T in the two-beam approximation
$\$(\alpha)$	shift matrix
T	amplitude of transmitted beam
T_c	transition temperature
t_g, t_{-g}	extinction distance of reflection g, $-g$
T^-	amplitude of transmitted beam with $s \Rightarrow -s$
u	distance between the foil center and an interface
$U(x,y)$	truncated Fourier transform of the diffraction pattern
$u(r_n)$	displacement of the scattering center at r_n
U_g, U_{-g}	reduced Fourier coefficient of lattice potential
$u_k(r)$	function with the periodicity of the crystal lattice (Bloch waves)
$V(r)$	lattice potential
$V(z,s)$	vacuum matrix
V_C	unit cell volume
V_g	Fourier coefficient of lattice potential
$V_j(r)$	electrostatic potential of scattering unit j
V_0	constant part of the lattice potential: mean inner potential
$V_{def,\,deformed}$	potential of a deformed lattice
V^*	volume of the reciprocal unit cell
$W(r)$	imaginary part of the lattice potential
x_{max}	position of the peak in the amplitude phase diagram
z_0, t	total crystal thickness

α	phase difference introduced by translation interfaces
α_g	$\equiv 2\pi \, \boldsymbol{g} \cdot \boldsymbol{R}$, phase difference introduced by planar defect
α_i	Cowley-Warren short-range-order parameter
β	Takagi deviation parameter
β_A	angle due to the Abbe limit
β_i	pair correlation coefficient
γ	stacking fault energy
$\delta(u, v)$	Dirac function
δ	$= s_1 t_{g_1} - s_2 t_{g_2}$, excitation difference at domain boundary
Δ	Laplace operator
$\boldsymbol{\delta}_L(r)$	vector describing the deviation of scattering units from r_L
ε	additional displacement field
ε	$= e\,E$, electron energy
ε_w	half-width of the half-maximum of a Lorentzian peak
θ_g	phase angle of Fourier coefficient V_g
θ_H	Bragg angle
θ_n	Bragg angle of order n
λ	wavelength of the electrons (or another radiation used)
Λ	moiré wavelength
Λ_p	parallel moiré wavelength
Λ_r	rotation wavelength
μ	shear modulus
$\mu(x, y)$	absorption coefficient
ν	Poisson's ratio
ϱ	radius of the final disc of confusion
$\varrho_{A,C,S}$	radius of different discs of confusion in the object space
σ_g	reduced dynamic deviation parameter of reflection $\boldsymbol{g}$
$\bar{\sigma}_i$	Flinn occupation operator of site i
σ_i	imaginary part of σ_g
σ_i	occupation operator
σ_r	real part of σ_g
$\sigma_{g,i},\ \sigma_{g,r}$	imaginary, real part of σ_g
$\sigma_j^{A,B}$	operator describing the occupation of the site j by an A, B atom
τ_g	absorption length of reflection $\boldsymbol{g}$
$\varphi(x, y)$	scalar potential
φ_g	amplitude of ψ_g
$\phi_{0,g}$	amplitude of the incident, diffracted beam
χ	phase shift due to lens aberrations
Ψ	total electron wave function
ψ	electron wave function
ψ_g	wave function of electrons in diffracted beam
$\psi_{S,T}$	ψ referring to plane waves with the amplitudes S, T
ψ_0	wave function of electrons in incident beam
$\Psi_1,\ \Psi_2$	Bloch wave fields in crystal
ω_j	coefficients of the cluster relation in the transition state

∇ Nabla operator

ANNNI   axial next-nearest neighbor Ising (model)
B.F.   bright field
BZ   Brillouin zone (boundary)
D.F.   dark field
f.c.c.   face-centered cubic
FS   final start
h.c.p.   hexagonal close-packed
HREM   high-resolution electron microscopy
RH   right handed
TEM   transmission electron microscopy

1.1 Introduction

In this chapter a survey is given of the theoretical background of electron diffraction and imaging techniques used in materials science. The use of mathematics is kept to a minimum, and whenever possible intuitive considerations are also advanced.

A discussion of both electron optics and instrumentation is considered to be beyond the scope of this contribution, especially since these topics depend on the particular instrument under consideration. We refer to other chapters of this Volume (Chap. 2, 3) and to the literature (Hirsch et al., 1965; Spence, 1981; Thomas, 1962) for a presentation of the instrumental aspects. In particular, the chemical aspects are treated in a separate contribution to this Volume (Chap. 2).

Since electron imaging is a diffraction phenomenon, diffraction theory is an essential basis for the interpretation of images. We therefore treat in detail kinematic and dynamic diffraction theory for perfect and for faulted crystals.

These aspects of electron optics, which are needed for a proper understanding of image formation under high resolution conditions, are treated as briefly as possible.

A number of special diffraction phenomena which are often met in electron microscopic studies are discussed as well.

A number of concrete case-studies, illustrating the described phenomena in specific materials, are presented in the final section.

1.2 Geometry of Diffraction

1.2.1 Bragg's Law and the Ewald Construction

The diffraction condition is in its simplest form expressed as Bragg's law (Bragg, 1929):

$$2 d_H \sin \theta_n = n \lambda \qquad (1\text{-}1)$$

where attention is focused on a set of lattice planes with Miller indices $H(h_1, h_2, h_3)$; d_H is the interplanar spacing, θ_n is the Bragg angle of order n and λ the wavelength of the radiation used. In the case of monokinetic electrons of 100 kV, λ is about 0.04 Å, and the Bragg angles are as a consequence very small, of the order of 1°. Bragg's law states that the path difference between beams (or waves) "reflected" by successive lattice planes is an integer number of wavelengths. This path difference Δ, indicated in Fig. 1-1 by a thick line, is on the left-hand side of Eq. (1-1). Note that we can only have diffraction for real values of θ_n, i.e., for $\lambda \leq 2 d_{H,\max}$, where $d_{H,\max}$ is the largest interplanar spacing occurring in the crystal.

The Bragg condition can also be expressed in vector form as

$$k - k_0 = g \qquad (1\text{-}2)$$

where the vector g is oriented along the unit normal e_n to the set of lattice planes H and has a length $1/d_H$; $g = e_n/d_H$; k and k_0 represent the wave vectors of the diffracted

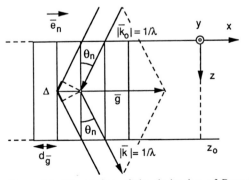

Figure 1-1. Illustration of the derivation of Bragg's law and the notations used. The path difference Δ is indicated by a thicker line segment.

and incident beams, respectively, where $|\boldsymbol{k}| = |\boldsymbol{k}_0| = 1/\lambda$.

The equivalence of the statements in Eqs. (1-1) and (1-2) can be seen as follows: Squaring Eq. (1-2) yields $k^2 + k_0^2 - 2\boldsymbol{k} \cdot \boldsymbol{k}_0 = g^2$ and since $\boldsymbol{k} \cdot \boldsymbol{k}_0 = \cos 2\theta_n / \lambda^2$ and $g = n/d_H$, one finds $(4/\lambda^2) \sin^2 \theta_n = (n/d_H)^2$, which is equivalent to Eq. (1-1). The relation Eq. (1-2) leads to the Ewald construction (Ewald, 1917) represented in Fig. 1-2. Let C be the center of a sphere with radius $k_0 = 1/\lambda$ through the origin of reciprocal space and through the reciprocal lattice node G. This clearly leads to $\boldsymbol{k}_0 + \boldsymbol{g} = \boldsymbol{k}_g$. The wave vector of the diffracted beam is thus obtained by joining the center C with the node G.

Let us consider the orders of magnitude of the parameters involved: for 100 kV electrons $\lambda = 3.7 \times 10^{-3}$ nm and $k_0 = 25$ nm^{-1}, whereas the mesh size of the reciprocal lattice is of the order $g \simeq 0.1$ nm^{-1}. It is therefore clear that the Bragg angles θ_n are small since they are of the order $\theta_n \sim \sin \theta_n \sim g/(2 k_0)$, i.e., at most a few degrees. Note also that since $|\boldsymbol{k}_0| = |\boldsymbol{k}_0 + \boldsymbol{g}| \gg |\boldsymbol{g}|$, $\boldsymbol{g}$ must be approximately perpendicular to $\boldsymbol{k}_0$. Since the radius $|\boldsymbol{k}_0|$ of the Ewald sphere is large as compared to $|\boldsymbol{g}|$, the diffraction pattern is an almost planar section of reciprocal space, which makes indexing of the diffraction pattern very simple and unambiguous. It is therefore often simpler to use single crystal electron diffraction patterns to determine a unit cell, rather than an X-ray powder diffraction pattern, the *precise* lattice parameters being obtained by X-ray methods.

In transmission electron microscopy, one almost invariable works in the so-called Laue geometry in which the transmitted and the scattered beams transverse the foil (Fig. 1-3). In our further description, we shall use the reference system of

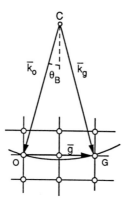

Figure 1-2. Ewald construction for the exact Bragg orientation. C is the center of Ewald's sphere, O is the origin of the reciprocal lattice, and G a reciprocal lattice node.

Fig. 1-3. The small size of the Bragg angles causes large sensitivity of the diffracted intensity to small local orientation differences of the lattice, as we shall see. This is the reason why defects such as dislocations can be revealed in transmission electron microscopy. In order to understand image formation at defects it is necessary to develop the theory of electron diffraction to a sufficient extent. This is the objective of the next sections.

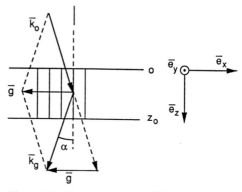

Figure 1-3. The geometry of diffraction in the Laue case, illustrating the notations used.

1.2.2 Atomic Scattering Factor – Interaction Strength

Electron scattering by a crystal is a co-operative phenomenon due to the scattering by the many individual atoms and the subsequent interference of the scattered waves.

As negatively charged particles, electrons interact by Coulomb forces with the positively charged nuclei as well as with the negatively charged atomic electron clouds in the structure. X-rays, on the other hand, practically only interact with the electron cloud, and neutrons only with the nucleus. For these reasons the strength of the interaction is largest for electrons. A scattering theory for electrons will have to take into account the strength of this interaction.

The atomic scattering factor for electrons thus contains two contributions (Mott and Masey, 1949):

$$f_e(\theta) = \frac{m e^2 \lambda}{2 h^2} \frac{Z - f_X(\theta)}{\sin^2 \theta} \qquad (1-3)$$

The first term is due to the interaction with the nucleus with atomic number Z, whereas the second term is due to the electron cloud, and it is proportional to $f_X(\theta)$, the scattering factor for X-rays. The other symbols have their usual meaning (m is the mass of the electron; e is the charge of the electron; λ is the wavelength of the electron; h is the Planck's constant). $f_e(\theta)$ is of the order of 10^4 times larger than f_X, but it decreases more rapidly with θ than $f_X(\theta)$. The amplitude of a spherical wavelet scattered by an isolated atom is then given by

$$a(r) = \frac{1}{r} f_e(\theta) \exp(2\pi i k r) \qquad (1-4)$$

Multiple scattering events cannot be neglected in electron diffraction, as is possible for X-rays and neutrons. The appropriate theory is thus the dynamic theory of diffraction, where such multiple scattering events are taken into account. Nevertheless, we shall see that for certain applications the kinematic theory can be used as a first approximation. Since the kinematic theory can in a number of cases be formulated analytically, whereas most problems in dynamic scattering theory require numerical treatment, it is of interest to devote some attention first to the kinematic theory. In most cases, the geometrical features of the kinematic theory fit the experiment, but significant deviations are found in the intensities of the spots.

1.3 Kinematic Diffraction Theory

1.3.1 Introduction

The specimen is modeled by its lattice potential, which contains the potentials due to the nuclei and contributions due to the electron clouds. Since the atomic arrangement is periodic in three dimensions, the lattice potential will have this property as well. We can thus write

$$V(r) = \sum V_g \exp(2\pi i g \cdot r) \qquad (1-5)$$

as a Fourier expansion, where g represents reciprocal lattice vectors. The constant part of the potential V_0, corresponding with $g \equiv 0$ can be separated out. We note that $V_0 \simeq 10-20$ V and that V_0 is small as compared to the accelerating potential E of the imaging electrons ($E \sim 100-400$ kV). The periodic variation of the lattice potential, superposed on this constant component V_0, is described by the various Fourier components V_g.

Real specimens have to be thin foils (50–100 Å for high resolution observations; 1000–3000 Å for diffraction contrast experiments) in order to be transparent for electrons of 100–400 kV. In such thin foils, Bragg's condition is somewhat relaxed since the number of interfering beams cannot be considered as infinite when the number of unit cells in the foil thickness is finite. This relaxation can be represented in reciprocal space by assuming the mathematical points to have become rods (so-called relrods) perpendicular to the foil plane, as we shall see.

As long as Ewald's sphere intersects this relrod, "reflection" will take place. The deviation from Bragg's exact interference condition is characterized by a vector s_g which joins the endpoint G of g with a point on Ewald's sphere along the direction perpendicular to the entrance face of the foil and oriented in the direction of propagation of the electrons (Fig. 1-4). With this sign convention, $s_g > 0$ if G is inside Ewald's sphere, and $s_g < 0$ if G is outside Ewald's sphere; Bragg's condition corresponds to $s_g = 0$.

In the case of symmetrical incidence, i.e., with $\overline{OC}$ along a zone axis, the magnitude

of s_g for reflections belonging to this zone is given by $s_g \simeq g^2/(2k_0)$. In the general case, we have the following relation:

$$|\boldsymbol{k}_0 + \boldsymbol{g} + s_g \boldsymbol{e}_z|^2 = k_0^2$$

that is

$$|\boldsymbol{k}_0 + \boldsymbol{g}|^2 + 2 s_g |\boldsymbol{k}_0 + \boldsymbol{g}| \cos \alpha + s_g^2 = k_0^2$$

Neglecting s_g^2 with respect to the other quantities we can solve for s_g:

$$s_g = \frac{k_0^2 - |\boldsymbol{k}_0 + \boldsymbol{g}|^2}{2 |\boldsymbol{k}_0 + \boldsymbol{g}| \cos \alpha} \tag{1-6}$$

We shall need this relation further below.

1.3.2 Kikuchi Lines

In sufficiently thick and rather perfect specimens spot patterns are no longer observed; instead a diffraction phenomena, first discovered by Kikuchi (1928), is produced. It usually consists in the occurrence of pairs of bright and dark straight lines in the diffraction pattern as shown in Fig. 1-5. In foils of intermediate thickness one can observe the Kikuchi pattern superposed on the spot pattern. The geometry of the Kikuchi pattern can be satisfactorily explained by assuming that electrons are not only Bragg scattered but that a substantial fraction, especially in thick foils, is scattered inelastically and incoherently by the crystal, the energy loss being small as compared to the energy of the incident electron. In this way, the electron wavelength is not appreciably changed. Inside the crystal these randomly scattered electrons impinge on the lattice planes from all directions, but preferentially in the forward direction, and can subsequently give rise to Bragg scattering.

Let us consider for instance the symmetrical situation with respect to the set of lattice planes H, with spacing d_H, as shown in Fig. 1-6a. Bragg scattering out of the

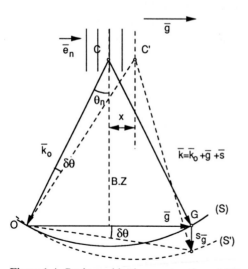

Figure 1-4. Reciprocal lattice construction of diffraction by a thin foil in which the Bragg condition is not exactly satisfied. Definition of the notations and of the reference system used, illustrating also the relation between the two deviation parameters x and s_g.

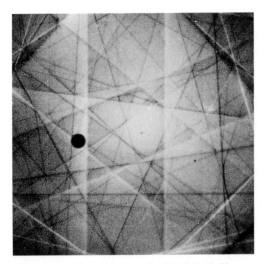

Figure 1-5. Kikuchi pattern in thin foil of silicon.

incident beam is then weak since the Bragg condition is not satisfied. However a fraction of the randomly scattered electrons will have the correct direction of incidence to give rise to Bragg scattering by the considered set of lattice planes. The geometrical locus of these Bragg scattered electron beams is a double cone of revolution with an opening angle $(\pi/2) - \theta_H$ and with its axis along H ($\theta_H = $ Bragg angle). These cones are therefore rather "flat", and the intersection lines of the two sheets of this double cone with the photographic plate P looks like two parallel straight lines, although in actual fact they are two branches of a hyperbolic conic section. It is also evident that the angular separation of these two lines is $2\theta_H$; the separation Δ observed on the plate is thus $\Delta = 2L\theta_H$, where L is the camera length, i.e., the distance of the specimen to the plate. This angular separation does not depend on the crystal orientation.

It should be noted that the geometry of this cone (i.e., the axis of revolution and the opening angle) is entirely fixed by the crystal lattice and independent of the incident beam direction. This implies that tilting the specimen over a small angle will lead to an equal tilt of the double cone, but it would leave the geometry of the spot diffraction pattern unchanged provided the same reflections remain excited, i.e., as long as the same "relrods" are intersected by Ewald's sphere. The relative position of the spot pattern and of the Kikuchi line pattern is thus very orientation sensitive, and as a consequence, it carries very useful information which can only otherwise be obtained with difficulty as we shall see.

We now consider the situation which arises when the specimen is tilted in such a way that the set of lattice planes g satisfies the Bragg condition. The situation with respect to the incident beam is then no longer symmetrical (Fig. 1-6 b). The elastically Bragg-scattered beam, which produces the spot G, is now one of the generators of the cone. One of the Kikuchi lines thus passes through the Bragg spot. It appears bright (B) on a positive print, i.e., it corresponds with an excess of electrons above the background. The other line (D), which appears dark due to a deficiency of electrons, passes through the origin. These features will now be explained. The dark line passing through the origin is produced against a high background caused by the predominantly forward, inelastically scattered electrons. Among these electrons, those which satisfy the Bragg condition are scattered elastically out of this background onto the sheet of the cone which passes through the Bragg spot. Along the parallel line through the origin, which is the locus of the electrons satisfying Bragg's condition, there is as a consequence a deficiency of electrons compared to the background. On the other hand, the same electrons which by their absence cause the dark line through the origin, cause an excess, compared to a lower background along the part of the

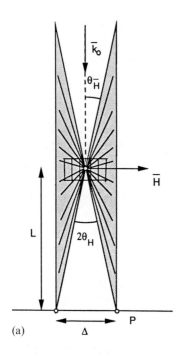

(a)

cone containing the coherently scattered Bragg beam. This background is somewhat smaller since the scattering angle is larger. Therefore the excess electrons produce a bright line through the Bragg spot. The angular separation of the bright line – dark line pair, is clearly the same as in the symmetrical orientation; the linear separation measured on the plate may slightly depend on the tilt angle, however. The symmetrical situation is represented schematically in Fig. 1-6a. In the symmetrical orientation the Kikuchi lines often form the limiting lines of "Kikuchi-bands", the inside of which exhibits a somewhat smaller brightness than the outside (Fig. 1-7) (Thomas, 1978).

In this particular orientation, the Kikuchi lines can be considered as images

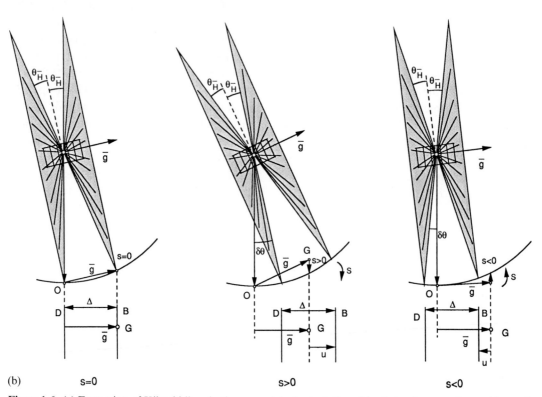

(b)   s=0   s>0   s<0

Figure 1-6. (a) Formation of Kikuchi lines in the symmetrical orientation; (b) relation between the positions of the Bragg spots and the associated Kikuchi lines for $s = 0$, $s > 0$ and $s < 0$.

of the Brillouin zone boundaries belonging to the different reflections.

1.3.3 Determination of the Sign and Magnitude of the Deviation Parameter s

Starting with a foil in the exact Bragg orientation for the reflection G, the bright Kikuchi line passes through G (Fig. 1-6 b), whereas the dark line passes through the origin of the reciprocal lattice. Tilting the specimen over a small angle $\delta\theta$, in the clockwise direction, i.e., toward $s < 0$ about an axis in the foil plane, normal to the g vector, the position of the bright Kikuchi line moves toward the origin over $u = L \delta\theta$ (Fig. 1-6 b). The vector g is then rotated over the same angle $\delta\theta$, and hence s becomes negative and equal to $s = g \delta\theta$; the relation between u and s is thus

$$u = \frac{L}{g} s \quad \text{and also}$$

$$\Delta u = \frac{L}{g} \Delta s \tag{1-7}$$

This relation allows one to determine the sign and the magnitude of s from the relative position of a diffraction spot and its associated Kikuchi line (Fig. 1-6 b). This relation also allows one to determine the orientation difference between two crystal parts. For practical applications we refer to Thomas (1978). The sign of s is required for a number of applications such as the determination of the sign of the Burgers vector of a dislocation, the vacancy or interstitial character of a dislocation loop, the orientation difference across a domain boundary, etc., as will be discussed below. The magnitude of s is needed when applying the weak-beam method (Sec. 1.6.3.8).

A different deviation parameter x is sometimes used; this is illustrated in Fig. 1-4. Let C' be the centre of Ewald's sphere in the deviated position and let BZ be the

Figure 1-7. Kikuchi bands in silicon.

Brillouin zone boundary belonging to g. Then x is the distance from C' to BZ. If the small angle between g and $g + s$ is called $\delta\theta$, one clearly obtains $s_g = g \delta\theta$. The normal from C' on $g + s$ encloses the same angle $\delta\theta$ with the trace of the BZ boundary. Since the radius of Ewald's sphere is $K \sim k_0$, one also obtains $\delta\theta \simeq x/K \simeq x/k_0$, and hence $s = (g/k_0) x$ is the relation between the two deviation parameters x and s.

1.3.4 Refraction of Electrons

Refraction of the incident electron beam takes place at the interface of the vacuum – crystal foil, because the lengths of the wave vectors are different in the two media:

$$K_0 \equiv K_{\text{vacuum}} = \frac{(2 m e E)^{1/2}}{h} \tag{1-8}$$

$$K \equiv K_{\text{crystal}} = \frac{[2 m e (E + V_0)]^{1/2}}{h} \tag{1-9}$$

But the tangential components have to be conserved at the interface. Figure 1-8 shows the relation between the two wave vectors; one obtains

$$n = \frac{\sin i}{\sin r} = \frac{K_{t, \text{vac}}/K}{K_{t, \text{cryst}}/K_{\text{cryst}}} =$$

$$= \frac{K_{\text{cryst}}}{K_{\text{vac}}} = \left(\frac{E + V_0}{E}\right)^{1/2} \quad (1\text{-}10)$$

The refractive index is thus

$$n = \left(1 + \frac{V_0}{E}\right)^{1/2} \quad (1\text{-}11)$$

Since $V_0 \ll E$, n is only slightly larger than 1, and the angle of refraction is very small, especially for quasi-normal incidence, as is the case in most observations. Refraction nevertheless produces an observable effect for grazing incidence. Small polyhedral particles may produce diffraction spots consisting of a number of components corresponding with the number of crystal wedges crossed by the beam.

1.3.5 The Basic Equations of the Kinematic Theory

The basic equation describing the interaction of the imaging electrons with

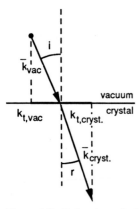

Figure 1-8. Refraction of electron beam at the entrance face of a foil; the tangential component of the wave vectors is conserved.

the periodic crystal potential is the Schrödinger equation (Gevers, 1978):

$$H \psi = \varepsilon \psi$$

with $H = (-h^2/8\pi^2 m)\Delta - e V(r)$ and $\varepsilon = e E$, where E is the accelerating potential.

We assume $e > 0$, $-e$ is then the electron charge and $E > 0$ ($\sim 100 - 400\,\text{kV}$).

Electrons used in electron microscopy have energies exceeding a hundred keV, and as a result they are to a non-negligible extent relativistic, since they travel at speeds larger than half the speed of light. Strictly speaking, the Dirac equations should be used rather than the Schrödinger equation. Fortunately since $V(r) \ll E$, the terms in V^2 and in EV can be neglected when compared to E^2 in the relativistic expression for the impulse, and the corrections with respect to the non-relativistic Schrödinger equation are then limited to the use of the relativistic mass correction and the replacement of the accelerating potential E by $E[1 + e E/(2 m_0 c^2)]$. The latter correction also leads to a slightly modified expression for the de Broglie wavelength of the electron:

$$\lambda = h \left[2 m_0 e E \left(1 + \frac{e E}{2 m_0 c^2}\right)\right]^{-1/2}$$

We introduce in the Schrödinger equation the quantity $E' = E + V_0$ and put $e E' = h^2 k_0^2/(2m)$ with $k_0 = 1/\lambda$; k_0 is then the wave vector of the incident electrons corrected for refraction by the constant part of the potential V_0; $|k_0| = |K_{\text{crystal}}|$.

Setting $V' = V - V_0$, the Schrödinger equation can be rewritten as:

$$\Delta\psi + 4\pi^2 k_0^2 \psi = -(8\pi^2 m e/h^2) V'(r)\psi$$

For abbreviation we further introduce:

$$U(r) = (2 m e/h^2) V'(r); \quad U_0 = 0$$

We then have:

$$U(r) = \sum U_g \exp(2\pi i \, g \cdot r)$$

and

$$\Delta\psi + 4\pi^2 k_0^2 \psi = -4\pi^2 U(r)\psi \qquad (1\text{-}12)$$

The right-hand side of this equation can be considered to be a perturbation, caused by the lattice, on the incident beam with wave vector k_0. The first Born approximation, which is equivalent to the kinematic approximation, now consists in using for ψ the unperturbed expression, i.e., the solution of Eq. (1-12) without expression on the right-hand side. This solution is the simple plane wave with unit amplitude

$$\psi_0 = \exp(2\pi i \, k_0 \cdot r) \qquad (1\text{-}13)$$

Substituting this expression on the right-hand side, one obtains

$$\Delta\psi + 4\pi^2 k_0^2 \psi =$$
$$= -4\pi^2 \sum_g U_g \exp[2\pi i(k_0+g)\cdot r]$$

This is an inhomogeneous linear differential equation whose solution is of the general form

$$\psi = \psi_0 + \sum_g \psi_g$$

Since ψ_0 is already a solution of Eq. (1-12) without the expression on the right-hand side, the term ψ_g has to satisfy the equations

$$\Delta\psi_g + 4\pi^2 k_0^2 \psi_g =$$
$$= -4\pi^2 U_g \exp[2\pi i(k_0+g)\cdot r] \qquad (1\text{-}14)$$

with the boundary condition $\psi_g = 0$ at the entrance face $z = 0$.

The mutual independence of the different beams is an inherent characteristic of the kinematical approximation. We shall call φ_g the amplitude of ψ_g and write

$$\psi_g(r) = \varphi_g(r) \exp[2\pi i(k_0+g)\cdot r]$$

We then substitute this in Eq. (1-14) and make use of the identity

$$\Delta(uv) = u\Delta v + v\Delta u + 2\nabla u \cdot \nabla v$$

to transform the Eq. (1-14), noting that

$$\Delta\{\varphi_g \exp[2\pi i(k_0+g)\cdot r]\} =$$
$$= \Delta\varphi_g \{\exp[2\pi i(k_0+g)\cdot r]\} +$$
$$+ \varphi_g \{\Delta\exp[2\pi i(k_0+g)\cdot r]\} +$$
$$+ 2\nabla\varphi_g \cdot \nabla[2\pi i(k_0+g)\cdot r]$$

After leaving out a common factor on both sides of the equation, one finds

$$\Delta\varphi_g + 4\pi i(k_0+g)\cdot\nabla\varphi_g +$$
$$+ 4\pi^2[k_0^2 + (k_0+g)^2]\varphi_g =$$
$$= -4\pi^2 U_g \qquad (1\text{-}15)$$

With the particular choice of the reference system shown in Fig. 1-9, we obtain

$$(k_0+g)\cdot\nabla\varphi_g =$$
$$= [(k_0+g)\cdot e_z]\frac{\partial\varphi_g}{\partial z} + [(k_0+g)\cdot e_x]\frac{\partial\varphi_g}{\partial x}$$

and with the definition of the angle α shown in Fig. 1-3, this expression becomes:

$$(k_0+g)\left(\cos\alpha\cdot\frac{\partial\varphi_g}{\partial z} + \sin\alpha\cdot\frac{\partial\varphi_g}{\partial x}\right)$$

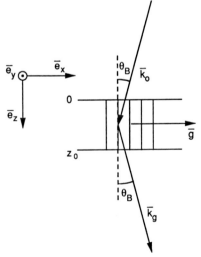

Figure 1-9. Reference system used in deriving the basic equations of kinematic diffraction.

In Eq. (1-15) we can neglect the term $\Delta\varphi_g$ since it is small compared to the second term; after dividing by $4\pi i (k_0 + g)\cos\alpha$, we then obtain

$$\frac{\partial\varphi_g}{\partial z} + \tan\alpha \cdot \frac{\partial\varphi_g}{\partial x} - 2\pi i s_g \varphi_g = \frac{\pi i U_g}{|k_0 + g|\cos\alpha}$$

The right-hand side of this equation has the dimensions of an inverse length and depends on the active reflection g. We shall set

$$\frac{1}{t_g} \equiv \frac{U_g}{|k_0 + g|\cos\alpha} \quad \text{with} \quad \alpha = \theta_B$$

and $\cos\theta_B = 1$ \hfill (1-16)

The equation then finally becomes

$$\frac{\partial\varphi_g}{\partial z} + \tan\alpha \cdot \frac{\partial\varphi_g}{\partial x} - 2\pi i s_g \varphi_g = \frac{\pi i}{t_g} \quad (1\text{-}17)$$

The expression for the *extinction distance* t_g is often transformed by making use of the approximation $|k_0 + g| \simeq k_0$ (since $k_0 \gg g$) and using the definition of U_g and noting that g is approximately perpendicular to k_0:

$$t_g = \frac{h^2 k_0 \cos\alpha}{2 m e |V_g|} \quad (1\text{-}18)$$

Then reintroducing the accelerating voltage E (neglecting V_0)

$$t_g = \frac{\lambda E \cos\alpha}{|V_g|} \quad \text{with} \quad \cos\alpha \simeq 1 \quad (1\text{-}19)$$

t_g is in fact a measure for the "strength" of the reflection. A small extinction distance corresponds to a strong reflection (i.e., larger $|V_g|$), whereas weak reflections have a long extinction distance. Low order reflections in elementary metals have extinction distances ranging from $100–1000\ \text{Å}$.

The extinction distance can be computed from the structure factor F_g by means of the relation $t_g = (\pi V_c m_0 v \cos\alpha)/(h F_g)$, where V_c is the unit cell volume, v is the speed of the electrons, and F_g is the structure factor (with $m = m_0$) with $\cos\alpha \simeq 1$ (Hirsch et al., 1965).

The amplitude of the kinematically scattered beam can now be obtained by integrating the Eq. (1-17). We note that V_g is in general complex and can be written as $V_g = |V_g| \exp(i\theta_g)$. The equation then becomes

$$\frac{\partial\varphi_g}{\partial z} + \tan\alpha \cdot \frac{\partial\varphi_g}{\partial x} - 2\pi i s_g \varphi_g = \frac{\pi i}{t_g} \exp(i\theta_g) \tag{1-20}$$

For a perfect foil φ_g cannot depend on x or y but only on z, and thus the partial derivative can be replaced by the ordinary derivative, and we finally obtain

$$\frac{d\varphi_g}{dz} - 2\pi i s_g \varphi_g = \frac{\pi i}{t_g} \exp(i\theta_g) \quad (1\text{-}21)$$

We shall call S_g the modulus of the amplitude of φ_g:

$$\varphi_g = S_g \exp(2\pi i s_g z)$$

where $S_g = 0$ for $z = 0$. Equation (1-21) then reads

$$\frac{dS_g}{dz} = \frac{\pi i}{t_g} \exp(i\theta_g) \exp(-2\pi i s_g z)$$

which after integration over the depth of the foil leads to

$$S_g = \frac{\pi i}{t_g} \exp(i\theta_g) \int_0^{z_0} \exp(-2\pi i s_g z)\, dz \tag{1-22}$$

where z_0 is the foil thickness. After performing the integration explicitly, one finds for the amplitude

$$\varphi_g = \frac{i \sin(\pi s_g z_0)}{s_g t_g} \exp(i\theta_g) \exp(\pi i s_g z_0)$$

and for the intensity (Fig. 1-10)

$$I_g = \varphi_g \varphi_g^* = \frac{\sin^2(\pi s_g z_0)}{(s_g t_g)^2} = \frac{I_0 \sin^2(\pi s_g z_0)}{(\pi s_g z_0)^2} \tag{1-23}$$

with $I_0 = \pi^2 z_0^2/t_g^2$.

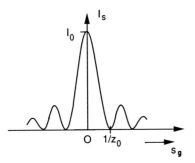

Figure 1-10. Rocking curve according to the kinematic theory.

Equation (1-23) describes the length profile of the "relrods" introduced in Sec. 1.3.1.

At constant s_g, the depth period is $\Delta z_0 = 1/s_g$. Equation (1-23) thus describes a periodic variation of the scattered intensity as a function of the thickness z_0; it has maxima for thicknesses $z_0 = (n + 1/2)/s_g$ and minima for $z_0 = n/s_g$. A wedge shaped specimen will thus exhibit thickness fringes or "Pendellösung" fringes, two successive fringes corresponding with parts of the wedge differing $1/s_g$ in thickness (Fig. 1-11).

This theory clearly ignores absorption and violates the conservation of electrons since each scattering center, irrespective of its depth in the foil, is assumed to see an incident electron beam with unit amplitude. This will be corrected in the dynamic theory.

It is worth pointing out that the thickness fringes result in fact from the beating of two waves in the crystal. To illustrate this we rewrite the expression for ψ_g

$$\psi_g = \varphi_g \exp[2\pi i (k_0 + g) \cdot r]$$

as

$$\varphi_g = \exp(i\theta_g) \exp(\pi i s_g z) \cdot$$
$$\cdot \frac{\exp(\pi i s_g z) - \exp(-\pi i s_g z)}{2 s_g t_g}$$

The two interfering waves have slightly different wave vectors:

$$k_0 + g + s_g e_z \quad \text{and} \quad k_0 + g$$

since ψ_g can be written as:

$$\psi_g = \exp(i\theta_g) \frac{\exp[2\pi i (k_0 + g + s e_z) \cdot r] - \exp[2\pi i (k_0 + g) \cdot r]}{2 s_g t_g}$$

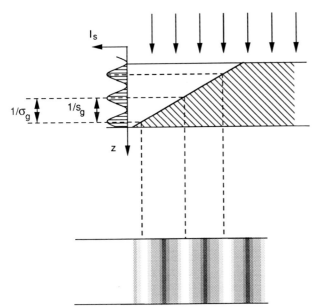

Figure 1-11. Schematic illustration of the formation of wedge fringes (thickness extinction contours).

1.3.6 Kinematic Rocking Curve

The function in Eq. (1-23) also describes the dependence of I_g on s_g, i.e., the rocking curve (Fig. 1-10). The rocking curve has a main maximum at $s_g = 0$ and its magnitude is $I_{g,\max} = (\pi z_0/t_g)^2$. It has zeros for $s_g = n/z_0$ ($n = integer$). There are secondary maxima approximately halfway between the zeros.

The curve is clearly symmetrical in s_g. This will still be the case according to the dynamic theory for the scattered beam, but not for the transmitted beam.

The kinematic theory does not describe what happens to the transmitted beam. At best, one could assume that electrons are conserved and that the transmitted beam intensity I_T varies with thickness z_0 and with the deviation parameter s_g in a manner complementary to that of the scattered beam, i.e., $I_T = 1 - I_g$.

Under which conditions can the kinematic theory lead to useful results? Neglecting the depletion of the incident beam is clearly only acceptable for thin crystals such that $I_g \ll 1$; that is the thickness should be small as compared to the extinction distance. Another way to satisfy the condition $I_g \ll 1$ is to assume s_g to be large, i.e., if the foil orientation is far from the exact Bragg condition.

It should be noted that for $s = 0$ the thickness fringes acquire an infinite period z which is the inverse of s_g. Since in actual fact thickness extinction contours are always observed, the kinematic theory is not valid for small s values, not even qualitatively. It will nevertheless be useful, especially for large s values, as in the weak beam technique.

1.3.7 The Column Approximation

The formula Eq. (1-22) can be understood intuitively as being due to the summation of the contributions due to the scattering elements dz in "columns" parallel to the beam direction. The factor $\exp(-2\pi i s z)$ takes into account the "phase" of the element dz due to its depth z behind the entrance face. The factor $(\pi i/t_g)\exp(i\theta_g)$ describes the strength and the phase of the reflection due to the diffracting material in the element dz, while the factor i accounts for the phase jump on scattering. The size of the "column" need not be specified. In a perfect foil the whole foil surface can be considered to be the basis of a "column"; the result would in any case be independent of the choice of the size of the column and of the point (x, y) on which the column is centered.

It is helpful to recall at this stage that we are dealing with the Laue case, and that Bragg angles are very small in electron diffraction; moreover, electron scattering is strongly peaked in the forward direction. The maximum sideways displacement that electrons can undergo on entering a foil of thickness z_0 is $\Delta = z_0 \theta_n$ (Fig. 1-12a) which

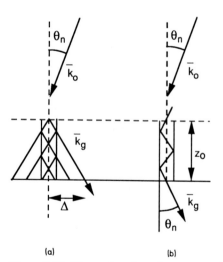

(a)   (b)

Figure 1-12. Illustrating the column approximation. (a) The maximum sideways displacement Δ in the case of single scattering; (b) the maximum sideways displacement Δ in the case of multiple scattering.

is very small since we are considering thin
foils. Moreover we must realize that multiple scattering takes place; this will be discussed explicitly in the dynamic theory.
This will confine the electrons to very narrow columns on traversing the foil (Fig.
1-12 b). It is therefore a good approximation to assume that electrons propagate
along narrow columns centered on the
point of impact.

The intensity of the beam, either scattered or transmitted, computed at the exit
surface of the column will then be representative for the site on which the column is
located. The columns can in a sense be
considered the picture elements (pixels) of
the map of the intensity distribution in a
given beam, at the back surface of the foil.
For a defect-free perfectly flat foil of constant thickness, this map will exhibit uniform intensity.

In a wedge-shaped defect-free crystal we
shall observe the thickness fringes discussed above. In the dark field image the
dark fringes are the geometrical loci of the
columns for which the emerging scattered
intensity is a minimum.

For a cylindrically bent crystal of uniform thickness the s value becomes spatially variable, the loci of constant s being
parallel to the axis of the cylinder (Figs.
1-13 and 1-14). Those columns for which
the s value is such as to produce a maximum of I_g at the exit face will produce a
bright pixel. The loci of bright pixels are
lines parallel to the axis of the cylinder;
they are lines of equal inclination or *bent
contours* (Heidenreich, 1949).

The bent contours in a cylindrically deformed thin cleavage foil of graphite are
shown in Fig. 1-14; they image directly the
rocking curve.

More detailed informations about the
column approximation are given by Takagi (1962).

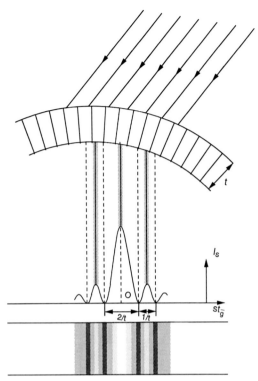

Figure 1-13. Cylindrically bent foil producing equi-
inclination contours (schematic).

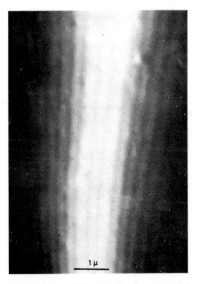

Figure 1-14. Cylindrically bent graphite foil of constant thickness illustrating equi-inclination contours
(Courtesy Delavignette).

1.3.8 Description of Deformed Crystals

The diffraction by a deformed crystal must be treated by introducing the deformation in the model via the lattice potential, since this determines the scattering. A crystal defect can be characterized by its displacement function $R(r)$, which is a vector field describing for each point with position vector r the local displacement of a volume element away from its position in the undeformed crystal.

The displacement field of a screw dislocation along the z-axis can for instance be written as:

$$R_x = 0, \quad R_y = 0, \quad R_z = b \, \varphi/(2\pi),$$
$$\varphi = \arctan(y/x) \tag{1-24}$$

where φ is the azimuth about the z-axis; that is, all displacements are parallel to the axis of the screw dislocation, i.e., with b.

The displacement field for a stacking fault or an anti-phase boundary, parallel to the foil plane at level $z = z_1$ behind the entrance face, is a step function; for instance (Fig. 1-15a), $R = 0$ for $z < z_1$ and $R_x = R_0$, $R_y = 0$, $R_z = 0$ for $z \geq z_1$.

The sense of R is defined as the translation of the exit part with respect to the front part of the foil. Changing the sense of

inclination of the fault plane with respect to the sense of g is equivalent to interchanging front and exit part and hence to changing the sign of R and also of $\alpha_g = 2\pi g \cdot R$.

A twin domain boundary parallel with the foil plane at level $z = z_1$ can likewise be represented by (Fig. 1-15b)

$$R = 0 \quad \text{for} \quad z < z_1 \quad \text{and}$$
$$R_y = R_z = 0, \quad R_x = kz;$$
$$k = 2\tan(\alpha/2) \tag{1-25}$$

1.3.9 Diffraction Equations for Deformed Crystals

One usually assumes that a displaced volume element carries along the lattice potential it had before the deformation; this approximation is called the *deformable ion approximation* (Hirsch et al., 1956):

$$V_{\text{deformed}}(r) = V_{\text{undeformed}}(r - R) \tag{1-26}$$

or

$$V_{\text{def}} = V_0 + \sum_g V_g \exp[2\pi i g \cdot (r - R)] =$$
$$= V_0 + \sum_g V_g \exp(-2\pi i g \cdot R) \cdot$$
$$\cdot \exp(2\pi i g \cdot r) \tag{1-27}$$

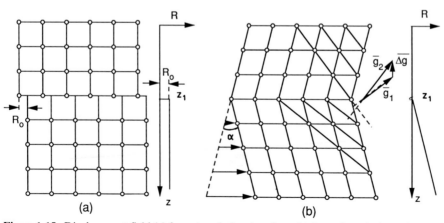

Figure 1-15. Displacement field (a) for a translation interface; (b) for a domain boundary or twin.

The Fourier coefficients now become functions of r. Strictly speaking, Eq. (1-27) is no longer a Fourier series; this is consistent with the fact that the crystal is no longer periodic. However for small values of R the reasoning used above can still be applied. The presence of the defect can thus be taken into account by the following substitution: $U_g \rightarrow U_g \exp(-2\pi i g \cdot R)$. Taking into account the expression for t_g [Eq. (1-16) or (1-19)] it can also be formulated as replacing $1/t_g$ by $(1/t_g)\exp(-i\alpha_g)$ with $\alpha_g = 2\pi g \cdot R(r)$.

Equation (1-17) then becomes

$$\frac{\partial \Phi_g}{\partial z} + \tan\alpha \frac{\partial \Phi_g}{\partial x} - 2\pi i s_g \Phi_g =$$
$$= \frac{\pi i}{t_g} \exp(i\theta_g)\exp(-i\alpha_g) \qquad (1-28)$$

Note that $\alpha_{-g}(r) = -\alpha_g(r)$.

Taking into account the column approximation and assuming the incident beam to be nearly normal to the entrance face (Laue case!) of the foil, we can put $\tan\alpha \simeq 0$. Equation (1-28) then becomes

$$\frac{d\Phi_g}{dz} - 2\pi i s_g \Phi_g =$$
$$= \frac{\pi i}{t_g} \exp(i\theta_g)\exp(-i\alpha_g) \qquad (1-29)$$

The physical meaning of this equation can be made clearer by rewriting it in terms of φ_g by the relation

$$\varphi_g = \Phi_g \exp(-i\alpha_g)$$

One then obtains

$$\frac{d\varphi_g}{dz} - 2\pi i \left(s_g + g \cdot \frac{dR}{dz}\right)\varphi_g = \frac{\pi i}{t_g}\exp(i\theta_g) \qquad (1-30)$$

This equation has the same form as Eq. (1-21) for the perfect crystal, except that the deviation parameter s_g, which is a constant in a defect-free foil, has to be replaced by an effective local value $s_{eff} = s_g + g \cdot (dR/dz)$

in the deformed crystal; s_{eff} is now generally a function of x, y and z and thus varies along an integration column, i.e., along z in a way which moreover depends on the chosen column at x, y. The local value of s_{eff} can now become significantly different from the s_g value in the perfect part of the foil. The scattered intensity leaving a column at the back surface of the crystal is a function of x and y since $R(r)$ depends on these coordinates. The intensity distribution at the exit surface of a foil, which is what we call an *image*, will thus be a function of x and y.

1.3.10 Stacking Fault Fringes

The integration of Eq. (1-22) along a column now leads to

$$S_g = \frac{\pi i}{t_g}\exp(i\theta_g)\int_0^{z_0}\exp[2\pi i s_g z + i\alpha_g(r)]\,dz \qquad (1-31)$$

In the simple case of a translation interface, parallel with the foil surfaces at $z = z_1$ behind the entrance face, this expression becomes:

$$S_g = \frac{\pi i}{t_g}\exp(i\theta_g)\left[\int_0^{z_1}\exp(2\pi i s_g z)\,dz + \right.$$
$$\left. + \exp(i\alpha_g)\int_{z_1}^{z_0}\exp(2\pi i s_g z)\,dz\right]$$

After performing the integrations and multiplying with the complex conjugate, one finds

$$I_S(s,z_0) = \left(\frac{1}{t_g s_g}\right)^2 \cdot$$
$$\cdot \{1 - \cos(\alpha_g + \pi s_g z_0)\cos(\pi s_g z_0) +$$
$$+ \cos(2\pi s_g u)[\cos(\alpha_g + \pi s_g z_0) -$$
$$- \cos(\pi s_g z_0)]\} \qquad (1-32)$$

with $u = z_1 - (1/2)z_0$ being the distance from the foil center. The intensity at the exit face of a column (i.e., for a given x and y value) depends periodically on u, i.e., on the level where the fault plane intersects this column, as well as on α and on the total thickness z_0. At constant z_0 and α the intensity is a periodic function of u with period $1/s_g$. Along a planar fault, inclined with respect to the foil plane, u varies linearly; the projected area of the fault along the beam direction exhibits a periodic intensity variation consisting of a set of fringes parallel to the central plane $u = 0$. In this approximation, the pattern is symmetrical with respect to the projection of the intersection line of the plane $u = 0$ with the fault plane, because I_S is an even function of u. A more refined dynamic theory is necessary to explain the details of the observed fringe patterns (Hirsch et al., 1960; Amelinckx, 1964).

1.3.11 Dislocation Displacement Fields

For a screw dislocation, parallel to the y-axis at $z = 0$, the foil being parallel to the (x, y) plane and the z-axis being perpendicular to the foil plane, the function α_g is

$$\alpha_g = (g \cdot b) \arctan\left(\frac{z}{x}\right) \tag{1-33}$$

The amplitude scattered by a column at (x, y) is then

$$A(x, y) = \frac{\pi i}{t_g} \exp(i\theta_g) \int_{-z_1}^{+z_2} \cdot \tag{1-34}$$

$$\cdot \exp\left[g \cdot b \arctan\left(\frac{z}{x}\right)\right] \exp(2\pi i s_g z)\, dz$$

where s_g refers to the perfect part of the foil.

For an edge dislocation along the y-axis and with a slip plane parallel to the foil plane the displacement function is (Friedel, 1964)

$$R = \frac{b}{2\pi}\left[\Phi + \frac{\sin 2\Phi}{4(1-v)}\right] e_x -\tag{1-35}$$

$$- \frac{b}{2\pi}\left[\frac{1-2v}{2(1-v)}\ln r + \frac{\cos 2\Phi}{4(1-v)}\right] e_z$$

where $\Phi = \arctan(z/x)$; e_x and e_z are the unit vectors along the x- and z-axis. For diffraction vectors g along the foil plane $g \cdot e_z = 0$ and α_g reduces to

$$\alpha_g = (g \cdot b)\left[\Phi + \frac{\sin(2\Phi)}{4(1-v)}\right] \tag{1-36}$$

It is clear that if $\alpha_g = 0$, i.e., if $g \cdot b = 0$, the displacement field of the dislocation does not perturb the diffraction phenomena, and as a result the dislocation will not produce an image. This condition is the usual extinction criterium for dislocations. However, if the slip plane of an edge dislocation is perpendicular to the foil plane, or stated otherwise, if the supplementary half plane is parallel to the foil plane, this simple extinction criterium is no longer valid since the slip plane of an edge dislocation exhibits a "bump" at the dislocation line, described by the term in e_z in Eq. (1-35).

1.4 Two-Beam Dynamic Theory

1.4.1 Derivation of the Basic Set of Equations: The Darwin Approach

The fundamental equations of the dynamic theory can be derived directly from the Schrödinger equation (Bethe, 1928). However a method similar to one initially attributed to Darwin (1914) can also be used to derive the dynamic equations for the experimentally important two-beam case (Amelinckx, 1964). It provides moreover direct physical insight into the diffraction process. We shall hereby make use of the kinematic theory, which can be applied at its limit to a very thin slice. We represent the electrons traveling along a column by the wave function

$$\psi(r) = \Phi_0(z)\exp(2\pi i k_0 \cdot r) +$$
$$+ \Phi_g(z)\exp(2\pi i k \cdot r)$$

where k_0 and k are the wave vectors of the incident and the diffracted beam, respectively; their amplitudes Φ_0 and Φ_g now depend on z.

The amplitude of the transmitted beam after interaction with the thin slice dz of the column, at depth z behind the entrance face can then be represented as (Fig. 1-16)

$$\Phi_0(x + \tan\theta\, dx, z + dz) =$$
$$= \Phi_0(x, z)\,\Phi_0(dz) + \Phi_g(x, z)\,\Phi_{-g}(dz) \quad (1\text{-}37)$$

This relation states that the transmitted beam results from the interference between the twice-transmitted beam of which the amplitude is $\Phi_0(x, z) \cdot \Phi_0(dz)$ and the twice-scattered beam with amplitude $\Phi_g(x, z) \cdot \Phi_{-g}(dz)$. For the expressions of $\Phi_0(dz)$ and $\Phi_g(dz)$, the use of the kinematic theory is justified since dz is arbitrarily thin. We can thus write [from Eq. (1-22)]

$$\Phi_0(dz) = 1\,;$$
$$\Phi_g(dz) = \frac{\pi i}{t_g}\exp(-2\pi i s_g z)\,dz \quad (1\text{-}38)$$

and since $\tan\theta \simeq \theta$, Eq. (1-37) becomes

$$\Phi_0(x + \theta\, dx, z + dz) - \Phi_0(x, z) =$$
$$= \frac{\pi i}{t_g}\exp(2\pi i s_g z)\,\Phi_g(x, z)\,dz$$

Limiting the left hand side to the term of the first order in a Taylor expansion leads to

$$\frac{\partial\Phi_0}{\partial z} + \theta\,\frac{\partial\Phi_0}{\partial x} = \frac{\pi i}{t_g}\exp(2\pi i s_g z)\,\Phi_g \quad (1\text{-}39)$$

A similar equation is valid for the scattered beam

$$\Phi_g(x - \theta\, dx, z + dz) =$$
$$= \Phi_0(x, z)\,\Phi_g(dz) + \Phi_g(x, z)\,\Phi_0(dz) \quad (1\text{-}40)$$

The factor $\Phi_{-g}(dz)$ refers to scattering by the slice dz, with the diffraction vector $-g$ acting in the reverse sense of that considered in the first equation. Changing g into $-g$ results in interchanging O and G in Fig. 1-16: i.e., G becomes the origin of the reciprocal lattice and as a consequence

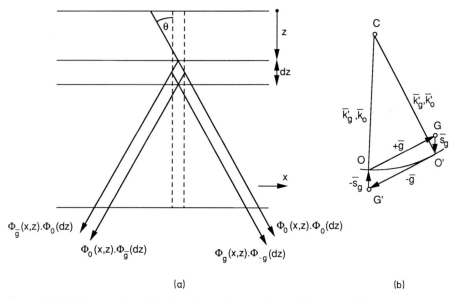

$\Phi_{\bar{g}}(x,z).\Phi_0(dz)$

$\Phi_0(x,z).\Phi_{\bar{g}}(dz)$

$\Phi_0(x,z).\Phi_0(dz)$

$\Phi_g(x,z).\Phi_{-g}(dz)$

(a)   (b)

Figure 1-16. (a) Derivation of the fundamental equations of dynamic diffraction; (b) changing $+g$ into $-g$ changes the sign of s_g.

also s_g has to be changed into $-s_g$. Finally, we obtain

$$\frac{\partial \Phi_g}{\partial z} - \theta \frac{\partial \Phi_g}{\partial x} = \frac{\pi i}{t_g} [\exp(-2\pi i s_g z)] \Phi_0 \quad (1\text{-}41)$$

Applying the column approximation, we can neglect the terms in θ and obtain the Darwin–Howie–Whelan (Howie and Whelan, 1961, 1962) set of coupled differential equations, describing the interplay between the transmitted and the scattered beam along a column of crystal. The basic form of this set of equations is thus

$$\frac{d\Phi_0}{dz} = \frac{\pi i}{t_{-g}} [\exp(2\pi i s_g z)] \Phi_g \quad (1\text{-}42)$$

$$\frac{d\Phi_g}{dz} = \frac{\pi i}{t_g} [\exp(-2\pi i s_g z)] \Phi_0 \quad (1\text{-}43)$$

1.4.2 Alternative Forms of the Diffraction Equations

Several alternative forms can be obtained for the set of equations, Eq. (1-42) and Eq. (1-43), by introducing phase factors in the amplitudes Φ_0 and Φ_g; such phase factors do not change the intensities.

One can for instance separate out the phase factor due to the depth z in the foil by setting

$$\Phi_0 = T' \exp(\pi i s_g z);$$

$$\Phi_g = S' \exp(-\pi i s_g z) \quad (1\text{-}44)$$

The set of equations then becomes

$$\frac{dT'}{dz} + \pi i s_g T' = \frac{\pi i}{t_{-g}} S' \quad (1\text{-}45)$$

$$\frac{dS'}{dz} - \pi i s_g S' = \frac{\pi i}{t_g} T' \quad (1\text{-}46)$$

Or one can also substitute

$$\Phi_0 = T; \quad \Phi_g = S \exp(-2\pi i s_g z)$$

and obtain the following set of equations:

$$\frac{dT}{dz} = \frac{\pi i}{t_{-g}} S \quad (1\text{-}47)$$

$$\frac{dS}{dz} = 2\pi i s_g S + \frac{\pi i}{t_g} T \quad (1\text{-}48)$$

Depending on the problem to be treated, the symmetrical form in Eqs. (1-45) and (1-46) or the asymmetrical form in Eqs. (1-47) and (1-48) will be used.

We note that the kinematic approximation is recovered by setting $\Phi_0 = 1$ or $T = 1$. One then obtains a single equation for the scattered beam equivalent to Eqs. (1-21) and (1-22).

Equations (1-45) and (1-46) are similar to those describing the motion of two coupled pendulums or of two coupled oscillating LCR circuits. Energy is periodically transferred from one pendulum to the other, or from one circuit to the coupled circuit. For $s = 0$ the two pendulums have equal lengths; for $s \neq 0$ they have unequal lengths, and the pendulum representing the incident beam is the longest one. Similarly, electrons are periodically transferred from the incident beam into the scattered beam and vice-versa. Therefore the Eqs. (1-46) and (1-47) are often referred to as the "Pendellösung" equations.

1.4.3 Dynamic Equations for Deformed Crystals

We have seen above (Sec. 1.3.9) that the deformation of the crystal lattice can be described by introducing a local effective deviation parameter, s_{eff} which is a function of r

$$s_{g,e} = s_g + \frac{d}{dz} (g \cdot R) = s_g + \frac{d\alpha}{dz} \quad (1\text{-}49)$$

where $\alpha = \alpha_g/(2\pi) = g \cdot R$ and which replaces s_g.

The substitution $s_g \rightarrow s_{g,e}$ allows one to easily adapt the sets of Eqs. (1-45), (1-46) and Eqs. (1-47) and (1-48) to the case of a deformed crystal.

In the case of a domain boundary with a small twinning vector, $R = k z e_\tau$ in the rear part of the crystal, and $s_2 = s_1 + \Delta s$; that is the deviation parameter in the rear part is slightly different from that in the front part.

The presence of a defect, e.g., a dislocation or a domain boundary, can also be described as a local change of g in direction, as well as in length. We have shown that at a domain boundary Δg changes abruptly from g to $g + \Delta g$ with Δg constant and perpendicular to the interface for a coherent twin.

For a dislocation we can define a "local" diffraction vector that would cause the same phase shift as the displacement field $R(r)$, i.e.

$$g \cdot [r - R(r)] = (g + \Delta g) \cdot r$$

i.e.

$$-g \cdot R(r) = \Delta g \cdot r$$

The expression for α can thus also be written as

$$\alpha_g = 2 \pi g \cdot R = -2 \pi \Delta g \cdot r \qquad (1\text{-}50)$$

where Δg now depends in general on r.

1.4.4 Solution of the Dynamic Equations for the Perfect Crystal

The system of Eqs. (1-47) and (1-48) can be uncoupled by eliminating S and T, respectively between the two equations. One obtains for S

$$\frac{d^2 S}{dz^2} - 2 \pi i s_g \frac{dS}{dz} + \frac{\pi^2}{t_g^2} S = 0 \qquad (1\text{-}51)$$

and a similar equation is obtained for T. These are linear second-order differential equations with constant coefficients for which solutions of the form

$$S = C \exp(2 \pi i \alpha z)$$

$(C = $ arbitrary constant$)$

exist, where α must satisfy the characteristic quadratic equation

$$\alpha^2 - s_g \alpha - (4 t_g^2)^{-1} = 0 \qquad (1\text{-}52)$$

This equation has two roots:

$$\alpha_{1,2} = 1/2 (s_g \pm \sigma_g) \quad \text{where}$$

$$\sigma_g = \frac{[1 + (s_g t_g)^2]^{1/2}}{t_g} \qquad (1\text{-}53)$$

The general solution of Eq. (1-51) is then

$$S = C_g^{(1)} \exp[\pi i (s_g + \sigma_g) z] + \\ + C_g^{(2)} \exp[\pi i (s_g - \sigma_g) z] \qquad (1\text{-}54)$$

and similarly

$$T = C_0^{(1)} \exp[\pi i (s_g + \sigma_g) z] + \\ + C_0^{(2)} \exp[\pi i (s_g - \sigma_g) z] \qquad (1\text{-}55)$$

Taking into account that $T = 1$ and $S = 0$ for $z = 0$ and that the Eqs. (1-47) and (1-48) must be satisfied for all values of z one finds that the coefficients $C_0^{(1)}$, $C_0^{(2)}$, $C_g^{(1)}$ and $C_g^{(2)}$ must satisfy the equations:

$$C_0^{(1)} + C_0^{(2)} = 1; \quad C_g^{(1)} + C_g^{(2)} = 0$$

$$C_0^{(1)} (s_g + \sigma_g) = (1/t_g) C_g^{(1)} \qquad (1\text{-}56)$$

$$C_0^{(2)} (s_g - \sigma_g) = (1/t_g) C_g^{(2)}$$

and thus, solving this set of linear equations

$$C_0^{(1)} = (1/2) [1 - (s_g/\sigma_g)] = \sin^2 (\beta/2)$$

$$C_0^{(2)} = (1/2) [1 + (s_g/\sigma_g)] = \cos^2 (\beta/2)$$

$$C_g^{(1)} = -C_g^{(2)} = (1/2) \sigma_g t_g = \\ = \sin (\beta/2) \cos (\beta/2)$$

where the Takagi (Takagi, 1962) parameter β defined as $\cotan \beta = s_g t_g$ was introduced. Substituting in Eqs. (1-54) and (1-55) and transforming the expressions so obtained leads to the following relations

$$T = \exp(\pi i s_g z) \cdot \tag{1-57}$$
$$\cdot [\cos(\pi \sigma_g z) - i(s_g/\sigma_g) \sin(\pi \sigma_g z)]$$

and

$$S = \frac{i}{\sigma_g t_g} \sin(\pi \sigma_g z) \exp(\pi i s_g z) \tag{1-58}$$

One can similarly show from Eqs. (1-45) and (1-46) that T' and S' satisfy the equations with the initial values $T'=1$ and $S'=0$ for $z=0$.

$$\frac{d^2 T'}{dz^2} + (\pi \sigma_g)^2 T' = 0; \quad \frac{d^2 S'}{dz^2} + (\pi \sigma_g)^2 S' = 0 \tag{1-59}$$

which lead to expressions for T' and S' which are the same as those for T and S apart from the phase factor $\exp(\pi i s_g z)$.

The intensities I_S and I_T can easily be obtained from Eqs. (1-57) and (1-58) $I_S = S S^*$ and $I_T = T T^*$, i.e.

$$I_S = \left(\frac{1}{\sigma_g t_g}\right)^2 \sin^2(\pi \sigma_g z) \tag{1-60}$$

and

$$I_T = 1 - I_S \tag{1-61}$$

since up to now we have neglected absorption. The depth period is now

$$1/\sigma_g = \frac{t_g}{[1 + (s_g t_g)^2]^{1/2}} \tag{1-62}$$

as compared to $1/s_g$ in the kinematic theory; it no longer diverges as $s_g \to 0$. In the latter case $(s_g=0)$ I_S becomes

$$I_{S,0} = \sin^2\left(\frac{\pi z}{t_g}\right) \tag{1-63}$$

and the transmitted intensity

$$I_{T,0} = \cos^2\frac{\pi z}{t_g} \tag{1-64}$$

which shows that $I_T + I_S = 1$.

Note that in the kinematic approximation, i.e., for large s_g, $(s_g t_g)^2 \gg 1$, σ_g reduces to s_g.

The expressions Eq. (1-63) and Eq. (1-64) show that as the electrons propagate into the crystal there is a continuous transfer of electrons from the transmitted to the scattered beam and vice-versa, without loss of electrons, as schematically represented in Fig. 1-17.

As in the kinematic case, the *Pendellösung* effect leads to the formation of thickness fringes in a wedge-shaped crystal. The depth period is now $1/\sigma_g$ and the Eqs. (1-63) and (1-64) would predict a strictly periodic variation. Figure 1-18 shows the thickness fringes of a wedge-shaped silicon crystal; it is evident that the contrast of the fringes (i.e., their amplitude) decreases with increasing thickness; that is, absorption takes place, as will be discussed below.

In view of the discussion of the wave fields to be treated further below, we can write the solution corresponding with the two roots of Eq. (1-52) separately.

Consider first the root $\alpha_1 = (1/2)(s+\sigma)$ (omitting for simplicity the index g). From the fact that the solutions

$$T = {'C_0^{(1)}} \exp[\pi i(s+\sigma)z] \quad \text{and}$$
$$S = {'C_g^{(1)}} \exp[\pi i(s+\sigma)z] \tag{1-65}$$

must satisfy Eq. (1-47), we deduce that

$$\frac{{'C_g^{(1)}}}{{'C_0^{(1)}}} = (s+\sigma)t \tag{1-66}$$

Requiring that $|{'C_0^{(1)}}|^2 + |{'C_g^{(1)}}|^2 = 1$, i.e. that in the wave field associated with the root α_1 the intensity be conserved, we obtain (recall that $\sigma^2 t^2 = 1 + s^2 t^2$)

$$'C_0^{(1)} = \{(1/2)[1 + (s/\sigma)]\}^{1/2};$$
$$'C_g^{(1)} = -\{(1/2)[1 - (s/\sigma)]\}^{1/2} \tag{1-67}$$

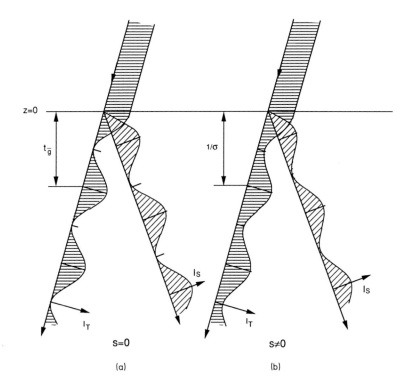

Figure 1-17. *Pendellösung* effect according to the dynamic diffraction theory: (a) $s = 0$; (b) $s \neq 0$.

By following a similar reasoning for the second root $\alpha_2 = (1/2)(s - \sigma)$, we find the expressions

$$'C_0^{(2)} = \{(1/2)[1 - (s/\sigma)]\}^{1/2} \, ;$$
$$'C_g^{(2)} = \{(1/2)[1 + (s/\sigma)]\}^{1/2} \qquad (1\text{-}68)$$

The first *Bloch field* can thus be represented by:

$$B^{(1)}(\boldsymbol{r}) = {'C_0^{(1)}} \exp(2\pi i \, \boldsymbol{k}_0^{(1)} \cdot \boldsymbol{r}) +$$
$$+ {'C_g^{(1)}} \exp[2\pi i (\boldsymbol{k}_0^{(1)} + \boldsymbol{g}) \cdot \boldsymbol{r}] \qquad (1\text{-}69)$$

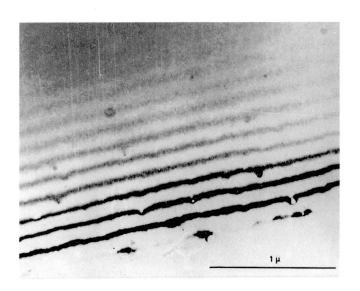

Figure 1-18. Thickness fringes in a wedge-shaped silicon crystal. Note the attenuation of the contrast with increasing thickness (Courtesy Delavignette).

with $k_0^{(1)} = K + \alpha_1^{(1)} e$, where K is the incident wave vector, corrected for refraction by the (mean) inner potential, and e is the normal to the entrance face (see Sec. 1.4.5).

The second Bloch field is obtained mutatis mutandis i.e., by changing the indices 1 to 2.

The expressions for the coefficients $'C_{0,g}^{(1,2)}$ can be simplified by introducing, after Takagi (1962), the dimensionless deviation parameter $st = \cotan\beta$, which leads to $s/\sigma = \cos\beta$ and hence

$$'C_0^{(1)} = 'C_g^{(2)} = \cos(\beta/2);$$
$$'C_0^{(2)} = -'C_g^{(1)} = \sin(\beta/2) \qquad (1\text{-}70)$$

1.4.5 Dispersion Surface and Wave Fields

1.4.5.1 Two-Beam Case

The dynamic theory can also be developed directly from the Schrödinger equation. This approach can easily be generalized to many-beam situations. Moreover it leads to a simple geometrical construction of the wave vectors of the different beams propagating in the crystal. We will call K the wave vector of the incident wave in the crystal, given by the relation $2me(E + V_0) = h^2 K^2$; i.e., K is corrected for refraction by maintaining the constant part of the lattice potential, but ignoring the periodic part. The periodic part of the lattice potential changes the wave vector of the incident wave to k_0. We further assume that only one scattered wave with wave vector $k_g = k_0 + g$ is excited, a part from the incident beam with wave vector k_0.

The wave function for the electrons in the periodic potential of the crystal is of the Bloch form; under the assumptions made, it can be written as

$$\Psi(r) = \psi_0 \exp(2\pi i k_0 \cdot r) + \psi_g \exp(2\pi i k_g \cdot r) \qquad (1\text{-}71)$$

Substituting this "ansatz" in Schrödinger's equation and taking into account that

$$V' = V - V_0 = \frac{h^2}{2me} \sum_g U_g \exp(2\pi i g \cdot r) \qquad (1\text{-}72)$$

leads to the following expression

$$(K^2 - k_0^2)\psi_0 \exp(2\pi i k_0 \cdot r) +$$
$$+ (K^2 - k_g^2)\psi_0 \exp(2\pi i k_g \cdot r) +$$
$$+ [\psi_0 \exp(2\pi i k_0 \cdot r) +$$
$$+ \psi_g \exp(2\pi i k_g \cdot r)] \sum_g U_g \exp(2\pi i g \cdot r) = 0$$

Setting the coefficients of the exponential functions in k_0 and k_g separately equal to zero, one obtains

$$(K^2 - k_0^2)\psi_0 + U_{-g}\psi_g = 0 \qquad (1\text{-}73)$$
$$U_g \psi_0 + (K^2 - k_g^2)\psi_g = 0 \qquad (1\text{-}74)$$

This system of homogeneous linear equations allows one to determine the ratio R_g of the amplitudes ψ_g/ψ_0. There will only be a non-trivial solution provided the determinant of the system vanishes, i.e.

$$(K^2 - k_0^2)(K^2 - k_g^2) =$$
$$= U_g \cdot U_{-g} = |U_g|^2 \qquad (1\text{-}75)$$

in a centro-symmetric crystal where $U_g = U_{-g}$.

This equation represents a complicated revolution surface with g as a rotation axis. Bearing in mind that the periodic term of the lattice potential has an amplitude much smaller than the constant part, one has $|K| \sim |k_0| \sim |k_g|$ and the relation in Eq. (1-75) can therefore be approximated fairly well by putting $K + k_0 \simeq 2K$ and $K + k_g \simeq 2K$; Eq. (1-75) then becomes

$$(K - k_0)(K - k_g) = \frac{|U_g|^2}{4K^2} \qquad (1\text{-}76)$$

Since $K^2 \gg |U_g|^2$, the right-hand side is small and hence the surface does not differ very much from $(K - k_0)(K - k_g) = 0$; i.e., in the asymptotic approximation the geometric locus of the endpoints of allowed vectors k_0 and k_g are two spheres with radius K and with centers O and G, respectively. Figure 1-19 represents the intersection of this surface with the plane determined by k_0 and k_g. The differences $K - k_0$ and $K - k_g$ must either be both positive or both negative. The two branches of the curve are thus situated in the obtuse "angles" of the intersecting asymptotic circles in Fig. 1-19.

This geometric locus, consisting of two sheets, is called the *dispersion surface*. Only the part of the surface close to the Brillouin zone boundary of g is of interest for our discussion. A planar section of this part of the surface is represented in Fig. 1-19.

The vectors connecting a point of the dispersion surface with the origin O and with the point G are wave vectors which are compatible with the periodic potential of the crystal, i.e., are Bloch waves. Among these potential waves only those which satisfy the boundary conditions will be actually excited. These conditions require the continuity of the wave functions and of their derivatives at the entrance face. It can

be shown (Bethe, 1928) that both conditions are satisfied if the tangential components of the wave vectors are conserved at the interface, i.e., if $k_{0,t}^{(i)} = K_t$ $(i = 1, 2)$, where (1) and (2) refer to points on the two sheets of the dispersion surface. Graphically, this means that the starting points of the wave vectors of the excited waves must be situated on the same normal to the entrance face as the starting point of K, which is itself unambiguously defined by the direction of the incident electron beam. These geometrical relations are represented in Fig. 1-19.

The knowledge of the dispersion surface makes it possible to obtain graphically the wave vectors of the Bloch waves for any given incident wave K corrected for refraction. The geometry of the dispersion surface belonging to g is determined by the reciprocal lattice nodes O and G, which are the centers of the asymptotic spheres with radius $|K|$. (In passing, it may be noted that these asymptotic spheres are the degenerate form of the dispersion surface in the free electron case, i.e., for $U_0 = 0$.)

One first constructs the incident wave vector K with its endpoint in O. Through the starting point C of K, one draws a normal to the entrance face of the foil. This normal intersects the surface in the points D and D′, which are the starting points of the vectors $\overline{DO}$, $\overline{D'O}$, $\overline{DG}$ and $\overline{D'G}$.

The construction produces for a given incident beam the wave vectors of the crystal waves along incident and diffracted directions. The periodic part of the crystal apparently causes *birefrigence*. The transmitted as well as the scattered beam both consist of two waves with slightly different wave vectors, as already shown in Sec. 1.3.5 for the kinematic case. It is the interference of these pairs of waves that causes "beating" leading to the periodic depth variation in intensity of transmitted and scattered

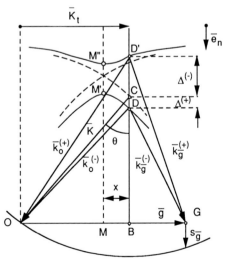

Figure 1-19. Dispersion surface construction in the two-beam case of dynamic diffraction.

wave. We shall show that the two approaches are in fact equivalent.

We now need to determine the wave vectors k_0^+ and k_0^- as well as k_g^+ and k_g^- corresponding with the two intersection points D and D' with the dispersion surface.

The geometry of Fig. 1-19 shows that

$$K^2 - k_0^2 = \overline{OC}^2 - \overline{OD}^2 =$$
$$= (\overline{OB}^2 + \overline{BC}^2) - (\overline{OB}^2 + \overline{BD}^2) =$$
$$= \overline{BC}^2 - \overline{BD}^2 =$$
$$= (\overline{BC} + \overline{BD})(\overline{BC} - \overline{BD}) =$$
$$= (\overline{BC} + \overline{BD})\,\overline{CD} \simeq 2\,\overline{BC} \cdot \overline{CD}$$

since $\overline{BD} \simeq \overline{BC}$.

Introducing the vector $\overline{CD} = \Delta e_n$, where e_n is the unit normal on the foil surface, since $K \cos\theta = \overline{BC}$ (from Fig. 1-19) we obtain finally

$$K^2 - k_0^2 = 2K\Delta\cos\theta \qquad (1\text{-}77)$$

On the other hand

$$k_g^2 - k_0^2 = (\overline{BG}^2 + \overline{BD}^2) - (\overline{DB}^2 + \overline{BD}^2) =$$
$$= \overline{BG}^2 - \overline{DB}^2 =$$
$$= [(1/2)g - x]^2 - [(1/2)g + x]^2 =$$
$$= -2gx$$

and since

$$K^2 - k_g^2 = (K^2 - k_0^2) - (k_g^2 - k_0^2) =$$
$$= 2K\Delta\cos\theta + 2\,gx \qquad (1\text{-}78)$$

Taking into account Eq. (1-77) and Eq. (1-78) and introducing the extinction distance t_g [Eq. (1-18)], we can write the relation in Eq. (1-75) as

$$\Delta^2 + s\Delta - \frac{1}{4t_g^2} = 0 \qquad (1\text{-}79)$$

The roots of this quadratic equation determine the values of Δ corresponding with the intersection points with the dispersion surface; one finds

$$\Delta^\pm = \frac{1}{2}(-s \pm \sigma) \qquad (1\text{-}80)$$

where σ was defined as Eq. (1-53). The separation of the two branches of the dispersion surface as measured along e is given by $\Delta^+ - \Delta^- = \sigma$. For $s = 0$ this separation $\overline{MM'}$ is minimal and it becomes $1/t_g$. This gives a simple geometric interpretation of t_g in reciprocal space.

The relation in Eq. (1-80) leads to two wave vectors k_0^+ and k_0^- for the incident wave, which are compatible with the *periodic* potential:

$$k_0^\pm = K - \Delta^\pm e \qquad (1\text{-}81)$$

The ratio of the amplitudes R_g can be obtained from the set of homogeneous equations, Eqs. (1-73) and (1-74):

$$R_g = \frac{K^2 - k_0^2}{|U_{-g}|} \equiv \frac{|U_g|}{K^2 - k_g^2} \qquad (1\text{-}82)$$

The two expressions are equivalent provided the relation in Eq. (1-75) is satisfied; from Eq. (1-80) one finds

$$R_g^\pm = \frac{2K\Delta^\pm\cos\theta}{|U_{-g}|} = 2t_g\Delta^\pm =$$
$$= t_g(-s \pm \sigma) \qquad (1\text{-}83)$$

The wave functions belonging to the two branches of the dispersion surface, i.e., to the two values of $\Delta^\pm$ or of $R_g^\pm$ are thus

$$\psi^\pm = \exp(2\pi i\,k_0^\pm \cdot r) +$$
$$+ R_g^\pm \exp[2\pi i(k_0^\pm + g) \cdot r] \qquad (1\text{-}84)$$

and the total wave function is then a linear combination of these:

$$\Psi = C^+\psi^+ + C^-\psi^- \qquad (1\text{-}85)$$

The arbitrary constants C^+ and C^- have to be chosen so that the initial conditions are met, i.e., so that

$$\psi_T = 1 \quad \text{and} \quad \psi_S = 0$$
$$\text{at} \quad z \equiv e \cdot r = 0 \qquad (1\text{-}86)$$

This leads to the following system of linear equations:

$$C^+ + C^- = 1; \quad C^+ R_g^+ + C^- R_g^- = 0 \quad (1\text{-}87)$$

which has the solution

$$C^{\pm} = \frac{1}{2}\left(1 \pm \frac{s}{\sigma}\right) \qquad (1\text{-}88)$$

and thus

$$\psi_T = C^+ \exp(2\pi i k_0^+ \cdot r) +$$
$$+ C^- \exp(2\pi i k_0^- \cdot r) \qquad (1\text{-}89)$$

Using Eqs. (1-81), (1-83) and (1-88), one obtains

$$\psi_T = \exp(2\pi i K \cdot r) \cdot \qquad (1\text{-}90)$$
$$\cdot \exp(\pi i s z)\left[\cos(\pi \sigma z) - i \frac{s}{\sigma}\sin(\pi \sigma z)\right]$$

Similarly, one finds

$$\psi_S = R_g^+ C^+ \exp(2\pi i k_g^+ \cdot r) +$$
$$+ R_g^- C^- \exp(2\pi i k_g^- \cdot r) \qquad (1\text{-}91)$$

and after the same transformation as for ψ_T, using Eqs. (1-81), (1-83) and (1-88),

$$\psi_S = \frac{i}{\sigma t_g}\exp[2\pi i (K + g)\cdot r] \cdot$$
$$\cdot \exp(\pi i s z)\sin(\pi \sigma z) \qquad (1\text{-}92)$$

The total wave function then becomes [from Eqs. (1-57) and (1-58)]

$$\Psi = [T + S \exp(2\pi i g \cdot r)]\exp(2\pi i K \cdot r) \qquad (1\text{-}93)$$

These expressions are clearly identical with those derived previously Sec. 1.4.4 from the Darwin–Howie–Whelan equations.

In terms of the expressions Eqs. (1-65) and (1-66), the total wave function in the crystal can also be expressed in terms of the "Bloch waves"

$$\Psi = A^{(1)} B^{(1)}(r) + A^{(2)} B^{(2)}(r) \qquad (1\text{-}94)$$

where the $A^{(1)}$ and $A^{(2)}$ represent the amplitudes of the Bloch waves $B^{(1)}$ and $B^{(2)}$

excited in the crystal. Their values have to be determined by the boundary conditions. Explicitly, we have:

$$\Psi = A^{(1)\prime} C_0^{(1)} \exp(2\pi i k_0^{(1)} \cdot r) +$$
$$+ A^{(2)\prime} C_0^{(2)} \exp(2\pi i k_0^{(2)} \cdot r) +$$
$$+ A^{(1)\prime} C_g^{(1)} \exp[2\pi i (k_0^{(1)} + g)\cdot r] +$$
$$+ A^{(2)\prime} C_g^{(2)} \exp[2\pi i (k_0^{(2)} + g)\cdot r] \qquad (1\text{-}95)$$

The first two terms refer to the transmitted beam T, whereas the last two terms refer to the scattered beam S. Expressing the boundary conditions $T = 1$, $S = 0$ at the entrance face one can determine the amplitude $A^{(1)}$ and $A^{(2)}$ of the Bloch waves from the system of linear equations

$$A^{(1)\prime} C_0^{(1)} + A^{(2)\prime} C_0^{(2)} = 1;$$
$$A^{(1)\prime} C_g^{(1)} + A^{(2)\prime} C_g^{(2)} = 0$$

this leads to

$$A^{(1)} = \cos(\beta/2); \quad A^{(2)} = \sin(\beta/2) \qquad (1\text{-}96)$$

It is easy to verify that the two expressions for Ψ, Eq. (1-95) and Eqs. (1-54), (1-55), are in fact identical when noting that

$$A^{(1)\prime} C_0^{(1)} = C_0^{(1)}; \qquad A^{(2)\prime} C_0^{(2)} = C_0^{(2)};$$
$$A^{(1)\prime} C_g^{(1)} = C_g^{(1)} \text{ and } A^{(2)\prime} C_g^{(2)} = C_g^{(2)}.$$

1.4.5.2 Many-Beam Case

We shall now discuss briefly how the n-beam case can be treated. An analytical treatment is only possible in high-symmetry situations. The solutions of the Schrödinger equation with a periodic potential are known to consist of so-called Bloch waves, i.e., plane waves with amplitudes having the periodicity of the lattice of the form

$$\psi(r) = u_k(r)\exp(2\pi i k \cdot r) \qquad (1\text{-}97)$$

where $u_k(r)$ has the periodicity of the crystal; it can thus be expanded in a Fourier series based on the reciprocal lattice of the crystal. The wave function is then a linear combination of such Bloch waves:

$$\Psi(r) = \sum_g C_g \exp[2\pi i(k+g) \cdot r] \qquad (1\text{-}98)$$

where the coefficients C_g have to be determined from the boundary conditions.

Recall the abbreviations introduced in Sec. 1.3.5, i.e., $U_g = (2me/h^2)V_g$; $K^2 = 2me(E+V_0)/h^2$. Using the expansion

$$V(r) = \sum_g V_g \exp(2\pi i g \cdot r)$$

we can substitute these expressions in the Schrödinger equation in its reduced form (Sec. 1.3.5) and then obtain

$$\sum_g [(K^2 - |k+g|^2) C_g + \sum_{h \neq g} U_{g-h} C_h] \cdot$$
$$\cdot \exp[2\pi i(k+g) \cdot r] = 0$$

This equation has to be satisfied for any value of r; this is only possible if the coefficients of the different exponentials are separately zero.

$$(K^2 - |k+g|^2) C_g + \sum_{h \neq g} U_{g-h} C_h = 0 \quad (1\text{-}99)$$

One linear equation is then obtained for each g vector. This set of homogeneous linear equations determines in principle the unknown coefficients C_g, a part from a proportionality factor. In practice, the number of equations in this set is limited to the number of beams for which the interaction is being studied. In the two-beam case, considered above in detail, it reduces to a set of two homogeneous equations, the solution of which has been discussed in detail in Sec. 1.4.4.

1.4.6 Wave Function at the Exit Face

The total wave function Ψ in the two-beam case can be imaged directly in the microscope by collecting the two interfering beams ψ_T and ψ_S in the objective aperture and focusing on the exit surface of the

specimen. The intensity distribution in this image is then given by

$$I = \Psi \Psi^* = TT^* + SS^* +$$
$$+ TS^* \exp(-2\pi i g \cdot r) +$$
$$+ T^*S \exp(2\pi i g \cdot r) \qquad (1\text{-}100)$$
$$= I_T + I_S + 2\sqrt{I_T I_S} \sin(2\pi i g \cdot r + \varphi)$$

where

$$\tan \varphi = \frac{s}{\sigma} \tan(\pi \sigma z_0) \qquad (1\text{-}101)$$

(z_0: foil thickness). It consists of sinusoidal fringes with a period equal to the separation $1/|g|$ of the lattice planes normal to g; moreover, they are parallel to this set of lattice planes. For $s=0$ one simply obtains

$$I = 1 + \sin\left(\frac{2\pi z_0}{t_g}\right) \sin(2\pi g x) \qquad (1\text{-}102)$$

with the x-axis chosen along g. The contrast of the fringe pattern depends on the thickness z_0. The localization of the fringes depends in turn on s since this determines φ. If $s=0$ the fringes can be said to coincide with the lattice planes: they are therefore called two-beam *lattice* fringes. Their formation can easily be understood on a purely geometric basis as shown in the next section.

1.4.7 Intuitive Model for the Origin of Lattice Fringes

In Fig. 1-20, let d be the distance between the lattice planes g (represented as full lines), for which the Bragg condition is assumed to be exactly satisfied. Rather than concentrating on the beams, we focus attention on the wave fronts, whose maxima are represented as solid lines with a spacing λ. We assume them to be plane waves. The superposition of the two systems of plane waves T and S behind the crystal will produce an interference pattern whose maxima are represented by black

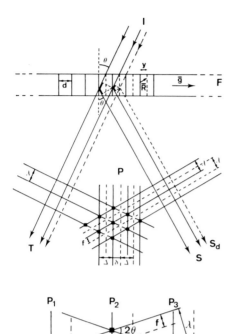

Figure 1-20. Intuitive derivation of the mechanism for the formation of lattice fringes (Amelinckx, 1986a).

dots (which are in fact straight lines viewed end on!) in Fig. 1-20. As the waves T and S propagate under the Bragg angle these maxima will move along planes parallel to the lattice planes and separated by the same interplanar spacing $\Delta \equiv d$. This follows directly from the symmetry of Fig. 1-20.

These maxima thus form a stationary pattern which can be observed behind the crystal and magnified, and which is directly related to the set of lattice planes.

If in one part of the foil the lattice planes are displaced over y, occupying the dotted positions, for instance because of the presence of a stacking fault with displacement vector R such that $g \cdot R = y$, this will not

affect the transmitted wave, but rather the reflected wave will suffer a phase shift. The new positions of the reflected wave fronts are indicated by dotted lines. The interference pattern will as a result be shifted over δ into the dotted positions. From elementary geometrical considerations it follows directly that $\delta/\Delta = y/d = g \cdot R/d$. From a measurement of the fractional shift δ/Δ, one can thus deduce the projection of R on g.

We conclude that it is justified to consider the observed fringes as an image of the family of lattice planes g, i.e., as a rudimentary image of the lattice of the structure present in the foil. In practice, the actual localization of the fringes depends on s, but their relative positions in different crystal parts depend only on the presence of lattice shifts.

The *lattice fringes* generated in this way are in fact the analogue of the optical interference fringes formed by the Fresnel biprism.

1.4.8 Absorption

1.4.8.1 Normal Absorption

Normal absorption is phenomenologically described by a complex refractive index, the absorption coefficient being related to the imaginary part of the refractive index. This can easily be seen quite generally by writing the expression for a plane wave as $\psi = \psi_0 \exp[i(kz - \omega t)]$ with $k = \omega/v = \omega n/c$ with $n = c/v$ ($v =$ velocity in medium, $c =$ velocity in vacuum).

For n complex $n_c = n + i\mu$ the complex wave vector becomes $k_c = (\omega/c)n + i(\omega/c)\mu$ and the wave is then represented by:

$$\psi = \psi_0 \exp[i(k_c z - \omega t)] = \qquad (1\text{-}103)$$
$$= \psi_0 \exp[i(kz - \omega t)] \exp\left(-\frac{\omega \mu z}{c}\right)$$

i.e., the amplitude absorption coefficient is $\mu \omega/c$. "Normal" absorption, i.e., absorption only depending on the path length, can be accounted for by an exponential damping factor. Such a factor emerges automatically when making the constant term V_0 of the lattice potential complex, replacing it by $(V_0 + i\,W_0)$. In view of the relation in Eq. (1-18) between the extinction distances t_g and the Fourier coefficients V_g of the lattice potential, this is equivalent to assuming that the extinction distance becomes complex.

The refractive index for electrons is related to the constant part V_0 of the lattice potential. The magnitude of the wave vector K_0 for electrons in vacuum is given by

$$h^2 K_0^2/(2\,m\,e) = E \qquad (1\text{-}104)$$

whereas the magnitude of the wave vector K in the crystal becomes

$$h^2 K^2/(2\,m\,e) = E + V_0 \qquad (1\text{-}105)$$

The boundary condition at the entrance face requires continuity of the tangential components $K_{0,\,t} = K_t$; this leads to refraction (Fig. 1-8) by the constant part of the potential V_0. The periodic part causes birefringence as we have seen for the two beam case, causing the incident beam to split in two beams having wave vectors with the same tangential component as K.

1.4.8.2 Anomalous Absorption

The number of electrons is, of course, ultimately conserved in any experiment and "absorption" in the original sense does not take place on electron diffraction. However electrons may be "diverted" by being scattered, inelastically or even elastically in such a way that they no longer contribute to the image formation. Phenomenologically this can be considered as some kind of absorption called "anomalous absorption". That such absorption oc-

curs can be deduced from images such as Fig. 1-18 representing the two-beam "Pendellösung" or thickness fringes in a silicon wedge. The contrast of the fringes decreases clearly with increasing thickness, but even at thicknesses where the fringes are no longer visible, electrons are still transmitted since the intensity is not zero in these areas. The strength of the absorption is found to depend on the active reflection and on the deviation parameter s_g; i.e., it not only depends on the path length as "normal" absorption does. Phenomenologically, it can be described in a way similar to that in the description of normal absorption, by assuming a complex lattice potential:

$$V(r) + i\,W(r) \qquad (1\text{-}106)$$

The physical basis for such a phenomenological approach was laid down by Yoshioka (1957). Taking into account anomalous absorption in the two-beam case then consists in assuming that the extinction distance, which is directly related to the Fourier coefficient, is complex; i.e., one substitutes

$$\frac{1}{t_g} \rightarrow \frac{1}{t_g} + \frac{i}{\tau_g} \qquad (1\text{-}107)$$

where τ_g is the absorption *length* belonging to the reflection g in the relevant equations.

The effect of including an imaginary part in V_0, i.e., replacing it by $V_0 + i\,W_0$, is to make the wave vector K complex; it becomes K^*. This can be seen as follows:

$$K^* = \left[\frac{2\,m\,e\,(E + V_0 + i\,W_0)}{h^2}\right]^{1/2} =$$
$$= (A + i\,B)^{1/2}$$

with

$$A = \frac{2\,m\,e\,(E + V_0)}{h^2}; \qquad B = \frac{2\,m\,e\,W_0}{h^2};$$

$$K = A^{1/2}$$

Now $(A + iB)^{1/2}$ can be approximated as $[1 + (i/2)(B/A)]A^{1/2}$ since $B/A \ll 1$ and hence

$$K^* = K + \frac{i}{2\tau_0} \qquad (1\text{-}108)$$

where τ_0 is defined as

$$\frac{1}{\tau_0} = \frac{2 m e W_0}{h^2 K} \qquad (1\text{-}109)$$

by analogy with $1/t_0 = 2 m e V_0/(h^2 K)$. Since at the interface $z = 0$, the tangential component of the wave vector must be conserved, the imaginary part must be oriented along e_n, since there is no tangential component for $z < 0$, i.e., one obtains

$$K^* = K + \frac{i}{2\tau_0} e_n \qquad (1\text{-}110)$$

Replacing K by its complex homologue K^* introduces an exponential attenuation factor in the expression $\exp(2\pi i K \cdot r)$, which now becomes

$$\exp(2\pi i K \cdot r)\exp\left(-\frac{\pi z}{\tau_0}\right) \qquad (1\text{-}111)$$

since $e_n \cdot r = z$.

The absorption coefficient for amplitudes is thus $\mu = \pi/\tau_0$; it affects to the same extent all the component waves in ψ_T and ψ_S.

In a similar way, anomalous absorption is introduced by the above-mentioned substitution in Eq. (1-107): $(1/t_g) \rightarrow (1/t_g) + (i/\tau_g)$. We note that σ, defined in Eq. (1-53), now also becomes complex:

$$\sigma = \sigma_r + i\sigma_i \qquad (1\text{-}112)$$

Since $\tau_g \gg t_g$, we can neglect the quantities $\sigma_i^2 \ll \sigma_r^2$ and $1/\tau_g^2 \ll 1/t_g^2$ and write to a good approximation:

$$\sigma_r = \frac{1}{t_g}[1 + (s\,t_g)^2]^{1/2}; \quad \sigma_i = \frac{1}{\sigma_r t_g \tau_g} \qquad (1\text{-}113)$$

with $\sigma_i \cdot r = \sigma_i z$.

1.4.9 Dynamic Equations Including Anomalous Absorption

The set of equations, taking into account anomalous absorption, is then obtained by substituting

$$\sigma_g = \sigma_{g,r} + i\sigma_{g,i} \qquad (1\text{-}114)$$

omitting the index g in the two-beam case. The solution of this set of equations is then obtained by performing the same substitution directly in the solutions ψ_T and ψ_S for the non-absorbing case.

Interpretable analytical expressions are obtained when adopting the following approximation $\sigma_g \rightarrow \sigma_r$ in the coefficients of the expressions for ψ_T and ψ_S but $\sigma_g \rightarrow \sigma_r + i\sigma_i$ in the exponentials. For a perfect crystal the solution for the two-beam case, including anomalous absorption, can thus be written from Eqs. (1-90) and (1-92):

$$\psi_T = \frac{1}{2}\left(1 - \frac{s}{\sigma_r}\right)\exp\left[2\pi i\left(K + \frac{s + \sigma_r}{2}e\right)\cdot r\right]$$
$$\cdot \exp(-\pi\sigma_i z) +$$
$$+ \frac{1}{2}\left(1 + \frac{s}{\sigma_r}\right)\exp\left[2\pi i\left(K + \frac{s - \sigma_r}{2}e\right)\cdot r\right]$$
$$\cdot \exp(\pi\sigma_i z) \qquad (1\text{-}115)$$

and similarly (neglecting in the coefficients $1/\tau_g \ll 1/t_g$)

$$\psi_S = \frac{1}{2\sigma_r t_g}\exp\left[2\pi i\left(K + g + \frac{s + \sigma_r}{2}e\right)\cdot r\right]\cdot$$
$$\cdot \exp(-\pi\sigma_i z) -$$
$$- \frac{1}{2\sigma_r t_g}\exp\left[2\pi i\left(K + g + \frac{s - \sigma_r}{2}e\right)\cdot r\right]\cdot$$
$$\cdot \exp(\pi\sigma_i z) \qquad (1\text{-}116)$$

It is clear from these expression that ψ_T and ψ_S both result from the interference between two beams differing slightly in wave vector; for the transmitted beam these are

$$K + \frac{s + \sigma_r}{2} e \quad \text{and} \quad K + \frac{s - \sigma_r}{2} e \qquad (1\text{-}117)$$

and for the scattered beam

$$K + g + \frac{s + \sigma_r}{2} e \quad \text{and}$$

$$K + g + \frac{s - \sigma_r}{2} e \qquad (1\text{-}118)$$

The beating of these waves causes the periodic variation in depth of the intensities of the transmitted and scattered beam.

1.4.10 Wave Fields

The four waves of the total wave function belong to two *wave fields*, corresponding with the two branches of the dispersion surface. For the first wave field σ_r has a positive sign (upper branch), for the second, a negative one (lower branch).

The first wave field can thus be represented as

$$\Psi_1 = \left\{ \frac{1}{2} \left(1 - \frac{s}{\sigma_r}\right) \cdot \right.$$

$$\cdot \exp\left[2\pi i \left(K + \frac{s + \sigma_r}{2} e\right) \cdot r\right] +$$

$$\left. + \frac{1}{2\sigma_r t_g} \exp\left[2\pi i \left(K + g + \frac{s + \sigma_r}{2} e\right) \cdot r\right] \right\} \cdot$$

$$\cdot \exp\left(-\pi \sigma_i z\right) \qquad (1\text{-}119)$$

and the second as

$$\Psi_2 = \left\{ \frac{1}{2} \left(1 + \frac{s}{\sigma_r}\right) \exp\left[2\pi i \left(K + \frac{s - \sigma_r}{2} e\right) \cdot r\right] - \right.$$

$$\left. - \frac{1}{2\sigma_r t_g} \exp\left[2\pi i \left(K + g + \frac{s - \sigma_r}{2} e\right) \cdot r\right] \right\} \cdot$$

$$\cdot \exp\left(\pi \sigma_i z\right) \qquad (1\text{-}120)$$

The first of these wave fields Ψ_1 is attenuated with increasing thickness z as a result of anomalous absorption, whereas the sec-

ond Ψ_2 is enhanced as described by the last factors. However, normal absorption introduces an attenuation factor $\exp(-\mu z)$, which affects both wave fields equally and ensures that the second wave field does not "diverge".

We have discussed in some detail the case of $s = 0$, i.e., $\sigma_r = 1/t_g$; one now finds

$$\Psi_1 = \exp\left[2\pi i (K + \tfrac{1}{2} g + \tfrac{1}{2}\sigma_r e) \cdot r\right] \cos(\pi g r) \cdot$$

$$\cdot \exp\left(-\pi \sigma_i z\right) \quad \text{and}$$

$$\Psi_2 = -i \exp\left[2\pi i (K + \tfrac{1}{2} g - \tfrac{1}{2}\sigma_r e) \cdot r\right] \cdot$$

$$\cdot \sin(\pi g r) \exp(\pi \sigma_i z) \qquad (1\text{-}121)$$

The wave vectors $K + (1/2)(g \pm \sigma_r e)$ are parallel with the set of lattice planes g, which are exactly in the Bragg orientation. The relation $g \cdot r = n$ ($n = integer$) represents these lattice planes in direct space; they are assumed to coincide with the planes containing the atoms in a primitive lattice. The equation $g \cdot r = n + (1/2)$ then represents the planes midway between the atomic planes.

The wave functions Ψ_1 and Ψ_2 thus represent waves propagating along these lattice planes with amplitudes described by the last factors (Fig. 1-21).

The first wave field Ψ_1 has maximum elongation along the planes $g \cdot r = n$ since then $\cos(\pi g \cdot r) = \pm 1$; i.e., its extrema coincide with the atomic planes. This field is attenuated by anomalous absorption.

The second wave field Ψ_2 on the other hand has maximum elongation between the atomic planes; it is enhanced by anomalous absorption.

These results are physically meaningful. It is to be expected that the electrons which propagate close to "atomic" cores, i.e., whose wave function is peaked at the atomic positions, will have an appreciably higher probability to excite X-rays, and hence to be "absorbed", than those passing between atomic planes.

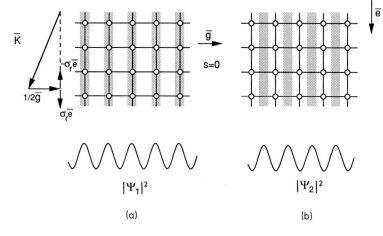

Figure 1-21. Propagation through a crystal foil of the two-wave fields formed in the two-beam case ($s = 0$). (a) The amplitude of the absorbed wave field ψ_1 reaches a maximum on the atomic planes; (b) the amplitude of the passing wave field ψ_2 reaches a maximum between the atomic planes.

The rapid damping of thickness fringes in a wedge shaped crystal can now easily be understood. The thickness fringes are due to the depth variation of I_T or I_S, which is itself related to the "beating" of two waves, one from each branch of the dispersion surface. Since one of these waves is much more attenuated than the other, the beating envelope, which has maximum amplitude when the two beating waves have equal amplitude, decreases with depth in the crystal.

The asymmetry in s_g of ψ_T, known as the Bormann effect (Bormann, 1941, 1950), can also be understood on the basis of this model. For $s > 0$ the amplitude $\frac{1}{2}[1 - (s/\sigma_r)]$ of the rapidly attenuated wave in ψ_T [Eq. (1-115)] is smaller than 1, whereas the amplitude $\frac{1}{2}[1 + (s/\sigma_r)]$ of the passing wave is enhanced. The opposite is true for $s < 0$; now the amplitude of the passing wave is smaller than 1. As a result, ψ_T will have a larger amplitude for $s > 0$ than for $s < 0$ for the same absolute value of s. A similar asymmetry in s does not occur for ψ_S [Eq. (1-116)] since the coefficients determining the amplitudes of the waves are the same: They are both $1/(2\,\sigma_r t_g)$, which, moreover, only depend on s^2.

1.4.11 Rocking Curves for Perfect Crystals

Explicit expressions for I_S and I_T can be obtained by computing $I_S = \psi_S \psi_S^*$ and $I_T = \psi_T \psi_T^*$. Remembering that $\sigma_i \ll \sigma_r$ and $\tau_g \gg t_g$, and neglecting higher order terms, one obtains after rather lengthy but straightforward calculations:

$$I_T = \left(\cosh u + \frac{s}{\sigma_r} \sinh u\right)^2 - \frac{1}{(\sigma_r t_g)^2} \sin^2 v \tag{1-122}$$

with

$$u \equiv \pi \, \sigma_i \, z; \quad v \equiv \pi \, \sigma_r \, z; \quad \sigma_i = 1/(\sigma_r t_g \sigma_g);$$
$$\sigma_r = [1 + (s\,t_g)^2]^{1/2}/t_g$$

and similarly

$$I_S = \frac{\sinh^2 u + \sin^2 v}{(\sigma_r t_g)^2} \tag{1-123}$$

These expressions are represented in Fig. 1-22. In the limit of $\sigma_i \to 0$ these expressions reduce to those for the non-absorption case.

In particular for $s = 0$ one obtains

$$I_T = \cosh^2 u - \sin^2 v \quad \text{and}$$
$$I_S = \sinh^2 u + \sin^2 v$$

Note that $I_T + I_S = \sinh^2 u + \cosh^2 u$ instead of 1 in the non-absorption case. This

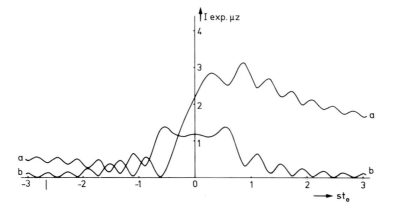

Figure 1-22. Rocking curves according to the dynamic theory taking anomalous absorption into account ($z_0 = 3\,t_g$; $\tau_g = 10\,t_g$). (a) Transmitted beam; (b) scattered beam.

apparently contradictory result is due to the fact that normal absorption was neglected: the latter would reduce this sum to a value smaller than 1.

1.5 Contrast at Planar Interfaces: Two-Beam Dynamic Theory

1.5.1 Introduction

We shall consider two types of interfaces: (i) interfaces for which the displacement field is a step function (stacking fault, antiphase boundaries, shear planes); (ii) interfaces for which the displacement function can be approximated by a linear function (twins, domain boundaries).

For the first type of interface, the s_g values are the same in both parts of the crystal, i.e., $s_{g_1} = s_{g_2}$ and the same g vector is active in both parts.

For the second type of interface the simultaneously excited g vectors are slightly different; in the case of a *coherent* twin, $g_1 = g_2 + \Delta g$, with Δg perpendicular to the coherent interface. As a result, the deviation parameters in the two crystal parts will generally be different, i.e., $s_{g_1} = s_{g_2} + \Delta s$.

1.5.2 Matrix Formulation

1.5.2.1 Perfect Crystal Matrix

First, a matrix notation will be introduced describing transmission and scattering by a slab of defect-free crystal. We will call T and S the transmitted and scattered amplitudes for an incident wave with unit amplitude. The initial values of T and S at the entrace face are $T = 1$ and $S = 0$ for $z = 0$. These initial values can be represented by a column vector $\begin{pmatrix} 1 \\ 0 \end{pmatrix}$. Similarly, we represent the transmitted and scattered amplitude at level z by a column vector $\begin{pmatrix} T \\ S \end{pmatrix}$. From the results of Sec. 1.4.4 we conclude – ignoring a common phase factor – that

$$T(s, z) = \cos(\pi \sigma z) - i\frac{S}{\sigma}\sin(\pi \sigma z) \quad (1\text{-}124)$$

$$S(s, z) = \frac{i}{\sigma t}\sin(\pi \sigma z) \quad (1\text{-}125)$$

where

$$\sigma = \sigma_r + i\sigma_i; \quad \sigma_r = [1 + (s\,t)^2]^{1/2}/t;$$

$$\sigma_i = (\sigma_r\, t\, \tau)^{-1}$$

For simplicity, we have dropped the indices g in s_g, σ_g, t_g and τ_g. In general, for an arbitrary incoming wave with initial values $\begin{pmatrix} T \\ S \end{pmatrix}_{in}$ we can write

$$\begin{pmatrix} T \\ S \end{pmatrix}_{out} = \begin{pmatrix} A & C \\ B & D \end{pmatrix} \begin{pmatrix} T \\ S \end{pmatrix}_{in} \qquad (1\text{-}126)$$

since the outcoming amplitudes depend linearly on the incoming ones. The elements A, B, C and D of the 2×2 matrix must still be determined. From

$$\begin{pmatrix} T \\ S \end{pmatrix} = \begin{pmatrix} A & C \\ B & D \end{pmatrix} \begin{pmatrix} 1 \\ 0 \end{pmatrix} \qquad (1\text{-}127)$$

we can conclude that $A \equiv T$ and $B \equiv S$. We now exploit the symmetry of the system of coupled Eqs. (1-45) and (1-46), which is satisfied for S and T. This system is mapped onto itself by the substitution $T \rightarrow S$; $S \rightarrow T$, $s \rightarrow -s$. This means that the solution for the initial values $\begin{pmatrix} 0 \\ 1 \end{pmatrix}$ is given by S^- and T^-; where $S^- = S(-s)$ and $T^- = T(-s)$, i.e., we obtain

$$\begin{pmatrix} S^- \\ T^- \end{pmatrix} = \begin{pmatrix} A & C \\ B & D \end{pmatrix} \begin{pmatrix} 0 \\ 1 \end{pmatrix} \qquad (1\text{-}128)$$

and thus $C \equiv S^-$ and $D \equiv T^-$ or finally

$$\begin{pmatrix} T \\ S \end{pmatrix}_{out} = \begin{pmatrix} T & S^- \\ S & T^- \end{pmatrix} \begin{pmatrix} T \\ S \end{pmatrix}_{in} \qquad (1\text{-}129)$$

where "in" and "out" refer to the incoming and outgoing waves in the transmitted and scattered directions, respectively, for a slab of defect-free material. The 2×2 matrix is called the *response* matrix, and it depends on two parameters: the deviation parameter s and the thickness z of the slab:

$$\begin{pmatrix} T & S^- \\ S & T^- \end{pmatrix} \equiv M(s, z) \qquad (1\text{-}130)$$

It is clear that this matrix must have the property

$$M(s, z_1 + z_2 + \ldots + z_n) = \qquad (1\text{-}131)$$
$$= M(s, z_n) M(s, z_{n-1}) \ldots M(s, z_2) M(s, z_1)$$

which can be shown in a straightforward manner by matrix calculus. It expresses that the response of a perfect crystal slab must be independent of whether or not we imagine the slab to be divided in thinner slabs with the same total thickness.

1.5.2.2 Matrices for Faulted Crystals, Shift Matrix

We have shown that in the framework of the *deformable ion approximation* lattice deformations can be accounted for by replacing the Fourier coefficient U_g by $U_g \exp(-2\pi i \boldsymbol{g} \cdot \boldsymbol{R})$, or alternatively, $1/t_g$ by $(1/t_g) \exp(-i\alpha_g)$ with $\alpha_g = 2\pi \boldsymbol{g} \cdot \boldsymbol{R}(r)$. Noting that $\alpha_{-g} = -\alpha_g$ we must also make the substitution $1/t_{-g} \rightarrow (1/t_{-g}) \exp(+i\alpha_g)$.

The set of equations describing diffraction by a faulted crystal, according to the two-beam approximation thus becomes

$$\frac{dT'}{dz} + \pi i s_g T' = \frac{\pi i}{t_{-g}} S' \exp(i\alpha_g) \qquad (1\text{-}132)$$

$$\frac{dS'}{dz} - \pi i s_g S' = \frac{\pi i}{t_g} T' \exp(-i\alpha_g) \qquad (1\text{-}133)$$

or the equivalent set obtained after the substitution:

$$T'' = T' \exp(\pi i \alpha);$$
$$S'' = S' \exp(-\pi i \alpha) \qquad (1\text{-}134)$$

with $\alpha \equiv \alpha_g/(2\pi)$; $s \equiv s_g$

$$\frac{dT''}{dz} + \pi i \left(s + \frac{d\alpha}{dz}\right) T'' = \frac{\pi i}{t_{-g}} S'' \qquad (1\text{-}135)$$

$$\frac{dS''}{dz} - \pi i \left(s + \frac{d\alpha}{dz}\right) S'' = \frac{\pi i}{t_g} T'' \qquad (1\text{-}136)$$

The effect of the deformation is thus modeled by replacing s by a local effective value $s + (d\alpha/dz)$, as already pointed out above.

Planar interfaces

For a stacking fault, $\alpha = $ constant, and these equations thus reduce to those for a perfect foil. This simply confirms the fact that the amplitudes of transmitted and scattered beams do not change in magnitude (but rather in phase) as a result of a parallel shift of a crystal slab.

Equations (1-135) and (1-136) are thus not suitable for a discussion of diffraction effects due to pure translation interfaces.

For the case of a domain boundary, however, with $\alpha = kz$, one finds that $(d\alpha/dz) = k$, and the equations for the rear part are obtained from those of the front part by substituting $s \rightarrow s + k$; i.e., the front and rear parts differ by the deviation parameter only, assuming that the same reflection g is excited in both parts.

For a discussion of the contrast at translation interfaces we consider the set of Eqs. (1-132) and (1-133); moreover, we generally assume that the deviation parameters s_1 and s_2 may also be different in the front and rear parts of the crystal, simulating the presence of a "mixed" interface; the thicknesses of the two parts are called z_1 and z_2 (Fig. 1-23).

The transmission matrix for the front part is clearly $M(z_1, s_1)$. The transmitted and scattered beams emerging from this front part are now incident on the second

part with thickness z_2. We shall represent the transmission matrix for this second part by $\begin{pmatrix} X & U \\ Y & V \end{pmatrix}$ and determine the expressions for the elements X, Y, U and V. We must now use Eqs. (1-132) and (1-133) with $s = s_2$. We note that this set of equations reduces to that for a perfect slab of crystal by means of the following substitution:

$$T' = T, \quad S' = S \exp(-i\alpha)$$

where $\alpha = $ constant, which does not affect the initial values. The solution for the initial values

$$\begin{pmatrix} T \\ S \end{pmatrix}_{z=0} = \begin{pmatrix} 1 \\ 0 \end{pmatrix} \text{ is thus } T = T(z_2, s_2) \quad (1\text{-}137)$$

and $S = S(z_2, s_2)$.

For the original system Eqs. (1-132) and (1-133) the solution is

$$T' = T(z_2, s_2) \quad \text{and}$$
$$S' = S(z_2, s_2) \exp(-i\alpha) \quad (1\text{-}138)$$

and the elements X and Y of the matrix are thus

$$X = T(z_2, s_2) \quad \text{and}$$
$$Y = S(z_2, s_2) \exp(-i\alpha) \quad (1\text{-}139)$$

In order to find U and V we look for the solution of Eqs. (1-132) and (1-133) with the initial values $\begin{pmatrix} T' \\ S' \end{pmatrix}_{z=0} = \begin{pmatrix} 0 \\ 1 \end{pmatrix}$. These initial values reduce to the previous ones if we interchange S' and T'. Making this interchange and changing $\alpha \rightarrow -\alpha$ and $s \rightarrow -s$ as well leaves the system of equations in Eqs. (1-132) and (1-133) invariant, and hence the solutions also remain the same as those of the initial set. We thus find

$$U = S(z_2, -s_2) \exp(i\alpha) \quad \text{and}$$
$$V = T(z_2, -s_2) \quad (1\text{-}140)$$

Introducing the notations T^- and S^- for T and S in which s has been replaced by

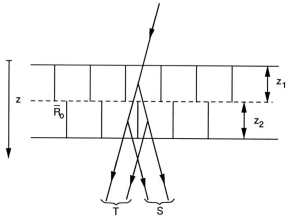

Figure 1-23. Intuitive interpretation of the expressions for the transmitted and scattered amplitude by a crystal slab containing a planar fault.

$-s$, we can write for the transmission matrix of the rear part

$$\begin{pmatrix} T_2 & S_2^- \exp(i\,\alpha) \\ S_2 \exp(-i\,\alpha) & T_2^- \end{pmatrix} \qquad (1\text{-}141)$$

with

$$T_2 = T(z_2, s_2); \quad S_2 = S(z_2, s_2);$$
$$T_2^- = T(z_2, -s_2); \quad S_2^- = S(z_2, -s_2)$$

The matrix Eq. (1-141) can conveniently be written as the product of three matrices:

$$\begin{pmatrix} 1 & 0 \\ 0 & \exp(-i\,\alpha) \end{pmatrix} \begin{pmatrix} T_2 & S_2^- \\ S_2 & T_2^- \end{pmatrix} \begin{pmatrix} 1 & 0 \\ 0 & \exp(i\,\alpha) \end{pmatrix}$$
$$(1\text{-}142)$$

Introducing the shorthand

$$\$(\alpha) \equiv \begin{pmatrix} 1 & 0 \\ 0 & e^{i\alpha} \end{pmatrix} \qquad (1\text{-}143)$$

called *shift matrix* and the response matrix

$$M_j(z_j, s_j) \equiv \begin{pmatrix} T_j & S_j^- \\ S_j & T_j^- \end{pmatrix} \qquad (1\text{-}144)$$

The final result can be written as

$$\begin{pmatrix} T \\ S \end{pmatrix} = \$(-\alpha)\, M_2\, \$(\alpha)\, M_1 \begin{pmatrix} 1 \\ 0 \end{pmatrix} \qquad (1\text{-}145)$$

This result can be generalized to a succession of interfaces in which the phases α_j' are all referred to the entrance part of the crystal; the deviation parameters in the successive parts are $s_1, s_2 \ldots s_j$. We then obtain

$$\begin{pmatrix} T \\ S \end{pmatrix} = \ldots \$(-\alpha_2')\, M_3\, \$(\alpha_2')\, \$(-\alpha_1')\, M_2\, \$(\alpha_1')\, M_1 \begin{pmatrix} 1 \\ 0 \end{pmatrix} \qquad (1\text{-}146)$$

Note that $\$(\alpha_1)\,\$(\alpha_2) = \$(\alpha_1 + \alpha_2)$. We can introduce the phase angles $\alpha_j = \alpha_j' - \alpha_{j-1}'$, which are now characteristic of the *relative* displacements of two successive lamellae, the rear lamella being displaced with respect to the front lamella. We then obtain

$$\begin{pmatrix} T \\ S \end{pmatrix} = \ldots M_3\, \$(\alpha_2)\, M_2\, \$(\alpha_1)\, M_1 \begin{pmatrix} 1 \\ 0 \end{pmatrix} \qquad (1\text{-}147)$$

1.5.2.3 Vacuum Matrix

Certain crystal slabs may be non-reflecting under the diffraction conditions prevailing in the rest of the foil; they then behave as a *vacuum lamella*. This behaviour is apparent for instance for a thin lamella of twinned material if a non-common reflection is active in the matrix. It is literally the case for a void in the crystal. It may also apply to thin precipitate lamellae which have a lattice different from that of the matrix. In such a non-reflecting slab of thickness z the extinction distance is infinite since no diffraction occurs out of the transmitted or scattered beam. The system of Eqs. (1-132) and (1-133) then becomes

$$\frac{dT}{dz} + \pi i s T = 0; \quad \frac{dS}{dz} - \pi i s S = 0 \quad (1\text{-}148)$$

which integrates to

$$T = T_0 \exp(-i\pi s z); \quad S = S_0 \exp(\pi i s z)$$

where T_0 and S_0 are the values of T and S at the entrance face of the vacuum lamellae, or in matrix form:

$$\begin{pmatrix} T \\ S \end{pmatrix}_{out} = \begin{pmatrix} \exp(-\pi i s z) & 0 \\ 0 & \exp(\pi i s z) \end{pmatrix} \begin{pmatrix} T \\ S \end{pmatrix}_{in}$$
$$(1\text{-}149)$$

We call this matrix the *vacuum matrix* and it is represented by $V(z, s)$, where z is the

thickness of the non-reflecting part measured along the integration column and s is the deviation parameter in the crystal part *preceding* the vacuum lamella.

1.5.2.4 Overlapping Interfaces

The contrast for a combination of overlapping interfaces in parallel planes can be described by the product of the appropriate sequence of matrices (Amelinckx and Van Landuyt, 1978). For instance, the diffraction contrast due to a cavity can be simulated by the product of matrices, where z_2 is the thickness of the cavity:

$$\binom{T}{S} = M_3(z_3, s_1) V(z_2, s_1) M_1(z_1, s_1) \binom{1}{0}$$

$$(1\text{-}150)$$

which gives the amplitudes of transmitted and scattered beam in the two-beam case. The normal absorption contrast due to mass–thickness differences has obviously to be taken into account as well in this case, but for small cavities the diffraction contrast effect predominates.

The contrast due to a microtwin in an f.c.c. crystal, assuming the microtwin to be non-reflecting, can be simulated with the following matrix multiplication:

$$\binom{T}{S \exp(i\alpha)} = \qquad (1\text{-}151)$$

$$= M_3(z_3, s) V(z_2, s) \$(\alpha) M_1(z_1, s) \binom{1}{0}$$

The vacuum matrix simulates the non reflecting microtwin lamella with thickness z_2. The shift matrix takes care of the phase shift due to the presence of the microtwin lamella between front and rear parts with thicknesses z_1 and z_3. This shift may be $\alpha = 0$ or $\alpha = \pm 2\pi/3$ depending on the number of atomic layers in the microtwin. Note that the matrices V and $\$$ commute. Contrast at overlapping stacking faults is simulated by the product of matrices:

1.5.3 Fringe Profiles for Planar Interfaces Including Anomalous Absorption

1.5.3.1 Introduction

Explicit expressions can be obtained for the depth variation, and thus also for the fringe profiles at inclined faults, by performing the matrix multiplication Eq. (1-147). One obtains in short hand notation

$$T = T_1 T_2 + S_1 S_2^- \exp(i\alpha);$$
$$S = T_1 S_2 \exp(-i\alpha) + S_1 T_2^- \qquad (1\text{-}154)$$

These expressions have a simple intuitive interpretation. The amplitude of the transmitted beam T results from the interference between the doubly transmitted beam $T_1 T_2$ and the doubly scattered beam $S_1 S_2^- \exp(i\alpha)$. The minus sign in S_2^- means that scattering takes place from the $-g$ side, and hence that the deviation parameter changes sign, since the primary scattered beam now functions as the incident beam for the second part. The scattering by the second part is accompanied by a phase shift α which is taken care of by the factor $\exp(i\alpha)$. A similar interpretation can be given to the expression for S (Fig. 1-23).

Introducing the explicit expressions for T_j and S_j given in Eqs. (1-57) and (1-58) and noting $\sigma = \sigma_r + i\sigma_i$, analytical expressions can be obtained for $I_T = T T^*$ and $I_S = S S^*$. These expressions can be written as the sums of three terms:

$$I_{T,S} = I_{T,S}^{(1)} + I_{T,S}^{(2)} + I_{T,S}^{(3)}$$

$$\binom{T}{S} = M_3(z_3, s) \$(\alpha_2) M_2(z_2, s) \$(\alpha_1) M_1(z_1, s) \binom{1}{0} \qquad (1\text{-}152)$$

If the fault separation is small, i.e., $z_2 \ll t_g$, one can obtain to a good approximation

$$\$(\alpha_2) M_2(z_2, s) \$(\alpha_1) \cong \$(\alpha_1 + \alpha_2) \qquad (1\text{-}153)$$

More cases are considered by Amelinckx and Van Landuyt (1978).

In the most general case the deviation parameters s_1 and s_2 are also assumed to be different in the front and rear parts: this case is treated by Gevers et al. (1965) and Gevers et al. (1964 b).

1.5.3.2 Translation Interfaces

Here we shall only consider the situation $s_1 \equiv s_2 \equiv s$, i.e., a pure translation interface. For $s = 0$ the expressions then become

$$I_{T,S}^{(1)} = \qquad (1\text{-}155)$$
$$= \frac{1}{2}\cos^2\left(\frac{\alpha}{2}\right)[\cosh(2\pi\sigma_i z_0) \pm \cos(2\pi\sigma_r z_0)]$$

$$I_{T,S}^{(2)} = \qquad (1\text{-}156)$$
$$= \frac{1}{2}\sin^2\left(\frac{\alpha}{2}\right)[\cosh(4\pi\sigma_i u) \pm \cos(4\pi\sigma_r u)]$$

$$I_{T,S}^{(3)} = \frac{1}{2}\sin\alpha\,[\sin(2\pi\sigma_r z_1)\sinh(2\pi\sigma_i z_2) \pm$$
$$\pm \sin(2\pi\sigma_r z_2)\sinh(2\pi\sigma_i z_i)] \quad (1\text{-}157)$$

where the upper sign applies to I_T, and the lower sign to I_S; $z_0 = z_1 + z_2$ is the total thickness; z_1 and z_2 are the thicknesses of the front and rear part, respectively; and $2u = z_1 - z_2$. In fact, u is the distance of the interface from the foil's central plane. In the case of $\alpha = n \cdot 2\pi$ ($n = integer$), $\sin\alpha = 0$ and $\sin(\alpha/2) = 0$, and as a result only the first term remains. This term represents the contribution of the perfect crystal since it depends on z_0 only; it describes the background on which the fringes are superposed, and these are described by the two remaining terms. The second term represents a function which depends periodically on u with depth period $1/(2\sigma_r)$ and exhibits an extreme for $u = 0$ which is a maximum for $I_T^{(2)}$ but a minimum for $I_S^{(2)}$. This term represents fringes which are parallel with the central line of the pattern.

In sufficiently thick foils i.e. for $2\sinh(2\pi\sigma_i z_0) \gg \tan(\alpha/2)$ and $\alpha = 2\pi/3$, the dominant behavior of the fringe pattern is described by the third terms which are pseudo-periodic with depth period $1/\sigma_r$ (Fig. 1-24).

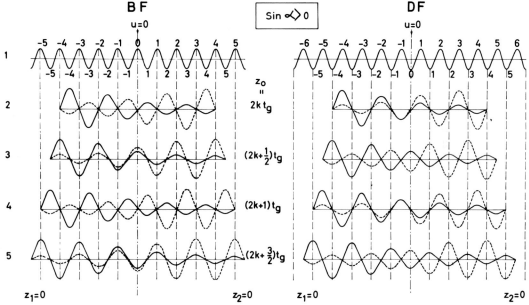

Figure 1-24. Schematic representation of the different terms occurring in the expressions for I_T and I_S of a foil containing a planar fault (Gevers et al., 1965).

At the entrance face z_1 is small and $z_2 \simeq z_0$ and the factor $\sinh(2\pi\sigma_i z_2)$ is large, and hence the first term in $I_{T,S}^{(3)}$ is dominant. It represents a damped sinusoid with depth period $1/\sigma_r$; the envelope disappears at the back surface where $z_2 \simeq 0$. If $\sin\alpha > 0$, the first extremum will be a maximum, and hence the first fringe will be bright. For $\sin\alpha < 0$ the first fringe will be a dark fringe.

At the exit face the term in $\sinh(2\pi\sigma_i z_1)$ is large since now $z_1 \simeq z_0$ and $z_2 \simeq 0$. This term again represents a damped sinusoid. The first extremum, which now corresponds with the last fringe, is either a maximum or a minimum depending on the sign of $\sin\alpha$, but it is different for I_T and I_S. This discussion is summarized in Fig. 1-24. As to the nature of the outer fringes [bright (B) or dark (D)], the results are represented in Table 1-1. These results are only valid close to the surfaces in sufficiently thick foils where anomalous absorption is sufficiently strong.

In a more detailed discussion of the fringe pattern in the central part of the foil, i.e., close to $u = 0$, the contribution from $I_{T,S}^{(1)}$ cannot be ignored, especially for those thicknesses (z_0 values) where the two terms in $I_{T,S}^{(3)}$ are opposite in sign for small values of u.

1.5.3.3 Properties of Fringes Due to Translation Interfaces (α Fringes)

A discussion of the complete expressions leads to the following properties of stacking fault fringes for the important case $\alpha = \pm 2\pi/3$, provided the foil is sufficiently thick (i.e., several extinction distances):

(i) The fringes are parallel to the closest surface.

Table 1-1. Survey of the properties of α and δ fringe patterns.

	$\alpha = 2\pi\,\boldsymbol{g}\cdot\boldsymbol{R}$					$\delta = s_1\,t_{g_1} - s_2\,t_{g_2}$			
	B.F		D.F.			B.F.		D.F.	
	F	L	F	L		F	L	F	L
$\sin\alpha > 0$	B	B	B	D	$\delta > 0$	B	D	B	B
$\alpha \neq \pi$	–	–	–	–		–	–	–	–
$\sin\alpha < 0$	D	D	D	B	$\delta < 0$	D	B	D	D

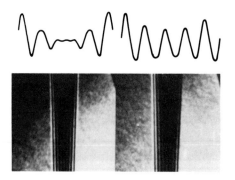
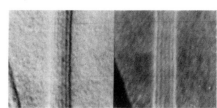

The nature of the edge fringes F (first) and L (last) is indicated by B (bright) or D (dark). A schematic profile, as well as an observed pattern, is given for the two types of fringes.

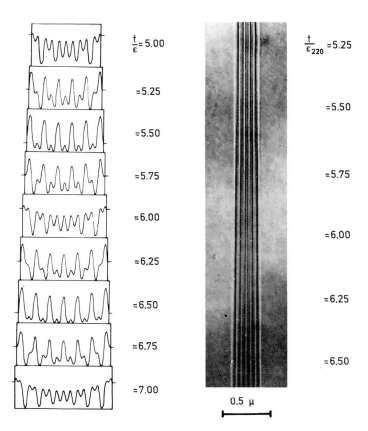

$\dfrac{t}{\varepsilon} = 5.00$

$= 5.25$

$= 5.50$

$= 5.75$

$= 6.00$

$= 6.25$

$= 6.50$

$= 6.75$

$= 7.00$

$\dfrac{t}{\varepsilon_{220}} = 5.25$

$= 5.50$

$= 5.75$

$= 6.00$

$= 6.25$

$= 6.50$

0.5 μ

Figure 1-25. Computed α fringe profiles ($\alpha = \pm\, 2\pi/3$) for a foil of increasing thickness ($s = 0$) and the corresponding image in a silicon wedge.

(ii) The bright field (B.F.) fringe pattern is symmetrical with respect to the foil center ($u = 0$).

(iii) The dark field (D.F.) pattern is antisymmetrical with respect to the foil center.

(iv) Close to the entrance face, the B.F. and D.F. patterns are similar but close to the exit face they are complementary. This is in fact a general feature of diffraction contrast images of defects in thick foils; it is also particularly true for dislocation images.

(v) With increasing thickness, new fringes are created in the center close to $u = 0$. The set of computed profiles for foils of increasing thickness compared with the observed pattern, shown in Fig. 1-25, allows these properties to be verified, whereas Fig. 1-26 shows an observed ex-

ample of a stacking fault in a wedge-shaped specimen of stainless steel.

The case $\alpha = \pi$, which usually occurs at antiphase boundaries, gives rise to fringes whose characteristics are singular and different from those due to faults with $\alpha = \pm\, 2\pi/3$ (Van Landuyt et al., 1964; Drum and Whelan, 1965). Since now $\cos(\alpha/2) = 0$ and $\sin\alpha = 0$, the fringe pattern is represented completely by the term

$$I_{T,S}^{(2)} = \tfrac{1}{2}\left[\cosh\left(4\pi\,\sigma_i\,u\right) \pm \cos\left(4\pi\,\sigma_r\,u\right)\right]$$

The following properties can immediately be deduced from this expression:

(i) B.F. and D.F. images are complementary with respect to the background described by $(1/2)\cosh\left(4\pi\,\sigma_i\,u\right)$.

(ii) The central fringe ($u = 0$) is bright in the B.F. image and dark in the D.F. image.

Figure 1-26. Observed stacking fault ($\alpha = \pm\,2\pi/3$) in a wedge shaped foil of stainless steel. The foil also contains twin bands, revealed as wedge fringes as well as dislocations (Courtesy Van Landuyt).

(iii) The fringes are parallel to the central line $u = 0$; with increasing thickness, new fringes are created at the surfaces.

(iv) The depth period is $1/(2\,\sigma_r)$, i.e., half of the period for fringes $\alpha = \pm\,2\pi/3$.

Some of these properties can be verified on the observed example of Fig. 1-27.

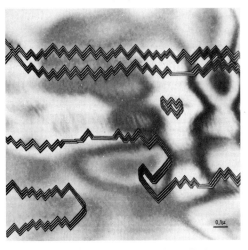

Figure 1-27. Fault fringes ($\alpha = \pi$) in rutile (Van Landuyt et al., 1964).

1.5.3.4 Domain Boundary Fringes (δ Fringes)

We assume that the boundary can be modeled as the interface between two juxtaposed crystal parts which have the same crystal structure, but which are slightly misoriented as a result of coherent twinning. Such boundaries often occur in crystals fragmented as a result of a phase transformation, whereby rotation symmetry is lost. The deviation parameters s_1 and s_2 as well as the extinction distances $1/\sigma_1$ and $1/\sigma_2$ in the two crystal parts are assumed to be slightly different (Gevers et al., 1965; Gevers et al., 1964a; Goringe and Valdré, 1966). The expressions for the amplitude T and S are then given by

$$T = T_1\,T_2 + S_1\,S_2^{-}\,;$$
$$S = T_1\,S_2 + S_1\,T_2^{-} \qquad (1\text{-}158)$$

and the intensities, by $I_T = T\,T^*$ and $I_S = S\,S^*$. Inserting the explicit expressions for S and T, one can show that I_T and I_S are again sums of three contributions similar to those in Sec. 1.5.3.1. For simplicity, we shall limit ourselves here to the symmetrical case $s_1 = -\,s_2 = s$. In thick foils, the dominant terms are then of the same form as in Sec. 1.5.3.1:

$$w^4\,I_{T,S}^{(3)} = -\frac{1}{2}\,\delta\,\{\cos\,(2\,\pi\,\sigma_{r,\,1}\,z_1)\,\cdot$$
$$\cdot\,\sinh\,[2\,(\pi\,\sigma_{i,\,2} \pm \varphi_2)] \mp \qquad (1\text{-}159)$$
$$\mp\,\cos\,(2\,\pi\,\sigma_{r,\,2}\,z_2)\,\sinh\,[2\,(\pi\,\sigma_{i,\,1}\,z_1 \pm \varphi_1)]\}$$

where

$$w^2 = 1 + (s\,t_g)^2\,; \quad \delta = s_1\,t_{g,\,1} - s_2\,t_{g,\,2}\,;$$
$$2\,\varphi_j = \operatorname{arcsinh}\,(s\,t_{g,\,j})$$

The upper sign applies to I_T and the lower to I_S. Close to the surfaces, one term is dominant and determines the nature of the outer fringes. Close to the entrance face $z_2 \simeq z_0$, the first term is the important one;

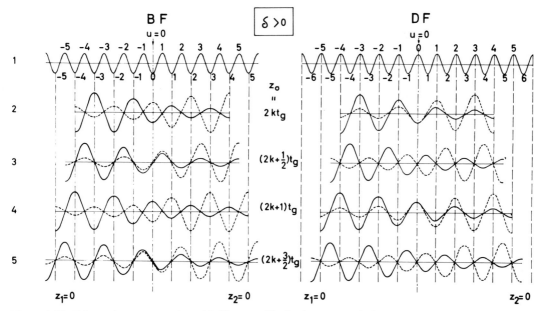

Figure 1-28. Schematic representation of δ fringe profiles in the symmetrical case $s_1 = -s_2$.

close to the exit face $z_1 \simeq z_0$, it is the second term. Close to the front surface, B.F. and D.F. images are thus again similar, whereas close to the exit face they are pseudo-complementary (Fig. 1-28).

1.5.3.5 Properties of Domain Boundary Fringes

The following properties can be deduced from the complete explicit expressions given in Gevers et al. (1965):

(i) The depth period may be different close to the front and rear surface if $t_{g,1}$ is significantly different from $t_{g,2}$.

(ii) If $t_{g,1} = t_{g,2}$, the image is symmetrical in the D.F., the outer fringes have the same nature. It is quasi-antisymmetrical in the B.F., i.e., the outer fringes have opposite nature.

(iii) The fringes are parallel to the closest surface; new fringes are added in the center.

(iv) The nature of the outer fringes depends on the sign of δ in the manner shown

in Table 1-1; the fringes lose contrast if $\delta = 0$.

(v) The domain contrast on either side of the boundary is in general different in the B.F. image; it may be the same in the D.F. image, in particular in the symmetrical orientation $s_1 = -s_2$.

These properties can be verified on the pair of images of Fig. 1-29. The complete

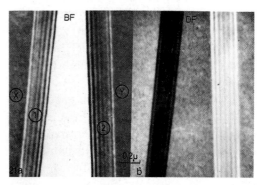

Figure 1-29. δ fringes at two domain boundaries in Al-Zr alloy (Zr_3Al_4): (a) bright field; (b) dark field (Courtesy Delavignette and Nandedkar).

expressions for I_T and I_S in Gevers et al. (1965) allow one to understand why in the dark field image the average level of intensity is different for the fringe patterns along the interfaces (1) and (2) in Fig. 1-29. These interfaces are inclined in the same sense with respect to the electron beam and hence s_1 and s_2 are interchanged along 1 and 2. The average intensity should be the same in the B.F. if $s_1 = -s_2$.

It is clear that the fringes associated with a pure *translation* interface (α fringes) have symmetry properties which are different from those of pure *inclination* interface (δ fringes). In particular, the nature of the outer fringes in sufficiently thick foils is a useful characteristic. These properties allow the two classes of interfaces to be distinguished by means of contrast experiments.

The symmetry properties of these fringe patterns for translation interfaces can also be deduced directly from the expressions in Eq. (1-154):

$$I_T(z_1, z_2, s, \alpha) = I_T(z_2, z_1, s, \alpha) \qquad (1\text{-}160)$$

i.e., the bright field image is symmetrical with respect to the central line $u = 0$. On the other hand

$$I_S(z_1, z_2, s, \alpha) = I_S(z_2, z_1, -s, -\alpha) \qquad (1\text{-}161)$$

It is possible that along certain planar interfaces $\alpha = 2\pi g \cdot R$ as well as $\delta = s_2 t_{g,2} - s_1 t_{g,1}$ differ from zero. The fringe patterns produced by such interfaces have characteristics intermediate between those of α and δ fringes.

If a diffraction vector which is common to both crystal parts (i.e., which is unsplit in the diffraction pattern) is operative, the δ component becomes inoperative and the α component is revealed separately. It is also possible to eliminate the α component selectively from the image by the use of a diffraction vector g for which $g \cdot R =$ integer. One can, for instance, image the lattice relaxation along stacking faults and the antiphase boundaries with displacement vector $R = R_0 + \varepsilon$ by using only a systematic row of reflections for which $g \cdot R = integer$. The presence of residual fringes in such images reveals the presence of additional displacements, which are different from R_0, e.g., $R_0 + \varepsilon$. The residual fringes are then due to $\alpha = 2\pi g \cdot \varepsilon$ (Fig. 1-30).

1.5.4 Applications

1.5.4.1 Determination of the Type of Stacking Fault

Close-packed layers of atoms can be stacked in an infinite number of ways which differ very little in free energy. In the cubic-face-centered stacking ... ABCABC ..., two essentially different types of stacking faults may occur on the (111) close-packed layers.

If a single layer plane is extracted, and the gap closed by a displacement $R_0 = (1/3)[111]$, one obtains the sequence (*intrinsic fault*)

$$\text{ABCA}\overset{\uparrow}{\text{C}}\text{ABCABC} \qquad (a)$$

The same sequence is obtained if a Shockley partial dislocation with Burgers vector $(1/6)[11\bar{2}]$ sweeps the crystal; it also results from the precipitation of a layer of vacancies. The fault contains two triplets in the hexagonal stacking.

The second important type of fault results from the insertion of one layer of atoms leading to the sequence (*extrinsic fault*)

$$\text{ABCA}\underset{\uparrow}{\text{C}}\text{BCABCABC} \qquad (b)$$

which contains again two triplets in the hexagonal stacking but in a configuration different from the sequence (a). The dis-

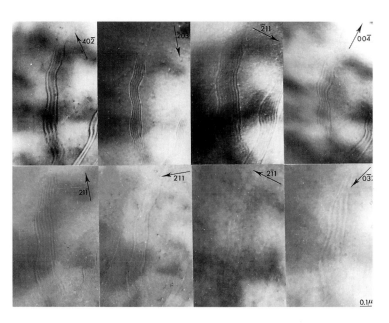

Figure 1-30. Residual fringe contrast due to relaxation along anti-phase boundaries in Ni$_3$Mo (Van Tendeloo and Amelinckx, 1974a).

placement vector is now $-(1/3)[111]$, i.e., the opposite of the previous one. This type of fault cannot be generated by a single glide motion; but it may result from the precipitation of a layer of interstitial atoms.

For some fundamental problems, the destinction between extrinsic and intrinsic faults is of importance (Hashimoto et al., 1962; Hashimoto and Whelan 1963; Gevers et al., 1963; Art et al., 1963). This information can be obtained from a single dark field image made in a well-defined reflection $g = [hkl]$. We then have $\alpha = 2\pi g \cdot R_0 = (2\pi/3)(h+k+l)$. Three kinds of reflections can be considered in the face-centered-cubic structure depending on whether or not $h+k+l = 3$ fold, 3 fold ± 1. It is clear that for $h+k+l = 3$ fold the fault contrast disappears, since $\alpha = 2\pi$. Reflections for which $h+k+l = 3$ fold -1 such as $\{200\}$, $\{2\bar{2}2\}$ and $\{440\}$ will be called "type A", whereas reflections with $h+k+l = 3$ fold $+1$, such as $\{1\bar{1}1\}$, $\{220\}$ and $\{400\}$ will be called "type B". A system-

atic analysis of all possible combinations for a given g vector pointing to the right is represented in Table 1-2. It becomes clear that for a given fault type and a given g vector the nature of the edge fringes in the dark field image is independent of the sense of inclination of the fault plane. A simple rule can be derived based on the dark field image: if the g vector, its origin being placed in the center of the fault image, points towards a bright fringe and the operative reflection is of type A, the fault is intrinsic. The conclusion changes if either the nature of the edge fringe or the class of the reflection changes. If the sense of inclination of the fault plane is also required, one needs in addition a bright field image, and the answer then follows directly from Table 1-2. In applying this method, one should be aware of the fact that the foil must be sufficiently thick so as to make sure that anomalous absorption is significant enough to determine the nature of the edge fringes.

Examples of application are shown in Figs. 1-31 and 1-32.

Table 1-2. Determination of the nature [intrinsic (I) or extrinsic (E)] of stacking faults in an f.c.c. crystal.

$\overrightarrow{g}$	B.F. A	B.F. B	D.F. A	D.F. B
E	D D	B B	B D	D B
	B B	D D	B D	D B
I	B B	D D	D B	B D
	D D	B B	D B	B D

The reflections giving rise to stacking fault fringes belong to two classes: A $\{200\}$, $\{2\bar{2}2\}$, $\{440\}$ and B $\{1\bar{1}1\}$, $\{220\}$, $\{400\}$. For each of these classes, for each slope and for a given sense of g, the nature of the edge fringes B (bright) and D (dark) is indicated. The table allows the nature of the SF to be determined from the dark field (D.F.) image only. For instance, if g belongs to class B and points to a B fringe in the D.F. image the fault is extrinsic (E), irrespective of the slope of the fault plane.

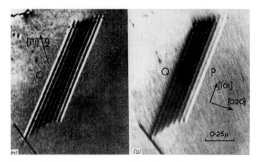

Figure 1-31. Intrinsic stacking fault in a low stacking fault energy Cu-Al alloy. Bright (a) and dark (b) field images are shown.

1.5.4.2 Domain Textures

Phase transformations accompanied by a decrease in space-group symmetry usually give rise to fragmentation of the crystal in domains whose structures are related by the symmetry elements lost during the

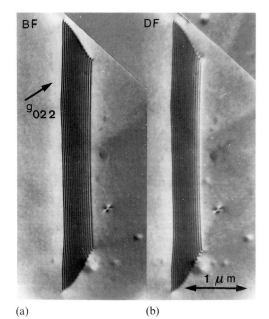

(a)   (b)

Figure 1-32. Extrinsic stacking fault in silicon formed by the precipitation of self-interstitials. (a) Bright field; (b) dark field (Courtesy de Veirman).

transformation. This is particularly true if the space group of the low-temperature product phase is a subgroup of that of the high-temperature parent phase (Van Tendeloo and Amelinckx, 1974b; Van Tendeloo et al., 1976).

The lost symmetry elements can either be *translations;* or *rotations.* In the first case the corresponding interfaces are called *translation interfaces;* they are characterized by a constant translation vector R_0 relating the structures in the two domains. Such interfaces are for instance out-of-phase boundaries and discommensuration walls. Stacking faults and crystallographic shear planes also belong to this category.

The interfaces corresponding with rotation symmetry elements are called *twins* or *domain boundaries.* It is convenient from the diffraction point of view to reserve the term "twin" to those cases where the lattices of the two components are so different that the corresponding g vectors of the two domains differ so much in orientation and/ or magnitude that they are not excited simultaneously.

The use of the term "domain boundary" will be restricted to the case where the lattices in the two parts are only slightly different so that the reciprocal lattice nodes belonging to the two domains are sufficiently close to be simultaneously excited, i.e., $g_2 - g_1 = \Delta g$ with $|\Delta g| \ll |g_1|$ or $|g_2|$. The distinction between twins and domain boundaries is obviously not strictly defined.

Finally, it is possible that the domains have a common lattice but differ nevertheless in their structure. In non-centrosymmetric crystals this is for instance the case for inversion domains in which the structures are related by an inversion operation. Also Dauphiné twins in non-centrosymmetrical α quartz, which are 180° rotation twins, belong to this family.

Different contrast phenomena can be exploited for imaging purposes. Domain textures can be revealed by means of interface contrast, by means of domain contrast or possibly by both simultaneously.

Domain contrast can arise from different origins. In the majority of the cases it is a slight misorientation of the lattices in the different domains that causes a Δg and hence also a difference in deviation parameter Δs for simultaneously excited reflections g_1 and g_2, which through the rocking curve (Fig. 1-22) correspond in general with a difference in scattered and transmitted intensity. Those reflections for which $\Delta g = g_2 - g_1$ (or $\Delta s = s_2 - s_1$) vanishes do not give rise to domain contrast.

Contrast may also arise from a difference in structure factor for the simultaneously excited reflections in the two domains. This may even be the case if the two domains have a common lattice, as for instance with Dauphiné twins (i.e., 180° rotation twin) in α quartz (Fig. 1-33).

Finally, in non-centrosymmetrical crystals inversion domains are revealed as a result of the violation of Friedel's law (which states that $I_{hkl} = I_{\overline{hkl}}$). In multiple beam situations, this law no longer strictly holds (Serneels et al., 1973). Dark field images made along a zone axis which in projection does *not* lead to a center of symme-

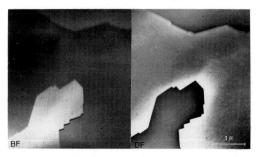

Figure 1-33. Domain contrast at Dauphiné twin boundaries in α quartz (Van Tendeloo et al., 1976).

try (i.e., along a non-centrosymmetric zone) will exhibit domain contrast if several reflections are excited simultaneously. The bright field image will *not* exhibit domain contrast as a result of this effect.

For translation interfaces, the lattices and structures in the two domains are related by a constant displacement vector R_0. As a result there is no domain contrast regardless of the diffraction conditions since the deviation parameters as well as the structure factors are always the same in the two domains. Even in multiple beam situations, no contrast is expected.

1.5.4.3 Diffraction Patterns of Domain Textures

Domain textures produce a composite diffraction pattern consisting of the superposition of the diffraction patterns of the separate domains; this is often sufficient to identify the lost symmetry elements using the fine structure of the diffraction spots.

In the superposition region of two domains, i.e., along an inclined interface, double diffraction phenomena often occur and may complicate the diffraction pattern. A dark field image made in such a double diffraction spot reveals the area of overlap, i.e., the inclined interface, as a bright area.

In order to simplify the interpretation, diffraction patterns can be made successively across single interfaces, separating two domains (Boulesteix et al., 1976; Manolikas et al., 1980 a, b).

The most commonly occurring interfaces are reflection twins and 180° twins. In the first case reciprocal space is as represented in Fig. 1-34 a; i.e., it exhibits a row of unsplit nodes through the origin and perpendicular to the mirror plane in real space. All other nodes are split along a direction parallel with the unsplit row, the

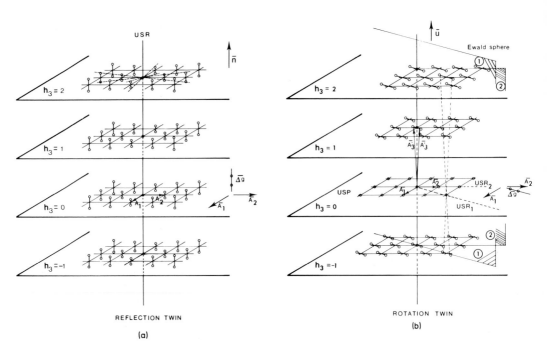

Figure 1-34. Reciprocal space of two types of twins in a centro-symmetrical crystal. (a) Reflection twin; (b) 180°-rotation twin (Boulesteix et al., 1976).

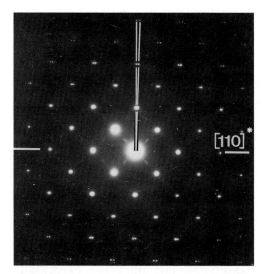

Figure 1-35. Diffraction pattern due to a reflection twin in $YBa_2Cu_3O_{7-\delta}$. Note the row of unsplit spots along the [110] direction. The weak elongated spots are due to two variants of the $2a_0$ structure (Zandbergen et al., 1987).

magnitude of the splitting being directly proportional with the distance to the unsplit row. The spot splitting is a direct measure for the twinning vector. Exploring reciprocal space by rotating the specimen around the unsplit row reveals diffraction patterns such as the one shown in Fig. 1-35.

The reciprocal space associated with a 180° rotation twin is represented in Fig.

1-34 b; there is an unsplit lattice plane perpendicular to the 180° rotation axis. All other nodes are split along the same direction parallel to the unsplit plane, the magnitude of the splitting increasing with the distance from the unsplit plane. Many sections of reciprocal space will look similar in the two cases. In order to differentiate the two cases, tilting experiments around two different rotation axes are required to make sure that an unsplit plane exists. Such an experiment is represented in Fig. 1-36 for the case of $MoTe_2$.

Special cases arise if the two domains have a common lattice, but differently oriented structures. Such a situation may occur if the symmetry of the structure is lower than that of the lattice; it exists for instance in δ-NiMo where the Bravais lattice is tetragonal, but the structure has a non-centrosymmetrical orthorhombic structure (Van Tendeloo and Amelinckx, 1973). As a result, 90° rotation twins around the pseudo-fourfold axis are possible as well as domains related by an inversion operation. The identification of the symmetry operations relating the different domains requires an "ad hoc" combination of bright and dark field images in various diffraction vectors. For inversion domains, the characteristic features, which can be used for identification purposes, have been de-

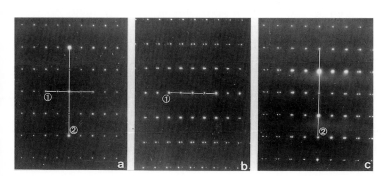

Figure 1-36. Tilting experiment in a $MoTe_2$ foil, proving the presence of an unsplit plane. (a) Untilted position; (b) tilted around the axis 1; (c) tilted around the axis 2 (Manolikas et al., 1979).

Figure 1-37. Domain contrast at inversion domains in the χ phase alloy Fe-Cr-Mo-Ti (Snijkers et al., 1972).

1.5.4.4 Inversion Boundaries in Non-Centro-Symmetrical Crystals

Planar interfaces separating non-centro symmetrical crystal parts related by an inversion operation require special consideration (Serneels et al., 1973). The two crystal parts have a common lattice and therefore the reflections g in one crystal part and $-g$ in the inversion related part are always simultaneously excited with the same s value. Assuming the validity of Friedel's law, which states that the modulus of the structure factors $|F_g|$ and $|F_{-g}|$ are the same even in a non-centrosymmetrical crystal, one does not expect any brightness contrast between inversion domains. According to the kinematic theory, this is strictly the case. Such a contrast difference is nevertheless observed (Fig. 1-37) under the appropriate experimental conditions, in particular in dark field images made under multiple beam situations along a zone axis which is not parallel with a twofold axis. Along a twofold axis (and *a fortiori* along a 4- or 6-fold axis) the projected structure is centro-symmetric and no contrast is observed. The contrast is due to the violation of Friedel's law under multiple-beam conditions.

For the identification of inversion boundaries, one can make use of these characteristics. If domain contrast is systematically absent in the B.F. image under all diffraction conditions, whereas domain contrast results in the D.F. image for multiple beam situations, one can conclude that the boundary is an inversion boundary.

Inversion boundaries are also revealed by interface contrast as α fringes, even in two-beam situations. The value of α is then equal to twice the phase angle of the active reflection. Since the crystal is non-centrosymmetrical, this phase angle can have

scribed in Serneels et al. (1973); they are summarized in Sec. 1.5.4.4. Inversion domains in the χ phase alloy in the system Fe–Cr–Mo–Ti are shown in Fig. 1-37.

The phase transition $\beta \rightarrow \alpha$ in quartz leads to a fragmentation of the crystal in α_1 and α_2 domains. The non-centrosymmetric pointgroup of β being *622* and that of α being *32,* the two α variants, called Dauphiné twins, are related by the lost symmetry element, i.e., by the 180° rotation about the threefold axis. The two variants have a common lattice and their reflections are therefore excited simultaneously. However, a number of common reflections have structure factors which differ for α_1 and α_2 in modulus, in phase, or in both. If the modulus is different, the domains exhibit structure factor contrast in bright as well as in dark field images. If the structure factors corresponding with the two reflections have the same modulus but a different phase α with respect to the same origin, the interfaces are imaged as α fringes and no domain contrast occurs. The non-centro symmetric character cannot be exploited in this case since the two variants α_1 and α_2 are not related by an inversion.

Figure 1-38. α fringes produced at the interfaces between inversion domains in the χ phase (Snijkers et al., 1972).

A concrete example is the contrast between Dauphiné twins in α quartz, which are related by a 180° rotation about the threefold axis of α quartz (Fig. 1-39).

A second example is the domain structure in the δ phase of NiMo, which has a tetragonal lattice but an orthorhombic structure. As a result, 90° rotation domains with a common lattice are formed;

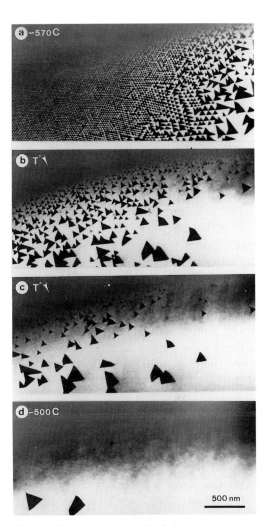

Figure 1-39. Regular network of triangular columnar domains of α_1 and α_2 Dauphiné twin variants in Quartz. At around 570 °C they form an incommensurately modulated structure (Van Tendeloo et al., 1976).

any value, and in particular it need not be 0 or π as in centro-symmetrical crystals (Fig. 1-38).

1.5.4.5 Structure Factor Contrast

The structure factors of simultaneously excited reflections in twin related crystal parts may be significantly different. Either the modulus or the phase of the two structure factors may be different. In fact, both may even be different. Structure factor domain contrast will result for all reflections for which the structure factors have a different modulus. If the structure factors have only a phase difference, the interfaces can be imaged as α fringes, but no domain contrast occurs under two-beam conditions. In order to derive the phase angle α, one writes the structure factors of the two domains with respect to the *same* origin. An expression of the form $F_2 = F_1 \exp(i\alpha)$ is then obtained.

they can be revealed by structure factor contrast (Van Tendeloo and Amelinckx, 1973).

Coherent precipitates may produce reflections which coincide with those of the matrix but the structure factor may be different for precipitate and matrix. Dark field images in such reflections will then reveal the precipitate by structure factor contrast.

1.6 Dislocation Contrast

1.6.1 Intuitive Considerations

Experimentally, it is found that in diffraction, contrast dislocations are imaged usually as dark lines on a brighter, uniform background. Using the weak beam method they appear as bright lines on a darker background.

The diffraction contrast at dislocations can qualitatively be understood on intuitive grounds (Amelinckx, 1964). Consider the foil of Fig. 1-40; it contains an edge dislocation in E. The set of lattice planes used for imaging has been represented. Let the foil be oriented in such a way that the Bragg condition for reflection by the set of lattice planes shown in Fig. 1-40 is only approximately satisfied in the perfect part of the foil, far away from the dislocation. Such a situation can easily be envisaged since the specimen is a thin foil, and as a consequence Bragg's condition is appreciably relaxed, the node points in reciprocal space having become *relrods* perpendicular to the foil plane. It is therefore possible for diffraction to occur over a certain angular range around the exact Bragg position. Imagine a situation where the reciprocal lattice node corresponding with the set of lattice planes is outside Ewald's sphere ($s < 0$). The local rotation of the lattice planes on the left of the dislocation at E_1 will then bring this set of planes closer to the exact Bragg orientation, and therefore, the locally diffracted beam will be more intense than in the perfect part of the foil.

On the contrary, to the right of the dislocation at E_2, the local rotation of the lattice planes does not allow Bragg's condition to be satisfied to the same degree, and hence, the diffracted beam will be weaker than in the perfect part of the foil. The relative intensities of the diffracted beam in different parts of the foil are represented schematically in Fig. 1-40 by lines of different thicknesses.

The part of the beam which is not diffracted is clearly transmitted since electrons are assumed not to be lost and hence the transmitted beam exhibits a complementary variation in intensity to that of the scattered beam.

Selecting the diffracted beam in the microscope and magnifying the corresponding diffraction spot will produce a map of the intensity distribution in this beam. This map will show a lack of intensity to the right of the dislocation and an excess of intensity over the background in E_1. In this approximation, the dislocation will thus be imaged as a bright–dark line pair against an intermediate background. This image is called a *dark field* image. Selecting the transmitted beam and magnifying the direct beam spot, a similar intensity map allows to be produced. The dislocation will now be imaged in a complementary way, bright and dark lines being interchanged as compared to the dark field image. This image is now called a *bright field* image. This is clearly an approximation which is only qualitatively correct for large s and thin foils. This will be corrected in the dynamic theory. The model is nevertheless useful for determining the image side, as we shall see below. Bright and dark field images are

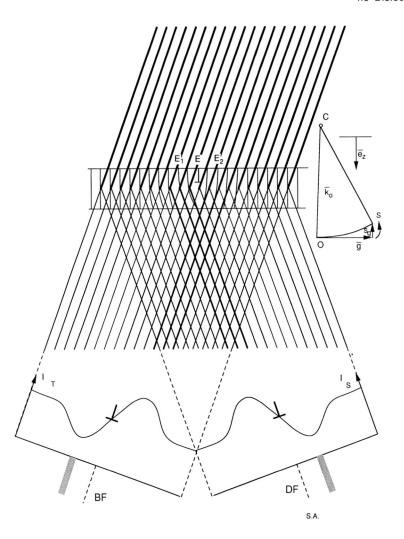

Figure 1-40. Intuitive model illustrating the image formation at an edge dislocation E, according to the kinematic theory (Amelinckx, 1964).

thus strongly magnified diffraction spots, the intensity distribution being the "image".

The formation of an "image" is possible because electron diffraction is largely a "local" phenomena. Electrons only sense a narrow column of material because the foil is thin, the Bragg angles are small and electron scattering is strongly peaked in the forward direction.

It is clear that a defect image will only be produced if diffraction occurs by a set of lattice planes deformed by the presence of the defect. Displacements parallel to the diffracting lattice planes do not give rise to contrast effects. Since to a first approximation all displacements around a dislocation are parallel to the Burgers vector, the criterium for the absence of contrast is $\boldsymbol{g} \cdot \boldsymbol{b} = 0$.

The same type of reasoning can be used to show that screw dislocations also produce a line image. The presence of a screw dislocation transforms the succession of lattice planes perpendicular to the dislocation line into a helical surface. To the left and to right of the dislocation line, the lattice planes are thus slightly inclined in opposite senses. The same type of reasoning

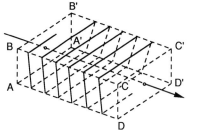
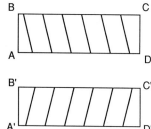

Figure 1-41. Displacement field around a screw dislocation.

as applied in the case of the edge dislocation allows one to conclude that also a screw dislocation can be imaged as a bright–dark line (Fig. 1-41).

Note that the dark line that images the dislocation line in the bright field image is not produced at the position of the dislocation line, but slightly displaced away from it. We shall call this the *image side.*

Changing the orientation of the foil so as to make $s > 0$ in the part of the foil far away from the dislocation core and using the same reasoning as we made above it is found that the image side has changed. This also occurs if, for the same sign of s, g is changed into $-g$, i.e., if reflection takes place from the other side of the lattice planes. The image side also changes if the dislocation changes sign, i.e., if $b \rightarrow -b$.

It turns out that the image side is determined by the sign of the quantity $p = (g \cdot b)s$, $(s \neq 0!)$ involving all three parameters. Provided the Burgers vector of a dislocation line is defined by the FS/RH convention (Bilby et al., 1955), the image is at the right on a positive print – as viewed from below the specimen – when $p > 0$, and it is at the left if $p < 0$, the dislocation line being oriented from bottom to top on the print. When using this criterion, one must be sure that all operations between the diffraction by the foil and the final positive print are properly taken into account. It is often more straightforward to use the intuitive reasoning sketched above.

An edge dislocation parallel to the foil surfaces in a plan parallel foil causes a slight misorientation of the two crystal parts separated by the dislocation, which can be measured by the displacement of the Kikuchi lines. This orientation difference increases as the foil thickness decreases. Its magnitude, which depends moreover on the position of the dislocation in the foil, is at a maximum if the dislocation is in the central plane of the foil (Siems et al., 1962 a, b). In that case $\theta \simeq b/D$ ($D = $ foil thickness). This effect produces a small intensity difference between the two crystal parts separated by the dislocation. The sense of the "buckling" depends on the sign of the dislocation (Fig. 1-42 a). This therefore enables one to determine the sign of the dislocation. The effect is often observed in thin specimens of layer structures.

Dislocation lines parallel to the electron beam produce diffraction contrast images as well (Ruedl et al., 1962 a, b). For a pure screw dislocation parallel to the incident beam, one would not expect any contrast since all displacements are parallel to b, and thus for the active g vectors $g \cdot b = 0$. However, it was shown by Eshelby and Stroh (1951), that close to the emergence points of the dislocation in the foil surfaces, elastic relaxation takes place, and the lattice planes parallel to the screw dislocation line do not remain flat but become slightly helically twisted close to the surface. This

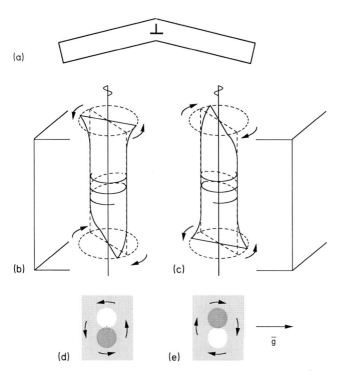

Figure 1-42. Surface relaxation effects at dislocations: (a) The presence of an edge dislocation causes buckling of the foil; (b), (c) helical surface relaxation around the emergence points of a screw dislocation; (d), (e) black–white dot contrast at the emergence points of screw dislocations.

deformation produces an image consisting of a black–white dot pair even if the extinction criterion $g \cdot b = 0$ is fulfilled. The line joining the dots is perpendicular to g.

Depending on the sign of the screw dislocation the surface relaxation produces a right- or left-handed helical twist and the dot pair is black–white or white–black (Fig. 1-42 b–e). As a result, this effect allows the sign of the screw dislocation to be deduced. Figure 1-43 shows screw dislocations in this configuration.

In addition an edge dislocation along the z-axis viewed end-on produces diffrac-

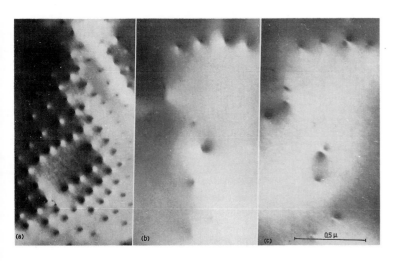

Figure 1-43. Dislocations viewed end on in (a) platinum; (b), (c) uranium dioxide.

tion contrast, because in the vicinity of the dislocation the local interplanar spacing is modified, i.e., g is changed in length as well as in orientation, which in turn causes a change in the local deviation parameter s. Along a column parallel to the dislocation line the s parameter remains constant (i.e., independent of z), but it becomes a function of x and y, and hence the scattered and transmitted intensities along a column depend on the location of the column, i.e., an image is produced. The contours of equal s value or of equal intensity are shown in Fig. 1-44; they reflect the symmetry of the strain field around an edge dislocation (Mannami, 1960, 1962).

An edge dislocation with its Burgers vector parallel to the incident beam would not produce contrast according to the simple $g \cdot b = 0$ criterion, the actual extinction criterion being $g \cdot R = 0$. However, this is an oversimplification since the displacement field of an edge dislocation is not uniaxial along b but also contains a component perpendicular to the glide plane, i.e., to b, producing a "bump" in the glide plane extending towards the supplementary half plane (Fig. 1-45). For columns close to the dislocation core, the s value thus varies along the column, and contrast is produced (Howie and Whelan, 1960).

Edge dislocations occur in this configuration in prismatic loops. The displacement field $R(r)$ now has a radial component $R_{\parallel}$, which is either inward or outward, depending on whether the loop is of the vacancy or of the interstitial type (Fig. 1-46).

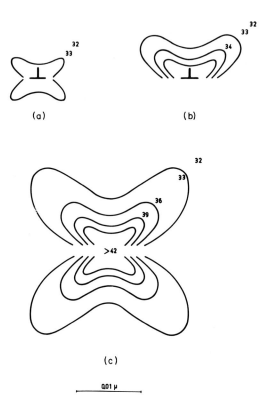

(a)   (b)

(c)

0.01 μ

Figure 1-44. Strain pattern around an edge dislocation viewed end on (Mannami, 1960).

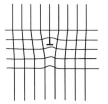

Figure 1-45. Schematic representation of the strain field around an edge dislocation. Note the "bump" in the glide plane.

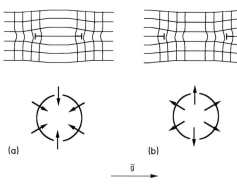

(a)   (b)

$\bar{g}$

Figure 1-46. Schematic representation of the radial strain field around prismatic loops and of the corresponding images: (a) Interstitial loop; (b) vacancy loop. Note the *line of no contrast* perpendicular to the active g vector.

For a diffraction vector g in the foil plane, the dot product $g \cdot R$ with the normal component $R_\perp$ will be zero all along the loop for any g parallel with the loop plane. On the other hand, $g \cdot R_\parallel$ will vary along the loop and it will be zero only along two diametrically opposed segments, as represented in Fig. 1-46 and observed in Fig. 1-47. The contrast will only disappear along these two segments; the *line of no contrast*, joining these two segments is perpendicular to the active g vector. The rest of the loop will exhibit contrast even though $g \cdot b = 0$.

Somewhat surprisingly, it is found that parallel dislocations with the same Burgers vector do not necessarily exhibit the same contrast if they are sufficiently close; one of them usually has a higher contrast (i.e., is marked by a darker line) than the other(s) (Delavignette et al., 1966). Which one will show more contrast depends on the sign of s and g. This behavior is due to the overlap of the strain fields of the two (or more) dislocations; it is the *combined* strain field that determines the image. This effect is especially striking in triple ribbons of partials in face-centered-cubic metals with low stacking fault energy and in graphite. An analytical theory, based on the kinematic approximation sufficiently accounts for these observations (Delavignette et al., 1966).

1.6.2 Image Profiles: Kinematic Theory

Image profiles can be calculated quantitatively in the framework of the kinematic theory (Hirsch et al., 1956, 1960; Whelan, 1958–1959) by inserting the adequate expression for the displacement field $R(r)$ in Eq. (1-31) and integrating along columns situated along lines normal to the dislocation line).

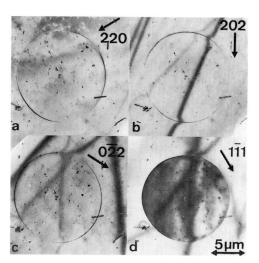

Figure 1-47. Frank dislocation loops in silicon. A *line of no contrast* perpendicular to g is present. Note the deformation of the extinction contour on crossing the dislocation.

A screw dislocation is assumed to be oriented along the y direction parallel to the foil plane and at a depth d behind the entrance face of the foil with thickness t (Fig. 1-48). According to isotropic elasticity theory, the displacement field is then of the type given by Eq. (1-24); for the particular geometry adopted here

$$\varphi = \arctan\left[(z - d)/x\right] \qquad (1\text{-}162)$$

The image profile is then obtained by integrating Eq. (1-31) from 0 to t along columns corresponding with various values of x. The integrals can be obtained analytically in terms of non-elementary func-

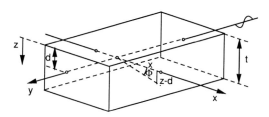

Figure 1-48. Geometry of screw dislocation in foil; notations used.

tions, also for edge dislocations (Gevers, 1962 a–c). However, the results are of qualitative value only since the foils are usually too thick for the kinematic theory to apply quantitatively or even qualitatively; moreover, best contrast is obtained for small values of s, where the kinematic theory breaks down.

1.6.3 Image Profiles: Dynamic Theory

According to the dynamic theory, including anomalous absorption, the set of coupled linear differential equations in Eqs. (1-135) and (1-136), where s_g is replaced by $s_{eff} = s_g + (d\alpha/dz)$, must be integrated with the adequate expression for α, given by (Fig. 1-48)

$$\alpha = (\boldsymbol{g} \cdot \boldsymbol{b}) \arctan [(z-d)/x] \qquad (1\text{-}163)$$

for a screw. $\boldsymbol{g} \cdot \boldsymbol{b} = n$ is called the order of the image; it is an integer for perfect dislocations since $\boldsymbol{b}$ is a lattice vector and $\boldsymbol{g}$ is a reciprocal lattice vector. In most cases, analytical solutions are difficult, if not impos-

sible, to obtain, but numerical results for a number of representative dislocation configurations are available (Howie and Whelan, 1961, 1962; Hirsch et al., 1962; Whelan and Hirsch, 1957; Whelan, 1978).

When the dislocation is inclined with respect to the foil surface, as is often the case, the parameter d in Eq. (1-163) is an additional variable, and profiles have to be computed for a set of d values. Fast methods, based on the linearity of the set of equations, Eq. (1-135) and Eq. (1-136), have been developed to obtain many such profiles so that a 2D map of the intensity distribution comparable to that recorded on a micrograph, can be printed by using characters of different darkness (Fig. 1-49) or by simulating the results on a TV monitor. For a description of these numerical methods we refer to Humble (1978) and Head (1967).

We shall now survey the main results that may be relevant for the interpretation of micrographs.

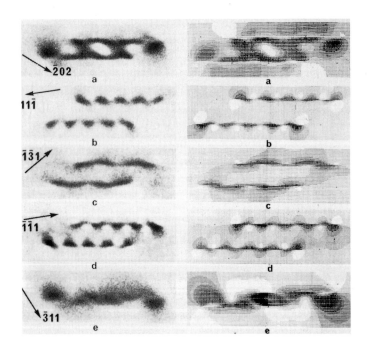

Figure 1-49. Example of computed dislocation images (right) compared with experimental images (left) (Humble, 1978).

1.6.3.1 Images of Screw Dislocations

The image for $n = 1$ and $s = 0$ with the screw dislocation parallel to the foil plane and in the foil center, exhibits a single dark peak, located very close to the position of the dislocation, in the bright field image as well as in the dark field image (Howie and Whelan, 1961, 1962). This is clearly in contradiction with the intuitive result, which is essentially based on kinematic diffraction and leads to complementary images. The non-complementarity or rather similarity of bright and dark field images is a result of anomalous absorption in thick crystals ($t = 5$ to $10\,t_g$). The peak width is of the order of $0.3-0.4\,t_g$. For $s \neq 0$ and large values of s, the image peak is displaced away from the dislocation position in a sense which is independent of the depth position, but which changes with the sign of s. The sense of the image shift is the same as that predicted by the intuitive reasoning of Sec. 1.6.1 for the B.F. image.

Since along an inclination extinction contour, s changes its sign, a dislocation image with $n = 1$ will change from one side of the line to the other on crossing an extinction contour in the manner represented schematically in Fig. 1-50. For $s = 0$ and for small values of s the image shift de-

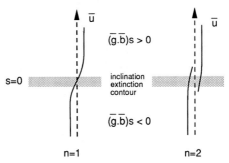

Figure 1-50. Behavior of the dislocation image on crossing a bent extinction contour: (a) $n = 1$; (b) $n = 2$.

pends on d, and especially close to the surfaces the image side switches with every $(1/2)\,t_g$ change in depth; close to the center of a thick foil ($\sim 8-10\,t_g$), the image shift is small (Fig. 1-51). As a result, inclined dislocations in thick foils will exhibit *"oscillating or alternating contrast"* close to foil surfaces (b) but not in the central part (a) (Fig. 1-52). As a consequence of anomalous absorption, the oscillations will be *in phase* close to the entrance face in corresponding bright and dark field images but they will be in *anti-phase* close to the exit face of the foil. This effect allows one to determine which end of the dislocation image corresponds to the vicinity of which foil surface. Since the oscillation period corresponds

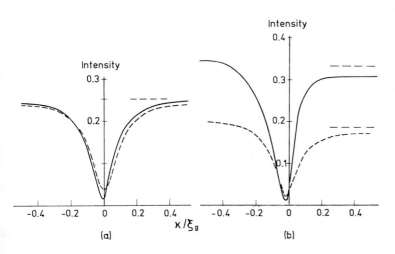

Figure 1-51. Computed image profile for a screw dislocation in the middle of the foil $t = 8\,t_g$, $n = 1$, $\tau_g/t_g = 10$: (a) $s = 0$; (b) $s\,t_g = 0.3$. The full lines represent the bright field image, whereas the broken curves represent the dark field image (Howie and Whelan, 1961, 1962).

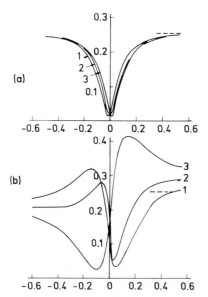

Figure 1-52. Computed bright field images for a screw dislocation parallel to the foil plane for $s = 0$, $n = 1$, $t = 8\,t_g$, $\tau_g = 10\,t_g$. (a) The curves 1, 2, 3 correspond to $d = 4, 4.25, 4.50\,t_g$; (b) the curves 1, 2, 3 correspond to $d = 7.25, 7.50, 7.75\,t_g$. From Howie and Whelan (1961, 1962).

with a depth change of one effective extinction distance, such images allow one to deduce the foil thickness.

In Fig. 1-53 the dislocations exhibit a pronounced zig-zag contrast close to the surface of a wedge shaped crystal of the layer structure SnS_2. In the lower right corner the foil thickness is uniform, and the dislocations exhibit a simple line contrast. The relation between the thickness extinction contours and the oscillation period of the dislocation contrast can be observed directly in the wedge-shaped part. The occurrence of "dotted" images at inclined screw dislocations also for $s = 0$ and $n = 1$ can be understood intuitively by noting that the top and bottom parts of a column passing through the dislocation core are related by a phase jump over $n\pi$ at the level of the core. Along these columns, the same variation in intensity will take place for $n = 1$ as for an inclined stacking fault with $\alpha = \pi$. The central part of the dislocation image will thus fluctuate in intensity with a period t_g.

This reasoning can be generalized by considering columns to the right and to the left of the projected dislocation line (Fig. 1-54). Depending on their distance to this line the phase change, which occurs at the level of the dislocation core, will vary for instance from 0 far to the left to 2π far to the right. At some position close to the projection of the dislocation, the phase change will then be $2\pi/3$ on the left and $4\pi/3 = -2\pi/3$ (mod 2π) on the right. For these columns, the analytical expressions for the fringe profiles of stacking faults with $\alpha = \pm 2\pi/3$ can be applied to explain dislocation contrast qualitatively. It is clear for instance that in the B.F. image the intensity profile along a line left of the dislocation will be inverted on the right side since the sign of α is changed; as a result *alternating* contrast results.

Whether *dotted* or *alternating* contrast will occur in foils of moderate thickness

Figure 1-53. Oscillating contrast at dislocations emerging in the surface in a wedge shaped crystal of the layer structure SnS_2.

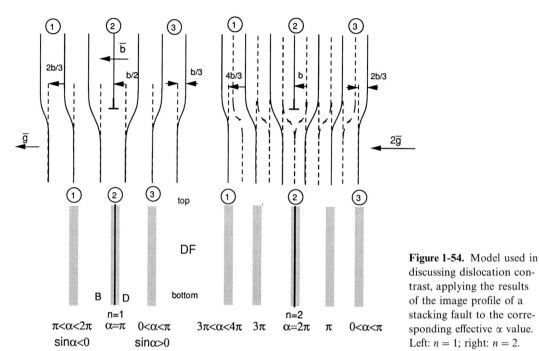

Figure 1-54. Model used in discussing dislocation contrast, applying the results of the image profile of a stacking fault to the corresponding effective α value. Left: $n = 1$; right: $n = 2$.

depends on the thickness of the foil (Howie and Whelan, 1961, 1962). For instance, in a foil with $t = 3t_g$, $s = 0$ and $\tau_g = 10 t_g$ the B.F. image is dotted and the dark field image alternating, whereas for a foil with $t = 3.5 t_g$ the reverse is true.

For $n = 2$ and $s = 0$, the image consists of two dark peaks, one on each side of the actual dislocation position. The two peaks differ in strength, their relative strength alternating with the depth in the foil with a period equal to t_g. These features are exhibited by the curves of Fig. 1-55 a: they lead to zig-zagging, or *oscillating*, images at inclined dislocations. If $s \neq 0$, the two peaks become strongly asymmetrical (Fig. 1-55 b). The type of asymmetry depends again on the sign of s; in practice one dark line is often produced, except for $s \simeq 0$. As a result, a dislocation image with $n = 2$ will, on intersecting an inclination extinction contour, behave as represented in Fig. 1-50. The image behavior on intersecting

an extinction contour thus allows n to be determined, which in turn allows one to deduce the projection of b on the active diffraction vector g and hence to determine the length of b, provided its direction is known. Note that for $n = 2$ the intensity of the bright central peak is in each case equal to the intensity prevailing in the perfect part of the foil. This can be understood by noting that for the column at the origin, i.e., passing through the dislocation core, the displacement just below the core will be a lattice vector (i.e., $\alpha = 2\pi$). Hence the column behaves as perfect material.

For $n = 1$ we have shown that this column behaves as if it contained a stacking fault with $\alpha = \pi$ at the position of the dislocation core.

1.6.3.2 Images of Edge and Mixed Dislocations

For a mixed dislocation along a direction given by its unit vector u and parallel

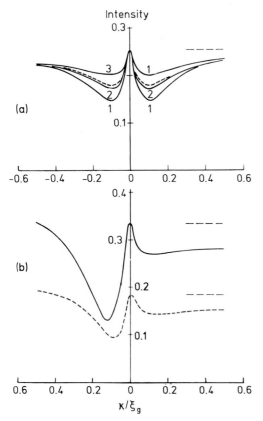

Figure 1-55. Computed image profiles of a screw dislocation for $n = 2$, $t = 8\,t_g$: (a) $s = 0$; (b) $s\,t_g = 0.3$. The profiles (1), (2) and (3) in (a) refer to $d = 4$, 4.25 and $4.50\,t_g$, respectively. Full lines are bright field images and dotted lines dark field images (Howie and Whelan (1961, 1962)).

to the foil plane, i.e., for the geometry represented in Fig. 1-56 the displacement field of the mixed dislocation, as given by isotropic elasticity theory, can be written as (Friedel, 1964)

$$R = \frac{1}{2\pi}\left\{b\,\varphi + \frac{b_e}{4(1-v)}\sin 2\varphi + \right.$$
(1-164)
$$\left. + \left[\frac{1-2v}{2(1-v)}\ln|r| + \frac{\cos 2\varphi}{4(1-v)}\right](b \times u)\right\}$$

v is Poisson's ratio ($v \simeq 1/3$), $\varphi = \alpha - \gamma$ (Fig. 1-56), b is the total Burgers vector and b_e the edge component of b. It is clear that for

the pure screw $b \times u = 0$ and $b_e = 0$, and the expression reduces to Eq. (1-163) for a pure screw. The term in $b \times u$ describes a displacement perpendicular to the slip plane, which is the plane determined by b and u, and which forms an angle γ with the foil plane. The parameter $p = (g \cdot b_e)/n$ describes the extent of edge character of the dislocation: $p = 0$ for pure screws and 1 for pure edges.

Images were computed for a mixed dislocation with slip plane parallel to the foil plane (i.e., for $\gamma = 0$) for $n = 1$, $s = 0$, $t = 8\,t_g$, $d = 4\,t_g$; i.e., the dislocation occupies the center of the foil and for different values of p. It turns out that the image of a pure edge dislocation is wider than that of a pure screw. The narrowest image corresponds with $p = -1/2$ and the widest with $p = 1/2$, i.e., with 45° mixed dislocations. The full widths vary between 0.3 and $0.8\,t_g$ (Fig. 1-57) (Whelan, 1978). The pure edge dislocation for which $g \cdot b = 0$ may nevertheless produce contrast as a consequence of the "bump" in the glide plane, i.e., because of the term in $b \times u$ in the displacement function. Such a situation arises for instance when a prismatic Frank loop is situated in the foil plane and g vectors parallel to the loop plane are used for imaging. Total extinction will only occur if $(b \times u) \cdot g = 0$, which is the case if u is parallel to the g vector. Closed Frank loops will thus produce contrast except for those dislocation segments which are parallel to the

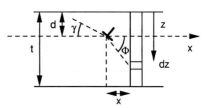

Figure 1-56. Illustration of the notation used in describing the displacement field of an edge dislocation in a general orientation.

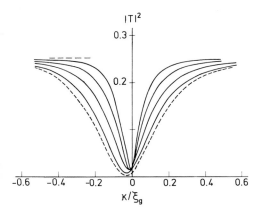

Figure 1-57. Computed bright field images of mixed dislocations for different values of p. In order of increasing width, the p values are -0.5, 0, 0.5 and 1.0. $g \cdot b = 1$, $t = 8\, t_g$, $d/t_g = 4$, $s = 0$, $\tau_g = 10\, t_g$. The broken curve is for $p = 1$ and $\tan \gamma = 1$ (Howie and Whelan, 1961).

active g vector, i.e., there is a *line of no contrast* perpendicular to g connecting the two loop segments which are parallel to g and hence out of contrast (Fig. 1-47). If the loop is a Frank loop, i.e., with b perpendicular to the loop plane, a line of no contrast is formed for *all* g vectors parallel to the loop plane. If the loop has an inclined Burgers vector, there will only be one g vector (and its negative) in the loop plane for which there is a *line of no contrast*. If on the other hand a loop exhibits *no* line of no-contrast for even a single g vector in the loop plane it is *not* a Frank loop. These considerations are used when determining the character of dislocation loops.

Changing the sign of x in Eq. (1-163) changes the sign of φ; however in Eq. (1-164) for the displacement field, the term in $b \times u$ only depends on x through $\cos 2\varphi$, which is an even function of φ, and on $\ln |r|$, which is thus also an even function of x; we can conclude that the image profile is symmetrical in x.

From the computed profiles of Fig. 1-58 one can conclude that for certain values of d the image may exhibit two peaks.

1.6.3.3 Partial Dislocations

For partial dislocations the Burgers vector is not a lattice vector, and therefore $g \cdot b$ is no longer necessarily an integer. For Schockley partials in the cubic-close-packed structure $b = (1/6)[\bar{2}11]$, the value of $n = (1/6)[-2h + k + l]$ and hence n becomes a multiple of $1/3$.

A partial dislocation forms the border of a stacking fault; the line image therefore separates two areas which generally have a different brightness. Images have been computed for $n = \pm 1/3$; $\pm 2/3$ and $\pm 4/3$ (Fig. 1-59). For $n = \pm 1/3$ the image is not observable; it would consist of a broad weak bright line, constituting the continuous transition between the brightness level on the fault side and that on the perfect side. For $n = \pm 2/3$ the image would be a dark line, but its visibility is still low. Images with $n = \pm 4/3$ should be as visible as ordinary dislocations. Partial dislocations

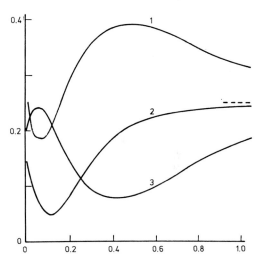

Figure 1-58. Image profiles for an edge-dislocation parallel to the foil plane and with Burgers vector perpendicular to it. $t = 8\, t_g$, $\tau_g = 10\, t_g$ (numbered 1, 2, 3 from top to bottom). Curve (1) $d/t_g = 7.75$, $m = \frac{1}{4}$; curve (2) $d/t_g = 4.00$, $m = \frac{1}{4}$; curve (3) $d/t_g = 7.75$, $m = -\frac{1}{4}$; $m = (1 - 2\nu)/4(1 - \nu)$ (ν: Poisson's ratio). From Howie and Whelan (1961, 1962).

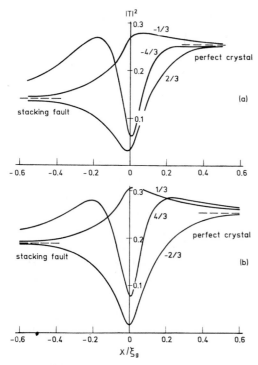

Figure 1-59. Computed bright field image profiles for partial dislocations: (a) $t = 6t_g$, $s = 0$, $\tau_g = 10t_g$, $d = 1.7t_g$; (b) $t = 6t_g$, $s = 0$, $\tau_g = 10t_g$, $d = 1.2t_g$. The fractions indicate the n values (Howie and Whelan (1961, 1962)).

of the Frank type behave as the loops discussed in Sec. 1.6.3.2.

1.6.3.4 Small Defects

Generally speaking the symmetry properties of images are directly related to the symmetry properties of the displacement field combined with the diffraction symmetry following from anomalous absorption.

For instance, spherical inclusions with elastic properties (Ashby and Brown, 1963), which differ from those of the surrounding material, give rise in an isotropic matrix to a radially symmetric displacement field, which can be represented as $R = \varepsilon r_0^3 r/r^3$, with $r \geq r_0$ as the radius of the precipitate. Inside the precipitate, $R = \varepsilon r$.

In an isotropic matrix $\varepsilon = (2/3)\delta$, where δ is the misfit between the lattices of inclusion and matrix. As in the case of a closed prismatic loop, there will be a *line of no contrast*, which is the geometrical locus of the points for which g is perpendicular to R (Fig. 1-60). The image consists of a pair of lobes separated by this line of no contrast, which is a symmetry line in the B.F. image, but a line of antisymmetry in the D.F. image. From the symmetry properties of the dark field image in sufficiently thick foils, and for inclusions close to the foil surfaces, the sign of ε, and even an estimate of its magnitude can be derived by means of a simple rule.

We first note that bright (and dark) field images of defects situated close to the entrance face are similar, whereas the images are pseudo-complementary when situated close to the exit face. The dark field images are similar whether close to the entrance face or close to the exit face; that is, there is a bright and a dark lobe, but the relative orientation of the vector l connecting the dark with the bright lobe with respect to the sense of g depends on the sign of ε. If $\varepsilon > 0$ (interstitial type), the g vector points toward the dark lobe (on a positive print!) for a defect close to any surface, but toward

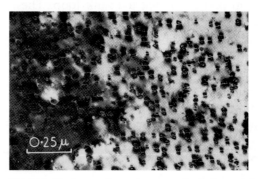

Figure 1-60. Image formation at spherical precipitates of cobalt in a copper-cobalt alloy. The image consists of a pair of lobes separated by *a line of no contrast* perpendicular to g (Ashby and Brown, 1963).

the bright lobe for $\varepsilon < 0$ (vacancy type). For a detailed discussion, we refer to Essmann and Wilkens (1964), Rühle et al. (1965), Rühle (1967), and Chik et al. (1967).

1.6.3.5 Qualitative Image Profiles

It is often useful to be able to predict even qualitatively the bright–dark characteristics of defects knowing their displacement field and the active reflections. Such intuitive considerations are only valid for the *dark field image* of defects situated *close to the exit surface* of the foil. The reasons for this restriction are the following: The intuitive considerations are essentially kinematic and are thus only valid for a very thin foil. If the defect or the relevant part of the defect is located close to the exit surface, it sees an incident beam produced by dynamic scattering in the preceding, perfect part of the foil, and it then scatters this beam kinematically, provided the bottom layer behind the defect it sufficiently thin [e.g., $(1/2)t_g$]. The kinematic image which is essentially formed by the scattered beam can be deduced in fact as a *dark field image*.

Intuitive reasoning serves to locate the columns for which $\sin\alpha > 0$ and those for which $\sin\alpha < 0$, where α is the integrated phase shift introduced by the defect between top and bottom part of the column. Referring to the symmetry properties of stacking fault profiles in sufficiently thick foils, we can then conclude which areas of the image will be bright and which ones dark with respect to the background brightness. Of course the details of the image, especially in the center of the foil, cannot be deduced in this way, but for a preliminary interpretation this type of reasoning may be quite useful.

A simple principle allows a qualitative discussion of defect images. The scattered intensity leaving the exit face of a column is determined by the thickness of the foil, i.e., by the length of the column and by the integrated phase difference, introduced by the presence of the defect between the top and bottom end of the column. Even for a complicated displacement field, one can approximate this phase difference from the geometry of the lattice planes in the vicinity of the defect or from the dot product $g \cdot R$. We apply this method to two examples.

(i) Spherical Inclusions

We consider as a first example the image predicted for a spherical inclusion with the spherically symmetrical displacement field presented in Fig. 1-61 ($\varepsilon > 0$). The *line of no contrast*, along which $g \cdot R = 0$ separates two regions one in which $g \cdot R > 0$ and one in which $g \cdot R < 0$. Since $|R|$ is small compared to a lattice vector, we will have $\sin\alpha > 0$ if $g \cdot R > 0$ and $\sin\alpha < 0$ if $g \cdot R < 0$. The sign of $\sin\alpha$ for a given column can be deduced by considering the

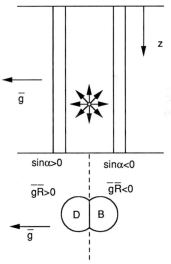

Figure 1-61. Schematic representation of the image formation at a spherical inclusion.

amplitude–phase diagram representing the kinematic amplitude due to this column:

$$\int_0^{z_0} \exp\left[2\pi i(sz + \boldsymbol{g}\cdot\boldsymbol{R})\right]dz$$

If $\boldsymbol{g}\cdot\boldsymbol{R}$ is positive for all values of z along the column, there is a net positive phase difference smaller than π between the top and bottom of the column, and hence $\sin\alpha > 0$. The fastest phase shift occurs at the level of the inclusion. The contrast at the exit end of the column is then of the same nature as that of a stacking fault by assuming the net phase change α_{eff} to occur at the level of the inclusion. The value of α_{eff} decreases with increasing distance from the inclusion. Using the symmetry properties of stacking fault fringes in sufficiently thick foils allows one to deduce the nature of the D.F. images of this kind of defect. The last fringe in the dark field image of a stacking fault is dark for $\sin\alpha > 0$. We thus find that $\boldsymbol{g}$ points toward the dark lobe for an inclusion close to the back surface, in agreement with the numerical results of the dynamic theory (Wilkens, 1978). The model also explains the periodic interchange of bright and dark lobes with the depth position of the inclusion.

(ii) Inclined Edge Dislocation

For an inclined dislocation one can approximate the effect of the displacement field by that due to a set of narrow strips of stacking fault situated at the same level in the foil as the dislocation core and parallel with the inclined dislocation line (Fig. 1-54).

The strips have a width equal to the size of an integration column. Along a given strip, the integrated phase change that is supposed to occur abruptly at the level of the stacking fault (i.e., of the core) is assumed to be a constant. An effective α_{eff}

value can be deduced from the integrated phase shift by the following expression valid in the kinematic approximation:

$$\int_{-z_1}^{z_2} \exp\left[i\,\alpha(z)\right]\exp\left(2\pi i\,s_g z\right)dz =$$
$$= \int_{-z_1}^{0} \exp\left(2\pi i\,s_g z\right)dz +$$
$$+ \exp\left(i\,\alpha_{eff}\right)\int_{0}^{z_2} \exp\left(2\pi i\,s_g z\right)dz$$

Geometrical considerations often allow one to find α_{eff} for a few special points, which will generally make it possible to have a qualitative idea of the profile.

This is a consequence of the invariance of the displacement field of a dislocation for a translation parallel to the line. For instance, for the strip coinciding with the dislocation core and for a $\boldsymbol{g}$ vector such that $n = 1$, the effective α value is π and if $n = 2$, it is $\alpha = 2\pi$. When $n = 1$, a strip close to the dislocation core on one side will have $0 < \alpha < \pi$ (i.e., $\sin\alpha > 0$), and on the other side, $\pi < \alpha < 2\pi$ (i.e., $\sin\alpha < 0$).

The qualitative behavior of the image profiles along these strips will resemble that deduced from the analytical expressions for a stacking fault Sec. 1.5.3. The profiles vary from strip to strip since the effective α value varies, and they thus describe a two-dimensional map, which is the image of the inclined dislocation. The use of an effective α value, which is the integrated phase change along a column, is justified by the fact that the intensity at the column exit is mainly determined by the phase difference between the top and bottom part and does not depend sensitively on the detail of its variation along the column. Rapid variation is in any case limited to the neighborhood of the dislocation core. As an example, we consider the inclined pure edge dislocation in an f.c.c. metal (Fig. 1-54). We assume $\boldsymbol{g}$ as well as $\boldsymbol{b} = (1/2)[110]$ to point to the left, and we further assume that $g \cdot b \equiv n = 1$; this means that $1/2(h+k) = 1$ and the active reflection g is 200 (or 020). Column (2) through the origin then suffers a lateral shift of $b/2$ at the level of the dislocation core, i.e., $\alpha = 2\pi g \cdot b = \pi$. The brightness variation of the core image will thus follow the same profile as that for a stacking fault with $\alpha = \pi$. Along column (3) the lateral shift is $(1/3)b$ and hence $\alpha = 2\pi/3 (\sin\alpha > 0)$, whereas along columns (1) the lateral shift is $(2/3)b$ and hence $\alpha = 4\pi/3 \equiv -(2\pi/3)(\mathrm{mod}\,2\pi)$ (i.e., $\sin\alpha < 0$). With reference to the stacking fault image profiles we then find that in the D.F. image close to the exit face there is a bright area to the left and a dark area to the right. At a distance of $(1/2)t_g$ from the exit surface, dark and bright are interchanged, i.e., the contrast alternates. The behavior close to the other surface as well as the B.F. profile can be deduced, as in the first example, by using the symmetry properties of stacking fault fringes.

With the vector $2g$ active, $n = 2$, as seen in Fig. 1-54. There is a phase shift $\alpha = 2\pi$ for column 2, and hence this column produces background intensity at the dislocation core. At column (3) the lateral displacement is now $2/3$ of the new interplanar spacing and hence $\alpha = 4\pi/3 \equiv -2\pi/3 (\mathrm{mod}\,2\pi)$, i.e., $\sin\alpha < 0$. At column (1) the lateral shift is now $(4/3)b$ $[(1/3)b(\mathrm{mod}\,b)]$ and hence $\alpha = 2\pi/3$ and $\sin\alpha > 0$. The bright and dark areas are thus interchanged as compared to the case of $n = 1$.

1.6.3.6 The Image Side of Dislocations

From the dynamic image simulations discussed in Secs. 1.6.3.1 and 1.6.3.2 we concluded that the black line image of a dislocation in the B.F. image is systematically one-sided provided s is large enough. This is true for $g \cdot b = 1$ as well as for $g \cdot b = 2$. This behavior is different for $n = 1$ and $n = 2$ if $s \simeq 0$, as discussed in Secs. 1.6.3.1 and 1.6.3.2.

The image side, i.e., the position of the black line in the B.F. image on a positive print with respect to the dislocation position, is correctly given by the intuitive kinematic considerations of Sec. 1.6.1. According to this theory, the image side is on that side of the dislocation core where locally the lattice planes normal to g are rotated toward the exact Bragg orientation; in Fig. 1-40 this is in the sense S. Finding S requires that the sign of s is known; this can be determined by means of the Kikuchi line pattern as shown in Sec. 1.3.3. For the edge dislocation in Fig. 1-62a, the positive sense, i.e., unit vector u was chosen as entering the plane of the drawing. The Burgers vector b is determined according to the FS/RH convention (Bilby et al., 1955). A right-handed closed Burgers circuit viewed along u is constructed in the real crystal (Fig. 1-62a). In the perfect reference crystal (Fig. 1-62b) the corresponding circuit is constructed and the Burgers vector b is found as the closure failure of this circuit, joining the final point with the starting point $b = FS$. For the concrete situation in Fig. 1-62, $s > 0$ and $(g \cdot b)s < 0$; the image is indicated by a solid line, and the dislocation line by a dotted line. The rule can be formulated as follows: the image side is to the right when viewed in the positive sense if $(g \cdot b)s < 0$. Changing the sign of one of the three parameters, $g \cdot b$ or s, changes the image side. It should be noted that the descriptions given in different reviews are sometimes confusing and do not always agree, because some authors refer to the image as seen along the incident electron beam, whereas other formulations refer to the image as seen from below (Paseman and Scheerschmidt, 1982;

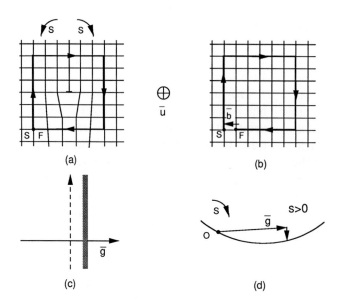

(a)

(b)

(c)

(d)

Figure 1-62. Derivation of the image side of edge dislocations using the FS/RH convention. (a) Burgers circuit in real crystal; (b) Burgers circuit in perfect crystal; (c) relation of the dislocation line (dotted line) and the image (wide dark band); (d) reciprocal lattice situation: $s > 0$.

Whelan, 1978). The sense of u depends on whether the first or the second viewpoint is adopted but this changes the sign of $p = (g \cdot b) s$. The most direct way is to apply intuitive reasoning.

1.6.3.7 Characterizing Dislocations

The full description of a dislocation line implies the determination of its core geometry and its Burgers vector, i.e., direction, magnitude, and sense of b. Methods are available to obtain all of these elements. The precise *position* of the dislocation can be found by making two images leading to opposite image sides either for active diffraction vectors $+g$ and $-g$ for the same sign of s, or for $+s$ and $-s$ for the same g vector.

The *direction* of the Burgers vector is determined by looking for two diffraction vectors g_1 and g_2 for which the dislocation is out of contrast or for which a residual contrast characteristic of $g \cdot b = 0$, is produced. The Burgers vector has then a direction parallel with $g_1 \times g_2$. An example of the application of this method to a hexagonal network of dislocations in graphite is

shown in Fig. 1-63. In this particular case a single "extinction" is in fact sufficient since the dislocations were known to be glide dislocations and thus have their Burgers vectors in the c plane. Since the foil is prepared by cleavage, it is also limited by

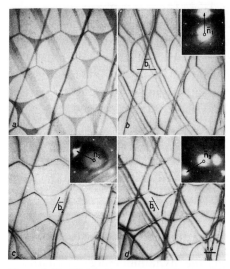

Figure 1-63. Dislocation network in graphite: The same area is shown under four different diffraction conditions. The Burgers vectors of all partials in the network were deduced from image extinctions (Delavignette and Amelinckx, 1962).

c planes. The three families of partial dislocations are seen to be successively brought to extinction using the indicated *g* vectors. Note also the simultaneous extinction of the three partials in the triple ribbon, which shows that they have the same Burgers vector. Their contrast is nevertheless different, as discussed in Sec. 1.6.1.

The *magnitude* of the Burgers vector for a perfect dislocation can be determined, once its direction is known, by looking for diffraction vectors for which $g \cdot b = 2$. In this case, use is made of the typical contrast effect that occurs where the dislocation crosses a bent extinction contour (see Sec. 1.6.3.1). If such a diffraction vector is identified, we then know the length of the projection of *b* on *g*. With the knowledge of the direction of *b* and of its projected length on *g*, the length of *b* can be found.

Finally the sense of *b* is found from the image side, which defines the sign of $(g \cdot b)\, s$. Knowing the sign of *s* from the Kikuchi pattern, one can then use the image side to find the sign of $g \cdot b$. The knowledge of *g* then leads to the sense of *b*.

An important application of the sign determination of the Burgers vector consists in determining whether a Frank loop is due to the precipitation of vacancies or of interstitials, i.e., whether *b* is either $+(1/3)[111]$ or $-(1/3)[111]$ (Edmondson, 1963; Groves and Kelly, 1961, 1963; Ruedl et al., 1962 a, b). Applying the relation determining the image side to the loop represented in Fig. 1-64, it follows that for a loop the image is either completely inside (a) or completely outside (b) the dislocation ring, depending on the sign of $(g \cdot b)\, s$. Since *b* is different for a vacancy loop and an interstitial loop, the image side will also differ for the same *g* and *s*. The type of contrast experiment required for an analysis of the nature of loops is illustrated in Fig. 1-65.

A difficulty with the application of this method arises with the need to know the sense of inclination of the loop plane. If the loops are known to be parallel to the foil surface, a known slope can be imposed by mounting the sample on a wedge. However this method is not always applicable.

Assuming the sense of the slope to be represented in Fig. 1-64 and *g* and *s* as shown, it is evident that the image is inside for interstitial loops but outside for vacancy loops. Changing the sign of *s* by tilting allows one to find the image side and hence to distinguish the two cases.

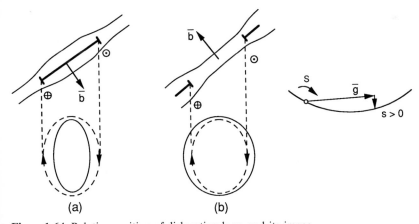

(a)   (b)

Figure 1-64. Relative position of dislocation loop and its image.

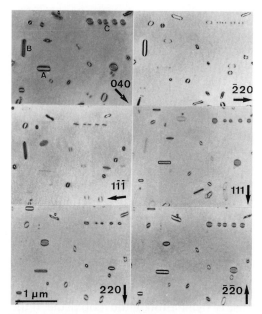

Figure 1-65. Contrast experiment on interstitial loops in silicon. Note the change in image side when changing $+g$ into $-g$.

An alternative application of the same principles consists in rotating the specimen through the exact Bragg orientation from $s > 0$ to $s < 0$ for a given g. It is then found that an *interstitial* loop will grow in size because of two effects: (i) The projected size increases and (ii) the image goes from inside for $s > 0$ to outside for $s < 0$. A *vacancy* loop will grow as long as $s > 0$ because of the geometrical effect; but beyond $s = 0$ the image side changes and the image size shrinks. The experiment must clearly be performed starting with loops which are steeply inclined. One can also make use of the asymmetrical image contrast, consisting of a line of no contrast; separating a bright and a dark lobe (or crescent), characteristic of Frank loops seen edge on, which moreover are close to the surface (Sec. 1.6.3.4). In the dark field image the asymmetry is the same at the top and the bottom of the foil as a result of anomalous absorption. If the diffraction vector g is

parallel with b and points from the bright to the dark lobe in the image, the loop has interstitial character. If g points from the dark to the bright lobe, the loop is a vacancy loop. A restriction is that the loop must be close to the surface [i.e., within $(1/2)t_g$].

1.6.3.8 The Weak-Beam Method

It is found empirically that dark field dislocation line images become narrower as the deviation parameter of the operating reflection increases in magnitude (Cockayne et al., 1969, 1971). Unfortunately the image contrast then decreases also, and long exposure times are required to record the image. This *sharpening* effect is systematically exploited in the weak-beam method of image formation; it is essentially due to a decrease in the effective extinction distance for increasing s as described by Eq. (1-53).

The method is especially useful for the study of the fine structure of dislocations, i.e., their dissociation in sequences of parallel partial dislocations connected by narrow stacking fault ribbons. Such configurations are used in one of the few direct methods for the measurement of stacking fault energies. This application requires the precise measurement of the separation of closely spaced parallel partial dislocations; sharp, precisely located dislocation images are thus a prerequisite for the application of this method.

Usually, a high-order reflection $3g$ or $4g$ is brought in the exact Bragg position and the reflection g is selected for the actual dark field imaging. It is also possible to excite a lower-order reflection, say g or $2g$ and use $-g$ for imaging (Fig. 1-66). A deviation parameter s of the order of $0.2\,\mathrm{nm}^{-1}$ for $100\,\mathrm{kV}$ electrons is a good trade-off between image sharpness and exposure time.

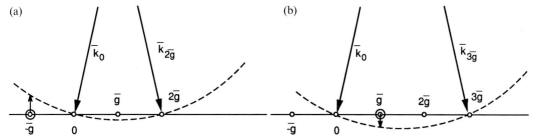

Figure 1-66. Weak-beam diffraction conditions. (a) $s = 0$ for $2\,\boldsymbol{g}$, image is made in $-\boldsymbol{g}$; (b) $s = 0$ for $3\,\boldsymbol{g}$, image is made in $\boldsymbol{g}$.

The described diffraction geometries allow one to achieve such values in most crystals that have a small lattice parameter of the order of $\sim 4\,\text{Å}$. The Kikuchi pattern is of great help in achieving the correct specimen orientation and in measuring the s value, using Eq. (1-7).

In weak-beam images the depth period of stacking fault fringes decreases significantly with increasing s_g [Eq. (1-53); this period is given approximately by its kinematic value $1/s_\text{g}$]. As a result, antiphase boundaries in ordered alloys, for instance, in which the extinction distances of the superlattice reflections are larger than the foil thickness, can still be imaged by fringe patterns when using the weak-beam method. We will now discuss semi-quantitatively the weak-beam image of an edge dislocation.

The columns close to the dislocation core can be considered as consisting of three parts (De Ridder and Amelinckx, 1971). The central lamella contains the dislocation core. The lattice planes in this part are inclined with respect to their orientation in the other two perfect parts (Fig. 1-67) in such a way that the local deviation parameter is much smaller in the core region than in the perfect parts where s is large. The scattered intensity will then mainly originate from this region, producing a bright peak on a darker background.

The kinematic theory allows an approximate expression for the peak position and peak width of the image to be derived. The amplitude scattered by a column at x is proportional to the kinematic integral

$$A = \int_0^{z_0} \exp\{2\pi i [s_\text{g} z + \boldsymbol{g} \cdot \boldsymbol{R}(x,z)]\}\, dz \quad (1\text{-}165)$$

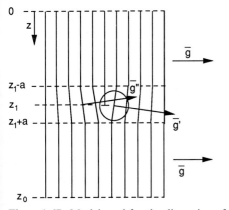

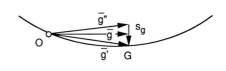

Figure 1-67. Model used for the discussion of weak beam image formation at an edge dislocation.

where $s_g z$ is the phase due to the depth position along the column of the scattering element dz, and $g \cdot R$ is the additional phase shift due to the local displacement field R of the dislocation at depth z. We can split the integration along the column into three parts:

$$A = \int_0^{z_1-a} \exp\left(2\pi i s_g z\right) dz +$$
$$+ \int_{z_1-a}^{z_1+a} \exp\left[2\pi i (s_g z + g \cdot R)\right] dz +$$
$$+ \int_{z_1+a}^{z_0} \exp\left(2\pi i s_g z\right) dz \qquad (1\text{-}166)$$

The first and third integral correspond with perfect crystal lamellae; their contributions are small since s is larger. Moreover, they do not depend on the presence of the defect.

In a complex-plane phase-amplitude diagram, the first and third integral are represented as arcs of circles with a small radius, $1/s_g$. They are connected by a circular arc with a much larger radius (Fig. 1-68): $1/s_{eff}$, where s_{eff} is the local s value close to the dislocation core. The final amplitude scattered by the column is then given roughly by the length of the line segment connecting to centers of the small circles; it is given to a good approximation by the second integral in Eq. (1-166).

The displacement function $R(x, z)$ can be expanded in the vicinity of $z = z_1$ in a Taylor expansion:

$$R = R(z_1) + (z - z_1)\left(\frac{\partial R}{\partial z}\right)_{z_1} +$$
$$+ \frac{1}{2}(z - z_1)^2 \left(\frac{\partial^2 R}{\partial z^2}\right)_{z_1} + \dots \qquad (1\text{-}167)$$

Limiting the expansion to the first two terms in the expression for the second integral, one obtains:

$$\exp\left\{2\pi i\left[R(z_1) - z_1\left(\frac{\partial R}{\partial z}\right)_{z_1}\right] \cdot g\right\} \times$$
$$\times \int_{z-a}^{z+a} \exp\left[2\pi i\left(s_g + g \cdot \frac{\partial R}{\partial z}\right) z\right] dz \qquad (1\text{-}168)$$

This expression will be maximum if the exponential in the integrand is unity, i.e., if x is such that

$$s_g + g \cdot \frac{\partial R}{\partial z} = 0 \qquad (1\text{-}169)$$

This expression also enters in the diffraction equation for a deformed crystal as the local effective parameter $s_{eff} = s_g + g \cdot \left(\frac{\partial R}{\partial z}\right)$. The peak position then corresponds with the x value x_{max}, which causes s_{eff} to become zero.

It is clear that for $s_g = 0$ the amplitude phase diagram reduces to a straight line segment connecting points on the two small circles representing the perfect parts of the foil; it is approximately equal to the distance between the centers of these small circles.

Introducing in the relation Eq. (1-169) the expressions for the displacement fields $R(x, z)$ for edge and screw dislocations, adopting the FS/RH convention, leads to the solution x_{max} for the position of the peak:

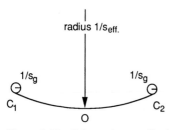

radius $1/s_{eff.}$

$1/s_g$ $1/s_g$

C_1 C_2

O

Figure 1-68. Schematic amplitude-phase diagram used in discussing the image formation according to the weak beam method.

$$x_{max} = -\frac{g \cdot b}{2\pi s_g}\left[1 + \frac{K}{2(1-\nu)}\right] \qquad (1\text{-}170)$$

with $K = 1$ for an edge dislocation and $K = 0$ for a screw dislocation (v = Poisson's ratio). In this approximation the peak position is independent of the foil thickness and of the depth position of the dislocation in the foil.

We note that this expression shows that the image side, i.e., the sign of x_{max} is determined by the sign of the product $(g \cdot b) s_g$.

Similarly, the image width at half maximum can be deduced from the kinematic theory; for $g \cdot b = 2$ one finds (De Ridder and Amelinckx, 1971)

$$\Delta x = \frac{0.28}{|s_g|} \left[1 + \frac{K}{\varepsilon(1-v)} \right]$$

where $v = 1/3$, $|s_g| = 0.2\,\text{nm}^{-1}$. One also finds that $\Delta z \sim 2.5\,\text{nm}$ for an edge dislocation.

The image peak lies further from the core position for increasing values of $g \cdot b$, as predicted also by the kinematic theory (Hirsch et al., 1956). In order to achieve the same precision in the image position, larger values of $|s_g|$ are thus needed for larger values of $g \cdot b$; this limits in practice $g \cdot b \le 2$.

An example of a weak beam image of dislocation ribbons in $RuSe_2$ is reproduced in Fig. 1-69.

1.7 Moiré Patterns

1.7.1 Intuitive Considerations

An electron microscopic specimen consisting of two similar thin films superposed with a small orientation difference produces an interference pattern consisting of parallel fringes when a g vector parallel to the films is active. In the bright field image this fringe pattern results from the interference between the doubly transmitted and the doubly scattered beam, which encloses

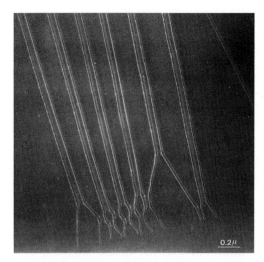

Figure 1-69. Example of weak beam image in $RuSe_2$. The dislocations are dissociated in partials; they are parallel to the foil plane.

a small angle. This angle is usually revealed in the diffraction pattern by a doubling of the spots. The doubly transmitted beam is the beam transmitted through the first thin film and subsequently through the second; the doubly scattered beam is formed in a similar way (Pashley et al., 1957; Bassett et al., 1958).

We shall discuss the geometric features of such fringes since they provide useful information in a number of cases. A geometric analogue consisting of the superposition of two line patterns is shown in Fig. 1-70. The lines represent lattice planes, one of which contains a dislocation. In (a) the two patterns have the same line spacing, but the directions of the lines enclose a small angle. In (b) the directions of the lines are the same, but their spacing is slightly different. The *moiré pattern,* or the superposition pattern, clearly shows a magnified representation of a dislocation. Moiré patterns can thus provide *geometrical* magnification, which was especially useful at a time when atomic resolution was not possible. With the development of atomic res-

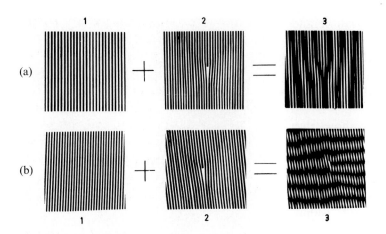

Figure 1-70. Formation of moiré patterns by two superposed line patterns, one of which contains a supplementary half-line (simulating a dislocation). (a) Rotation moiré pattern; (b) parallel moiré pattern.

olution microscopy, moiré imaging lost some of its importance; the geometric features are still useful however.

1.7.2 Theoretical Considerations

Consider a "sandwich" crystal consisting of two plan parallel slabs I and II. Let part I of a column be derived from part II by the displacement field $u(r)$. The phase shift between the waves diffracted locally by the two parts of the crystal is then $\alpha = 2\pi(g+s) \cdot u \simeq 2\pi g \cdot u$. This expression is of the same form as the phase shift introduced by a stacking fault, the main difference being that u is *not* a constant vector R but now depends on r, and hence α is also a function of r. The transmitted and scattered amplitudes are then given by Eqs. (1-132) and (1-133), in which α enters through the periodic factor $\exp(i\alpha)$. Without solving the system of equations, it is clear that the loci of the points of equal intensity (i.e., the fringes) are given by $\exp(i\alpha) = constant$ [(i.e., by $\alpha = constant + k \cdot 2\pi$ ($k = integer$)].

Assuming r to be a lattice vector, $g \cdot r = integer$ and hence for small difference vectors

$$\Delta g \cdot r + g \cdot \Delta r = 0 \quad \text{with} \quad \Delta r = u(r) \quad (1\text{-}171)$$

and thus $\alpha \equiv 2\pi g \cdot u = -2\pi \Delta g \cdot r$.

Provided $u(r)$ is such that Δg does not depend on r, which is true for moiré patterns, the lines of equal intensity are given by $\Delta g \cdot r = constant + k$ ($k = integer$). This equation represents a set of parallel straight lines perpendicular to $K = -\Delta g$, where K can be considered as the wave vector of the fringe system, with wavelength $\Lambda = 1/K$.

In the case of a *rotation moiré*, $K = 2g \sin(\theta/2) \simeq g\theta$ (for small θ); or expressed in terms of the interplanar spacing d_g of the active reflection

$$\Lambda_r = \frac{d_g}{\theta} \quad (1\text{-}172)$$

The fringes are parallel to g for small values of θ.

For *parallel moiré* patterns, $\Delta g = g_2 - g_1$ with g_2 parallel to g_1, or in terms of interplanar spacings

$$\Delta g = \frac{1}{d_2} - \frac{1}{d_1} = \frac{d_1 - d_2}{d_1 d_2} \quad (1\text{-}173)$$

and

$$\Lambda_p = \frac{d_1 d_2}{d_1 - d_2} \quad (1\text{-}174)$$

The fringes are again perpendicular to Δg, i.e., also perpendicular to g_1 and g_2.

If an orientation difference as well as a spacing difference are present, mixed moiré patterns are found. One can always decompose Δg into a component perpendicular to Δg and a component parallel to g:

$$\Delta g = \Delta g_{\parallel} + \Delta g_{\perp}$$

with the fringes still perpendicular to Δg, they enclose an angle β with the direction of g given by $\tan \beta = \Delta g_{\perp}/\Delta g_{\parallel}$.

One obtains further

$$|\Delta g|^2 = |\Delta g_{\perp}|^2 + |\Delta g_{\parallel}|^2$$

and hence

$$\frac{1}{\Lambda^2} = \frac{1}{\Lambda_{\parallel}^2} + \frac{1}{\Lambda_{\perp}^2} \qquad (1\text{-}175)$$

The intensity variation of the fringe system can be found in a similar way as for stacking faults.

For a quantitative theory of the intensity profiles we refer to (Gevers, 1963a, 1963b; Hashimoto and Uyeda, 1957).

As we have seen for coherent domain boundaries, Δg is perpendicular to the interface. When the interface is perpendicular to the incident beam, which is the usual geometry for moiré patterns, the projection of Δg on the interface thus vanishes and no moiré fringes are formed. The fringe pattern imaging this type of interface therefore has a different origin. The image for an inclined interface consists of the δ fringes discussed above, which are perpendicular to the projection of Δg on the foil plane (Sec. 1.6.3.4), i.e., perpendicular to the intersection lines of the interface with the foil surfaces.

If Δg has an arbitrary orientation with respect to the contact plane between the two crystal parts, Δg has a perpendicular component as well as a parallel component with respect to the interface, and the image can be a complicated mixture of both types of images. The parallel component gives rise to moiré-type fringes, the perpendicular component to δ type fringes.

An important application of parallel moiré fringes consists in the determination of the lattice parameter of one of the two components in a sandwich, the lattice parameter of the other one being known. This can be of interest for the identification of plate-like coherent precipitates in a matrix with a known lattice parameter. Moiré fringes formed at the interface between *voidite* and diamond are shown in Fig. 1-71, whereas Fig. 1-72 shows the moiré

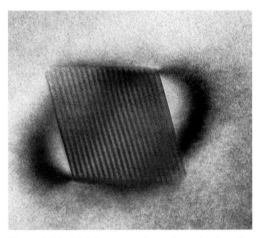

Figure 1-71. Moiré pattern formed at the interface between voidite and diamond (Courtesy Van Tendeloo and Luyten).

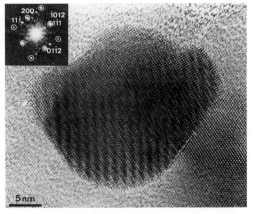

Figure 1-72. One-dimensional moiré pattern between a SiC particle and a silicon matrix superposed on a high-resolution dot pattern (Courtesy De Veirman).

fringes at the interface between silicon and SiC precipitate particles.

Moiré fringes have also been used as a tool in the study of dislocations. Ending moiré fringes reveal the emergence points of dislocations in one of the two components of the sandwich. The number N of supplementary half-fringes depends on the reflection used to produce the dislocation image; it is given by $N = \boldsymbol{g} \cdot \boldsymbol{b}$. This number is independent of the character of the dislocations. Supplementary half-fringes therefore can not be interpreted as meaning necessarily that the corresponding dislocation has edge character. Partial dislocations bordering stacking faults are revealed by a *fractional* number of supplementary half-planes, i.e., along the trace of the stacking fault, the moiré fringes are shifted discontinuously over a function $\boldsymbol{g} \cdot \boldsymbol{b}$ such as 1/3, 2/3 of the interfringe distance.

The moiré fringes are also shifted by a surface step in one of the components. The fringe shift is not only a function of the step height but also of the deviation parameter, and hence of the specimen orientation.

If two or more diffraction vectors are active, a crossed grid of moiré fringes is formed, which has the rotation symmetry of the two films.

1.8 Two-Beam Lattice Fringes

1.8.1 Theoretical Considerations

We have seen in Sec. 1.4.6 that in the two-beam case the wave function at the exit surface of the foil is of the form

$$\Psi(r) = \exp(2\pi i \boldsymbol{K} \cdot \boldsymbol{r}) [T + S \exp(2\pi i \boldsymbol{g} \cdot \boldsymbol{r})]$$
$$(1\text{-}176)$$

Only the transmitted beam with amplitude T and one diffracted beam $\boldsymbol{g}$ with amplitude S are considered. The expressions for T and S are given by Eqs. (1-57) and (1-58).

When these two beams are admitted through the selector aperture they produce an interference pattern with an intensity distribution given by $I = \Psi\Psi^*$, or explicitly

$$I = TT^* + SS^* - 2(TT^*SS^*)^{1/2} \cdot$$
$$\cdot \sin(2\pi g x + \varphi)$$

or

$$I = I_T + I_S - 2(I_T I_S)^{1/2} \sin(2\pi g x + \varphi)$$
$$(1\text{-}177)$$

where e_x is parallel to $\boldsymbol{g}$, i.e., $g x = (\boldsymbol{g} \cdot \boldsymbol{e}_z) x$; φ is given by

$$\tan\varphi = \frac{s}{\sigma}\tan(\pi\sigma t) \qquad (1\text{-}178)$$

The expression in Eq. (1-177) represents sinusoidal fringes with a period $1/g = d_g$ superposed on a background $I_T + I_S$. The contrast C defined as $C = (I_{max} - I_{max})/(I_{max} + I_{min})$ is then given by

$$C = \frac{2(I_T I_S)^{1/2}}{I_T + I_S}$$

It will be optimal for $I_S = I_T$; if $s_g = 0$, this occurs for $t = (1/4) t_g$.

The intensity distribution in Eq. (1-177) will be modified on imaging by the microscope because an additional phase difference χ will be introduced between T and S due to defocusing and lens aberrations.

Furthermore, the presence of an interface with displacement vector $\boldsymbol{R}$ will change the relative phase of T and S by $2\pi\boldsymbol{g} \cdot \boldsymbol{R}$. In the part of the foil containing such a planar interface, the phase angle will thus become $\varphi \rightarrow \varphi + \chi + 2\pi\boldsymbol{g} \cdot \boldsymbol{R}$, i.e., the fringes will thus be shifted.

1.8.2 Properties of Lattice Fringes

Although the fringes have the same spacing as the lattice planes corresponding with $\boldsymbol{g}$, they cannot be said to represent these lattice planes, since the fringes are not localized with respect to the crystal lattice; their positions depend on several factors and in particular on s [through the relation in Eq. (1-178)] and on $g \cdot R$, if an interface is present. On crossing the borderline between a crystal area with and without a stacking fault with displacement vector R, the fringes will thus suffer a parallel shift over a fraction of the interfringe spacing given by $g \cdot R$. Lattice fringes can thus be used to determine R.

For $s = 0$ (i.e., $T = T^*$; $S = -S^*$, $\varphi = 0$) and no absorption ($I_T + I_S = 1$), the expression in Eq. (1-177) reduces to a particularly simple form corresponding with optimum contrast: It allows the main features of lattice fringes to be deduced:

$$I = 1 + \sin\left(2\pi \frac{t}{t_g}\right) \sin(2\pi g x) \qquad (1\text{-}179)$$

The fringe contrast clearly depends in a periodic way on the foil thickness t. The fringes in a wedge-shaped crystal periodically change in contrast, reversal taking place after a thickness change of $(1/2) t_g$, passing through a thickness where the contrast disappears (for $t = (1/2) k t_g$, $k =$ integer). The fringes remain perpendicular to g in a wedge as well, but if $s \neq 0$, the fringe position (i.e., φ) also becomes a function of the thickness, and the fringes then form a small angle with the direction of the lattice planes.

In practice, use is often made of the equi-inclination method represented in Fig. 1-73, where the incident beam is tilted over the Bragg angle with respect to the lattice planes. In this case, $s = 0$ for g, and the phase shift χ due to angle-dependent aberrations is eliminated, since the two beams T and S now enclose the same angle with the optical axis of the microscope. Fringes with a spacing smaller than the point reso-

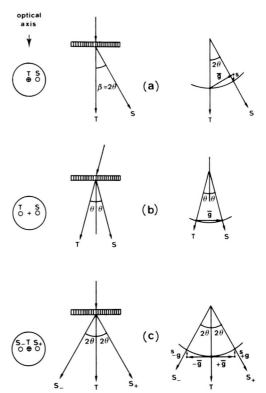

Figure 1-73. Different beam configurations used in producing lattice fringes. (a) Two-beam, normal incidence; (b) two-beam, tilted incidence; (c) symmetrical three beam case. For each mode, the geometry is shown in direct space (center) and in reciprocal space (right). The positions of the objective selector aperture are represented schematically for each case (left).

lution of the microscope can thus be produced.

One-dimensional multiple-beam fringes can be formed by admitting more beams of the linear sequence: $-ng, \ldots, -2g, -g, 0, +g, +2g, \ldots, ng$ in the selector aperture. We shall now discuss qualitatively the case of $-g$, 0, g. The intensity profile of the fringe pattern is now of the form

$$I(x) = I_0 + A \sin(2\pi g x + \varphi_1) + \\ + B \sin(4\pi g x + \varphi_2)$$

where A, B, φ_1 and φ_2 are complicated expressions of the deviation parameters, of the crystal thickness and of the Fourier coefficients V_g and V_{-g}. The most important feature is the occurrence of a second harmonic in the fringe pattern; that is, main and subsidiary fringes occur in an alternating fashion, the separation of successive fringes being $(1/2)d_g$. By choosing a proper defocus value the two sets of fringes become equally intense, and fringes with a spacing of $(1/2)d_g$ are observed.

Including higher-order reflections clearly introduces higher order harmonics in the fringe profile. The image formed by non-colinear reflections is the subject of the chapter on high resolution methods.

A general review on two-beam lattice fringes is given by Menter (1956), Spence (1981) and Hashimoto et al. (1961).

1.9 High-Resolution Electron Microscopy (HREM)

1.9.1 Introduction

We have seen in Sec. 1.8 that sinusoidal fringes are formed by the interference between two beams, the transmitted beam and one diffracted beam g. Such lattice fringes give only a very rudimentary image of the crystal structure, i.e., one Fourier component g is imaged. Such images are nevertheless useful for certain applications in simple structures. Their formation principle can easily be generalized by admitting many beams, i.e., many Fourier components, corresponding with a two-dimensional array of diffraction spots, through the selector aperture. In such cases, the resulting image depends sensitively on the lens system. We will therefore briefly discuss the necessary elements of electron optics.

The formation under multiple beam conditions of *structure* images can be un-

derstood intuitively by extending the considerations of Sec. 1.8, which emphasized the wave nature of electrons. An intuitive picture, emphasizing the particle nature of electrons, is equally possible; it is given further below.

Analytical expressions for the image contrast can only be obtained in certain simple situations, these will be reviewed below. The images of concrete structures can only be obtained by computer simulations (Cowley and Moodie, 1957).

More detailed informations about high-resolution electron microscopy are given in: Spence (1981), Van Dyck (1978) and Amelinckx (1986 b).

1.9.2 Image Formation in an Ideal Microscope

The image formation in an ideal microscope is shown schematically in Fig. 1-74. Let the incident electron beam be described by a plane wave of amplitude 1 and let the object be characterized by a two-dimensional transmission function $q(x, y)$.

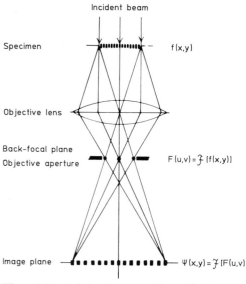

Figure 1-74. Relation between object, diffraction pattern and image in an ideal microscope.

Diffraction occurs in the object, and electrons emerge from the exit face; the back surface of the object can thus be considered as being a planar assembly of point sources of spherical wavelets in the sense of Huyghens. The interference between these wavelets generates the diffracted beams in the case of a crystalline specimen, and produces a diffraction pattern in the back focal plane of the objective lens. This diffraction pattern can to a good approximation be described by Fraunhofer diffraction, because of the relative dimensions of the lenses and of the electron wavelength and because of the paraxial nature of most of the diffracted beams; this is again a consequence of the fact that in electron diffraction Bragg angles are very small because of the short wavelength of electron beams (~ 0.04 Å at 100 kV). The diffraction amplitude is then the Fourier transform of the function $q(x, y)$ (Fig. 1-74). In turn the diffraction pattern in this back focal plane acts as a source of Huyghens spherical wavelets, which interfere to produce an enlarged image of the transmission function. This image is again the Fourier transform of the diffraction pattern. We can thus conclude that the ideal microscope acts as an analogue computer and performs a double Fourier transformation and thus reproduces the object apart from a linear magnification. Unfortunately, the actual situation is somewhat more complicated.

1.9.3 Image Formation in a Real Microscope

Real microscopes are subject to a number of limitations which induce deviations from the ideal imaging conditions described above.

1.9.3.1 Spherical Aberration

In real magnetic lenses the paraxial approximation which leads to point-to-point representation in the Gaussian focal plane (i.e., the focal plane considered in geometrical optics) breaks down. This is due to the fact that the value of $\sin \beta$ (β = angle between v and H), which enters in the expression for the Lorentz force on a moving charge, can no longer be approximated by the angle β, but higher order terms up to β^3 are required. This is analogous to the limitations introduced by approximating $\sin \beta = \beta$ in Snell's law which is only valid for paraxial rays in ordinary optics.

The radius of the disc of confusion in object space resulting from this lens aberration is then given by (Glaser, 1952)

$$\varrho_s = C_s \beta^3 \tag{1-180}$$

where C_s is the spherical aberration constant, which has a value somewhere between 1 and 10 mm. A typical high-resolution microscope operating at 200 kV has a value $C_s = 1.2$ mm.

As a result of spherical aberration, electron beams inclined at an angle with the optical axis suffer a phase shift with respect to the axial beam at $\beta = 0$, given by

$$\chi_s = 2 \pi \frac{\Delta}{\lambda} \tag{1-181}$$

where Δ is the path difference caused by the fact that the beam does not pass along the axis. From Fig. 1-75, it can be concluded that $\Delta = \varrho_s \sin \beta \simeq \varrho_s \beta$ and hence $d\Delta = \varrho_s d\beta$ and $d\chi_s = 2 \pi \varrho_s d\beta/\lambda = 2 \pi C_s \beta^3 d\beta/\lambda$. After integration from 0 up to the angle β (Scherzer, 1949)

$$\chi_s = \frac{\pi C_s \beta^4}{2 \lambda} \tag{1-182}$$

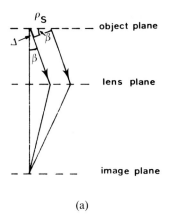

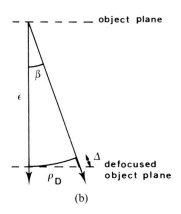

Figure 1-75. Derivation of the path difference caused by spherical aberration (a) and defocusing (b).

(a)   (b)

1.9.3.2 Aperture

The microscope contains an objective aperture which eliminates beams which enclose an angle β with the optical axis, exceeding a limiting angle β_A, in order to reduce the spherical aberration. This imposes a limit to the theoretically achievable resolution called the Abbe limit. A geometric point is imaged as a circle (the disc of confusion) with the radius

$$\varrho_A = 0.61 \frac{\lambda}{\beta_A} \qquad (1\text{-}183)$$

which means in practice that only points separated by at least this distance in the object can be observed as separate points in the final image. The aperture also determines the beams which are allowed to contribute to the image.

1.9.3.3 Defocus

Most high resolution images are automatically made under conditions where visual contrast is best. It turns out that in the exact Gaussian focal plane the contrast is smallest, at least for a phase object, which is the case for most specimens. One therefore is usually forced to work under somewhat defocused conditions, which ensure the *best* contrast and definition. Also defo-

cusing causes "phase shifts" and a disc of confusion which we can estimate with reference to Fig. 1-75 b.

Defocusing the electron microscope by an amount ε but leaving the plane of observation unchanged – the object being situated near the first focal plane – results in an apparent displacement ε of the object plane (Fig. 1-75 b). One clearly obtains

$$\varrho_D = \varepsilon \sin \beta \simeq \varepsilon \beta$$

and for the value of $\varDelta$

$$\varDelta = \frac{\varepsilon}{\cos \beta} - \varepsilon = \varepsilon \left(\frac{1}{1 - \beta^2/2} - 1 \right) \simeq \varepsilon \frac{\beta^2}{2}$$

hence

$$\chi_D = 2\pi \frac{\varDelta}{\lambda} = \pi \varepsilon \frac{\beta^2}{\lambda} \qquad (1\text{-}184)$$

$\varepsilon > 0$ results in lens stengthening, and $\varepsilon < 0$ in lens weakening.

1.9.3.4 Chromatic Aberration

As a result of instabilities in the high voltage of the microscope, the incident electron beam exhibits a wavelength spread, since λ is related to the acceleration potential E by the non-relativistic approximate relation

$$\lambda = \frac{h}{(2 m e E)^{1/2}} \qquad (1\text{-}185)$$

Moreover, variations in the lens currents $\Delta I/I$ also cause aberrations which are of the same nature.

A third origin of aberration is the inelastic scattering in the specimen which is equivalent to a change in energy of the electrons entering the lens system: $\Delta E/E$. The net effect of all these phenoma on the image formation is to cause a spread of Δf on the focal distance f of the objective lens. The latter is proportional to $E I^{-2}$. Hence, assuming ΔE and ΔI to be uncorrelated, Δf is given by

$$\Delta f = C_{\mathrm{C}} \left[\left(\frac{\Delta E}{E} \right)^2 + 4 \left(\frac{\Delta I}{I} \right)^2 \right]^{1/2} \quad (1\text{-}186)$$

The corresponding disc of confusion in object space has a radius

$$\varrho_{\mathrm{C}} = \beta \, \Delta f$$

The constant C_{C} in Eq. (1-186) is called the chromatic aberration constant. Since the instabilities in high voltage and in lens current can be reduced to a magnitude smaller than 10^{-6}, Δf takes a typical value of 100 Å, which corresponds to a C_{C} value smaller than 10 mm.

1.9.3.5 Beam divergence

Because of the finite dimensions of the electron source and of the condensor lens aperture, the incident beam is somewhat divergent. Under the intense illumination conditions used in high resolution imaging, the apex angle of the illumination cone may reach a value of the order of $\approx 10^{-3}$ rad.

The influence of incoherent beam divergence on the image can be described as being due to the superposition of independent images (i.e., intensities) corresponding with different incident directions within the divergence cone (Hibi, 1962; Spence, 1981).

1.9.3.6 Ultimate Resolution

Apart from the lens imperfections discussed above, which lead to image blurring and phase shifts, a number of other imperfections occur; but these are unimportant relative to the ones we have already discussed. Furthermore, resolution also depends on mechanical stability (vibration, drift, ...). These effects can be eliminated to a large extent by a proper microscope design; they will not be discussed any further here.

The ultimate resolution is thus currently limited mainly by three inevitable phenomena: finite aperture, spherical aberration and chromatic aberration.

The final disc of confusion has a radius given by

$$\varrho = (\varrho_{\mathrm{A}}^2 + \varrho_{\mathrm{S}}^2 + \varrho_{\mathrm{C}}^2)^{1/2}$$

Because of the difference in angular dependence of the different aberrations, it turns out that in present day high-resolution electron microscopes, chromatic aberration has only a relatively small influence. However, in recently built microscopes with an optimized lens design, the chromatic aberration has again become more

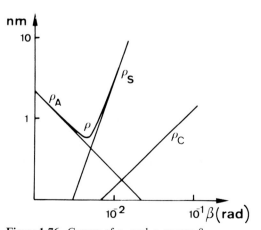

Figure 1-76. Curves of ϱ_{A} and ϱ_{s} versus β.

important and is for such instruments the resolution-limiting parameter.

The limiting factor at small angles ($\beta < 5 \times 10^{-3}$ rad) is the aperture, whereas in the range $\beta > 5 \times 10^{-3}$ rad, the limiting factor is the spherical aberration. (This is true for $E = 100$ kV; $C_S = 8.2$ mm and $C_C = 3.9$ mm).

The curves ϱ_A and ϱ_S versus β have opposite slopes (Fig. 1-77). There is therefore

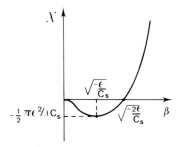

Figure 1-77. Dependence of $\chi(\beta)$ on β.

a minimum value for ϱ which occurs for $\partial\varrho/\partial\beta = 0$ where $\varrho = (\varrho_A^2 + \varrho_S^2)^{1/2}$. This minimum, which corresponds to the optimum compromise β_0 between spherical aberration and aperture effects, occurs for

$$\beta_0 = \left(\frac{0.61\,\lambda}{\sqrt{3}\,C_S}\right)^{1/4} \tag{1-187}$$

The corresponding radius of the confusion disc is then

$$\varrho_0 = 0.9\,\lambda^{3/4}\,C_S^{1/2} \tag{1-188}$$

(Representative values are $\beta_0 = 5 \times 10^{-3}$ rad; $\varrho_0 \simeq 0.5$ nm). This expression makes it clear that one gains more in resolution by decreasing the wavelength (i.e., by increasing the accelerating voltage) than by decreasing C_S. Unfortunately, increasing the accelerating voltage also increases radiation damage to the specimen. A compromise is therefore sought by using intermediate voltages (e.g., 400 kV) combined with optimized lens designs.

1.9.3.7 Phase Shift

We have pointed out that spherical aberration and defocus cause phase shifts of the non-axial electron beams with respect to the axial beam. These phase shifts depend on β in the following manner, from Eqs. (1-182) and (1-184):

$$\chi(\beta) = \frac{\pi\,C_S\,\beta^4}{2\,\lambda} + \frac{\pi\,\varepsilon\,\beta^2}{\lambda} \tag{1-189}$$

Note that $\chi = 0$ for the defocus value

$$\varepsilon = -\frac{1}{2}\,C_S\,\beta^2 \tag{1-190}$$

This value still depends on β and can thus only be satisfied approximately in a limited range of β values. It is thus necessary to take the phase shifts into account when performing image calculations. For negative ε values the function $\chi(\beta)$ presents stationary points where $\partial\chi/\partial\beta = 0$; there is a minimum for $\beta^2 = -\varepsilon/C_S$ and a maximum for $\beta = 0$. The corresponding values of $\chi(\beta)$ are respectively $\chi_{min} = -(1/2)\,\pi\,\varepsilon^2/(\lambda\,C_S)$ and $\chi_{max} = 0$. The general aspect of the curve is as presented in Fig. 1-77.

The phase shift can also be expressed in terms of a coordinate system (u, v) in the back focal plane of the objective lens. The axes u and v are parallel with the x- and y-axis of the object plane, respectively, ignoring the image rotation, which is inherent to magnetic lenses (but is corrected in recent designs). We obtain the relations

$$u = \frac{\beta_x}{\lambda} \quad \text{and} \quad v = \frac{\beta_y}{\lambda}$$

where β_x and β_y are the angles enclosed with the optical axis (the diffraction angles) in the x and y directions, respectively. One then obtains the relation

$$\beta^2 = \beta_x^2 + \beta_y^2 = (u^2 + v^2)\,\lambda^2 \tag{1-191}$$

1.9.4 Mathematical Formulation of Image Formation

1.9.4.1 General

The amplitude distribution in the back focal plane of the objective lens is given by the Fourier transform (F) of the object function. In the case of a crystalline specimen, the object function is in fact the electron wave function at the exit face of the thin foil, i.e., $q(x, y) = \Psi(x, y)$. The amplitude distribution in the diffraction pattern is then

$$Q(u, v) = \underset{u, v}{F} [q(x, y)]$$

The final image amplitude is in turn the Fourier transform of the diffraction amplitude. However, the electrons are now moving in a lens system and thus undergo the phase shifts $\chi(u, v)$ discussed above. Moreover, an aperture is limiting the number of beams transmitted through the system. This can be taken care of by introducing an aperture function $A(u, v)$ in the plane of the diffraction pattern. This function is 1 over the surface of the aperture and zero outside of this. The amplitude of the final image can then be represented by the truncated Fourier transform of the diffraction pattern:

$$U(x, y) = \underset{x, y}{F} \{A(u, v) Q(u, v) \exp[-i \chi(u, v)]\}$$

in which the expression for the phase shift χ is given above as

$$\chi(u, v) = \frac{1}{2} \pi C_s \lambda^3 (u^2 + v^2)^2 + \pi \varepsilon \lambda (u^2 + v^2) \tag{1-192}$$

The intensity in the image is then $I = U U^*$.

It has been shown that the blurring effect due to beam divergence and chromatic aberration can to a first approximation be simulated by using a smaller effective radius of the aperture in the calculation, thereby eliminating the beams with the largest Bragg angles (Fejes, 1977).

1.9.4.2 Phase Grating Approximation

We shall now illustrate the image calculations in the simple case where the specimen can be assimilated with a phase grating; i.e., the specimen only causes phase changes in the electron beam (Grinton and Cowley, 1971).

The phase change introduced by a local potential $V(r)$ in a slice of thickness dz can be estimated as follows:

$$d\chi(r) = 2\pi \left(\frac{1}{\lambda'} - \frac{1}{\lambda}\right) dz =$$
$$= \frac{2\pi}{\lambda} \left(\frac{\lambda}{\lambda'} - 1\right) dz \tag{1-193}$$

where λ and λ' are the electron wavelengths in a field-free region and in regions with electrostatic potential $V(r)$, respectively. One obtains

$$\lambda'(r) = \frac{h}{\{2 m e [E + V(r)]\}^{1/2}}$$

whereas

$$\lambda(r) = \frac{h}{(2 m e E)^{1/2}}$$

The phase shift is thus given by

$$d\chi(r) = \frac{2\pi}{\lambda} \left\{\frac{[E + V(r)]^{1/2}}{E^{1/2}} - 1\right\} dz$$

Since V/E is a small quantity (V of the order of volts; $E \simeq 100$ kV), it is a good approximation to write

$$d\chi(r) = \frac{2\pi}{\lambda} \left[\left(1 + \frac{V}{E}\right)^{1/2} - 1\right] dz \simeq$$
$$\simeq \frac{2\pi}{\lambda} \frac{V}{2E} dz = \sigma V dz \tag{1-194}$$

with $\sigma = \pi/(\lambda E)$. The total phase shift on propagating through the slab is then obtained by integration over the thickness of the sample:

$$\chi(x, y) = \sigma \underset{\text{thickness}}{\int} V(x, y, z) dz = \sigma \varphi(x, y) \tag{1-195}$$

where $\varphi(x,y)$ represents the projected potential along the z direction. Assuming that the object function represents a pure phase grating, it can be written as

$$q(x,y) = \exp[i\,\sigma\,\varphi(x,y)] \qquad (1\text{-}196)$$

In reality, the sample also causes some absorption, which can be accounted for by introducing an exponential damping factor. The same factor takes into account a position dependent apparent "loss" of electrons due to all kinds of processes which cause electrons to be intercepted by the finite aperture. The object function then becomes:

$$q(x,y) = \exp[i\,\sigma\,\varphi(x,y) - \mu(x,y)] \qquad (1\text{-}197)$$

Two extreme models which allow a simple analytical discussion, but which nevertheless illustrate the main features will now be discussed: the weak phase object model and the *thick* object model; ε is assumed to be small.

1.9.4.3 Weak-Phase Object

In very thin foils, i.e., in weakly scattering objects, this exponential can be expanded up to the first order:

$$q(x,y) = 1 + i\,\sigma\,\varphi(x,y) - \mu(x,y) \qquad (1\text{-}198)$$

The Fourier transform then becomes

$$Q(u,v) =$$
$$= \delta(u,v) + i\,\sigma\,\Phi(u,v) - M(u,v) \qquad (1\text{-}199)$$

where $\delta(u,v)$ is the Dirac function.

The image amplitude then becomes

$$U(x,y) = \mathop{F}_{x,y}\{Q(u,v)\exp[-i\,\chi(u,v)]\} =$$

$$= \mathop{F}_{x,y}[\delta(u,v) + i\,\sigma\,\Phi(u,v) - M(u,v)][\cos\chi(u,v) - i\sin\chi(u,v)] =$$

$$= \mathop{F}_{x,y}\{[\delta(u,v) + \sigma\,\Phi(u,v)\sin\chi(u,v) - M(u,v)\cos\chi(u,v)] + i[\sigma\,\Phi(u,v)\cos\chi(u,v) +$$

$$+ M(u,v)\sin\chi(u,v)]\}$$

since $\chi(0,0) = 0$

$$(1\text{-}200)$$

We note that an image which is directly related to the object will be obtained if $\sin\chi(u,v) = \pm 1$ for all u and v; in this case, $\cos\chi = 0$. Let us consider in particular the case $\sin\chi = -1$. One then obtains:

$$U(x,y) = \mathop{F}_{x,y}\{\delta(u,v) - \sigma\,\Phi(u,v) - i\,M(u,v)\}$$
$$= 1 - \sigma\,\varphi(x,y) - i\,\mu(x,y)$$

The image intensity then reads up to first-order terms

$$I(x,y) = U\,U^* = [1 - \sigma\,\varphi(x,y) - i\,\mu(x,y)]\cdot$$
$$\cdot[1 - \sigma\,\varphi(x,y) + i\,\mu(x,y)] \simeq$$
$$\simeq 1 - 2\,\sigma\,\varphi(x,y) \qquad (1\text{-}201)$$

The image intensity thus clearly has a direct relation with the object represented by $\varphi(x,y)$. The intensity will be smaller, the larger the projected potential. The image contrast, i.e., $(I - I_0)/I_0 = -2\,\sigma\,\varphi(x,y)$, where $I_0 = 1$, is directly proportional with the projected potential φ.

If it were possible to make $\sin\chi = +1$, one would obtain $I(x,y) = 1 + 2\,\sigma\,\varphi(x,y)$.

The intensity is now larger for a larger projected potential. The situation is somewhat similar to positive and negative phase contrast. The lenses have introduced phase shifts of $\pi/2$, similar to what the quarter wavelength ring does in optical phase contrast microscopy. The lens aberrations are in fact exploited to produce *phase* contrast, which would be absent otherwise (Cowley and Ilijima, 1972).

1.9.4.4 Optimum Defocus Images

If we wish the image to be a "faithful" representation of the projected potential we must thus require that $\sin \chi \approx \pm 1$, not just for a single beam but for as many diffracted beams contributing to the image as possible.

The value of $\sin \chi$ will not vary rapidly in the vicinity of a stationary point of χ, i.e., in a minimum or a maximum. We have shown above that χ adopts the stationary value $\chi = -\pi \varepsilon^2/(2 \lambda C_S)$ for $\beta = (-2\varepsilon/C_S)^{1/2}$. Since this value of χ is essentially negative, we cannot satisfy simultaneously the requirement $\sin \chi = +1$ (i.e., $\chi = \pi/2$), but we can do so for $\sin \chi = -1$ (i.e., $\chi = -\pi/2$). It is sufficient to choose the defocus ε in such a way that $-\pi \varepsilon^2/(2 \lambda C_S) = -\pi/2$, that is,

$$\varepsilon_S = -(\lambda C_S)^{1/2} \tag{1-202}$$

Since the $\sin \chi$ function is in turn stationary around $\chi = \pm \pi/2$, the $\sin \chi$ versus β curve will present a flat part in the region of $\beta = (-\varepsilon/C_S)^{1/2}$ provided $\varepsilon = \varepsilon_S$.

The defocus value $\varepsilon_S = -(\lambda C_S)^{1/2}$ which corresponds to the optimum imaging condition of a phase grating, is called the *Scherzer defocus* (Scherzer, 1949) and the quantity $(\lambda C_S)^{1/2}$ is used as a unit of defocus called the *Scherzer*. A more complete expression in $\varepsilon_S = -[(4/3)\lambda C_S]^{1/2}$ (see, e.g., Thomas and Goringe, 1979).

The dependence of $\sin \chi (u,v)$ on the diffraction angle $\beta = \lambda(u^2 + v^2)^{1/2}$ is represented in Fig. 1-78 for a typical situation close to the Scherzer defocus. It is a rapidly oscillating function; it is thus not possible to fullfil the required condition for all β values. The curve of Fig. 1-78, which was drawn for the Scherzer defocus value of -2100 Å, does exhibit in this particular case, a region where $\sin \chi$ is approximately -1 as expected. Beams which are diffracted in this angular range thus give rise to an image which is a direct representation of the object. If not enough beams can be passed through this "window" in the $\sin \chi$ versus β curve, the image may be rudimentary – of course in the sense that it can only give true detail up to some maximum spatial frequency. This limiting frequency corresponds roughly with the β angle for which $\sin \chi$ goes the first time through zero at Scherzer focus, that is, $\beta_{\text{max}} = (-2\varepsilon/C_S)^{1/2}$ for $\chi(\beta_{\text{max}}) = 0$; with $\varepsilon = \varepsilon_S$ this becomes $\beta_{\text{max}} = 2[\lambda/(3 C_S)]^{1/4}$, and the radius of the corresponding disc of confusion is $\varrho = (\lambda/3)^{3/4} C_S^{1/4}$, to be compared with the expression given above [Eq. (1-189)].

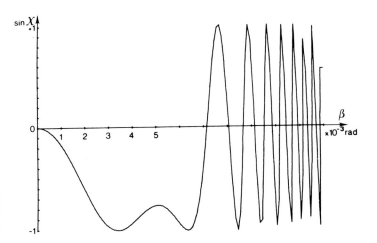

Figure 1-78. Dependence of $\sin \chi (u,v)$ on β: i.e., image transfer function without damping.

Figure 1-79a and b show two image transfer functions for two different instruments; the advantage of high voltage becomes quite apparent from the width of the window.

1.9.4.5 Projected Charge Density Approximation or *Thick*-Phase Object

Let us now consider a situation where the spherical aberration and the effect of the aperture can be neglected. The image amplitude is then given by

$$U(x, y) = \underset{x, y}{F} \left[\exp\left(-i\pi \frac{\varepsilon \beta^2}{\lambda} \right) Q(u, v) \right]$$

$$(1\text{-}203)$$

For small defocus values of ε, the exponential can be expanded, retaining only the first-order term, and with $\beta^2 = \lambda^2 (u^2 + v^2)$, one finds

$$U(x, y) \simeq \underset{x, y}{F} \{[1 - i\pi\varepsilon\lambda(u^2 + v^2)] Q(u, v)\} =$$

$$= q(x, y) - i\pi\varepsilon\lambda \underset{x, y}{F} [(u^2 + v^2) Q(u, v)]$$

The Fourier transform can be written explicitly as

$$\underset{x, y}{F} [(u^2 + v^2) Q(u, v)] =$$

$$= \int (u^2 + v^2) Q(u, v) \exp[2\pi i(ux + vy)] \, du \, dv$$

It is easy to show that

$$\frac{\partial^2}{\partial x^2} \int Q(u, v) \exp[2\pi i(ux + vy)] \, du \, dv =$$

$$= -4\pi^2 \int u^2 Q(u, v) \cdot$$

$$\cdot \exp[2\pi i(ux + vy)] \, du \, dv$$

and a similar expression is valid for the second partial derivative with respect to y. We can thus write

$$U(x, y) = q(x, y) + \frac{i\pi\varepsilon\lambda}{4\pi^2} \left(\frac{\partial^2}{\partial x^2} + \frac{\partial^2}{\partial y^2} \right) \times$$

$$\times \int Q(u, v) \exp[2\pi i(ux + vy)] \, du \, dv =$$

$$= \left[1 + \frac{i\varepsilon\lambda}{4\pi} \Delta \right] q(x, y)$$

$$(1\text{-}204)$$

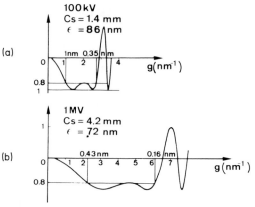

Figure 1-79. Image transfer function for two different microscopes. (a) 100 kV; (b) 1000 kV.

where Δ is the Laplace operator in two dimensions $(\partial^2/\partial x^2) + (\partial^2/\partial y^2)$.

We note that in the case of the phase-grating approximation $q(x, y) = \exp[i\sigma\varphi(x, y)]$ and that

$$\Delta \exp(i\sigma\varphi) =$$

$$= \left\{ i\sigma\Delta\varphi - \sigma^2 \left[\left(\frac{\partial\varphi}{\partial x} \right)^2 + \left(\frac{d\varphi}{\partial y} \right)^2 \right] \right\} \exp(i\sigma\varphi)$$

In this special case one thus obtains for the image amplitude

$$U(x, y) =$$

$$= \left\{ 1 - \frac{\varepsilon\lambda\sigma}{4\pi} \Delta\varphi - \frac{i\varepsilon\lambda\sigma^2}{4\pi} \left[\left(\frac{\partial\varphi}{\partial x} \right)^2 + \left(\frac{\partial\varphi}{\partial y} \right)^2 \right] \right\}$$

$$\times \exp(i\sigma\varphi)$$

$$(1\text{-}205)$$

Up to the first order in ε, the image intensity is thus given by

$$I(x, y) = UU^* = 1 - \frac{\varepsilon\lambda\sigma}{2\pi} \Delta\varphi$$

$$(1\text{-}206)$$

We know from electrostatics that the Laplacian of the potential φ is related to the charge density ϱ by Poisson's equation

$$\Delta\varphi = -4\pi\varrho$$

We thus finally obtain

$$I(x, y) = 1 + 2\varepsilon\lambda\sigma\varrho(x, y)$$

$$(1\text{-}207)$$

From this expression it becomes clear that the image contrast is now proportional to the projected charge density $\varrho(x, y)$ and to the defocus ε. In the Gaussian image plane $\varepsilon = 0$ and all contrast disappears; the contrast reverses with the sign of the defocus ε, i.e., bright becomes dark and vice versa. The same phase object, e.g., a column of atoms, may thus appear bright or dark depending on the defocus.

Details about the projected charge density approximation are given by Lynch and O'Keefe (1972), O'Keefe (1973) and Lynch et al. (1975).

1.9.5 Two-Dimensional Multiple-Beam Images

1.9.5.1 Wave Approach (Fourier Synthesis)

In Sec. 1.4.7 we presented an intuitive picture for the generation of one-dimensional lattice fringes based on a wave approach. This approach can easily be generalized to the case where more than two beams interfere. Each beam admitted through the selector aperture contributes one (or several) Fourier component(s), i.e., planar sinusoidal image waves or intensity waves. The electron microscopic image can be considered to be due to the superposition of these different *image waves,* one corresponding with each diffraction vector.

The wave vectors of these waves are given by the vectors joining the origin with the diffraction spots, corresponding with the different admitted beams, which are interfering also among each other as a consequence of multiple scattering. An image wave has an amplitude proportional with the intensity of the diffraction spot corresponding with the considered wave vector. In the simple case where only the central beam O, one first-order g(OA) and one second-order $2g$ reflection (OB) in a linear arrangement are admitted, one obtains

straight fringes with a periodic intensity distribution containing two Fourier terms, one with period $1/g$ and the second with period $1/(2g)$, corresponding with the vectors $g = \overline{OA} = \overline{AB}$ and $2g = \overline{OB}$. The more beams that are used in the linear arrangement, the more Fourier components contribute, and the more detailed fringes will be obtained.

The generalization to two dimensions is obvious. For the simple diffraction pattern where apart from the direct beam only four scattered beams forming a square are admitted one obtains the superposition of four waves, i.e., four sets of fringes with wave vectors g_1, g_2, g_3, and g_4 (Fig. 1-80). Moreover, the higher harmonics, i.e., $g_5 = \overline{AC}$ and $g_6 = \overline{BD}$ are also inevitably present. Finally, $g_7 = \overline{AB}$, $g_8 = \overline{BC}$, $g_9 = \overline{CD}$ and $g_{10} = \overline{DA}$ are also produced. The superposition of all these waves, i.e., their interference, is sufficient to produce a rudimentary image revealing the lattice, but without structural details on a sub-unit cell level, unless the structure is very simple, e.g., a face-centered-cubic element. In the particular case presented in Fig. 1-80 the number of beams, i.e., Fourier components, is sufficient to represent the structure. The more beams admitted, the finer the structural detail that can be revealed in the image. There is of course a limit to the detail that can be represented imposed by the width of the window through which the beams can be transmitted through the optical system in the correct phase relationship, and this in turn is determined by the resolution of the microscope. The considerations developed above implicitly assume that all considered waves interfere "in phase", i.e., in the correct phase relationship with the incident beam and among themselves. Unfortunately, increasing order of the Fourier components corresponds to beams enclosing angles of in-

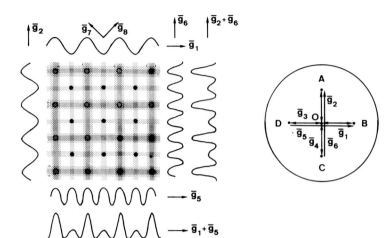

Figure 1-80. Image formation by the superposition of different Fourier components.

creasing magnitude with the optical axis. As long as we only make use of the beams which pass through the *window*, or *plateau*, in the image transfer function, the different components interfere with approximately the correct phase relationship and hence produce a directly interpretable image for a properly chosen defocus value (Scherzer defocus). For high-resolution studies it is thus important to have an instrument with a wide plateau in the image transfer function and to eliminate the beams outside of it by an aperture.

From Fig. 1-79 the advantages of the use of high voltage become apparent in this respect.

One can use also beams corresponding with β angles outside of the window, but then the resulting image is in general no longer directly interpretable except in very special cases, such as in the *aberration free focus method* (Hashimoto et al., 1978–79). For simple crystals (e.g., Si) with a small unit cell and using a small number of beams, it is possible, by choosing an adequate defocus and an appropriate C_s value, that a number of reflections outside of the plateau still keep the correct phase by passing through small high-order windows cor-

responding with approximately the same phase shift as the plateau. An image is then obtained which represents the true structure with a resolution which may exceed the point resolution.

1.9.5.2 Particle Approach (Channelling)

In the intuitive theory described above the incident electrons are modeled as plane waves. It is also possible to conceive an intuitive theory based on the particle nature of the incident electrons. An atom column parallel with the incident beam constitutes a potential well with cylindrical symmetry; it acts on the electron beam like a succession of convergent divergent lenses. This behavior is due to the periodic transformation of potential energy into the kinetic energy of the electron and vice versa. The incident electron flux is uniform on entering the foil. The focusing effect of the atom columns creates first a higher density of electrons, i.e., a maximum, close to the center of the atom strings; subsequent defocusing first causes the distribution to become more uniform again and finally to exhibit a minimum at the atom column positions. Periodically, focusing and defocusing occur along the atom column (Fig. 1-81). Each type of column, depending on the chemical nature of its constituent atoms, corresponds to a characteristic length Δ of the order of $4-10$ nm. Depending on the foil thickness, certain atom columns will thus give rise to a maximum in electron density; whereas others may give rise to a minimum at the exit face of the foil. The electron distribution at the exit face of the foil acts as the object. Each point of the exit face is a source of spherical Huyghens wavelets which interfere to form the diffraction pattern in the backfocal plane of the objective lens and subsequently form the magnified image of the object on the screen or photographic plate.

In this process, the electrons propagate through the microscope lenses and they undergo the effects of the lens aberrations. At Scherzer focus, i.e., at a suitable negative defocus, it can be shown that a bright area at the exit face of the foil will be imaged as a dark dot in the final image and vice-versa. Depending on the foil thickness, a given atom column will thus be imaged either as a relative maximum in brightness or as a relative minimum, i.e., as a bright or a dark dot. In very thin foils (5–8 nm), one usually finds that most heavy atom columns will be images as dark dots, whereas channels in the structure or light atom columns, will appear bright. The situation reverses in somewhat thicker foils where the heavy atoms appear as bright dots. In foils of 20 nm thickness or more, usually only a fraction of the atom columns is imaged as bright dots. This often results in images which only reveal the unit cell but no sub-unit cell detail. In long period modulated structures, the modulation period is usually best visible in the thick parts of the specimen, because the satellite spots corresponding with the modulation only become sufficiently intense to influence the image in the thicker parts of the specimen.

Superstructures in ordered binary alloys can sometimes be imaged by selecting only superstructure diffraction spots, excluding the basic spots. Since the superstructure spots clearly carry the information on the superstructure, they preferentially image the atoms responsible for the superstructure, i.e., the minority atom columns; in a binary alloy this is sufficient to determine the structure and its defects.

From these considerations it is clear that one should not necessarily attempt to achieve the ultimate resolution. Depending on the information searched for the use of images made with selectively chosen beams is often more relevant.

1.9.5.3 Image Interpretation

The interpretation usually proceeds by the *trial and error* method, which consists in comparing the calculated image for a given structural model with the observed image. In this calculation the phase shifts introduced by the lens system have to be properly taken into account. Usually the

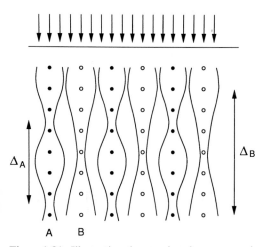

Figure 1-81. Illustrating the atomic column approximation.

calculation is made for different defocus values and for different specimen thicknesses; since we know that these parameters affect strongly and differently the phase of the different Fourier components and hence the final image.

A number of computational methods are used (e.g., Cowley and Moodie, 1957); for a survey of these methods we refer to Spence (1981) and to Van Dyck (1978). A fully dynamic direct space method has been formulated by Van Dyck and Coene (1984). Examples of the application of this method have been reproduced in Fig. 1-82. The calculated images are plotted on a cathode ray screen, simulating in this way images of the same nature as the ones observed in the microscope. Direct retrieval methods are likely to gradually replace the trial and error method used at present for the interpretation of high resolution images (Kirkland, 1984). It seems possible to reconstruct the projected lattice potential and hence to retrieve the structure from a sequence of digitalized images of the same area, made at a number of close-spaced defocus values (Van Dyck and Op de Beeck, 1990).

1.9.5.4 Defects Revealed by High-Resolution Electron Microscopy

High-resolution images reveal atom columns which are parallel with the viewing direction and are periodic along this direction. Only two-dimensional configurations consisting of such columns can adequately be represented by a dot pattern. As a result, only the core of dislocation lines parallel to the viewing direction can be imaged, and even in this case only the edge component is revealed (Fig. 1-83). The fact that non-periodic defects such as dislocations can be imaged suggests that electron diffraction is a highly localized phenomenon in a sufficiently thin foil, and provides a justification for the use of the atomic column approximation.

Dislocations which are parallel to the foil plane will perturb the shape of the atom columns close to the core by displacements along b. Lattice planes parallel to b will nevertheless remain flat and their spacing is unchanged by the presence of the dislocations. As a result, rows of bright dots, which are due to atom columns along lines parallel with b will degenerate into

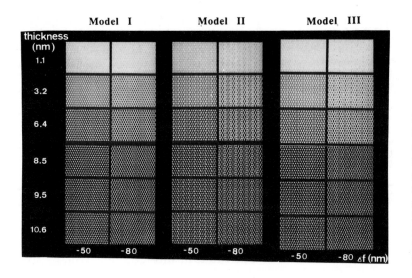

Figure 1-82. Example of image of the incommensurate structure in $Ca_{0.85}CuO_2$, simulated by means of the direct space method. Three different Ca arrangements were simulated; the best fit with the observed image is found for model I (see Fig. 1-144a) (Milat et al., 1991).

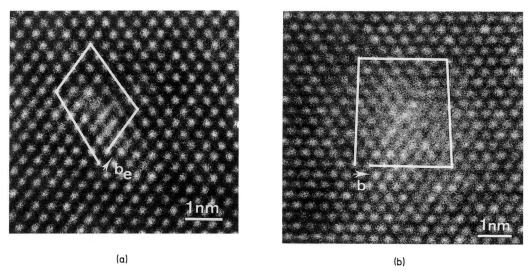

(a)   (b)

Figure 1-83. High-resolution images of dislocations in silicon. (a) 60° dislocation; (b) 90° dislocation.

continuous bright lines parallel with **b**. If the unit vector along the normal to these lines is **n**, the condition $\boldsymbol{b} \cdot \boldsymbol{n} = 0$ is satisfied. This condition is in fact the same as the $\boldsymbol{g} \cdot \boldsymbol{b} = 0$ criterion for the extinction of dislocation images in diffraction contrast. The

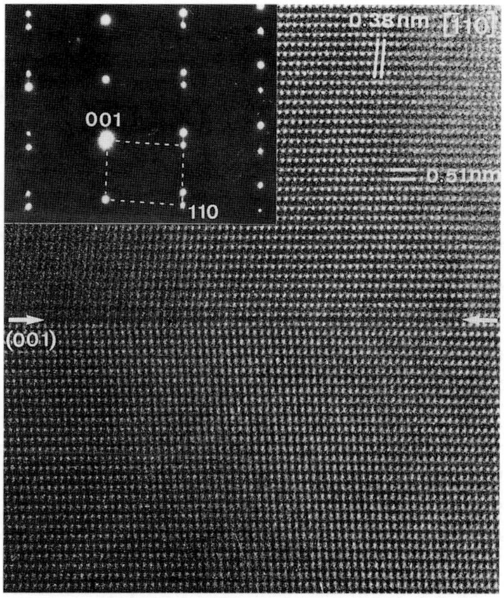

Figure 1-84. Twin interface along (001) in YTaO$_4$ imaged along a common lattice row (Courtesy G. Van Tendeloo).

direction of the Burgers vector can in principle be deduced from such images, but not the core configuration.

Antiphase boundaries and stacking faults can adequately be imaged by a proper choice of the imaging zones parallel to the planar interface and along a close-packed direction. Twins are adequately imaged along a direction which is common to the lattices of both components (Fig. 1-84). The same applies to interfaces between an epitaxial layer and a substrate. Other zones will of course also produce images, but they will not be directly interpretable. Examples of high-resolution images of structures and of defects are presented in Sec. 1.12.

1.10 Geometric Diffraction Theory

1.10.1 Introduction

Rather than starting from Schrödinger's equation, the kinematic theory can be based on the direct summation of the contributions of scattering units, which may be single atoms, atom clusters or unit cells, taking properly into account the phase differences between the waves scattered by these units due to their geometrical configuration. This approach will be introduced here in order to allow for a discussion of a number of special diffraction phenomena.

Let the scattering units be placed at positions r_j and let the electrostatic potential responsible for the scattering be represented by $V_j(r)$.

The total potential is then

$$V(r) = \sum_j V_j(r - r_j) \qquad (1\text{-}208)$$

The phase difference with respect to the origin, on diffraction in the direction h due to the volume element dr at position r, is $2\pi h \cdot r$, and the total amplitude of the scattered beam is then obtained by summation over all scattering units

$$A(h) = \int V(r) \exp(-2\pi i h \cdot r) dr = \qquad (1\text{-}209)$$
$$= \int \sum_j V_j(r - r_j) \exp(-2\pi i h \cdot r) dr$$

and after interchanging the integration and the summation

$$A(h) = \sum_j f_j(h) \exp(-2\pi i h \cdot r_j) \qquad (1\text{-}210)$$

with

$$f_j = \int_{\text{unit}} V_j(r) \exp(-2\pi i h \cdot r) dr \qquad (1\text{-}211)$$

where f_j is now the scattering amplitude of unit j. The relation in Eq. (1-209) expresses that $A(h)$ is the Fourier transform of $V(r)$.

In a periodic crystal it is possible to group the atomic scatterers in larger units which are all identical apart from a lattice translation $r_L = \sum_i l_i a_i$ (l_i: integers; a_i: base vectors of the lattice). These units are called *unit cells*.

The scattering amplitude due to one unit cell then becomes

$$F(h) = \sum_k f_k(h) \exp(-2\pi i h \cdot \varrho_k) \qquad (1\text{-}212)$$

where ϱ_k describes the positions within the unit cell of the individual atoms with scattering factor f_k referred to the origin; i.e., we have set $r_k = r_L + \varrho_k$. The scattered amplitude can then be written

$$A(h) = \sum_k \sum_L f_k \exp(-2\pi i h \cdot r_L) \cdot$$
$$\cdot \exp(-2\pi i h \cdot \varrho_k) =$$
$$= F(h) \sum_L \exp(-2\pi i h \cdot r_L) \qquad (1\text{-}213)$$

The summation over the lattice yields delta functions at the nodes g of the reciprocal lattice based on the vectors b_i defined as $b_i \cdot b_j = \delta_{ij}$ ($i, j = 1, 2, 3$), i.e., $g = \sum_j h_j b_j$ and finally (N is the number of unit cells)

$$A(h) = N F(h) \delta(g - h) \qquad (1\text{-}214)$$

where the Dirac δ expresses the Bragg condition.

If the diffraction condition is not exactly satisfied $h = g + s$, we obtain [from Eq. (1-213)]

$$A(h) = F(h) \sum_L \exp[-2\pi i(g + s) \cdot r_L] \simeq$$
$$\simeq F(g) \sum_L \exp(-2\pi i s \cdot r_L) \qquad (1\text{-}215)$$

since $g \cdot r_L = integer$ and $F(h) \simeq F(g)$. In a continuum approximation of the summation, applying moreover the column approximation and putting $s \cdot r_L = sz$, where the z-axis is parallel with the column

$$A(h) = F(g) \int_{\text{column}} \exp(-2\pi i s z) dz \qquad (1\text{-}216)$$

This relation has the same physical content as Eq. (1-22) and both expressions become identical provided $(\pi i/t_g) \exp(i\theta_g) = F(g)$, i.e., $F(g) \propto 1/t_g$.

Deformation can be introduced in this formalism by describing the displacement $R(r)$ of the unit cells along a column $[R(r) \ll r_L]$. The relation in Eq. (1-215) then becomes

$$A(h) = \qquad (1\text{-}217)$$
$$= F(g) \sum_L \exp\{-2\pi i(g + s) \cdot [r_L + R(r_L)]\}$$

Applying the column approximation and neglecting small quantities such as $s \cdot R$, this reduces to

$$A(h) = \tag{1-218}$$
$$= F(g) \sum_{column} \exp\{-2\pi i[s \cdot r_L + g \cdot R(r_L)]\}$$

Replacing the summation by an integration along the column, one obtains an expression which is equivalent with Eq. (1-31)

$$A(h, x, y) = \tag{1-219}$$
$$= F(g) \int_{column} \exp\{-2\pi i[sz + g \cdot R(x, y, z)]\}\, dz$$

The lattice potential can be introduced in this formalism by noting that

$$V(r) = \sum_g V_g \exp(2\pi i g \cdot r) \tag{1-220}$$

and applying the relation in Eq. (1-208) to obtain $A(h)$

$$A(h) = \int \sum_g V_g \exp[2\pi i(g-h) \cdot r]\, dr =$$
$$= \sum_g V_g \delta(g-h) = N V_h \quad \text{with} \quad g = h \tag{1-221}$$

(N is the number of unit cells).

The Fourier coefficients V_g are obtained by Fourier inversion

$$V_g = \frac{1}{V} \int V(r) \exp(-2\pi i g \cdot r)\, dr \tag{1-222}$$

$$V_g = \frac{F_g}{V}$$

The relations derived in this section will be used to describe a number of particular diffraction phenomena observable in electron diffraction patterns.

1.10.2 Diffraction by Modulated Structures

The recent years it has become clear that a number of structures produce electron diffraction patterns consisting of a lattice of strong spots, so-called *basic spots,* exhibiting moreover arrays of much weaker satellite spots associated with these basic spots. Often these arrays are linear, but they may also be two- or even three-dimensional arrays. Such weak spots are much easier to observe in electron diffraction than in X-ray or neutron diffraction. In the past these satellite spots were often ignored in structure determinations, either intentionally or because they were too weak to be observed in X-ray or neutron diffraction patterns. The structures obtained in this manner are thus only average structures, as we shall see. It has now become clear that such satellites reveal long-period superstructures called *modulated structures* (Amelinckx et al., 1989 a, b; Janssen and Janner, 1987).

Different types of modulated structures are distinguishable:

(i) Composition Modulation

The scattering units which are associated with the nodes of the basic lattice may vary periodically in chemical nature or in occupancy, the periodicity of this variation being described by a lattice with a larger unit cell than that of the basic lattice, but not necessarily related rationally to this lattice.

(ii) Deformation or Displacement Modulation

The scattering units may be displaced away from the nodes of the average or basic lattice according to a periodic pattern, the periodicity of the displacement pattern being larger than that of the basic lattice.

(iii) Interface Modulation

The long period structures may consist of the periodic juxtaposition of "modules" (i.e., slabs or blocks) of a simpler, basic structure. Two successive modules are separated by planar interfaces: e.g., stacking faults (in polytypes), out-of-phase boundaries (in long period anti-phase boundary superstructures), twin boundaries (in polysynthetic twinning), inversion boundaries, or discommensuration walls.

We shall discuss the characteristic features of the diffraction patterns due to these different types of modulated structures.

It should be noted that in all cases the modulation period can either be commensurate or incommensurate with the period of the basic lattice.

Incommensurate diffraction patterns may be due either to incommensurately modulated structures or to *uniform sequences* of commensurate modules. A uniform sequence is one which at any point in space approximates best a given period, which need not be an integer number of unit cells.

Uniform sequences consisting of modules containing an integer number of unit cells, e.g., $n \cdot a$ and $m \cdot a$ (n, $m = integer$, $n < m$) generating a given average period Λ ($n \cdot a < \Lambda < m \cdot a$) can be obtained by one of the several algorithms, such as the Fujiwara square wave method (Fujiwara, 1957) (limited to the case $m = n + 1$) or the more general cut and projection method (Elser, 1985, 1986; Levine and Steinhardt, 1984, 1986; Duneau and Katz, 1985; Frangis et al., 1990; Van Tendeloo et al., 1989 a, b). The modulation period Λ is obtained directly from the diffraction pattern $\Lambda = 1/q$, where q is the satellite spacing.

1.10.3 Composition Modulation

We make use of the expression in Eq. (1-213). The structure amplitude of the scattering units now becomes a periodic function of their position r, i.e., it becomes

$F(h, r)$: this function clearly only has values at discrete points $r = r_L$. The scattering amplitude is thus given by

$$A(h) = \sum_L F(h, r_L) \exp(-2\pi i h \cdot r_L) \quad (1\text{-}223)$$

Let $\{G\}$ represent the set of reciprocal lattice vectors corresponding with the periodicity of the composition variations. We can then expand $F(h, r)$ in a Fourier series:

$$F(h, r) = \sum_G a_G(h) \exp(2\pi i G \cdot r) \quad (1\text{-}224)$$

and the scattered amplitude then becomes

$$A(h) = \sum_G \sum_L a_G(h) \exp[2\pi i (G - h) \cdot r_L] =$$
$$= \sum_G a_G(h) \sum_L \exp[2\pi i (G - h) \cdot r_L] \quad (1\text{-}225)$$

The second sum leads to delta-like maxima at the nodes g of the reciprocal lattice of the basic structure, i.e., $G - h = g$, and hence $h = g + G$. Each Bragg spot g of the unmodulated structure gives rise to an array of satellites of which the positions are given by the vectors G. Each satellite is thus characterized by two reciprocal lattice vectors g and G.

1.10.4 Displacement Modulation

We now assume all scattering units to be the same, but their positions are assumed to deviate from their average lattice position r_L by a vector δ_L which is a periodic function of the position vector r: i.e., $\delta_L(r)$ (Van Landuyt et al., 1974; Overhauser, 1962).

Let the periodicity of the displacement field be described by a lattice and the associated reciprocal lattice, different from the basic lattice.

The scattered amplitude then becomes

$$A(h) = F(h) \sum_L \exp[-2\pi i h \cdot (r_L + \delta_L)] =$$
$$= \sum_L F(h) \exp(-2\pi i h \cdot \delta_L) \cdot$$
$$\cdot \exp(-2\pi i h \cdot r_L) \quad (1\text{-}226)$$

This expression is formally equivalent with Eq. (1-213). It is possible to interpret it as being due to the average lattice occupied by scattering units with site-dependent structure amplitudes. The general conclusion of the previous paragraph thus remains valid.

We can be somewhat more explicit, however, and note that since the displacement field $\delta(r)$ is periodic, then $\exp(-2\pi i h \cdot \delta_L)$ is also periodic with the same period and can thus be expanded in a Fourier series.

$$\exp(-2\pi i h \cdot \delta_L) =$$
$$= \sum_G a_G(h) \exp(2\pi i G \cdot r) \qquad (1-227)$$

and hence

$$A(h) = F(h) \sum_L \sum_G a_G(h) \exp[2\pi i (G-h)] \qquad (1-228)$$

which leads again to delta-like peaks at $G - h = g$ or $h = g + G$ describing the diffraction peaks. For $|\delta_L| \ll 1/|g|$ it is possible to approximate

$$\exp(-2\pi i h \cdot \delta_L) \simeq 1 - 2\pi i h \cdot \delta_L \qquad (1-229)$$

and since δ_L can also be expanded in a Fourier series with vectorial coefficients A_G, which are the polarization vectors of the various harmonics:

$$\delta_L = \sum_G A_G \exp(2\pi i G \cdot r_L) \qquad (1-230)$$

Within this approximation one thus obtains by comparing Eqs. (1-227), (1-229) and (1-230)

$$a_G(h) = -2\pi i h \cdot A_G \quad \text{for} \quad G \neq 0 \qquad (1-231)$$

and

$$a_G(h) = 1 \quad \text{for} \quad G = 0$$

From this one can conclude for instance that satellites will be extinct if $h \cdot A_G = 0$, i.e., if the polarization vector A_G of the Fourier component G is perpendicular to the diffraction vector h. The extinction conditions of satellites can be used to derive the symmetry properties of the displacement field δ_L (Yamamoto, 1980, 1982 a, b, 1985; De Wolff, 1974, 1984; De Wolff et al., 1981; Bird and Withers, 1986; Perez-Mato et al., 1986, 1984).

In a number of cases the displacements are large enough to lead to visible displacements of the atom columns in high resolution images. Reasonable models of the modulated structure can then be deduced from such images. This is for instance the case in the superconducting material $Bi_2Sr_2CaCu_2O_8$. Figure 1-85 shows a diffraction pattern exhibiting satellite reflections, and Fig. 1-86 shows the corresponding high-resolution image. The layer planes are clearly wavy. The inset is the simulated image based on a model inspired by the high-resolution image (Zandbergen et al., 1988).

1.10.5 Composition Versus Displacement Modulation

If the composition modulation is assumed to be sinusoidal, the Fourier expan-

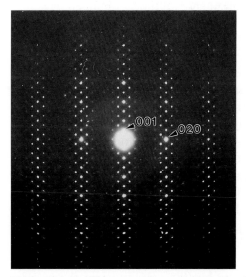

Figure 1-85. Satellite arrays due to displacive modulation of the structure of $Bi_2Sr_2CaCu_2O_8$.

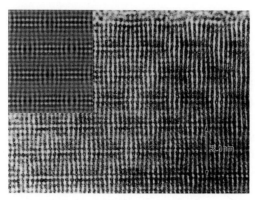

Figure 1-86. High-resolution image of the modulated structure in $Bi_2Sr_2Ca_2CuO_8$, as viewed along the same zone as the diffraction pattern of Fig. 1-85, i.e., along the layer planes. The inset is a simulated image based on a model derived from the high-resolution image (Zandbergen et al., 1988).

sion of the structure factor only contains two terms, and as a result the basic spots have only two satellites, one on each side, so-called side-bands.

The intensity of the satellites decreases with increasing order of the basic reflections because of the decrease in atomic scattering amplitude with increasing scattering angle.

On the other hand, in the case of deformation modulation the intensity of the satellites belonging to higher-order basic reflections is relative to that of the basic spot larger than that of satellites associated with low-order basic reflections. This is a consequence of the fact that the amplitude of the satellite G belonging to the basic reflection h is proportional with $F(h)$ but also with $h \cdot A_G$, i.e., with the length of h.

If the deformation modulation is sinusoidal, the intensity of the satellites decreases rapidly with increasing order; in practice only one satellite on each side of the basic spot will have an observable intensity, but the amplitudes of the higher-order satellites do not vanish. The number

of visible satellites increases as the deformation wave becomes more square.

In practice, a composition modulated structure will also be displacement modulated with the same period; a clear distinction is thus neither possible nor meaningful.

1.10.6 Translation Interface Modulation

We will limit ourselves to the most commonly occurring case of periodic translation interfaces. The basic structure is assumed to be divided into slabs (i.e., moduli) separated by a set of planar interfaces normal to e and separated by a constant spacing Δ. The interfaces between slabs of basic structure all have the same displacement vector R_0. The wave vector of the modulation is then $q = (1/\Delta) e$. Let $u(r_n)$ represent the displacement of the scattering center at r_n, with respect to a scattering center in the origin. This function has the property

$$u(r_n + N \Delta e) = u(r_n) + N R_0$$

$(N = integer)$ \hfill (1-232)

This relation follows from the cumulative nature of the displacements R_0 at each interface. We now define a function

$$T(r) = u(r) - (q \cdot r) R_0 \tag{1-233}$$

which is periodic if Eq. (1-232) is satisfied; one obviously finds

$$T(r + \Delta e) = T(r) \tag{1-234}$$

the period of T along e is thus $\Delta = 1/q$. The function $\exp[-2\pi i h \cdot T(r)]$ is then also periodic with the same period as $T(r)$ and can thus be expanded as

$$\exp[-2\pi i h \cdot T(r)] =$$
$$= \sum_m A_m(h) \exp(2\pi i m q \cdot r) \tag{1-235}$$

The amplitude of the diffracted beam can now be obtained as the following sum over n, i.e., over the scattering units:

$$A(h) = \sum_n f_n(h) \exp\{-2\pi i h \cdot [r_n + u(r_n)]\} \quad (1\text{-}236)$$

or taking into account Eqs. (1-233), (1-235) and (1-236):

$$A(h) = \sum_n \sum_m f_n A_m(h) \cdot \quad (1\text{-}237)$$
$$\cdot \exp\{-2\pi i [h + m q + (h \cdot R_0) q] \cdot r_n\}$$

The summation over n leads to peaks at positions given by

$$H = h + (m + h \cdot R_0) q \quad (m = integer) \quad (1\text{-}238)$$

The long period structure thus exhibits a diffraction pattern consisting of main peaks at h. With each basic spot h a linear sequence of equidistant satellites $m q \equiv m/\Delta$ ($m = integer$) is associated; it is perpendicular to the interfaces. The linear sequences are shifted with respect to the positions of the basic spots over a fraction $h \cdot R_0$ of the interspot distance. Provided the basic structure is known, the interface modulated structure can be deduced from the geometry of the diffraction pattern. The orientation of the interfaces is normal to the sequences of satellites; the spacing of the interfaces is the inverse of the satellite separation and finally the projection of R_0 on the diffraction vectors can be deduced from the observed fractional shifts $h \cdot R_0$. Examples of the application of this method are shown in Sec. 1.12.11.

A review on the translation interface modulation method is given by Van Landuyt et al. (1970) and Amelinckx et al. (1984).

1.11 Short-Range Order in Binary Systems

1.11.1 Introduction

Although diffuse scattering can be due to numerous phenomena, a discussion of all of these is outside the scope of this contri-

bution. One can distinguish two main types: thermal diffuse scattering due to phonons (i.e., dynamic disorder) and diffuse scattering due to static disorder (e.g., short-range order). In principle, the two types can be distinguished by making a high-resolution image and revealing at least the unit cell. If the optical diffraction pattern of the high-resolution negative exhibits the same type of diffuse scattering as the electron diffraction pattern, the disorder is static. If on the other hand the disorder is dynamic, the average structure, which is the one which is recorded in the high resolution image, remains undeformed and the optical diffraction pattern will then only reveal sharp spots (Van Tendeloo and Amelinckx, 1986).

In addition, the temperature dependence may allow the two cases to be distinguished. Thermal diffuse scattering is appreciably temperature dependent and is gradually enhanced by an increase in temperature and suppressed by cooling; static disorder can also be temperature dependent, but the change from diffuse scattering to sharp spots is more abrupt and occurs usually below a rather well defined temperature.

In the short-range-order state the diffuse intensity is spread out over the whole of reciprocal space; it has a three-dimensional distribution. Nevertheless more or less pronounced intensity maxima are usually present at well-defined crystallographic positions. In binary alloys based on a face-centered-cubic alloy this is mostly the case at "special positions" such as $\{100\}$ $\{1/2\ 1/2\ 1/2\}$ $\{1\ 1/2\ 0\}$ ..., corresponding with the intersection points of symmetry elements. Upon ordering, the diffuse intensity distribution gradually evolves into the configuration of sharp superstructure spots characteristic of the long-range-ordered structure.

For intermediate situations it often happens that the diffuse intensity first becomes concentrated along a surface or along a curve, usually passing through the positions of the superstructure spots, which are usually prefigured by local enhancements of the intensity. This intensity distribution corresponds to an intermediate state of order, termed "transition state", which is in a sense a precursor of the long-range-ordered state (Fig. 1-87). The diffuse intensity pattern has the translation symmetry of the reciprocal lattice.

We shall show that in the transition state local atom configurations, called "clusters" are formed, which also occur in the long-range-ordered state. The geometry of the diffuse intensity surface restricts the possible configurations of atoms. These have to satisfy a "cluster relation" which imposes conditions on the shape and composition of the clusters, i.e., on the occupation of the cluster sites, but nevertheless leaves enough freedom to allow a disordered structure to be generated.

Figure 1-87. Example of diffuse scattering due to a transition state in nickel-intercalated NbS_2.

This type of diffuse scattering has mainly been studied by electron diffraction because this technique produces an undistorted image of sections of reciprocal space. We shall therefore give a brief survey of the theory that allows electron diffraction patterns of *transition state* structures to be interpreted.

1.11.2 Transition State Diffuse Scattering

We consider for simplicity a binary system of A and B atoms located on a lattice with a primitive unit cell. The atomic fractions of A and B atoms are respectively m_A and m_B with $m_A + m_B = 1$. The occupation of the site (j), located at r_j is described by the occupation operators σ_j^A and σ_j^B. The operator σ_j^A takes the value 1 when the site j is indeed occupied by an A atom and the value 0 when occupied by a B atom; if this is not the case $\sigma_j^B = 1$ and $\sigma_j^A = 0$. We thus clearly obtain $\sigma_j^A + \sigma_j^B = 1$.

The scattered amplitude, according to the geometrical theory, can then be written as

$$A(h) \equiv \sum_j (f_A \sigma_j^A + f_B \sigma_j^B) \exp(2\pi i\, h \cdot r_j) \tag{1-239}$$

where the sum extends over all lattice sites; f_A and f_B are the atomic scattering amplitudes for A and B atoms, respectively. We can now introduce the Flinn Operator $\bar{\sigma}_j$ which describes essentially the deviation from the average occupancy:

$$\tag{1-240}$$
$$\bar{\sigma}_j = m_A - \sigma_j^A = \sigma_j^B - m_B \text{ with } m_A \geq m_B$$

Substituting these expressions in Eq. (1-239) allows one to separate the Bragg diffracted peak $A_B(h)$ from the diffuse scattering $A_D(h)$.

One then finds

$$A(h) = A_B(h) + A_D(h) \tag{1-241}$$

with
$$\tag{1-242}$$
$$A_B(h) = (m_A f_A + m_B f_B) \sum_j \exp(2\pi i\, h \cdot r_j)$$

and

$$A_D(h) = (f_B - f_A) \sum_j \bar{\sigma}_j \exp(2\pi i h \cdot r_j)$$

The first expression $A_B(h)$ can be understood by noting that identical atoms, all having as a scattering amplitude the weighted average, are assumed to be placed at the node points r_k of the lattice; this clearly leads to the Bragg spots since

$$\sum_j \exp(2\pi i h \cdot r_j) = \delta(h-g) \qquad (1\text{-}243)$$

The second term describes the scattering by the deviations from this averaged structure; it represents diffuse scattering.

We note that $\bar{\sigma}_j = m_A$ when the site j is occupied by a B atom whereas $\bar{\sigma}_j = -m_B$ when the site is occupied by an A atom. The average over all lattice sites of $\bar{\sigma}_j = \langle \bar{\sigma}_j \rangle = 0$, which is obvious from the definition of $\bar{\sigma}_j$. We also note that $\langle \sigma_j^A \rangle = m_A$ and $\langle \sigma_j^B \rangle = m_B$ and hence

$$\langle \bar{\sigma}_j^2 \rangle = \langle \bar{\sigma}_j \bar{\sigma}_j \rangle =$$
$$= \langle (m_A - \sigma_j^A)(\sigma_j^B - m_B) \rangle = m_A m_B \qquad (1\text{-}244)$$

The diffuse scattering disappears in the Bragg spots, i.e., for $h = g_i$, because $g_i \cdot r_j = integer$ and hence $\exp(2\pi i g_i \cdot r_j) = 1$ and thus

$$A_D(g) = (f_B - f_A) N \langle \bar{\sigma}_j \rangle = 0 \qquad (1\text{-}245)$$

(N is the number of lattice sites).

The reduced diffuse scattering amplitude, defined as $A_D(h)/(f_B - f_A)$ thus finally becomes

$$a(h) = \sum_j \bar{\sigma}_j \exp(2\pi i h \cdot r_j) \qquad (1\text{-}246)$$

i.e., the diffuse intensity can be expressed as a Fourier sum, of which the Flinn operators are the Fourier coefficients corresponding with the lattice sites r_j. The expression $a(h)$ has the translation symmetry of the reciprocal lattice; since the r_j are lattice vectors one has:

$$\sum_j \bar{\sigma}_j \exp[2\pi i (h+g) \cdot r_j] =$$
$$= \sum_j \bar{\sigma}_j \exp(2\pi i h \cdot r_j) \qquad (1\text{-}247)$$

since $g \cdot r_j = integer$.

Since also the diffuse intensity contours have the translation symmetry of the reciprocal lattice, the equation of the locus can be expressed by means of a Fourier representation (De Ridder et al., 1976, 1977; De Ridder, 1978):

$$f(h) = \sum_k \omega_k \exp(-2\pi i h \cdot r_k) \qquad (1\text{-}248)$$

We now express analytically that $a(h)$ can only differ from zero along the locus $f(h) = 0$; this leads to

$$a(h) f(h) \equiv 0 \qquad (1\text{-}249)$$

This does *not* mean though that intensity has to be present all along $f(h)$.

Explicitly, this becomes

$$\sum_j \sum_k \omega_k \bar{\sigma}_j \exp[2\pi i h(r_j - r_k)] \equiv 0 \qquad (1\text{-}250)$$

Since $r_j - r_k$ is again a lattice vector, we rewrite Eq. (1-250) as $r_j - r_k \to r_j$; that is, we make a substitution on the indices $j - k \to j$ and thus also $j \to j + k$, and thus Eq. (1-250) becomes

$$\sum_j \left(\sum_k \omega_k \bar{\sigma}_{j+k} \right) \exp(2\pi i h \cdot r_j) \equiv 0 \qquad (1\text{-}251)$$

Since the exponentials have a modulus equal to unity, this expression can only be identical to zero if all coefficients are zero, i.e., if

$$\sum_j \omega_k \bar{\sigma}_{j+k} = 0 \qquad (1\text{-}252)$$

This is a homogeneous linear equation among Flinn operators, which imposes the occupation conditions on the cluster sites described by the set $\{r_k\}$ used in Eq. (1-248).

We now demonstrate its use in a simple example which is relevant for a number of binary systems based on a cubic lattice. In a number of such systems the diffuse surface of Fig. 1-88 was observed. Its shape can to a good approximation be represented analytically by the equation

$$\cos \pi h + \cos \pi k + \cos \pi l = 0 \qquad (1\text{-}253)$$

in h, k, l space, or expressed in exponential form

$$\exp(\pi i h) + \exp(-\pi i h) + \exp(\pi i k) +$$
$$+ \exp(-\pi i k) + \exp(\pi i l) +$$
$$+ \exp(-\pi i l) = 0 \qquad (1\text{-}254)$$

Comparing this expression with Eq. (1-248), we conclude that the $\omega_k = 1$ and that the cluster vectors r_k are $\pm (1/2)[100]$; $\pm (1/2)[010]$ and $\pm (1/2)[001]$.

This can be seen by writing, for instance, $\pi i k = 2\pi i [hkl] \cdot (1/2)[010]$. The set of cluster vectors describes an octahedron of sites around each reference site. The cluster relation states that the sum of the Flinn operators $\Sigma \bar{\sigma}_j = 0$ must be zero, which means that all such octahedra, centered on an arbitrary lattice site, must have the macroscopic composition which is consistent with the fact that the diffuse intensity contours pass through the positions of the superstructure spots. This condition is of course satisfied particularly in the long-range-ordered structure, since the unit cell has the macroscopic composition. The condition becomes more stringent with smaller clusters.

At the limit, when the cluster is as large as the whole crystal, the condition $\sum_j \bar{\sigma}_j = 0$ is obviously satisfied. This can also be seen by applying the condition from Eq. (1-249) to this case:

$$a(\boldsymbol{h}) \sum_j \delta(\boldsymbol{h} - \boldsymbol{g}_j) \equiv 0$$

The corresponding cluster relation is then $\sum_j \bar{\sigma}_j = 0$. This relation states that diffuse scattering can occur everywhere in reciprocal space except in the reciprocal lattice nodes.

1.11.3 Examples of Applications

1.11.3.1 Planes of Diffuse Scattering

The simplest non-trivial case is that of parallel planes of diffuse scattering perpendicular to [100] (Fig. 1-90a and b). They can be represented by the equation $\sin \pi h = 0$ or $\exp(\pi i h) - \exp(-\pi i h) = 0$, which leads to the cluster relation $\bar{\sigma}_{100} = \bar{\sigma}_{\bar{1}00}$.

This means that order is imposed along the [100] direction only; the relative positions of the equally spaced linear arrangements along [100] remain random, however.

1.11.3.2 Lines of Diffuse Scattering

A second example is concerned with one-dimensional stacking disorder of (001) layers (Fig. 1-91c and d). In reciprocal space, lines of diffuse scattering are present along [001]* for instance. The geometry of the diffuse scattering can now be represented by the intersection of two loci

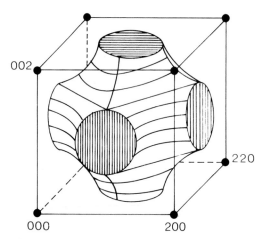

Figure 1-88. Diffuse intensity surface observed in a number of systems in which the clusters are octahedra.

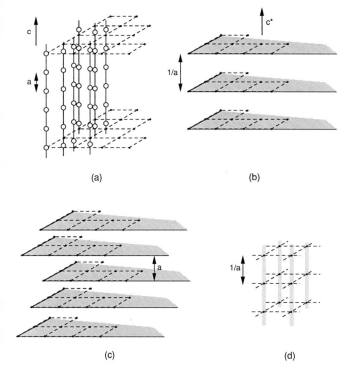

(a)   (b)

(c)   (d)

Figure 1-89. Diffuse scattering due to various forms of disorder. (a) Disordered linear arrays of equally spaced scattering centers in direct space; (b) corresponding planes of diffuse scattering in reciprocal space; (c) one-dimensional disorder of equidistant layers in direct space; (d) lines of diffuse scattering in reciprocal space.

$\sin \pi h = 0$, $\sin \pi k = 0$; these equations must be satisfied simultaneously; similarly, the corresponding cluster relations $\bar{\sigma}_{100} = \bar{\sigma}_{\bar{1}00}$; $\bar{\sigma}_{010} = \bar{\sigma}_{0\bar{1}0}$ must also be satisfied simultaneously. These relations impose order within the (001) layers, but leave the relation between successive layers random.

A well-known concrete example is the faulted stacking of close-packed layers (111) in an f.c.c. crystal. Certain reflections which are common to the cubic and to the hexagonal close-packed stacking remain sharp, but other reflections which are different in the h.c.p. and f.c.c. stacking become streaked in a direction perpendicular to the layer planes. A simple rule allows one to deduce which reflections will be become streaked when a stacking is faulted. If the vector relating the correct stacking position to the faulted position in a given stacking sequence is R_0, referred to the

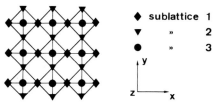

Figure 1-90. Vertex sharing octahedra, illustrating the notation used, in particular the sublattices 1, 2 and 3.

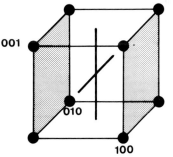

Figure 1-91. Locus of diffuse intensity resulting from orientational disorder in perovskite-like structures for one sublattice. The actual pattern is the superposition of three orientation variants of this type.

perfect lattice, those reflections (referred to the same perfect lattice) for which $g \cdot R_0 \neq integer$ will become streaked. This is a consequence of the fact that a shift R_0 affects the structure factor $F(g)$ only if $g \cdot R \neq integer$. For stacking faults in a cubic close-packed arrangement $R_0 = (1/6)[11\bar{2}]$; i.e., the reflections for which $h + k - 2l = 6\text{-fold}$ remain sharp; the others become streaked.

1.11.4 Short-Range-Order Parameters

From Eq. (1-254) we obtain by Fourier inversion

$$\sigma_j = (1/V^*) \int_{V^*} a(h) \exp(-2\pi i h \cdot r_j) dh \tag{1-255}$$

where the integral extends over the volume V^* of the reciprocal unit cell. For the reduced intensity of the diffuse scattering one obtains

$$I(h) = \sum_l \sum_{l'} \bar{\sigma}_l \bar{\sigma}_{l'} \exp[2\pi i h \cdot (r_l - r_{l'})] =$$
$$= N \sum_j \langle \bar{\sigma}_l \bar{\sigma}_{l+j} \rangle \exp(2\pi i h \cdot r_j) \tag{1-256}$$

The averages of the products of Flinn operators are related to the Warren-Cowley short-range-order parameters α_{0j} (Cowley, 1950) by

$$\alpha_{0j} = \frac{\langle \bar{\sigma}_l \bar{\sigma}_{l+j} \rangle}{\langle \bar{\sigma}_l^2 \rangle} = \frac{\langle \bar{\sigma}_l \sigma_{l+j} \rangle}{m_A m_B} \tag{1-257}$$

Fourier inversion of Eq. (1-256) leads to

$$\alpha_{0j} = \frac{1}{m_A m_B N V^*} \int_{V^*} I(h) \exp(-2\pi i h \cdot r) dh \tag{1-258}$$

This expression allows in principle the short-range-order-parameters to be obtained from measurements of the diffuse intensity distribution.

1.11.5 Diffuse Scattering due to Orientational Disorder

Orientational disorder may also give rise to diffuse scattering concentrated on simple geometric loci. This is the case, for instance, at high temperature in certain perovskites, where corner-linked rigid oxygen octahedra may give rise to orientationally disordered arrangements revealed in reciprocal space as lines and/or planes of diffuse scattering.

Consider the configuration of vertex-sharing oxygen octahedra represented in Fig. 1-90. Small tilts of the octahedra about an arbitrary axis can be decomposed in tilts about the three mutually perpendicular tetrad axis x, y and z. Tilting about one of these axis leaves the atoms situated on that axis unchanged but displaces the others. It is therefore meaningful to subdivide the lattice of oxygen atoms in three sublattices 1, 2 and 3 corresponding directly to oxygen atoms situated on the fourfold axis along x, y and z (Fig. 1-90). In particular the displacement of atom j on sublattice 1 can be written as:

$$\delta r_j = \sigma_j^{(y)} \delta y \, e_y + \sigma_j^{(z)} \delta z \, e_z \tag{1-259}$$

where e_z and e_y are the unit vectors along the y and z axis, and δy and δz are the magnitudes of the displacements in these direction. Since the displacements result from rotations of rigid interlinked octahedra it is reasonable to assume that δy and δz have the same magnitude for all atoms. The sense of the displacement can of course either be positive or negative, i.e., the operators $\sigma_j^{(y)}$ and $\sigma_j^{(z)}$ in Eq. (1-259) can only adopt the values $+1$ or -1.

It is sufficient to make the calculation for one sublattice; the result is similar, mutatis mutandis, for the others. The scattered amplitude in the direction h produced by sublattice 1 is obtained by summing the contributions of all atoms, at positions $r_j ...$, on this sublattice

$$a(h) = f_o(h) \sum_j \exp\left[2\pi i\, h \cdot (r_j + \delta r_j)\right] \quad (1\text{-}260)$$

$f_o(h)$ is the atomic scattering amplitude of oxygen; h is a position vector in reciprocal space, defining $h \cdot e_x = h$; $h \cdot e_y = k$ and $h \cdot e_z = l$ can be rewritten Eq. (1-260) as

$$a(h) = f_o(h) \sum_j \exp\left(2\pi i\, h \cdot r_j\right) \cdot \quad (2\text{-}261)$$
$$\cdot \exp\left[2\pi i (\sigma_j^{(y)} k\, \delta y + \sigma_j^{(z)} l\, dz)\right]$$

Similar expressions apply for the sublattices 2 and 3. We note that

$$\exp\left(2\pi i\, \sigma_j^{(y)} k\, \delta y\right) = \quad (1\text{-}262)$$
$$= \cos\left(2\pi\, \sigma_j^{(y)} k\, \delta y\right) + i \sin\left(2\pi\, \sigma_j^{(y)} k\, \delta y\right)$$

Since the cosine is an even function, whereas the sine is an odd function, one also finds

$$\exp\left(2\pi i\, \sigma_j^{(y)} k\, \delta y\right) = \quad$$
$$= \cos\left(2\pi k\, \delta y\right) + i\, \sigma_j^{(y)} \sin\left(2\pi k\, \delta y\right) \quad (1\text{-}263)$$

A similar expression can be written for the term in δz; it is obtained by the substitution $y \to z$ and $k \to l$. The scattered amplitude is thus

$$a(h) = f_o(h) \sum_j \exp\left(2\pi i\, h \cdot r_j\right) \times$$
$$\times \left[\cos\left(2\pi k\, \delta y\right) + i\, \sigma_j^{(y)} \sin\left(2\pi k\, \delta y\right)\right] \times$$
$$\times \left[\cos\left(2\pi l\, \delta z\right) + i\, \sigma_j^{(z)} \sin\left(2\pi l\, \delta z\right)\right] \quad (1\text{-}264)$$

This expression consists of four terms:

$$a(h)/f_o(h) = c_k c_l \sum_j \exp\left(2\pi i\, h \cdot r_j\right) \quad (1\text{-}265\,a)$$

$$+ i c_k s_l \sum_j \sigma_j^{(z)} \exp\left(2\pi i\, h \cdot r_j\right) \quad (1\text{-}265\,b)$$

$$+ i c_l s_k \sum_j \sigma_j^{(y)} \exp\left(2\pi i\, h \cdot r_j\right) \quad (1\text{-}265\,c)$$

$$- s_k s_l \sum_j \sigma_j^{(y)} \sigma_j^{(z)} \exp\left(2\pi i\, h \cdot r_j\right) \quad (1\text{-}265\,d)$$

where c_k, c_l, s_k, s_l represent the following:

$$c_k = \cos\left(2\pi k\, \delta y\right); \quad c_l = \cos\left(2\pi l\, \delta z\right)$$
$$s_k = \sin\left(2\pi k\, \delta y\right); \quad s_l = \sin\left(2\pi l\, \delta z\right)$$

The term (a) in Eq. (1-265) represents the Bragg peaks of the undeformed structure, the amplitude being reduced by a factor $c_k c_l$, which is close to 1 since the tilts are small.

The other terms in Eq. (1-265) represent the diffuse scattering; note that each of these terms is of the same form as term (a); each term therefore corresponds with a different part of the locus of diffuse intensity. We shall comment on the interpretation of these terms and deduce the loci of diffuse scattering from these expressions. Rather than following the direct method illustrated above, which consists in deriving the cluster relation from the geometry of the diffuse scattering locus, we now derive the loci from the knowledge of the occupation parameters.

The term (b) in Eq. (1-265) refers to the displacements in the z direction resulting from a rotation about the y direction, whereas the term (c) refers to the displacements in the y direction as a result of a rotation about the z direction. The last term is discussed separately.

The displacements in the z direction, resulting from a rotation about the y-axis are opposite in sign for adjacent oxygen atoms along the x direction as a result of the vertex-sharing and of the rigid character of the octahedra; this is expressed by the relation $\sigma_{100}^{(z)} = - \sigma_{000}^{(z)}$. A similar reasoning applies to the displacements in the z direction resulting from a rotation about the x-axis; for successive O-atoms along the z-axis, $\sigma_{001}^{(z)} = - \sigma_{000}^{(z)}$.

The following relations thus exist between the occupation parameters:

$$\sigma_{100}^{(z)} + \sigma_{000}^{(z)} = 0 \quad \text{and}$$
$$\sigma_{001}^{(z)} + \sigma_{000}^{(z)} = 0 \quad (1\text{-}266)$$

they are of the form of Eq. (1-242) with $\omega_{100} = \omega_{000} = 1$ and $\omega_{001} = \omega_{000} = 1$.

Knowing the ω's, the equations of the geometric loci can be written in the same form as Eq. (1-248), i.e., as

$$\exp(-2\pi i h) + 1 = 0 \quad \text{and}$$
$$\exp(-2\pi i l) + 1 = 0$$

or

$$h = n + 1/2 \quad \text{and} \quad l = n + 1/2 \quad (n = integer)$$

which represent planes parallel with the cube planes $x = 0$ and $z = 0$. Both relations have to be satisfied simultaneously; they thus represent a set of straight lines parallel to the y direction passing through the cube centers.

The third term refers to the diffuse intensity associated with displacements along y resulting from tilting about the z direction. Following the same type of reasoning as above, one now has $\sigma^{(y)}_{100} + \sigma^{(y)}_{000} = 0$. Similarly, from tilting about the y direction $\sigma^{(y)}_{010} + \sigma^{(y)}_{000} = 0$ follows, and thus for the geometrical locus

$$h = n + 1/2; \quad k = n + 1/2$$

i.e., a set of straight lines along the z direction through the centers of the cubes.

The last term (d) in Eq. (1-265) requires some more discussion since the coefficients are now products of occupation parameters.

From the rotation of rigid corner-linked octahedra, it follows in the same way as explained above that

$$\sigma^{(z)}_{100} = -\sigma^{(z)}_{000}; \quad \sigma^{(z)}_{001} = -\sigma^{(z)}_{000};$$
$$\sigma^{(y)}_{100} = -\sigma^{(y)}_{000}; \quad \sigma^{(y)}_{010} = -\sigma^{(y)}_{000}$$

One can conclude from these relations that

$$\sigma^{(y)}_{100}\,\sigma^{(z)}_{100} = \sigma^{(y)}_{000}\,\sigma^{(z)}_{000}$$

and thus

$$\omega_{100} = 1; \quad \omega_{000} = 1$$

The corresponding locus is clearly $\exp(-2\pi i h) - 1 = 0$ or $h = n$ ($n = integer$).

This equation represents a set of planes parallel to $h = 0$. The complete locus of diffuse intensity resulting from the orientational disorder on sublattice 1 is represented in Fig. 1-91. Similar loci, mutatis mutandis, result from the orientational disorder on the sublattices 2 and 3. We can then conclude that planes of diffuse intensity passing through the basic reflections are present along the three families of cube planes. Moreover lines of diffuse intensity parallel to the three cube directions pass through the cube centers. This is the type of diffuse scattering observed in a number of perovskite structures in the high-temperature phase.

The basic literature on Sec. 1.11.5 is: Verwerft et al. (1988, 1989).

1.11.6 Displacive Disorder

The method summarized here can easily be generalized to systems in which diffuse scattering is due to displacive disorder. In the precursor phase of the ω phase, for instance, the diffuse scattering can be related to the presence of a random distribution of nuclei of all orientation and translation variants of the ω phase. The cluster relation is then implicitly determined *a priori* from the model. It states that within the smallest cluster compatible with the body-centered structure, i.e., within a centered cube, the sum of all the displacements of the type $(1/12)\langle 111\rangle$ should be zero. This relation is in fact satisfied in each variant of the long-range-ordered ω structure; it leads to the ω coefficients, via the cluster relation of Eq. (1-252); one obtains $\omega_j = 1$. Using as cluster vectors $\mathbf{0}$ and all symmetry-related $(1/2)\langle 111\rangle$ vectors, the equation for the diffuse scattering surface becomes $1 + \Sigma \exp[-\pi i(h + k + l)] = 0$ from Eq. (1-248), where the sum contains eight terms corresponding with all sign combinations of h, k and l. This can be rewritten as (Van Dyck et al., 1989)

$$\cos \pi h \cdot \cos \pi k \cdot \cos \pi l = -\tfrac{1}{8} \qquad (1\text{-}267)$$

This locus represents *spheroidal* surfaces centered on the midpoints of the edges of the cube as well as on the cube center: the diameter of the surfaces measured along the edges is about $0.92\,a^{-1}$. The surfaces are not spheres however. This locus fits well with observation.

1.11.7 Non-Periodic Arrays of Scattering Centers

1.11.7.1 Introduction

From the viewpoint of diffraction theory, it is useful to distinguish two essentially different types of non-periodic arrays of scattering centers (e.g., atomic planes). Certain arrays are called *uniform;* they approximate as closely as possible a given average spacing $\varDelta$ at each point in the sequence, by mixing two block sizes L and S, such that $S < \varDelta < L$, where L and S may be mutually incommensurate. The resulting sequence may be non-periodic but it nevertheless has long-range order, since it is completely defined up to infinity by a deterministic algorithm, such as the Fujiwara square wave (Fujiwara, 1957), the *cut and projection method* or the *slit method* (Elser, 1985, 1986; Duneau and Katz, 1985; Frangis et al., 1990).

On the other hand, certain sequences are only determined in stochastic terms, for instance by the probability of finding a scattering center at the site $x_n = nd$ ($n = integer$), provided there is one in the origin at $x = 0$, i.e., by a pair correlation function. In this case, there is no long-range order. Whereas in the first case sharp diffraction maxima are still produced, possibly at incommensurate positions, the second case gives rise to broadened diffraction peaks, the higher the order of the satellite reflection, the broader the peak.

Uniformly spaced non-periodic sequences of atomic planes exhibiting two different interplanar distances occur, for instance, in icosahedral quasi-crystals. This is the fundamental reason why quasi-crystals still produce sharp diffraction spots, even though they are not built on a lattice, i.e. (see Sec. 1.12.5), even though the parallel atomic planes do not form equidistant sequences. We shall illustrate the diffraction effects associated with both situations by means of simple one-dimensional models.

1.11.7.2 Uniform Sequences

Let $\varDelta$ be the average spacing and m the fraction of L (long) spacings and $1 - m$ the fraction of S (short) spacings in a sufficiently long sequence; we then must have

$$\varDelta = (1 - m)\,S + m\,L$$

from which we deduce

$$m = \frac{S - \varDelta}{S - L} \quad \text{and} \quad 1 - m = \frac{\varDelta - L}{S - L} \qquad (1\text{-}268)$$

The total length of a sequence of n ($n = integer$) average spacings $\varDelta$ would be $n\varDelta$. The closest number of L spacings in such a length is $n \cdot m$, and that of S spacings is $n(1 - m)$. However the numbers $n \cdot m$ and $n(1 - m)$ are not necessarily integers and the actual length x_n formed by the sum of these segments is the integer part of this sum. Or stated explicitly

$$x_n = n\varDelta - [L\{nm\} + S\{n(1-m)\}] \quad (1\text{-}269)$$

where $\{z\}$ denotes the fractional part of z. Since $\{n\} = 0$, we find that

$$x_n = n\varDelta + (S - L)\{nm\} \qquad (1\text{-}270)$$

x_n is in general not periodic with n, even though the expression $\{zm\}$ is periodic in z

with period $1/m$; this can be seen as follows:

$$\left\{ \left(z + \frac{1}{m} \right) m \right\} = \{z\,m + 1\} = \{z\,m\}$$

Since $\{z\,m\}$ is periodic with period $1/m$, the exponential function $\exp(2\pi i \alpha \{z\,m\})$ is also periodic with the same period. We shall need this expression further below. This exponential can thus be expanded in a Fourier series of the form

$$\exp(2\pi i \alpha \{z\,m\}) =$$
$$= \sum_p A_p \exp(-2\pi i m p z) \tag{1-271}$$

The Fourier coefficients can be obtained by the usual procedure. We multiply both sides with $\exp(2\pi i m q z)\,dz$ and we integrate over one period, i.e., from $-1/(2m)$ to $+1/(2m)$

$$\int_{-1/(2m)}^{+1/(2m)} \exp[2\pi i (\alpha \{z\,m\} + m q z)]\,dz =$$
$$= \sum_p \int_{-1/(2m)}^{1/(2m)} A_p \exp[2\pi i (q-p) m z]\,dz$$

The integral on the right-hand side is zero unless $p = q$; it then reduces to A_p/m. The left-hand side integrates to $\sin[\pi(p+\alpha)]/$ $[m\pi(p+\alpha)]$ with $q \to p$. Use was made here of the relation $\{z\,m\} = z\,m - [\![z\,m]\!]$, where $[\![z\,m]\!]$ denotes the integral part of $z\,m$. The Fourier expansion is then

$$\exp(2\pi i \alpha \{z\,m\}) =$$
$$= \sum_p \frac{\sin[\pi(\alpha+p)]}{\pi(\alpha+p)} \exp(-2\pi i p z\,m) \tag{1-272}$$

The amplitude diffracted by a linear array of scattering centers with scattering factor $f(\mathbf{h})$ at positions x_n is then [from Eqs. (1-210) and (1-270)]

$$A(\mathbf{h}) = f(\mathbf{h}) \sum_n \exp(2\pi i h n \Delta) \cdot$$
$$\cdot \exp[-2\pi i h (L-S)\{n\,m\}]$$

or making use of the Fourier expansion in Eq. (1-272), and with $\alpha = -h(L-S)$ and $z \equiv n$

$$A(\mathbf{h}) = f(\mathbf{h}) \sum_p \frac{\sin[\pi(p+\alpha)]}{\pi(p+\alpha)} \times$$
$$\times \sum_n \exp[-2\pi i n(-h\Delta + p\,m)]$$

The sum over n produces sharp peaks at positions $-h\Delta + p\,m = l$, where l is an integer. We thus find reflections at positions $h = -(l/\Delta) + p/(\Delta/m)$ or from Eq. (1-268)

$$h = -\frac{l}{\Delta} + p \frac{S-\Delta}{\Delta(S-L)} \tag{1-273}$$

The diffraction pattern thus consists of spots (with $p = 0$) at $h = -l/\Delta$, i.e., with a spacing $1/\Delta$ equal to the inverse of the *average* spacing; these spots only depend on one index l.

Their amplitude is given by

$$A(h) = f(h) \frac{\sin \pi \alpha}{\pi \alpha} \sum_n \exp(2\pi i n h \Delta) \tag{1-274}$$

with $h = -l/\Delta$ this leads to

$$A(h) = \sum_l N f(h) \frac{\sin[\pi h(L-S)]}{\pi h(L-S)} \delta\left(h + \frac{l}{\Delta}\right) \tag{1-275}$$

where N is the number of scattering centers. If $L = S$, this reduces to the well-known result $A(h) = N f(h) \sum_l \delta(h + l/\Delta)$.

Moreover there are *satellite* spots at the positions given by Eq. (1-273), characterized by two indices l and p. The amplitudes of the satellites are given by:

$$A(h) = \sum_{l,p} N f(h) \frac{\sin\{\pi[p - h(L-S)]\}}{\pi[p - h(L-S)]} \cdot$$
$$\cdot \delta\left\{ h - \left[-\frac{l}{\Delta} + \frac{p}{(\Delta/m)} \right] \right\} \tag{1-276}$$

The satellites have a spacing of m/Δ, i.e., smaller than the spacing between basic spots; their positions are determined by the positions of the *basic* spots (this justifies their name!).

If m is a rational number, the diffraction pattern is commensurate; if m is an irrational number, it is incommensurate.

It is very often the case that $L = (k+1)a$, $S = ka$ (k is an integer; a is the lattice parameter of the basic structure). One then finds for

$$\frac{S-\Delta}{(S-L)\Delta} = \frac{\Delta - ak}{a\Delta}$$

and

$$h = -\frac{l}{\Delta} + p \left[\frac{1}{a} - \frac{k}{\Delta} \right]$$

$$h = \frac{p}{a} + \frac{-l - kp}{\Delta}$$

Introducing two new integral indices r and s defined as $p \equiv r$ and $-l - kp \equiv s$ one finds

$$h = \frac{r}{a} + \frac{s}{\Delta} \qquad (1\text{-}277)$$

In this case one can thus also describe the diffraction pattern as consisting of the spots represented by r/a due to the basic structure present within the blocks, and of arrays of satellite spots with a separation $1/\Delta$, determined by the average spacing, and given by s/Δ.

In the particular case of Figs. 1-131 and 1-132: $\Delta = (9/4)a$, $k = 2$, $m = 1/4$, $1 - m = 3/4$, $L = 3a$ and $S = 2a$. The block sequence is thus $2\bar{2}2\bar{3}$. The central row of the diffraction pattern is then given by $h = (r/a) + (4/9)(s/a)$; the spots with $s = 0$ are the basic spots.

1.11.7.3 Probabilistic Sequences

We consider a linear array of potential sites for scattering centers at $x_n = nd$, i.e., the sites are equidistant. Each site can either be occupied or vacant (Van Dyck et al., 1979). The occupation of site i is described by the occupation operator σ_i,

which is 1 if x_i is occupied, and zero if x_i is vacant. The pair correlation is described by β_i which represents the probability of finding the first occupied site at x_i to the right of the origin, provided the origin x_0 is occupied. We assume the pair correlation functions for first, second, third ... neighbors to be independent. The probability of finding the second occupied site at x_i to the right of the origin is then given by $\sum_j \beta_j \beta_{i-j}$. The probability of finding the third occupied site at x_i is similarly given by $\sum_{j,h} \beta_j \beta_h \beta_{i-j-h}$. The total probability of finding the i-th site to the right of the origin occupied by a scattering center is then given by

$$\alpha_i^+ = \beta_i = \sum_j \beta_j \beta_{i-j} + \sum_{j,h} \beta_j \beta_h \beta_{i-j-h} + \cdots \qquad (1\text{-}278)$$

This expression follows from the possibility that between the origin and the site i there may be 0, 1, 2, ... occupied sites.

It was shown above [Eq. (1-246)] that the normalized amplitude of the scattered beam is given

$$\frac{A(h)}{f(h)} \equiv a(h) = \sum_i \sigma_i \exp(2\pi i h x_i) \qquad (1\text{-}279)$$

The normalized intensity is then given by

$$I(h) = \sum_j \sum_k \sigma_j \sigma_k \exp[2\pi i h (x_j - x_k)]$$

and noting that we can put $x_k = x_j + x_i$, i.e., $k = j + i$

$$I(h) = N \sum_i \alpha_i \exp(-2\pi i h x_i) \qquad (1\text{-}280)$$

with

$$\alpha_i = \frac{1}{N} \sum_j \sigma_j \sigma_{j+i} = \langle \sigma_0 \sigma_i \rangle \qquad (1\text{-}281)$$

where N is the number of sites in the array. Note that $\alpha_0 = 1$.

It is useful to consider first the two extreme situations:

(i) Complete disorder: A fraction m of the sites is randomly occupied by scattering centers.

(ii) Complete order with a period $X = N_0 d$.

In the first case $\alpha_i = m$ for all $i \neq 0$ and $\alpha_0 = 1$; one finds [from Eq. (1-280)] with $x_i = n d$

$$I(h) = N(1 - m) + N m \sum_n \exp(-2\pi i h n d) \tag{1-282}$$

The first term results from the terms with $\alpha_0 = 1$; it represents a constant background, whereas the second term leads to Bragg peaks of the basic lattice situated at $n h$ with $h = 1/d$.

For the second case we have $\sigma_i = \sigma_{i+N_0 p}$ ($p = integer$) and the normalized amplitude now becomes

$$a(h) = \sum_i \sigma_i \exp(-2\pi i h x_i)$$

or with $X = N_0 d$

$$= \sum_{i=0}^{N_0-1} \sum_p \sigma_{i+pN} \exp[-2\pi i h (x_i + p X)] =$$

$$= \left[\sum_{i=0}^{N_0-1} \sigma_i \exp(-2\pi i h x_i) \right] \cdot$$

$$\cdot \sum_p \exp(-2\pi i h p X) \tag{1-283}$$

The factor between brackets is the structure factor of the superlattice unit cell, whereas the second factor leads to the Bragg peaks of the superstructure at $p h$ with $h = 1/(N_0 d) = 1/X$ ($p = integer$).

We shall now discuss disordered arrays of scattering centers, which can either be vacancies in an otherwise occupied row or atoms in a nearly vacant row. As a result of Babinet's theorem of complementary screens, the resulting diffraction patterns are similar for these two cases. The α_i in Eq. (1-280) represents the probability of finding a scattering center at site i, provided the origin is occupied, i.e., α_i is the sum of

the expression given above for α_i^+ in Eq. (1-278) and of a similar expression for

$$\alpha_i^- = \alpha_{-i}^+ \tag{1-284}$$

$$\alpha_i^- = \sum_j \beta_j \beta_{-i-j} + \sum_{j,h} \beta_j \beta_h \beta_{-i-j-h} + \dots$$

Equation (1-280) for the intensity can now be written as

$$I(h) = N \sum_i \left(\beta_i + \sum_j \beta_j \beta_{i-j} + \right. \tag{1-285}$$

$$\left. + \sum_{j,k} \beta_j \beta_k \beta_{i-j-k} + \dots \right) \exp(-2\pi i h x_i)$$

We consider in particular the term

$$\sum_i \sum_j \beta_i \beta_{i-j} \exp(-2\pi i h x_i) \tag{1-286}$$

Taking into account that $k = i - j$ and $x_k = x_i - x_j$, one can factorize Eq. (1-286)

$$\sum_j \beta_j \exp(-2\pi i h x_j) \sum_k \beta_k \exp(-2\pi i h x_k) \tag{1-287}$$

Since the two factors only differ by dummy indices, they are in fact identical. We shall write as an abbreviation

$$f \equiv f(h) \equiv \sum_j \beta_j \exp(-2\pi i h x_j)$$

Replacing the summation by an integration, i.e., $f(h) = \int \beta(x) \exp(-2\pi i h x) \, dx$, it is clear that $f(h)$ is approximated well by the Fourier transform of $\beta(x)$. We then obtain from Eq. (1-285)

$$\sum_i \alpha_i^+ \exp(-2\pi i h x_i) = \tag{1-288}$$

$$= f(h) + f^2(h) + f^3(h) + \dots = \frac{f(h)}{1 - f(h)}$$

since the series is a geometrical progression. We have a similar contribution from the sites left of the origin:

$$\sum_i \alpha_i^- \exp(-2\pi i h x_i) =$$

$$= \sum_i \alpha_i^+ \exp(2\pi i h x_i) \tag{1-289}$$

This last sum can be interpreted as the complex conjugate of Eq. (1-288) and it is therefore equal to $f^*(h)/[1 - f^*(h)]$. For the total intensity of the diffracted beam one must sum the two contributions for $x_i > 0$ and $x_i < 0$ as well as the contribution due to the origin at $x_i = 0$:

$$\frac{I(h)}{N} = 1 + \frac{f}{1-f} + \frac{f^*}{1-f^*} \qquad (1\text{-}290)$$

or after transformation

$$\frac{I(h)}{N} = \frac{1 - |f|^2}{(1-f)(1-f^*)} \qquad (1\text{-}291)$$

We assume that the pair correlation is described by a continuous function $\beta(x)$ peaked at the average distance Δ between occupied sites, such that $\beta_i = \beta(x_i)$. We then introduce the shifted distribution $p(x)$ such that $p_i = p(x_i)$, which has the same shape but is peaked at the origin. We can then write $\beta_i = p_{i-\Delta}$ and

$$\begin{aligned} f(h) &= \sum_i \beta_i \exp(-2\pi i h x_i) = \\ &= \sum_i p_i \exp[-2\pi i h(x_i - \Delta)] = \\ &= \exp(2\pi i h \Delta) \sum_i p_i \exp(-2\pi i h x_i) = \\ &= g(h) \exp(2\pi i h \Delta) \qquad (1\text{-}292) \end{aligned}$$

where $g(h)$ is the Fourier transform of $p(x)$:

$$g(h) = \int p(x) \exp(-2\pi i h x) \, dx$$

Substituting $f(h)$ by its expression, Eq. (1-292), in Eq. (1-291) we find

$$\frac{I(h)}{N} = \frac{1 - g^2(h)}{1 + g^2(h) - 2g(h)\cos(2\pi h \Delta)} \qquad (1\text{-}293)$$

In many cases of interest, $p(x)$ is sharply peaked; the Fourier transform $g(h)$ is then a broad peak and the complementary function $\delta(h) = 1 - g(h)$ is then much smaller than $1: \delta(h) \ll 1$, for not exceedingly large values of h. It increases with increasing h but is slowly varying in the vicinity of the peaks. Introducing $\delta(h)$ in Eq. (1-293) then leads to

$$\frac{I(h)}{N} = \frac{\delta(h)}{1 - \cos(2\pi h \Delta) + \delta^2(h)} \qquad (1\text{-}294)$$

We have hereby neglected $\delta^2(h)$ with respect to $\delta(h)$ in the numerator, but we have kept $\delta^2(h)$ as compared to $1 - \cos(2\pi h \Delta)$ in the denominator since $\cos(2\pi h \Delta)$ could become close to 1 in the vicinity of the basic peaks where $h \simeq 1/\Delta$.

Let us consider the shape of $I(h)$ in the vicinity of the peaks which correspond with $\cos(2\pi h \Delta) = 1$ or with $h \simeq n/\Delta$, i.e., for $h_n = (n + \varepsilon)/\Delta$ ($n = integer$; $\varepsilon \ll 1$), where ε is the deviation from the expected position of the peak. We can then write $\cos(2\pi h \Delta) \simeq \cos[2\pi(n + \varepsilon)] \simeq \cos(2\pi\varepsilon) \simeq 1 - 2\pi^2 \varepsilon^2$ and hence

$$I(h) = \frac{\delta(h)}{\delta^2(h) + 2\pi^2 \varepsilon^2} \qquad (1\text{-}295)$$

Since $I_{max} = 1/\delta(h)$ we can also write

$$I(h) = \frac{I_{max}}{1 + \dfrac{2\pi^2 \varepsilon^2}{\delta^2(h)}} = \frac{I_{max}}{1 + 2\pi^2 \varepsilon^2 I^2_{max}} \qquad (1\text{-}296)$$

The half-width at the half-maximum of the Lorentzian peak is $\varepsilon_w = \delta(h)/\pi\sqrt{2} = 1/I_{max}\pi\sqrt{2}$, where we have treated $\delta(h)$ as a constant in the vicinity of the peaks.

One thus finds that ε_w increases with h since $\delta(h)$ increases with h. The sharper the peak function $p(x)$ is, the wider will be the fourier transform $g(h)$, and the slower will be the decrease of I_{max} with h. The diffraction pattern thus consists of *satellite* peaks at $h = n/\Delta$ associated with each of the basic peaks. With increasing n (or h), the heights of these peaks decrease, while their widths increase.

1.12 Applications and Case Studies

1.12.1 Introduction

It is obvious that the number of applications of electron microscopy has become so large that an exhaustive review of all applications, even when restricted to materials science, is an almost impossible task. We shall therefore limit this review to a few typical examples of those applications for which electron microscopy and electron diffraction have made an essential contribution, leading to conclusions which could not have been obtained by other techniques.

The interpretation of the images is only meaningful in terms of the underlying solid-state phenomena. We shall therefore briefly sketch for each application the framework in which the images acquire their significance and interpretation. The choice of the examples is mainly motivated by the availability of suitable photographs.

1.12.2 The Fine Structure of Dislocations

1.12.2.1 Measuring the Stacking Fault Energy

In most materials the dislocations are not simple line defects but in fact consist either of two or more partial dislocations connected by strips of stacking fault or of out-of-phase boundary. The simplest situation arises when glide takes place between two close-packed layers of identical "spherical" atoms in an elemental face-centered-cubic crystal. The glide motion along the (111) plane in the $[\bar{1}10]$ direction then follows the valleys; that is, it takes place in two steps, each performed by the motion of a partial, the first with Burgers vector $b_1 = (1/6)[\bar{2}11]$, the second with Burgers vector $b_2 = (1/6)[\bar{1}2\bar{1}]$, which encloses an angle of 60° and leads to a symmetry translation $(1/2)[\bar{1}10]$ along the (111) glide plane. Between the two partials, a stacking fault ribbon with a displacement vector is present, given by one of the Burgers vectors of the partials.

The two partial dislocations repel one another because their Burgers vectors enclose an acute angle. In an infinite solid, this repulsion is proportional with $1/d$ (d = partial separation), and its magnitude is a function of the orientation of the partials, i.e., of their character (screw or edge). The presence of the stacking fault ribbon causes an effective attractive force per unit length between the two dislocations, which is independent of their separation and numerically equal to the stacking fault energy γ. An equilibrium separation is thus established. Assuming that the repulsive force is known, it is then possible to deduce the stacking fault energy from the measured equilibrium separation of the partials. Dislocation ribbons are thus sensitive probes for the stacking fault energy, a quantity that is difficult to measure in any other way. The following relations are valid in an infinite isotropic solid:

$$d = d_0 \left(1 - \frac{2\nu}{2-\nu} \cos 2\phi \right)$$

with

$$d_0 = \frac{\mu b^2 (2-\nu)}{8\pi\gamma(1-\nu)}$$

where ϕ is the angle between the total Burgers vector and the ribbon direction; μ is the shear modulus, and ν Poisson's ratio.

The orientation dependence of the ribbon width could be verified on an image such as that in Fig. 1-92, which represents a curved dislocation in a graphite foil. Plotting d as a function of $\cos 2\phi$, the slope of the straight line obtained gives the effective value of the Poisson ratio as well as the intercept d_0 to be used in the second relation, which then yields a value for the stacking fault energy.

Using this method, it is implicitly assumed that the repulsive force between dislocations is proportional to $1/d$, which is only the case in an infinite solid. In a thin

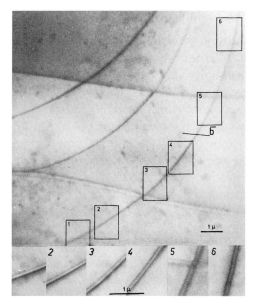

Figure 1-92. Curved dislocation ribbon in the (0001) plane of graphite. Several segments are reproduced as magnified insets. The direction **b** of the total Burgers vector, as determined by extinction experiments, is indicated. Note the systematic change in width with orientation (Delavignette and Amelinckx, 1962).

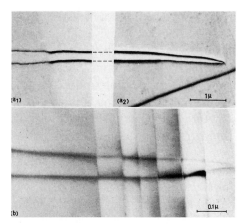

Figure 1-93. A wide ribbon in SnS$_2$ gradually approaching the surface. As the ribbon crosses surface steps, it becomes discontinuously narrower. The ribbon closes where it emerges in the surface (Siems et al., 1962a, b).

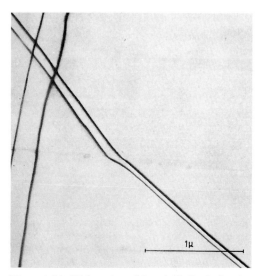

Figure 1-94. Dislocation ribbon in SnS$_2$. Refraction, accompanied by a width change, occurs on passing underneath a surface step (Siems et al., 1962a, b).

foil, the repulsive force between dislocations parallel to the foil surfaces decreases with decreasing distance to the specimen surfaces. This behavior can be visualized, as in Fig. 1-93, where a ribbon gradually approaching the surface in a wedge shaped lamella of thin disulphide closes as it emerges to the surface.

The energy of a dislocation ribbon also depends on its distance to the surface. As a result, the shape of minimum energy of a dislocation ribbon crossing a surface step is not a straight line; *refraction* of the ribbon as well as a change in width occur on passing underneath the surface step (Fig. 1-94). The *index of refraction* is the ratio of the total energies of the ribbon in the two parts of the foil on either side of the surface step.

These images prove that these surface effects are not negligible. When measuring stacking fault energies, care should thus be taken to use foils of maximum thickness and moreover to take the widest ribbon as the most representative. The width of narrow stacking fault ribbons can best be determined by imaging in the weak-beam mode (Cockayne, 1969, 1971).

An alternative method of measuring ribbon widths consists in the use of high-resolution images with the viewing direction along the partial dislocations. Such images are shown in Fig. 1-95 for ribbons in Silicon; they reveal the stacking fault as well as the structure of the partial dislocations. In this geometry, the ribbon width is not as sensitive to the foil thickness.

Other geometrical configurations involving stacking faults can be used, such as the partial separation in triple ribbons in graphite (Fig. 1-96) and in close packed structures, or the radius of curvature of partials in a network of extended nodes (Fig. 1-97). In the latter case one has approximately $\gamma = (1/2)\mu b^2/R$ (R: radius of curvature, b: Burgers vector of partial). More accurate relations are discussed in (Amelinckx, 1979).

1.12.2.2 Multiribbons

Ordering in alloys based on close-packed structures leads to long symmetry translations along the glide directions in the close-packed glide planes. As a result, ribbons consisting of several partials, separated either by stacking faults or by out-of-phase boundaries result. The equilibrium separation of superdislocations (i.e., perfect dislocations with respect to the basic lattice, but partial dislocations with respect to the ordered structure) can be used to derive values of the anti-phase boundary energy in the same way as described above for stacking faults. In Ni_4Mo as many as ten partial dislocations are connected by faults and anti-phase boundaries.

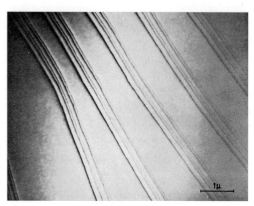

Figure 1-97. Sixfold ribbons of partials in the (0001) plane of $CrCl_3$. The Burgers vectors of the partials form a zigzag glide path (Amelinckx and Delavignette, 1962).

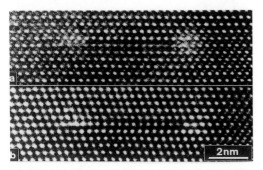

Figure 1-95. High-resolution images of dislocation ribbons in silicon as viewed along the close-packed direction parallel with the partial dislocation lines (Courtesy Bender).

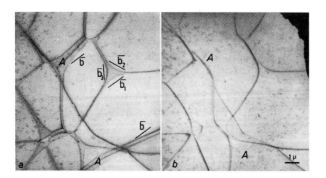

Figure 1-96. (a) Widely extended dislocation node of partial dislocations in graphite. In "A", a triple ribbon is present; the three partials have the same Burgers vector as follows from the contrast experiment in (b), where the three partials are simultaneously out of contrast. Nevertheless the contrast at the three partials is different in (a) (Delavignette and Amelinckx, 1962).

The dislocations involved in glide between the close-packed layers of anions (X) in layered ionic sandwich crystals of the CdI_2 type AX_2 (XAXXAX ...) or AX_3 are of particular interest. The glide motion takes place between the two close-packed anion layers, which are weakly bonded by Van der Waals forces. Dislocations can thus dissociate into two or more Shockley partials. Although in the close-packed layers, between which glide takes place, all X atoms are equivalent, the A cations in the adjacent central layers of the sandwiches may form configurations which lead to a large unit mesh in the glide plane, either because not all octahedral cation sites are occupied (e.g., in $CrCl_3$, $CrBr_3$) or because the cations form metal-metal bonded clusters (e.g., in $NbTe_2$, $TaTe_2$).

In the chromium trihalides (Amelinckx and Delavignette, 1962) multiribbons containing either four or six partials are observed. Assuming, that glide takes place along the close-packed anion layers by the propagation of Shockley partials, two types of stacking faults occurring between successive partials can be distinguished:

(i) faults violating only the chromium stacking, i.e., involving only third neighbors;

(ii) faults violating the stacking of the chromium ions as well as that of the anions, i.e., involving next-nearest neighbors.

Intuitively, it is clear that the type (ii) faults will have a larger energy than those of type (i). The sixfold ribbons correspond with a "straight" zigzag glide path along the close-packed directions in the (0001) glide plane of the anion sublattice; they contain the two types of faults in an alternating fashion, the outer ribbons corresponding with high-energy faults. Diffraction contrast images of such ribbons are reproduced in Fig. 1-97. The structure of the fourfold ribbons can similarly be related to the structure.

In $NbTe_2$ (J. Van Landuyt et al., 1970), which has a deformed CdI_2 structure, sixfold ribbons also occur. In this structure, the Nb ions form clusters of three parallel close-packed niobium rows, having a somewhat smaller separation than in the ideal hexagonal structure which probably occurs only in the temperature range in which the crystal is grown. The resulting structure then becomes monoclinic on cooling. The unit mesh in the glide plane is now a centered rectangle, which can adopt three different but equally probable orientations differing by 60°. As a consequence, the room temperature structure is fragmented in domains corresponding with the three possible orientations of the clustered niobium rows. The monoclinic symmetry causes the glide paths along the three close-packed directions within the same domain to become non-equivalent.

The zigzag glide paths in the direction enclosing an angle of 30° with the long side of the rectangular mesh consist of six partials, whereas the glide path along the other close-packed direction (i.e., along the short side of the rectangle) repeats after two partials. The Burgers vector is conserved all along the dislocation lines. Hence, when a sixfold ribbon passes through a domain wall, the glide path changes its orientation relative to the underlying structure. A sixfold ribbon in one domain is thus transformed into three separate twofold ribbons in the adjacent domain. Whereas in the sixfold ribbon the six partials are held together by stacking faults, this is no longer the case with the three twofold ribbons which repel one another. The image of Fig. 1-98 illustrates the described behavior of a sixfold ribbon intersecting a set of parallel domain boundaries in $NbTe_2$.

Figure 1-98. Sixfold ribbon of partials in NbTe₂ intersecting domain boundaries along which the underlying structure changes 60° in orientation. In half of the domains, the sixfold ribbons separate into three twofold ribbons, which form bulges as a result of repulsive forces (Van Landuyt et al., 1970 b).

1.12.2.3 Plastic Deformation: Glide Dislocations

Plastic deformation is a subject that has been intensely studied at an early stage by means of diffraction contrast (see, e.g., Hirsch et al., 1965; and Behtge and Heydenreich, 1982). High-voltage electron microscopy (∼1000 kV) has been of considerable interest in this respect, because it is possible to study thicker foils, which are more representative of the bulk material than the thin foils required at 100 kV. Figure 1-99 shows a procession of glide dislocations in f.c.c. stainless steel, confined to their (111) glide plane, as observed in high-voltage electron microscopy. The strictly planar arrangement illustrates that the dislocations are dissociated, and therefore that cross-glide is a difficult process. The dissociation is too small to be directly observable at this resolution, but it has been found from other images that the stacking fault energy is rather small in stainless steel. Note the periodic contrast of the dislocations in the vicinity of their emergence points in the foil surfaces and the absence of such contrast in the central part of the foil.

Figure 1-100 shows a network of intersecting glide dislocations confined to the (111) glide plane in an f.c.c. copper alloy (Cu-Ga) with low stacking fault energy. One set of dislocation nodes is dissociated and gives rise to the dark triangular areas; the other one is contracted. Such nodes allow the stacking fault energy to be deduced from the curvature of the partial dislocations forming the extended nodes.

The image of Fig. 1-101 shows glide dislocations in the layer plane (001) of NbTe₂, which is parallel with the foil plane, the specimen having been obtained by cleavage. In every other domain the dislocation multiribbons consist of six partials. In the remaining domains the dislocations are simple ribbons, as described above in Sec. 1.12.2.2.

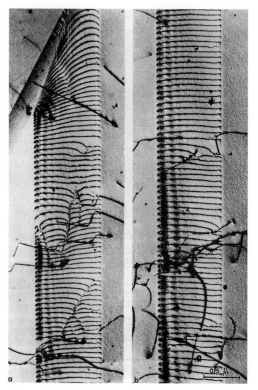

Figure 1-99. Procession of dislocations confined to a glide plane in stainless steel. Note the wavy contrast close to the surfaces and its absence in the central part of the foil (high-voltage electron micrograph).

Figure 1-100. Network of dissociated dislocations in a Cu-Ga alloy with low stacking fault energy (Courtesy Art).

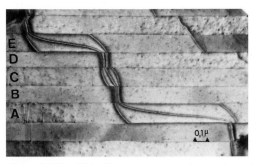

Figure 1-101. Glide dislocations in the layer plane of NbTe$_2$. Note the interaction between dislocations and twin domain walls (Van Landuyt et al., 1970b).

The image illustrate the strong interactions of the glide dislocations with the domain walls. On entering a domain in which the sixfold ribbons would have to be formed, the single ribbons line up with the domain wall, because in forming the multiribbon stacking faults would have to be generated. This leads to an effective interaction between dislocation ribbons and domain walls.

The reader interested in more details on Sec. 1.12.2 is referred to: Amelinckx (1979).

1.12.3 Small Particles

Certain materials can only be obtained as small crystals because only a limited quantity of substance is available. Other particles have to be very small because this is essential for their use, such as catalysts, silver halides in photographic emulsions, or magnetic particles for recording. The crystallographic study of such particles can only be performed by means of electron microscopy.

An interesting example is Fullerite, a material consisting of large C_{60} molecules (Van Tendeloo et al., 1991). These molecules form spherical shells of carbon atoms interconnected as hexagons and pentagons adopting the topology of a soccer ball. In as grown microcrystals, from organic solutions the layers of hollow spheres are stacked in a hexagonal fashion. It could be shown that under electron irradiation in the microscope vacuum the stacking transforms in situ into the stable cubic stacking, i.e., the microcrystal undergoes a phase transformation. Figure 1-102a shows the evolution of the diffraction pattern whereas Fig. 1-102b shows the high resolution image of a Fullerite crystallite.

Very small silver particles obtained by evaporation in vacuum on an amorphous carbon substrate exhibit pentagonal symmetry due to multiple twinning on {111} planes. Since the angle between two {111} planes in the f.c.c. lattice is 70° 1/4, five repeat twins on {111} around the same axis leave an angular gap of 8° 3/4; i.e., a grain boundary should be present. Most particles seem to contain five equivalent coher-

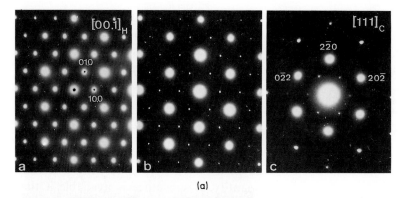

(a)

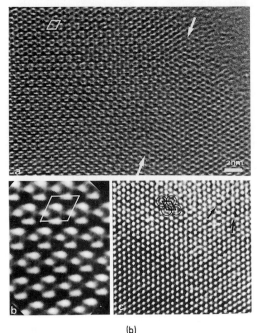

(b)

Figure 1-102. (a) Evolution under electron irradiation of the diffraction pattern of Fullerite (C_{60}) from hexagonal, "a", to cubic, "c"; (b) high-resolution images of a Fullerite microcrystal consisting mainly of C_{60} molecules (Van Tendeloo et al., 1991).

ent twin interfaces (Fig. 1-103); this suggests that in such cases the fivefold axis is in fact a line discontinuity accommodating the angular misfit.

Although the silver halides are very radiation sensitive, it is nevertheless possible to observe sufficiently small microcrystals under high-resolution conditions for short periods of time. The print-out process could in this way be followed in situ. Surface layers of AgCl are peeled off, and simultaneously epitaxial layers of metallic silver are formed. Figure 1-104 shows the superposition pattern of photolytic silver on a cube plane of AgCl.

Small silicon carbide precipitate particles, imbedded in a silicon matrix, are imaged in HREM in Fig. 1-104. The orientation relationship between the two lattices is evident from this figure. From the moiré pattern formed between matrix and precipitate particle, it is possible to deduce the lattice parameters of the precipitate and hence to identify the particle.

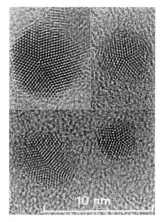

Figure 1-103. High-resolution image of multiply twinned silver microcrystals obtained by evaporation in vacuum on an amorphous substrate (Courtesy Coessens).

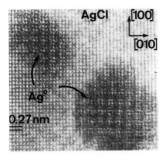

Figure 1-104. Silver chloride microcrystal exposed to electron irradiation. An epitaxial layer of metallic silver has been formed on the surface and gives rise to a coincidence pattern (Courtesy Coessens).

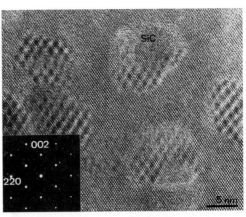

Figure 1-105. High-resolution image of a coincidence pattern between a silicon carbide particle and the surrounding silicon matrix, after heat treatment (Courtesy De Veirman).

1.12.4 Point Defect Clusters

Vacancies in quenched metals form disc shaped agglomerates in $(111)_{f.c.c.}$ or $(0001)_{h.c.p.}$ layers, limited by Frank-type dislocation loops. If the stacking fault energy is large enough, the loop is *unfaulted*, since energy is gained by nucleating a Shockley partial and sweeping the loop, transforming the sessile Frank loop into a perfect glissile loop. Such unfaulted loops in quenched aluminium are shown in Fig. 1-106.

If the stacking fault energy is small enough, which is true, for instance, in gold and in Ni-Co alloys, the Frank loop is transformed into a stacking fault tetrahedron consisting of four intersecting triangular stacking faults in {111} planes, limited along their intersecton lines by edge-type stair rod dislocations with a Burgers vector of the type (1/6)[110]. For intermediate values of the stacking fault energy, the Frank loops may remain faulted. In Fig. 1-107 stacking fault tetrahedra in gold are imaged in diffraction contrast, whereas in Fig. 1-108 HREM is used to image a vacancy type stacking fault tetrahedron in phosphorus ion-implanted and subsequently annealed silicon, as viewed along a [110] zone. The nature of

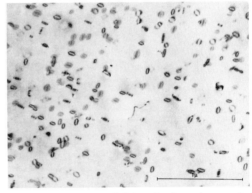

Figure 1-106. Unfaulted dislocation loops in quenched aluminum (Hirsch et al., 1958).

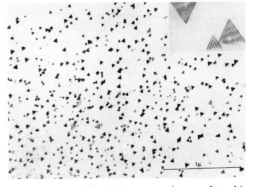

Figure 1-107. Diffraction contrast image of stacking fault tetrahedra in quenched gold. The inset shows a magnified image (Hirsch and Silcox, 1958).

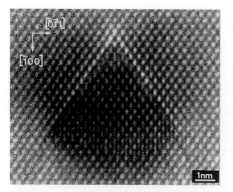

Figure 1-108. High-resolution image of a vacancy type tetrahedron in ion-irradiated and annealed silicon (Coene et al., 1985a).

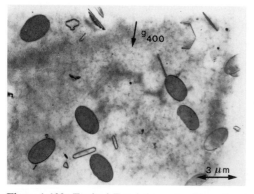

Figure 1-109. Faulted Frank loops in silicon due to interstitials (Courtesy Bender).

Figure 1-110. Diffraction pattern of a quasi-crystal of Al-Mn exhibiting fivefold symmetry (Courtesy Van Tendeloo).

the tetrahedron in silicon could be determined by image simulation of tetrahedra due to vacancies and interstitials, and by comparing computed and observed images.

Faulted Frank loops in silicon are imaged in diffraction contrast in Fig. 1-109. The presence of the stacking fault causes contrast inside the loop. Figure 1-47 shows a contrast experiment on an extrinsic Frank-type dislocation loop in silicon. Note the presence of a line of no contrast perpendicular to the active g vector. Note also the deformation of the extinction contours where they cross the dislocation loop. For $g = [1\bar{1}1]$ the loop exhibits stacking fault contrast as do the loops in Fig. 1-109.

1.12.5 Quasi-Crystals

This new class of materials, characterized by the presence of *non-crystallographic* symmetry elements, such as 5, 8, 10, and 12-fold rotation axes and the simultaneous absence of translation symmetry, was first observed by electron diffraction and electron microscopy (Schechtman et al., 1984). Figure 1-110 shows the diffraction pattern of a quenched Al-Mn alloy along the fivefold zone axis, and the corresponding high-resolution image is reproduced in Fig. 1-111: Note the absence of translation symmetry. (See also Vol. 1, Chap. 1.)

1.12.6 Mixed Layer Compounds

High-resolution electron microscopy is particularly suitable for the study of homologous series of mixed layer compounds such as $As_2Te_3 (GeTe)_n$. These compounds consist of a variable number n of GeTe layers, which have a slightly deformed sodium chloride structure which alternates with single five-layered lamellae of As_2Te_3.

The (111) layers of GeTe fit perfectly on to the close-packed layers of the As_2Te_3 lamellae, giving rise to hexagonal or rhombohedral layer structures depending on the n value. High-resolution images made along close-packed rows of the layers reveal directly the stacking. In particular, the As_2Te_3 lamellae can clearly be distinguished from the blocks of GeTe layers, and the number n of GeTe layers can be observed directly. In Fig. 1-112 the n values are, for instance, $n = 5$ and $n = 9$ (Kuypers et al., 1988).

In the Y–Ba–Cu–O system of superconducting compounds, a number of layered structures with various compositions were identified and imaged (Krekels et al., 1991). In orthorhombic (quasi-tetragonal) $YBa_2Cu_3O_{7-\delta}$ the succession of layers is ... $CuO–BaO–CuO_2–Y–CuO_2–BaO–$ ... The CuO layer consists of $Cu–O–Cu–O$ chains parallel with the b_0 direction. This single chain layer can be replaced by a layer of double chains $(CuO)_2$ along b_0. If each CuO layer is replaced by a double layer $(CuO)_2$, the composition is $YBa_2Cu_4O_8$. A view along the b_0 direction shows the open channels in the double chain layers as rows of elongated bright dots, whereas along the a_0 direction the

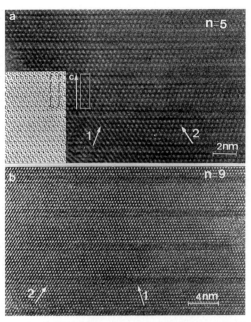

Figure 1-112. HREM images of (As_2Te_3) $(GeTe)_n$ with $n = 5$ and $n = 9$ as viewed along the close-packed atom rows (Kuypers et al., 1988).

double chain layer is revealed as a staggered double row of bright dots. In $Y_2Ba_4Cu_7O_{15}$ only every other CuO layer is replaced by a double chain layer. The structure images of these different phases are compared in Fig. 1-113.

1.12.7 MTS₃ Compounds

The MTS_3 compounds (M = Rb, Sn, ...; T = Nb, Ta, ...) form a remarkable series of mixed layer compounds. They consist both of a regular alternation of TS_2 layers having a trigonal prismatic NbS_2 structure, and of two-layer lamellae MS having a slightly deformed sodium chloride structure. The striking feature is that successive layers have different rotation symmetries. The TS_2 layers have hexagonal symmetry, whereas the MS layers have fourfold symmetry. The distance between successive atom rows is the same for the close-packed

Figure 1-111. High-resolution image of an icosahedral quasi-crystal of Al-Mn; note the absence of translation symmetry and the presence of pentagonal dot arrangements (Courtesy Van Tendeloo).

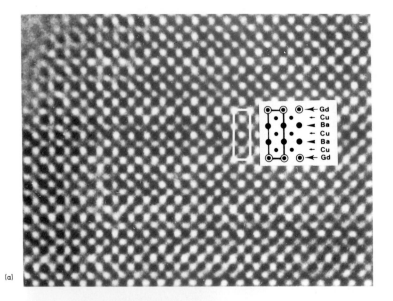

(a)

Figure 1-113. Comparison of the structures of the 1−2−3, 1−2−4 and 1−2−3½ compounds in the Y−Ba−Cu−O system. (a) $YBa_2Cu_3O_{7-\delta}$; (b) $YBa_2Cu_4O_8$; (c) $Y_2Ba_4Cu_7O_{15}$ (Courtesy of Krekels).

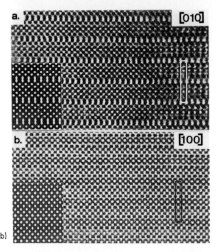

(b)

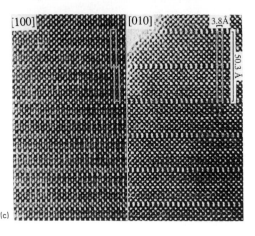

(c)

rows in the TS_2 layers as for the $\langle 100 \rangle$ rows in the sodium chloride layer. One family of atom rows of one layer fits in the "grooves" between the atom rows in the adjacent layer, and hence the lattice parameter along the direction normal to these rows is common for both types of lamellae. On the other hand, the a_0 parameter along the rows is very different for the two types of lamellae. Along this direction, the two types of layers "modulate" one another. The diffraction pattern reveals this mutual modulation; along the zone normal to the layers it exhibits the superposition of a hexagonal grid of strong spots due to the TS_2 layers and a square grid of strong spots due to the MS layers. Moreover, each spot of one grid is surrounded by an array of weaker satellite spots having the geometry of the other grid. Such a diffraction pattern is reproduced in Fig. 1-114 together with the corresponding high resolution image (Kuypers et al., 1989).

The high-resolution image exhibits a quasi-periodic variation of the brightness of the dots which image the atom columns. The brightest dots form a quasi-periodic

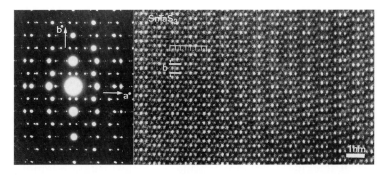

Figure 1-114. Diffraction pattern of SnTaS$_3$ along the [001] zone and the corresponding high-resolution image (Kuypers et al., 1989).

pattern based on a centered elongated rectangular unit mesh. Strictly speaking, the pattern is not periodic since there is no simple relation between the a_0 parameters of the two types of lamellae.

1.12.8 Planar Interfaces

It is well known that two simple types of stacking faults can occur in the face-centered cubic structure. The *intrinsic* fault, formed either by the extraction of a layer, or by glide of a Shockley partial, is represented by the stacking symbol abcabca-cabc ... The *extrinsic* fault, formed for instance by the precipitation of interstitials in a Frank loop corresponds with the stacking symbol abcabacabc ... The two types of faults have comparable energies in certain materials. In a network of dissociated dislocations all nodes are then dissociated; this is the case, for instance, in silicon (Fig. 1-115) and in certain alloys (Ag-Sn). The two kinds of faults have opposite displacement vectors of the type $(a/3)[111]$ and can thus be distinguished by the characteristic fringe pattern which they produce when situated in inclined planes (see Sec. 1.5.4.1). They can be identified directly in high-resolution images, as in Figs. 1-95 and 1-96, where intrinsic stacking faults are present between the partials.

A stacking fault of the type ... ababcacac ... in 2H wurtzite is imaged in Fig. 1-116 by means of HREM.

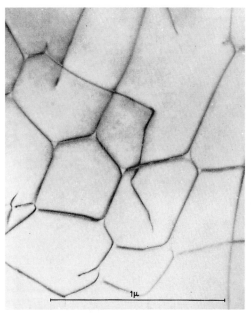

Figure 1-115. Network of extended dislocations in silicon; all nodes are dissociated (Aerts et al., 1962).

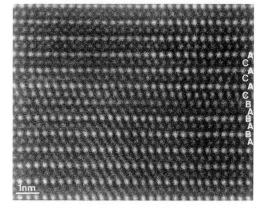

Figure 1-116. Stacking fault in 2H wurtzite in HREM (Coene et al., 1985 b).

Figure 1-117a is a lattice image of a crystal of the 15R polytype of SiC with stacking sequence abcbc ... Lattice fringes with a spacing equal to the thickness of the five layered lamellae exhibit different lattice spacings at the level of two stacking faults. In Fig. 1-117b all single atom layers are imaged as fringes. Finally, in Fig. 1-117c enough reflections were collected to produce atomic resolution of the 15R stacking.

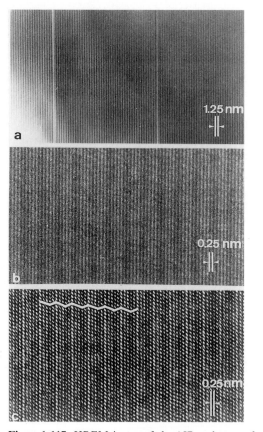

Figure 1-117. HREM image of the 15R polytype of SiC (abcbc ...). (a) Only two successive spots in the 000 *l* row are used in the image formation; the period corresponding with the five layer lamellae is revealed. Two stacking faults are revealed by fringes with a different spacing. (b) A cluster of five spots in 000 *l* row is used to image the structure; individual atom layers are now visible within the five-layer lamellae. (c) Several parallel rows of spots are used. The stacking sequence can now be observed (Van Tendeloo et al., 1982).

1.12.9 Domain Structures

Phase transformations are usually accompanied by a decrease in symmetry with decreasing temperature. As a result, a single crystal of a higher symmetric phase becomes fragmented into domains whose structures are related by the symmetry elements lost in the transition to the lower symmetry phase. The lost rotation symmetry elements give rise to orientation variants of the low-temperature phase, whose number is given by the ratio of the order of the point group of the high-temperature phase and the order of the point group of the low-temperature phase. The loss of translation symmetry gives rise to translation variants related by displacement vectors given by the lost lattice translations. Their number is determined by the ratio of the volumes of the primitive unit cells of the low- and high-temperature phases (Van Tendeloo and Amelinckx, 1974b).

Orientation variants are separated by domain boundaries, whereas translation variants are separated by out-of-phase boundaries. The orientation of the domain boundaries is determined by the requirement that the strain energy should be a minimum. This will be the case for strain-free interfaces. As a result, the orientation of certain interfaces (W) follows entirely from symmetry, whereas others (W') have orientations which depend on the lattice parameters of the two phases involved, at the transition temperature.

For example, in the α−β transition of quartz, referred to above, the α phase has the point group *32* (order 6) and the β phase the point group *622* (order 12). The number of orientation variants in the α phase is thus $12 \div 6 = 2$ (α_1 and α_2; Dauphiné twins), and they are related by a 180° rotation about the threefold axis. There is no change in translation symme-

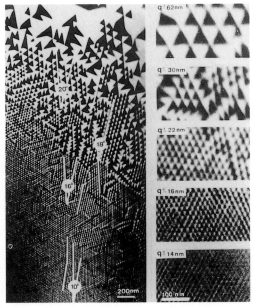

Figure 1-118. Domain fragmentation in quartz as a function of temperature. A temperature gradient is present across the specimen. At the highest temperature, the incommensurately modulated phase is observed (Van Tendeloo et al., 1976).

try. Images of the domain-fragmented α phase are shown in Fig. 1-118. In the case of quartz, the situation is actually somewhat more complicated by the occurrence of an intermediate incommensurate phase between α and β, which is only stable in a narrow temperature interval (~ 1.5 K). This phase was discovered by diffraction contrast electron microscopy (Van Tendeloo et al., 1976).

Quite striking domain structures were studied by diffraction contrast in the monoclinic room-temperature phase of ferroelastic lead orthovanadate [$Pb_3(VO_4)_2$] (Manolikas and Amelinckx, 1980a, 1990b). The structure is rhombohedral at high temperature (γ phase), but on cooling it transforms at 120°C into a monoclinic structure (β phase) which is stable at room temperature. The rhombohedral parent phase is fragmented in domain patterns which minimize the strain energy. They consist of combinations of completely symmetry-determined walls (W) and of walls (W′) whose strain-free orientation depends on the lattice parameters below and above the $\gamma \leftrightarrow \beta$ transition temperature (i.e., on the spontaneous strain tensor). The most striking configuration is the pattern of Fig. 1-119, which contains concentric *stars* of decreasing size. The pattern in (a) of Fig. 1-120 contains a central triangle of metastable γ phase, surrounded by areas consisting of three different variants of the β phase. On cooling, it transforms further in situ into the configuration (b), where the γ triangle has become smaller and is rotated over 180°.

Similar patterns occur in other domain textures resulting from a phase transformation between parent and product phases belonging to the same point groups as γ and β lead orthovanadate, respectively.

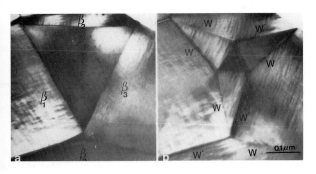

Figure 1-119. Domain pattern in lead orthovanadate resulting from the $\gamma \rightarrow \beta$ phase transition. The central triangle of the star pattern is still in the γ phase. The two photographs refer to the same area; in (b) the temperature was somewhat lowered with respect to that in (a) (Manolikas and Amelinckx, 1980).

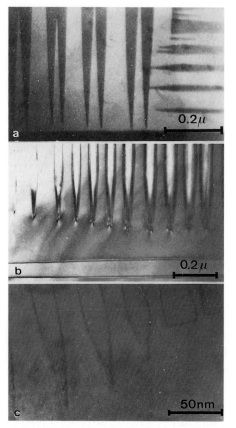

Figure 1-120. Orthorhombic twins in $YBa_2Cu_3O_{7-\delta}$ (Zandbergen et al., 1987), revealed using three different contrast modes: (a) Domain contrast; (b) interface contrast; (c) high-resolution imaging.

The compound $YBa_2Cu_3O_7$ is tetragonal at high temperature, where the … O–Cu–O–Cu … chains in the CuO layers are disordered. Below the transition temperature, which depends on the oxygen content, the chains order in any given area along one out of two mutually perpendicular, equally probable, orientations, which then becomes the b_0 direction of the orthorhombic structure. The disorder–order transition thus produces two structural variants, which have roughly perpendicular b_0 axes, and which are twin related by a mirror operation with respect to (110) or ($1\bar{1}0$). These two orientation variants are

revealed using different imaging modes in Fig. 1-120 (Zandbergen et al., 1987).

1.12.10 The Structure of Ordered Alloys

The structure determination of ordered alloys by means of X-ray diffraction is hampered by the occurrence of numerous orientation variants in *single crystals* of the ordered phase. As many as 12 orientation variants may occur, e.g., in Au_5Mn_2. X-ray diffraction is therefore mostly limited to the use of powder diffraction methods with their inherent limitations. Although electron diffraction, even when complemented by HREM, is usually far inferior to X-ray diffraction for the solution of structure determination problems, it is particularly convenient, however, for the structure determination of binary alloys, especially when HREM is also applied to the same specimen area (Amelinckx, 1978–1979).

The structures of binary alloys are usually superstructures based on a face-centered cubic lattice. It is therefore sufficient to determine the positions of the minority atoms to deduce the complete structure. The minority atoms define the superlattice and produce the superstructure spots in the diffraction pattern. These spots are located within the unit meshes of the f.c.c. reciprocal lattice. When producing a high-resolution image along the cube direction by selecting all superstructure spots within the square formed by {200} reflections, a sufficient number of Fourier components is excited to produce an adequate representation of the minority atom sublattice by bright dots for a suitably chosen focus. This imaging mode is called *bright field superlattice imaging*.

As an alternative, applicable to the simpler superstructures, one can select all superstructure spots within the square of f.c.c. basic reflections 000, 200, 020, 220 and produce a *dark field superlattice image*. In this case, the incident beam must be tilted in order to ensure that the center of Ewald's sphere projects into the center of the square formed by the four above-mentioned reflections. Under these conditions, the phase differences resulting from the angle dependent lens aberrations are minimized for the set of selected beams. In the particular case of the tetragonal Au_4Mn superstructure, viewed along [001], these aberrations are completely eliminated when applying this imaging mode (Van Tendeloo and Amelinckx, 1978).

Figure 1-121 reproduces a number of bright field superlattice images of simple superstructures derived from a basic f.c.c. lattice, viewed along a cube direction. In all images, the bright dots represent the minority atom columns. The relations between the images and the juxtaposed structure models is evident. In Fig. 1-122 a dark field superlattice image of a domain structure in Au_4Mn is reproduced. Two orientation variants with a common [001] axis, as well as a number of out-of-phase boundaries, are revealed by the bright dot pattern, which indicates the positions of the manganese columns.

1.12.11 Long-Period Alloy Superstructures

Electron microscopy and electron diffraction have been particularly successful in the study of long-period superstructures in general and in particular in alloys. Particularly in this field, X-ray diffraction and electron diffraction are complementary, the former method being used to determine the basic structure, while the structure of the long-period derivative is deduced from electron diffraction patterns or from HREM images. The interpretation of the electron diffraction pattern is based on the

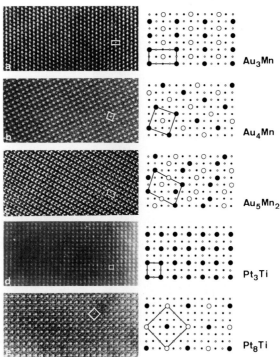

Figure 1-121. Ordered structures of a number of alloys as viewed in HREM along a cube direction of the f.c.c. basic structure (Amelinckx, 1978–79). (a) Au_3Mn [100]; (b) Au_4Mn along [001]; (c) Au_5Mn_2 along [010]; (d) Pt_3Ti along [100]; (e) Pt_8Ti along [001]. All images were made using the bright field superlattice imaging mode. The bright dots represent columns of minority atoms.

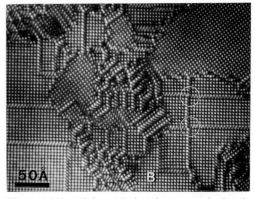

Figure 1-122. High-resolution image made in the dark field superlattice imaging mode, of the microstructure of Au_4Mn. Two orientation variants with a common [001] axis are present. Several out-of-phase boundaries are visible as well; the bright dots represent manganese columns (Van Tendeloo and Amelinckx, 1978).

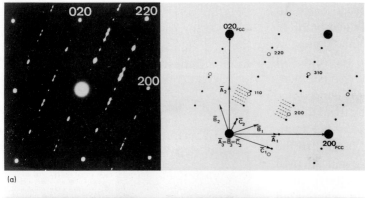

(a)

(b)

Figure 1-123. One-dimensional long-period superstructure of Au_4Mn. (a) Diffraction pattern and (b) corresponding high-resolution image (Van Tendeloo and Amelinckx, 1977).

geometric features, as described in Sec. 1.10.6.

The diffraction pattern of a one-dimensional superstructure of the Au_4Mn phase and the corresponding high-resolution image are reproduced in Fig. 1-123 a and b. The structure model of Fig. 1-124 follows directly from the high-resolution image. It is derived from the Au_4Mn structure by introducing periodic non-conservative out-of-phase boundaries with a displacement vector $(1/10)[135]$ when referred to the Au_4Mn lattice. Its theoretical composition is $Au_{22}Mn_6$. This model can be shown to be in agreement with the fractional shifts, which can be observed directly in Fig. 1-123 since the diffraction pattern of the Au_4Mn phase is superposed on that of the long-period structure (Van Tendeloo and Amelinckx, 1977).

In slightly more manganese deficient material, a two-dimensional superstruc-

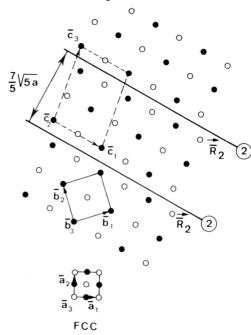

Figure 1-124. Structural model for the one-dimensional long-period superstructure of Fig. 1-123. Only manganese atoms are represented (Van Tendeloo and Amelinckx, 1977).

ture with composition $Au_{31}Mn_9$ is formed by introducing two mutually perpendicular families of out-of-phase boundaries in $(301)_{f.c.c.}$ planes, of the same type as in the one-dimensional superstructure. The diffraction pattern of this superstructure is superposed on that of the basic structure in Fig. 1-125; the corresponding high-resolution image is shown in Fig. 1-126. The superstructure clearly consists of square islands of Au_4Mn structure containing 3×3 manganese columns. Such a model leads to fractional shifts in two directions in agreement with the diffraction pattern of Fig. 1-125 (Van Tendeloo and Amelinckx, 1981).

The first long-period alloy structure was discovered in 1925 in the equiatomic alloy CuAuII (Johansson and Linde, 1925). The period is about 10 basic unit cells long, but its exact length depends on the composition and on alloying elements. The basic structure is a tetragonal superstructure of the f.c.c. structure; it is derived from this structure by having alternating (001) atomic layers occupied by gold and by copper. The conservative anti-phase boundaries are situated in cube planes (100) or (010), and the corresponding displacement vectors are $(1/2)[011]$ and $(1/2)[101]$; i.e., a given (001) plane of the long period structure is alternatingly occupied by copper and by gold. An HREM image of the structure as viewed along a [001] cube direction is reproduced in Fig. 1-127. The interfaces are not atomically flat.

A unique long-period alloy structure, with an exceptionally long c parameter ($c = 8.02$ nm) occurs in Nb_5Ga_{13}, which has a structure derived from the tetragonal DO_{22} structure $NbGa_3$ (Fig. 1-129 a) (Takeda et al., 1988). The superstructure results from the introduction of periodic non-conservative Nb-rich out-of-phase

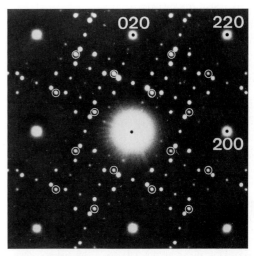

Figure 1-125. Diffraction pattern of two-dimensional long-period superstructure of the Au_4Mn structure. The spots due to the basic Au_4Mn structure are surrounded by circles. Note the fractional shifts in two directions (Van Tendeloo and Amelinckx, 1981).

boundaries on (001) planes of the DO_{22} structure. Whereas in all long-period alloy structures all parallel out-of-phase boundaries are identical, i.e., all have the same displacement vector, this is not the case here; out-of-phase boundaries with two different displacement vectors $(1/2)[011]$ and $(1/2)[101]$ (with respect to

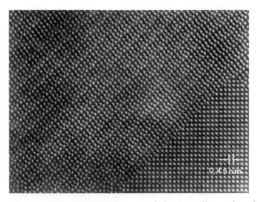

Figure 1-126. HREM image of the two-dimensional long-period superstructure of the Au_4Mn structure of which the diffraction pattern is shown in Fig. 1-125 (Van Tendeloo and Amelinckx, 1981).

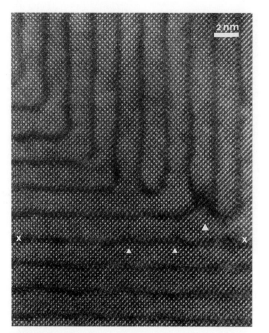

Figure 1-127. HREM image along [001] of the CuAu (II) structure. Note the somewhat wavy shape of the anti-phase boundaries (Yasuda et al., 1987).

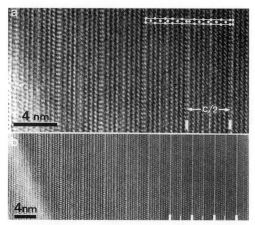

Figure 1-128. HREM image of the long-period superstructure of the alloy Nb_5Ga_{13} which is derived from the DO_{22} structure, and which contains alternatingly two different types of out-of-phase boundaries (Takeda et al., 1988).

the basic f.c.c. structure) alternate, doubling in this way the repeat distance along the c direction. A model of the structure, which can be compared with the HREM image of Fig. 1-128 is shown in Fig. 1-129; only Niobium atoms are represented.

The alloy $Au_{3+}Zn$ forms long-period superstructures derived from the $L1_2$ superstructure by the creation of conservative anti-phase boundaries in (001) cube planes, with a displacement vector $(1/2)[110]$. Complicated sequences of the type predicted by the ANNNI model (Elliott, 1961) are observed; they are formed by domain slabs consisting of double and triple layers of $L1_2$ unit cells. A high-resolution image of a $2\bar{2}2\bar{3}$ sequence viewed along a cube direction is reproduced in Fig. 1-130 where the bright dots represent zinc columns. These structures have been shown to be transient structures the ground state structure being $2\bar{2}$. The "three

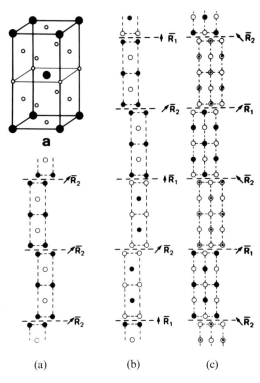

Figure 1-129. Model of the Nb_5Ga_{13} structure to be compared with Fig. 1-128 (Takeda et al., 1988).

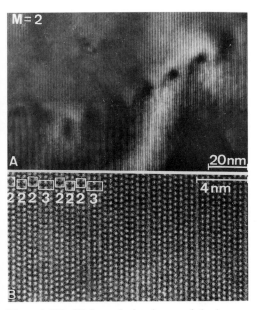

Figure 1-130. High-resolution image of the long period superstructure of $Au_{3+}Zn$, with stacking symbol $2\bar{2}2\bar{3}$. The bright dots represent zinc columns. The part (A) represents the contact between the transient $2\bar{2}2\bar{3}$ structure and the stable $2\bar{2}$ structure (Broddin et al., 1990).

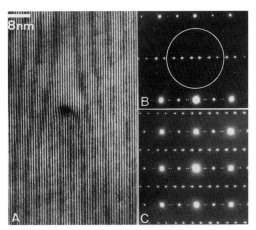

Figure 1-131. Lattice fringes along [001], imaging the "three" bands in $Au_{3+}Zn$ as darker fringes. Note the formation of discommensuration nodes by the fusion of four discommensuration walls. The circle indicates the reflections collected to make the image (Broddin et al., 1990).

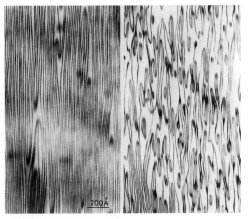

Figure 1-132. Fourfold discommensuration nodes in Ni_3Mo, revealed by diffraction contrast (Van Tendeloo et al., 1974c, d).

bands" are in fact discommensuration walls in this structure, which are removed on annealing by the formation and elimination of fourfold discommensuration nodes, as imaged in Fig. 1-131 by means of (001) lattice fringes (Broddin et al., 1990).

The first observations of *discommensurations* and of *discommensuration nodes* were performed on the alloy $Ni_{3+}Mo$, using diffraction contrast, at a time when the term "discommensuration" had not yet been coined (McMillan, 1976). The interfaces of Fig. 1-132 were described as "out-of-phase boundaries", with a displacement vector equal to one quarter of a lattice vector (Van Tendeloo et al., 1974a, b). Although in alloys there is no essential difference between out-of-phase boundaries and discommensuration walls, the defects of Fig. 1-132 would at present presumably be termed "discommensurations" by most authors.

Conservative anti-phase boundaries in the alloy Cu_3Pd with an $L1_2$ structure revealed by diffraction contrast are shown in Fig. 1-133; they represent the first stage in the formation of a one-dimensional long-period anti-phase boundary structure from the disordered phase. A number of nonconservative anti-phase boundaries be-

Figure 1-133. Diffraction contrast image of the first stage in the formation of a one-dimensional long-period structure in Cu_3Pd. Note the "meandering" of the anti-phase boundaries (Broddin et al., 1989).

come unstable and start "meandering", forming in this way parallel sets of conservative anti-phase boundaries (Broddin et al., 1989).

The same alloy system also exhibits a two-dimensional long-period superstructure, based on a rectangular grid of out-of-phase boundaries in the $L1_2$ basic structure. One family of anti-phase boundaries is conservative; the second one is non-conservative. An HREM image together with the corresponding model is shown in Fig. 1-134. HREM studies have shed some light on the formation mechanism of such superstructures (Broddin et al., 1988).

1.12.12 Minerals

Anorthite ($CaAl_2Si_2O_8$) is a complicated silicate which has a primitive triclinic Bravais lattice (space group $P\bar{1}$) at room temperature. Above $T_c = 514$ K the same unit cell becomes body-centered ($I\bar{1}$). This can be concluded from the diffraction pattern, since the spots of the type $h + k + l = odd$ gradually disappear above T_c. On cooling the crystal from the high-temperature phase to room temperature, it breaks up into two translation variants separated by very "raggy" anti-phase boundaries with a $(1/2)[111]$ displacement vector. No orientation variants are formed. The domain boundaries are revealed by diffraction contrast dark field imaging in reflections for which $h + k + l = odd$. On heating above 514 K the boundaries disappear, but on cooling they reappear at exactly the same place and with the same shape as before; i.e., there is a pronounced memory effect presumably due to impurity pinning. This is illustrated by the heating-cooling cycle in Fig. 1-135, whereas the corresponding diffraction patterns along [101] are reproduced in Fig. 1-136 (Van Tendeloo et al., 1989 a).

1.12.13 Fabrication-Induced Defects in Semiconductors

Semiconductor single crystal chips often undergo a long sequence of fabrication

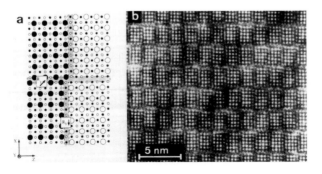

Figure 1-134. HREM image of the two-dimensional long-period superstructure of Cu_3Pd, compared with a model of the structure. The horizontal set of anti-phase boundaries is non-conservative, whereas the vertical set is conservative (Broddin et al., 1989).

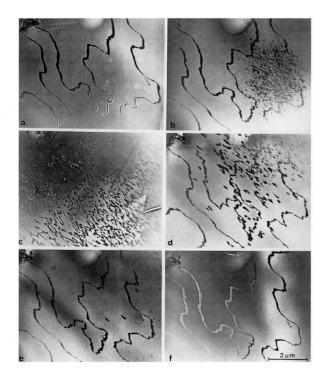

Figure 1-135. Evolution of anti-phase boundaries in anorthite ($CaAl_2Si_2O_8$) during a heating–cooling cycle from room temperature up to above 514 K. All images refer to the same area. Note the memory effect (Van Tendeloo et al., 1989).

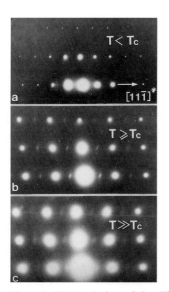

Figure 1-136. Evolution of the diffraction pattern of anorthite during the same heating–cooling cycle as in Fig. 1-135 (Van Tendeloo et al., 1989a).

steps (thermal treatment, oxidation, etching, ...), some of which can be accompanied by a deterioration of the crystal's physical properties, a process affecting the performance of the final device. The microminiaturization of the electronic devices makes detailed control of the crystal perfection strongly dependent on electron microscope techniques, both on high-resolution images of cross sectional specimens of devices, and on high-voltage electron microscopy for the study of thick specimens at low resolution and low magnification.

Figure 1-137 shows a TEM image of processions of dislocations observed end-on in a cross section view of a field-effect device. At the edge of the constriction in the silicon oxide layer sources have generated dislocations along the glide planes of maximum resolved shear stress, in order to relieve the stresses generated by the oxidation process. The dislocations apparently

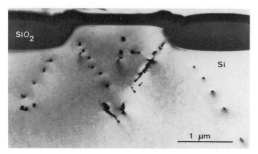

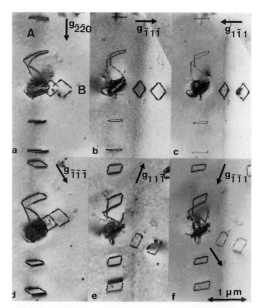

Figure 1-137. TEM image of a cross section of a field-effect device. Dislocations are emitted from the edges of the constriction in the silicon oxide layer; the dislocations are seen end-on (Vanhellemont and Amelinckx, 1987).

form "inverse" pile-ups, their spacing being smallest close to the source (Vanhellemont and Amelinckx, 1987).

In Fig. 1-138 finger-shaped gate areas formed in a field oxide layer on a silicon chip have similarly generated stresses which were relieved by dislocation generation. In this case the dislocations are imaged in a plane view.

Oxide or other precipitate particles may put the surrounding silicon matrix under a compressive stress. This stress is often large enough to give rise to *prismatic punching*, whereby discs of self interstitials surrounded by a loop of perfect dislocation are emitted. Such loops are glissile on a cylindrical surface. The cross section is determined by the precipitate's shape, and the direction of the generators by the Burgers vector of the dislocations, i.e., $(1/2)\langle 110 \rangle$ (Fig. 1-139).

Figure 1-139. *Prismatic punching* around a precipitate particle in a silicon matrix (Courtesy of Bender).

With sophisticated crystal growth techniques, such as molecular beam epitaxy, it has recently become possible to fabricate artificial periodic layer structures consisting of various combinations of semiconducting materials. Cross sections of samples examined by electron microscopy have revealed the precise geometry of the obtained products, in particular, the quality of the interfaces and the thickness of the successive layers. Since electron microscopy is destructive, it is not likely to become a routine production control

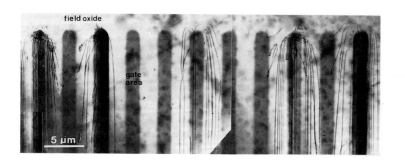

Figure 1-138. Finger-shaped gate areas in a field oxide. Dislocations are generated along the edges; they are observed in a plane view (Courtesy of Vanhellemont).

method, but it is important for the calibration of non-destructive, but less direct, methods. In particular it allows one to check the quality of the interfaces with atomic resolution.

At the interfaces between different epitaxial layers, misfit dislocations are often formed; these can be imaged directly in HREM, as, for example, in the interface between InSb and GaAs in Fig. 1-140, from which it is clear that the supplementary half planes of the edge dislocations are formed in the GaAs layer, whose lattice parameter is smaller than that of the InSb layer.

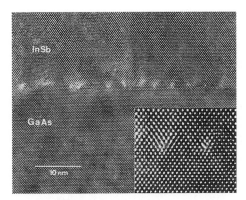

Figure 1-140. Misfit dislocations along the interface between InSb and GaSb. The inset shows two dislocations; the supplementary halfplanes are formed in the GaAs layer (Courtesy of Luyten).

1.12.14 Superstructures Due to Non-Stoichiometry

In the high T_c superconductors, the oxygen content, as well as the way in which it is accommodated in the crystal, is of considerable importance, since it determines to a large extent the value of T_c. In the $YBa_2Cu_3O_{7-\delta}$ compound, oxygen deficiency is mainly associated with the CuO layers in the structure. These layers consist of chains of ... $O-Cu-O-Cu-O$ ... parallel with the b_0 direction and separated by a_0. It is this feature of the structure that breaks the tetragonal symmetry and reduces it to an orthorhombic one. In oxygen-deficient crystals there is a strong tendency to deplete complete chains in the CuO layer, rather than breaking up chains into short segments by oxygen vacancies. The average separation of chains then increases with increasing oxygen deficiency. In the particular case of $YBa_2Cu_3O_{6.5}$, the average separation of the chains is $2a_0$. This was first shown by means of electron microscopy (Zandbergen et al., 1987). Subsequently, evidence was found for more complicated arrangements such as $3a_0$ and for mixed sequences of $2a_0$ and $3a_0$

(Reyes-Gasga et al., 1989). Since oxygen is a low Z element, it does not produce pronounced dot contrast, but nevertheless, evidence could be obtained from images, as well as from the diffraction pattern (Fig. 1-141) for the presence of the $2a_0$ structures and for the spatial arrangement of the chains in these structures.

In the compound $1-2-3\frac{1}{2}$ similar $2a_0$ and $3a_0$ structures due to oxygen deficiency in the single CuO layers were found. The relaxation of the heavy atoms around the rows of vacancies allows the vacancy chains to become visible, as in Fig. 1-142, which visualizes the vacant chains as viewed end-on in the $3a_0$ structure in the $1-2-3\frac{1}{2}$ phase (Krekels et al., 1991).

In $La_2CuO_{4-\delta}$ the oxygen deficiency is apparently accommodated along crystallographic shear planes formed by edge-sharing CuO_6 octahedra. The shear planes have a strong tendency to be uniformly spaced, giving rise to crystallographic shear structures. Such a shear structure is imaged at high resolution in Fig. 1-143. The inset shows the well-resolved atom columns in the thinnest part of the specimen (Van Tendeloo and Amelinckx, 1991).

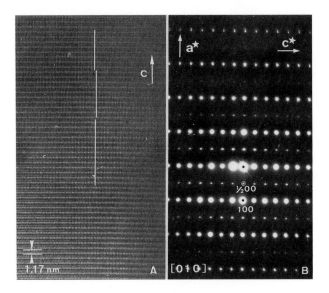

Figure 1-141. The vacancy-ordered $2a_0$ structure in $YBa_2Cu_3O_{7-\delta}$ (Reyes-Gasga et al., 1989). (A) HREM image; (B) diffraction pattern.

Figure 1-142. The vacancy-ordered $3a_0$ structure in the phase $Y_2Ba_4Cu_7O_{15}$. The high-resolution images are compared with simulated images, based on a model that allows for relaxation of the barium ions around the vacancy rows (Courtesy of Krekels).

The structure of $Ca_{0.85}CuO_2$ is remarkable for the way in which the complicated stoichiometry is accommodated in a rather simple structure (Milat et al., 1991). The structure can be considered to consist of an orthorhombic framework of ribbons of edge-sharing planar CuO_4 groups "stuffed" with calcium atoms. The ribbons are positioned in a centered arrangement when viewed along the ribbons, i.e., parallel to the b_0 direction. When viewed along the a_0 direction, i.e., along the normal to the CuO_4 planes, the copper sublattice is centered as well. This framework gives rise to "tunnels" along the b_0 direction, consisting of interpenetrating deformed oxygen octahedra. The calcium atoms occupy positions along these tunnels. Coulomb repulsion tends to space them uniformly, but on the other hand, the centers of the oxygen octahedra are preferred sites for inserted atoms such as calcium. The actual configuration is a trade-off between a uniformly spaced array and an arrangement in which arrays of five octahedral sites are occupied along the tunnels, regularly alternating with a vacant site, so that a ratio

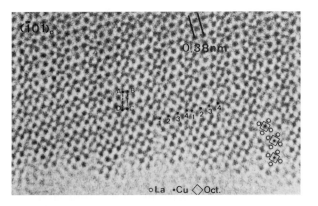

Figure 1-143. HREM image of a shear structure attributed to oxygen deficiency in $La_2CuO_{4-\delta}$ (Van Tendeloo and Amelinckx, 1992).

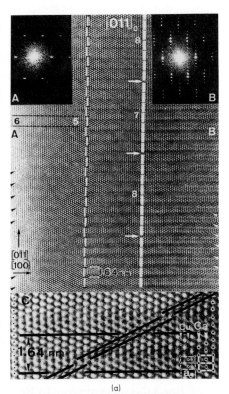

Cu/Ca $\approx 6/5$ is achieved. Imaging along a suitable zone electron microscopy made it possible to reveal the copper atoms of the framework simultaneously with the calcium atoms in the tunnels as parallel, alternating bright dot rows (Fig. 1-144). It is then clear that seven dots along the copper row (i.e., six spacings b_0) correspond with six dots (i.e., five spacings) along the calcium rows. The superstructure therefore has a periodicity amounting to the smallest common multiple of the copper and the calcium spacing, i.e., $6b_0$, the average calcium spacing along the tunnels being $(6/5)\,b_0$.

1.12.15 Various Applications

1.12.15.1 In Situ Studies

The availability of cooling and heating specimen holders allows for the in situ study of the phenomena accompanying phase transitions. When a specimen goes through a disorder–order transition, different phases of the domain fragmentation can be followed. The creation and elimination of discommensuration walls is directly

Figure 1-144. HREM image of the incommensurate structure in $Ca_{0.85}CuO_2$ (Milat et al., 1991). (a) $[011]_0$ zone to be compared with Fig. 1-82; (b) $[010]_0$ zone. Note the relation between the rows of calcium and copper atoms.

observable in dark field images made in clusters of incommensurate reflections.

Static images referring to the phase transitions in lead orthovanadate and in quartz are reproduced respectively in Fig. 1-119 and in Fig. 1-118. When performing such observations, one should be aware of the effect of the electron beam on the specimen, which results in an increase in temperature depending on the thermal conductivity of the foil, and which may also cause some radiation effects, which in turn may interfere with the transition.

1.12.15.2 Radiation Damage

Electron microscopy, in particular high-voltage electron microscopy, has been used extensively to study in situ radiation effects as well as post irradiation defect configurations. The point defects, precipitates, and small dislocation loop can be characterized using the methods mentioned above (Sec. 1.6.3.7).

1.12.15.3 Radiation Ordering

Some surprising results were found by in situ studies of ordering alloys, exhibiting a short-range-order state, such as Ni_4Mo. When irradiated with 1 MV electrons at low-temperature, ordered Ni_4Mo becomes completely disordered. When irradiating in a temperature range below, but close to, the order-disorder transition temperature, the irradiation causes the alloy to order up to a certain degree. The order parameters can be determined by following the evolution of the intensity of the order diffraction spots. These phenomena result from the competition between the ordering effect due to radiation-enhanced diffusion at the irradiation temperature and the disordering effect of the irradiation as a result of atomic collisions. In a certain temperature range, the short-range-order state will be produced by irradiation. Certain alloy phases, which could not be ordered by thermal treatment, were found to order under electron irradiation. This kind of behavior was reviewed by Russell (1985).

1.12.15.4 Magnetic Domain Structures

Magnetic domain walls, both Néel walls and Bloch walls, can be studied making use of Lorentz contrast. The magnetization vector changes in orientation and (or) sense across a domain wall. As a result, the electron beam is deflected differently, because of the Lorentz force, in the domains on both sides of the wall. Under defocused conditions, the domain wall can thus be revealed either as a bright or as a dark line, depending on whether the change in the magnetization across the boundary, produces an excess or a deficiency of electrons behind the specimen at the position of the walls.

1.13 Acknowledgements

The author is grateful to a number of colleagues for the use of photographs as illustrations in this survey; their names are mentioned either in the text or in the figure captions. I am grateful to my colleagues Professor J. Van Landuyt and Professor G. Van Tendeloo for useful discussions and for providing the facilities to prepare this manuscript. Special thanks are due to Professor D. Van Dyck for the illuminating discussions we had on a number of subtle points of theory. The photographic illustrations were mostly provided by Professor G. Van Tendeloo with the skillful printing of F. Schallenberg and A. De Muynck. The manuscript was patiently typed by A. De Belder and H. Evans, whereas the line drawings were drafted carefully by M. Schrijnemakers.

1.14 References

Aerts, E., Delavignette, P., Siems, R., Amelinckx, S. (1962), *J. Appl. Phys. 33*, 3078.

Amelinckx, S. (1964), *The Direct Observation of Dislocations,* Suppl. 6 in: *Solid State Physics:* Seitz, F., Turnbull, D. (Eds.). London: Academic Press.

Amelinckx, S. (1978–79), *Chimica Scripta 14,* 197.

Amelinckx, S. (1979), in: *Dislocation in Solids,* Vol. 2, Chap. 6: Nabarro, F. R. N. (Ed.). Amsterdam: North Holland, p. 68.

Amelinckx, S. (1986a), *J. Electron. Microscop. Techn. 3,* 131.

Amelinckx, S. (1986b), in: *Examining the Submicron World:* Nato-ASI, Series B, Vol. 137. New York: Plenum Press, p. 71.

Amelinckx, S., Delavignette, P. (1962), *J. Appl. Phys. 33,* 1458.

Amelinckx, S., Van Landuyt, J. (1978), in: *Diffraction and Images in Material Science:* Amelinckx, S., Gevers, R., Van Landuyt, J. (Eds.). Amsterdam, New York, Oxford: North Holland, p. 107.

Amelinckx, S., Van Landuyt, J. (1989), in: *Encyclopedia of Physical Science and Technology, 1989 Yearbook:* Meyers, R. (Ed.). London: Academic Press, p. 617.

Amelinckx, S., Van Landuyt, J., Van Tendeloo, G. (1984), in: *Modulated Structure Materials:* Tsakalakos, T. (Ed.). Dordrecht: Martinus Nijhoff Publ., p. 183.

Amelinckx, S., Van Tendeloo, G., Van Dyck, D., Van Landuyt, J. (1989a), *Phase Transitions, Vol. 16/17,* 3.

Amelinckx, S., Van Heurck, C., Van Tendeloo, G. (1989b), in: *Quasi-crystals and Incommensurate Structures in Condensed Matter, Proc. Third Intern. Meeting on Quasi-crystals (Mexico):* Yacaman, M. J., Romen, D., Castaño, V., Gomez, A. (Eds.). Singapore: World Scientific, p. 300.

Art, A., Gevers, R., Amelinckx, S. (1963), *Phys. Stat. Sol. 3,* 967.

Ashby, M. F., Brown, L. M. (1963), *Phil. Mag. 8,* 1083, 1649.

Bassett, G. A., Menter, J. W., Pashley, D. W. (1958), *Proc. Roy. Soc. A 246,* 345.

Bethe, H. A. (1928), *Ann. Physik [4] 87,* 55.

Bethge, H., Heydenreich, J. (Eds.) (1982), *Elektronenmikroskopie in der Festkörperphysik.* Berlin, Heidelberg, New York: Springer Verlag.

Bilby, B. A., Bullough, R., Smith, E. (1955), *Proc. Roy. Soc. A 231,* 263.

Bird, D. M., Withers, R. L. (1986), *J. Phys. C 19,* 3497, 3507.

Bormann, G. (1941), *Physik Z. 42,* 157.

Bormann, G. (1950), *Z. Physik 127,* 297.

Boulesteix, C., Van Landuyt, J., Amelinckx, S. (1976), *Phys. Stat. Sol. (a) 33,* 595.

Bragg, W. L. (1929), *Nature 124,* 125.

Broddin, D., Van Tendeloo, G., Van Landuyt, J., Amelinckx, S., Loiseau, A. (1988), *Phil. Mag. B 57, Nr. 1,* 31.

Broddin, D., Van Tendeloo, G., Van Landuyt, J., Amelinckx, S. (1989), *Phil. Mag. 59, Nr. 1,* 47.

Broddin, D., Van Tendeloo, G., Amelinckx, S. (1990), *J. Phys. Condens. Matter 2,* 3459.

Chik, K. P., Wilkens, M., Rühle, M. (1967), *Phys. Stat. Sol. 23,* 113.

Cockayne, D. J. H., Ray, I. L. E., Whelan, M. J. (1969), *Phil. Mag. 20,* 1265.

Cockayne, D. J. H., Jenkins, M. J., Ray, I. L. E. (1971), *Phil. Mag. 24,* 1383.

Coene, W., Bender, H., Amelinckx, S. (1985a), *Phil. Mag. A 52,* 369.

Coene, W., Bender, H., Lovey, F. C., Van Dyck, D., Amelinckx, S. (1985b), *Phys. Stat. Sol. (a) 87,* 483.

Cowley, J. M. (1950), *J. Appl. Phys. 24,* 24.

Cowley, J. M., Iijima, S. (1972), *Z. Naturf. A 27,* 445.

Cowley, J. M., Moodie, A. F. (1957), *Acta Cryst. 10,* 609.

Darwin, C. G. (1914), *Phil. Mag. 27,* 315, 675.

Delavignette, P., Amelinckx, S. (1962), *J. Nucl. Mater. 5,* 17.

De Ridder, R. (1978), in: *Diffraction and Imaging Techniques in Material Science:* Amelinckx, S., Gevers, R., Van Landuyt, J. (Eds.). Amsterdam, New York, Oxford: North Holland, p. 429.

De Ridder, R., Amelinckx, S. (1971), *Phys. Stat. Sol. (b) 43,* 541.

De Ridder, R., Van Tendeloo, G., Van Dyck, D., Amelinckx, S. (1976), *Phys. Stat. Sol. (a) 38,* 663.

De Ridder, R., Van Tendeloo, G., Van Dyck, D., Amelinckx, S. (1977), *Phys. Stat. Sol. (a) 40,* 669.

De Wolff, P. M. (1974), *Acta Cryst. A 30,* 777.

De Wolff, P. M. (1984), *Acta Cryst. A 40,* 34.

De Wolff, P. M., Janssen, T., Janner, A. (1981), *Acta Cryst. A 37,* 625.

Drum, C. M., Whelan, M. J. (1965), *Phil. Mag. 11,* 205.

Duneau, M., Katz, A. (1985), *Phys. Rev. Lett. 54,* 2688.

Edmondson, B. (1963), *Proc. Joint Conf. on Inorganic and Intermetallic Crystals,* Univ. of Birmingham (U.K.).

Elliott, R. J. (1961), *Phys. Rev. 124,* 346.

Elser, V. (1985), *Phys. Rev. B 32,* 4892.

Elser, V. (1986), *Acta Cryst. A 42,* 36.

Eshelby, J. D., Stroh, A. N. (1951), *Phil. Mag. [7], 42,* 1401.

Essmann, U., Wilkens, M. (1964), *Phys. Stat. Sol. 4,* K53.

Ewald, P. P. (1917), *Ann. der Phys. 54,* 519.

Fejes, P. L. (1977), *Acta Cryst. A 33,* 109.

Frangis, N., Kuypers, S., Manolikas, C., Van Tendeloo, G., Van Landuyt, J., Amelinckx, S. (1990), *Journal of Solid State Chem. 84,* 314.

Friedel, J. (1964), *Dislocations.* London: Pergamon.

Fujiwara, K. (1957), *J. Phys. Soc. Japan 12,* 7.

Gevers, R. (1962a), *Phil. Mag. 7,* 1681.

Gevers, R. (1962b), *Phil. Mag. 7,* 59.

Gevers, R. (1962c), *Phil. Mag. 7,* 651.

Gevers, R. (1963a), *Phys. Stat. Sol. 3,* 2289.

Gevers, R. (1963b), *Phil. Mag. 7,* 769.

Gevers, R. (1978), in: *Diffraction and Imaging Techniques in Material Science:* Amelinckx, S., Gevers R., Van Landuyt, J. (Eds.). Amsterdam, New York, Oxford: North Holland, p. 9.

Gevers, R., Art, A., Amelinckx, S. (1963), *Phys. Stat. Sol. 3,* 1563.

Gevers, R., Delavignette, P., Blank, H., Van Landuyt, J., Amelinckx, S. (1964a), *Phys. Stat. Sol. 5,* 595.

Gevers, R., Art, A., Amelinckx, S. (1964b), *Phys. Stat. Sol. 7,* 605.

Gevers, R., Van Landuyt, J., Amelinckx, S. (1965), *Phys. Stat. Sol. 11,* 689.

Glaser, W. (1952), *Grundlagen der Elektronoptik.* Wien: Springer Verlag.

Goringe, M. J., Valdré, U. (1966), *Proc. Roy. Soc. A295,* 192.

Grinton, G. R., Cowley, J. M. (1971), *Optik 34,* 221.

Groves, G. W., Kelly, A. (1961), *Phil. Mag. [8] 6,* 1527.

Groves, G. W., Kelly, A. (1963), *Proc. Joint Conf. on Inorganic and Intermetallic Crystals,* Univ. of Birmingham (U.K.).

Groves, G. W., Whelan, M. J. (1962), *Phil. Mag. [8] 7,* 1603.

Hashimoto, H., Whelan, M. J. (1963), *J. Phys. Soc. Japan 18,* 1706.

Hashimoto, H., Uyeda, R. (1957), *Acta Cryst. 10,* 143.

Hashimoto, H., Mannami, M., Naiki, T. (1961), *Phil. Trans. Royal Soc. 253,* 459.

Hashimoto, H., Howie, A., Whelan, M. J. (1962), *Proc. Roy. Soc. A269,* 80.

Hashimoto, H., Endo, H., Takai, Y., Tomita, H., Yokota, Y. (1978–79), *Chemica Scripta 14,* 23.

Head, A. K. (1967), *Aust. J. Phys. 20,* 557.

Heidenreich, R. D. (1949), *J. Appl. Phys. 20,* 993.

Hibi, T. (1962), in: *Fifth Int. Congress on Electron Microscopy.* New York: Academic Press, p. KK1.

Hirsch, P. B., Horne, R. W., Whelan, M. J. (1956), *Phil. Mag. 1,* 667.

Hirsch, P. B., Silcox, J. (1958), in: *Growth and Perfection of Crystals:* Doremus, R. H., Roberts, B. W., Turnbull, D. (Eds.). New York: Wiley, p. 262.

Hirsch, P. B., Silcox, J., Smallmann, R., Westmacott, K. (1958), *Phil. Mag. [8] 3,* 897.

Hirsch, P. B., Howie, A., Whelan, M. J. (1960), *Phil. Trans. Roy. Soc. A252,* 499.

Hirsch, P. B., Howie, A., Whelan, M. J. (1962), *Phil. Mag. [8] 7,* 2095.

Hirsch, P. B., Nicholson, R. B., Howie, A., Pashley, D. W., Whelan, M. J. (1965), *Electron Microscopy of Thin Crystals.* London: Butterworths.

Howie, A., Whelan, M. J. (1960), *Proc. European Reg. Conf. on Electron Microscopy,* Delft, Vol. 1, p. 194.

Howie, A., Whelan, M. J. (1961), *Proc. Roy. Soc. A263,* 217.

Howie, A., Whelan, M. J. (1962), *Proc. Roy. Soc. A267,* 206.

Humble, P. (1978), in: *Diffraction and Imaging Techniques in Material Science:* Amelinckx, S., Gevers, R., Van Landuyt, J. (Eds.). Amsterdam, New York, Oxford: North Holland, p. 315.

Janssen, T., Janner, A. (1987), *Advances in Physics 36,* 519.

Johansson, C. H., Linde, J. O. (1925), *Ann. Physik [4] 78,* 439.

Kikuchi, S. (1928), *Japan J. Phys. 5,* 23.

Kirkland, E. J. (1984), *Ultramicroscopy 15,* 151–157.

Krekels, T., Van Tendeloo, G., Amelinckx, S., Karpinski, J., Kaldis, E., Rusiecki, S. (1991), *Appl. Phys. Lett. 59,* 23.

Kuypers, S., Van Tendeloo, G., Van Landuyt, J., Amelinckx, S., Shu, H. W., Jaulmes, S., Flahaut, J., Laruelle, P. (1988), *J. of Solid State Chemistry 73,* 192.

Kuypers, S., Van Tendeloo, G., Van Landuyt, J., Amelinckx, S. (1989), *Acta Cryst. A45,* 291.

Levine, D., Steinhardt, P. J. (1984), *Phys. Rev. Lett. 53,* 2477.

Levine, D., Steinhardt, P. J. (1986), *Phys. Rev. B34,* 596.

Lynch, D. F., O'Keefe, M. A. (1972), *Acta Cryst. A28,* 536.

Lynch, D. F., Moodie, A. F., O'Keefe, M. A. (1975), *Acta Cryst. A29,* 537.

Mannami, M. (1960), *Acta Cryst. 13,* 363.

Mannami, M. (1962), *J. Phys. Soc. Japan 17,* 1160.

Manolikas, C., Amelinckx, S. (1980a), *Phys. Stat. Sol. (a) 60,* 607.

Manolikas, C., Amelinckx, S. (1980b), *Phys. Stat. Sol. 61,* 179.

Manolikas, C., Van Landuyt, J., Amelinckx, S. (1979), *Phys. Stat. Sol. (a) 53,* 327.

McGillavry, C. H. (1940), *Physica 7,* 329.

McMillan, W. L. (1976), *Phys. Rev. B14,* 1496.

Menter, J. W. (1956), *Proc. Roy. Soc. A236,* 119.

Milat, O., Van Tendeloo, G., Amelinckx, S., Babu, T. G. N., Greaves, C. (1992), *Journal of Solid State Chemistry 97,* 405.

Mott, N. F., Massey, H. S. W. (1949), *The Theory of Atomic Collisions.* Oxford: Clarendon Press.

O'Keefe, M. A. (1973), *Acta Cryst. A29,* 389.

Overhauser, A. W. (1962), *Phys. Rev. B3,* 1373.

Pasemann, M., Scheerschmidt, K. (1982), in: *Elektronenmikroskopie in der Festkörperphysik:* Bethge, H., Heydenreich, J. (Eds.). Berlin, Heidelberg, New York: Springer Verlag, p. 257.

Pashley, D. W., Menter, J. W., Bassett, G. A. (1957), *Nature 179,* 752.

Pérez-Mato, T. M., Madariaga, G., Tillo, M. J. (1984), *Phys. Rev. B30,* 1534.

Pérez-Mato, T. M., Madariaga, G., Tillo, M. J. (1986), *J. Phys. C19,* 2613.

Reyes-Gasga, J., Krekels, T., Van Tendeloo, G., Van Landuyt, J., Amelinckx, S., Bruggink, W. H. M., Verweij, H. (1989), *Physcia C159,* 831.

Ruedl, E., Delavignette, P., Amelinckx, S. (1962a), *Proc. I.A.E.A. Symposium on Radiation Damage in*

Solids and Reactor Materials, Venice 1962, Vol. 1, p. 363.

Ruedl, E., Delavignette, P., Amelinckx, S. (1962 b), *J. Nucl. Mater. 6*, 46.

Rühle, M. (1967), *Phys. Stat. Sol. 19*, 263, 279.

Rühle, M., Wilkens, M., Essmann, U. (1965), *Phys. Stat. Sol. 11*, 819.

Russell, K. C. (1985), *Progr. Mat. Sci. 28*, 229.

Schechtman, D., Blech, I., Gratias, D., Cahn, J. W. (1984), *Phys. Rev. Lett. 53*, 1951.

Scherzer, O. (1949), *J. Appl. Phys. 20*, 20.

Serneels, R., Snijkers, M., Delavignette, P., Gevers, R., Amelinckx, S. (1973), *Phys. Stat. Sol. (b) 58*, 277.

Siems, R., Delavignette, P., Amelinckx, S. (1962 a), *Phys. Stat. Sol. 2*, 421.

Siems, R., Delavignette, P., Amelinckx, S. (1962 b), *Phys. Stat. Sol. 2*, 636.

Snijkers, M., Serneels, R., Delavignette, P., Gevers, R., Amelinckx, S. (1972), *Crystal Lattice Defects 3*, 99.

Spence, J. C. H. (1981), *Experimental High Resolution Electron Microscopy*, in: *Monographs on the Physics and Chemistry of Materials, Oxford Science Publications.* Oxford: Clarendon.

Takagi, S. (1962), *Acta Cryst. 15*, 1311.

Takeda, M., Van Tendeloo, G., Amelinckx, S. (1988), *Acta Cryst. A 44*, 938.

Thomas, G. (1962), *Transmission Electron Microscopy of Metals.* New York: J. Wiley.

Thomas, G. (1978), in: *Diffraction and Imaging Techniques in Materials Science:* Amelinckx, S., Gevers, R., Van Landuyt, J. (Eds.). Amsterdam, New York, Oxford: North Holland, p. 217.

Thomas, G., Goringe, M. J. (1979), *Transmission Electron Microscopy of Materials.* New York: J. Wiley.

Van Dyck, D. (1978), in: *Diffraction and Imaging Techniques in Material Science:* Amelinckx, S., Gevers, R., Van Landuyt, J. (Eds.). Amsterdam, New York, Oxford: North Holland, p. 355.

Van Dyck, D., Coene, W. (1989), *Ultramicroscopy 15*, 29.

Van Dyck, D., Op de Beeck, M. (1990), *Proceedings of the XIIth. International Congress for Electron Microscopy.* San Francisco: San Francisco Press Inc., p. 26.

Van Dyck, D., De Ridder, R., Van Tendeloo, G., Amelinckx, S. (1977), *Phys. Stat. Sol. (a) 43*, 541.

Van Dyck, D., Conde, C., Amelinckx, S. (1979), *Phys. Stat. Sol. (b) 56*, 327.

Van Dyck, D., Danckaert, J., Coene, W., Selderslaghs, E., Broddin, D., Van Landuyt, J., Amelinckx, S. (1989), in: *Computer Simulation of Electron Microscope Diffraction and Images:* Krakow, W., O'Keefe, M. (Eds.). Warrendale: The Minerals, Metals and Materials Society, p. 107.

Vanhellemont, J., Amelinckx, S. (1987), *J. Appl. Phys. 61 (b)*, 2176.

Van Landuyt, J., Gevers, R., Amelinckx, S. (1964), *Phys. Stat. Sol. 7*, 519.

Van Landuyt, J., De Ridder, R., Gevers, R., Amelinckx, S. (1970 a), *Mat. Res. Bull. 5*, 353.

Van Landuyt, J., Remaut, G., Amelinckx, S. (1970 b), *Phys. Stat. Sol. 41*, 271.

Van Landuyt, J., Van Tendeloo, G., Amelinckx, S. (1974), *Phys. Stat. Sol. (a) 26*, K9.

Van Tendeloo, G., Amelinckx, S. (1973), *Mat. Res. Bull. 8*, 721.

Van Tendeloo, G., Amelinckx, S. (1974 a), *Phys. Stat. Sol. (a) 22*, 621.

Van Tendeloo, G., Amelinckx, S. (1974 b), *Acta Cryst. A 30*, 431.

Van Tendeloo, G., Amelinckx, S. (1977), *Phys. Stat. Sol. (a) 43*, 553.

Van Tendeloo, G., Amelinckx, S. (1978), *Phys. Stat. Sol. (a) 49*, 337.

Van Tendeloo, G., Amelinckx, S. (1981), *Phys. Stat. Sol. (a) 65*, 431.

Van Tendeloo, G., Amelinckx, S. (1986), *Scripta Met. 20*, 335.

Van Tendeloo, G., Amelinckx, S. (1991), *Physica C 176*, 575.

Van Tendeloo, G., Van Landuyt, J., Delavignette, P., Amelinckx, S. (1974 a), *Phys. Stat. Sol. (a) 25*, 697.

Van Tendeloo, G., Delavignette, P., Van Landuyt, J., Amelinckx, S. (1974 b), *Phys. Stat. Sol. 26*, 299.

Van Tendeloo, G., Van Landuyt, J., Amelinckx, S. (1976), *Phys. Stat. Sol. (a) 33*, 723.

Van Tendeloo, G., Van Landuyt, J., Amelinckx, S. (1982), *40th Annual Proc. Electron Microscopy Society of America:* Bailey, G. W. (Ed.), p. 340.

Van Tendeloo, G., Ghose, S., Amelinckx, S. (1989 a), *Phys. Chem. Minerals 16*, 311.

Van Tendeloo, G., Van Heurck, C., Amelinckx, S. (1989 b), *Solid State Commun. 71*, 705.

Van Tendeloo, G., Op de Beeck, M., Amelinckx, S., Bohr, J., Krätschmer, W. (1991), *Europhys. Lett. 15 (3)*, 215.

Verwerft, M., Van Tendeloo, G., Van Landuyt, J., Coene, W., Amelinckx, S. (1988), *Phys. Stat. Sol. (a) 109*, 67.

Verwerft, M., Van Dyck, D., Brabers, W. A. M., Van Landuyt, J., Amelinckx, S. (1989), *Phys. Stat. Sol. (a) 112*, 451.

Whelan, M. J. (1958–59), *J. Inst. Met. 87*, 392.

Whelan, M. J. (1978), in: *Diffraction and Imaging Techniques in Materials Science:* Amelinckx, S., Gevers, R., Van Landuyt, J. (Eds.). Amsterdam, New York, Oxford: North Holland, p. 43.

Whelan, M. J., Hirsch, P. B. (1957), *Phil. Mag. 2*, 1121, 1303.

Wilkens, M. (1978), in: *Diffraction and Imaging Techniques in Materials Science:* Amelinckx, S., Gevers, R., Van Landuyt, J. (Eds.). Amsterdam, New York, Oxford: North Holland, p. 185.

Yamamoto, A. (1980), *Phys. Rev. B 22*, 373.

Yamamoto, A. (1982 a), *Acta Cryst. A 37*, 838.

Yamamoto, A. (1982 b), *Acta Cryst. B 38*, 1446, 1451.

Yamamoto, A. (1985), *Acta Cryst. A 38*, 87.

Yasuda, K., Nakagawa, M., Van Tendeloo, G., Amelinckx, S. (1987), *Journal of Less-Common Metals 135,* 169.

Yoshioka, H. (1957), *J. Phys. Soc. Japan 12,* 628.

Zandbergen, H. W., Van Tendeloo, G., Okabe, T., Amelinckx, S. (1987), *Phys. Stat. Sol. (a) 103,* 45.

Zandbergen, H. W., Groen, W. A., Mylhoff, F. C., Van Tendeloo, G., Amelinckx, S. (1988), *Physica C 156,* 325.

General Reading

Amelinckx, S. (1964), *The Direct Observation of Dislocations,* Supplement 6 in: *Solid State Physics:* Seitz, F., Turnbull, D. (Eds.). London: Academic Press.

Diffraction and Imaging Techniques in Material Science (1970, 1978): Amelinckx, S., Gevers, R., Van Landuyt, J. (Eds.). Amsterdam, New York, Oxford: North Holland, Publishing Company.

Dislocation in Solids (1979): Nabarro, F. R. N. (Ed.). Amsterdam, New York, Oxford: North Holland, Publishing Company.

Hirsch, P. B., Nicholson, R. B., Howie, A., Pashley, D. W., Whelan, M. J. (1965), *Electron Microscopy of Thin Crystals.* London: Butterworths.

Elektronenmikroskopie in der Festkörperphysik (1982): Bethge, H., Heydenreich, J. (Eds.). Berlin, Heidelberg, New York: Springer Verlag.

Spence, J. C. H. (1981), *Experimental High Resolution Electron Microscopy, Monographs on the Physics and Chemistry of Materials, Oxford Science Publications.* Oxford: Clarendon Press.

Thomas, G. (1962), *Transmission Electron Microscopy of Metals.* New York: John Wiley and Sons Inc.

2 Analytical Electron Microscopy

Ernest L. Hall

GE Corporate Research and Development, Schenectady, NY, U.S.A.

List of Symbols and Abbreviations

A_d	area of the solid-state X-ray detector
A	atomic weight
A, B, C	elements
B	electron gun brightness
b	broadening of the electron beam
C_A, C_B	weight fractions of elements A and B
C_s	spherical aberration
C_1, C_2	first and second condenser lenses
c	diameter of the CBED discs
d_p	probe diameter
d	interplanar spacing
d_i	diameter of demagnified image
d_{sa}	diameter of the probe due to spherical aberration
d_0	diameter of the electron source
E, E_0	energy, initial electron energy
ΔE	energy loss
E_e	edge energy
E_M	mean electron energy
$G, G_x, G_1, G_2, \ldots$	diameters of the HOLZ rings
g	spacing of the diffraction discs
$H, H_x, H_1, H_2, \ldots$	spacings along the electron beam direction
I	beam current
I_A, I_B	X-ray intensities from element A, B
$I_K, (I_{AK}, I_{BK})$	intensity of the K-edge window of width Δ (of element A, B)
I_p	intensity in the plasmon loss peak
$I_{0\Delta}$	intensity in a window from $0\,\mathrm{eV}$ to $\Delta\,\mathrm{eV}$
I_0	intensity in the zero-loss peak
k_{AB}	proportionality factor for elements A, B in Cliff-Lorimer equation
k_{AB}^{TF}	k_{AB} for an infinitely thin foil
L	camera length (sample to film distance)
l	distance between X-ray detector and sample
m	mass of an (incident) electron
$N, (N_A, N_B)$	number of atoms (of element A, B)
n_i	integer
Q	ionization cross section
R	spatial resolution
R_A	atomic radius
r, r_A, r_B	absorption edge jump ratio
s, s_i	parameter describing the deviation from θ_B
t	specimen (or foil) thickness
U	overvoltage
V	velocity of an (incident) electron
V_0	microscope high voltage

X	fluorescence correction factor
Z	atomic number
α	generalized semi-angle of convergence of the probe
α_p	convergence angle at specimen
α_x	X-ray takeoff angle
α_a	angular range of the electrons entering probe-forming lens
α_c	illumination angle of the TEM
α_i	convergence of the beam formed by the first condenser lens
α_s	illumination angle of the AEM
β_c	objective angle of the TEM
β_s	detector angle of the AEM
β_0	EEL spectrometer acceptance angle
Δ	energy window
ε	detector absorption parameter
θ_e	elastic scattering angle
θ_B	Bragg angle
θ_i	inelastic scattering angle
λ	electron wavelength
$(\mu/\varrho)^X_{SPEC}$	mass absorption coefficient for X-rays of element X
ξ_g	extinction distance
ϱ	specimen density
σ	parameter of the Gaussian distribution function
$\sigma_K, (\sigma_{AK}, \sigma_{BK})$	ionization cross-section for the K-shell excitation (of element A, B)
Ω	solid angle for X-ray collection
ω	fluorescent yield
ADF	annular dark field
AEM	analytical electron microscope
ALCHEMI	atom location by channelling-enhanced microanalysis
BSE	backscattered electron
BN	boron nitride
CBED	convergent-beam electron diffraction
CRT	cathode-ray tube
EDS	energy dispersive spectrometer
EELS	electron energy loss spectrometer
EXAFS	extended X-ray absorption fine structure
f.c.c.	face-centered cubic
FWHM	full-width-at-half-maximum
HOLZ	higher-order Laue zones
SAED	selected area electron diffraction
SE	secondary electron
SEM	scanning electron microscope
SIGMAK	method for calculating K-edge ionization cross section
SIGMAL	method for calculating L-edge ionization cross section

ST	scanning transmission
STEM	scanning transmission electron microscope
TEM	transmission electron microscope
UTW	ultra-thin window
YAG	yttria alumina garnet
ZOLZ	zero-order Laue zone

2.1 General Introduction

Chapters 1 and 3 in this volume describe the development, operation, and capabilities of the transmission electron microscope (TEM) and the scanning electron microscope (SEM). These two instruments developed, starting in the 1940s, along parallel but nonconverging paths. Despite the fact that these two microscopes have optical columns which are fundamentally similar, the mode of operation and information provided by TEMs and SEMs remained distinct. In the TEM, a parallel beam of high-energy electrons is used to irradiate a thin, electron-transparent sample. Post-specimen optics are then used to produce either a high-magnification, high-resolution image of the internal structure of the sample, or an electron diffraction pattern from a selected area which contains crystallographic or crystal structure information. The images or electron diffraction patterns are viewed on a fluorescent screen, and hence the information is provided in "parallel" or simultaneous mode. Alternatively, the SEM uses a focussed beam of lower-energy electrons, which is scanned across the surface of a bulk sample. In this instrument, various products of the electron-specimen interaction are used for imaging or microanalysis: secondary electrons reveal surface topography, backscattered electrons contain composition or orientation information, and emitted X-rays are collected for chemical analysis. The images are produced in "serial" or sequential fashion on a cathode-ray tube, and only surface or near-surface data are provided.

Around 1970, several different groups of researchers and instrument manufacturers recognized the value of combining the capabilities of these two types of instrument, signalling the birth of a new hybrid instrument. This hybrid instrument was origi-

nally referred to as a scanning transmission electron microscope (STEM), but this name has been largely supplanted by the term analytical electron microscope (AEM). The stage had been set for the development of this instrument by initial work in the 1960s by Peter Duncumb at Tube Investment Research Laboratories in England (Duncumb, 1968) and by Albert Crewe at the University of Chicago in the United States (Crewe, 1970), among others. In the early 1970s, research proceeded rapidly in both the areas of equipment development and of the development of theoretical treatments to aid in the understanding of imaging or analytical data. By the mid-1970s, AEM instruments were available from a number of commercial manufacturers, launching an era which has revolutionized both the fields of electron microscopy and of materials science, in which detailed, quantitative, high-spatial resolution chemical, structural, and crystallographic information is available from a single specimen in a single instrument.

The term "analytical electron microscopy" undoubtedly has different meanings to different researchers, and so a definition of the term at this point is useful. For the purposes of this chapter, we will define an analytical electron microscope as an instrument which uses specimens in the form of thin, electron transparent wafers, and which uses various products of the interaction of a focussed probe of high-energy electrons (usually $100-400$ keV) with these specimens for imaging, diffraction, or microanalysis. This is a working definition only, and is not intended to be completely comprehensive or exclusive. For example, most AEMs are capable of operation in modes other than that described above. However, this definition helps to describe the scope of the subjects which will be considered in this chapter.

The development of AEM instruments proceeded along two distinct paths, which remain separate today and describe two different classes of instrument. The first of these is the TEM/STEM hybrid instrument, which developed by adding scanning capabilities and analytical equipment such as X-ray detectors and electron energy loss spectrometers to conventional TEMs. These microscopes are thus capable of operating in either AEM, TEM, and SEM modes, and are by far the most common type of AEM in use today. The second AEM type is often referred to as a "dedicated" AEM, and its optical column is similar to an SEM. This instrument has no conventional TEM capabilities (in other words, it does not use post-specimen optics to form an image or diffraction pattern) and hence operates only in an AEM or SEM mode. A cross-sectional view of the electron optical column of each of these instrument types is shown in Fig. 2-1.

2.1.1 Basic Principles of Analytical Electron Microscope (AEM) Operation

Regardless of the type of AEM, the basic principle of operation in the AEM mode is the same: pre-specimen optics are used to produce a small, focussed probe of electrons which can be scanned over the specimen to produce an image or held stationary on a feature of interest for microanalysis. Thus, this operating mode is similar to the operation of an SEM. However, the use of a thin specimen in the AEM results in some fundamental differences in the electron-specimen interaction when compared with a bulk specimen in an SEM. Figure 2-2 shows schematically the principal products of the electron-specimen interaction in an AEM. Each of these products has a use for imaging, crystallo-graphic analysis, or chemical analysis, as described below.

(a) Forward-scattered electrons: The vast majority of the electrons which pass through the sample are forward-scattered or transmitted, or, on other words, lose no energy and do not change direction. These electrons, when collected by an electron detector, can be used to form a scanning transmission (ST) image of the internal microstructure of the specimen. This ST image is similar in characteristics to a conventional bright-field TEM image.

(b) Inelastically-scattered electrons: These electrons have lost energy during transit through the specimen and have been scattered to small angles close to the forward direction. These electrons are used in electron energy loss spectroscopy to provide compositional and chemical information about the specimen. They can also be used for certain specialized imaging modes.

(c) Elastically-scattered electrons: In crystalline materials, the elastically-scattered electrons are produced primarily by diffraction processes. These electrons lose no energy but are scattered to relatively high angles, related to the spacing of the diffracting planes by Bragg's law. In the AEM mode, the elastically-scattered electrons produce a convergent-beam electron diffraction (CBED) pattern, which contains a wealth of information about the crystal structure, symmetry, orientation, and thickness of the sample.

(d) Secondary electrons (SE): The secondary electrons, which are loosely-bound outer-shell electrons from the atoms in the sample, are the primary electron signal collected for imaging in the SEM. In the AEM, these electrons can also be collected and provide topographical information about the top surface of the sample.

(e) Backscattered electrons (BSE): These electrons are also used extensively in SEM imaging, and are primary beam electrons which exit the top foil surface after having been scattered in the sample by elastic and inelastic scattering processes. These electrons can be collected in the AEM and can be used to form an image of the top surface of the sample in which regions of different orientation or different average atomic number can be distinguished.

(f) Auger electrons (AE): Auger electrons are one of the by-products, along with X-rays, which result from the inner-shell ionizations of the atoms in the sample caused by primary beam electrons. The energies of the emitted Auger electrons can be used to measure the chemistry of the sample, and this information is specific to the first few atomic layers beneath the sample surface. In general, AEM instruments are not able to

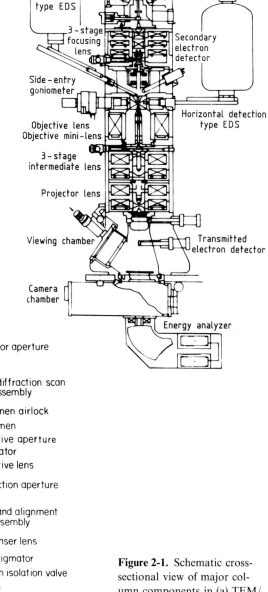

Figure 2-1. Schematic cross-sectional view of major column components in (a) TEM/STEM-type AEM (courtesy JEOL USA, Inc.) and (b) "dedicated" AEM (adapted from VG Microscopes Ltd.).

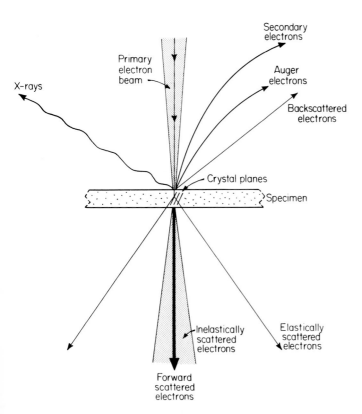

Figure 2-2. Schematic diagram of principle electron-specimen interactions in an AEM.

collect the Auger electrons. However, in recent years, a few specially-designed instruments have become available which are specifically configured to study the structure and chemistry of surfaces in an AEM sample, while simultaneously having available all of the usual AEM modes. Auger electron spectroscopy is used in these instruments for surface studies.

(g) X-rays: X-ray spectroscopy is the most widely-used and most powerful of the analytical capabilities available in the AEM. X-rays emitted from the irradiated area of the AEM specimen are collected and used to provide quantitative information about the composition of the internal constituents of the area.

From this very brief overview, it should be clear that there are a wide variety of imaging, diffraction, and microanalysis modes available in the AEM. Some of these modes are used in other types of instruments: X-ray spectroscopy, secondary electron and backscattered electron detection are used in the SEM and electron microprobe, for example, and Auger electron spectrometers are important surface analysis tools. The power of the AEM is that all of these modes are available in a single instrument, using a single sample. In addition, some AEM capabilities are available only in this instrument, such as transmitted electron imaging, convergent beam electron diffraction, and electron energy loss spectroscopy.

Each of the imaging and analysis modes described above have unique capabilities and limitations, and these will be the subject of the remainder of this chapter. Central to all of these modes is the production of a focussed beam of electrons which can be used to probe the sample, and this is the topic which we will consider first. Following that, we will discuss in detail imaging, X-ray microanalysis, electron energy loss spectroscopy, and convergent-beam electron diffraction in the AEM. The treatments of these subjects in this chapter are necessarily brief; for a more complete discussion of any topic, the reader is referred to the text by Williams (1984) or the anthologies edited by Hren et al. (1979) or by Joy et al. (1986).

2.2 Probe Formation in the AEM

In the analytical electron microscope, a great deal of attention must be paid to the formation of a small, well-defined, focussed beam of electrons which contains sufficient current to generate the analytical signal of interest. This is of particular concern in the AEM for several reasons. First, many AEM analyses require the maximum in spatial resolution, and hence it is important that the focussed probe be as small as possible, and that all of the incident electrons be confined to that probe. Second, many of the signals generated in the AEM are inherently weak – for example, the number of X-ray quanta generated is very low because the sample is thin – and it is thus important to have as much current as possible in a given size probe. Finally, the high-energy electrons associated with AEM instruments are very effective at generating high-energy X-rays from column components, and therefore careful collimation of the electrons is essential. The diameter of and current in the probe are controlled by the electron gun and the probe-forming pre-specimen optics, as described below.

2.2.1 The Electron Gun

Most TEM/STEM hybrid microscopes in use today employ a thermionic emitter to generate electrons, and tungsten and lanthanum hexaboride (LaB_6) are the emitters currently in use. Conversely, most "dedicated" AEMs use field-emission tungsten emitters, which require a very high vacuum for reliable operation. These three types of electron gun differ primarily in their brightness (B), which describes the current density in the probe per solid angle, and is thus the most important figure of merit for comparison of electron guns. For a given probe, the brightness is given by

$$B = 4 I/(\pi d_p \alpha)^2 \quad A/cm^2/sr \qquad (2\text{-}1)$$

where I is the beam current, d_p is the probe diameter, and α is the semi-angle of convergence of the probe. Table 2-1 gives some typical gun parameters for thermionic W and LaB_6 and field-emission W. It can be seen that field-emission W has the highest brightness values, followed by thermionic LaB_6 and W, in that order. Thus, microscopes with field-emission W filaments are capable of higher spatial resolution microanalysis, since more current is available in a small (~ 2.0 nm) probe. However, the stringent vacuum requirements associated with field emission W guns ($\sim 10^{-10}$ torr for room-temperature field emission) have limited the use of these guns, especially in hybrid TEM/STEMs. In addition, the total electron current produced by a field-emission gun is low relative to a thermionic emitter due to the much smaller virtual source size in the field emitter (Table 2-1), and thus these guns may be less suitable in cases were lower resolution, high elemental sensitivity analysis is desired. In some cases, thermally-assisted field emission guns have been used, which have less severe vac-

Table 2-1. Electron gun characteristics at 100 keV (adapted from Geiss and Romig, 1986).

Characteristic	W	LaB$_6$	Heated field emission	Cold field emission
Filament temp. in K	2700	2000	1200	300
Brightness in A/cm^2 sr	2.5×10^5	1.0×10^7	1.0×10^8	2.0×10^8
Total fil. current in μA	100	200	100	20
Filament life in h	100	500	500	1000
Vacuum required in torr	10^{-5}	10^{-7}	10^{-9}	10^{-10}
Probe size on sample, FWHM, in nm	25	10	5	2
Current in probe on sample, in nA	0.05	0.05	1.00	1.00

uum requirements ($\sim 10^{-9}$ Torr), but these guns are less bright and less stable than room-temperature field emitters (Table 2-1). At the present time, a number of different filament materials and gun types are being actively investigated, including carbides and metal-oxide-coated W for field emission, as well as extended Schottky emission (Orloff, 1989).

2.2.2 Condenser (Probe-Forming) Lens System

The role of the pre-specimen condenser lens system is to form a small, well-collimated probe of electrons which is focussed at the specimen position and used for imaging and microanalysis. In modern instruments this lens system can be quite complex, consisting of as many as five lenses. However, through microprocessor control of the lenses, this complexity is often transparent to the user, and these systems, as well as the less complex older systems, can be regarded as consisting of only three components: two condenser lenses and a condenser aperture. Figure 2-3a shows the electron ray path through the pre-specimen optics and the role of each of these three components. The first condenser lens forms a demagnified image (d_i) of the electron source (d_0), and increasing the strength of that lens demagnifies the

image further, or, in other words, creates a smaller diameter focussed beam of electrons. Increasing the first condenser lens strength also increases the convergence of the beam (α_i) entering the second condenser lens. Because of the relationship between probe size and current in the probe [Eq. (2-1)], increasing the first condenser lens setting will also decrease the probe current. The relationship between probe size, probe current, and first condenser lens current is shown schematically in Fig. 2-4a. Eq. (2-1) effectively limits the minimum probe size, depending upon the type of analysis desired. For high spatial resolution studies, the minimum probe size which contains sufficient current for signal detection is chosen by varying the first condenser lens setting.

The condenser aperture is located between the first condenser and probe-forming lenses. It is used to select the angular range of the electrons (α_a) which enter the probe-forming lens. If the magnetic lenses in this system were perfect, the condenser aperture would not be needed and the probe-forming lens could accept the maximum angular range possible. However, the aberrations in the condenser lenses, in particular spherical aberration, increase with increasing α, and these aberrations increase the diameter of the final probe. For example, the increase in the diameter of the

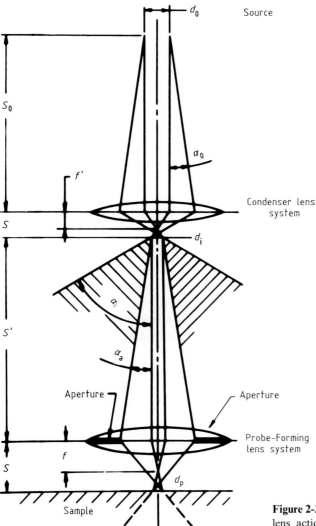

Figure 2-3 a. Schematic diagrams of probe-forming lens action in the AEM. Ray traces through two probe-forming lenses plus condenser aperture (from Goldstein et al., 1981).

probe due to spherical aberration is given by

$$d_{sa} = (1/2) C_s \alpha_p^3 \qquad (2\text{-}2)$$

where C_s is the spherical aberration coefficient of the lens and α_p is the convergence angle at the specimen, as shown in Fig. 2-3 a. Cliff and Kenway (1982) have shown that the electron distribution in the final probe can be represented schematically as

shown in Fig. 2-5, with a narrow full-width-at-half-maximum (FWHM) but with a long "tail" due to spherical aberration. In order to minimize the extent of this tail, a small condenser aperture is necessary to limit α. However, as the aperture size decreases, the current in the probe also decreases, as shown schematically in Fig. 2-4 b. It should be stressed that the extent of the electron tails due to using a

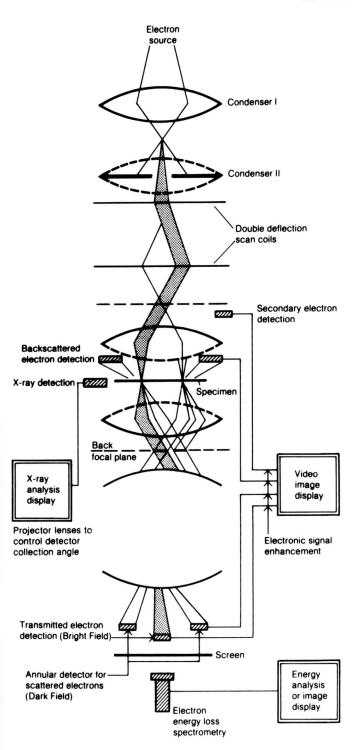

Figure 2-3 b. Ray traces through entire optical column of AEM (originally from Geiss, 1979; modified by Williams, 1984).

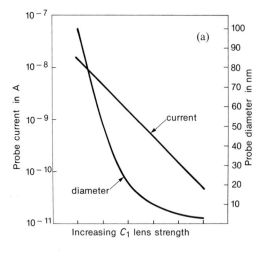

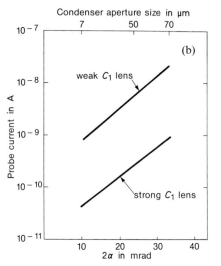

Figure 2-4. Schematic plots of (a) variation in probe diameter and current as a function of first condenser lens strength and (b) variation in probe current with condenser aperture size and first condenser lens strength; data taken from Williams (1984).

large condenser aperture and thus enhanced spherical abberation can be very sizeable; Fig. 2-6 shows an image of the electron probes produced using the same first condenser lens excitation but different condenser apertures. For imaging, the effect of the large electron tails is small, because most of the intensity is in the focussed spot and the FWHM of this spot is

primarily a function of first condenser lens setting, as Fig. 2-6 shows. However, for applications such as high spatial resolution microanalysis, the electron tails will produce X-rays far from the intended point of analysis and can lead to erroneous results.

The second condenser lens, sometimes referred to as a condenser-objective lens since it usually consists of the pre-field of the objective lens, is used to focus the probe onto the specimen. In addition, as Fig. 2-3 b shows, a set of double-deflection scan coils are used to move the beam for imaging, microanalysis, or microdiffraction.

One additional method of improving the performance of the probe-forming lens sys-

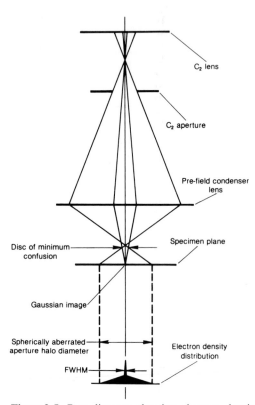

Figure 2-5. Ray diagram showing electron density distribution in probe (from Cliff and Kenway, 1982). Diameter of spherically-aberrated tail increases with increasing C_2 aperture size.

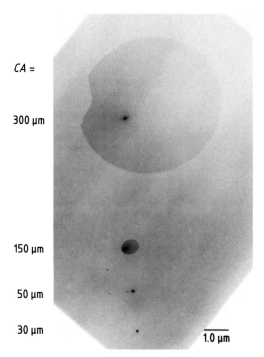

CA =

300 µm

150 µm

50 µm

30 µm

1.0 µm

Figure 2-6. Depiction of effect of condenser aperture size on electron distribution on probe. Electron distribution is imaged here as contamination pattern on a thin carbon substrate. Asymmetry of the halo around the probe and incomplete halo are a result of miscentered aperture and dirt on the aperture, respectively.

tem is to lower the inherent aberration coefficients in the condenser lenses. Recent microscopes have been produced with values of C_s as low as 0.4 mm in the condenser-objective lens (Yanaka et al., 1989), compared with typical values of 2–3 mm in previous instruments. These low values of C_s reduce the probe enlargement due to spherical aberration, and can lead to a decrease in probe size for a given current level of a factor of 2 or more.

The size of the final probe and the current in the probe are thus a function of the brightness of the gun, the first condenser lens setting, the size of the condenser aperture, and the optical perfection of the condenser lenses. As mentioned before, in

AEM applications, particular care must be paid to optimizing all of the components of the probe-forming system, since most AEM studies require high spatial resolution and thus very small probes. The current in these small probes is inherently low and other factors, such as the small thickness of material probed, may contribute to further limiting the signal produced by the electron-specimen interaction. Thus, work is constantly continuing on methods to maximize the electron current in the probe.

2.3 AEM Imaging

The AEM image formation process is fundamentally different from the image formation process in the TEM mode, despite the fact that a single sample and a single microscope column can be used for both AEM and TEM imaging. In the AEM mode, all images are formed by rastering a focussed probe of electrons over an area of the sample, and collecting some resultant signal from the electron-specimen interaction using an appropriate detector. The image is thus formed sequentially and displayed on a cathode-ray tube. No lenses, other than the probe-forming optics, are involved in this process. The magnification is changed by altering the size of the raster area on the sample. In the TEM mode, the sample is flooded by a parallel beam of electrons, and lenses are used to form and magnify the image. The information is produced in parallel fashion and observed on a fluorescent screen.

These fundamental differences in the methods of producing and recording information in the AEM and TEM result in some important advantages and disadvantages associated with each mode. The resolution of the TEM image is related only to the wavelength of the electrons and the

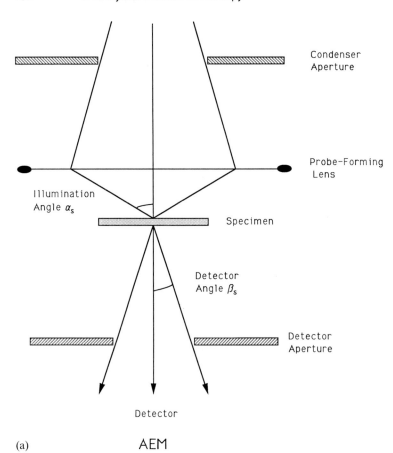

Condenser
Aperture

Probe-Forming
Lens

Illumination
Angle α_s

Specimen

Detector
Angle β_s

Detector
Aperture

Detector

(a)   AEM

Figure 2-7. Illustration of the principle of reciprocity for the equivalence of TEM (a) and AEM (b) bright-field images.

aberrations in the optical system, and this results in a point-to-point resolution between 0.2 and 0.3 nm in most cases. The viewing and recording media for the TEM images, a fluorescent screen and photographic film, respectively, have a large number of pixels and are capable of producing very high quality images which can then be enlarged $20 \times$ or higher.

On the other hand, the resolution in AEM images is related to the area from which the signal is obtained, which in general is approximately the probe size. As we have seen, typical probe sizes in the AEM range from 1 nm for a field-emission gun instrument to 10 nm for an AEM with a thermionic emitter, and so the fundamental resolution in AEM images is inferior to

TEM images. In addition, the number of pixels associated with a CRT is on the order of 3% of those available in photographic film (Brown, 1977), and so the quality of the recording medium and the potential for additional enlargement is greatly reduced. However, these disadvantages of AEM images are to a great extent mitigated by the advantage that a large variety of image types exists in the AEM mode, most of which are unavailable in the TEM mode. These image types include AEM bright field, annular dark field, energy-filtered, secondary, and backscattered electron images, and compositional images using X-rays or electron energy loss electrons. Since all of these images are captured electronically for display on a CRT,

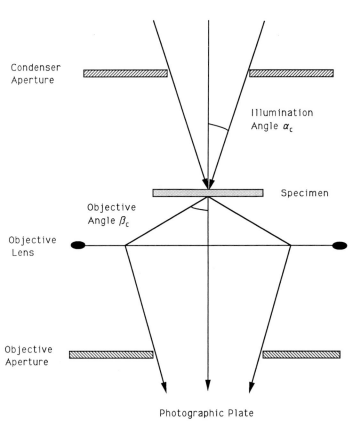

Condenser
Aperture

Illumination
Angle α_c

Specimen

Objective
Angle β_c

Objective
Lens

Objective
Aperture

Photographic Plate

Figure 2-7 b.   TEM

they can be manipulated in sophisticated ways, including processing to reduce or enhance contrast, or signal mixing to produce new images with unique information.

In the subsequent subsections, we will briefly consider each of the imaging modes available in the AEM.

2.3.1 AEM Bright-Field Imaging

AEM bright-field images are formed by rastering the focussed electron probe over a thin sample, and collecting the electrons which have either been forward-scattered (transmitted) or inelastically-scattered to small angles near to the forward-scattered beam. These electrons are usually collected

using a scintillator-photomultiplier electron detector. The electrons collected by this bright-field detector are similar in type to those that pass through the objective aperture during TEM bright-field imaging, and so the TEM and AEM bright-field images are in this way similar. In fact, early studies (Cowley 1969, 1979; Crewe and Wall, 1970) showed that a "principle of reciprocity" could be formulated which would describe the conditions under which the TEM and AEM bright-field images would be identical. Figure 2-7 shows the geometry of image formation in both modes, and the principle of reciprocity states that if

$$2\alpha_s = 2\beta_c \qquad (2\text{-}3)$$

and

$$2\beta_s = 2\alpha_c \qquad (2\text{-}4)$$

with these angles as defined in Fig. 2-7, then the two types of bright-field image are identical. In practice, in the TEM the incident illumination is generally near parallel and so $2\alpha_c$ is very small compared with $2\beta_s$, and so reciprocity is generally not achieved. The effect of a large $2\beta_s$ compared with $2\alpha_c$ is that several diffracted beams are collected by the bright-field detector, and thus the AEM image will have reduced diffraction contrast compared with the TEM image (Booker et al., 1974; Maher and Joy, 1976). This is equivalent to using a large objective aperture in TEM imaging.

In practice, this reduced contrast in the AEM bright-field image is of little consequence, since detailed image studies will generally be done in TEM mode, where contrast and resolution are improved. The AEM bright-field image is generally used only to locate areas for microanalysis or other AEM operations, and the contrast and resolution in this image is sufficient for this purpose. In addition, the true resolution of crystalline defects in either AEM or TEM bright-field images is often related to the defect image width, which is usually much larger than the resolution capabilities of the microscope. The bright-field imaging capabilities of the AEM are of critical importance only for dedicated AEMs, which have no TEM capabilities. However, these machines are equipped with field-emission guns, and the smaller probe sizes and higher currents in these AEMs result in improvements in image resolution and allow for the imaging conditions to be set up to give enhanced diffraction contrast.

2.3.2 Annular Dark-Field Imaging (ADF)

Annular dark-field AEM images are produced using a ring-shaped detector, again often of the scintillator-photomultiplier type, to collect the majority of the electrons which are elastically and inelastically scattered to large angles. This type of detector is shown schematically in Fig. 2-8. This type of detector is most easily incorporated in the dedicated AEM, because of restrictions imposed by the physical construction of the microscopes, and nearly all of the applications of ADF imaging in the AEM have come from these machines. The use of ADF imaging originated with Crewe et al. (1970) and his pioneering dedicated AEM. Crewe et al. showed that the electrons collected by the ADF detector were primarily

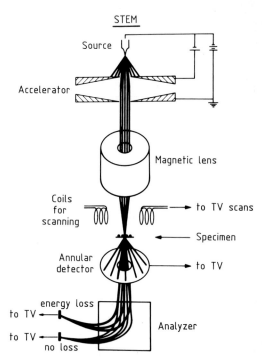

Figure 2-8. Schematic diagram of collection of annular dark field signal (from Isaacson et al., 1979). Collection of inelastically-scattered signal ("energy loss"), used for Z-contrast imaging of individual atoms, is also shown.

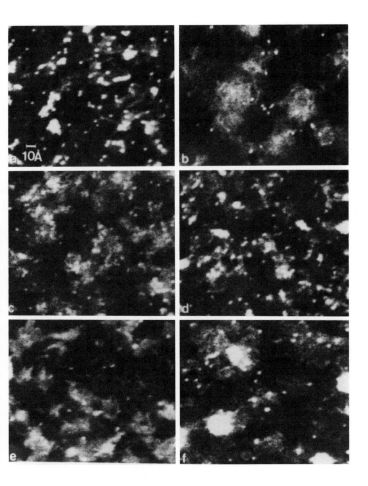

Figure 2-9. Examples of single atom imaging, from Isaacson et al. (1979). (a) Uranium; (b) gold; (c) cadmium; (d) indium; (e) silver and (f) 80% platinum-20% palladium.

elastically scattered electrons, with a scattering probability based on the cross-section proportional to $Z^{1/2}$, where Z is the atomic number. Hence, ADF images will display some atomic number contrast. Crewe enhanced this contrast by using another signal, the inelastically-scattered electron signal, which he obtained from a bright-field detector after using an electron energy-loss spectrometer to filter out the unscattered or transmitted electrons. The inelastic scattering probability is proportional to $Z^{3/2}$, and so by electronically ratioing these two signals, an image is produced in which the contrast is proportional to Z. Using the very small probes produced by his field-emission gun dedicated

AEM, Crewe was able to use Z-contrast imaging to view individual atoms of high atomic number materials such as gold on this carbon substrates (Fig. 2-9). For a review of this technique, see Isaacson et al. (1979).

A more recent application of ADF Z-contrast techniques has been developed by Pennycook and Boatner (1988). They have used a specially-designed ADF detector to collect high-angle scattered electrons, which have a Z-dependence which approaches the full Z^2-dependence of Rutherford scattering. Combining this detector with a field-emission gun dedicated AEM which has a probe-forming lens capable of producing a 0.22 nm diameter probe (Pennycook, 1989) has enabled them to produce high-resolution Z-contrast images of crystalline materials (Fig. 2-10). These images have the potential of allowing direct assessment of chemical identity and compositional changes on an atomic level.

On a more conventional resolution level, ADF imaging has found a number of applications, particularly in low-contrast materials such as unstained polymers. The ADF detector is very efficient, collecting as many as 90% of the scattered electrons. The images produced are thus much more sensitive to scattering power than conventional dark-field images, which can be produced in both the TEM and AEM but use only a fraction of scattered electrons for image formation. For materials which scatter weakly, or for which differences in scattering power between various constituents is low, the ADF images can prove valuable, particularly after additional electronic image processing.

2.3.3 Secondary, Backscattered Electron Images

In principle, the AEM is also capable of producing both backscattered and secondary electron images of the surface of

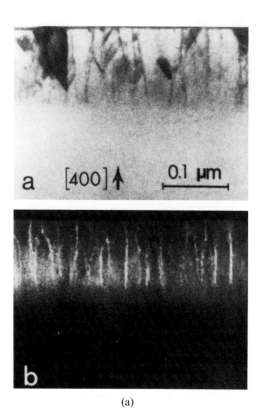

(a)

Figure 2-10. Examples of Z-contrast imaging in crystalline materials. (a) Cross-section of Sb-implanted Si showing conventional image and Z-contrast image of Sb distribution. (b) High-resolution Z-contrast image of $YBa_2Cu_3O_{7-x}$ superconductor, showing planes of Ba and Y atoms. Examples from Pennycook (1989).

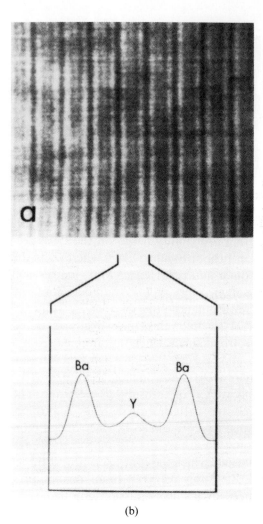

(b)

the thin foil facing the electron beam. These images are described in great detail in the chapter on SEM (Chap. 3). Secondary images show the topography of the foil surface. They can be a useful complement to bright-field images in the AEM, since they can reveal if a feature is located on the surface or within the foil. They can also show phase distribution if one of the phases has etched differently during specimen preparation, and essentially the entire foil area can be studied, rather than just the limited thin area available in the transmitted electron images. The backscattered images are sensitive to atomic number and crystallograpic orientation, and can again be a useful tool for examining large areas of the sample.

In practice, two factors limit the use of these images in the AEM. The first is the limited area available for the positioning of these detectors, and the general configuration of the AEM where the sample sits deep within the probe-forming lens. The secondary electrons are easier to collect, since these electrons are extracted from the sample region using a high potential, but limitations still exist. The backscattered electrons are much more problematical, since backscattered detectors are line-of-sight large-area detectors, and few AEMs can accommodate these detectors. The second factor is that the yield, or signal strength, decreases for the higher accelerating voltages used in AEMs, particularly for secondary electrons.

On the positive side, the use of thin specimens and very small probe sizes obtainable in the AEM results in very high resolution for SE and BSE images in AEMs equipped to collect these signals.

2.3.4 Compositional Imaging Using X-Rays or Electron Energy Loss Electrons

In the AEM it is possible to produce compositional maps of an area of interest when in the AEM mode. X-ray compositional images are widely used in SEM and microprobe studies of bulk samples, and the AEM compositional imaging modes are similar. In this method, certain elements of interest are chosen and the peak intensities from these elements in either the X-ray (see Sec. 2.4) or electron energy loss spectrum (see Sec. 2.5) are used to modulate the intensity of the display CRT, resulting in an intensity map where large concentrations of a given element are shown as higher brightness on the CRT.

In the case of X-ray imaging, the gross counts in a peak are generally used for the map, and it is possible to map a large number of elements simultaneously, usually as different color windows on a portion of the CRT. Figure 2-11 shows an example of such a map. The resolution of the compositional map is usually much less than that of an image of the same area, because in order to reduce the analysis time the screen is divided into a smaller number of pixels than is used for imaging. Even so, the time to create a map can be lengthy, since at each pixel a statistically significant X-ray spectrum must be acquired. The typical time to produce a series of X-ray maps from a single area is about 0.5 h.

X-ray maps have had only limited application to thin specimens compared to their more extensive use for bulk samples. The primary reason for this is that most AEM analyses are conducted at high magnifications and analysis of small features is required, and compositional maps are poorly suited for this purpose. Since many fewer X-rays are produced by thin samples, and the areas of interest are smaller, much

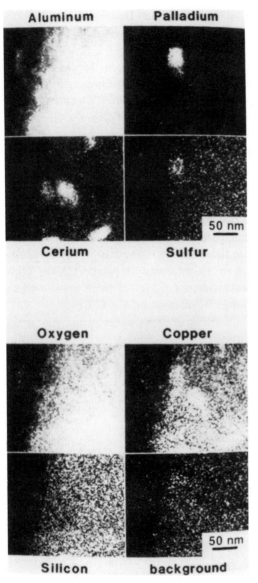

Figure 2-11. Example of high-resolution X-ray mapping in an AEM. Map shows distribution of various elements in sample of catalyst particles on Al_2O_3 flake. From Lyman et al. (1987).

sample preparation may affect the results. Nevertheless, X-ray maps can be useful in complex multiphase samples as a convenient visual display of elemental distribution.

Many of the capabilities and limitations of X-ray maps also apply to electron energy loss spectroscopy (EELS) maps, and some additional complications are present in this case. As will be discussed (Sec. 2.5), the EELS spectrum tends to be more complex than the EDS spectrum, with a rapidly decreasing background, severe peak overlaps, and large peak intensity variations as a result of thickness changes, atomic number, and transition type. Consequently, raw peak intensities cannot be used for mapping, but as a minimum background subtraction must be performed. More correct maps will be obtained if the background-subtracted intensities are then also corrected for differences in cross-section, so that the intensities are more closely related to composition. This results in much more lengthy analysis times, even when compared to X-ray maps. However, data acquisition times have been greatly accelerated by modern parallel EELS spectrometers, and these times can be significantly less than required for X-ray spectrum acquisition. Thus, quantitative EELS maps have become a reality in recent years. An example is shown in Fig. 2-12. Once again, the major limitations to this technique are analysis time and sample thickness variations, the latter having a much greater effect on the EELS maps than on X-ray maps. However, the EELS maps tend to be well-suited for very thin specimens (< 100 nm thick), where count rate limitations are severe for X-ray maps.

2.3.5 Energy-Filtered Images

One final imaging mode available in AEM instruments is that of energy-filtered

longer analysis times are required at higher magnifications, which necessitates a much more stable microscope and specimen, or, alternatively, some on-line drift-correction routine. In addition, thickness differences between phases and regions caused by the

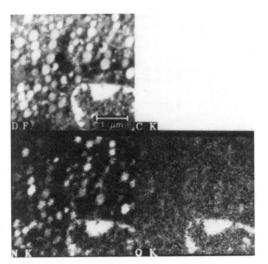

Figure 2-12. Example of EELS map. The dark-field image and the C, N, and O maps are of unstained Lowicryl-embedded freeze-substituted preparation of chromaffin cells. The carbon distribution is nearly uniform. From Leapman and Ornberg (1988).

images. This mode is somewhat related to both annular dark-field images and to EELS maps. In this mode, an EELS spectrometer is used to filter out all electrons except those that have undergone a specific energy loss upon passing through the sample. These particular loss electrons are then used to form the image. If a Omega-filter type of EELS is used (see Sec. 2.5), then the image is produced in parallel on a fluorescent screen. If a magnetic-sector type of EELS is used, the image is produced serially on a CRT. If the energy window used for imaging corresponds to a strong peak in the EELS spectrum, then the image has the characteristics of a compositional map, with the bright areas rich in the element corresponding to the peak in the EELS spectrum. However, this map has the advantage that the resolution of the TEM or AEM image is maintained. If the energy window used for imaging corresponds to a portion of the EELS spectrum without major peaks, and excludes the zero loss peak

(transmitted beam), then the image produced will be similar to an ADF image. This is because the electrons used to form the image correspond to a portion of the inelastically-scattered signal, which are also used in the ADF image. However, the ADF detector also collects the elastically-scattered electrons, and so differences will also exist, especially for crystalline materials in a strong diffracting condition. Energy-filtered images of this second type have been shown to have enhanced contrast and reduced chromatic aberration compared to normal bright-field images (Reimer et al., 1988, 1989). As discussed in the section of ADF imaging, samples with weak scattering contrast, such as unstained biological or polymeric materials, this imaging mode can be useful owing to the enhanced contrast. Reduced chromatic aberration will generally result in the ability to produce higher-quality images of thick specimens.

2.4 Quantitative X-Ray Spectroscopy

2.4.1 Fundamentals of X-Ray Collection in the AEM

The ability to collect X-rays emitted from a small electron-irradiated volume in a thin AEM sample is probably the most widely-used of the AEM techniques, and is the capability which has led to the rapid development of the field of analytical electron microscopy. In modern-day instruments, X-ray collection is done almost exclusively using energy-dispersive X-ray spectrometers, and the configuration and important components of the collection system are shown in Fig. 2-13. In the AEM, the thin, electron-transparent sample sits immersed in the objective lens, which is a

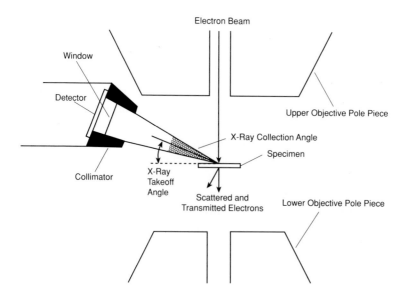

Figure 2-13. Schematic diagram of a typical energy-dispersive X-ray detector mounted on an AEM column.

high-excitation lens with a very narrow gap if high-resolution capabilities are required. The energy-dispersive X-ray spectrometer consists of a doped Si crystal, which is cooled by liquid nitrogen. The important spectrometer components through which the X-rays from the sample must pass in order to reach the Si crystal are a thin window, which protects the Si crystal from light and may serve to isolate the crystal from the microscope vacuum, and thin layers of gold and electrically inactive Si ("Si dead layer") on the surface of the detector crystal. In the microscope, the detector is brought as close as permitted by the design of the lens to the sample, in order to maximize the solid angle for X-ray collection, given by

$$\Omega = A_d/l^2 \qquad (2\text{-}5)$$

where A_d is the area of the detector and l is the distance between the detector and the sample. The angle that the detector makes with the horizontal sample plane, called the X-ray takeoff angle, varies depending upon the microscope from $0°$ (horizontal) to $70°$. The lower the takeoff angle, the farther the X-rays must travel through the sample before exiting the foil, and, as we will see in a later section, as this path length increases, X-ray absorption effects increase. Thus, from the standpoint of absorption, higher takeoff angles are preferable. However, in general, the higher takeoff angle designs tend to have poorer solid angles, since the detector must be positioned above the objective lens, whereas the lower takeoff angle designs allow the detector to enter through the pole-piece gap. In addition, the higher takeoff angle detectors tend to be affected more severely by high-energy backscattered electrons, whose flux is higher at higher angles.

One area of active research in recent years has been the development of new detector window materials. Traditionally, detector manufacturers used a thin layer of Be as a window, which protected the Si crystal from the external atmosphere but absorbed X-rays from elements with atomic number less than 9, excluding detection of C, N, and O. More recently, manufacturers have introduced detectors which are windowless or which have only a thin aluminum film as a light shield. These detectors can detect Be in some cases, and all elements of higher atomic number, but must have a valving system to avoid exposure to water vapor or hydrocarbons in air or in a poor microscope vacuum. Other window materials include boron films, thin diamond-like carbon films, and aluminized polymer sheets supported on grids. These windows will allow the detection of boron and higher atomic numbers. Figure 2-14 shows a plot of relative transmission of X-rays through various window materials.

Manufacturers are also experimenting with other materials besides Si as a detector material, and high-purity intrinsic Ge detectors are now commercially available. Ge has better X-ray stopping power than Si, and this translates in a better ability of Ge to detect X-rays with energy greater than 20 keV. This becomes important for applications where mixtures of high atomic number elements may be present in the sample. The Ge detector enables well-separated, high-energy ($E > 40$ keV) K lines to be used for analysis, rather than L or M lines, which may be convoluted. In addition, spectral resolution is generally higher for Ge detectors. The main disadvantages of Ge detectors are: lack of experience in manufacture and use of these detectors, particularly for quantitative analysis; inherent problems with peak tailing and numerous escape peaks (related to the loss of deposited X-ray energy from the detector through the excitation of Ge X-rays); and the effect of a strong Ge L absorption effect at around 1.2 keV on the collection of low-energy X-rays.

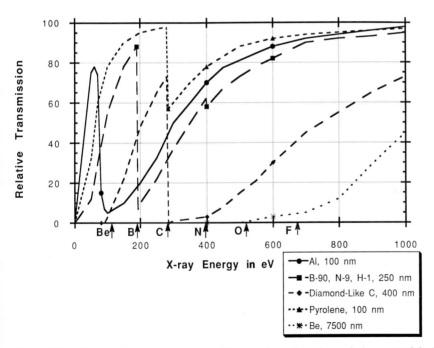

Figure 2-14. Plot of relative transmission of X-rays through various window materials. The window materials shown are Be ($\sim 8 \mu m$), $B_{90}N_9H_1$ (Kevex patented Quantum), diamond-like C, Al, and pyrolene. Note that the X-rays below 0.6 keV (below F) are not transmitted by Be window. Other windows show transmission to lower energies.

2.4.2 Spectrum Processing and Quantitation

A typical energy-dispersive X-ray spectrum obtained using an AEM, in this case from a YBaCuO superconductor, is shown in Fig. 2-15. The detector used was of the windowless type with a Si analyzing crystal. In this case, the spectrum displays many strong peaks with good peak-to-background ratio and no major overlaps or convolutions with peaks of different elements. The X-ray continuum background is low and relatively flat compared with spectra taken from bulk specimens using an SEM since the thin sample generates less background intensity. In viewing a spectrum of this sort, information is obviously available about the presence of elements in the electron-irradiated volume,

but in most cases very little semi-quantitative information can be extracted by simple visual inspection of the spectrum. The relative peak heights will be related to the relative concentrations of the elements only in cases where peaks of the same family are compared (e.g., K-lines) and where the peaks lie in a portion of the spectrum where the detector response is approximately flat (e.g., 3–8 keV). For example, in stainless steel, it is possible to make a qualitative assessment of the relative amounts of Cr, Fe, and Ni present by comparing peak heights of the K lines, but for a spectrum such as the one in Fig. 2-15, no such information is available since peaks of different families and of widely-different energies are present.

The first step in the quantitative analysis of an X-ray spectrum is to extract the peak intensities corresponding to each element. This involves subtracting the background and deconvoluting any peak overlaps. Modern X-ray analysis systems are capable of performing each of these functions, although the approach may differ depending upon the manufacturer. For background subtraction, a mathematical model is often used which depends primarily on the incident beam energy and the average

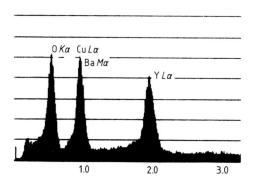

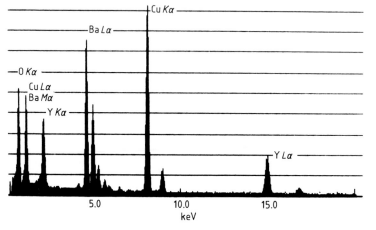

Figure 2-15. Typical X-ray spectrum from AEM of $YBa_2Cu_3O_7$ superconductor, obtained using a windowless detector. Inset shows low-energy region in more detail.

atomic number of the specimen (Kramers, 1923); regions where there is poor fit between the model and the observed background, often in the region of low X-ray energies, for example, can then be adjusted manually. Alternatively, digital filtering can be used to remove the background, in which a filter function is applied to each spectrum channel and the net result is to transform the slowly-varying background into a horizontal line of zero intensity, with the rapidly-varying peaks still present and unaffected. Both of these approaches have been used successfully and have their own minor merits and flaws. After background subtraction, peak deconvolution, if necessary, is performed by generating a Gaussian model for each of the peaks in the spectrum, or by comparing each peak in the spectrum to one obtained for a nonconvoluted standard. After background subtraction and peak deconvolution, the intensity in each peak is obtained, and it is generally preferable to use as much of the peak structure as possible for this measurement since the X-ray intensities from thin specimens are low and failure to include some of the peak area will result in lowered statistical accuracy.

The measured X-ray intensities can then be converted to compositions by following a procedure originally described by Cliff and Lorimer (1975). Cliff and Lorimer proposed that the X-ray intensities could be related to the amount of the element present using the relationships

$$\frac{C_A}{C_B} = k_{AB}\frac{I_A}{I_B} \qquad (2\text{-}6)$$

and

$$C_A + C_B = 1 \qquad (2\text{-}7)$$

where C_A and C_B are the weight fractions of elements A and B in the sample, I_A and I_B are the X-ray intensities from these two

elements, and k_{AB} is a constant of proportionality. In Eq. (2-6), the factor k_{AB} is not a function of the relative amounts of elements A and B in the sample, or of foil thickness, but is a function of accelerating voltage. Equation (2-6) is valid only in cases were absorption and fluorescence, which are major factors in the quantitation of X-ray spectra from bulk specimens, are negligible. For thin films, the importance of absorption and fluorescence corrections has been studied by many researchers, and it has been shown that in some cases an absorption correction needs to be added to Eq. (2-6). This modification of Eq. (2-6), originally proposed by Goldstein et al. (1977), is

$$\frac{C_A}{C_B} = k_{AB}^{TF}\frac{I_A}{I_B}\left[\frac{(\mu/\varrho)_{SPEC}^A}{(\mu/\varrho)_{SPEC}^B}\right] \cdot \qquad (2\text{-}8)$$
$$\cdot \left\{\frac{1-\exp\left[-(\mu/\varrho)_{SPEC}^B\,\varrho t\,\mathrm{cosec}\,\alpha_x\right]}{1-\exp\left[-(\mu/\varrho)_{SPEC}^A\,\varrho t\,\mathrm{cosec}\,\alpha_x\right]}\right\}$$

where k_{AB}^{TF} is the k_{AB} factor for an infinitely thin foil, $(\mu/\varrho)_{SPEC}^X$ is the mass absorption coefficient for X-rays of element X in the specimen, ϱ is the specimen density, α_x is the X-ray takeoff angle, and t is the specimen thickness. Equation (2-8) shows that the magnitude of the absorption correction increases as ϱ or t increase or α_x decreases. This is because as t increases or α_x decreases, the path length that the X-rays travel before exiting the sample increases, and as ϱ increases, the effective mass-thickness increases. Equation (2-8) also shows that it is the difference between the mass absorption coefficients of the X-rays from elements A and B that is important, with the magnitude of the absorption correction increasing as this difference increases.

From Eq. (2-8) it is possible to estimate the conditions under which an absorption correction is required. The critical parameters are obviously $\varrho, \alpha_x, t, [(\mu/\varrho)^A - (\mu/\varrho)^B]$, and the magnitude of the error due to absorption which is acceptable. In Table 2-2 is listed the maximum foil thickness t for which absorption can be ignored and Eq. (2-6) used, as a function of takeoff angle and acceptable error for a variety of elements in different samples. The table shows that for medium and high energy X-rays, such as Fe, Cr, and Mo K_α, in medium atomic number materials, the absorption correction is small for the foil thicknesses generally encountered in AEM samples (100–300 nm). This is because the difference in mass absorption coefficients for these X-rays is small. However, for lower energy X-rays such as Al K_α and Mo L_α in medium atomic number materials such as Ni or Fe, the mass absorption coefficient difference is large and so the absorption correction is large. For example, an absorption correction is necessary for NiAl in all but the thinnest samples. Table 2-2 also illustrates the advantage of a higher X-ray takeoff angle for cases where absorption is a problem. Finally, the worst-case situation with respect to absorption is often encountered when light elements such as C,

N, or O are analyzed; Table 2-2 shows that very large absorption effects are possible for the X-rays from these elements.

Having now established that Eq. (2-6) or, in the case where absorption is important, Eq. (2-8) can be used for quantification of X-ray data, we shall briefly discuss how the various factors in Eqs. (2-6) and (2-8) can be determined.

2.4.2.1 k_{AB} Factor Determination

Regardless of whether an absorption correction is necessary or not, an accurate measure of k_{AB} is essential for quantitative analysis. There are three methods for determining a value of k_{AB}.

Measurement from standards: The most accurate method for the determination of k_{AB} is to measure this value experimentally from a standard of known composition. The requirements for a standard are that the material be homogeneous, stable during the analysis, that it contain A and B in sufficient quantity that statistically meaningful data can be obtained, and that it not contain other elements that cannot be easily deconvoluted from the X-ray lines of A and B. The material must also be able to be prepared as a thin foil specimen without altering the composition of the specimen. Once the sample is prepared, I_A and I_B can be measured, and since C_A and C_B are known, Eq. (2-6) can be used to calculate k_{AB}. This assumes, however, that the absorption correction for A and B is small. If it is significant, then Eq. (2-8) must be used to calculate k_{AB}^{TF}. One can test for the importance of absorption in the standard either by calculating the size of the absorption factor in Eq. (2-8) or empirically by examining how the measured k_{AB} factor from the standard changes with foil thickness. An example of this type of study in both non-absorbing and absorbing sys-

Table 2-2. Maximum thickness allowable for no absorption correction.

Material	Thickness in nm		
	10% error $\alpha_x = 20°$	10% error $\alpha_x = 68°$	3% error $\alpha_x = 68°$
Fe–Cr (stainless steel)	4000	11 000	3300
Fe–Mo(L) (stainless steel)	88	238	72
Fe–Mo(K) (stainless steel)	831	2 250	674
NiAl	34	91	27
$Cr_{23}C_6$	10	26	8
SiC	8	23	7

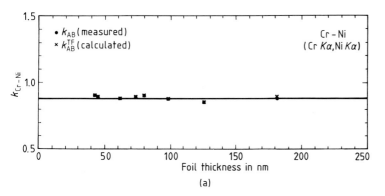

(a)

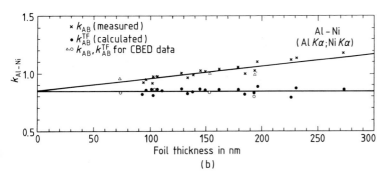

(b)

Figure 2-16. Examples of experimental k factor determination from binary standards. (a) Cr-Ni, (b) Al-Ni. Both experimentally measured and absorption-corrected data are shown. From Kouh and Hall (1982).

tems is shown in Fig. 2-16. In Fig. 2-16a, data from a Cr-Ni standard is shown. At each analysis point, I_{Cr}, I_{Ni}, and the foil thickness t were measured. Then, Eq. (2-6) was used to calculate k_{AB} (measured), shown as filled circles, and Eq. (2-8) was also used to calculate k_{AB}^{TF}, shown as x's. It can be seen that k_{AB} and k_{AB}^{TF} are equivalent throughout the thickness range studied, and that k_{AB} does not vary with foil thickness, both indicators of a very small absorption correction. Thus, each data point is an accurate measurement of $k_{AB} = k_{AB}^{TF}$. Figure 2-16b shows a similar analysis for and Al-Ni standard. Here, the measured k_{AB} data from Eq. (2-6) show a variation with thickness, corresponding to relatively more Al K_α absorption as the foil thickness increases. From the k_{AB} data, the value of k_{AB}^{TF} can be determined in two ways: the k_{AB} data can be extrapolated to zero thickness, or each point can be corrected using Eq. (2-

8). Figure 2-16b shows that each of these methods yields the same result. The important point here is that one must ensure that experimentally-measured k_{AB} values are not affected by absorption.

Calculation from first principles: It is also possible to calculate k_{AB}, using the expression (Goldstein et al., 1977; Zaluzec, 1979)

$$k_{AB} = [Q\,\omega\,a\,\varepsilon/A]_B/[Q\,\omega\,a\,\varepsilon/A]_A \qquad (2-9)$$

where Q is the ionization cross section, ω is the fluorescent yield, a is the ratio $K_\alpha/(K_\alpha + K_\beta)$, A is the atomic weight, and ε is a detector absorption parameter. In this calculation, the most significant problem is uncertainty in the value of Q. Table 2-3 gives k_{AB} values based on a variety of models for Q, and it can be seen that significant variation in k_{AB} can result. This problem becomes more severe when different families of lines are used in the quantitation,

Table 2-3. Theoretical k_{A-Fe} factors for K and L X-ray lines at 120 kV as a function of ionization cross section model (from Williams, 1984).

Element	k_{MM}[a]	k_{GC}[a]	k_P[a]	k_{BP}[a]	k_{SW}[a]	k_Z[a]
			K lines:			
Na	1.420	1.340	1.260	1.450	1.170	1.180
Mg	1.043	0.954	0.898	1.030	0.836	0.850
Al	0.893	0.822	0.777	0.877	0.723	0.742
Si	0.781	0.723	0.687	0.769	0.638	0.713
P	0.813	0.759	0.723	0.803	0.671	0.699
S	0.827	0.776	0.743	0.817	0.688	0.722
K	0.814	0.779	0.755	0.807	0.701	0.745
Ca	0.804	0.774	0.753	0.798	0.702	0.747
Ti	0.892	0.869	0.853	0.888	0.807	0.850
Cr	0.938	0.925	0.917	0.936	0.887	0.917
Mn	0.980	0.974	0.970	0.979	0.953	0.969
Fe	1.000	1.000	1.000	1.000	1.000	1.000
Co	1.063	1.069	1.074	1.066	1.096	1.075
Ni	1.071	1.085	1.096	1.074	1.143	1.220
Cu	1.185	1.209	1.227	1.190	1.310	1.225
Zn	1.245	1.278	1.305	1.255	1.440	1.299
Mo	3.130	3.520	3.880	3.270	3.840	3.721
Ag	4.580	5.410	6.230	4.910	5.930	5.725

Element	k_{MM}[a]	k_P[a]	k_{BP}[a]	k_{SW}[a]	k_Z[a]
		L lines:			
Sr[b]	1.730	1.330	1.320	1.640	1.474
Zr[b]	1.620	1.260	1.240	1.510	1.335
Nb[b]	1.540	1.210	1.180	1.430	1.344
Ag[b]	1.430	1.160	1.090	1.260	1.305
Sn	2.550	2.090	1.930	2.210	2.360
Ba	2.970	2.520	2.250	2.490	2.870
W	3.590	3.370	2.680	2.800	3.807
Au	3.940	3.840	2.940	3.050	4.308
Pb	4.340	4.310	3.250	3.340	4.809

[a] Code for models, MM: Mott and Massey, GC: Green and Cosslett, P: Powell, BP: Brown and Powell, SW: Schreiber and Wims, Z: Zaluzec (for full references, see Williams, 1984); [b] factors are the ratio of (L_α and L_β)/Fe K_α.

e.g., Fe K_α and Mo L_α. In addition, in order to calculate ε accurately, the thickness and composition of the various detector components through which the X-rays pass, such as the window and the gold layer and

Si dead layer on the detector crystal must be know accurately. Often, these values are not well known, but in some cases can be determined empirically. These uncertainties combine to result in a large possible error associated with calculated k_{AB}, generally estimated to be about $\pm 20\%$. Modern X-ray analysis packages are capable of quickly calculating k_{AB} factors, and for analyses where standards are not available, these factors may be the only alternative. However, the microscopist must realize the inherently greater error associated with calculated factors.

Calculated/measured literature values: A number of studies have been performed which measure or calculate k_{AB} factors for a wide variety of elements and X-ray lines. An early study is shown in Fig. 2-17a, which compares calculated and measured k_{X-Si} for K_α lines, for a detector with a beryllium window. A more complete study has been published by Schreiber and Wims (1981), which gives k_{AB} factors for most K, L, and M lines. These factors are generally comparable in accuracy to the calculated values discussed above. One important caveat, however, is associated with light element analysis. As shown in Fig. 2-17a, the calculated k_{X-Si} values from this study begin to increase rapidly below 2 keV. This is primarily due to enhanced absorption of X-rays with energy less than 2 keV by the detector, and the most important component associated with this absorption is the window. Nearly all published studies of calculated or measured k_{AB} factors were done for beryllium window detectors, and these values will be inaccurate for other window materials for X-ray lines with energy below 2 keV. Figure 2-17b shows a recent compilation of k_{AB} factors for a UTW (aluminized polymer) window, and can be compared with Fig. 2-17a. Above 2 keV, the window material has little effect

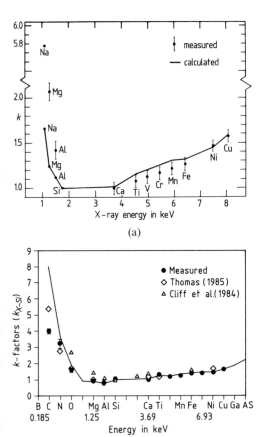

(a)

(b)

Figure 2-17. Compilation of experimental k_{AB} factors (B = Si) and comparison with calculation. (a) Be window detector (Mehta et al., 1979). (b) UTW (aluminized parylene) window detector (200 kV) (Krishnan and Echer, 1987; including data of Cliff et al., 1984; and Thomas, 1985). The detector parameters are: parylene window: 0.1 μm; Al coating: 0.15 μm; Au contact: 20.00 nm; Si dead layer 0.125 μm.

on the k_{AB} factors. Below 2 keV, major differences are seen. Thus, below 2 keV, differences in window type, or, even for the same window type, differences in window thickness may lower the accuracy associated with using calculated or measured k_{AB} from other investigators. For accurate quantitation of X-rays below 2 keV in energy, measurement of k_{AB} factors from standards is essential.

2.4.2.2 Foil Thickness

In order to use the absorption correction in Eq. (2-8), it is necessary to measure the foil thickness at each analysis point. There are a variety of different methods which can be used for thickness measurement, including: extinction (thickness) fringes; projected widths of crystallographic features; convergent-beam electron diffraction (see Sec. 2.6); X-ray intensity; transmitted electron intensity; intensity in EELS spectra; and separation of tilted contamination spots. The first three of these can be used for accurate thickness measurement, but are applicable only to crystalline specimens. The last four can be used with any specimen. The contamination spot method is the most straightforward and frequently-used technique. In this method, a focussed electron probe is positioned at or near the analysis point. Hydrocarbons on the specimen surface will migrate by surface diffusion to the analysis point and form visible spots of carbonaceous material on the top and bottom surfaces of the foil. When the foil is then tilted by a known amount, the separation of the contamination spots gives the foil thickness. Care must be taken to accurately determine the proper measurement distance, and uncertainties in locating the base of the contamination mounds makes this technique relatively inexact. In addition, modern clean vacuum technology in AEMs and in specimen preparation equipment has made it more difficult in recent years to produce contamination spots in a reliable and reproducible fashion. More research is needed into generally-applicable, fast, and accurate methods for thickness determination, and methods based upon X-ray or EELS intensity seem most promising.

2.4.2.3 Other Factors

The other factors which must be evaluated in order to use Eq. (2-8) for quantitation are the density ϱ, the mass absorption coefficients, and the X-ray takeoff angle. The latter parameter is a combination of the microscope geometry and the tilt of the specimen along the axis normal to the detector. The density, if not known, can be calculated for crystalline materials from knowledge of the composition and the unit cell parameters. The mass absorption coefficient can be calculated from the equation

$$(\mu/\varrho)_{SPEC}^A = C_A (\mu/\varrho)_A^A + C_B (\mu/\varrho)_B^A +$$
$$+ C_C (\mu/\varrho)_C^A + \dots \qquad (2\text{-}10)$$

where $(\mu/\varrho)_i^A$ is the mass absorption coefficient of X-rays from element A in a matrix of pure element i. These values are available in tabulated form (Henke and Ebisu, 1974). Calculation of both the density and the mass absorption coefficient require knowledge of the sample composition, and thus must be evaluated iteratively according to Eq. (2-8).

2.4.3 Other Aspects of X-Ray Spectrocsopy

There are a number of other factors which need to be considered when assessing the accuracy and spatial resolution of X-ray spectroscopy in the analytical electron microscope. We will consider those factors in this section.

2.4.3.1 Artifacts in Spectrum Acquisition

The use of high-energy electrons and a thin sample which may be strongly diffracting leads to a number of possible artifacts which can occur in the EDS X-ray spectrum generated in an AEM. It is convenient to divide these artifacts into two distinct types: microscope-related and specimen-related.

The microscope-related artifacts are generally caused by uncollimated high-energy electrons or X-rays striking column components or parts of the sample far from the point of analysis. This type of artifact was common in early AEM instruments, and much work has been done by manufacturers and users of modern instruments in order to eliminate these artifacts. Microscope-related artifacts can be minimized by ensuring that column components that can be irradiated by electrons or X-rays are made of or coated with materials with low X-ray emission, and by proper collimation for both incident electrons and excited X-rays. In an AEM, there are simple procedures which can be used to test for the presence of these types of artifacts. The first of these is a "hole count". It makes use of the fact that most AEM specimens consist of thin regions of material which impinge on a hole in the specimen. If the electron beam is placed in the hole near the edge of the specimen, no X-rays characteristic of the specimen should be produced. If characteristic X-rays are seen, then either electrons which have not been focussed into the probe or high-energy X-rays from column components are striking the specimen. This test can also give a crude approximation of probe size if the spectrum is monitored as the probe is brought closer to the edge of the specimen, and can also check for spherically-aberrated tails on the probe as described in Sec. 2.2.

A second test consists of using a standard of high atomic number material, such as gold or tantalum, to generate high-energy characteristic X-rays. If additional peaks, such as Cu or Fe, are seen in the spectrum, then the characteristic X-rays are due to fluorescing column components.

The second type of artifact is specimen-related, and usually results from or can be enhanced by an improperly-oriented sample. The most common of these problems is the excitation of the sample far from the point of analysis by backscattered electrons or by continuum or characteristic X-rays. Figure 2-18 shows schematic examples of these effects. In the case of backscattered electrons (Fig. 2-18 a), these electrons tend to spiral back up the optic axis due to the action of the lens field. If the sample is tilted, they can strike the sample. A solution to this problem is to tilt the sample as close to the horizontal plane as possible. A similar effect occurs for spurious X-rays, as shown in Fig. 2-18 b. In this case, the continuum intensity is peaked in the forward direction at an angle of approximately 45° to the forward beam direction, and a tilted sample will intersect this strong continuum signal and will fluoresce. Again, a horizontal sample will show the smallest effect.

In practice, it is impossible to eliminate this latter effect, namely, the fluorescence of the sample away from the point of analysis by continuum and characteristic X-rays. This effect occurs even when the electron probe is well collimated and is contained completely within the phase from which the information is desired. This will lead to the generation of spurious X-rays, and result in quantitation inaccuracies. This problem will be greatest when the X-rays generated by the electron probe from the area of interest have an energy which is just above the absorption edge for one of the major elements in the matrix of the sample, and thus maximum fluorescence will occur.

Two examples of this spurious effect are given in Table 2-4. The first is for TaC particles in a Ni-base matrix (Koch et al.,

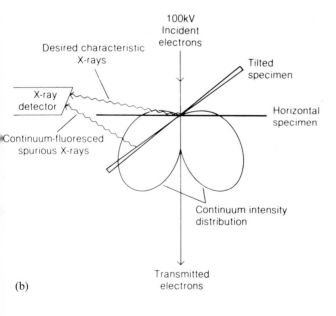

(a)

(b)

Figure 2-18. Schematic depiction of the origin of spurious X-ray signals in the AEM. (a) High-energy backscattered electron excitation of the sample. (b) Continuum and characteristic X-ray fluorescence of the sample. From Williams (1984).

Table 2-4. Results of X-ray microanalysis of second phases, demonstrating secondary fluorescence effects (Koch et al., 1983; Hall, 1990).

Example 1: TaC particles in Ni-base matrix, in at.%

Particle type:	Ta	Ni	Cr
TaC particles in matrix	66	30	5
TaC particles, extracted	91	0	9

Example 2: Cr-rich $M_{23}C_6$ particles in Fe-base matrix, in wt.%

Particle type:	Fe	Cr	Ni	Mo
$M_{23}C_6$ particles in matrix	24	62	3	11
$M_{23}C_6$ particles, extracted	18	67	2	13

1983). In this sample, the TaC particles were large (~ 2 μm) compared to the electron probe size (~ 10 nm). Comparing the composition of the particles measured while embedded in the foil to that measured after the matrix had been removed and the particles were on a carbon support film shows that a large spurious nickel signal is measured when the particles are in the matrix. This is a particularly large effect because the Ta L_α X-rays strongly fluoresce Ni. Another, more typical, example is also given, that of Cr-rich $M_{23}C_6$ particles in a Fe-base matrix (Hall, 1990). In this case, enhanced levels of Fe are found if the particles are measured while embedded in the Fe-base matrix.

Since this effect is unavoidable, it is a strong limitation to the accurate analysis of second phases embedded in a matrix of different chemistry. All in-situ analyses will be affected by this phenomenon. If very accurate quantitation of second phases is required, it is necessary to extract the particles or phases of interest from the matrix, using the well-known carbon-extraction techniques (von Heimendahl, 1980).

There are other artifacts possible which are related to the diffracting condition of the sample, and occur only for crystalline samples. If the sample is in a strong diffracting condition, then anomalously high electron penetration may occur (often described as the "Borrmann effect") and affect microanalysis (Cherns and Howie, 1973; Bourdillon et al., 1981; Bourdillon, 1982). In practice, this effect tends to be less of a problem for thin foils than it is in bulk samples. Another artifact associated with strong diffraction is "coherent Bremsstrahlung", which can give rise to small peaks in the X-ray spectrum which can be confused with elemental peaks (Reese and Spence, 1984). This effect is strongest in zone axis orientations, and is due to the coherent production of continuum intensity by planes perpendicular to the electron beam. The location of the peaks can be predicted from the electron energy and the spacing of the planes (Spence and Tafto, 1983). These peaks can be avoided by avoiding zone axis orientations during microanalysis; an additional test is to change the accelerating voltage and examine if the peak energies change.

The last specimen-related "artifact" can actually be used as an important analysis tool, and that is ALCHEMI (atom location by channelling-enhanced microanalysis) (Spence and Tafto, 1983). This effect occurs only for ordered crystalline materials which are oriented in a manner to produce alternating planes of atoms of different types. Figure 2-19 shows a schematic example. In this case, changing the diffracting condition can cause the localization of the electron beam either on rows of A atoms or rows of B atoms. This can greatly affect the ratio of A/B intensity in the X-ray spectrum. Large errors in microanalysis can thus occur if this condition is inadvertently achieved. However, this effect can be

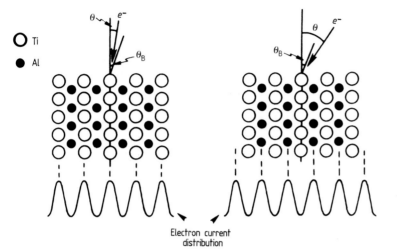

○ Ti
● Al

Electron current
distribution

Figure 2-19. Schematic representation of the principle of ALCHEMI. In an ordered material such as TiAl, it is possible to localize the electron wave on either the Ti or Al atoms by adjusting the diffracting condition. From Otten (1989).

used to determine the sublattice on which a third element is located, and is thus very useful in some studies. To avoid the ALCHEMI conditions, a weakly-diffracting condition is again best.

2.4.3.2 Beam Broadening and Spatial Resolution

Very few subjects in the area of AEM analysis have received as much attention and have been so vigorously investigated as that of electron probe broadening in the sample and its effect on spatial resolution of microanalysis. This is quite appropriate, since the principal reason for using an AEM to perform X-ray microanalysis of thin specimens is to obtain better spatial resolution than is possible using bulk specimens in an SEM or microprobe. Many AEM analyses are also performed near the limits of resolution for the instrument. It is thus important to examine the true spatial resolution for accurate X-ray microanalysis in the AEM, and to determine what parameters limit this spatial resolution.

It has been known for many years from the work on bulk samples that inelastic and elastic scattering processes cause the electrons in the incident high-energy beam to deviate from a linear path while passing through the sample. This results in broadening of the electron probe. Convoluted with this broadening is the normal divergence of the electrons in the probe due to the highly convergent nature of the probe-forming optics, but this effect tends to be small relative to the beam broadening caused by scattering and will be subsequently ignored. The scattering of the electron beam has been modelled for many years for bulk samples using Monte Carlo approaches (Newbury and Myklebust, 1982), where the electron trajectories in the sample are computed and a statistical treatment yields the electron interaction volume in the specimen. It is a simple process to truncate these models at a small specimen thickness and apply the results to thin foil samples. Figure 2-20 shows typical results for Si, Cu, and Au. These plots show that a significant amount of beam spreading has occurred in the sample, and that the probe diameter at the foil exit surface varies from about 13 nm for Si to 52 nm in gold for a beam energy of 100 keV. This can be compared with an incident probe diameter of ∼ 5 nm. However, this spread-

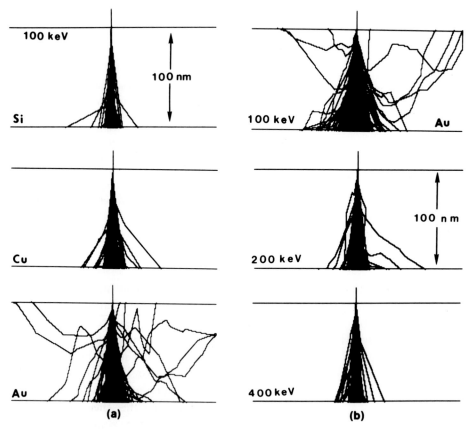

Figure 2-20. Examples of Monte Carlo electron trajectory plots for thin films. (a) 100 nm thick foils of Si, Cu, and Au for a beam energy of 100 keV; (b) a 100 nm thick foil of Au at beam energies of 100, 200, and 400 keV. From Newbury (1986).

ing is still much less than would be encountered in a bulk sample, where typical interaction volumes are $1-5 \, \mu m$.

A useful analytical model has been developed by Reed (1982) to describe probe broadening, and is helpful because it shows the effect of various experimental parameters on the amount of broadening. This equation assumes a single scattering event at the mid-thickness of the foil, and hence is only approximate and only applies to relatively thin foils of low atomic number material. Nevertheless, comparison with Monte Carlo results have shown that this model is a good approximation for many

situations. Reed's equation gives the broadening, b, as

$$b = 7.21 \times 10^5 \, (\varrho/A)^{1/2} \, (Z/E) \, t^{3/2} \qquad (2\text{-}11)$$

where ϱ, A, Z, and t are the density, atomic weight, atomic number, and thickness (in cm) of the specimen and E is the electron beam energy in eV. This equation shows that broadening increases as the density, atomic number, or thickness of the specimen increase, or as the accelerating voltage decreases. Table 2-5 gives broadening values for a point probe of 100 kV electrons as predicted by Eq. (2-11) and also from a Monte Carlo calculation (Goldstein et al.,

Table 2-5. Broadening, in nm, of a point probe of electrons as predicted by the single scattering model [Eq. (2-11)] and by Monte Carlo calculations (in parentheses) (Newbury and Myklebust, 1979; Kyser, 1979).

Element	Foil thickness in nm				
	10	50	100	300	500
Carbon	0.19 (0.22)	2.1 (1.9)	5.9 (4.1)	31.0 (16.0)	66.0 (33.0)
Aluminum	0.3 (0.41)	3.3 (3.0)	9.3 (7.6)	49.0 (30.0)	105.0 (66.0)
Copper	0.79 (0.78)	8.8 (5.8)	25.0 (18.0)	129.0 (97.0)	278.0 (244.0)
Gold	1.8 (1.7)	20.0 (15.0)	56.0 (52.0)	293.0 (599.0)	630.0 (1725.0)

1977). It can be seen that the two approaches give similar results. For samples in the typical range of thicknesses for AEM specimens (100–300 nm), and for medium atomic number materials, Table 2-5 shows that the beam broadening limits the spatial resolution for microanalysis to 10–100 nm.

In order to calculate the actual spatial resolution for a given analysis, it is necessary to take into account the initial probe diameter as well as the broadening, and sum these in quadrature, as

$$R = (d_p^2 + b^2)^{1/2} \qquad (2\text{-}12)$$

where R is the spatial resolution and d_p is the initial probe diameter.

Two types of microanalysis are most affected by the issue of beam broadening and spatial resolution. These are shown schematically in Fig. 2-21. The first of these is the case of small particles, in size near to the spatial resolution, embedded in a foil. In this case, several factors will affect the results of quantitative X-ray microanalysis. As Fig. 2-21 shows, the position of the particle in the foil may be critical, because the particle will intersect more of the incident beam of electrons if it is near the bottom surface of the foil due to beam broadening. The position of the particle within the foil is generally not known, and so this effect cannot be accounted for. In any case, the situation in Fig. 2-21 will lead to a large matrix contribution to the X-ray signal,

and will also be subject to the spurious fluorescence errors described previously. Thus, for small particles embedded in a foil, for the situation where the particle size is less than the foil thickness and near the spatial resolution, the analytical solution is untenable and only qualitative analysis is possible. Extraction of the particles is the only method for quantitative analysis.

A more tenable situation is shown on the right side in Fig. 2-21. In this case, the composition at a thin interface is of interest, where either an enrichment or depletion of a solute is present. The zone containing the enrichment or depletion is thinner than the spatial resolution, and the

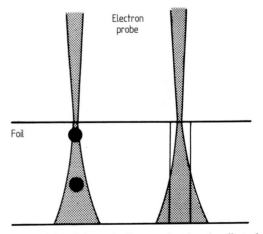

Figure 2-21. Schematic diagram showing the effect of beam broadening on the analysis of small particles and of interfacial layers. From Doig et al. (1981).

composition may be varying in this zone. This situation has been studied extensively experimentally. For very thin interface layers, such as a single monolayer of solute at a grain boundary, the effect of beam broadening is large. Figure 2-22 shows both experimental and calculated profiles for the case of a monolayer of Fe segregated to a grain boundary in MgO (Hall et al., 1981). The main effect of increasing sample thickness and hence increasing beam broadening is a decrease in the amount of Fe measured at the grain boundary, as more of the beam is scattered into the surrounding matrix. From results such as those in Fig. 2-22, the amount of solute at the grain boundary at zero thickness can be estimated. This geometry also lends itself to various analytical treatments. The electron distribution in the probe can be calculated from Monte Carlo results, the solute distribution at the grain boundary can be modelled, and these two profiles can be mathematically convoluted and the results compared to the experimental results. This approach has been applied successfully to a number of interfacial studies (Hall et al., 1981; Doig and Flewitt, 1977; Doig et al., 1981; Baumann and Williams, 1981; Hall and Briant, 1984).

Studies of thin segregation layers and other high spatial resolution investigations can be simplified if the spatial resolution for X-ray microanalysis is enhanced, and considerable effort has been expended over the past few years to improve this parameter. As shown in Eq. (2-11), for a given sample, the only two methods of decreasing beam broadening and increasing spatial resolution is to decrease specimen thickness or to increase the acceleration voltage. Each of these methods has its own set of limitations. Most AEM X-ray analyses are performed at the maximum accelerating voltages that the microscope is capable of.

A recent trend has been to manufacture AEMs which have higher maximum accelerating voltages, such as 300 and 400 kV. However, the higher voltages can create a new set of problems with gun stability and spurious X-rays from column components. Also, the increases in spatial resolution predicted by Eq. (2-11) associated with increased accelerating voltages have not been confirmed experimentally (Michael et al., 1990). Moving to thinner regions of the sample has a tendency to increase problems with surface films (discussed subsequently), and to also decrease the number of X-rays emitted from the sample. Another trend in recent years has been to design higher voltage microscopes with higher brightness guns. The higher brightness guns, generally of the field-emission type, produce smaller initial probes with large current, and thus are very effective at increasing spatial resolution, since thinner regions can be analyzed and both terms in Eq. (2-12) will be reduced. As shown in Table 2-5, the ultimate achievable spatial resolution, for a 1 nm probe of electrons, a 50 nm thick foil, a 300 kV accelerating voltage, and a medium atomic number material such as Fe, Ni, or Cu, is on the order of 5 nm.

2.4.3.3 Fluorescence Effects

In addition to absorption effects playing a role in accurate quantitation of X-ray intensity data, it is possible to have effects due to fluorescence. Fluorescence is related to absorption, and occurs when an X-ray of element A is absorbed by element B and results in the emission of an X-ray of element B. Fluorescence effects are most large when the energy of the X-ray from element A is just greater than the absorption edge for element B, and in addition when there is a small amount of element B in a matrix

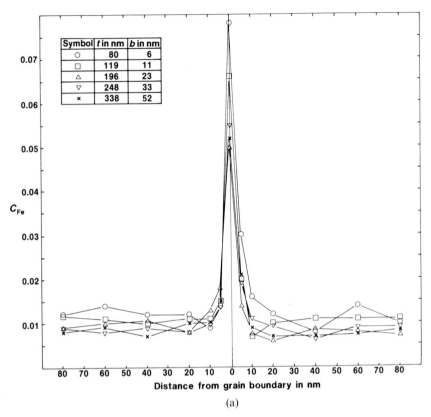

Symbol	t in nm	b in nm
○	80	6
□	119	11
△	196	23
▽	248	33
×	338	52

(a)

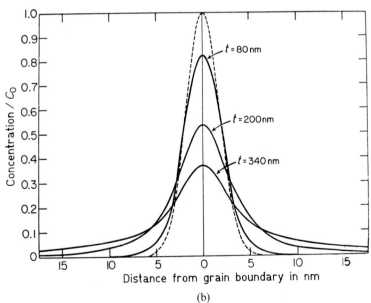

(b)

Figure 2-22. (a) Experimental and (b) calculated profiles as a function of foil thickness for the case of Fe segregated to a grain boundary in MgO. The model assumes a Gaussian probe of diameter 2.5 nm, and an Fe distribution at the grain boundary as shown by dashed curve. From Hall et al. (1981).

of A. In bulk samples, due to the large mass thickness of the sample, fluorescence effects are significant and a correction is routinely applied. Like absorption, fluorescence is lessened as the sample thickness decreases. In general, fluorescence is only a minor effect in AEM samples. This is due to the fact that absorption is strongest for low energy X-rays, and the subsequent secondarily-emitted lower-energy X-rays are affected by low fluorescence yields for low energy X-rays. The net result is that a noticeable fluorescence effect has been found only in the most extreme cases, namely, for small amount of element B in a matrix of A where A fluoresces B strongly. An example of this is Fe-5%Cr. Nockolds et al. (1980) have developed a model which can be used to correct for fluorescence in these cases, and which can be used to modify the simple Cliff-Lorimer relationship in a manner similar to that used to correct for absorption. In the case of fluorescence,

$$C_A/C_B = k_{AB}(I_A/I_B)(1 + X) \qquad (2\text{-}13)$$

where X is the fluorescence factor given by

$$X = C_A \omega_A \frac{(r-1)_B}{r_B} \frac{A_B}{A_A} \left(\frac{\mu}{\varrho}\right)_B^A \frac{U_A \ln U_A}{U_B \ln U_B} \cdot$$
$$\cdot \left(\frac{\varrho t}{2}\right) \left[0.923 - \ln\left(\frac{\mu}{\varrho}\right)_{SPEC}^A \varrho t \sec \alpha_x \right] \quad (2\text{-}14)$$

where ω is the fluorescent yield, r is the absorption edge jump ratio, U is the overvoltage, and the other terms have the same meaning as in the absorption correction. This correction is directly proportional to t, the sample thickness. The data reduction packages available with modern X-ray spectrometers can evaluate this correction easily using the same information input to assess absorption. As an example, the value of X for Fe-5%Cr for a foil thickness of 100 nm is 0.077, or a 7.7% quantitation error.

2.4.3.4 Contamination

Another factor which can affect the accuracy of quantitation, and also the spatial resolution, of X-ray spectroscopy in the AEM is the buildup of contamination on the foil during analysis. As discussed in the section on foil thickness determinations, it is possible to develop large hemispherical mounds of carbonaceous compounds on the top and bottom surfaces of the foil during microanalysis. These compounds exist on the sample surfaces or in the microscope vacuum. If these mounds become large enough, they can have two effects: they can absorb X-rays exiting the foil surface, and they can cause the beam to spread before entering the foil. The first of these effects can be estimated using the approach described in the section on the absorption correction in thin films. In an A-B alloy, the relative absorption of the X-rays from elements A and B by carbon can be calculated, and the magnitude of the absorption difference determined as a function of carbon thickness. Table 2-6 gives the maximum thickness of carbon contamination permitted in order to keep the absorption difference less than 5%. The calculation assumes an X-ray takeoff angle of 20°. It is clear from this table that for medium or high energy X-rays, little effect of contamination will be seen on the accuracy of quantitation. However, a significant effect will be seen in the case of lowenergy X-ray detection.

A second effect of carbon contamination is to cause spreading of the electron probe before it enters the specimen. This will degrade the spatial resolution of the analysis. The topic of probe spreading was considered previously, and the single scattering model presented there can be used to calculate the magnitude of probe spreading as a function of the thickness of the carbon

Table 2-6. Maximum carbon thickness for 5% error in quantitation.

A-B	Thickness
$Fe\,K_\alpha - Cr\,K_\alpha$	28 µm
$Fe\,K_\alpha - Mo\,K_\alpha$	37 µm
$Ni\,K_\alpha - Al\,K_\alpha$	270 nm
$Fe\,K_\alpha - O\,K_\alpha$	14 nm
$Fe\,K_\alpha - N\,K_\alpha$	6 nm
$Fe\,K_\alpha - C\,K_\alpha$	67 nm

Table 2-7. Beam spreading due to contamination, 100 kV electrons.

Carbon thickness in nm	Spreading in nm
10	0.2
100	5.0
500	57.0

contamination layer. Table 2-7 gives the results of these calculations for a point probe of electrons. This effect will only be significant in the case of an older AEM with a relatively dirty vacuum system, a dirty sample, a very high spatial resolution analysis, or some combination of these. Most AEMs, even with poor vacuum systems, will not produce contamination layers thicker than 100 nm in the usual time span for microanalysis. In newer machines with oil-free vacuum systems, little or no contamination is seen during microanalysis.

2.4.3.5 Accuracy of Quantitation

The subject of the statistical accuracy of X-ray microanalysis in the AEM has been considered by Romig and Goldstein (1979). The total error in an analysis using the Cliff-Lorimer method is the sum of the errors in k_{AB}, I_A, and I_B. The error in I_A and I_B is given, at the 2σ (95%) confidence level, as

$$Error\,(\%) = 2\,(\sqrt{N}/N) \times 100 \qquad (2\text{-}15)$$

The error associated with k_{AB} can be measured for an experimental determination but must be estimated if k_{AB} is calculated or taken from other work. In the case of an experimental measurement, the error is simply the standard deviation of the experimental values from the standard after any absorption effects have been accounted for, and is generally about 5%. As discussed previously, calculated or textbook values may be in error by 20% or more.

In the calculation of total error, therefore, in a case where k_{AB} is 1.0 ± 0.5, $I_A = 50\,000$ counts (0.9% error), and $I_B = 5000$ counts (2.8% error), the total error is $5 + 0.9 + 2.8 = 8.7\%$. From this calculation, the value of accurate assessment of k_{AB}, and of large numbers of counts in the characteristic peaks, is obvious.

2.4.3.6 Surface Films

It has been shown in the past that in some cases the accuracy of quantitation using thin samples in the AEM can be affected by the presence of surface films. In general, these films are found on metals which contain an oxide-forming element as a solute, and which have been prepared for microscopy by electropolishing. Examples include Al-Cu, Al-Zn, and Al-Mg alloys, which form solute-rich films (Thompson et al., 1977; Pountney and Loretto, 1980), and Ni-Al alloys, which form films rich in Al (Fraser and McCartney, 1982). The effect of these films is most noticeable in the thinnest regions of the sample, where enhanced levels of the solute are seen, as shown in Fig. 2-23. It is generally recommended that the thinnest regions of any AEM sample be avoided, since the highest density of artifacts exist at the foil edge. However, in high spatial resolution microanalysis, or when small particles embedded in a matrix are of interest, it is often impor-

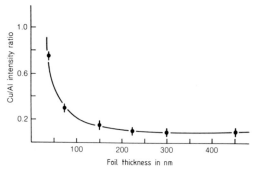

Figure 2-23. Effect of a Cu-rich surface layer on an Al foil on the Cu/Al X-ray intensity ratio as a function of foil thickness. From Thompson et al. (1977).

tant to work in the thinnest regions possible. In these cases, care should be taken to check for the presence of surface films by monitoring I_A/I_B as a function of foil thickness in the sample matrix.

If such films are present, they can often be easily removed or avoided. Removal can often be accomplished by ion milling the sample for a few minutes, or by altering the electropolishing conditions. In some cases, an alternative method of sample preparation, such as microtomy, may be necessary.

2.5 Electron Energy Loss Spectroscopy (EELS)

Electron energy loss spectrometers have been an important component of analytical electron microscopes since the inception of the dedicated-type of AEM. The original conception of an EELS unit dates back to Hillier and Baker (1944) in the 1940s, but this technique was not developed further for many years until being rediscovered by Wittry et al. (1969) and Crewe (1970). At the present time, high-quality EEL spectrometers are available

commercially for attachment to virtually any AEM, and these devices have sophisticated user interfaces and a great deal of software for data acquisition, display, and quantitation. In this section, we will discuss the basic principles of the technique, EELS spectra acquisition and interpretation, quantitation, and the specific capabilities and limitations of this technique. It is worth noting that compositional analysis by EELS is a unique capability of the AEM; because it requires the use of very thin samples, it is not available in other electron-beam instruments. For a detailed description of EELS, the text by Egerton (1986) is recommended.

2.5.1 Basic Principles of EELS

The EELS technique is based on the fact that when a high-energy beam of incident electrons pass through a thin, electron-transparent sample, the electrons in the incident beam lose energy through a number of inelastic scattering processes. If a spectrometer is used to acquire and display a plot of the number of these high-energy electrons exiting the sample as a function of their energy, then information can be obtained from this spectrum about the sample composition, thickness, and the local chemical environment of the atoms, among other things.

To begin, the fundamental event which needs to be considered is the electron-specimen interaction. When a high-energy beam of electrons passes through a sample, the exiting electrons can be divided into three categories:

(a) unscattered (forward-scattered or transmitted) electrons, which have lost no energy and have not changed direction;

(b) elastically scattered electrons, which experience negligible energy loss but a significant change in direction; and

(c) inelastically scattered electrons, which experience both energy loss and a change in direction.

Elastic interactions generally result from scattering from the nucleus (billiard-ball type collisions), and the most well-known example of elastic scattering is Bragg diffraction. Elastic scattering angles are typically $1-2°$ ($20-40$ mr), and can be calculated for amorphous or crystalline materials using the formulas:

amorphous: $\theta_e = \lambda/(2\pi R_A)$ (2-16)

crystalline: $\theta_e = \lambda/(2d)$ (2-17)

where θ_e is the elastic scattering angle, λ is the electron wavelength, A is the atomic radius and d is the interplanar spacing. In comparison, there are a large number of inelastic scattering processes. These include: phonon excitation, where the energy loss is on the order of 0.02 eV and thus can be considered as quasi-elastic; electron excitation, associated with the loosely-bound valence electrons, where losses are on the order of $1-150$ eV; inner-shell ionization of the strongly-bound valence electrons, which results in losses from about 50 eV to several thousand eV; and plasmon excitation, with losses in the range of $10-50$ eV. It is these last two loss mechanisms that are of importance in EELS, and will be considered in detail below.

The scattering angles associated with inelastic scattering are on average much less than those due to elastic scattering. The inelastic scattering angle θ_i is given by

$\theta_i = \Delta E/mV$ (2-18)

where ΔE is the energy loss and m and V are the mass and velocity of the incident electron. For an energy loss of 250 eV, the scattering angle is 1.37 mr. Thus, the inelastically-scattered electrons are localized near to the transmitted beam.

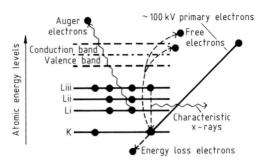

Figure 2-24. Processes occurring within an atom during high-energy electron irradiation. From Williams (1984).

It is the inner-shell ionization process which is most importance to EELS, and this is shown schematically in Fig. 2-24. A high-energy incident electron is capable of ejecting an inner-shell electron from an atom in the sample, and in doing so will lose a characteristic amount of energy. The relaxation of the atom to its ground state, in which the inner-shell vacancy is filled by a higher-shell electron, results in the emission of an X-ray from the atom. Thus, the energy-loss and X-ray emission processes are complimentary, and for every X-ray produced, there is an associated energy-loss electron. As in X-ray spectroscopy, there are a number of possible ionization events associated with an atom, and thus a multiplicity of characteristic losses from a given element. The notation used to describe various energy-loss events, and the corresponding peaks in the EELS spectrum, is shown in Fig. 2-25.

It is the physics of the collection process of energy-loss electrons, when compared with X-rays, that led to the initial interest in EELS. Let us first consider X-ray collection. X-rays are isotropically emitted from atoms over 4π steradians. As described in Sec. 2.4.1, and shown in Fig. 2-13, the solid-state X-ray detector intercepts only a small fraction of these X-rays, given by the

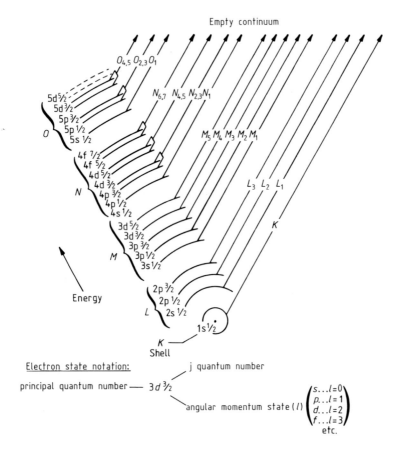

Empty continuum

Electron state notation:

principal quantum number — $3d\,\frac{3}{2}$

j quantum number

angular momentum state (l)

$$\begin{pmatrix} s...l=0 \\ p...l=1 \\ d...l=2 \\ f...l=3 \\ etc. \end{pmatrix}$$

Figure 2-25. Notations of ionization events associated with EELS peaks. From Ahn and Krivanek (1983).

solid angle [Eq. (2-5)], a function of the detector area and its proximity to the sample. For the typical AEM, the solid angle is about 0.1 steradian, leading to a collection efficiency of 0.8%. However, there are additional X-ray losses. The most fundamental of these is the fluorescent yield, which is the number of X-rays emitted from the atom relative to the number produced. This yield is low for low atomic number materials since low-energy X-rays tend to be absorbed within the atom to produce an Auger electron (Fig. 2-24). This competing process reduces the fluorescent yield for an X-ray such as carbon K_α to 0.3%. There are additional losses, particularly for low-energy X-rays, due to absorption in the sample and in the materials in the detector between the sample and the analyzing crys-

tal. Thus, in the best case of high-energy X-rays, where the main limitation is the solid angle, the collection efficiency for X-rays is about 1%, and this can drop several orders of magnitude for low-energy X-rays.

The collection efficiency associated with EELS is significantly different. The collection geometry is shown in Fig. 2-26. The EELS spectrometer collects a certain angular range of electrons scattered from the specimen, defined by the spectrometer acceptance aperture. Since the inelastically-scattered electrons are scattered to small angles, the spectrometer is capable of collecting the majority of these electrons. Figure 2-27 shows how the EELS collection efficiency varies with aperture size and energy loss. For a typical value of $\beta_0 = 20\,\mathrm{mr}$, Fig. 2-27 shows that $\sim 50\%$ of all electrons

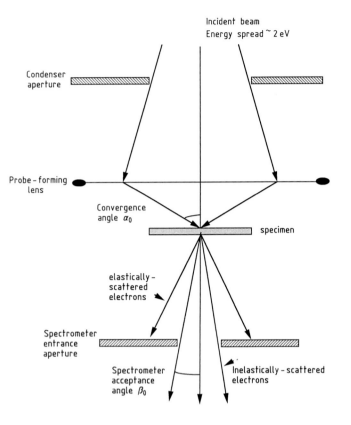

Incident beam
Energy spread ~ 2 eV

Condenser
aperture

Probe - forming
lens

Convergence
angle α_0

specimen

elastically -
scattered
electrons

Spectrometer
entrance
aperture

Spectrometer
acceptance
angle β_0

Inelastically - scattered
electrons

Figure 2-26. Schematic diagram of EELS collection process.

which have lost less than 1000 eV are collected. Therefore, since there are equal numbers of energy-loss electrons and X-rays, this greatly-increased collection efficiency is a major advantage of EELS. It translates directly into predictions of greatly enhanced elemental sensitivity, or reduced analysis time, for EELS (Isaacson and Johnson, 1975). However, other factors need to be considered, described in subsequent sections, in order to fully evaluate this comparison of analysis techniques.

2.5.2 EELS Spectrum Acquisition and Characteristics

The microscope setup for EELS spectrum acquisition is virtually identical to that for X-ray spectroscopy; in fact, EDS

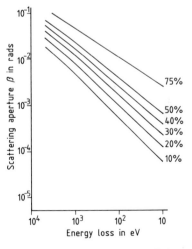

Figure 2-27. EELS collection efficiency as a function of spectrometer acceptance angle and energy loss. From Isaacson (1978).

and EELS spectra can be obtained simultaneously. However, the EELS spectrum is much less affected by spurious contributions, and so essentially no artifacts need be considered. The important parameters for spectrum acquisition are shown in Fig. 2-26, and consist primarily of the convergence angle α_0 and the spectrometer acceptance angle β_0. As in EDS spectroscopy, a well-collimated probe of electrons is focussed onto the point of interest for analysis. The EELS spectrum is relatively insensitive to values of α_0, but the conver-

gence angle will be convoluted with the scattering angle θ in the exiting beam and so in practice it is preferable to make α_0 as small as possible, and in particular it should be a small fraction of β_0. The choice of β_0 will affect the spectrum, and the same principles apply that govern optical systems. As β_0 is decreased, the spectrometer aberrations, which increase with distance from the optic axis, will decrease. However, at the same time, the total signal admitted to the spectrometer will decrease, and, in particular, the higher energy-loss electrons, which are scattered farthest from the optic axis [Eq. (2-18)], will be excluded. Thus, β_0 is generally chosen to be just large enough to admit the entire spectrum of interest, generally from 0 to about 2000 eV loss.

A typical EELS spectrometer attached to an AEM is shown schematically in Fig. 2-28. The great majority of EELS spectrometers on AEMs at present are of the magnetic sector type, as shown, in which the electron beam is bent through $90°$ by the action of a magnetic field. The action of the spectrometer is to convert the dispersion in energy in the electron beam into a dispersion in position at the spectrometer focus plane at the exit of the magnet. Thus, the spectrum exists in two dimensions in this plane, with electrons of energy E_0 focussed to one point and electrons of energy $E_0 - \Delta E$ focussed to a different point. Two distinct types of detection system are then used to collect and display the information. Early spectrometers, prior to about 1988, exclusively used a scintillator-photomultiplier electron detector for this purpose, as shown in Fig. 2-29a. This detector uses a mechanical slit of variable width to exclude all electrons except those of a particular energy from the detector. The current in the magnet is the ramped so that the entire spectrum can be collected by the detector, and the result is

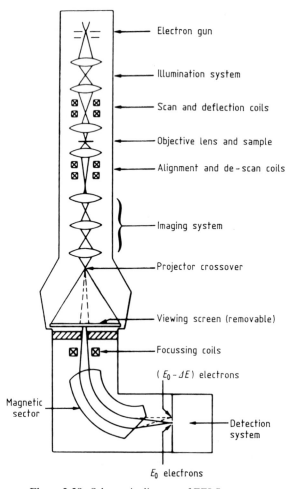

Figure 2-28. Schematic diagram of EELS spectrometer attached to a TEM/STEM-type AEM. From Ahn and Krivanek (1983).

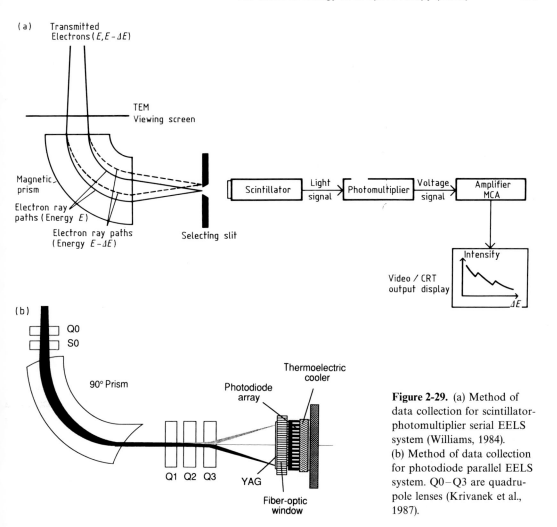

Figure 2-29. (a) Method of data collection for scintillator-photomultiplier serial EELS system (Williams, 1984). (b) Method of data collection for photodiode parallel EELS system. Q0–Q3 are quadrupole lenses (Krivanek et al., 1987).

displayed on a CRT, generally using the same display package as used by the EDS system. The spectrum is thus collected serially, and this requires a substantial amount of time, for example 100 s for a 1000 eV spectrum, to acquire. The energy resolution of the spectrum is a function of the slit size, with resolution increasing but signal strength decreasing as the slit size is reduced. A compromise of 3–5 eV for the resolution is generally selected. A second, superior type of detector has been available since 1988, and is now the only detector type commercially available. This is

a parallel detector, shown in Fig. 2-29 b, which uses a photodiode array to collect the spectrum in parallel. In this case, no slit is used, and the energy resolution is related to the energy spread in the incident beam, which is usually about 0.5 eV for a field-emission gun and about 2 eV for a thermionic emitter such as LaB_6. Since the spectrum is acquired in parallel, the acquisition time is reduced by a factor of 10^3 if a 1000 channel spectrum is compared. This results in the capability for studies of dynamic changes, since a spectrum can be acquired in as little as 0.1 s. It also represents a significant improvement compared with EDS collection times, and has allowed EELS composition maps to become possible (see Sec. 2.3.4).

Other types of electron energy loss spectrometer are possible, and a good summary of these is provided by Egerton (1986). Two of the more common types are the Wien filter (Fig. 2-30a) and the omega filter (Fig. 2-30b). The Wien filter employs both magnetic and electrostatic components to produce a spectrum at the recording slit. This type of spectrometer has been shown to be capable of very high energy resolution (Geiger et al., 1970) but, since it is connected to the microscope high

voltage V_0, requires very stable high-voltage supplies. The omega filter is now available on some commercial AEMs, and is a purely magnetic device. It has the advantage of producing energy-filtered images if a slit is positioned at D_2 and the intermediate lens focuses on the plane O_3. Alternatively, if the intermediate lens is focussed on D_2, a spectrum is recorded.

We will now consider the elements of an EELS spectrum. A typical test spectrum, from a BN flake on a very thin carbon support film, is shown in Fig. 2-31. Six regions are marked, namely: (1) zero-loss peak; (2) low loss region; (3) gain change; (4) background; (5) B peak area; and (6) N

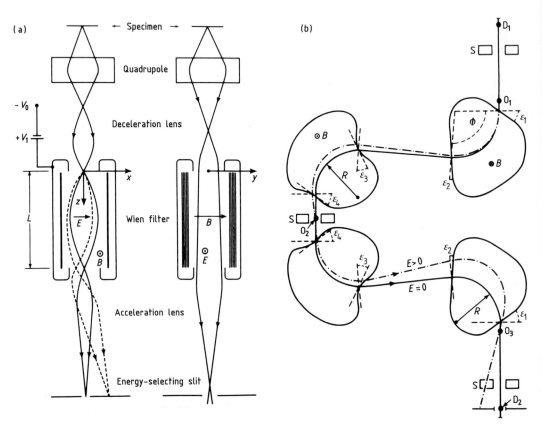

Figure 2-30. (a) Wien-filter spectrometer. The EELS spectrum is formed at the energy-selecting slit. From Batson (1985). (b) Optics of an aberration-corrected omega filter. The object is at D_1. An energy-filtered image is formed at O_3 in conjunction with an energy-filtering slit at D_2; the EELS spectrum is formed at D_2. From Pejas and Rose (1978).

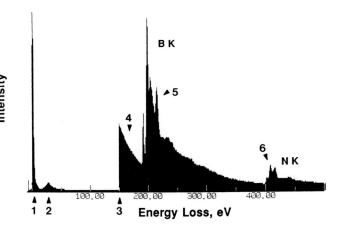

Figure 2-31. EELS spectrum from BN flake on thin carbon support film, showing: (1) zero-loss peak; (2) low-loss region; (3) gain change; (4) decreasing background; (5) B peak; and (6) N peak.

peak area. Each of these regions will be described below.

Region 1, shown in detail in Fig. 2-32a, contains the zero-loss peak, corresponding to the unscattered or transmitted electrons. The vast majority of the incident electrons reside in this peak. It should be noted that, by convention, the positive energy scale refers to the amount of energy lost. The width of the zero-loss peaks reflects both the true spread in energy of the electrons emitted (nominally ∼ 2 eV for thermionic emission) convoluted with any additional effects due to the spectrometer. The zero-loss peak centroid is defined as zero energy loss. The signal rate in this peak is on the order of 1–2 billion electrons per second, and so the electron detectors are used to generate an analog signal in this range. The zero-loss peak contains information about the scattering power of the specimen, related to its density and thickness and the scattering cross section of the atoms in the sample. The zero-loss peak has two primary uses. The full-width at half-maximum of the peak defines the energy resolution of the EELS spectrum, which is equal to or greater than the true energy spread in the probe. The peak can also be used as a standard intensity for quantitative analysis.

Region 2 (Fig. 2-32b), extending from the zero loss peak to about 50 eV energy loss, is the low-loss region, which contains the so-called "plasmon" loss peak. This peak, seen in all samples, is due to the transfer of energy to the loosely bound conduction or valence electrons in the sample. In metals and other good conductors, the loosely bound electrons can be excited collectively, resulting in a sharp plasmon loss peak at a specific energy (e.g., 15 eV in Al). In materials with lower conductivity, which have more tightly bound outer electrons, a broad peak is seen in the vicinity of 20–30 eV. For a given specimen, the relative height of the plasmon or low-loss peak is a sensitive function of the specimen thickness. EELS analysis tends to be very dependent upon specimen thickness, and thus this internal indication is quite useful. The plasmon loss peaks in metals have also been used for chemical analysis, and a review of this technique has been published by Williams and Edington (1976).

Region 3 (Fig. 2-32c) shows a large gain change at approximately 150 eV. This is necessary because the intensity in the EELS spectrum falls off very rapidly with energy loss, due to the behavior of the inelastic scattering cross section. The gain change is accomplished by switching the

(a)

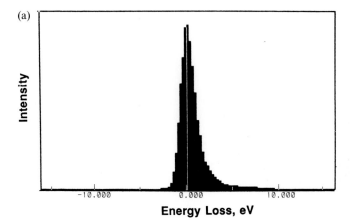

(b)

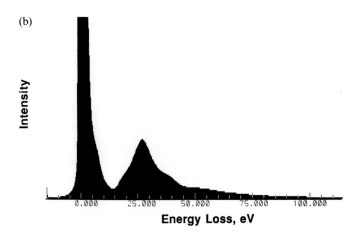

(c)

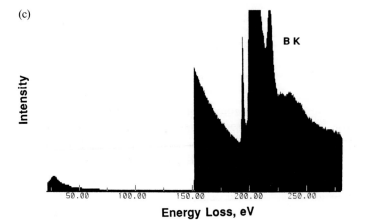

Figure 2-32. Enlarged views of spectrum in Fig. 2-31. (a) Zero-loss peak; (b) low-loss region; (c) gain change and decreasing background; (d) B peak; and (e) N peak.

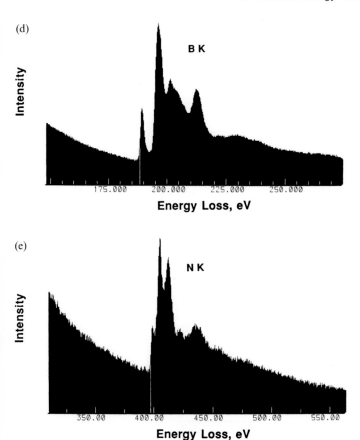

detector from analog to electron-counting modes, which results in this case in a gain change of about 3000 ×.

Region 4 (Fig. 2-32 c) shows the rapidly-decreasing background which is most clearly visible after the gain change. The background in the EELS spectrum is due to electrons which have experienced multiple or random energy losses, and the background intensity will increase with increasing sample thickness.

Region 5 (Fig. 2-32 d) shows the boron peak area. Its onset is at 188 eV, the minimum energy required to remove a K-shell electron from the boron atom. The peak area has multiple maxima and a long, extended tail. Several different processes contribute to this complex peak structure. The first of these is multiple scattering (for example, an electron which has ionized a boron atom and has experienced a plasmon loss event will contribute to a peak at 188 + 25 eV). Secondly, an electron can transition to a bound excited state in the atom, and this can also be affected by the local chemical (bonding) environment. The long tail is due to the fact that the incident electron can lose any energy $E_e + \Delta E$, where E_e is the edge energy (188 eV), in ejecting the inner-shell electron, although the probability of transferring the additional energy ΔE to the ejected electron decreases with ΔE. Finally, there is an oscillatory structure present in the tail far

Table 2-8. Principle EELS edges, onset energy in eV (from Ahn and Krivanek, 1983).

Z	Element	Edge	Energy
1	H	K	13.6
2	He	K	21.2
3	Li	K	54.7
4	Be	K	111.0
5	B	K	188.0
6	C	K	283.8
7	N	K	401.6
8	O	K	532.0
9	F	K	685.0
10	Ne	K	867.0
11	Na	K	1072.0
12	Mg	K	1305.0
13	Al	K	1560.0
		$L_{2,3}$	73.1
14	Si	K	1839.0
		$L_{2,3}$	99.2
15	P	$L_{2,3}$	132.2
16	S	$L_{2,3}$	164.8
17	Cl	$L_{2,3}$	200.0
18	Ar	$L_{2,3}$	245.2
19	K	$L_{2,3}$	293.6
20	Ca	$L_{2,3}$	346.4
21	Sc	$L_{2,3}$	402.2
22	Ti	$L_{2,3}$	455.5
23	V	$L_{2,3}$	513.0
24	Cr	$L_{2,3}$	574.0
25	Mn	$L_{2,3}$	640.0
26	Fe	$L_{2,3}$	708.0
27	Co	$L_{2,3}$	779.0
28	Ni	$L_{2,3}$	854.0
29	Cu	$L_{2,3}$	931.0
30	Zn	$L_{2,3}$	1020.0
31	Ga	$L_{2,3}$	1115.0
32	Ge	$L_{2,3}$	1217.0
33	As	$L_{2,3}$	1323.0
34	Se	$L_{2,3}$	1436.0
35	Br	$L_{2,3}$	1550.0
36	Kr	$L_{2,3}$	1674.0
		$M_{4,5}$	88.9
37	Rb	$L_{2,3}$	1804.0
		$M_{4,5}$	110.3

Table 2-8. (Continued).

Z	Element	Edge	Energy
38	Sr	$L_{2,3}$	1940.0
		$M_{4,5}$	133.1
39	Y	$L_{2,3}$	2080.0
		$M_{4,5}$	157.4
40	Zr	$L_{2,3}$	2222.0
		$M_{4,5}$	180.0
41	Nb	$L_{2,3}$	2370.0
		$M_{4,5}$	204.6
42	Mo	$L_{2,3}$	2520.0
		$M_{4,5}$	227.0
44	Ru	$M_{4,5}$	279.4
45	Rh	$M_{4,5}$	307.0
46	Pd	$M_{4,5}$	334.7
47	Ag	$M_{4,5}$	366.7
48	Cd	$M_{4,5}$	403.7
49	In	$M_{4,5}$	443.1
50	Sn	$M_{4,5}$	484.8
51	Sb	$M_{4,5}$	527.0
52	Te	$M_{4,5}$	577.0
53	I	$M_{4,5}$	619.0
54	Xe	$M_{4,5}$	672.0
55	Cs	$M_{4,5}$	725.0
56	Ba	$M_{4,5}$	781.0
57	La	$M_{4,5}$	832.0
58	Ce	$M_{4,5}$	883.0
59	Pr	$M_{4,5}$	931.0
60	Nd	$M_{4,5}$	978.0
62	Sm	$M_{4,5}$	1080.0
63	Eu	$M_{4,5}$	1131.0
64	Gd	$M_{4,5}$	1185.0
65	Tb	$M_{4,5}$	1241.0
66	Dy	$M_{4,5}$	1295.0
67	Ho	$M_{4,5}$	1351.0
68	Er	$M_{4,5}$	1409.0
69	Tm	$M_{4,5}$	1468.0
70	Yb	$M_{4,5}$	1528.0
71	Lu	$M_{4,5}$	1588.0
72	Hf	$M_{4,5}$	1662.0
73	Ta	$M_{4,5}$	1735.0
74	W	$M_{4,5}$	1809.0
75	Re	$M_{4,5}$	1883.0
76	Os	$M_{4,5}$	1960.0

Table 2-8 (Continued).

Z	Element	Edge	Energy
77	Ir	$M_{4,5}$	2040.0
78	Pt	$M_{4,5}$	2122.0
79	Au	$M_{4,5}$	2206.0
80	Hg	$M_{4,5}$	2295.0
81	Tl	$M_{4,5}$	2389.0
82	Pb	$M_{4,5}$	2484.0
83	Bi	$M_{4,5}$	2580.0
90	Th	$N_{6,7}$	335.2
		$O_{4,5}$	87.9
92	U	$N_{6,7}$	380.9
		$O_{4,5}$	96.3

from the edge, which is similar in origin to EXAFS (extended X-ray absorption fine structure) peaks in X-ray experiments in that it results from interactions of the ejected electron with neighboring atoms. Thus, the position of the peak structure in the EELS spectrum indicates the type of atom present, and the shape of the peak structure is related to bonding.

Finally, region 6 (Fig. 2-32e) contains the nitrogen peak, with onset at 401 eV. Its characteristics are similar to the boron peak.

We will now consider in more detail changes in peak shapes in the EELS spectra due to differences in transition type or chemical environment. These changes are both an advantage and a disadvantage; they contain additional information about the specimen, but they can complicate interpretation and in particular quantitation. Figure 2-33 illustrates the three most common edge types expected from various elements, as predicted from the behavior of the scattering cross-sections. From atomic numbers 1 to 14 (hydrogen to silicon), which spans a range from 13 to 1839 eV energy loss, the K edges are generally chosen for analysis, and these approach an ideal hydrogenic, step-like shape. From atomic numbers 13 to 36 (aluminum to krypton, losses 73 to 1727 eV), the $L_{2,3}$ edges are generally used, and these are characterized by a delayed onset of $\sim 20\,\text{eV}$. Finally, from atomic numbers 37 to 83 (rubidium to bismuth, losses 110 to 2688 eV), the $M_{4,5}$ edges are the strongest, and these have an extended delayed onset of $\sim 50\,\text{eV}$.

Table 2-8 summarizes the types of edges available from each element. Note that, unlike EDS, in principle the entire periodic table is available for analysis. The choice

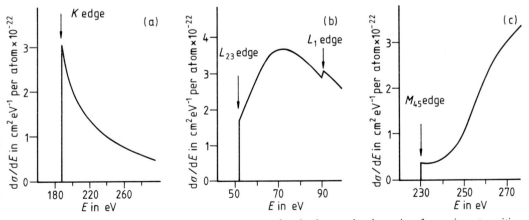

Figure 2-33. Edge shape expected in EELS spectrum after background subtraction for various transitions (Leapman et al., 1978). Shown are modeled edge profiles for (a) B K edge; (b) Mg L edges; (c) Mo M edges.

of edges given above assumes that most studies will be conducted between 50 and 2000 eV energy losses. Below 50 eV, the edge visibility is affected by the strong low-loss peaks and background in this region; above 2000 eV the edge visibility is affected by the very low scattering cross section for large losses, and by aberrations in the spectrometer which are larger for larger scattering angles (higher losses).

The predicted peak shapes in Fig. 2-33 and the onset energies in Table 2-8 can be strongly affected by the local chemical environment of the atoms in the sample. This is obvious when the ideal hydrogenic edge shape in Fig. 2-33 is compared with the true K-edge shape for B and N in Fig. 2-32. In addition to shape changes, edge onset energies can be shifted by about -2 eV to $+7$ eV due to chemical effects. The strong dependence of edge shape and energy on chemical state has led to the development of handbooks of EELS spectra, such as those by Ahn and Krivanek (1983) and by Zaluzec (1987) which show the expected

EELS edges for many element in different states. These handbook spectra can serve as "fingerprints" to identify elements and compounds. Results similar the these library spectra are shown in Figs. 2-34 and 2-35. In Fig. 2-34a, the $L_{2,3}$ edges from the transition elements are shown after background subtraction. The ratio of the L_2 to L_3 peaks is a characteristic of the element, and the predicted shape for the peak (Fig. 2-33) does not occur until Ge. Figure 2-34b shows a similar compilation for the $M_{4,5}$ peaks from Y to Ag. The predicted delayed onset is seen. Figure 2-34c shows the Mo $M_{4,5}$ prior to background subtraction. The delayed onset and long tail associated with the $M_{4,5}$ peaks is one of the problems associated with EELS analysis.

Figure 2-35 illustrates changes in peak shape associated with chemical state. Figure 2-35a shows the K-edge for boron in BN (see Figs. 2-31, 2-32) and in elemental B. Figure 2-35b shows a similar comparison for three forms of carbon, and Fig. 2-

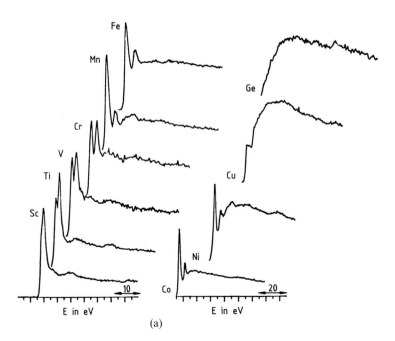

(a)

Figure 2-34. Comparison of (a) $L_{2,3}$ and (b) $M_{4,5}$ edges from various elements after background subtraction (Egerton, 1979). (c) Mo $M_{4,5}$ edge shown prior to background subtraction (Ahn and Krivanek, 1983).

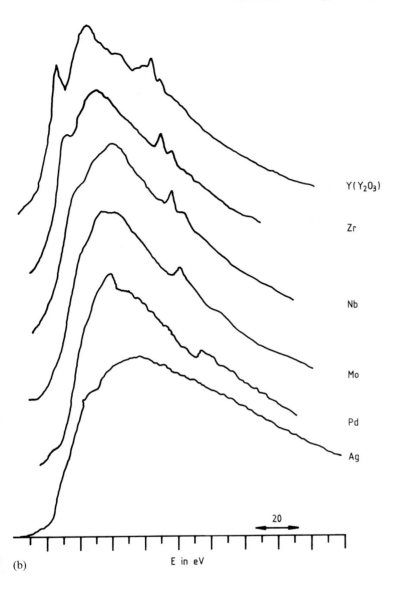

Y(Y$_2$O$_3$)

Zr

Nb

Mo

Pd

Ag

20

E in eV

(b)

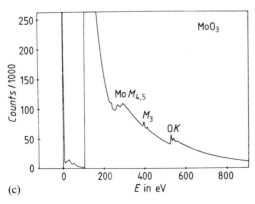

(c)

35c compares the Si $L_{2,3}$ edges in elemental Si versus SiO$_2$. These major changes in peak shape, such as that seen for Si, are both a useful tool for compound identification and a complicating factor for quantitative analysis of EELS spectra.

As mentioned previously, the extended fine structure on the tails of the EELS peaks are another important tool for elemental analysis. This extended fine structure can be analyzed using methods sim-

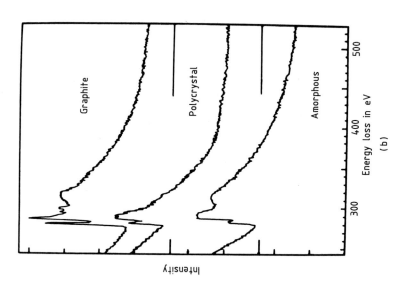

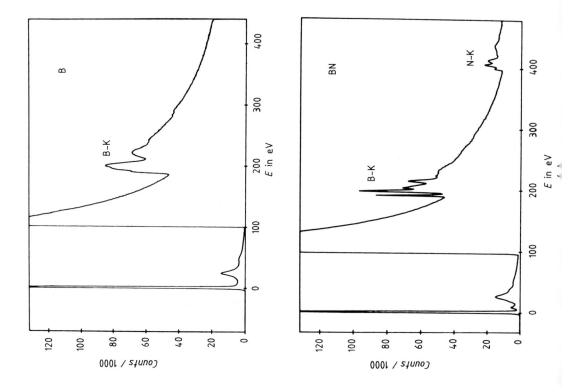

Figure 2-35. Comparison of peak shapes for same elements in different materials. (a) B in pure B vs. BN (Ahn and Krivanek, 1983); (b) C in three different forms (K-shell ionization-loss spectra) (Ichinokawa, 1982); (c) Si in pure Si, SiO$_2$ (Ahn and Krivanek, 1983).

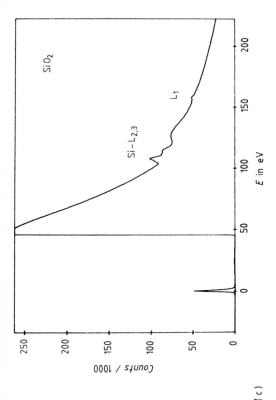

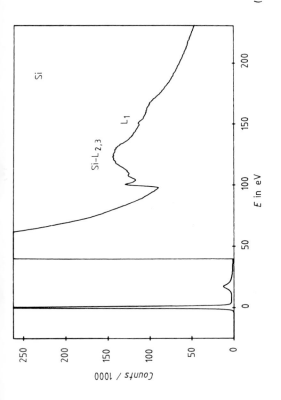

(c)

ilar to those developed for processing of EXAFS X-ray peaks, and can yield detailed information about local chemical environment. For description of these techniques, see Isaacson and Johnson (1975) or Egerton (1986).

2.5.3 Quantitation of EELS Spectra

The method for quantitative analysis of EELS spectra is similar to that for EDS spectra: background modelling and subtraction, resolution of peak overlaps, evaluation of peak intensities, and then use of an analytical expression to convert intensities into composition. However, the nature of the EELS spectrum results in some special procedures for these steps. We will illustrate the quantitation procedure in Fig. 2-36, using the BN EELS spectrum from Figs. 2-31 and 2-32.

Background subtraction and the deconvolution of peak overlaps is a particular problem in EELS quantitation. The background is not flat but rapidly decreasing with energy loss, and hence must be carefully modelled. Three empirical models have been found to work well with EELS background, namely:

(1) $I = A e^{-r}$ fit,

(2) $I = A x^{-2} + B x^{-1} + C$ polynomial fit,

(3) $I = A x + B$ log-polynomial fit.

The software package for EELS spectrum acquisition and display is capable of fitting each of these expressions to a selected background region just ahead of the peak (Fig. 2-36a), and, once the best fit is obtained, extrapolating the background beneath the edge (Fig. 2-33b) and subtracting the background (Fig. 2-36c). The same procedure can then be applied to subsequent peaks. However, as Fig. 2-36c shows, even for widely-separated peaks

(a)

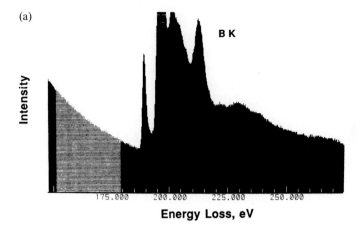

(b)

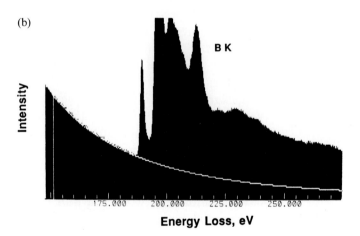

(c)

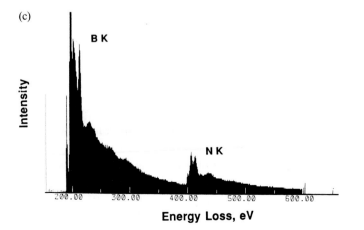

Figure 2-36. Method of background subtraction in EELS spectrum. (a) Background region ahead of peak is selected; (b) background is fit to analytical expression and extrapolated beneath peak; (c) spectrum after first background subtraction; (d) B tail structure is removed from underneath N peak in similar fashion.

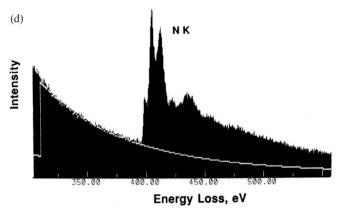

(d)

Intensity

N K

|350.00   400.00   450.00   500.00|

Energy Loss, eV

such as B (188 eV) and N (401 eV), the N peak structure sits on the tail of the B peak. Thus, the second background subtraction must be applied to the B tail (Fig. 2-36 d) in order to extract the true N intensity. For more closely-spaced peaks, or more extended edges, such as the $M_{4,5}$ edges, the separation of peaks is a particular problem and limitation to accurate quantitation. Since the peaks cannot be simply modeled, deconvolution depends upon the ability to visually separate the two peaks.

After background subtraction, the intensity in the peak is evaluated over a particular energy window Δ, which extends from the peak onset to some higher energy loss. A typical value for Δ is 50 eV. Two methods can then be used for quantitation, as shown in Fig. 2-37 (Egerton, 1978). Figure 2-37 a shows the "absolute" method, in which the edge intensity is ratioed to the intensity in a window of the same size containing the zero-loss and low-loss region. This method is used to calculate the number of atoms corresponding to a particular peak. In this case,

$$N = I_K/(I_{0\Delta}\,\sigma_K) \qquad (2\text{-}19)$$

where N is the number of atoms, I_K is the intensity in the K-edge window of width

Δ, $I_{0\Delta}$ is the intensity in a window from 0 eV to Δ eV, and σ_K is the ionization cross-section for K-shell excitation, which is a function of the spectrometer acceptance angle β_0 and of Δ. The ionization cross-section for K and L excitations can be modelled and computed using the SIGMAK or SIGMAL routines developed by Egerton (1979).

The second method, which is more widely used, is for situations where more than one type of atom is present in the spectrum. In this case, a ratio method is used for the elements A and B such that

$$N_A/N_B = (I_{AK}/I_{BK})(\sigma_{BK}/\sigma_{AK}) \qquad (2\text{-}20)$$

where N, I, and σ have the same meaning as in Eq. (2-19) for elements A and B.

2.5.4 Limitations to EELS Analysis

We have now considered many of the features of EELS analysis, and have discussed in particular the high collection efficiency and relatively straightforward quantitation procedures which make this technique so attractive. We will now consider some of the parameters which limit the applicability of this technique.

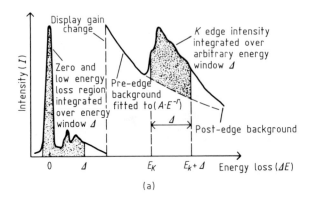

(a)

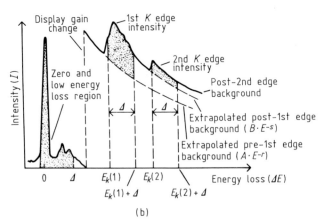

(b)

Figure 2-37. Schematic depiction of quantitation methods. (a) Absolute method, using window of width Δ at both zero-loss and elemental peaks; (b) ratio method, using windows of width Δ at two elemental peaks. From Williams (1984).

2.5.4.1 Sample Thickness

Perhaps the greatest limitation to EELS analysis is that imposed by specimen thickness. As the thickness of a specimen increases, the likelihood that an electron will undergo multiple inelastic scattering events, and thus contribute to the background rather than to a peak, increases. EELS analysis is thus limited to specimen thicknesses which are less than inelastic scattering mean free path (the distance that an electron travels on average between scattering events). The mean free path λ, is dependent upon several parameters, most importantly the scattering cross section, and is generally given as

$$\lambda = A/(N_0 \varrho Q) \tag{2-21}$$

where A and ϱ are the atomic weight and density of the sample, N_0 is Avogadro's number, and Q is the inelastic scattering cross section. Q will generally increase with increasing atomic number and decrease with increasing accelerating voltage and increasing edge energy. Thus, λ and the useable sample thickness for EELS analysis decreases as the atomic number increases or the accelerating voltage decreases. For medium atomic number materials and 100 kV electrons, the mean free path is 50–100 nm, which imposes severe sample thickness constraints for EELS analysis. Moreover, many samples are not uniformly thick, with particles or second phases often resistant to thinning during specimen preparation and hence quite thick.

Examples of the effect of sample thickness are shown in Fig. 2-38. Figure 2-38 a shows the change in the carbon K-edge as a function of foil thickness. As the thickness increases, the background intensity increases and the peak intensity decreases. This will, of course, have a disastrous affect on quantitation, as shown in Fig. 2-38 b, which shows the effect of thickness on the ratio of peak intensities from pairs of elements in various samples. The ratio changes rapidly with thickness, and thus will invalidate quantitation using Eq. (2-20). The thickness measure used in Fig. 2-38 b is the ratio of the intensity in the plasmon loss peak (I_p) to the intensity in the zero-loss peak (I_0), which was mentioned previously as a useful internal thickness measure. For EELS analysis, I_p should be $> 0.1\, I_0$, and this corresponds to a maximum thickness of near the inelastic scattering mean free path.

Ideally, samples for EELS analysis should be uniformly thin and less than 100 nm thick. It is interesting to note that the sample thickness dependence of EELS is in a sense opposite to that of EDS, where increased elemental sensitivity results from increased counts associated with increased thickness. The principle solution to the sample thickness problem in EELS is to use microscopes with increased accelerating voltages. Microscopes operating at 300 or 400 kV are capable of producing EELS spectra from proportionally thicker samples than those operating at only 100 kV.

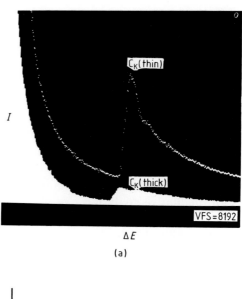

(a)

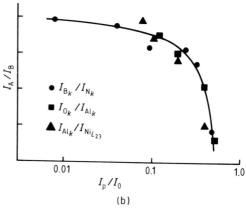

(b)

Figure 2-38. The effect of foil thickness on EELS spectra. (a) Effect on C peak from carbon film. Peak shape and peak/background ratio change (Williams, 1984). (b) Change in peak intensity ratios (I_A/I_B) for pairs of elements as a function of foil thickness. Thickness is given as ratio of plasmon loss intensity to zero-loss intensity (I_p/I_0) (Zaluzec, 1984).

2.5.4.2 Quantitation Uncertainties

A second limitation to EELS analysis is uncertainties in the values needed for quantitative analysis. We have already discussed the problems with background subtraction, peak overlap, and thickness effects on the peak intensities. In addition there can be large cross-section inaccuracies associated with the SIGMAK and SIGMAL calculations. These models in general are accurately only to about $\pm 25\%$, and so significant errors can result in quantitation.

Some significant progress has been made in recent years to solve the background/

overlap/quantitation problems, particularly in situations where minimum detectable mass is most important. Modern analysis systems make use of first and second difference spectra, and shift-averaging techniques, to reveal small peaks superimposed on large backgrounds. These techniques have been successful, when coupled with parallel recording systems, in advancing the state-of-the-art to near the theoretically predicted minimum detectability limits for qualitative analysis.

2.5.4.3 Stability Limitations

In the past, specimen and microscope stability problems were major limitations to EELS analysis. With serial recording systems, and analysis times of 100 s or longer, many factors interfered with spectrum acquisition. Drift of the spectrometer or sample could occur during analysis. Specimen contamination and carbon build-up could occur, especially during high-spatial resolution analyses with a small probe. Since EELS requires the use of very thin films, and is very sensitive to light elements, thin carbon layers are very visible in the spectrum. Mass loss from the specimen could also occur during analysis. Fortunately, all of these problems were solved by parallel recording. In fact, as mentioned above, the speed of parallel recording is now a significant advantage of EELS over EDS for beam-sensitive samples, and any sample changes can be observed dynamically in the parallel EELS spectrum.

2.6 Convergent-Beam Electron Diffraction (CBED)

One final mode of information retrieval in the analytical electron microscope is that of convergent-beam electron diffrac-

tion (CBED). In its simplest form, CBED consists of focussing a small probe of electrons onto a feature of interest and recording the electron diffraction pattern from that feature. The physics and geometry of the electron diffraction process and pattern are essentially identical to that described in Chap. 1 for more conventional selected area electron diffraction (SAED). In SAED, a small aperture is used to define an area from which the diffraction information is obtained. The minimum area associated with SAED is limited by both lens aberrations and by the construction of the aperture itself to approximately 0.2 μm in diameter. The advantage of CBED is that the only fundamental limitation to the minimum size of the area from which the information can be obtained is the size of the probe, which can be as small as 0.5 nm. Therefore, CBED has allowed the extension of electron diffraction to much smaller areas, and has become a very important tool for the analysis of crystalline materials.

In addition to these spatial resolution advantages, the CBED patterns can contain significant additional information about the sample compared with SAED patterns. We will consider in this section the formation of CBED patterns and the information available in them.

2.6.1 CBED Pattern Formation

There are a variety of ways in which a CBED pattern can be formed in the AEM, and these are described in detail in the reviews by Spence and Carpenter (1986) and Williams (1984). Some of these involve rocking the electron beam using electromagnetic coils before or after the sample, and recording the pattern using an electron detector for display on a CRT. These methods are specialized and will not be consid-

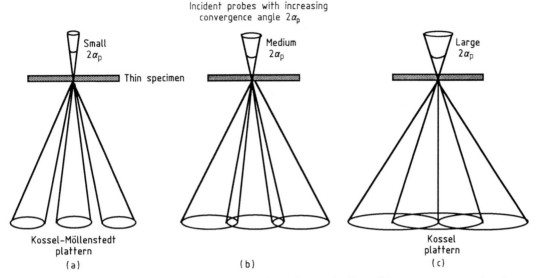

Figure 2-39. Schematic depiction of CBED pattern formation, and effect of beam convergence (condenser aperture size) on disc diameter.

ered in detail here. The results from these methods are essentially equivalent to the static-beam methods which are much more commonly used and will be described in this section.

Figure 2-39 shows the manner in which the CBED patterns are formed. A convergent electron beam is focussed on an area of interest, and a diffraction pattern is produced, which can be imaged on a fluorescent screen and recorded using photographic film. Since the electron beam in convergent, the resulting transmitted and diffracted beams are divergent, yielding a pattern consisting of discs rather than the discrete points typical of SAED patterns. The diameter of the discs is a function of the incident beam convergence, which in turn is controlled by the condenser aperture. A schematic CBED pattern is shown in Fig. 2-40. The CBED analogue to the conventional SAED pattern is labeled ZOLZ, for zero-order Laue zone. Also shown are first-order and second-order Laue zone rings (FOLZ and SOLZ, re-

spectively) which are unique to CBED patterns and will be described subsequently. The relationship between the incident beam convergence angle, α_p, the diameter of the discs c, and the spacing of the diffraction discs g is given by

$$2\alpha_p = c(2\theta_B)/g \qquad (2-22)$$

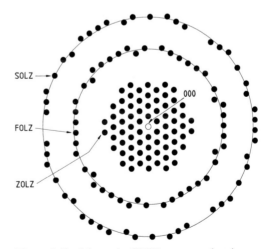

Figure 2-40. Schematic CBED pattern, showing zero-order, first-order, and second-order Laue zones (ZOLZ, FOLZ, and SOLZ, respectively).

where θ_B is the Bragg angle for the diffracting planes corresponding to the diffraction spot. As described in the previous chapter in the section on electron diffraction, Bragg's law gives $\lambda = 2\,d\sin\theta_B$, where λ is the electron wavelength and d is the spacing of the diffracting planes. Also, the relationship between d and g is $d = \lambda\,L/g$, where L is the camera length (sample to film distance). Equation 2-22 shows the direct relationship between α_s and the diameter of the diffraction discs. In the limit of very small α_p (small condenser aperture), the CBED pattern is similar in appearance to the SAED pattern.

2.6.2 Elements of the CBED Pattern

Another way of viewing the formation of any diffraction pattern is through the Ewald sphere construction, which describes the diffraction geometry and intensity as resulting from the intersection of a sphere of radius $1/\lambda$, which is attached to the incident electron beam, with a three-dimensional reciprocal lattice of the crystal, which is attached to the crystal (Chap. 1). This formulation is also useful in describing CBED patterns, and is shown schematically in Fig. 2-41. The points at which the sphere intersects the intensity points in the reciprocal lattice result in strong diffracted intensity (spots) in the electron diffraction pattern. Because the sphere is curved, and the reciprocal lattice exists in three dimensions, it is possible for the sphere to intersect points in the reciprocal lattice planes above the plane normally viewed in the two-dimensional SAED pattern (the ZOLZ plane). This leads to the FOLZ and SOLZ rings in Fig. 2-40. The major difference between the Ewald sphere interpretation in the case of CBED compared to SAED pattern is that since each CBED pattern spot is a disc of finite diameter, the CBED pattern spots show more of reciprocal space and thus contain more information that the SAED points. This, along with the higher spatial resolution associated with the focussed probe of electrons, is a key advantages of CBED over SAED.

A typical CBED pattern is shown in Fig. 2-42, which can be compared with the schematic pattern in Fig. 2-40. The pattern is conveniently divided into three distinct

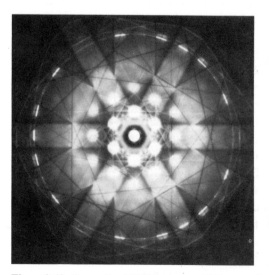

Figure 2-42. Example of CBED pattern, (111) zone of Ni (f.c.c.).

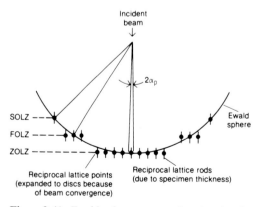

Figure 2-41. Ewald sphere construction showing formation of CBED pattern in Fig. 2-40. From Williams (1984).

regions, each containing unique information. The three regions are the transmitted beam or 000 disc; the zero-order Laue zone (ZOLZ); and the higher-order Laue zones (HOLZ). We will describe each of these in detail below.

2.6.2.1 The 000 Disc

Each intensity disc in the CBED pattern contains a wealth of information produced by elastic and inelastic scattering events within the crystal. The 000 disc contains some of the most readily-interpretable information, since the pattern symmetry is centered on this disc. Figure 2-43 shows a 000 disc from a crystalline material. Within the disc are broad, roughly circular minima and maxima bands, and these are related to the rocking curve, the plot of electron intensity versus electron beam incidence angle. Each point in the CBED disc corresponds to a different angle between the incident beam and the crystal, and intensity changes associated with diffraction are mapped out in these discs. The spacing of these minima and maxima are related to foil thickness for a given sample.

Also within the 000 disc are a series of lines which have a distinct symmetry. These are called HOLZ lines. For each dark (defect) line in the 000 disc, there is a corresponding bright (excess) line in the HOLZ rings. These lines are essentially equivalent to Kikuchi lines (see Chap. 1 in this volume) from HOLZ planes, but arise from elastic scattering only when observed within the diffraction maxima. These lines can be indexed in essentially the same way that Kikuchi lines can, and their arrangement can be used to assess crystal symmetry and lattice parameter, as described subsequently.

Three sample factors are important in controlling the visibility of CBED features

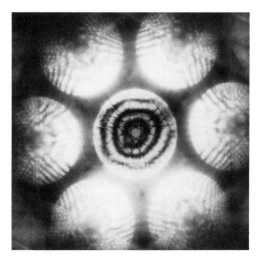

Figure 2-43. Example of HOLZ lines in 000 disc, 111 zone of Si (diamond cubic).

such as the HOLZ lines. The first of these is sample crystalline perfection. Since features such as HOLZ and Kikuchi lines are a sensitive function of the exact angular relationship between the electron beam and the sample, any defects which result in plane bending will obscure these features. In this sense, CBED patterns have an advantage over SAED patterns, since the area from which the pattern comes is smaller and hence more likely to be perfect. However, if the probe rests on a defect of some sort, the pattern will be affected. A second factor is sample thickness. Since HOLZ and Kikuchi lines are additive phenomena, a finite sample thickness is necessary for maximum visibility. However, inelastic scattering in very thick samples will obscure these features. Hence, there is an optimum range of intermediate thicknesses which are best for HOLZ and Kikuchi line observation. Finally, electron energy plays a role, HOLZ lines being most visible at lower electron voltages, in general < 200 kV.

2.6.2.2 The Zero-Order Laue Zone

The arrangement of discs in the ZOLZ is equivalent to the conventional SAED pattern. The spacings and angular relationships in this zone can be used to determine the spacings and angular relationships of atoms in the sample plane normal to the electron beam as described previously for SAED patterns, and thus the crystal structure of the sample can be deduced. All of the formulations used for SAED pattern analysis also apply to the spots in the ZOLZ. However, there is additional information available within the ZOLZ discs which is absent in the SAED patterns and which can be used for thickness and symmetry assessment, as described below.

2.6.2.3 Higher-Order Laue Zones

The HOLZs contain significant additional information which is missing in conventional SAED patterns. The diameter of the HOLZ rings gives information about the spacing of the crystalline planes in the sample which are perpendicular to the electron beam. This information is available from SAED patterns only by taking multiple patterns at a variety of sample tilts. In the CBED pattern, a single pattern can give complete three-dimensional information.

The analysis of the HOLZ spacing is done in the fashion illustrated in Fig. 2-44. The diameters of the HOLZ rings measured on the photograph of the pattern, G_1 and G_2, are related to the spacings along the electron beam direction H_1 and H_2 by (Steeds, 1979)

$$G_x = [2 H_x/(\lambda L)]^{1/2} \qquad (2\text{-}23)$$

The value of H can then be related to the unit cell parameters using the method described by Raghavan et al. (1984).

A second use of the HOLZ pattern is in the assessment of crystal symmetry. The arrangement of the intensity maxima in the HOLZ rings is an important clue to the true symmetry, as described below.

2.6.3 Spezialized Information Available in CBED Patterns

In this section, we will describe three key parameters which are available from the CBED patterns: the total crystal symmetry, lattice parameter, and the foil thickness.

2.6.3.1 3-D Crystal Symmetry

The most common method of extracting crystal symmetry information from CBED patterns involves the examination of the symmetry in the 000 HOLZ lines, the

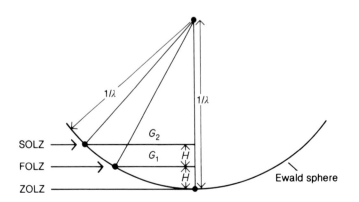

Figure 2-44. Method of measurement to get lattice spacing in beam direction from diameter of FOLZ or SOLZ. From Steeds (1979).

Table 2-9. Relationship between CBED pattern symmetries and diffraction groups (from Buxton et al., 1976).

| Diffraction group | Bright field | Whole pattern | Dark field | | $\pm G$ | | Projection diffraction group |
			General	Special	General	Special	
1	1	1	1	none	1	none	1_R
1_R	2	1	2	none	1	none	
2	2	2	1	none	2	none	21_R
2_R	1	1	1	none	2_R	none	
21_R	2	2	2	none	21_R	none	
m_R	m	1	1	m	1	m_R	$m1_R$
m	m	m	1	m	1	m	
$m1_R$	2mm	m	2	2mm	1	$m1_R$	
$2m_Rm_R$	2mm	2	1	m	2	–	$2mm1_R$
2mm	2mm	2mm	1	m	2	–	
2_Rmm_R	m	m	1	m	2_R	–	
$2mm1_R$	2mm	2mm	2	2mm	21_R	–	
4	4	4	1	none	2	none	41_R
4_R	4	2	1	none	2	none	
41_R	4	4	2	none	21_R	none	
$4m_Rm_R$	4mm	4	1	m	2	–	$4mm1_R$
4mm	4mm	4mm	1	m	2	–	
4_Rmm_R	4mm	2mm	1	m	2	–	
$4mm1_R$	4mm	4mm	2	2mm	21_R	–	
3	3	3	1	none	1	none	31_R
31_R	6	3	2	none	1	none	
$3m_R$	3m	3	1	m	1	m_R	$3m1_R$
3m	3m	3m	1	m	1	m	
$3m1_R$	6mm	3m	2	2mm	1	$m1_R$	
6	6	6	1	none	2	none	61_R
6_R	3	3	1	none	2_R	none	
61_R	6	6	2	none	21_R	none	
$6m_Rm_R$	6mm	6	1	m	2	–	$6mm1_R$
6mm	6mm	6mm	1	m	2	–	
6_Rmm_R	3m	3m	1	m	2_R	–	
$6mm1_R$	6mm	6mm	2	2mm	21_R	–	

ZOLZ spots, and the HOLZ rings. A complete description of the development of this technique is given by Gjonnes and Moody (1965), Goodman and Lempfuhl (1968), Buxton et al. (1976) and Steeds (1979). In this method, Tables 2-9 and 2-10 from Buxton et al. (1976) are used. In Table 2-9, the symmetry in the 000 disc ("Bright Field") and in the whole pattern (ZOLZ plus HOLZ) is determined and used to define the diffraction group. Ambiguities can be resolved by examining the symmetry in a dark-field disc when centered on the optic axis. "General" refers to an *hkl* disc not positioned on any of the major symmetry elements, whereas "special" corresponds to a disc on a major element such as a mirror plane. The $\pm G$ columns refer to the sym-

Table 2-10. Relationship between diffraction groups and crystal point groups (from Buxton et al., 1976).

Diffraction groups	1	$\bar{1}$	2	m	$2/m$	222	$mm2$	mmm	4	$\bar{4}$	$4/m$	422	$4mm$	$\bar{4}2m$	$4/mmm$	3	$\bar{3}$	32	$3m$	$\bar{3}m$	6	$\bar{6}$	$6/m$	622	$6mm$	$\bar{6}m2$	$6/mmm$	23	$m3$	432	$\bar{4}3m$	$m3m$
$6mm1_R$																											×					
$3m1_R$																										×						
$6mm$																									×							
$6m_Rm_R$																								×								
61_R																							×									
31_R																						×										
6																					×											
6_Rmm_R																				×												×
$3m$																			×												×	
$3m_R$																		×												×		
6_R																	×												×			
3																×												×				
$4mm1_R$															×																	×
4_Rmm_R														×																	×	
$4mm$													×																			
$4m_Rm_R$												×																		×		
41_R											×																					
4_R										×																						
4									×																							
$2mm1_R$								×							×												×		×			×
2_Rmm_R					×						×			×						×			×									
$2mm$							×																			×						
$2m_Rm_R$						×						×		×										×				×		×		
$m1_R$							×						×												×	×					×	
m				×															×			×										
m_R				×															×													
21_R					×																											
2_R			×						×	×											×											
2			×															×														
1_R		×			×			×			×				×		×			×			×				×		×			×
1	×		×	×		×	×		×	×		×	×	×		×		×	×		×	×		×	×	×		×		×	×	
	1	$\bar{1}$	2	m	$2/m$	222	$mm2$	mmm	4	$\bar{4}$	$4/m$	422	$4mm$	$\bar{4}2m$	$4/mmm$	3	$\bar{3}$	32	$3m$	$\bar{3}m$	6	$\bar{6}$	$6/m$	622	$6mm$	$\bar{6}m2$	$6/mmm$	23	$m3$	432	$\bar{4}3m$	$m3m$
																															Point groups	

metry found in pairs of $\pm g$ reflections. After the diffraction group is found, then Table 2-10 can be used to relate the diffraction group to the point group of the crystal.

The space group of the crystal can then be determined from a knowledge of the point group and by assessment of the "dynamic absences" in the CBED pattern (Steeds et al., 1978). Dynamic absences often show up in CBED discs as dark lines, and these can be used to determine the presence of screw axes or glide planes. The space group is deduced from the point group in conjunction with these additional symmetry elements.

There are several reviews of the use of CBED patterns to determine crystal symmetry that the reader is referred to, including those by Steeds and Evans (1980), Steeds (1979), and Williams (1984).

2.6.3.2 Lattice Parameter Measurements

A second specialized application of CBED patterns is the accurate measurement of lattice parameter from microareas. The normal SAED pattern, and the spacing of spots in the ZOLZ of a CBED, gives only a low-precision assessment of the lattice parameter of the sample, with an accuracy of about ± 0.005 nm. This is low accuracy compared with a technique such as X-ray diffraction, which has an accuracy close to 0.0001 nm. Use of the HOLZ lines in the 000 disc in a CBED pattern can, however, lead to lattice parameter accuracies approaching that of X-ray diffraction. This is because the position of the HOLZ

lines is a very sensitive function of the lattice parameter and the accelerating voltage. This technique requires the indexing and simulation of the HOLZ lines present in the 000 disc, which can be done by a number of available software packages (for a review of these techniques, see Williams, 1984). The simulated patterns are then compared with the experimental pattern, as shown in Fig. 2-45. Changes in the size of polygons in the pattern is the most sensitive method of pattern matching.

This method had the tremendous advantage that accurate lattice parameter information can be obtained from very small areas. However, it suffers from the disadvantage that absolute determination of the lattice parameter requires knowledge of the accelerating voltage very accurately, and very careful setup of the experimental conditions to insure reproducibility. Thus,

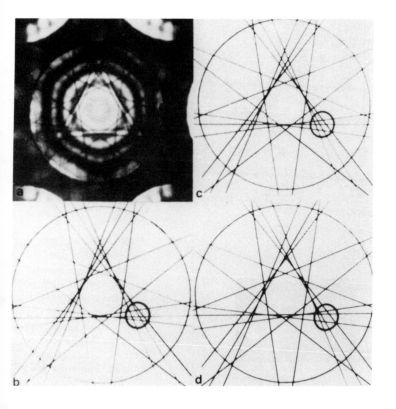

Figure 2-45. Example of precision lattice parameter measurement from HOLZ lines. (a) Experimental pattern from Udimet 720, f.c.c., (111) zone. (b)–(d) Simulated patterns for matching to (a). Lattice parameters correspond to (b) 0.355 nm; (c) 0.3556 nm; (d) 0.356 nm. Best match to (a) is (c). From Ecob et al. (1981).

the technique is better suited to observation of changes in lattice parameter, such as near a defect, within a single sample. The results can also be affected by imperfections in the sample.

2.6.3.3 Foil Thickness Determination

The final quantity available in the CBED pattern is the local foil thickness at the point of analysis. This quantity is useful for quantitative measurement of volume fractions of defects or phases, and is also needed for absorption correction in X-ray spectroscopy (Sec. 2.2.2). The method for foil thickness determination was originally developed by Kelly et al. (1975). In this method, the sample is tilted until there is only one strongly diffracting beam in the CBED pattern. In this case, fringes are seen in the diffracted beam, as shown in Fig. 2-43. The maxima and minima are due to the changes in incidence angle between the convergent electron beam and the sample, shown schematically in Fig. 2-46. As mentioned previously, the CBED discs are

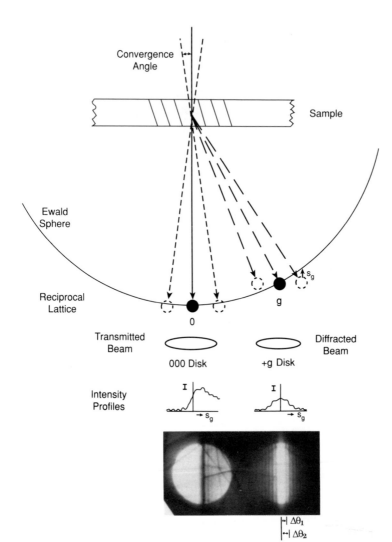

Figure 2-46. Schematic diagram showing origin of fringes used for thickness measurement in diffraction disc, and example of pattern. On the figure, "s" is the deviation parameter from the Bragg condition, and $\Delta\theta_i$ is the distance from the center maximum to each minimum.

essentially a map of the rocking curve. In the case shown in Fig. 2-46, the electrons which are parallel to the optic axis make the Bragg angle θ_B with the diffracting planes, but all of the other electrons are deviated by some amount from this angle. This deviation is described by the parameter s (Fig. 2-46). Kelly et al. showed that the relationship between the distance $\Delta\theta_i$ (Fig. 2-46) between the central extremum (maximum or minimum) and each subsequent minimum is related to s_i by

$$s_i = (\lambda/d^2)\,\Delta\theta_i/(2\,\theta_B) \qquad (2\text{-}24)$$

where d is the spacing of the diffracting planes. Thus, each CBED pattern gives several values of s_i, which are then related to foil thickness t by

$$(s_i/n_i)^2 + (1/n_i^2)(1/\xi_g^2) = (1/t)^2 \qquad (2\text{-}25)$$

where n_i is an integer and ξ_g is the extinction distance. Thus, a plot of $(s_i/n_i)^2$ versus $(1/n_i^2)$ results in a slope of $(1/\xi_g^2)$ and a y-intercept of $(1/t)^2$. The foil thickness can be determined from a single CBED pattern in this way.

This method of foil thickness determination has been shown to be accurate to about $\pm 3\%$ (Allen and Hall, 1982), which is much more accurate than most other methods. Its main limitation is that the range of thicknesses to which it applies is restricted by the visibility of the fringes, with thin samples giving too few and thick samples too closely-spaced fringes for accurate measurement. This is a particular problem for high atomic number materials (Allen, 1981).

2.7 Summary

In this chapter, we have attempted to show the strength and diversity of the analytical electron microscope as an analytical tool for the materials scientist. It is clear that an enormous amount of information about the structure, chemistry, and crystallography of a sample can be obtained in the AEM, with spatial resolution in the 0.2–10 nm range. Work is continuing at present on improvements to electron sources, optics, detectors, recording media, and computer simulation and processing of data. These advances will ultimately lead to the achievement of a goal which has been present since the advent of the AEM – that of atomic-scale imaging and microanalysis of materials.

2.8 References

Ahn, C. C., Krivanek, O. L. (1983), *EELS Atlas*. Warrendale, Pennsylvania: Gatan Inc.

Allen, S. M. (1981), *Phil. Mag. A 43*, 325.

Allen, S. M., Hall, E. L. (1982), *Phil. Mag. A 46*, 243.

Batson, P. E. (1985), *SEM 1985-Vol. 1*. Chicago: SEM Inc., p. 15.

Baumann, S. F., Williams, D. B. (1981), *J. Micros. 123*, 299.

Bethe, H. A. (1933), *Handbook of Physics, Vol. 24*. Berlin: Springer, p. 273.

Booker, G. R., Joy, D. C., Spencer, J. P., Graf von Harrach, J., Thompson, M. N. (1974), *SEM 1974, Vol. 1*. Chicago: IITRI, p. 225.

Bourdillon, A. J. (1982), *Microbeam Analysis 1982*. San Francisco: San Francisco Press, p. 84.

Bourdillon, A. J., Self, P. G., Stobbs, W. M. (1981), *Quantitative Microanalysis with High Spatial Resolution*. London: The Metals Society, p. 147.

Brown, L. M. (1977), *Developments in Electron Microscopy and Analysis*. Bristol and London: The Institute of Physics, p. 141.

Buxton, B. F., Eades, J. A., Steeds, J. W., Rackham, G. M. (1976), *Phil. Trans. Royal Soc. 281*, 181.

Cherns, D., Howie, A. (1973), *Z. für Naturforschung 28 a*, 565.

Cliff, G., Kenway, P. B. (1982), *Microbeam Analysis 1982*. San Francisco: San Francisco Press, p. 107.

Cliff, G., Lorimer, G. W. (1975), *J. Microsc. 103*, 203.

Cliff, G. Lorimer, G. W. (1981): *Quantitative Microanalysis with High Spatial Resolution*. London: The Metals Society, p. 47.

Cliff, G., Maher, D. M., Joy, D. C. (1984), *J. Micros. 133*, 255.

Cowley, J. M. (1969), *Appl. Phys. Letters 15*, 58.

Cowley, J. M. (1979), *Introduction to Analytical Electron Microscopy*. New York: Plenum Press, p. 1.

Crewe, A. (1970), *Science 168*, 3937.

Crewe, A., Wall, J. (1970), *Optik 30*, 461.

Crewe, A., Wall, J., Langmore, J. (1970), *Science 168*, 1338.

Doig, P., Flewitt, P. E. J. (1977), *J. Micros. 103*, 203.

Doig, P., Lonsdale, D., Flewitt, P. E. J. (1981), *Quantitative Microanalysis with High Spatial Resolution*. London: The Metals Society, p. 41.

Duncumb, P. (1968), *J. de Microscopie 7*, 581.

Ecob, R. C., Shaw, M. P., Porter, A. J., Ralph, B. (1981), *Phil. Mag. 44*, 1117.

Egerton, R. F. (1978), *SEM/1978/Vol. 1*. AMF O'Hare, Illinois: SEM Inc., p. 133.

Egerton, R. F. (1979), *Ultramicroscopy 4*, 169.

Egerton, R. F. (1986), *Electron Energy Loss Spectroscopy in the Electron Microscope*. New York: Plenum Press.

Fraser, H. L., McCarthy, J. P. (1982), *Microbeam Analysis 1982*. San Francisco: San Francisco Press, p. 93.

Geiger, J., Nolting, M., Schröder, B. (1970), *Electron Microscopy – 1970*. Paris: Soc. Fran. de Micro. Elect., p. 111.

Geiss, R. H. (1979), *Introduction to Analytical Electron Microscopy*. New York: Plenum Press, p. 43.

Geiss, R. H., Romig, A. D. (1986), *Principles of Analytical Electron Microscopy*. New York: Plenum Press, p. 29.

Gjonnes, J., Moodie, A. F. (1965), *Acta Cryst. 19*, 65.

Goldstein, J. I., Costley, J. L., Lorimer, G. W., Reed, S. J. B. (1977), *SEM 1977*, Vol. 1. Chicago: IITRI, p. 315.

Goldstein, J. I., Newbury, D. E., Echlin, P., Joy, D. C., Fiori, C., Lifshin, E. (1981), *Scanning Electron Microscopy and X-ray Microanalysis*. New York: Plenum Press, p. 53.

Goodman, P., Lempfuhl, G., (1968), *Acta Cryst. A 24*, 339.

Hall, E. L. (1990), *unpublished research*, GE Corp. Res. and Dev., Schenectady, New York.

Hall, E. L., Briant, C. L. (1984), *Met. Trans. A 14 A*, 1549.

Hall, E. L., Imeson, D., Vandersande, J. B. (1981), *Phil. Mag. A 43*, 1569.

Henke, B. L., Ebisu, E. S. (1974), *Advances in X-ray Analysis Vol. 17*. New York: Plenum Press, p. 150.

Hren, J. J., Goldstein, J. I., Joy, D. C. (Eds.) (1979), *Introduction to Analytical Electron Microscopy*. New York: Plenum Press.

Hillier, J., Baker, R. F. (1944), *J. Appl. Phys. 15*, 663.

Ichinokawa, H. (1982), *private communication*. Japan: Waseda University.

Isaacson, M. (1978), *SEM/1978/Vol. 1*. AMF O'Hare, Illinois: SEM Inc., p. 763.

Isaacson, M., Johnson, D. (1975), *Ultramicroscopy 1*, 33.

Isaacson, M., Ohtsuki, M., Utlaut, M. (1979), *Introduction to Analytical Electron Microscopy*. New York: Plenum Press, p. 343.

Joy, D. C., Romig, A. D., Goldstein, J. I. (Eds.) (1986), *Principles of Analytical Electron Microscopy*. New York: Plenum Press.

Kelly, P. M., Jostsons, A., Blake, R. G., Napier, J. G. (1975), *Phys. Stat. Sol. A 31*, 771.

Koch, E. F., Hall, E. L., Yang, S. W. (1983), *Proc. 41st Annual Meeting EMSA*. San Francisco: San Francisco Press, p. 250.

Kouh, Y. M., Hall, E. L. (1982), *General Electric Technical Information Series, Report No. 82CRD156*, Schenectady, New York.

Kramers, H. A. (1923), *Phil. Mag. 46*, 836.

Krishnan, K. M., Echer, C. J. (1987), *Analytical Electron Microscopy 1987*. San Francisco: San Francisco Press.

Krivanek, O. L., Ahn, C. C., Keeney, R. B. (1987), *Ultramicroscopy 22*, 103.

Kyser, D. F. (1979), *Introduction to Analytical Electron Microscopy*. New York: Plenum Press, p. 199.

Leapman, R. D., Ornberg, R. L. (1988), *Ultramicroscopy 24*, 251.

Leapman, R. D., Rez, P., Mayers, D. F. (1978), *Proc. 9th Int. Cong. on Elect. Mic.*, Toronto: Imperial Press, p. 526.

Lyman, C. E., Stenger, H. G., Michael, J. R. (1987), *Ultramicroscopy 22*, 129.

Maher, D., Joy, D. C. (1976), *Ultramicroscopy 1*, 239.

Mehta, S., Goldstein, J. I., Williams, D. B., Romig, A. D. (1979), *Microbeam Analysis 1979*. San Francisco: San Francisco Press, p. 119.

Michael, J. R., Williams, D. B., Klein, C. F., Ayer, R. (1990), *J. Micros. 160*, 41.

Newbury, D. E. (1986), *Principles of Analytical Electron Microscopy*. New York: Plenum Press, p. 1.

Newbury, D. E., Myklebust, R. L., (1982), *Ultramicroscopy 3*, 391.

Nockolds, C., Nasir, M. J., Cliff, G., Lorimer, G. W. (1980), *Electron Microscopy and Analysis 1979*. Bristol and London: The Institute of Physics, p. 417.

Orloff, J. (1989), *Ultramicroscopy 28*, 88.

Otten, M. T. (1989), *Philips Electron Optics Bulletin 126*, 21.

Pejas, W., Rose, H. (1978), *Electron Microscopy 1978*. Toronto: Micro. Soc. of Canada, p. 44.

Pennycook, S. J. (1989), *Ultramicroscopy 30*, 58.

Pennycook, S. J., Boatner, L. A. (1988), *Nature 336*, 565.

Poutney, J. M., Loretto, M. H. (1980), *Electron Microscopy 1980*. Leiden: 7th European Congress on Electron Microscopy Foundation, p. 180.

Raghavan, M., Scanlon, J. C., Steeds, J. W. (1984), *Met. Trans. A 15 A*, 1299.

Reed, S. M. (1982), *Ultramicroscopy 7*, 405.

Reese, G. M., Spence, J. C. H. (1984), *Phil. Mag. A 49*, 697.

Reimer, L., From, I., Rennekamp, R. (1988), *Ultramicroscopy 24*, 339.

Reimer, L., Rennekamp, R., Bakenfelder, A. (1989), *Proc. 47th Annual Meeting EMSA*. San Francisco: San Francisco Press, p. 412.

Romig, A. D., Goldstein, J. I. (1979), *Microbeam Analysis 1979*. San Francisco: San Francisco Press, p. 124.

Schreiber, T. P., Wims, A. M., (1981), *Ultramicroscopy 6*, 323.

Spence, J. C. H., Carpenter, R. W. (1986), *Principles of Analytical Electron Microscopy*. New York: Plenum Press, p. 301.

Spence, J. C. H., Tafto, J. (1983), *J. Microsc. 130*, 147.

Steeds, J. W. (1979), *Introduction to Analytical Electron Microscopy*. New York: Plenum Press, p. 387.

Steeds, J. W., Evans, N. S. (1980), *Proc. 38th Annual Meeting EMSA*. Baton Rouge, Los Angeles: Claitors, p. 188.

Steeds, J. W., Rackham, G. M., Shannon, M. D. (1978), *Developments in Electron Microscopy and Analysis*. London: Academic Press, p. 351.

Thomas, L. E. (1985), *Ultramicroscopy 18*, 173.

Thompson, M. N., Doig, P., Edington, J. W., Flewitt, P. E. J. (1977), *Phil. Mag. 35*, 1537.

von Heimendahl, M. (1980), *Electron Microscopy of Materials*. New York: Academic Press, p. 74.

Williams, D. B. (1984), *Practical Analytical Electron Microscopy in Materials Science*. Mahwah, New Jersey: Philips Electronic Instruments, Inc.

Williams, D. B., Edington, J. W. (1976), *J. Microsc. 108*, 113.

Wittry, D. B., Ferrier, R. P., Cosslett, V. E. (1969), *Brit. J. Appl. Phys. 2*, 1967.

Yanaka, T., Moriyama, K., Buchanan, R. (1989), *Mat. Res. Soc. Symp. Proc. 139*, 271.

Zaluzec, N. J. (1979), *Introduction to Analytical Electron Microscopy*. New York: Plenum Press, p. 121.

Zaluzec, N. J. (1984), in Williams, D. B.: *Practical Analytical Electron Microscopy in Materials Science*. Mahwah, New Jersey: Philips Electronic Instruments, Inc.

Zaluzec, N. J. (1987), *Library of Electron Energy Loss Spectra*. Available from the author (Argonne National Lab., Argonne, Illinois), or from EDAX, Mahwah, New Jersey.

General Reading

Egerton, R. F. (1986), *Electron Energy Loss Spectroscopy in the Electron Microscope*. New York: Plenum Press.

Goldstein, J. I., Newbury, D. E., Echlin, P., Joy, D. C., Fiori, C., Lifshin, E. (1981), *Scanning Electron Microscopy and X-Ray Microanalysis*. New York: Plenum Press.

Hren, J. J., Goldstein, J. I., Joy, D. C. (Eds.) (1979), *Introduction to Analytical Electron Microscopy*. New York: Plenum Press.

Joy, D. C., Romig, A. D., Goldstein, J. I. (Eds.) (1986), *Principles of Analytical Electron Microscopy*. New York: Plenum Press.

Williams, D. B. (1984), *Practical Analytical Electron Microscopy in Materials Science*. Mahwah (NJ): Philips Electronic Instruments, Inc.

3 Scanning Electron Microscopy

David C. Joy

University of Tennessee, Knoxville, TN, U.S.A.

List of Symbols and Abbreviations

A	atomic weight
C	contrast level
C_s	spherical aberration coefficient
D_f	depth of field
d	probe diameter
d_0	beam diameter
d_{hkl}	spacing of lattice planes $(h\,k\,l)$
E	incident beam energy
E_D	energy density
e_{eh}	energy of creation for an electron-hole pair
I	beam current
I_b	incident beam current
I_s	current flowing in the ground connection
M	demagnification of the lens
R_{BS}	escape depth and lateral spread
S	average signal level
Z	atomic number

α	beam convergence angle
β	brightness
δ	number of secondary electrons/number of incident electrons
θ	incident angle
λ	wavelength
ϱ	density
τ	recording time

BS	backscattered electron
CRT	cathode ray tube
EBIC	electron beam induced conductivity
ECP	electron channeling patterns
SACP	selected area channeling pattern
SE	secondary electron
SEM	scanning electron microscope

3.1 Introduction and Historical Summary

The scanning electron microscope (SEM) is the most widely used form of electron microscope in the field of the materials sciences. More than 10000 SEMs are installed worldwide and two new instruments are delivered each day. The SEM is popular because it uniquely combines some of the simplicity and ease of specimen preparation of the optical microscope with much of the performance capability and flexibility of the more expensive and complex transmission electron microscope.

The SEM was originally devised in Germany in the 1930's by Knoll and von Ardenne and can be considered as having had its origins in the early electrical facsimile machines then under development. Later, important developments were made by Zworykin, Hillier and Snyder at the RCA Research Laboratories in the USA in the 1940's. The design and performance of their instrument anticipated much that was found in later microscopes, but their success was ultimately limited by the poor vacuum conditions under which they had to work. The current form of the instrument is the result of the work of Oatley and his students at Cambridge University between 1948 and 1965. For a comprehensive historical survey see Oatley (1982). The first commercial SEM, the Cambridge "Stereoscan" was produced by Cambridge Instruments in the UK in 1965. Today nearly a dozen companies now manufacture scanning microscopes for the international market with instruments with prices varying from US $ 40000 to in excess of US $ 500000.

3.2 Principles of the Scanning Electron Microscope

The principle of the SEM is shown in Fig. 3-1. Two electron beams are used simultaneously. One strikes the specimen to be examined, the other strikes a cathode ray tube (CRT) viewed by the operator. As a result of the impact of the incident beam on the specimen a variety of electron and photon emissions are produced. The chosen signal is collected, detected, amplified and used to modulate the brightness of the second electron beam so that a big collected signal produces a bright spot on the CRT while a small signal produces a dimmer spot. The two beams are scanned synchronously so that for every point scanned on the specimen there is a corresponding point on the CRT. Typically the beams scan square patterns on both the specimen

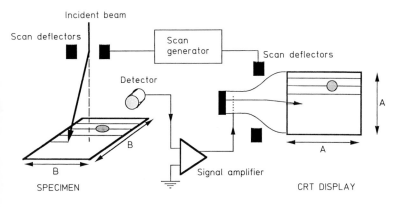

Figure 3-1. Schematic diagram showing the basic principles of the Scanning Electron Microscope (SEM).

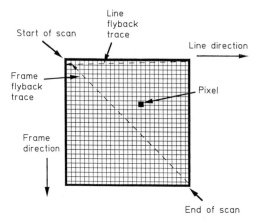

Figure 3-2. The SEM scan raster arrangement.

and the CRT. They start (Fig. 3-2) at the top left hand corner of the area, scan a *line* of points parallel to the top edge and then, when they reach the end of the line, they flyback to the starting edge and scan a second line and so on until the whole square area has been *rastered*. Each complete image is conventionally called a *frame*. If the display area of the CRT tube is $A \times A$ in size and the area scanned on the specimen is $B \times B$ in size, then variations in the signal from the specimen will be mapped on to the CRT as variations in brightness with a linear magnification of A/B. Thus a magnified map or image of the specimen is produced without the need for any imaging lenses.

This method of imaging offers several important advantages:

(1) Magnification is achieved in a purely geometric manner and can be varied by simply changing the dimensions of the area scanned on the specimen. It is therefore easy to change from a low ($\times 20$) to a high ($\times 100\,000$) magnification or to any intermediate value since it is not necessary to change any lenses or even to refocus the image.

(2) Any emission which can be stimulated from the specimen under the impact of the incident electron beam – e.g., electrons, X-rays, visible photons, heat, or sound – can be collected, detected and used to form an image. The SEM is therefore not restricted to imaging with radiations which can be focussed by lenses and can give many different views of the same specimen.

(3) Several different types of image can be produced and displayed simultaneously from the same area of the sample so enabling different types of information to be correlated. It is only necessary to provide a suitable detector, amplifier, and display screen for each signal of interest. Furthermore these signals can be mixed with each other to generate still more types of imaging information.

(4) Because the picture on the screen is formed from an electrical signal which varies with the position of the beam and hence with time the image can be electronically processed to control or enhance contrast, reduce noise, identify features etc.

A consequence of this arrangement is that a fundamental limit to the imaging performance is set by the display CRT screen. The smallest feature that can be discerned on the CRT is equal to the size of the electron spot on the display screen. Conventionally it is assumed that 1000 scan lines, each containing 1000 picture elements or *pixels*, make up each image frame scanned. Each picture is therefore composed of 1000×1000 i.e. 10^6 pixels. When the SEM is operating at a magnification of M then the resolution in the image i.e. the smallest detail on the specimen that can be observed, is equal to the pixel size divided by the magnification. Since the size of the spot on the CRT is typically 100 to 200 micrometers then for magnifications of a few hundred times the resolution is limited to a micrometer or so. Only at high magnifications is the resolution limited by

more fundamental electron-optical considerations. For a more detailed discussion
see Goldstein et al. (1981).

3.3 Components of the Scanning Electron Microscope

The main components of an SEM are
contained in two units, the *electron column*
which contains the electron beam scanning
the specimen, and the *display console*
which contains the second electron beam
which impinges on the CRT. The high energy electron beam incident on the specimen is generated by an electron *gun*, two
basic types of which are in current use. The
first (Fig. 3-3 a) is the thermionic gun in
which electrons are obtained by heating a
tungsten or lanthanum hexaboride cathode
of filament to between 1500 and 3000 K.
The cathode is held negative at the required accelerating voltage E_0 with respect
to the grounded anode of the gun so that
the negatively charged electrons are accelerated from the cathode and leave the anode with an energy E_0 kilo-electron volts
(keV). Thermionic guns are in very wide
use because they may safely by run in
vacuums of 10^{-5} Pa (i.e., 10^{-7} torr) or
worse. The alternative source (Fig. 3-3 b) is
the field emission gun in which a sharply
pointed wire of tungsten is held close to an
extraction anode to which is applied a potential of several thousand volts. Electrons
tunnel out of the tungsten wire, which need
not be heated and can be at room temperature, into the vacuum and are then accelerated as in the thermionic gun towards
the anode. Field emission guns depend on
an atomically clean emitter surface, thus
they must be operated under ultra-high
vacuum conditions, typically in a vacuum
of 10^{-7} Pa (i.e., 10^{-9} torr) or better. For
either emitter the entire length of the elec-

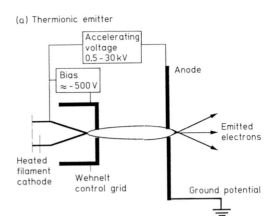

(a) Thermionic emitter

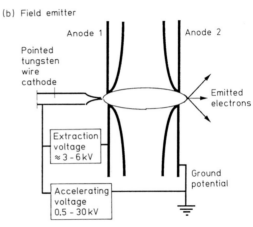

(b) Field emitter

Figure 3-3. (a) Schematic diagram of a thermionic
electron gun for the SEM. (b) Schematic diagram of
field emission electron gun for the SEM.

tron column traveled by the electron beam
from the gun to the specimen chamber
must also be pumped to an adequate vacuum using oil-diffusion, turbo-molecular
or ion pumps individually or in combination.

3.4 Performance Limits of the Scanning Electron Microscope

The performance of the SEM depends
on a number of related factors, perhaps the
most important of which is the output of

the electron source. The source is quantified by its brightness, β, which is the current density $(A\,m^{-2})$ it delivers into unit solid angle (steradian). The brightness increases linearly with the accelerating voltage of the microscope, but also varies greatly from one type of source to another. At a given energy a field emission gun is between ten and one hundred times as bright as an LaB_6 thermionic emitter which is, in turn, between three and ten times brighter than a tungsten thermionic emitter. Thus at 20 keV a field emission gun has a brightness in excess of $10^{12}\,A\,m^{-2}/sr$.

The diameter of the beam of electron from the gun is reduced or demagnified by passing it through two or more lenses before it reaches the sample surface (Fig. 3-4). An electron lens consists of a coil of wire carrying a current. It focuses the electron beam in exactly the same way as a glass lens focuses light, but it has the convenient property that the focal length can be varied by changing the magnitude of the current flowing through the coil. The electron

probe diameter d at the specimen is given by:

$$d = s\,M_1\,M_2 \qquad (3\text{-}1)$$

where s is the effective diameter of the electron source (about 20 to 50 μm for a thermionic source and 10 nm for a field emission source) and M_1 and M_2 are the demagnifications of the lenses. By varying the excitations of the lenses the beam diameter at the specimen can be set to any desired value from the source size downward.

The beam is usually scanned across the sample in two stages as shown. The electrons are initially deflected across the optical axis in one direction, then immediately deflected in the opposite sense through twice the angle. This arrangement ensures that all of the scanned rays pass through a single point making it possible to place an aperture between the specimen chamber and the electron-optical column to define the beam convergence angle α (where α = diameter of aperture/2 × working distance).

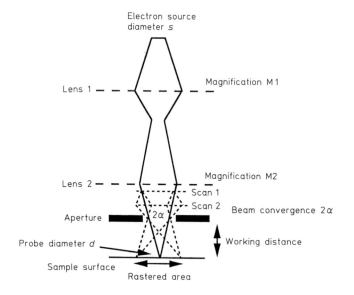

Figure 3-4. Simplified ray diagram for an SEM including the effects of demagnification and scanning.

The brightness, defined above, is constant throughout the electron optical system, and hence the value β measured for the focussed probe of electrons impinging on the specimen is the same the value that would be measured at the source. If the probe diameter is d, if the incident beam current is I_B, and if the convergence angle of the beam is α then from our definition the brightness β at the sample, and hence at the gun, is:

$$\beta = I_B/[(\pi d^2/4)(\pi \alpha^2)] \tag{3-2}$$

For typical operation α is fixed, and β is constant, thus;

$$I_B = (4\alpha^2 \beta/\pi^2) d^2 \tag{3-3}$$

which shows that as the diameter d of the probe is made smaller the current contained in the probe falls as d^2. The spatial resolution of the SEM, that is the smallest detail on the sample that can be observed, cannot be significantly less than the probe diameter d. The brightness equation [Eq. (3-2)] therefore indicates that the available source brightness will set a limit to the resolution of the microscope because, as discussed below, a certain minimum incident current is required to form an image.

There are also other limits to the resolution that can be achieved. In particular electron-optical lenses are not perfect but suffer from a variety of defects or *aberrations* (Joy et al., 1986). Firstly there are aberrations in the lenses that increase the probe diameter above the value predicted by Eq. (3-1). For a beam convergence angle α, defined from Fig. 3-4, the actual probe diameter d_0 is given as:

$$d_0^2 = d^2 + (0.5\,C_s\,\alpha^3)^2 + (\lambda/\alpha)^2 \tag{3-4}$$

where C_s is the spherical aberration coefficient of the lens and λ is the wavelength of the electrons equal to $1.226\,(E^{-1/2})$ nano-

meters where E is the incident energy of the beam in electron volts. The effect of these aberrations is to increase the diameter of the probe while reducing the amount of current that it contains. It is evident from Eq. (3-4) that both very large, and very small, values of α will lead to a probe whose size is dominated by aberrations. The value of α must therefore be chosen to maximize in some way the performance of the SEM. The most useful optimization is that which places the maximum current into a probe of a given diameter. From Eqs. (3-3) and (3-4) it can be shown that this maximum current I occurs for a value of α equal to $(d/C_s)^{1/3}$ which is an angle of the order of 5×10^{-3} radian i.e. about 0.1 of a degree. This angle is very small because electron-topical lenses have unacceptable aberrations for larger convergence angles. They are thus inefficient when compared to the glass lenses used for light microscopy where the maximum convergence angle can approach 1 radian. The maximum current I is then given by the relation

$$I = 1.88\,\beta\,\frac{d^{8/3}}{C_s^{2/3}} \tag{3-5}$$

where as before β is the brightness of the electron source. Since, for a modern SEM, the spherical aberration coefficient C_s is about 10 mm Eq. (3-5) shows that for a probe size of the order of 10 nm (100 Å) the beam will contain a current of between 10^{-12} and 10^{-9} A depending on the actual brightness of the electron gun.

The depth of field D_f of the image, defined as the vertical focussing range outside of which the image resolution is visibly degraded, is given as:

$$D_f = \frac{\text{pixel size at specimen}}{\alpha} \tag{3-6}$$

Since α is between 10^{-3} and 10^{-2} radians then the depth of field is typically several hundred times the pixel size. At low magnifications therefore the depth of field can be of the order of several millimeters giving the SEM an unrivaled ability to image complex surface topography and to produce images with a pronouned three-dimensional quality to them. This is demonstrated in the image of crystals of tungsten shown in Fig. 3-5. Here the vertical depth of the image is comparable with the lateral field of view, yet all of the features remain sharply in focus. It should be remembered, however, that this fortunate effect is the result of the fact that electron-optical lenses must be *stopped down* to very small apertures in order to work, and that the accompanying result of this is that very few of the electrons leaving the source actually reach the specimen i.e. the SEM is inefficient at using its radiation. An exact optical analogy would be the *pinhole* camera which also has a large depth of field but requires long period exposures.

Figure 3-5. Crystals of tungsten formed on the cold part of a lamp filament from material evaporated from a hot spot. Image courtesy Robertson, C. D.

3.5 Signal to Noise Limitations in the Scanning Electron Microscope Image

The image in the SEM is built up from an electrical signal which is varying with time. If the average signal level is S, and if as the beam scans across some feature the signal changes by some amount ∂S then the feature is said to have a "contrast" level C given by

$$C = \partial S / S \qquad (3\text{-}7)$$

Changes in the signal can also occur because of statistical fluctuations in the incident beam current and in the efficiency with which the various emission processes take place within the specimen. Thus repeated measurements across the same feature on a specimen will give signal intensities which will vary randomly around some mean value. These inherent statistical variations constitute a *noise* contribution to the image which therefore has a finite "signal to noise" ratio. For image information to be visible the magnitude of the signal change ∂S occurring at the specimen must exceed the magnitude of the random fluctuations by a factor of 5 times or so (Rose, 1948). This leads to the concept of threshold current I_{TH} which is the minimum incident beam current required to observe a feature of contrast level C. I_{TH} is given by the relationship:

$$I_{TH} = \frac{4 \times 10^{-12}}{C^2 \tau} \text{ (A)} \qquad (3\text{-}8)$$

where τ is the time in seconds required to record one frame of the image (assumed to contain 10^6 pixels). The observation of low contrast features therefore needs high beam currents or long exposure times. Typical visual images are produced in 1 second or less but images to be photographed are recorded for 30 to 100 seconds to improve the signal to noise ratio. The threshold current requirement sets a fundamental limit to the performance of the SEM in all modes of operation. A comparison of Eq. (3-8) with Eq. (3-5) shows that because the beam current varies rapidly with the beam diameter the resolution limit of the SEM will depend on the value of d_0 at which the current falls to I_{TH} for the feature of interest. SEMs with higher brightness guns (large β) or high quality lenses (small C_s) will perform the best.

3.6 Sample Preparation

Because the SEM chamber can readily accommodate big samples, and because the beam can focus samples which are rough or irregularly shaped, the preparation of specimens for SEM examination is relatively straightforward. In an ideal world no prior preparation at all would be necessary, but in practice it is always desirable to ensure that the surface(s) to be examined are free from oil or grease, such as that carried by the fingers, which can cause contamination. Samples in which water plays a substantial role, such as animal or plant tissue, cannot be examined directly because the evaporation of the water into the vacuum system of the microscope will lead to extensive shrinkage of the sample and even to total structural collapse. Instead the water must be removed by substituting it with alcohol and the structure then stabilized chemically. Alternatively the water can be *cryo-fixed* by rapid freezing to, and subsequent maintenance at, liquid nitrogen temperatures.

If the specimen is not a good electrical conductor then it is often desirable to provide some conductivity by evaporating a thin metal layer (typically 3 to 10 nm of gold) on to the surface of interest and electrically grounding this. The sample can then be stably examined at any desired beam energy. The drawback with this procedure is that real surface detail may be obscured by the film, and artefactual detail may be introduced. The current trend is therefore to try and operate at an incident beam energy, E_2, where the sample is in an dynamic charge balance (see below). Since this implies low voltage operation however this solution is not available if a high beam energy, for example for X-ray microanalysis, must be employed. An interesting alternative is to use an *environmental* SEM (Lane, 1970) in which, by means of a differentially pumped vacuum system, the specimen can be examined while being held in moist, low pressure, air. The water vapor drains surface charge from the specimen making it possible to observe uncoated insulators stably even at high beam energies. Figure 3-6 shows this technique in use to observe the interface between a sticky peel-off label and its backing strip. By using a pressure of 3 torr in the vicinity of the specimen the uncoated sample could be examined at 20 keV without any sign of charging.

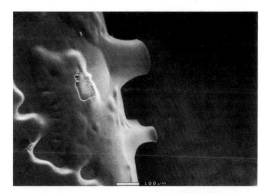

Figure 3-6. The interface of a peel-off label and its backing strip, as the two are being pulled apart. The sample is uncoated and is being observed in an environmental SEM at a pressure of 3 torr. The beam energy is 20 keV. Image courtesy Robertson, C. D.

3.7 Electron-Solid Interactions and Imaging Modes

The interaction of the electron beam with a solid specimen produces a wide variety of emissions, all of which are potentially useful for imaging (Fig. 3-7). Each of these signals is produced with a different efficiency, comes from a different volume of the sample, and carries different information about the specimen. Figure 3-8 shows a schematic plot of the energy distribution of the electrons produced by an incident beam of energy E_0. It can be seen that the distribution displays two peaks, one at an energy close to that of the incident beam and a second peak at a much lower energy. The high energy peak is made up of electrons which have been *backscattered* by the sample – that is it consists of incident electrons which have been scattered through such large angles within the sample that they emerge traveling in the opposite direction to the one from which they arrived. The fractional yield of backscattered electrons i.e. number of backscattered electrons/number of incident electrons, is called the backscatter yield and usually identified with the Greek letter η. The average energy of these electrons is about 0.5 to 0.6 of the incident energy E_0 and the backscatter yield η is typically 0.2 to 0.4.

The lower energy peak is made up of electrons produced by a variety of inelastic scattering processes within the sample rather than by backscattered incident electrons. The majority of these electrons, forming what is usually called the *secondary* electron signal, lie within the energy range 0 to 50 eV. As before we can define a relative secondary yield i.e. number of secondary electrons/number of incident electrons, identified by the Greek letter δ. The average energy of the secondary electrons is about 4 eV, independent of the incident beam energy, but the secondary yield δ varies rapidly with accelerating energy being of the order of 0.1 or less for most materials at 30 keV, but of the order of unity for energies of one or two keV. Because the secondaries are low in energy they cannot travel more than a few nanometers through the sample to reach the surface and escape, thus the secondary signal is surface specific coming only from the *escape depth* region beneath the surface.

The total current flowing into and out of the specimen must balance to zero thus:

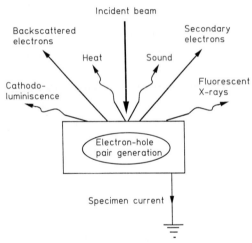

Figure 3-7. The available imaging modes on the SEM.

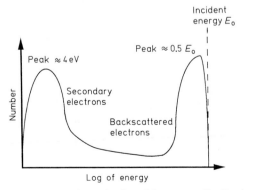

Figure 3-8. Schematic plot of the energy distribution of electrons emitted from a specimen in the SEM.

$$- I_\text{b} + \delta I_\text{b} + \eta I_\text{b} + I_\text{SE} = 0 \qquad (3\text{-}9)$$

where I_b is the incident beam current and I_{SE} is the current flowing in the ground connection to the sample. This *specimen current* contains information about both the secondary and backscatter signals and can form the basis of an important imaging mode. It can be seen from Eq. (3-9) that if $(\delta + \eta)$ is unity then no current flows to earth. At this condition the incident electron beam is neither injecting charge into the specimen, or extracting charge from it. When the sample being examined is an electrical conductor this situation is not of much significance except for studies of integrated circuits using the voltage contrast technique discussed below, but when the specimen is not a good electrical conductor then no specimen current can flow to earth, so any excess (or deficit) charge is retained in the specimen. At high incident electron energies (greater than a few keV) the total yield $(\delta + \eta)$ is less than unity, so charge is injected into the specimen which therefore charges negatively. This reduces the effective incident energy of the beam so $(\delta + \eta)$ increases, but charging will continue until the effective incident energy reaches a value E_2 at which $(\delta + \eta)$ becomes unity. At this energy each incident electron produces on average one exiting electron so charge balance is obtained, and no further charge is deposited and the specimen potential stabilizes. At this E_2 energy therefore it is possible to form an image from insulating or poorly conducting materials.

Table 3-1 gives E_2 values for some materials of interest (Joy, 1989).

Thus typically E_2 is of the order of a few keV. Because many materials of current interest (polymers, ceramics, composites) are poor electrical conductors modern SEMs are optimized to operate in this low energy range so that samples can be examined without the need to make them electrically conductive by the application of a thick metallic coating.

3.7.1 Secondary Electron Imaging Modes

Secondary electrons are the most popular choice of interaction with which to form an image. There are two main reasons for this:

(1) Because they are low in energy it is possible to collect a high fraction of all of the secondaries produced by the sample by biasing the detector to a modest positive potential so that it attracts electrons to itself. Efficient collection is therefore possible even when the detector is not in line of sight of the sample.

(2) The main contrast mechanism associated with secondary electrons produces images which are readily interpretable by analogy with reflected light images in the macroscopic world.

Figure 3-9 shows a schematic drawing of the detector, first discussed by Everhart and Thornley, which is now standard for secondary electron detection. The detector is based on a disc of scintillator which emits light under the impact of electrons. The light travels along a light pipe, through a vacuum window, and into a photomultiplier. Because the amount of light produced by the scintillator depends directly on the energy of the electron which strikes it, secondary electrons which have only a few eV of energy would produce only a very little signal. To increase the efficiency

Table 3-1. E_2 values for some materials of interest.

Material	E_2 (keV)
Photo-resist	0.6
Amorphous carbon	0.8
Teflon	1.9
Quartz	3.0
Alumina	3.5

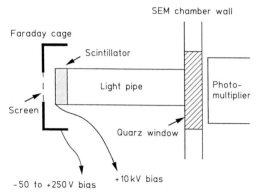

Figure 3-9. The Everhart-Thornley detector for secondary electrons.

therefore a bias of $+ 10 \, \text{kV}$ is applied to the front face of the scintillator so as to accelerate all incoming electrons to at least 10 keV energy before impact. This bias can however deflect the incident beam so to prevent this the scintillator is screened by a Faraday cage, made of wide-spaced mesh, held at about $+ 200 \, \text{V}$ relative to the sample. The field produced by this is sufficient to ensure that 50% or more of the secondaries emitted from the specimen are collected. The Everhart-Thornley detector satisfies all of the requirements for an imaging detector:

(1) High sensitivity – capable of imaging with currents below 10^{-14} A.

(2) Wide dynamic range – the logarithmic characteristic of the photomultiplier permits satisfactory operation for detected currents over 7 or 8 orders of magnitude.

(3) Good bandwidth – the ability of the detector to respond to signals changing with time is limited by the decay of the Scintillator and the behavior of the photomultiplier. Typically operation is possible with scan rates up to TV speeds.

(4) Stable to vacuum conditions – the Everhart-Thornley detector can survive repeated exposure to atmosphere without any degradation of properties or variation in gain or sensitivity.

3.7.1.1 Topographic Contrast

The dominant imaging mechanism for secondary electrons is *topographic contrast*. It has been estimated that in excess of 90% of all scanning micrographs rely on this mode. The effect arises because an increase in θ the angle of incidence between the beam and the surface normal will lead to an increase in the yield of secondary electrons as shown in Fig. 3-10. This can be understood by noting that as θ is increased the fraction of secondary electrons produced within the escape region of the surface also increases. The secondary signal $I(\theta)$ at an angle of incidence θ is related to the signal $I(0)$ at normal incidence by the approximate relationship:

$$I(\theta) = I(0) \sec \theta \tag{3-10}$$

If an electron beam moves over a surface which has topography then the local angle of incidence between the beam and the surface normal will change and produce a corresponding change in the secondary signal. As shown in Figs. 3-11 and 3-12 when the sample has a rough surface or significant surface topography then this results in an image containing pronounced light and shadow effects. What is, perhaps, surprising is the ease with which images of this

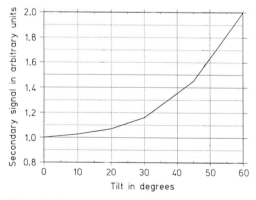

Figure 3-10. Variation of yield of secondary electrons with angle of incident beam.

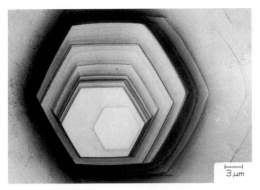

Figure 3-11. Secondary electron image of graphite flake subjected to oxygen and heat showing the formation of hexagonal etch pits. Beam energy is 25 keV. Image courtesy Robertson, C. D.

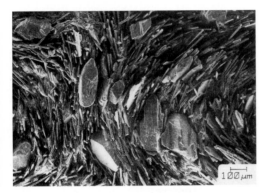

Figure 3-12. Secondary electron image of plasma etched polymer revealing the mica filler. The mica crystals and the flow of the extruded polymer can be seen. Micrograph recorded at 20 keV. Courtesy Posta, M. L.

type can be understood and interpreted. The reason can be understood by the application of the so-called *Principal of Reciprocity* which states that in any optical or electron-optical device if the illumination source and detector are interchanged than the picture produced will remain the same. This is simply a consequence of that fact that the properties of light or electrons are independent of the direction in which they are traveling. In the SEM the specimen is illuminated from above (i.e., the gun) by a beam of electrons and the signal is collected from the detector placed to one side of the specimen and normal to the line joining the specimen to the source. By reciprocity the signal is thus the same as if the detector was placed above the specimen, and the source was placed at the detector. By extension we can also replace our electron source by a light source and our detector by a light detector e.g., human eyes. In this case we know from Lamberts' cosine law that a surface will reflect into our detector (i.e., the eyes) a light signal which varies as the cosine of the angle made with the light source – so that faces at glancing incidence to the light will reflect only a small signal, while faces almost normal to the light source will reflect a large signal i.e. if the angle between the surface and the light source is φ then

$$I(\varphi) = I(0)\cos\varphi \qquad (3\text{-}11)$$

But $\varphi = 90° - \theta$ so an application of the principle of reciprocity and a comparison of Eqs. (3-10) and (3-11) shows that the secondary electron signal varies in exactly the same way as would a light signal from the same surface. Thus the view of a surface obtained in the SEM is very similar to that which could be obtained if the observer were to look down the column at the specimen illuminated by a light source placed at the Everhart-Thornley detector. Faces at a high angle of inclination to the beam and facing the detector will be bright, while surfaces normal to the beam will be darker. We also always instinctively expect the illumination to come from the top of a picture, since the sun shines in the sky rather than at our feet, so when observing a secondary electron micrograph our invariable first step is to rotate it so as to place the illumination at the top of the picture. Then all bright surfaces are tilted up and facing towards the detector, while darker surfaces are horizontal or facing away from the detector. Our brain, based on a lifetime of experience in using such clues, can then reconstruct the surface topography with a high degree of confidence. It is this convenient result which makes it possible for an observer to intuitively interpret secondary electron images and produce the correct deductions about the nature of the surface. This, coupled with the three-dimensional quality which comes from the high depth of field, explains much of the reason for the popularity of the SEM and secondary electron imaging. Fortunately this same principle can be applied up to even the highest magnifications, see for example Fig. 3-13, allowing images to be readily interpreted even when the feature size is approaching a few nanometers.

Secondary electrons are produced by both the incident primary electron as it passes through the escape region, and by exiting backscattered electrons as they again pass through the escape region as

they travel back to the surface. The ratio of these two contributions varies with the material of the sample, but typically only 20 to 30% of the detected secondary signal arises from the direct interaction with the primary beam. The secondary electrons produced by the incident beam, the so-called SE1 contribution, are produced within a few nanometers of the beam impact point and within a few nm of the surface and so carry high spatial resolution information, about the surface while secondaries produced by backscattered electrons, the SE2 contribution, can result from backscattered electrons which emerge from points microns away from the beam and from depths of a micron or more into the specimen. Thus the SE2 contribution carries only low resolution information and so reduce the contrast of the SE1 component. From Eq. (3-8) this implies that the incident beam current must be increased in order to achieve an adequate signal-to-noise ratio and this in turn implies from Eq. (3-5) that a larger probe diameter, and hence a worsened spatial resolution, must be used. High resolution imaging in the secondary electron mode therefore requires a high brightness electron source. We can also predict that high resolution imaging, relying on the SE1 electrons, will only be possible when the pixel size in the image is comparable with the size of the volume over which SE1's are produced i.e. a few nm. Thus, not surprisingly, high resolution imaging requires a high magnification (typically greater than $\times 20\,000$). The current best commercial instruments such as the Hitachi S-900 can achieve resolutions at 30 keV of the order of 1 nm or below, while more conventional SEMs can deliver 5 to 10 nm or suitable specimens.

Figure 3-13. Secondary electron image of surface of high performance magnetic recording disc showing topographic contrast. Recorded at 25 keV, original magnification $\times 300\,000$ on Hitachi S-900 SEM.

3.7.1.2 Voltage Contrast

It has been known from the earliest days of scanning microscopy that surfaces at different potentials produced images of different brightness (Knoll, 1941). This effect, now known as *voltage contrast*, is explained by Fig. 3-14 which shows the typical layout of the SEM specimen chamber. For a specimen at ground potential (Fig. 3-14a) the field from the Everhart-Thornley detector is of the order of 10 V mm^{-1}, sufficient to collect 60% or more of all of the emitted secondaries. If, as in Fig. 3-14b, the potentials of points on the surface are changed to $+5$ volts and -5 volts with respect to earth, then two complementary effects will produce contrast. Firstly, the collection field from the detector to the negatively charged area is increased while the corresponding field to the positively charged area is reduced. Secondly, the region which has a negative potential will repel the secondary electrons emitted from it but the positively charged area will attract and recollect some of the secondaries emitted from it. Together these effects mean that the positively biased strip will appear dark while the negatively biased strip will be bright. Figure 3-15 shows how this effect can be utilized to display visually the potentials on an operating integrated circuit, and arranged so as to operate inside the SEM. Since in modern high density integrated circuits the individual interconnect lines which carry the signals and operating potentials are only fractions of a micrometer in width the SEM provides the only feasible way of monitoring the operation of the circuit. The spatial resolution of this mode will be the same as for normal topographic observation.

Voltage contrast observations are invariably carried out at low beam energies. This is primarily because of the charge balance condition discussed earlier. If the incident energy is above E_2 then the beam injects electrons into the circuit under examination and this may influence its operation. Equally if the incident energy is be-

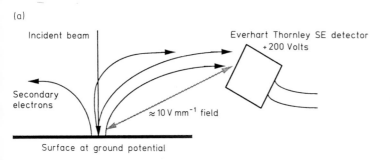

(a)

Incident beam Everhart Thornley SE detector
+200 Volts

Secondary electrons

≈ 10 V mm^{-1} field

Surface at ground potential

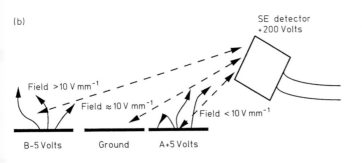

(b)

SE detector
+200 Volts

Field >10 V mm^{-1}
Field ≈ 10 V mm^{-1}
Field <10 V mm^{-1}

B-5 Volts   Ground   A+5 Volts

Figure 3-14. (a) Schematic layout of SEM specimen chamber showing sample and detector positions, and the electrostatic collection fields. (b) Corresponding situation when biases of $+5$ and -5 volts relative to earth are placed on features A and B.

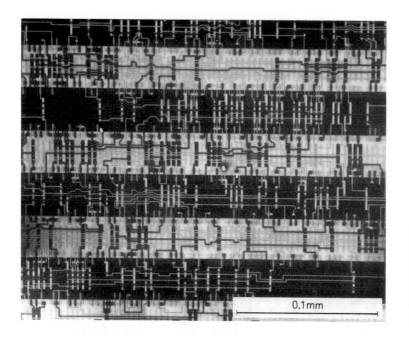

0.1mm

Figure 3-15. Voltage contrast image from an integrated circuit. Bright features are at a negative potential relative to ground while darker features are at a positive potential.

low E_2 then the beam extracts electrons from the circuit, so *loading* it in the same way as a conventional probe connected to a test meter or oscilloscope loads a circuit. The goal is therefore to choose an incident energy as close to E_2 as possible so that the minimum interference is caused to the circuit.

Considerable care must, however, be used in interpreting these voltage contrast results since the observed contrast arises from changes in the electrical field distribution in the specimen chamber and only indirectly from the potentials themselves. In addition to the fields from the detector to the sample there are also fields between regions on the chip which are at different potentials. For example the points labeled A and B in Fig. 3-14 b differ in potential by 10 volts while their spacing is a few microns, so the field between them is of the order of $10^3 \, \mathrm{V\,mm^{-1}}$ which is $100 \times$ greater than the field from the detector. This *local field* will modify the collection efficiency point to point on the surface so

that regions of constant potential will not display constant brightness. As a result the image should only be used as a qualitative guide to potentials. For a more detailed discussion of this topic and its quantitative aspects see Holt and Joy (1990).

3.7.1.3 Magnetic Contrast

Many materials, such as magnetic recording tape or floppy discs, recording heads, or naturally occurring substances such as cobalt, have magnetic fields above their surfaces. A secondary electron leaving the surface will travel through this field and will be deflected by the Lorentz force that is produced. Since this deflection will be normal to both the direction of travel of the electron and to the magnetic field, a leakage field into (or out of) the plane containing the specimen and the secondary detector will produce a deflection in such as sense as to add to, or subtract from, the collection efficiency of the detector. This variation in the collection efficiency for the secondary electrons produces *magnetic contrast* (Joy and Jakubovics, 1968). Figure 3-16 shows an example of such contrast from a single crystal of cobalt. The pattern which is visible comes from the stray magnetic field above the surface, which in turn reflects the underlying magnetic domain structure of the material. If the crystal were to be rotated in its own plane by 180° then the fields, and hence the contrast, would reverse. In this mode the spatial resolution of the contrast will be limited by the scale of the domain structure that produced it. On bulk materials this may be of the order of a micron or so, while for thin foils the corresponding value might be as low as 0.1 μm since the domain size is typically equal to or less than the sample thickness. As for other secondary electron modes the performance is best at low beam energies although on strongly magnetic materials severe astigmatism may result if too low an energy is selected. Other classes of magnetic materials (e.g., *cubic* materials such as iron) do not exhibit leakage fields above their surfaces and

their domain structures must be imaged using alternative techniques discussed below.

3.7.2 Backscattered Imaging Modes

Backscattered electrons are those incident electrons which have been scattered through an angle in excess of 90 degrees within the sample and so can leave it again. They typically have an energy which is of the order of 0.5 of their original incident energy and so, for normal operating energies, their energy is much greater than that of the secondary electrons discussed above. Consequently the backscattered electrons can emerge from considerable depths within the specimen, an estimate for this escape depth R_{BS} being

$$R_{BS} = 0.0083 \, A \, E^{1.67}/(Z^{0.889} \, \varrho) \, (\mu m) \quad (3\text{-}12)$$

where A is the atomic weight of the sample in g mol^{-1}, Z is the atomic number, ϱ is the density, and E is the beam energy in keV. For a beam energy of 20 keV backscattered electrons can therefore carry information about regions several micrometers below the sample surface compared with a few nanometers for the secondary electrons. However because the lateral area over which the backscattered electrons emerge from the surface is also of the same order as R_{BS} then the spatial resolution of the image will be worse than that of the secondary signal.

Backscattered electron imaging modes are complementary to secondary modes and produce unique information of their own. However they have so far attracted somewhat less attention because of the problem of signal collection. While the low energy secondary electrons are readily collected by the application of a small bias field, the high energy backscattered electrons travel in straight lines from the spec-

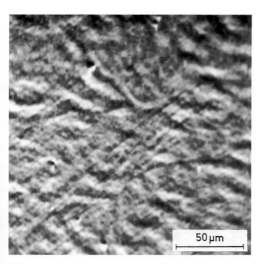

Figure 3-16. Magnetic contrast from a single crystal of cobalt. The large scale structures are the bulk magnetic domains within the crystal.

imen and must therefore be collected by placing a suitable detector in the path of the electrons. The simplest, but least efficient, technique is to use the Everhart-Thornley detector described above but to place bias of -50 volts on the Faraday cage. This is sufficient to reject any secondary electrons but allows those backscattered electrons moving towards the detector to be collected. However the solid angle presented by the detector at the sample is small so the collection efficiency is low (below 5%) and the asymmetrical arrangement produces heavy shadows in the image. A better arrangement (Fig. 3-17 a) again uses the Everhart-Thornley approach but places the scintillator concentrically around the beam and directly above the specimen. No bias is applied to the scintillator so only high energy backscattered electrons produce an output. For beam energies of above a few keV this type of arrangement is highly efficient ($> 50\%$) and has sufficient bandwidth to permit TV-rate imaging. A popular alternative (Fig. 3-17 b) is to use a solid-state detector which is simply a large P-N diode. A high energy electron passing through the diode will produce an avalanche of electrons to give a substantial current gain. The efficiency of this process rises linearly with the energy of the electrons and there is typically a cut-off energy of a few keV below which no output is produced. Solid state detectors are cheap to manufacture and occupy little space in the specimen chamber, however a detector large enough to be efficient has a large capacitance which limits the available bandwidth and restricts operation to slow-scan speeds. Finally electron micro-channel plates (Fig. 3-17c) are becoming important. These devices consist of a thin plate containing many thousands of micron-sized tubes each of which acts as an electron-multiplier. This can be mounted con-

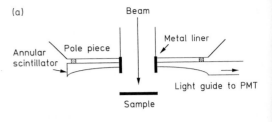

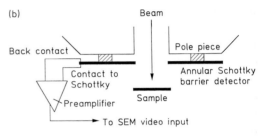

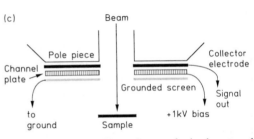

Figure 3-17. (a) Scintillator detector for backscattered electron imaging. (b) Solid state detector for backscattered imaging. (c) Micro-channel plate detector for backscattered electron imaging.

centric with the beam and symmetrically above the sample. These collectors can be made both sensitive and efficient and can function for incident beam energies down to 1 keV. Their only drawback is that they are fragile and have a limited lifetime.

3.7.2.1 Atomic Number Contrast and Topography

It has been known for nearly one hundred years (Starke, 1898) that the yield η of backscattered electrons varies with the atomic number of the target. The backscatter yield is also somewhat dependent on the energy of the incident electrons although for energies above 5 keV the backscattering yield is almost independent of the accelerating voltage. In this case η can be approximated by the function:

$$\eta = -0.0254 + 0.016\,Z - 0.000186\,Z^2 \tag{3-13}$$

where Z is the atomic number of the target, or the weighted mean value of the atomic number if the target is a compound. η varies between about 0.05 for carbon ($Z = 6$) and about 0.5 for gold ($Z = 79$). Since the yield is a monotonic function of Z a backscattered image can display contrast which is directly related to the atomic number of the area sampled by the electron probe. Thus in a multiphase alloy regions of the specimen with different atomic numbers will appear with different brightnesses, the level of the contrast depending on the difference in Z. Under normal imaging conditions regions with an effective difference of only about 0.5 in Z units can be distinguished. Figure 3-18 shows an example from the cathode plate of a space satellite Ni-Cd battery which has failed. While

the secondary electron image shows only topography, the atomic number image shows how the cadmium (which appears bright because of its high atomic number) has been redeposited on the nickel cathode (which appears relatively dark because of its lower atomic number). In principle this technique can be made quantitative by calibrating on suitable compounds and interpolating. The spatial resolution of the technique is limited by the lateral spread R_{BS} of the beam [Eq. (3-12)] and is thus of the order of fractions of a micron. When a feature is less than this size then its apparent atomic number will be influenced by the materials which surround it and consequently care must be employed in interpreting the data.

While this technique is of considerable value a difficulty arises if the specimen surface is not completely flat. This is because the backscatter yield also varies with the angle of incidence between the beam and the surface so producing topographic contrast. This variation is indistinguishable from that caused by changes by in the atomic number and so could lead to an error. In addition because backscattered electrons travel in straight lines from the specimen to the detector features on the surface can cast electron *shadows*. These problems can be minimized by using the arrangement shown in Fig. 3-19 in which the backscatter detector is divided into two halves. Using as an example the topography shown in the figure, it can be seen that the expected profiles from the A and B sections of the detector are quite different. If the signals from the two halves are added then, because the shadowing effects are opposite, the topographic components cancel leaving only the atomic number variation. If, on the other hand, the signals are subtracted then the atomic number variations which are the same for both detectors dis-

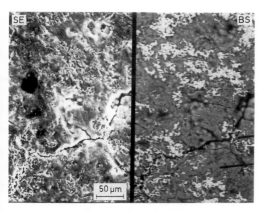

Figure 3-18. Secondary electron and backscattered electron images of the cathode plate of a space satellite battery. The SE image shows only the topography of the surface. The atomic number contrast in the BS image shows that Cadmium (appearing as bright particles) has redeposited across the nickel cathode (darker background).

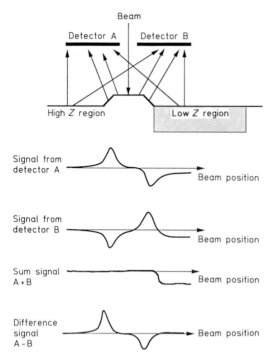

Figure 3-19. Schematic arrangement of split detectors for backscattered imaging allowing chemical and topographic contrasts to be separated.

appear leaving only the topographic variations. By comparing the images with the detectors added (A + B) and subtracted (A − B) it is possible to separate chemical contrast effects from topographic contrast. This scheme can be further extended by using four quadrants instead of two halves to give even more flexibility.

3.7.2.2 Magnetic Contrast

Common magnetic materials such as iron do not have significant leakage fields above their surfaces. This is because they can be magnetized along any of the three cubic axes and so can always arrange to close their flux internally. They therefore do not produce any *magnetic contrast* in the secondary electron mode described earlier. However their magnetic domain

structure can still be observed in the backscattered electron mode as shown schematically in Fig. 3-20. Here the sample surface is tilted at about 45 to 60 degrees to the incident beam. In this case contrast arises because the electron experiences a Lorentz force which deflects it as it travels through the internal flux. Depending on the direction of the flux the electron interaction volume is displaced either slightly towards or slightly away from the surface so modifying the backscattering coefficient. Contrast will only occur when the flux has a component parallel to the tilt axis, so rotation of the sample about its own axis will cause different magnetization components to come into contrast. Figure 3-21 shows an example of such contrast from a sample of Fe-Si transformer core material displaying the classic *fir-tree* domain pattern. The backscattered magnetic contrast effect is very weak, the signal variation between the brightest and the darkest area being of the order of 1% or less. From Eq. (3-8) it can be seen that, because of this, a large incident beam current will be required to achieve an adequate signal to noise ratio and hence the spatial resolution will be poor. Some improvement in contrast can be obtained by

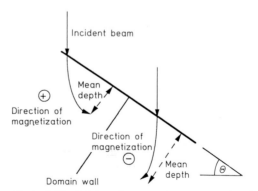

Figure 3-20. Origin of magnetic contrast in the backscattered mode of imaging from crystals with cubic magnetic anisotropy.

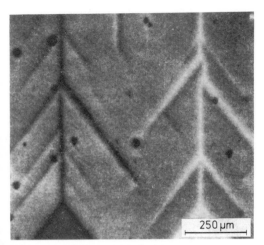

Figure 3-21. Magnetic contrast in the backscattered imaging mode showing characteristic *fir-tree* domain pattern. Sample is sheet of Fe-Si transformer core material observed at 30 keV. Tilt was 45°.

operating at high beam energies (i.e., greater than 30 keV) because the peak to peak magnitude of the contrast increases as $E^{3/2}$. This technique has been applied to the study of transformer core materials, magnetic recording discs, and to the study of thin film recording heads for computer disc storage.

3.7.2.3 Electron Channeling Patterns

The transmission electron microscope played a major role in advancing our knowledge of materials science because the techniques of electron diffraction made it possible to study the crystallography of the specimen and the crystallographic defects, such as dislocations and stacking faults, which determine its basic properties. The SEM can also provide this type of information through the technique of electron channeling patterns (ECP). For an amorphous material the yield of backscattered electrons depends only on the atomic number of the specimen as discussed above, but for a crystalline material the yield is also

found to depend on the angle made by the incident beam with the lattice. This is because, as shown in Fig. 3-22a, if the electron enters the crystal at an arbitrary angle there is a high probability that it will soon come close to a nucleus and be scattered back out of the lattice, but if the beam is traveling along one of the symmetry directions of the lattice (Fig. 3-22b) then the electron *channels* into the crystal and may travel to a considerable depth before coming close to a nucleus and hence has a lower back scattering probability (Joy et al., 1982). The angles θ at which this occurs are those which satisfy Bragg's law, i.e. $\sin \theta = \lambda/(2\,d_{hkl})$, where λ is the wavelength of the electron and d_{hkl} is the spacing of the lattice planes of Miller indices $(h\,k\,l)$. For 20 keV electron λ is of the order of 0.01 nm so for a typical lattice spacing of 0.3 nm the

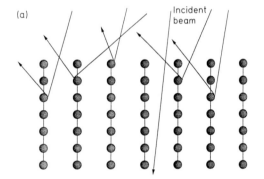

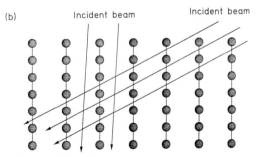

Figure 3-22. Origin of electron channeling contrast, (a) random angles of incidence lead to backscattering, (b) incident directions aligned with crystal symmetry directions channel into the specimen.

Bragg angle is one or two degrees. If the angle of incidence of the beam to the surface is changed by this sort of angle then the backscattering yield will vary, and if the angle of incidence is changed in two dimensions rather than just one then an image formed from the backscattered electrons will show a map of these variations which will hence display the crystallographic symmetry of the lattice about the beam direction. The necessary variation in beam direction is achieved by observing the specimen at low magnification, since as the beam scans from one side of the sample to the other its angle of incidence changes by several degrees. Because the channeling contrast comes only from the top twenty or thirty nanometers of the crystal surface care is required in the preparation of the specimen. Usually the sample is chemically cleaned to remove any contaminants, and the chemically or electro-chemically etched to remove any residual mechanical damage due to polishing or handling.

Figure 3-23 shows a channeling pattern recorded from a single crystal of the quaternary semiconductor compound indium phosphide gallium arsenide InPGaAs. The pattern is comprised of bands, the angular width of which is equal to twice the appropriate Bragg angle, which intersect at poles. In this example the pole is seen to have a four-fold axis of symmetry with mirror planes at $45°$ to each other. The beam is therefore traveling along the (011) axis of the crystal in this example. The indices of the bands are of the type $\{220\}$ and $\{400\}$, and in each case the bands are running parallel to the trace of the respectively indexed lattice planes on the surface of the crystal. If the sample is moved laterally then the pattern will not change since the symmetry of a crystal does not alter under translation, but if the crystal is rotated or tilted then the pattern will move as if

Figure 3-23. Electron channeling pattern recorded from a single crystal of the quaternary indium phosphide gallium arsenide. The symmetry is that of the $\{011\}$ axis of the crystal.

rigidly fixed to the lattice. By recording a sequence of pattern as the crystal is tilted a montage channeling map can be generated which shows all possible unique symmetry conditions for the crystal. It can be shown (Joy et al., 1982) that the poles in the ECP correspond to zone axes in the stereographic projection of the crystal, and thus by comparing the ECP from a crystal of unknown orientation with the montage the orientation can be determined by inspection.

While this technique is valuable it is applicable only to large single crystals, since the whole area scanned by the beam must be of the same orientation. To overcome this limitation a selected area channeling pattern (SACP) technique has been developed (van Essen et al., 1970). This allows the ECP to be obtained from micron sized regions of the sample. The beam no longer scans over the surface of the specimen but rocks about a fixed point on the surface. Since the ECP comes only from the change in the angle of incidence this procedure still produces a pattern but now from an area defined by the accuracy with which the beam can be held stationary on the surface. Figure 3-24 shows a simple application of this technique. The sample consists of small crystals of artificially grown diamonds. Working in the standard SE imaging mode described earlier one of the crystals with a well developed facet structure (Fig. 3-24a) is selected. The SEM is then switched to SACP mode to give the pattern shown (Fig. 3-24b) which in this case can be seen by its clear three-fold symmetry to be a {111} pole. To demonstrate this orientation by a conventional TEM or X-ray technique would require very considerable sample preparation and processing. This

technique is of special value in studies of fracture as it permits a direct determination of the orientations associated with each mode of fracture (Newbury et al., 1974; George et al., 1989).

Figure 3-25a shows the backscattered electron image of a 0.5 micrometer thick layer of amorphous silicon thermally regrown on a two micrometer thick layer of oxide which is, in turn, on a substrate of silicon. The image shows the *finger-like* grain structure produced as a consequence of the regrowth procedure in which a hot wire is moved across the surface. As the molten zone follows the heater the solid-liquid interface is pulled out parallel to the movement of the wire. The changes in contrast between the different grains arise because of the ECP effect, since depending on its exact orientation each grain will backscatter slightly more or slightly less signal. The image shows that the material is not a single crystal but gives no quantitative evidence as to how big the angular misorientations between grains might be. However an SACP examination (Fig. 3-25b) of this area produces a pattern which can be identified as a {110} pole. On closer examination it can be seen that the pole is actually the superimposition of several separate {110} poles with angular offsets of about 1° between them. This occurs because the selected area is straddling the boundary between two grains, each nominally of {110} orientation but with a random misalignment between them. Studies of this type demonstrate that this type of rapid regrowth produces a strongly textured crystal with a well defined surface normal orientation but with a random misorientation in the plane of the film. To obtain this same type of information using more conventional techniques such as transmission electron microscopy would require a lengthy preparation procedure to backthin

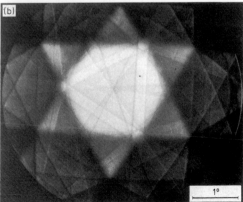

Figure 3-24. (a) Image of a small crystal of artificial diamond. (b) Corresponding selected area channeling pattern showing orientation of top facet to be {111}.

Figure 3-25. (a) Channeling contrast image of amorphous silicon regrown on silicon oxide. The brightness variation between adjacent grains indicates an angular misorientation. (b) Selected area channeling pattern from the same sample showing that the average orientation is {110}, but that adjacent grains have an offset of about 1 degree.

fect can also be imaged in channeling mode. Because the distortion is small in magnitude and spreads over only a small distance in the crystal the electron-optical conditions required to image it are severe, but with the latest field emission SEMs the desired parameters can be obtained (Joy, 1990). Figure 3-26 shows the contrast observed from individual dislocations laying at depths up to 0.25 μm from the surface of a solid sample of molybdenite MoS_2. The width of the dislocation image is determined by the extent of the strain field from the dislocation as well as by the physical size of the electron probe. This method of observing crystallographic defects is of importance because, unlike the standard techniques of transmission electron microscopy, the sample need not be thinned. The arrangement of the defects is therefore more representative of the structures that might be observed within the bulk of the specimen.

3.7.3 Other Contrast Modes

In addition to the electron imaging modes discussed above there are other types of information available in the SEM

the regrown film, and the result would be less statistically valid because the available field of view in a transmission micrograph is much smaller than that attainable in the SEM so far fewer grains can be sampled. X-ray techniques would also be less satisfactory because reflections would be obtained from both the film and the substrate, whereas the ECP contrast comes only from the top 100 nm or less of the surface and thus contains no complicating information from the substrate.

It can finally be noted that the distortion of the lattice around a crystallographic de-

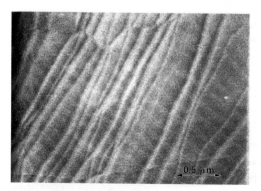

Figure 3-26. Channeling contrast image of individual crystallographic dislocations in a solid sample of molybdenite MoS_2. Imaged recorded in backscattered mode at 40 keV on JEOL JSM890 field emission SEM.

image which utilize other types of interactions between the incident beam and the sample.

3.7.3.1 Charge Collection Microscopy

There are special contrast modes available when the sample under examination is a semiconductor (or insulator) rather than a metal. These arise because of the special properties of semiconductor materials. When the high energy incident electron enters a semiconductor it creates electron-hole pairs by promoting electrons across the band-gap into the empty conduction band while leaving behind holes (which have the properties of a positively charge electron) in the previously filled valence band. The energy e_{eh} required to create one electron-hole pair is about three times the band-gap, for example e_{eh} in silicon is 3.6 eV, this a single 10 keV incident electron could generate nearly 3000 of these carrier-pairs. Under normal conditions the electrons and holes will recombine within a very short time, typically a microsecond or less, but if an electric bias field is imposed across the specimen then the electrons and holes will drift in opposite directions because they have opposite charge. The motion of these charges within the specimen will then produce a current flow in the external circuit which is supplying the bias. The effect of the incident electron has therefore been to produce localized conductivity within the semiconductor, or Electron Beam Induced Conductivity (EBIC). If the electric field is produced by an external voltage source then this effect is known as beta conductivity. However in the vicinity of a P-N junction there is a space-charge depleted region together with an associated electrical field. If the P-N junction is shorted by a connecting wire then a current, the charge collected current, will flow

in this circuit when the electron beam is in the junction region because the electron-hole pairs will be separated by the depletion field. If the beam is moved away from the junction region then the carriers must first diffuse back to the depleted region before they can be separated and produce a current. The measured charge collected signal will thus fall away at a rate determined by the minority carrier diffusion length.

If the charge collected current is amplified and used to form a picture then a powerful new imaging mode is produced, applicable to both semiconductor materials and to completed devices. Figure 3-27 shows a charge collected image produced from an wafer of GaAs. In this case the necessary collection field has been produced by forming a Schottky barrier (i.e., a thin gold film) on the top surface of the wafer. A depleted region then extends downwards into the

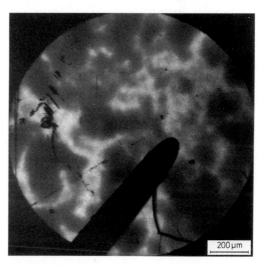

Figure 3-27. Charge collected image from a wafer of GaAs. The bright area indicates the limits of the Schottky barrier formed on the top surface of the crystal. The dark *cloudy* areas are regions of local impurity concentration and the dark lines are defects. The black pointed bar is the electrical contact to the Schottky.

materials for a depth of a few microns but laterally for the full extent of the barrier. The contrast in the image is associated with electrically active sites, such as impurity decorated dislocations and stacking faults. Such features will cause locally enhanced recombination rates for the carriers so leading to a fall in the externally collected current. The charge collection mode therefore provides a direct way of imaging individual crystallographic defects in a semiconductor material.

3.7.3.2 Cathodoluminescence

The recombination of the electron-hole pairs mentioned above results in the release of energy, some fraction of which may be radiated from the sample as in the visible or infrared portion of the electromagnetic spectrum. This emission of light is known as cathodoluminescence. If the light is collected and amplified then an image may be formed as for any other mode of operation. The major problem associated with cathodoluminescence is that the yield of photons is very small, only about 1 for each 10^6 incident electrons, so the light signal is weak and carefully optimized collection optics are required. Typically an ellipsoidal mirror is used with the sample placed at one of the foci and the detector, usually a cooled photomultiplier, at the other. In this way about 20 to 40% of the emitted radiation can be collected. When spectral analysis is required the detector is replaced by a light-pipe which directs the light to the entrance slit of a suitable spectrometer.

The spatial resolution of the cathodoluminescence signal is set by the beam interaction volume and is therefore of the order of R_{BS} [Eq. (3-12)]. It is not subject to the usual diffraction resolution limit of conventional optical microscopy because no imaging lenses are used. Because the cathodoluminescence signal is weak a high beam current is required and in practice the resolution is usually set by the probe diameter needed to produce acceptable signal levels, typically 1 μm diameter or more.

The intensity and spectral distribution of the cathodoluminescence signal depends on several factors, for a semiconductor the major spectral peak will be at the band gap energy, but this peak will shift and broaden as the sample temperature is increased. Spectral measurements are therefore usually made with the sample held at liquid nitrogen temperatures (77 K) or lower. The production of cathodoluminescence radiation is in competition with non-radiative modes of recombination such as those involving deep traps. The efficiency of radiative recombinations is enhanced by the presence of impurities however, so the total light intensity monotonically follows the doping level of a semiconductor over several orders of magnitude, providing an accurate and convenient tool for monitoring this in a material. Just as in the charge collection image case, electrically active defects will act as local centers for non-radiative recombination causing a fall in the emitted light output. Figure 3-28 shows a cathodoluminescence image from a GaAs crystal which contains strain induced dislocations, which appear as dark lines in the image. The information obtained is comparable to that from the charge collected image but no external field, P-N junctions, or Schottky barriers are required.

Cathodoluminescence is also observable from many naturally occurring minerals and biological materials. In such cases the spectrum is often very complex, each of the spectral peaks being associated with the excitation of a specific interatomic bond. In principle cathodoluminescence provides a tool for the examination of such bonds,

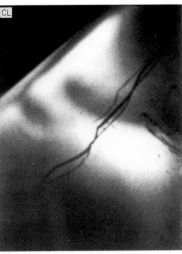

100 µm

Figure 3-28. Cathodoluminescence image from GaAs crystal produced using the total light emitted by the sample. The dark lines in the image are strain induced crystallographic defects.

but in many cases the radiation damage produced by the ionizing action of the incident beam leads to the destruction of the bonds before good data can be obtained.

3.7.3.3 Fluorescent X-Rays

A final beam interaction which must be mentioned is the production of fluorescent X-rays. These result from the decay back to its ground state of an atom which has been ionized by an incident electron. Since the energy carried away by the X-ray photon is a unique property of the atom ionized detection and measurement of the X-ray will permit a chemical identification to be made. Because the efficiency of the X-ray process is quite small, typically 1 photon for each 10^4 incident electrons, the most common procedure is to detect the X-ray signal with a solid state energy dispersive X-ray spectrometer. Such a device has a relatively high collection efficiency ($\approx 1\%$) leading to detected count rates of the order of a few thousand photons per seconds. When used in conjunction with a pulse height analyzer (multi-channel analyzer) a visual display of the X-ray spectrum is gen-

erated in real time. A rapid, quantitative elemental microanalysis of the sample is therefore possible from any area of the sample irradiated by the electron beam. The spatial resolution of X-ray microanalysis is determined again by the beam interaction volume since X-rays can be generated by all electrons with an energy greater than the critical energy for the line of interest. A typical resolution is of the order of 1 to 2 µm for an incident energy of 20 keV.

The combination of X-ray microanalysis and the various imaging modes described previously is a very powerful one in all areas of science and a high fraction of all new scanning microscopes are now equipped with some form of X-ray capability. This topic is considered in more comprehensive detail in another chapter of this volume.

3.8 Unwanted Beam Interactions

Although interactions between the sample and the beam of incident electrons provides the information carrying contrast of the image there are other interactions which are equally potent but less beneficial. The electron beam is a powerful source of ionizing radiation, it has been stated (Grubb and Keller, 1972) that the specimen in the SEM is subjected to as much radiation as an individual standing thirty meters from a ten megaton H-bomb. While this type of effect is of little concern when observing metals, ceramics and semiconductors, it is very harmful to polymers and biological materials because in such materials ionization can lead to the breaking and cross-linkage of bonds. This *radiation damage* is a severe limitation to the ability of the SEM to efficiently image such materials. The intensity of the radiation damage depends on the energy deposited per unit volume by the beam. If we assume that the interaction volume of the electron beam is approximately a sphere of diameter R_B [see Eq. (3-12)] then the energy density E_D is thus

$$E_D = E/R_B^3 \qquad (3\text{-}14)$$

which, using the result of Eq. (3-12) implies that the E_D varies as E^{-4}. Thus although the physical extent of radiation damage is larger at high beam energies, the energy deposited and hence to magnitude of the damage, rises rapidly as the beam energy is reduced.

The sample can also be modified by another indirect action of the beam. If the surface of the specimen irradiated by the beam is not clean but, for example, covered with a thin film of oil from finger contact or backstreaming from a vacuum pump, then as electrons pass through this layer they *crack* it to form a low-grade polymer. The poor electrical conductivity of this cracked layer results in charging and the production of a radially directed electrostatic field which attracts further contamination across the surface towards the region being examined. Within a few seconds from 5 to 50 nm of contamination can be formed, obscuring surface detail and impeding microanalysis. Direct sample damage due to heating of specimens is rare except in materials of very low thermal diffusivity under conditions where large diameter beams, and high currents, are used. For typical sub-micron diameter electron beams the temperature gradient in the vicinity of the beam is so high that, in almost all materials, there is adequate radial heat flow to prevent local heating. Finally it can be noted that in a few very high resistivity materials (such as glasses) mechanical damage can result from dielectric breakdown as a result of sample charge up under the beam.

3.9 Conclusion

The SEM is one of the most versatile forms of microscope available. Its combination of high spatial resolution, high depth of field, and its analytical power make it the ideal tool with which to examine and interpret the microstructure of materials.

3.10 References

George, E. P., Porter, W. D., Joy, D. C. (1989), *Mat. Res. Soc. Symp. Proc. 133*, 311.
Goldstein, J. I., Newbury, D. E., Echlin, P. E., Joy, D. C., Fiori, C. (1981). *Scanning Electron Microscopy and X-ray Microanalysis.* New York: Plenum Press.
Grubb, D. T., Keller, A. (1972), *Proc. 5th European Cong. on Electron Microscopy.* London: Institute of Physics, p. 554
Holt, D. B., Joy, D. C. (1990), *Application of SEM in Semiconductor Science.* London: Academic Press.
Joy, D. C. (1989), *Scanning 11,* 1
Joy, D. C. (1990), in: *High Resolution Electron Microscopy of Defects in Materials:* Sinclair, R. (Ed.). Pittsburgh (PA): Materials Research Society, Symposia Series No. 183, p. 199.

Joy, D. C., Jakubovics, J. P. (1968), *Phil. Mag. 17,* 61.
Joy, D. C., Newbury, D. E., Davidson, D. (1982), *J. Appl. Phys. 53,* R81.
Joy, D. C., Romig, A. D., Jr., Goldstein, J. I. (1986) *Principles of Analytical Electron Microscopy.* New York: Plenum Press, Chapter 2.
Knoll, M. (1941), *Naturwissenschaften 29,* 335.
Lane, W. C. (1970), *Proc. 7th Ann. SEM Symposium:* Johari, O. (Ed.). Chicago: IITRI, p. 42.
Newbury, D. E., Christ, B. W., Joy, D. C. (1974), *Met. Trans. 5,* 1505.
Oatley, C. (1982), *J. Appl. Phys. 53,* R1–13.
Rose, A. (1948), in: *Advances in Electronics:* Marton, A. (Ed.). New York: Academic Press, p. 131.
Starke, H. (1898), *Ann. Phys. 66,* 49.
van Essen, C. G., Schulson, E. M., Donaghay, R. H. (1970), *Nature 225,* 847.

General Reading

Chapman, J. N., Craven, A. J. (1983), *Quantitative Electron Microscopy.* Edinburgh: SUSSP.
Joy, D. C., Romig Jr., A. D., Goldstein, J. J. (1986), *Principles of Analytical Electron Microscopy.* New York: Plenum Press.
Kirschner, J., Murata, K., Venables, J. A. (Eds.) (1987), *Physical Aspects of Microscopic Characterization of Materials.* Chicago: Scanning Microscopy International.
Lyman, C. E. (1990), *Scanning Electron Microscopy, X-Ray Microanalysis and Analytical Electron Microscopy.* New York: Plenum Press.
Newbury, D. E., Joy, D. C., Echlin, P., Fiori, C. E., Goldstein, J. I. (1986), *Advanced Scanning Electron Microscopy and X-Ray Microanalysis.* New York: Plenum Press.
Ruska, E. (1980), *Early Development of Electron Lenses and Electron Microscopy.* Stuttgart: S. Hirzel Verlag.
Watt, I. M. (1985), *Principles and Practice of Electron Microscopy.* Cambridge University Press.

4 X-Ray Diffraction

Robert L. Snyder

Institute for Ceramic Superconductivity, New York State College of Ceramics, Alfred University, Alfred, NY, U.S.A.

List of Symbols and Abbreviations

a, b, c	unit cell edge translation vectors
a^*, b^*, c^*	reciprocal cell translation vectors
B, B_{ij}	Debye-Waller temperature factor and tensor components
c	speed of light
d_{hkl}	interplanar spacing vector
d_{hkl}^*	reciprocal cell interplanar spacings
E	energy
e	charge on the electron
F_{hkl}	structure factor
F_i	backscattering amplitude from neighboring atoms
F_N	Smith, Snyder figure of merit evaluated at line N
f, f_0	atomic scattering factor
$\Delta f', \Delta f''$	anomalous dispersion scattering components
$g(r)$	radial distribution function
G	Gaussian function
h	Planck's constant
$h k l$	Miller indices
I	intensity
I_0	incident intensity
$I_{i\alpha}$	intensity of reflection i from phase α
I^{rel}	relative intensity, usually on a scale of 100
$K_{\alpha 1}, K_{\alpha 2}, K_\beta$	characteristic X-ray emission lines
k	wave vector (magnitude: $2\pi/\lambda$)
L	Avagadro's number
L	Lorentzian function
Lp	Lorentz and polarization corrections
M	multiplicity of a plane
M_{20}	de Wolff figure of merit
m_e	mass of the electron
N_i	number of atoms in ith coordination shell
P	profile due to instrumental effects, the convolution of $W*G$
Q_{hkl}	$10^4 d_{hkl}^{*2}$
R, r	distance
$RIR_{\alpha, \beta}$	reference intensity ratio of phase α with respect to β
S	profile from diffraction by the sample
S_α	Rietveld scale factor for phase α
u	root mean square amplitude of vibration
V	unit cell volume
V	accelerating voltage
W	atomic weight
$W*G$	wavelength and instrumental profiles
X	weight fraction
x	thickness
x, y, z	atomic fractional coordinates

Z	number of asymmetric units per unit cell
z	charge on the nucleus
α, β, γ	interaxial angles
$\alpha^*, \beta^*, \gamma^*$	reciprocal cell interaxial angles
β	full width at half maximum of a diffraction peak
$\beta_\varepsilon, \beta_\tau$	peak broadening due to strain and size
ε	residual lattice stress
θ	Bragg diffraction angle
θ_m	diffraction angle of monochromator
λ	wavelength
λ_{SWL}	short wavelength limit from an X-ray tube
μ	linear X-ray absorption coefficient
μ_0	absorption of an atom in the absence of neighbors
μ/ϱ	mass absorption coefficient
ν	frequency
$\bar{\nu}$	wave number
ϱ	density
$\varrho(r), \varrho(x\,y\,z)$	electron density at location r or $x\,y\,z$
σ_i	displacement between absorbing atoms
τ	crystallite size
ϕ	phase angle
ψ	wave function
ω	fluorescent yield
χ	EXAFS interference function

BNL/NSLS	Brookhaven National Laboratory National Synchrotron Light Source
CD-ROM	compact disk read only memory
CVD	chemical vapor deposition
EDD	electron diffraction database
EDS	energy dispersive spectroscopy
EISI	elemental and interplanar spacings index
EXAFS	extended X-ray absorption fine structure
FET	field effect transistor
FOM	figure of merit
ICDD	international centre for diffraction data
IUPAC	international union of pure and applied chemistry
MCA	multichannel analyzer
PC	desktop computer
PDF	powder diffraction file
PHA	pulse height analyzer
PSD	position sensitive detector
RDF	radial distribution function
WDS	wavelength dispersive spectroscopy
XRD	X-ray diffraction
XRF	X-ray fluorescence
ZBH	zero background holder

4.1 Introduction

X-ray diffraction has acted as the cornerstone of twentieth century science. Its development has catalyzed the developments of all of the rest of solid state science and much of our understanding of chemical bonding. This article presents all of the necessary background to understand the applications of X-ray analysis to materials science. The applications of X-rays to materials characterization will be emphasized, with particular attention to the modern, computer assisted, approach to these methods.

4.2 The Nature of X-Rays

X-rays are relatively short wavelength, high energy electromagnetic radiation. When viewed as a wave we think of it as a sinusoidal oscillating electric field with, at right angles, a similar varying magnetic field changing with time. The other description is that of a particle of energy called a photon. All electromagnetic radiation is characterized by either its energy E, wavelength λ (i.e., the distance between peaks) or its frequency v (the number of peaks which pass a point per second). The following are useful relationships for interconverting the most common measures of radiation energy.

$$\lambda = \frac{c}{v} \qquad (4\text{-}1)$$

$$E = h v \qquad (4\text{-}2)$$

where c is the speed of light and h is Planck's constant. Spectroscopists commonly use wavenumbers particularly in the low energy regions of the electromagnetic spectrum, like the microwave and infrared. A wave number ($\bar{v}$) is frequency di-

vided by the speed of light

$$\bar{v} = \frac{v}{c} = \frac{c/\lambda}{c} = \frac{1}{\lambda} \qquad (4\text{-}3)$$

The angstrom (Å) unit, defined as 1×10^{-10} m, is the most common unit of measure for X-rays but the last IUPAC vention made the nanometer (1×10^{-9} m) a standard. However, here we will use the traditional angstrom unit. Energy in electron volts (eV) is related to angstroms through the formula,

$$E \,(\text{eV}) = \frac{h c}{\lambda_{\text{cm}}} = \frac{12\,396}{\lambda_{\text{Å}}} \qquad (4\text{-}4)$$

Electron volts are also not IUPAC approved in that the standard energy unit is the Joule which may be converted by

$$1 \,\text{eV} = 1.602 \times 10^{-19} \,\text{J} \qquad (4\text{-}5)$$

It should be noted that despite the IUPAC convention, Joules are never used by crystallographers or spectroscopists, while a few workers have adopted the nanometer in place of the angstrom. Table 4-1 lists the various measures across the electromagnetic spectrum.

4.3 The Production of X-Rays

There are four basic mechanisms in nature which generate X-rays. These are related to the four fundamental forces that exist in our universe. Any force when applied to an object is a potential source of energy. If the object moves kinetic energy is generated. The weak and strong nuclear forces combine to produce not very useful X-rays, along with many other wavelengths and subatomic particles, in high energy nuclear collisions. The force of gravitation also produces X-rays which are not useful in the materials characterization

Table 4-1. Values of common energy units across the electromagnetic spectrum.

Quantity	Units	IR	UV	Vacuum UV	Soft X-ray	X-ray	Hard X-ray	γ
Wavelength	Å	10 000	1000	100	10	1	0.1	0.01
Wavelength	nm	1 000	100	10	1	0.1	0.01	0.001
Wavenumber	cm^{-1}	10^4	10^5	10^6	10^7	10^8	10^9	10^{10}
Energy	eV	1.24	12.4	124	1239.6	12.4 keV	124 keV	1.24 MeV
Energy	J	2×10^{-19}	2×10^{-18}	2×10^{-17}	2×10^{-16}	2×10^{-15}	2×10^{-14}	2×10^{-13}

laboratory, by giving rise to neutron stars and black holes which, in the process of accreting matter, produce X-rays visible at astronomical distances. However, it is the Coulombic force which produces the X-rays we harness in the laboratory.

4.3.1 Synchrotron Radiation

Particle accelerators operate on the principle that as a charged particle passes through a magnetic field it will experience a force perpendicular to the direction of motion, in the direction of the field. This causes a particle to curve through a "bending magnetic" and accelerate. As long as energy is supplied to the magnets, a beam of particles can be continuously accelerated around a closed loop. Accelerating (and decelerating) charged particles will give off electromagnetic radiation. When the particles are accelerated into the GeV range, X-radiation will be produced. A synchrotron is a particle acceleration device which, through the use of bending magnetics, causes a charged particle beam to travel in a circular (actually polyhedral) path.

Today there are a number of synchrotron facilities around the world which are dedicated to the production of extremely intense sources of continuous (white) X-radiation ranging from hundredths to hundreds of angstroms in wavelength. In recent years there has been a burst of activity in the use of these sources. The wavelength tunability and very high brightness of these sources has opened a wide range of new characterization procedures to researchers. The addition of magnetic devices to make the particle beam wiggle up and down on its path between bending magnetics, called wigglers and undulators, have raised the intensity of X-rays available for experiments by as much as a factor of 10^{12}. In addition, since the X-rays are only produced as the charged particles fly by the experimenter's window every few nanoseconds, time-resolved studies in the nanosecond range have become accessible. See the chapter on synchrotron radiation by Sherwood et al. in this volume for more information on synchrotron techniques.

4.3.2 The Modern X-Ray Tube

The conventional method of producing X-rays in a laboratory is to use an evacuated tube invented by Coolidge (1913). Figure 4-1 shows a modern version of this tube whose function is illustrated in Fig. 4-2. This tube contains a tungsten cathode filament which is heated by an AC voltage ranging from 5 to 15 V. The anode is a water-cooled target made from a wide range of pure elements. Electrons are accelerated in vacuum under potentials of 5000 to 80 000 volts and produce a spectrum of the type shown in Fig. 4-3. As the accelerated electrons reach the target they are repelled by the electrons of the target atoms, causing a slowing down or breaking. To slow down an electron and conserve energy, the electron must lose its energy in the only manner available to it – radiation. The German word for breaking is brems and for radiation is strahlung. Most of the early discoveries concerning X-rays oc-

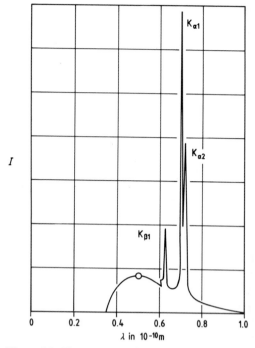

I

Figure 4-3. The spectrum from a Mo target X-ray tube.

Figure 4-1. The modern sealed X-ray tube (Courtesy of Siemens AG).

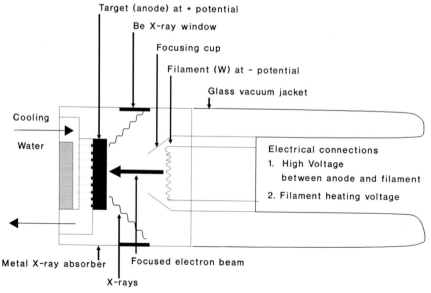

Figure 4-2. Schematic of an X-ray tube.

$(\lambda) = \mathrm{m}, \quad (v) = (c/\lambda) = \mathrm{s}^{-1}, \quad (hv) = (E) = \mathrm{eV},$

$(c) = \mathrm{m/s}, \quad (E) = \mathrm{J} \text{ or eV}, \quad \left(h\dfrac{c}{\lambda}\right) = \mathrm{eV}, \quad (hc) = \mathrm{eV} \cdot \mathrm{m},$

$(hc)/\mathrm{eV} = \mathrm{m} = (\lambda).$

curred in Germany; for example, Röntgen, the discoverer of X-rays, worked at the University of Munich (although the discovery was actually made in Würzburg), where others like von Laue and Ewald were to make dramatic advances. Since Germans seem to have a running competition to form the world's longest words, they called this continuous spectrum *bremsstrahlung*; which we have adopted as a rather odd sounding English word.

The maximum energy of a photon from such an X-ray tube would arise from a single dead stop collision of an accelerated electron with a target electron. The kinetic energy of the electron is the product of e and V, where e is the charge on the electron and V is the accelerating voltage. If this energy is completely converted to a photon of energy $h\nu$ then the *short wavelength limit* (λ_{SWL}) of the photons in the continuous spectrum will be

$$\lambda_{\text{SWL}}(\text{Å}) = h c/(e V) = 12\,398/V \qquad (4\text{-}6)$$

Superimposed on the white radiation from an X-ray tube are some very narrow spikes. The wavelengths of these lines were first shown by Moseley to be a function of the atomic number of the target material. They arise from billiard ball like collisions which eject inner shell electrons from the target atoms. This process is described more fully below. It should be noted that it makes no difference whether an inner shell "photoelectron" is ejected by an electron, as in an X-ray tube or, by a photon as in an X-ray spectrometer, the resulting emission lines will be the same. It is these nearly monochromatic emission lines which we employ for most of our X-ray experiments.

4.3.3 High Intensity Laboratory X-Ray Devices

The conventional modern X-ray tube uses a cup around the tungsten filament held at a potential of a few hundred volts more negative than the cathode so that the electrons are repelled and focused onto the target. The focal spot is actually a line about two centimeters in length, reflecting the length of the filament. Intensity is defined as the photon flux passing a unit area in unit time. Thus, focusing the electrons onto a smaller area increases the intensity. Various modifications of design parameters have produced "fine focus" and "long fine focus" X-ray tubes which take advantage of this fact to produce higher intensity. However, approximately 98% of the energy from the impacting electrons goes into producing heat. The limitation on the intensity which may be produced is the efficiency of the cooling system which prevents the target from melting.

Since the X-rays may be viewed from any of the four sides of the tube, two sides will produce X-rays from the line projection of the filament. The other two sides view the projection of the line from the end giving a focal spot (actually a rhombus) of about 1 mm^2 in size, when viewed from the usual take-off angle of from 3° to 6°. The take-off angle is the angle at which an experiment views the X-ray tube target. The higher the angle the more divergent X-rays will be present and the lower will be the resolution of any experiment. On the other hand, decreasing the angle decreases intensity but, by limiting the amount of angular divergence in the beam, increases the experimental resolution.

Microfocus tubes use the focusing cup to squeeze the electron beam down to a spot focus with a diameter as small as 10 μm. These units are used for experiments requiring extremely intense beams and can accept the small area of illumination. Such tubes usually have replaceable targets. Another, more popular, method to increase the intensity of an X-ray tube is to increase the power on the target and avoid melting it by rotating it. These rotating anode tubes continuously bring cool metal into the path of the focused electron beam. Such units can typically be run as high as 18 kW compared to about 1.8 kW for a sealed tube. They produce very intense X-ray beams. However, owing to the mechanical difficulties of a high speed motor drive which must feed through into the vacuum, there are difficulties in routine continuous operation. In recent years these units have become more common and more reliable.

The last laboratory method for generating X-rays is to charge a very large bank of capacitors and to dump the charge, in a very short time, to a target. These *flash X-ray* devices can reach peak currents of 5000 A in the hundreds of kV range. The extremely intense X-ray flash lasts for only a few nanoseconds but this has not stopped workers from performing some very clever experiments within this incredibly small time window.

4.4 Interaction of X-Rays with Matter

Consider the simple experimental arrangement shown in Fig. 4-4. Any mechanism which causes a photon, in the collimated incident X-ray beam, to miss the detector is called absorption. Most of the mechanisms of absorption involve the conversion of the photon's energy to another form; while some simply change the photon's direction. For the purposes of this discussion it is best to consider I_0 a monochromatic beam and that the detector is set only to detect X-rays of that energy. We may place the possible fates of an X-ray photon, as is passes through matter, in the following categories.

4.4.1 No Interaction

The fundamental reason for all X-ray—atom interactions is the acceleration experienced by an atom-bound electron from the oscillating electric field of the X-ray's electromagnetic wave. The probability of any interaction decreases as the energy of the wave increases. The probability of interaction is approximately proportional to the wavelength cubed. Thus, short wavelength photons are very penetrating while long wavelengths are readily absorbed. There is always a finite probability that an X-ray will pass through matter without interaction.

The simple cubic relationship of interaction probability is disturbed by the phenomenon of *resonance absorption*. When the energy of the incident radiation becomes exactly equal to the energy of a quantum allowed electron transition between two atomic states, a large increase is observed in the probability of a photon's being absorbed. The dramatic increase in absorption as photons reach the ionization potential of each of the electrons in an atom results in a series of absorption edges shown in Fig. 4-8.

4.4.2 Conversion To Heat

Heat is a measure of atomic motion. Heat may be stored in the quantum allowed translational, rotational and vibrational energy states of the atoms or molecules in a material. It also can be stored in the various excited electronic

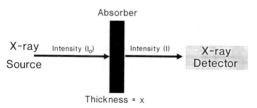

Absorber

X-ray Source   Intensity (I_0)   Intensity (I)   X-ray Detector

Thickness = x

Figure 4-4. A simple absorption experiment.

states allowed to an atom and in the motion of the relatively free electrons in metals. The principal mechanism for converting photons to heat in insulators is the stimulation of any of the modes of vibration of the lattice.

There are two classes of vibrational modes allowed to any lattice. One is the acoustic modes of vibration which may be stimulated by a mechanical force such as a blow or an incident sound wave. The other class is the optic modes of vibration. Optic modes are characterized by a change in dipole moment as the atoms vibrate. This change in electrical field in the lattice allows these modes to interact with the electric field of a photon. Thus, an X-ray photon may stimulate an optic lattice vibrational mode which we observe as heat. The efficiency of the coupling between the lattice vibrational modes, called phonons, and photons, depends both on the lattice itself and on the energy of the incident photon. Thus, we observe sample heating in an X-ray beam to be higher in some samples than in others.

In fact X-rays can also gain energy by absorbing a phonon. The energy of the lattice vibrational modes is on the order of 0.025 eV, while a Cu K_α photon has an energy of 8 keV. Thus, the modification of the incident X-ray beam is rather small, and of course can be studied to understand the phonon structures of solids. However, Raman spectroscopy and thermal neutron scattering are better for these types of studies. Photons whose energy has been modified by a phonon interaction contribute to experimental background as thermal diffuse scattering.

4.4.3 Photoelectric Effect

In a photon–electron interaction, if the photon's energy is equal to, or greater

than, the energy binding the electron to the nucleus, the electron may absorb all of the energy of the photon and become ionized as shown in Fig. 4-5. The free electron will leave the atom with a kinetic energy equal to the difference between the energy of the incident photon and the ionization potential of the electron. This high energy electron can, of course, go on to initiate a number of photon creating events. However, any secondary photon must have a lower energy. The experiment illustrated in Fig. 4-4 assumes that the detector is set only to count pulses of the same energy as in the incident monochromatic beam. Thus, these secondary or fluoresced photons of lower energy do not get included in the measurement of intensity.

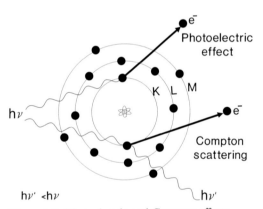

Figure 4-5. Photoelectric and Compton effects.

4.4.3.1 Fluorescence

An atom, ionized by having lost one of its innermost K or L shell electrons, is left in an extremely unstable energy state. If the vacancy has occurred in any orbital beneath the valence shell, then an immediate rearrangement of the electrons in all of the orbitals above the vacancy will occur. Electrons from higher orbitals will cascade down to fill in the hole. This process, illustrated in Fig. 4-6, causes the emission of

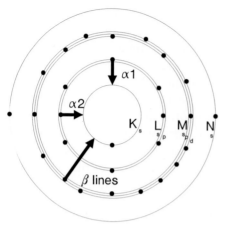

Figure 4-6. Fluorescence from an ionized atom.

secondary fluorescent photons. The energy gaps between the various electron orbitals are fixed by the laws of quantum mechanics. Thus, the photon emitted by an electron falling to lower energy (getting closer to the nucleus) will have a fixed energy, depending only on the number of protons in the nucleus. The photons fluoresced by any element will thus have X-ray wavelengths characteristic of that element.

If the ionized electron comes from the K shell, then there is a certain probability that an L_p, L_s or an M electron will fall in to replace it. The names of the resulting emitted photons are the $K_{\alpha 1}$, $K_{\alpha 2}$ and K_β, respectively. For a Cu atom the transition probabilities are roughly $5:2.5:1$, respectively. The energies of any of these lines must, of course, be less than that of the original incident X-ray which caused the ionization. The study of the fluoresced photons is called X-ray fluorescence spectroscopy (XRF). This technique allows the rapid qualitative analysis of the elements present in a material and with more work, the quantitative analysis of the elemental composition. See Chap. 9 of this Volume by Jenkins for a complete description of this method.

4.4.3.2 Auger Electron Production

There is a special tertiary effect of photoelectron production called the emission of Auger (pronounced oh-jay) electrons. Sometimes the removal of an inner-shell electron produces a photon which in turn gets absorbed by an outer-shell, valence electron. Thus, the incident X-ray gets absorbed by, for example, a K shell electron which leaves the atom. An L shell electron can fall to the K shell to fill in the hole and thereby causes the emission of a K_α X-ray photon. However, before this photon can leave the atom it gets absorbed by a valence electron which ionizes and flies off leaving a doubly charged ion behind. This process is illustrated in Fig. 4-7.

The kinetic energy of the Auger electron is not dependent on the energy of the initial X-ray photon which ionized the K electron. Any X-ray with sufficient energy to create the initial K hole can be responsible for the subsequent production of an Auger electron of fixed kinetic energy. This very specific kinetic energy is equal to the difference in energy between the fixed-energy K_α or K_β photon which ionized the Auger

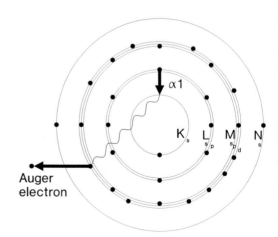

Figure 4-7. Auger electron emission from an ionized atom.

electron and the fixed binding energy of the valence electron to the nucleus. The study of these Auger electrons, called Auger Spectroscopy, allows us to measure the precise energy of the chemical bonds which involve the valence electrons. It also allows us to see subtle energy differences between chemical bonds. A full description of this characterization method may be found in Chapter 13 by Lou in Volume 2 B.

4.4.3.3 Fluorescent Yield

The efficiency of the production of characteristic X-rays is a function of the atomic numbers, z, of the elements in the absorber. We measure this efficiency with a quantity called the fluorescent yield (ω) which for a given spectral series is the ratio of the number of photons produced to the number of vacancies created

$$\omega = \frac{z^4}{A + z^4} \tag{4-7}$$

where A is a constant on the order of 10^6 for the K series and 10^8 for L series X-rays. The fluorescent yield becomes poorer approaching zero, as the atomic number decreases. Thus, traditional X-ray fluorescence was seldom used for elements lighter than sodium. The reason for this poor efficiency is that as the atomic number of the absorber decreases it is more and more likely that the characteristic X-ray photon will be reabsorbed and cause the emission of an Auger electron. Therefore the efficiency of the production of Auger electrons is inversely proportional to the efficiency of producing (fluorescent) characteristic X-rays as a function of atomic number. It should be noted however, that modern high vacuum wavelength dispersive spectrometers (WDS) are able to observe, and even quantitatively analyze, elements as low in atomic number as beryllium.

4.4.4 Compton Scattering

This phenomenon, illustrated in Fig. 4-5, amounts to an inelastic collision between a photon and an electron. Part of the energy of the incident photon is absorbed by an electron and the electron is thus excited. However, instead of the remaining energy of the original photon converting to kinetic energy of the excited photoelectron, some of it is re-emitted as an X-ray photon of lower energy. Not only has the energy of this Compton photon been lowered but it loses any phase relationship to the incident photon. For this reason the process is called *incoherent scattering*. The increase in wavelength of the scattered photon is dependent on the angle between the scattered photon and Compton electron,

$$\Delta\lambda = \frac{h}{m_e c}(1 - \cos 2\,\theta) \tag{4-8}$$

The intensity of incoherent radiation, $I_{Compton}$ is given by,

$$I_{Compton} = Z - \sum_{j=1}^{\substack{\text{number of atoms in unit cell}}} f_j^2 \tag{4-9}$$

where f is a measure of the X-ray scattering from an atom and related to the number of electrons on an atom as described in Sec. 4.7.7.2 below. f is equal to Z at zero degrees θ and falls off as θ increases. As Z increases, f^2 will increase more rapidly, thus Compton scattering decreases in intensity as the atomic number of the scatterer increases.

4.4.5 Coherent Scattering

The last important mechanism of X-ray absorption in matter is the one which leads to the phenomenon of diffraction. Coherent scattering is analogous to a perfectly elastic collision between a photon and an electron. The photon changes direction after colliding with the electron but transfers none of its energy to the electron. The result is that the scattered photon leaves in a new direction but with the same phase and energy as that of the incident photon. It is coherent scattering that leads to the phenomenon of diffraction. However, since the incident photon changes direction, and therefore will miss the detector in Fig. 4-4, it is considered an absorption mechanism.

A minor point could disturb the reader here. The rules of quantum mechanics appear to restrict the energy of an atom-bound electron to fixed values. How, therefore, can such an electron accelerate from the electric field of a photon of random energy and temporarily absorb its energy? The answer lies in the uncertainty principle which, in this context, states that an electron can have a large energy within a time interval approaching zero. For coherent scattering, since there is no phase shift, the electron must absorb and re-emit the photon's energy in exactly zero time. In zero time the electron may take on any value of energy.

4.4.6 Absorption

A great number of experiments of the type illustrated in Fig. 4-4 have shown that when electromagnetic waves, of any fixed wavelength, are absorbed by any form of matter, the following general equation holds

$$I = I_0 e^{-\mu x} \qquad (4\text{-}10)$$

or

$$\ln \frac{I}{I_0} = -\mu x \qquad (4\text{-}11)$$

where:

I_0 is the intensity of incident X-ray beam,
I is the intensity of beam at the detector,

μ is the linear absorption coefficient (in cm^{-1}) and
x is the thickness of material (in cm).

The difference between I and I_0 for a fixed wavelength is therefore dependent on the thickness of the absorber and on the linear absorption coefficient μ. μ is a constant related to the absorbing material. Since all of the absorption processes described above ultimately depend on the presence of electrons, then it seems clear that the ability of a material to absorb electromagnetic radiation will relate to the density of electrons. In turn, the electron density of a material will be determined by the types of atoms composing the material and the closeness of their packing. The linear absorption coefficient of a material will therefore depend on the types of atoms present and the density of the material. However, if we eliminate the functional dependence on density, which is determined by the type and strength of the chemical bonds in a material, we will obtain a true constant, (μ/ϱ), for each element at any specified wavelength.

The quantity (μ/ϱ) is called the *mass absorption coefficient*. The (μ/ϱ) of Zr is shown as a function of wavelength at the top of Fig. 4-8. The sharp discontinuity in the curve is called *absorption edge*. They correspond to the entrance of a new absorption mechanism. The highest energy edge corresponds to the ionization potential of a K electron and is called the K edge. Other edges correspond to the L and M electron ionizations. The exact location of each of the edges corresponds to the energy required for the removal of that electron. Thus, absorption edges yield fundamental information about the structure of the atom, its environment and the amount of energy required to excite that series of characteristic fluoresced radiation. The

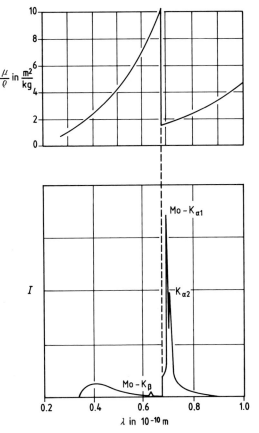

Figure 4-8. The absorption spectrum of Zr and the effect of a Zr filter on a Mo X-ray spectrum.

fine structure of these edges gives information on the exact energies of the electrons within the atom. These energies are slightly affected by the bonding environment around the atom. To explain the fine structure would require a full molecular orbital description of the atoms in the absorber and this fact dampened the spirits of many early researchers.

When it is necessary to find the mass absorption coefficient of a compound, solution, or mechanical mixture, containing more than one element, it may be computed by taking the weighted average of the mass absorption coefficients of the constituent elements

$$\left(\frac{\mu}{\varrho}\right)_{mixture} = X_1 \left(\frac{\mu}{\varrho}\right)_1 + X_2 \left(\frac{\mu}{\varrho}\right)_2 + \dots \quad (4\text{-}12)$$

or

$$\left(\frac{\mu}{\varrho}\right)_m = \sum_{j=1}^{number\ of\ elements} X_j \left(\frac{\mu}{\varrho}\right)_j \quad (4\text{-}13)$$

where ϱ is the density, $(\mu/\varrho)_j$ are the mass absorption coefficients of the constituent elements, and X_j are the weight fractions of the elements present.

An interesting application of the absorption edge is when the edge of one element (A) is located between the K_α and K_β lines for another element (B). When this occurs the K_β from the B atoms will be very strongly absorbed while the longer wavelength K_α will be only slightly absorbed. This means that a suitable thickness of element A can act as a *beta-filter* for characteristic radiation from element B. The bottom of Fig. 4-8 shows the effect of putting a Zr foil in a beam of radiation generated from a Mo X-ray tube. This method of emphasizing the K_α to make a pseudo-monochromatic beam for diffraction experiments has been used for many years. On most modern diffractometers a graphite crystal is used as a monochromator for the diffracted beam, eliminating the use of beta-filters.

4.5 The Detection of X-Rays

The nature of X-ray detectors and the electronics used with them is an important part of understanding the approaches to, and limitations of, X-ray applications. Beginning with the introduction of the Geiger-Müller counter (1928) and its use for the measurement of diffracted X-rays by LeGalley (1935), the modern X-ray diffractometer has developed primarily with the use of scintillation detectors, while fluorescence spectrometers more commonly use proportional counters. Most recently position sensitive proportional counters and high energy resolution solid state detectors have gained popularity. The principles of each of the methods used to detect X-rays will be described here, illustrating both the types of electronics and the physics employed.

4.5.1 Non-Electronic Detectors

4.5.1.1 Photographic Film

This is one of the oldest methods of detecting X-rays. When exposing the silver halide grains in a film so that they will become developable (i.e., reduced to pure silver on the plastic substrate) the grain size is important. Each particle in the emulsion which absorbs an X-ray photon becomes sensitized. Once sensitized the entire particle will reduce to silver in the development process. Thus, if a region of the film is exposed to an intense X-ray flux the same particles may be struck more than once. This causes a loss of information analogous to the dead time in electronic counters. The film darkening is only proportional to the intensity of the exposing X-rays over what is called the linear range of the film. To allow for this effect, when measuring intense sources, one must use a film of smaller grain size to extend the linear range or reduce the X-ray intensity with filters in front of the film or reduce the incident beam flux.

4.5.1.2 Fluorescent Screens

ZnS doped with Ni will emit a greenish visible light when struck by X-rays. Screens made of this material are useful in aligning X-ray cameras and intensifying X-rays to be recorded with a film sensitive to visible light.

4.5.1.3 Human Skin

Just to be complete we should consider this material which will redden and ulcerate under exposure to X-rays. Owing to its very slow response time and high dead time it is not a very useful detector. It is recommended that workers keep their skin out of the X-ray beam. In this context the phrase "dead time" has a more macabre meaning. The conventional meaning is the amount of time an electronic detector is inactive while processing a photon, before it can respond to another photon. The dead time of over-exposed human skin is infinite.

4.5.2 Gas-Ionization Detectors

The creation and collection of gas ions and electrons in an electrostatic field has led to the development of a number of different X-ray detectors:

(1) Ionization chamber.
(2) Sealed gas proportional counters.
(3) Flowing gas proportional counters.
(4) Geiger-Müller detectors.
(5) Position sensitive proportional counters.

All of these detectors are based on the same basic principle illustrated in Figs. 4-9 and 4-10. Figure 4-9 is a simple schematic

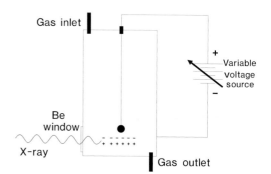

Figure 4-9. A schematic diagram of an ion collection device.

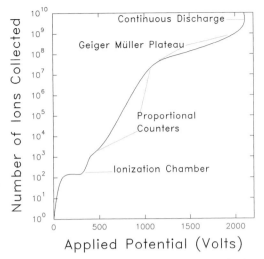

Figure 4-10. Gas amplification in ion collection devices.

of an ion collection device. The container has a beryllium window to allow the entrance of X-rays. The gas in the container is varied depending on the exact application the user has in mind however, P10 counting gas (10% methane and 90% Ar) is the most common gas used. The entering X-ray, on striking an argon atom, will, by the photoelectric effect, create an ion pair (an Ar^+ ion and an electron). The valence electron ejected from the Ar atom leaves with a kinetic energy equal to the energy of the ionizing X-ray minus its binding potential. The photon continues to lose its energy by colliding with other Ar atoms and creating a number of secondary ion pairs. The number of ion pairs created is proportional to the energy of the incident photon.

The fate of the ions and electrons will depend on the potential applied between the central wire and the outer cathode container. If, for example, no potential is applied, all of the electrons will recombine with ions and we have accomplished nothing very interesting! The five types of ion collection detectors mentioned above result from differing amounts of applied po-

tential to the circuit in Fig. 4-9 as illustrated in Fig. 4-10.

1. *Ionization chamber.* If the voltage applied to the ion collection chamber is on the order of 100 to 200 V, then the electrons and ions will drift to their respective electrodes and a current pulse will flow through the circuit. A $Cu K_\alpha$ photon (8040 eV) and an effective ionization energy of Ar of about 25 eV means that about 300 ion pairs can be produced and collected. This extremely small current must be amplified before a useful detector can be built. The ionization chamber is no longer in common use.

2. *Sealed gas proportional counters.* If the voltage applied to the ion collector is in the 700 to 800 V range then, the perhaps 300 electrons accelerate so rapidly to the anode that they collide with other Ar atoms, causing secondary ionizations. This effect is called gas amplification. At a constant applied potential the amount of current in the pulse that results will be proportional to the energy of the incident X-ray, hence the name proportional counter. The *energy resolution ($\Delta E/E$ of about 20%)* is quite good: about 3–4 times better than a scintillation detector. The dead time in these detectors is primarily the result of the time it takes for the relatively slow moving Ar^+ ions to move to the cathode and discharge. This process is slowed even more by the fact that the electrons move about 1000 times faster than the Ar^+ ions, leaving the gas with a net positive space charge. Due to mutual ion repulsions within this space charge, it takes a few micro seconds for discharge to complete, determining the dead time.

3. *Flowing gas proportional counters.* The principles are exactly as described for the sealed counter. The advantages are a nearly unlimited lifetime via occasional replacement of the anode wire, and much

wider range of sensitivity to X-ray energies made possible by the use of thinner X-ray detector windows.

4. *Geiger-Müller detectors.* When about 1000 V are applied to the sealed chamber a full Townsend avalanche occurs, where a single ionization causes every Ar atom in the chamber to ionize. This detector needs little external amplification due to the large current pulse which results from any X-ray entering the chamber. The principle disadvantage is a complete loss of information on the energy of the incident X-ray. In addition, since the Townsend avalanche is complete in this type of detector the space charge effect is extreme, producing a dead time of about 200 µs.

5. *Position sensitive proportional counters.* In recent years a number of designs have been developed which use the principles of the flow proportional counter in a number of clever ways which tell where on the anode wire the ionizing X-ray photon hit. The position sensitive detector (PSD) shown in Fig. 4-11 uses a graphitized wire to slow the current pulses which travel in both directions, to a sensitive time discriminator which looks at the time difference between the pulses arriving from each end

of the wire and determines the location where the photon struck. When X-rays scatter from a sample to a detector, the angle between the incident and scattered beams is called (2 θ). The wire of the PSD can be placed to sense the X-rays scattering from more than ten degrees of 2 θ simultaneously. The photon location information can be used to address the appropriate channels in a multichannel analyzer (MCA) where the full scattering (or diffraction) pattern can be stored, while the PSD scans across 2 θ (Göbel, 1982a). A number of PSD designs allow one to view a few to over a hundred degrees around an X-ray experiment.

4.5.3 Solid State Detectors

1. *Lithium-doped Si or Ge crystals.* This type of detector must be immersed in liquid nitrogen in order to reduce the diffusion of the Li out of the semiconductor and to reduce the electrical background noise so that a field effect transistor (FET) can amplify the delicate signal. The principle of operation is almost exactly like that of the ion collection devices described above. An

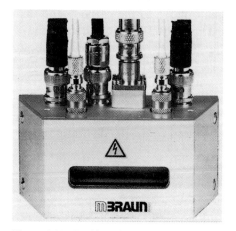

Figure 4-11. Position sensitive detector with its associated electronics (Courtesy of Braun).

X-ray photon enters the Si and is absorbed creating a photo electron whose energy is equal to that of the X-ray minus the binding energy of the electron. The energetic photoelectron then loses its energy in 3.8 eV increments by creating a string of electron hole pairs. Thus, the number of electrons created are proportional to the energy of the incoming photon. An electrical bias across the detector causes the electrons to be collected for amplification at the FET. The energy resolution (5%) is the best of any of the current detectors: about 15 times better than a scintillation detector. The dead time is determined primarily by the electronic amplification process and is on the order of 50 μs.

2. *Scintillation Detectors* are composed of a crystal of NaI with some Tl doped into it. When an X-ray photon hits the atoms in this crystal it get absorbed by producing a photo electron and a few Auger electrons. These electrons activate the production of fluoresced photons (scintilla) of blue light with a wavelength of 4100 Å, at the thallium sites. The total number of blue light photons produced are absorbed by a photoelectric material, typically a cesium-antimony intermetallic alloy. Approximately one electron is produced for each ten photons which strike the photocathode. The electrical signal in turn, is proportional to the energy of the original X-ray which entered the crystal. Thus, we have an energy discriminating detector. Unfortunately, owing to a rather large number of losses, the overall resolution is on the order of 75% of the energy of the incident photon. Thus, the resolution is not such as to be able to resolve the difference between Cu K_α (8.047 keV) and Cu K_β (8.9094 keV). This is made up to some extent by the high quantum efficiency (most entering photons are counted) and a low dead time of about 1 μs.

4.5.4 The Electronic Processing of X-Ray Signals

To operate a scintillation detector we need a high voltage source to put an electrical bias across the photocathode material and the very sensitive series of amplifiers (dynodes), which build the signal from the photon to a level that it can be amplified by a conventional amplifier and transported through a wire for processing. In a proportional counter we also have to apply a potential which will determine the gas amplification factor.

The preamplified signal brought back from most X-ray detectors is proportional to the energy of the X-ray photon which caused it. The only electronic detector still in occasional use for which this is not true is the Geiger-Müller detectors. The output from these detectors can be sent directly to a rate meter or scalar/timer for display. The electronics associated with the detection of X-rays on most modern X-ray equipment are illustrated in Fig. 4-12.

The initial preamplified voltage pulse arriving at the signal processing equipment is first sent to a linear amplifier to adjust its level to one appropriate for pulse height discrimination. An electrical circuit known as a pulse height analyzer (PHA), which is like a band pass filter, is the device we use to throw away extraneous pulses caused by any event other than a photon in the energy range of interest. Anyone not familiar with an electronic band pass filter should think of the very common graphic equalizer, or spectrum analyzer, attached to most modern stereo systems. This device breaks the music signal into a series of frequency bands which may be individually amplified. The PHA acts in an analogous manner, rejecting all pulses above and below voltages selected by the user.

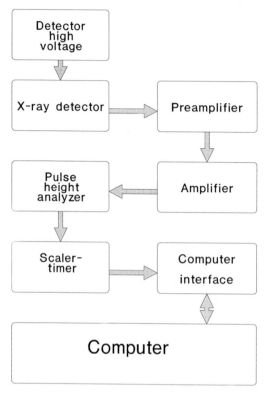

Figure 4-12. The electronics of signal processing.

The advantage of this device is clear. For example, cosmic rays send showers of very high energy particles into our detectors, also contamination of our tube target with another metal (like tungsten vaporized from the filament) will produce other characteristic radiation. In addition, there is a dark current, which is the phrase we use for electrons which spontaneously boil off of the photocathode in a scintillation detector, or spontaneous ionization in a proportional detector. These are a few of the many sources of background counts which will produce pulses of significantly higher or lower voltage than those caused by the radiation of interest. A pulse height analyzer can be set to reject all but the pulses from the radiation desired, within its and the detector's energy resolution.

After a pulse has been accepted by the PHA, it is passed on to two independent circuits. A scalar/timer, which allows us to count the number of pulses arriving in any time interval, and a rate meter. The rate meter is a circuit which takes in the random arrival of pulses and puts out an average signal, which can be displayed on a calibrated volt meter or a strip chart recorder.

In energy dispersive spectrometry (EDS) the high energy resolution of the Si(Li) crystal allows a different electronic approach. In place of the PHA and the scalar timer there is a multichannel analyzer (MCA). This device acts like an audio graphic equalizer with, from 4000 to 64 000 separate filters. The typically 4000 channels of an MCA each act as a separate scalar with its own PHA. An MCA allows us to store and then view photons of many different energies which strike the detector. Each energy will be counted into one of the scalars in the MCA. This is accomplished by converting the voltage associated with each ionization event to a digital number, using an analog to digital converter, and then using the resulting number to address a channel in the MCA. For wavelength dispersive spectrometry (WDS) and normal diffraction work we are only interested in photons of a single energy and therefore use the system shown in Fig. 4-12.

4.6 Crystallography

Symmetry can be thought of as an invisible motion of an object. If an object is hidden from view and moved in such a manner that when it reappears you cannot tell that it has been moved, then the object is said to possess *symmetry*. For example, if a perfect snowflake is covered and then rotated about its center by 60°, an observer will not be able to tell that any movement has occurred. The rotation of an object around an imaginary axis is the simplest symmetry element to visualize and is referred to as rotational symmetry. To get the complete set of all possible symmetry elements that an object may possess we must also allow for the possibility that, on removing the object from view, no movement at all was made. This of course, produces an effect identical to rotating the object about any of its symmetry axis and must also be a symmetry element. This element is called the identity operator by mathematicians but is more simply referred to as a 360° rotation by crystallographers who refer to it with the symbol 1.

Rotational symmetry is said to occur around a *proper axis* when the "handedness" of an object does not change. For example think of four right hand arranged in a circle each separated by 90°. The 4-fold axis relating them is proper, in that the hands remain "right" during each rotation. A rotation which changes the "handedness" after each operation is called *improper*. An improper rotation is a rotation followed by the operation of inversion through the origin of a coordinate system. Inversion causes an object located at x, y, z to change coordinates to $\bar{x}, \bar{y}, \bar{z}$. This, of course, turns a right hand into a left. Table 4-2 shows the complete set of crystallographic proper and improper axes of rotation.

The symbols used (n and $\bar{n}$) are called Hermann-Mauguin (Hermann, 1931; Mauguin, 1931) symbols and are uniformly used by crystallographers. The value of n refers to a rotation of $360/n$ degrees. Spectroscopists prefer a different set of symbols called Schönflies (1891) notation. Schönflies symbols only apply well to the shapes of objects or point groups and have serious ambiguity when applied to repeat-

Table 4-2. Crystallographic symmetry elements.

Degrees of rotation	Proper axis	Improper axis
$360°/n$	n	$\bar{n}$
360	1	$\bar{1}$
180	2	$m (= \bar{2})$
120	3	$\bar{3}$
90	4	$\bar{4}$
60	6	$\bar{6}$

ing arrays of objects or space groups which are described below.

The reader will note that the symmetry elements in Table 4-2 do not include a 5-fold rotation or any rotation greater than 6. In crystallography and mathematical group theory we always seek the *irreducible representation* of an object's symmetry (i.e., the simplest set of symmetry elements which describe the full symmetry of an object). The only classical irreducible rotations which can fill space are 1-, 2-, 3-, 4- and 6-fold rotations.

The absence of a 5-fold axis from Table 4-2 is due to the very old principle that objects with 5-fold symmetry cannot fill space, and therefore may not be observed as a crystallographic symmetry. The reader may easily verify this by trying to arrange cutouts of pentagons into a space filling array in two dimensions. As with most age-old principles, given sufficient time we must begin to add qualifications. Beginning in 1984, Shechtman et al. (1984) opened a whole new field of crystallography by discovering five fold symmetry in an electron diffraction pattern of a rapidly quenched Al_6Mn alloy. In their pattern there is not one repeating pentagon (which clearly cannot fill space) but two different sized pentagons in a pattern which, while filling space, does not produce any simple repetition. The ratio of the sides of these two pentagons is 1.6 ± 0.02, which is a number referred to as the "golden mean" of Hellenic architecture $(\sqrt{5} - 1)/2$. A two dimensional arrangement of two different 4-sided rhombs having the same, golden mean, ratio of their perimeters can also fill space when arranged in a pattern known as a *Penrose tile*. Penrose tilings have become the most widely studied of two dimensional space filling patterns which have no repeating unit cell. These patterns show local regions of five-fold symmetry. Figure 4-13 shows, previously thought to be impossible, ten fold symmetry. In fact examples of eight and twelve fold axis have now also been reported. None of these materials qualify as a crystal in the conventional sense in that, they are only quasi-periodic. Hence they have been called quasi-crystals. However, this name should not lead one to conclude that these materials are not real crystals, in that the best of these give diffraction patterns as sharp as any conventional crystal.

The full nature of quasicrystals, is still a subject of active investigation, with no clear structural model able to explain the growing number of examples. For further information see the recent comprehensive review by Steurer (1990) and the book by Hargittai (1990).

4.6.1 Point Groups

Now that we have examined the possible types of simple symmetry that an object, which can fill space may possess, we may next ask how many unique combinations of symmetry these space-filling objects may have. The simplest symmetry group is 1, or no symmetry at all. The addition of any other symmetry element to group 1 will, of course, raise the total symmetry and have to be classified as a new group. The addition of a center of symmetry puts us into group $\bar{1}$. Increasing the symmetry further, we must look at objects containing only a 2-fold axis (group 2) or a mirror (group *m*). At this point we see the first opportunity of combining two symmetries in an object to form a unique new group, 2/*m*, which refers to objects which have a 2-fold axis with a mirror perpendicular. The two axes in the plane of the mirror cannot meet at 90° or they would be forced to become 2-fold axes forcing the total symmetry into the group 222, *mm*2 or *mmm*.

Similar arguments lead to the conclusion that there are only 32 unique ways of combining symmetry elements in objects which can repeat in three-dimensions to fill space. These are called point groups. We should comment here on the type of coordinate systems required to specify these various symmetries. In the groups 1 and $\bar{1}$ there need be no relationship between the three coordinate axes or the angles between them. We refer to this coordinate system as *triclinic* or *anorthic*. For the groups 2, *m* and 2/*m*, there need be no relation between the lengths of the three axes

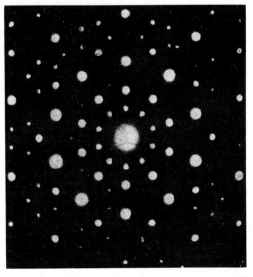

Figure 4-13. Electron diffraction pattern of a rapidly quenched Ni–Ti–V alloy (from Zhang et al., 1985).

but this symmetry requires that one axis meet the plane of the other two at right angles. This system with one unique axis is called *monoclinic*. The groups 222, *mm2* and *mmm* require that all three axes be orthogonal. This class is called *orthorhombic*. To increase the symmetry further will require that one or more of the axes must be equal in length. In the *hexagonal* and *tetragonal systems* the unique axis is called *c* by convention. In *cubic* all three orthogonal axes are equal. In the last crystal class *rhombohedral*, the three axes are equal in length but each are separated by a common non-90° angle α. Rhombohedral differs a bit from the crystal systems in that we can represent it on a hexagonal axis; thus, we really only need six coordinate systems to represent all symmetries. Table 4-3 shows the relationships among the axes and interaxial angles demanded by the symmetry of the various point groups. The seven different coordinate systems are called crystal systems.

4.6.2 Bravais Lattices

It is convenient for us to imagine that the atoms in a crystal are arranged in groups which repeat three-dimensionally to make up a crystal structure. If we look at each symmetrically unique group of atoms as if they were a single point, then we can imagine the entire crystal structure as a three-dimensional lattice of these points. This concept is called a space lattice and the points which define it are lattice points. It is important to understand that the lattice points which we use to abstractly think about a crystal structure are not atoms, but represent a collection of atoms (which are sometimes, but definitely not always, a molecule or a formula unit). The lattice point represents the *asymmetric unit* or *basis* of the crystal structure which is the smallest group of atoms upon which the crystallographic symmetry operates to produce the complete crystal structure.

We can recognize symmetry in the various possible ways of arranging points into three-dimensional arrays. These periodic arrays can be sorted into seven unique arrangements. These seven basic patterns are called crystal systems and define the coordinate systems described in Table 4-3. In a periodic array of points we can think of another type of symmetry: translational symmetry. Imagine that there are translation vectors extending between the lattice points along the three principal axes of the coordinate system. Connecting all of these vectors will produce a collection of boxes each called a unit cell. The three vectors which define the unit cell have lengths of *a*, *b* and *c*, making angles of α, β and γ with each other. The number of asymmetric units in each unit cell is called Z. In two dimensions we can recognize patterns cor-

Table 4-3. The seven crystal systems and fourteen Bravais lattices.

Crystal class	Axis system	Lattice symmetry	Bravais lattice
Cubic	$a = b = c$, $\alpha = \beta = \gamma = 90°$	$m3m$	**P, I, F**
Tetragonal	$a = b \neq c$, $\alpha = \beta = \gamma = 90°$	$4/mmm$	**P, I**
Hexagonal	$a = b \neq c$, $\alpha = \beta = 90°$, $\gamma = 120°$	$6/mmm$	**P**
Rhombohedral	$a = b = c$, $\alpha = \beta = \gamma \neq 90°$	$\bar{3}m$	**R**
Orthorhombic	$a \neq b \neq c$, $\alpha = \beta = \gamma = 90°$	mmm	**P, C, I, F**
Monoclinic	$a \neq b \neq c$, $\alpha = \gamma = 90°$, $\beta \neq 90°$	$2/m$	**P, C**
Triclinic	$a \neq b \neq c$, $\alpha \neq \beta \neq \gamma \neq 90°$	$\bar{1}$	**P**

responding to a general rhombus, $60°$ rhombus, square, rectangle and a rectangle with an additional lattice point in the center of its face or face centered. If we chose only the simple cell in the last case it would have no symmetry. The choice of the face centered cell allows us to specify *mm* symmetry.

When the unit cell, which repeats in space, is defined by one point at each corner, the unit cell is called *primitive* and given the symbol **P**. If there is an additional lattice point in the middle of each cell it is called *body centered* and given the symbol **I**. When the cell has points in each corner and in the center of each face, it is called *face centered* and the symbol **F** is used. The rather strange rhombohedral cell, in which the three interaxial angles are equal and acute, is given the symbol **R**. The last type of Bravais lattice is when only one face is centered. The symbol used to describe this type of cell depends on the name of the face that is centered. The plane perpendicular to the *a* and *b* directions of the lattice is called **C**. Likewise, the face perpendicular to the *a* and *c* directions is called **B**. On considering the seven basic lattice symmetries and the various types of unit cells allowed to each symmetry, there are fourteen Bravais lattices which describe the only ways of filling space in a periodic manner. These are also indicated in Table 4-3.

4.6.3 Space Groups

We have seen that there are 14 space lattices and 32 point groups. If we arrange the 32 point groups in the various patterns allowed by the 14 Bravais lattices we will be able to distinguish 230 unique three-dimensional patterns which we call space groups. Every crystal structure can be classified into one of these 230 symmetry groups. It has long been recognized that the notation we would need to describe these 230 space groups would need to be extremely complex. However, if we recognize that space groups can be thought to result from the combination of simple primitive symmetry elements (proper and improper rotations and reflection) with the translational symmetries of the Bravais lattices, we can recognize two new combinational symmetry elements: glide planes and screw axes. These non-primitive symmetry elements allow for the succinct space group notation used in the solid state sciences. This notation is most comprehensively described in the *International Tables for Crystallography*, Volume A (1983).

The combination of a translational movement with a rotation changes the rotational axis into a *screw axis*. The symbol for a screw axis is N_j, where N stands for the type of rotational axis (i.e., $360/N$ = the number of degrees in the rotation), and j tells the fraction of the cell translated (i.e., j/N). So 2_1 means a rotation of $180°$ and a translation of $1/2$ of the unit cell along the direction of the rotation axis. A 3_2 means a rotation of $120°$ followed by a translation of $2/3$ of the cell.

The combination of a translation with a mirror reflection is called a *glide plane*. The location of the symbol for the glide plane in the space group symbol tells which mirror the reflection occurs across, while the actual symbol (a, b, c, n or d) tells the direction of the translation. After reflection across a mirror, an "a" glide translates in the *a* direction. Similarly for "b" and "c" glides. An "n" glide translates along the face diagonal of the unit cell while the "d" glide moves along the body diagonal after reflection.

4.6.4 Space Group Notation

The direction of the mirror plane before the translation of a glide operation, introduces the subject of space group notation. Space group symbols are made of two parts: the first is a capital letter designating the Bravais lattice type [**P**, **C**, (**B** or **A**), **I**, **F** or **R**]. The second part of the symbol tells the type of symmetry the unit cell possesses. The symbols used conventionally only specify the minimum symmetry to uniquely identify the space group. All implied symmetry is not mentioned in the symbol. This fact along with a bit of arbitrariness in our choice of origin and definition of directions in a unit cell leads us to one of the great generalizations of modern crystallography: "Look up your space group in Volume A of the International Tables for Crystallography" (1983). This invaluable reference work lists each possible view of a unit cell and all of the implied symmetry along with a wealth of other information.

Since the only possible symmetry that can occur in triclinic is 1 and $\bar{1}$, there are only two space groups: **P**1 and **P**$\bar{1}$. In order to increase the symmetry from triclinic we must add a 2-fold axis or a mirror. To do this requires that we bend one of the axis so that it makes a 90° angle with the plane defined by the other two. This places us in the monoclinic crystal class. Since there can be only one symmetry plane or 2-fold axis, its name is purely arbitrary. We will adopt the most common **b** axis unique notation here.

The various monoclinic space groups are composed of all possible combinations of the point group symmetries 2, m, 2/m and the corresponding translational elements 2_1, c, $2_1/m$, $2/c$ and $2_1/c$ with the Bravais lattices **P** and **C**. These are shown in Table 4-4. Note that in Table 4-4 there are three blank entries for the groups **C**2_1,

Table 4-4. The 13 monoclinic space groups.

Point group	**P**	**C**
2 (C_2)	**P**2	**C**2
	P2_1	
m (C_s)	**P**m	**C**m
	Pc	**C**c
2/m (C_{2h})	**P**2/m	**C**2/m
	P2_1/m	
	P2/c	**C**2/c
	P2_1/c	

C2_1/m and **C**2_1/c. This is because these combinations of symmetry are contained by implication in the groups **C**2, **C**2/m and **C**2/c respectively. This means that there are only 13 unique monoclinic space groups, if any more symmetry is added to a cell the cell is forced into a higher symmetry crystal class and the rest of the 230 allowed space groups result.

4.6.5 Reduced Cells

Any lattice can be described by a large number of choices of unit cells. A cubic lattice may easily be described by three translation vectors which will give monoclinic or even triclinic symmetry. Since any three non-collinear reciprocal lattice vectors will define a unit cell which will repeat in space to give the space lattice, there in fact are an extremely large number of unit cells which may be chosen to describe any lattice (most of them non-primitive). Which is the correct cell? To answer this question we must examine what we mean by "correct". Since the recognition of symmetry saves a great deal of work in describing crystals we follow the usual crystallographic convention of choosing the unit cell with the highest symmetry. However, since we have a huge number of cells to choose from for any lattice, recognizing this cell may, at times, be difficult.

Niggli (see Mighell, 1976) has described a mathematical procedure which will always produce the same reduced cell from any of the large number of descriptions for a lattice. The reduced cell is defined as that cell whose axes are the three shortest noncoplanar translations in the lattice; consequently there is only one such cell in any lattice. It is, by convention, the standard choice for the triclinic cell. This Niggli cell is a good place to start, once three potential translation vectors have been found which describe the lattice. Sublattices and superlattices may sometimes be obtained by halving or centering the "correct" cell. The Niggli procedure will show that these cells also reduce to the same reduced cell. In fact, Niggli showed in 1928, that there are only 28 unique reduced cells. Mighell (1976) has taken advantage of this fact in designing the Crystal Data, database which classifies all materials with known unit cells according to their reduced cells. Program TRACER II of Lawton and Jacobson (1965), carries out the Niggli procedure. Program LATTICE, by Mighell and Himes (1986), will not only find the reduced cell but also will attempt to identify the cell in the Crystal Data computer database.

4.6.6 Miller Indices

The unit cell is a very useful concept and we use it not only to characterize the symmetry of a crystal, but also to specify crystallographic directions and even interatomic distances. To describe directions and distances, we imagine planes, with various orientations, intercepting the translation vectors at various points. We imagine that these planes are members of sets which cut all cells identically. Each plane may be characterized by the integers which result when the reciprocal of the axial intercepts

are taken. For example, the plane which passes through each unit cell intercepting the a axis at $1/2$ and parallel to the b and c axes has Miller indices of $1/2, 1/\infty, 1/\infty$ or (200). To find the Miller indices of any set of planes, only one trick is needed: we must use the mathematical relation that parallel planes will meet at infinity. Thus, a plane parallel to an axis has an intercept of infinity and the reciprocal of infinity gives a Miller index of zero. The following general rule will always, unambiguously, allow the determination of the Miller indices of any plane: Locate the first plane of the family away from the origin of the unit cell and take the reciprocal of its axial intercepts. Figure 4-14 illustrates a series of planes parallel to the c axis with their Miller indices.

Since the imaginary planes, characterized by Miller indices, are defined in terms of the lengths of the unit cell edges, then the perpendicular distances between the planes in any family are characteristic of the unit cell size and hence, the crystal structure. The distances (d_{hkl}) between these planes are on the same order as the distances between atoms. We should note that these **d**-values are vectors, having direction and magnitude. It should also be noted that Miller indices defining a plane are usually written in parenthesis, as $(h\,k\,l)$, while crystallographic directions which are the normal to any plane, are written in square brackets as $[h\,k\,l]$. The subject matter of Section 4.6 is also treated in Vol. 1, Chapter 1.

4.7 Diffraction

Diffraction occurs when waves scattering from an object constructively and destructively interfere with each other. Waves are characterized by their wavelength (λ)

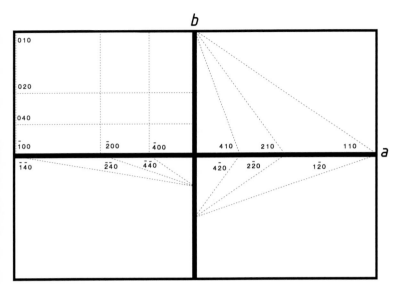

Figure 4-14. Planes, indicated by dashed lines, parallel to the **c** axis. The coordinate system origin is in the center with four unit cells shown about it.

which is defined as the distance between peaks. If a wave scatters from an object it will do so in all directions. If a second wave scatters from another object, displaced from the first by a distance on the order of the wavelength, then there will be some angle at which we can view the two scattered waves such that they will be in phase. This phenomenon is illustrated in Fig. 4-15.

Visible light is an electromagnetic wave with a wavelength of about 50 000 Å. If a surface is ruled with periodic grooves that are 50 000 Å apart and visible light falls on this surface, the light will scatter from each of the rulings and be diffracted. The reason for the diffraction is that the path difference between rays scattering from each adjacent

groove is on the order of λ which causes all of the scattered waves to be in phase at some angle. For a fixed scatterer spacing, the angle will depend on the wavelength, with short wavelengths diffracting at higher angles than longer ones. If the incident light is white (i.e., containing all wavelengths), an observer sees this effect as a rainbow; the different wavelengths of the visible light diffract from this ruled surface or grating at different angles, depending on their wavelength.

4.7.1 Bragg's Law

W. L. Bragg (1913) was the first to show that the scattering process which leads to diffraction can be equally visualized as if the X-rays were "reflecting" from the imaginary planes defined by Miller indices. This view pushes the reflection analogy rather hard; however, this rather strained analogy allows a simple derivation of the overall controlling law which describes the phenomena of diffraction. Of course, the same law can be derived without need of the reflection analogy but requires a bit more effort.

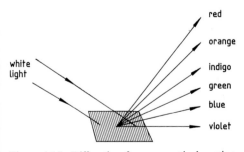

Figure 4-15. Diffraction from an optical grating.

In Fig. 4-16 X-rays impinge on a set of atomic planes, with indices $(h\,k\,l)$, from the left, making an angle θ with them. The distance between the planes is d_{hkl}. It helps if you consider atoms to be located on the planes acting as sources of scattering and thus, for these planes the d_{hkl} would correspond to an interatomic distance. However, all of the atoms in a unit cell will be bathed in the X-ray beam and, in fact, all will be scattering in all directions. It is not necessary that there be any atoms on the planes for diffraction to be thought of as occurring from them but, it makes our picture a little easier to understand. If we look only at beam 1 and beam 2 we see that beam 2 must travel the distance ABC farther than beam 1. If beam 1 and 2 start out in phase, meaning lined up peak to peak and valley to valley, then the extra distance ABC will cause beam 2 to be misaligned with 1 after reflection.

The degree of phase shift or misalignment is equal to the distance ABC. Constructive interference or diffraction will occur when the two waves (and all the waves scattering from deeper planes in the crystal) come out in phase. This of course will happen when the distance $ABC = 1\lambda$ or 2λ or 3λ or, in general, when

$$n\lambda = ABC \qquad (4\text{-}14)$$

where n is an integer. All that remains is to get the distance ABC in terms of the measurable angle θ. To do this we look at the right triangle ABO and notice that $d\sin\theta = AB$, or $2\,d\sin\theta = ABC$. Then the condition for diffraction to occur is

$$n\lambda = 2\,d\sin\theta \qquad (4\text{-}15)$$

This is the famous equation first derived by W. L. Bragg and called Bragg's law. This equation allows us to relate the distance between a set of planes in a crystal and the angle at which these planes will diffract X-rays of a particular wavelength. It is usually more convenient to divide both sides of the Bragg equation by n and to define d/n as d_{hkl}. This makes Bragg's law look like

$$\lambda = 2\,d_{hkl}\sin\theta \qquad (4\text{-}16)$$

In the original notation Eq. (4-15) we consider first order $(n = 1)$ or second order $(n = 2)$ diffraction from any plane like (111). With the modified Eq. (4-16), we consider all diffraction to be first order from each of the families of planes in the crystal [i.e., from (111) and (222) etc.]. An examination of the Bragg equation shows that when λ is known and θ is measured, we can calculate d_{hkl} and discover the dimensions of the unit cell and even of atoms and their bonds.

The artificial reflection analogy used in obtaining the Bragg equation does not, in any way, compromise its validity. It can be derived in the same form by a rigorous analysis of the Huygens scattering from each of the atoms in the crystal.

4.7.2 The Reciprocal Lattice

P. P. Ewald (1913) devloped what is by far the most useful method for describing and explaining diffraction phenomena. Its wide acceptance is an indication of its basic

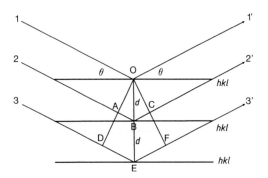

Figure 4-16. Diffraction of X-rays from the planes in a crystal.

simplicity. The concept introduced by Ewald has the name reciprocal lattice or reciprocal space.

When a particular arrangement of an X-ray source, sample and detector is proposed, we would like to be able to predict the various motions which will have to be applied to the three components in order to see particular diffraction effects. If, for example, we wish to see diffraction from the 100 planes of a LiF single crystal which has observable and identifiable cleavage, we simply use the Bragg reflection analogy and orient the crystal so that its 100 surface makes an equal angle with the beam and detector. The angle can be calculated from Bragg's law using the lattice parameter for LiF, which we must look up (i.e., in the Crystal Data database), and the wavelength of the X-rays to be used. However, if we wish to look at diffraction from the (246) planes, the visualization of the required orientation of the crystal becomes formidable and even the computation of the magnitude of d_{246} is non-trivial. Ultimately, the problem can be thought of as the difficulty of visualizing two dimensional planes intersecting a three dimensional unit cell.

In general, problems become easier to understand if we visualize them with fewer dimensions than are required in our three dimensional world. When one or more dimensions can be removed from an analysis the mathematics and conceptualization usually become much easier. For this reason many derivations begin in one dimension and only at the end expand to three.

Our problem, in diffraction theory, is that the Bragg planes in a space lattice are inherently three dimensional and restricting our analysis to a two dimensional plane within the lattice really doesn't simplify their three dimensional relationships with the other planes in the lattice. So we

use another method to try and remove a dimension from the problem; we will represent each two dimensional plane as a vector: d_{hkl} is defined as the vector pointing from the origin of the unit cell to a point perpendicular to the first plane in the family ($h k l$) as illustrated for the 110 plane in Fig. 4-17. If we plot all of the d_{hkl} vectors as points in the appropriate crystallographic coordinate system, we observe that these points begin on each axis at the fractional coordinate 1 and increase in density as we approach the origin. (Fractional coordinates are simply the distance expressed as a fraction of the unit cell edge.) Figure 4-18 shows the vectors representing the planes shown in Fig. 4-14. If the outermost points $(100, 010, 001, \bar{1}00, 0\bar{1}0, 00\bar{1})$ are connected, we will observe the three dimensional shape of the unit cell. Thus, this construction has the advantage of representing two dimensional planes as points while maintaining the symmetry relationships between planes. However, Fig. 4-18 shows a very serious hindrance: as we approach the origin of this vector diagram the density of points, representing planes, increases to infinity. Rather than simplifying our real space picture this vector space construction has definitely made it more complex.

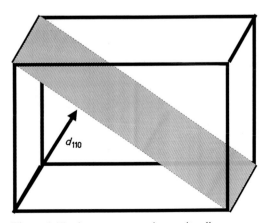

Figure 4-17. d_{110} as a vector in a unit cell.

The full three dimensional vector representation of the planes in a unit cell will be a sheaf of vectors projecting out from the origin in all directions, getting thicker and thicker as we approach the center. On examining Fig. 4-18 we see that the vectors are approaching the origin according to the reciprocals of the d_{hkl} values. Ewald, on noting this relationship, proposed that we plot not the d_{hkl} vectors, but instead the reciprocal of these vectors. We will define this reciprocal vector as

$$d_{hkl}^* = \frac{1}{d_{hkl}} \qquad (4\text{-}17)$$

Let us now redraw the rather complicated Fig. 4-18 but plotting d_{hkl}^* vectors instead of $d_{hkl}'^s$. Figure 4-19 shows this construction. The units are in reciprocal Ångströms and the space is therefore a reciprocal space. Notice that the points in this space repeat at perfectly periodic intervals defining another space lattice only this one we will call a reciprocal lattice. The repeating translation vectors in this lattice are called

a^*, b^* and c^*. The interaxial angles are α^*, β^* and γ^*, where the reciprocal of an angle is defined as the complement or 180° minus the angle.

The concept of the reciprocal lattice makes the visualization of the Bragg planes extremely easy. To establish the index of any point in the reciprocal lattice one simply has to count out the number of repeat units in the a^*, b^* and c^* directions. Figure 4-19 only shows the $h\,k\,0$ plane through the reciprocal lattice, however, the concept is of a full three dimensional lattice, which extends in all directions in reciprocal space. If the innermost points in this lattice are connected we will see the three dimensional shape of the reciprocal unit cell which is directly related to the shape of the real space unit cell (in fact, the surfaces of this figure define the well known Brillouin zone). Thus, the symmetry of the real space lattice propagates into the reciprocal lattice.

The reciprocal lattice has all of the properties of a real space lattice. Any vector in

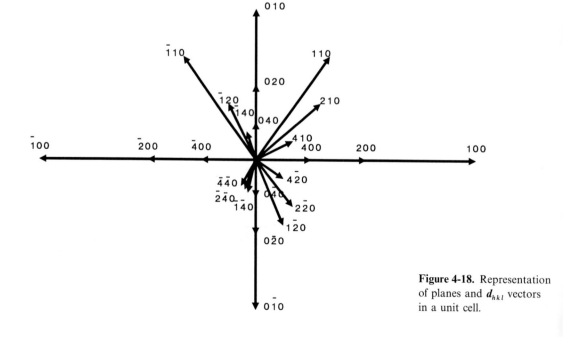

Figure 4-18. Representation of planes and d_{hkl} vectors in a unit cell.

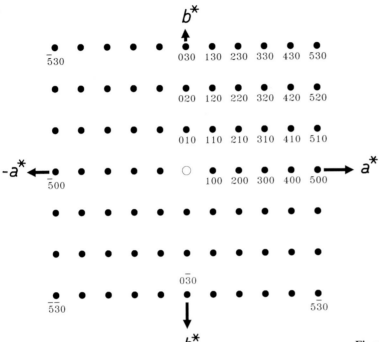

Figure 4-19. The reciprocal lattice.

this lattice, representing a set of Bragg planes, can be resolved into its components

$$d^*_{hkl} = h\,a^* + k\,b^* + l\,c^* \qquad (4\text{-}18)$$

An important point here is that the integers in Eq. (4-18) are in fact, equal to the Miller indices. We shall see in the next section that the process of X-ray diffraction can be thought of as recording the reciprocal lattice. It is this fundamental relation between the size, shape and symmetry of the real and reciprocal lattices which allows X-ray diffraction to determine the crystallographic information about a material and hence, gives rise to the name of this discipline as X-ray crystallography.

4.7.2.1 Relationship between d_{hkl}, hkl and Translation Vectors

There are an infinite number of sets of planes which can be thought of as inter-

secting a unit cell. We characterize these planes with their hkl values and the d_{hkl} interplanar spacings. The d_{hkl} values are a geometric function of the size and shape of the unit cell. The relationship between d_{hkl} and real unit cell is cumbersome and usually stated in a different form for each crystal system. However, the functional relation between the square of the reciprocal lattice vectors and the size and shape of the reciprocal unit cell is easily derived, has a simple form and applies easily to all crystal systems. The d^{*2}_{hkl} equation coupled with the relations given in Table 4-5 will allow simple computations relating d_{hkl} to the lattice parameters in any crystal system.

Table 4-5 shows the equations relating the real and reciprocal lattice translation vectors for the various crystal systems.

To derive d^{*2}_{hkl} we simply square Eq. (4-18). To get the square of a vector we must multiply it by itself, and since it is a

Table 4-5. Direct and reciprocal space relationships [a].

System	a^*	b^*	c^*
Orthogonal systems	$a^* = 1/a$	$b^* = 1/b$	$c^* = 1/c$
Hexagonal	$a^* = 1/a \sin \gamma$	$b^* = 1/b \sin \gamma$	$c^* = 1/c$
Monoclinic	$a^* = 1/a \sin \beta$	$b^* = 1/b$	$c^* = 1/c \sin \beta$
Triclinic	$a^* = \dfrac{b\,c \sin \alpha}{V}$	$b^* = \dfrac{a\,c \sin \beta}{V}$	$c^* = \dfrac{a\,b \sin \gamma}{V}$

[a] Note that the following relationships hold and are especially useful for triclinic:
$\cos \alpha^* = \cos \beta \cos \gamma - \cos \alpha/(\sin \beta \sin \gamma)$,
$\cos \beta^* = \cos \alpha \cos \gamma - \cos \beta/(\sin \alpha \sin \gamma)$,
$\cos \gamma^* = \cos \alpha \cos \beta - \cos \gamma/(\sin \alpha \sin \beta)$,
$V^* = a^* b^* c^* \sqrt{1 - \cos^2 \alpha^* - \cos^2 \beta^* - \cos^2 \gamma^* + 2 \cos \alpha^* \cos \beta^* \cos \gamma^*}$.

vector the multiplication takes the form of a dot product

$$d_{hkl}^{*2} = d_{hkl}^* \cdot d_{hkl}^* \qquad (4\text{-}19)$$

$$d_{hkl}^{*2} = h^2 a^{*2} + k^2 b^{*2} + l^2 c^{*2} +$$
$$+ 2hk\, a^* b^* \cos \gamma^* + 2hl\, a^* c^* \cos \beta^* +$$
$$+ 2kl\, b^* c^* \cos \alpha^* \qquad (4\text{-}20)$$

This expression relates the square of the inverse of d_{hkl} to the size and shape of the reciprocal unit cell for any plane in any crystal system. It is useful to point out that in orthogonal crystal systems the final three terms of the equation include a $\cos 90°$ term and therefore go to zero. For many applications then, Eq. (4-20) reduces to a particularly simple form. For example, Eq. (4-20) becomes for

cubic $\qquad d_{hkl}^{*2} = (h^2 + k^2 + l^2)\,a^{*2}$,

tetragonal $d_{hkl}^{*2} = (h^2 + k^2)\,a^{*2} + l^2 c^{*2}$,

hexagonal $d_{hkl}^{*2} = (h^2 + hk + k^2)\,a^{*2} +$
$$+ l^2 c^{*2},$$

orthorhombic $d_{hkl}^{*2} = h^2 a^{*2} + k^2 b^{*2} +$
$$+ l^2 c^{*2}.$$

4.7.2.2 The Ewald Sphere of Reflexion

In Fig. 4-20 we define a sphere with a radius of $1/\lambda$ and place it such that the origin of the reciprocal lattice is tangent to it. If we now imagine a real crystal at the center of this sphere and an X-ray beam entering from the right, we have all the components needed to visualize the diffraction process geometrically instead of mathematically, using the Bragg equation. It is clear that a rotation of the crystal (and its associated real space lattice), will also rotate the reciprocal lattice because the reciprocal lattice is defined in terms of the real lattice. We now examine this arrangement, at a specific time in the rotation of the crystal, when the 230 point in the reciprocal lattice is brought in contact with the sphere.

This orientation is illustrated in Fig. 4-21 where, $\overline{CO} = 1/\lambda$. From our definitions, $\overline{OA} = d_{230}^*/2$, hence

$$\sin \theta = \frac{\overline{OA}}{\overline{CO}} = \frac{d_{230}^* \lambda}{2}$$

or

$$\lambda = \frac{2 \sin \theta}{d_{230}^*}$$

but by definition,

$$d_{230} = \frac{1}{d_{230}^*}$$

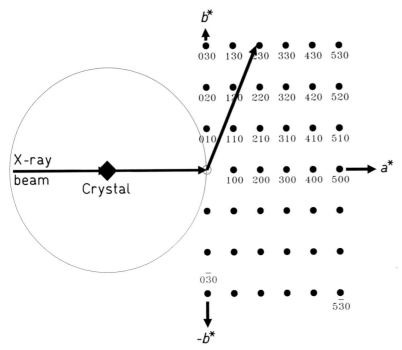

Figure 4-20. The Ewald sphere of reflection with a crystal in the center and its associated reciprocal lattice tangent to the sphere, at the point where the direct X-ray beam emerges.

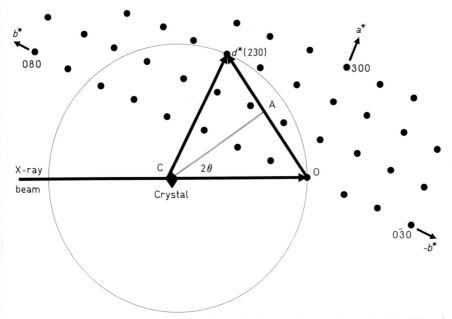

Figure 4-21. The Ewald sphere of reflection with the crystal rotated so that the 230 reciprocal lattice point touches it, permitting it to diffract.

therefore,

$$\lambda = 2\,d_{hkl}\sin\theta$$

Thus, we see that the reciprocal lattice and sphere of reflection concepts incorporate Bragg's law. As each lattice point, representing a d_{hkl}^* value, touches the sphere of reflection, the condition for diffraction is met and the corresponding real space lattice plane "reflects" in the direction $C-d_{hkl}^*$. If we look at the non-imaginary components of our construction, we see the sample and the incident and diffracted X-ray beams. Thus, the Ewald construction allows us to easily analyze an otherwise complex diffraction geometry, showing what motion must be applied to a crystal in order to produce a diffracted beam in a particular direction.

The Ewald sphere construction is very useful in explaining diffraction phenomena in any type of geometry. The principal advantage is that it avoids the need to do calculations to explain any phenomena and instead allows us to visualize an effect using a pictorial, mental model. In addi-

tion, it permits the simple analysis of the otherwise complex relationships among the crystallographic axis and planes.

4.7.3 Single Crystal Diffraction Techniques

There have been a number of techniques developed to record the diffraction pattern of single crystals. Each may be viewed as a method of visualizing the reciprocal lattice with various distortions introduced by a particular experimental geometry. For example, if we set up the experiment shown in Fig. 4-22 with a crystal rotating around its a axis, in an X-ray beam, and place a cylindrical film around the crystal, we will record each of the $0\,k\,l$ points of the reciprocal lattice, projected onto a line on the film. This technique, called an oscillation or rotation photograph permits rapid determination of the length of the axis being rotated about. Each of the resulting parallel lines of points contains all of the diffraction points in each $n\,k\,l$ plane of the reciprocal lattice, when n is constant in each line. The intensities of the reciprocal lattice

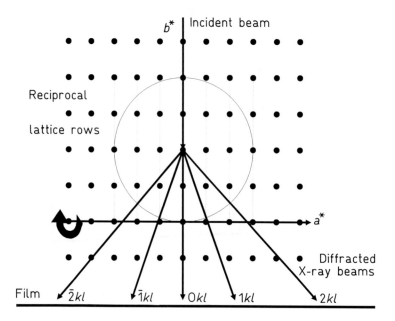

Figure 4-22. The experimental arrangement for a rotation photograph.

points on the film are related to the types and locations of the atoms in the unit cell and will be discussed later. This recorded pattern of spots is referred to as the intensity weighted reciprocal lattice.

Since all of the reflections in each reciprocal lattice plane are projected onto a line, the identification of individual reflections is quite difficult. One of the earliest procedures allowing the easy indexing of individual spots and subsequent measurement of their intensity was realized in the Weissenberg camera. This device allows the isolation of a single plane of the reciprocal lattice by placing an absorbing screen over all of the rows of points in the rotation photograph, other than the row to be measured. The film is then oscillated left and right as the crystal is rotated in the X-ray beam. This causes each diffraction spot in the row to be recorded in a different location on the film. Figure 4-23 shows a Weissenberg camera and Fig. 4-24 the $0\,k\,l$ zone of a crystal spread over the film. This distortion of the reciprocal lattice is readily analyzed permitting indexing.

Martin Buerger (1964) designed a single crystal X-ray diffraction camera, called the precession camera, which will take pictures of the reciprocal lattice exactly the way we imagine it. This camera records the diffraction pattern while the crystal is precessed around the X-ray beam. If the film is kept parallel to the reciprocal lattice plane being photographed, during this motion, an undistorted picture of the reciprocal lattice plane can be made. Figure 4-25 shows such a camera with the layer-line screen in place, allowing only the reciprocal lattice layer desired to pass through the annulus. The photograph of the $0\,k\,l$ layer of ammonium oxylate shown in Fig. 4-26 was taken in this manner. It is interesting to think of this procedure as taking a photograph of a mental construction.

Figure 4-23. The Weissenberg camera with the crystal hidden from view inside the film cassette on its movable mount (courtesy of Siemens AG).

Today most single crystal diffraction is done on automated single crystal diffractometers. These devices have four individual degrees of freedom allowing the positioning of each reciprocal lattice point on the sphere of reflection and then scanning an electronic detector past it to record its

Figure 4-24. A Weissenberg photograph of the $0\,k\,l$ reciprocal lattice net.

Figure 4-25. The Buerger precession camera with layer line screen behind the crystal and in front of the film cassette (Courtesy of Siemens AG).

intensity. In this manner a few thousand data points can be measured per day. For very large unit celled materials, two dimensional array detectors have been developed to permit rapid measurement of hundreds of thousands of diffraction intensities. The rapid recording of large quantities of diffraction intensities is currently the subject

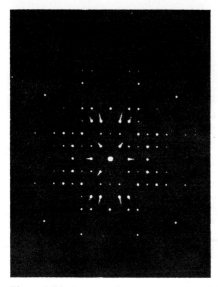

Figure 4-26. A precession photograph of the $0kl$ zone of ammonium oxalate.

of intensive development. An example of this thrust is the two dimensional array of charge coupled devices (CCD) which can record X-rays over a region of about 30 cm² with a resolution on the order of 100 μm.

Another thrust area of applied crystallography is the use of the high brightness of synchrotron sources to collect full single crystal intensity data-sets from tiny crystallites. Normally the crystals required for diffractometer measurement are on the order of 0.1–0.3 mm in dimension. Recent studies have obtained full diffraction data-sets from crystals with volumes on the order of 10–100 μm³. In a recent study by Newsom et al. (1991) single crystal intensity data was collected from a 7 μm zeolite crystal (about 2500 unit cells on an edge).

4.7.3.1 X-Ray Topography

X-ray topography is the imaging of single crystals from the viewpoint of a particular crystallographic plane. In the Lange camera procedure a narrowly collimated X-ray beam is directed at a sample oriented at the Bragg angle for, say, the (111) reflection. The diffracted beam is then directed in a narrow line across a piece of film. Then, both the crystal and film are moved in unison so that each band of the crystal, diffracting the (111) X-rays, is recorded on a corresponding band of the film. The resulting photograph, recorded on high resolution film or a nuclear emulsion plate, shows all of the crystal imperfections, grain boundaries and domain structure. The resolution of such a topograph is on the order of 1 μm which is three orders of magnitude less than electron microscopy. However, the X-ray topograph examines large areas of the crystal, is non-destructive and can be run in non-ambient conditions. Higher intensity synchrotron sources have allowed the use of full parallel beam geometry, eliminating the need to scan the crystal and film. These procedures are widely used in the materials industries involved with single crystals, silicon in particular.

4.7.3.2 Laue and Kossel Techniques for Orientation Determination

Two other single crystal techniques are particularly important to materials science and technology. The Laue technique is so named because it is the procedure used by Friedrich, Knipping and von Laue (1912) to record the very first X-ray diffraction pattern. The procedure is to place a single crystal in front of a white X-ray beam and allow the various planes, which happen to be in diffracting position, to pick out the wavelength, which meets the Bragg condition, and diffract it. The reciprocal space way of thinking about this technique is to picture the white X-ray beam as a series of Ewald spheres within each other, each showing a common surface point which is the origin of the reciprocal lattice. Each sphere corresponds to the $1/\lambda$ of the longest and shortest wavelengths in the beam. This solid Ewald sphere has individual onion-skin like layers, some of which will intersect each of the points in the reciprocal lattice within the solid sphere, putting them all into diffracting position. This situation is illustrated in Fig. 4-27. This geometry implies that the reflections in each crystallographic zone will lie on the edge of a cone surrounding the zone axis. Depending on the angle at which this cone intersects the film, we will observe zonal reflections to lie along various conic sections, like hyperbolas and ellipses.

This technique can have the film placed in transmission, behind the crystal or in back-reflection, bringing the incident X-ray beam through a hole in the film. The

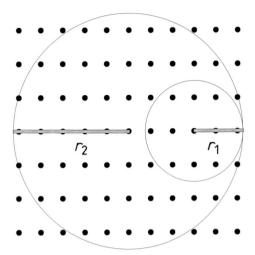

Figure 4-27. The reciprocal lattice model for the Laue method. All reciprocal lattice points between the two limiting Ewald spheres will diffract.

patterns obtained are rather difficult to interpret quantitatively in that the positions of the diffraction spots on a film will be completely dependent on, and extremely sensitive to, the orientation of the crystal. However, the pattern of spots will immediately show the symmetry of the reciprocal axis direction pointing at the film. Thus, this method has found wide popularity in both industry and research in aligning large single crystals along specific crystallographic directions. Since the Laue method is the only practical way of orienting large industrial crystals, the difficulty of interpreting the Laue pattern quantitatively has been solved by automating the procedure. A number of algorithms are available which interpret Laue patterns and enable this technique to be very widely used.

The other method of single crystal diffraction important to materials science is Kossel (1936) photography. Kossel diffraction is a single crystal technique but the crystal is usually part of a polycrystalline aggregate in a material. It results from the stimulation of characteristic X-ray emission by atoms within a tiny single crystal, which is part of a polycrystalline aggregate. This happens, for example, in a transmission electron microscope when the energy of the beam is higher than the K excitation potential of say, Ti atoms in the material being examined. The K_α radiation emitted by the Ti atoms in the few cubic micron volume of the electron beam, emerges as a spherical wave which will diffract from the various Bragg planes in the crystal below it. The spherical wave will diffract as a cone about each reciprocal lattice vector. A film will record circles, ellipses and the various conic sections depending on the angle each reciprocal lattice vector, and its associated cone of diffraction, makes with the film. The efficiency of this method is reflected in the short times (seconds) required to expose a Kossel photograph compared with those (hours) required for a Laue photograph. An example Kossel photo is shown in Fig. 4-28. The pseudo-Kossel method refers to a trick used to change the radiation in Kossel diffraction. For example, if Cu K_α is desired a thin layer of Cu can be deposited on the surface of the sample to produce the desired X-rays when the electron beam strikes it. For a good example of the use of this method see Bellier and Doherty (1977).

As with Laue diffraction, the Kossel technique is used both in transmission and reflection. However, this monochromatic diffraction method uses the very low intensity X-ray beam generated from the atoms in the small crystallite. Thus, samples for transmission Kossel diffraction usually need to be thinned to under 100 µm to permit the diffracted beam to pass. Like the Laue method, Kossel diffraction is most often applied to the study of orientation of crystallites, however here it is used in polycrystalline materials. In the Kossel technique

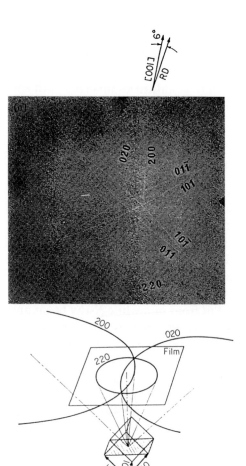

Figure 4-28. Kossel lines from a 10 µm thick single crystal of 3% Si-steel (100)[001] using the Fe K_α radiation stimulated from a 20 kV electron beam. Courtesy of Dr. Yukio Inokuti.

the monochromatic spherical wave is generated very close to the diffracting planes so the diffracted beam must travel a relatively long distance to the film. This very long lever arm, makes the method very sensitive to small orientation tilts or d_{hkl} value changes due to strain displacement and thus, can be used to evaluate residual stress in crystallites. A recent application by Inokuti et al. (1987) obtained approximately 1500 Kossel patterns from individual grains in a silicon containing transformer steel. The Goss texture was then displayed by coloring all grains with similar orientation.

4.7.4 Preferred Orientation

Preferred orientation of crystallites in bulk, polycrystalline materials, is a vital subject to many materials industries. A number of industrial materials are based on vector physical properties of the crystallites. For example the barium hexaferrite ceramic magnets, used commonly as seals on refrigerator doors, are polycrystalline materials in which only the $(0\,0\,l)$ crystallite directions have a magnetic moment. Thus, the fabrication and quality control procedures must involve manipulating and measuring the degree of preferred orientation in the ceramic. Extruded wires show a characteristic preferred orientation as do most pressed powder materials. The most common way of evaluating the type and extent of preferred orientation is to measure the pole figure for a particular crystallographic direction. The pole figure is simply the intensity of a particular Bragg diffraction line plotted as a function of the three dimensional orientation of the specimen. It is determined on a pole figure diffractometer which is essentially the same as a single crystal diffractometer, able to rotate the specimen through all orientations while monitoring the diffraction intensity of a reflection. The results are displayed as a pole figure which is a two dimensional stereographic projection. Bunge (1982) has written a comprehensive treatment of preferred orientation evaluation techniques. A non-conventional method which avoids the requirement of an expensive automated pole figure device while often providing the required information should be mentioned. All of the preferred orientation on crystallites lying within a materials surface can be obtained by comparing the observed powder pattern diffraction intensities to the "ideal" intensities which can be obtained by calculating the powder pattern from known atomic positions. The trick is to find a way to compare these two patterns which are on different scales. Snyder and Carr (1974) devised a scale independent function which allows this comparison and permits direct computation of a vector properties enhancement or disenhancement in a particular crystallographic direction. See the chapter on measurement and control of texture by Cahn (Chap. 10) in Volume 15 of this series for a modern summary of texture analysis.

4.7.5 Crystallite Size

In Fig. 4-16 we saw that when diffraction occurs, the distance between adjacent planes, ABC, must exactly equal $1\,\lambda$. Geometry requires that when this condition is met that the distance between three sets of planes, DEF, must be $2\,\lambda$ and all similar triangles from lower lying planes must also be an integral multiple of λ. If we increase the angle of incidence θ, so that the distance ABC becomes $1.1\,\lambda$ then DEF will be $2.2\,\lambda$. The scattered waves from the sixth deepest plane will be $6.6\,\lambda$ and will therefore be exactly out of phase $(0.5\,\lambda)$ with the waves scattered from the first plane. The scattering from the second and seventh planes will also be exactly out of phase. When all the planes from all of the unit cells are considered we see that no net scattering will occur; namely, there will be another unit cell at some depth within the crystal which will exactly cancel the scattering from any other cell except exactly at the Bragg angle where all scattering is in phase.

If θ is set closer to the Bragg angle so that the distance ABC is equal to $1.001\,\lambda$

then the scattering from the first plane will be cancelled by the scattering from the plane 500 layers deep in the crystal, with a phase shift of 500.5 λ. Similarly, if ABC is 1.00001 λ the scattering will be cancelled by a plane 50 000 layers deep in the crystal. Thus, it is clear that Bragg reflections should occur only exactly at the Bragg diffraction angle producing a sharp peak. However, if the crystal is only 1000 Å in size then the planes needed to cancel scattering from, for example the (100) plane, with an ABC distance of 1.0001 λ (i.e., the 5000th plane) are not present. Thus, the diffraction peak begins to show intensity at a lower θ and ends at an angle higher than the Bragg angle. This is the source of "particle size broadening" of diffraction lines. The observed broadening can be used to determine the crystallite size of materials less than one micrometer. Crystallites larger than one micron typically have a sufficient number of planes to allow the diffraction peak to display its inherent Darwin (1914) width (i.e., the width dictated by the uncertainty principle) additionally broadened by instrumental effects, with little contribution from size broadening (see Sec. 4.13).

The crystallite size broadening (β_τ) of a peak can usually be related to the crystallite size (τ) via the Scherrer equation (1918)

$$\tau = \lambda/(\beta_\tau \cos \theta) \tag{4-21}$$

The additional broadening in diffraction peaks beyond the inherent peak widths due to instrumental effects, can be used to measure crystallite sizes as low as 10 Å. However, a second cause of broadening, due to stress, can complicate the picture. The size of small particles dispersed in a matrix can be determined by small-angle scattering of X-rays (or neutrons). For details see Vol. 2 B, Chapter 12.

4.7.6 Residual Stress

There are two types of effects strain in a material can produce on diffraction patterns. If the stress is uniformly compressive or tensile, it is called a macro-stress and, the distances within the unit cell will either become smaller or larger, respectively. This will be observed as a shift in the location of the diffraction peaks. These macro-stresses are measured by an analysis of the lattice parameters which will be described later in Sec. 4.12.4.

When the residual stress in a material produces a distribution of both tensile and compressive forces, they are called micro-stress and, the observed diffraction profiles will be observed to broaden about the original position. See Langford et al. (1988) and Delhez et al. (1988) for recent discussions of these effects. Both of the size and strain broadening effects generally produce a symmetric broadening, the observed asymmetry in diffraction profiles is usually due to instrumental effects (see Sec. 4.13 below). Micro-stress in crystallites can come from a number of sources: vacancies, defects, shear planes, thermal expansions and contractions, etc. (see Warren, 1969). Whatever the cause of the residual stress in a crystallite, the effect will cause a distribution of d-values about the normal, unstrained, d_{hkl} value. Figure 4-29 illustrates the expansion and contraction of d_{hkl} values caused by compression and extension of unit cells in micro-stress.

The broadening in a peak due to stress has been shown to be related to the residual stress ε, by

$$\beta_\varepsilon = 4 \varepsilon \tan \theta \tag{4-22}$$

The fact, that stress induced diffraction peak broadening follows the tangent of θ while crystallite size broadening follows a $1/\cos \theta$ allows us to separate these effects.

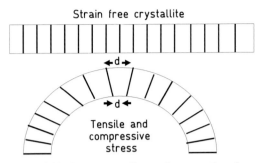

Strain free crystallite

Tensile and
compressive
stress

Figure 4-29. Stress expanding and contracting d's.

The most common procedure for accomplishing this was developed by Warren and Averbach (1950); this, along with a more recent procedure will be described later in Sec. 4.13 and 4.13.7. For a complete review of these and the preferred orientation methods mentioned above, and how preferred orientation affects determination of residual stresses, see Chap. 10 by Cahn in Volume 15 of this series.

4.7.7 The Intensities of Diffracted X-Ray Peaks

All of the diffraction theory discussed until now has looked at the metric aspects of a diffraction pattern. That is, the positions of the diffraction maxima and how they are related to the size and shape of the unit cell. We have not considered why the intensities of the diffraction lines are what we observe. In order to understand what determines the intensity of a diffraction peak we need to examine a number of independent phenomena which determine the intensity. First we must consider how much intensity a single electron will coherently scatter (the more massive nucleus is a very inefficient scatterer and may be ignored). Next we need to consider the interference effects which will occur due to the electrons being distributed in space around atoms and the fact that atoms are not sta-

tionary in a lattice but vibrate in an anisotropic manner. Then the interference effects of atoms scattering from different regions of the unit cell must be allowed for, and finally optical and absorption effects must be considered.

4.7.7.1 Scattering of X-Rays by a Bound Electron

X-rays are electromagnetic radiation and from a fixed point will be seen as an oscillating electric field. The field will cause a bound electron to also oscillate (i.e., accelerate and decelerate) and, therefore, reradiate the incident radiation through a solid angle of 360°. This process occurs coherently, with the scattered waves having the same phase as the incident waves.

J. J. Thomson (1906) first showed that the intensity scattered from an electron is

$$I = \frac{I_0 e^4}{m_e^2 r^2 c^4} \frac{1 + \cos^2 2\theta}{2} \qquad (4\text{-}23)$$

where I_0 is the intensity of the incident beam, e is the charge on the electron, m_e is the mass of the electron, c is the speed of light and r is the distance from the scattering electron to the detector. Note that the inverse square law is explicit in the Thomson equation. The final term involving the cosine function is called the *polarization factor* and results from the fact that the incident X-ray beam is unpolarized while the scattered beam will be decreasingly polarized as the angle of view is increased. The Thomson equation addresses only coherently scattered radiation; all other absorption mechanisms are ignored.

4.7.7.2 Scattering of X-Rays by an Atom

In Fig. 4-30 we see X-ray beams X and X′ coherently scatter from electrons at A and B around an atom. When viewed at an angle of 0°, scattered waves Y and Y′ are

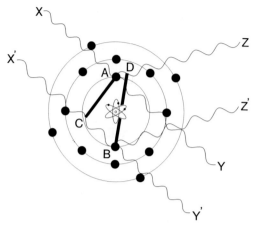

Figure 4-30. The scattering of X-rays from an atom.

where

$$k = 4\pi(\sin\theta)/\lambda \qquad (4\text{-}25)$$

The electron density function $\varrho(r)$ around the atom is related to the quantum mechanical wave function Ψ via

$$\varrho(r) = 4\pi r^2|\Psi|^2 \qquad (4\text{-}26)$$

Every few years a quantum mechanic will produce a better set of wave functions describing the electron densities of atoms and the atomic scattering factors are recomputed. The wave functions for all atoms except hydrogen must be obtained by approximate methods and this introduces a small amount of error into the atomic scattering factors. Experience in using these f_0's has shown that the error in them is less than the typical errors from other sources and for most work can be considered negligible. The preferred atomic scattering factors today were computed by Cromer and Waber (1965) using Dirac-Slater orbitals.

4.7.7.3 Anomalous Dispersion

There are two other factors which influence the intensity of the scattering from an atom and it is convenient to consider these

exactly in phase, but when viewed from any non zero angle, we find that wave Z' has to travel CB-AD farther than wave Z. Because atomic dimensions are on the same order as the wavelength of X-rays, this path difference causes partial destructive interference and lowers the resultant amplitude. Therefore, the intensity of scattered radiation decreases with angle of view (θ). This phenomenon is described by the quantity f_0 which is called the *atomic scattering factor*. The function f_0 is normalized in units of the amount of scattering from a single electron. At zero degrees f_0 will be equal to the number of electrons surrounding any atom or ion. Figure 4-31 shows typical atomic scattering factor curves. Since the phase difference CB-AD depends on the wavelength and the angle of view, the function f is specified as a function of $\sin\theta/\lambda$.

The actual shape of the function f_0 must be calculated by integrating the scattering over the electron distribution around an atom as shown in Eq. (4-24)

$$f_0 = \int_0^\infty \varrho(r)\frac{\sin(kr)}{kr}\,dr \qquad (4\text{-}24)$$

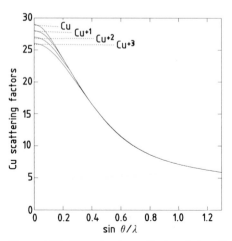

Figure 4-31. The atomic scattering factor f_0 for a copper atom.

as modifying the atomic scattering factor. The first is a phenomenon called *anomalous scattering*. In normal *Thomson scattering*, the electron acts as an oscillator under the stimulation of the oscillating electric field of the incident radiation. We can crudely picture this effect as a true movement of the electron back and forth away from the nucleus. When the frequency of the radiation gets high enough to cause the electron on its next oscillation away from the nucleus, to no longer feel the restoring attraction, ionization occurs. This is a wave description for picturing the photoelectric effect. Using his notation we can think of removal of an electron as a resonance phenomenon and the various absorption edges as natural frequencies of vibration.

Using this over-simplified model we can think of frequencies just short of the resonance frequency as lifting the electron from its inner orbital to the outer unfilled energy states of the atom. Since there is a finite quantum probability that the electron could exist in these states there will be a slight delay time as it oscillates back toward the nucleus. This delay will in turn cause a delay in the radiation being scattered by the oscillating electron. We will see the effects of this time delay as a phase shift in the scattered wave. Though a physicist might object to this simple argument it allows one to picture the origin of the observed phase lag. To properly describe anomalous scattering, we need to correct the normal scattering factor f_0 with a real ($\Delta f'$) and an imaginary ($\Delta f''$) term. The effective scattering from an atom will be

$$|f|^2 = (f_0 + \Delta f')^2 + (\Delta f'')^2 \qquad (4\text{-}27)$$

4.7.7.4 Thermal Motion

Figure 4-30 shows that the size of the atom causes some destructive interference from the scattering of electrons that are separated by distances on the order of a fraction of the wavelength of the X-rays. If the atom in question is vibrating about its lattice site then its effective size is larger and the interference effects are, in turn, larger. We have seen, in Fig. 4-31, that the interference in a stationary atom causes the atomic scattering factor to fall off somewhat exponentially as a function of $(\sin\theta)/\lambda$. In order to describe the thermal motion induced enhancement of this fall off in f, Debye (1913) and later Waller (1928) defined the parameter B, which is related to the vibrational amplitude of the atom

$$B = 8\pi^2 u^2 \qquad (4\text{-}28)$$

B is called the *Debye-Waller temperature factor*. It is directly related to u^2, the mean-square amplitude of vibration of an atom. The amount and direction of atomic vibration will depend on the temperature (i.e., the amount of kT, thermal energy available), the atomic mass and the direction and strength of the force constants holding the atom bonded in its location. Figure 4-32 shows the effect of increasing B on f

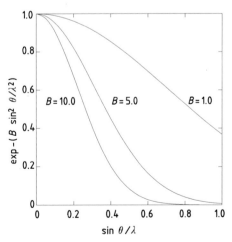

Figure 4-32. Effect of the temperature parameter B on f.

when B is used in the form

$$f = f_0 \exp\left(-\frac{B \sin^2 \theta}{\lambda^2}\right) \qquad (4\text{-}29)$$

The temperature factor B is the same for all directions of vibration of an atom and is therefore called the isotropic temperature factor. In fact, most atoms in solids will have special directions in which they can vibrate with higher amplitudes. Thus, a more accurate description of the thermal vibration in a solid is to use a tensor to describe the anisotropic motion. To do this we recognize that the $\sin^2 \theta/\lambda^2$ term in Eq. (4-29) is simply $d^{*2}/4$. We have already derived the equation for d^{*2} in terms of its vector components as Eq. (4-20). Thus, we can substitute Eq. (4-20) into (4-29) and break B into six B_{ij} anisotropic terms,

$$\exp\left(-\frac{B\,d^{*2}}{4}\right) = \exp[-\tfrac{1}{4}(B_{11}\,h^2\,a^{*2} +$$

$$+ B_{22}\,k^2\,b^{*2} + B_{33}\,l^2\,c^{*2} + 2B_{12}\,hk\,ab +$$

$$+ 2B_{13}\,hl\,ac + 2B_{23}\,kl\,bc)] \qquad (4\text{-}30)$$

In Eq. (4-30) the cosines of the reciprocal interaxial angles are included into the values of the B_{ij} cross terms.

4.7.7.5 Scattering of X-Rays by a Crystal

Since a crystal is made up of repeating identical unit cells, we need only consider diffraction from a single unit cell to see the effects from a whole crystal. Each atom in the unit cell will scatter an X-ray beam of intensity given by the atomic scattering factor f and with a phase depending on its location in the cell. If we consider an atom at the origin of the unit cell scattering with a phase angle of $0°$, then an atom half way along the a edge will scatter with a phase angle of $180°$ when we consider Bragg reflection from the 100 plane. However, for the 200 plane these two atoms will scatter

in phase, each having a phase angle of $0°$. Our task is to find a mathematical notation that will allow us to add up the scattering contributions from all of the atoms in the unit cell allowing for the interference effects caused by their different locations.

The resultant scattering from all of the atoms in the unit cell for a particular diffraction line is called the structure factor F_{hkl}. In order to conveniently sum up the scattering amplitudes from all of the atoms in the cell, allowing for their individual phase angles, we represent this scattering as a complex vector. The Euler coordinate system which has an imaginary axis normal to the real axis is ideal for our current needs. When vectors in an Euler coordinate system are added, the resultant vector will have an amplitude and phase angle which properly reflects the interference between the waves that the vectors represent. In order to represent the scattered wave from each atom in a unit cell as a vector, we simply need to find a direction and magnitude to associate with each wave. The amplitude of the wave is equal to f the atomic scattering factor and this we will define as the length of the atomic scattering vector. We may use the phase angle to define the direction of the vector in polar coordinates.

Any vector in an Euler coordinate system can be resolved into its components as $f \cos \phi + f \mathrm{i} \sin \phi$, where ϕ is the phase angle. The well known Euler relation allows us to express these vectors in exponential form

$$f \exp(\mathrm{i}\,\phi) = f \cos \phi + f \mathrm{i} \sin \phi \qquad (4\text{-}31)$$

To generalize this notation we will look at a real three dimensional unit cell containing N atoms. Each atom will scatter X-rays with a different amplitude equal to its atomic scattering factor evaluated at the diffraction angle of the hkl planes in question, and for the wavelength being used. The phase angle of the scattered wave from each atom, j, will be

$$\phi = 2\pi(h x_j + k y_j + l z_j) \qquad (4\text{-}32)$$

The problem of describing the scattering from a unit cell reduces to adding the waves scattered from each atom in the cell, represented by vectors in a complex coordinate system, each with its own amplitude and phase angle. The resultant diffracted wave for any set of Bragg planes is called the structure factor F_{hkl} and may now be written as

$$F_{hkl} = \sum_{j=1}^{\text{number of atoms}} f_j \cdot$$
$$\cdot \exp[2\pi i(h x_j + k y_j + l z_j)] \qquad (4\text{-}33)$$

Figure 4-33 illustrates the resultant scattering (F_{hkl}) form a unit cell containing two atoms, represented by their scattering vectors.

4.7.8 Calculated X-Ray Intensities

The intensity we observe for any Bragg reflection i, (i.e., $h k l$), is proportional to F_i^2.

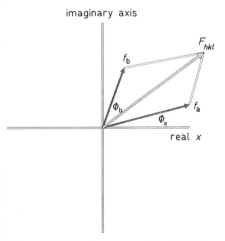

imaginary axis

F_{hkl}

f_b

Φ_b

Φ_a

f_a

real x

Figure 4-33. Resultant scattering (F_{hkl}) from a unit cell with two atoms whose scattering is represented by vectors f_a and f_b.

Therefore, if we know the position of the atoms in the unit cell (x, y, z) we can calculate the structure factor and relate this to the intensity. In addition to the Thomson Eq. (4-23), a few other terms are needed to complete this relationship. Two related to the sample itself and three to the type of diffraction experiment used to measure the intensities.

- The number of types of planes in the set i, called the plane *multiplicity factor* M_i, will directly affect the intensity. For example, the face diagonal reflection (110) and ($\bar{1}$10), (1$\bar{1}$0), ($\bar{1}\bar{1}$0) are equivalent in all unit cells where the a and b axis meet at 90°, making $M_{110} = 4$. When the a and b axis are not orthogonal the (110) and ($\bar{1}\bar{1}$0) lose their equivalence to the ($\bar{1}$10) and (1$\bar{1}$0), allowing two independent diffraction maxima each with $M_i = 2$.

- Dynamic scattering considerations require a scale factor of $1/(2 V)^2$ where V is the volume of the unit cell.

- The *Lorentz factor* is a measure of the amount of time that a point in the reciprocal lattice remains on the sphere of reflection. For example, the high angle (longest) reciprocal lattice vectors cut the Ewald sphere in a manner approaching a tangent. This causes the high angle reflections to remain in diffracting position longer and increase their intensity relative to the low angle reflections. For a powder diffractometer this effect can be eliminated, and all reflections put onto the same intensity scale, by including the term $1/(\sin^2\theta \cos\theta)$ in our expression for calculating diffraction intensities. Usually, the Lorentz factor is combined with the polarization term from the Thomson equation and called the Lp correction.

- The additional polarization of a monochromator crystal will also effect the intensity. For a diffracted beam mono-chromator the term is, $\cos^2 2\theta_m$. Where, θ_m is the Bragg angle of the monochromator crystal.

- The last term required to calculate a diffraction intensity allows for the differing path length that the X-ray beam takes in the sample, depending on diffraction angle and experimental geometry. The beam will be absorbed according to Eq. (4-10). For a powder diffractometer with a flat brickette sample, the volume of sample irradiated is independent of diffraction angle, so this absorption term reduces to a simple, $1/\mu$. For other geometries such as a cylindrical sample the absorption term will have a trigonometric component.

We are now ready to put all of the above considerations together and write the, rather formidable, equation for the diffraction intensity, for line i of phase α, for a powder diffractometer as

$$I_{i\alpha} = \frac{I_0 \lambda^3 e^4}{32 \pi r m_e^2 c^4} \frac{M_i}{2 V_\alpha^2 \mu} |F_{i\alpha}|^2 \cdot$$

$$\cdot \left(\frac{1 + \cos^2 2\theta_i \cos^2 2\theta_m}{\sin^2 \theta_i \cos \theta_i} \right) \quad (4\text{-}34)$$

where,

$$F_{hkl} = \overset{\text{number of atoms in cell}}{\underset{j=1}{\sum}} \left\{ (f_{j0} + \Delta f_j' + \Delta f_j'') \cdot \right.$$

$$\left. \cdot \exp\left[-B_j \sin^2\left(\frac{\theta}{\lambda}\right)^2 \right] \right\} \cdot$$

$$\cdot \exp[2\pi i(h x_j + k y_j + l z_j)] \quad (4\text{-}35)$$

It is a triumph of our understanding of physics that computations of diffraction intensities, using Eq. (4-34), produce the observed values. This triumph has been heavily exploited in determining the crystal structure of materials. This ability, first used by the Braggs in (1913), has estab-lished the basis of modern solid state science. To test that a particular model for a crystal structure is correct we use Bragg's law, i.e., Eq. (4-16), and the lattice parameters with Eq. (4-20), to calculate the positions of possible diffraction lines. Then Eq. (4-34) is used to compute the intensity of these lines. Even slight errors in the crystal structure model will be seen as discrepancies between the observed and calculated intensities. When each of the intensities can be independently measured, as in the case of single crystal diffractometry, they can be used in a least squares procedure to refine the atomic locations. When the reflections are difficult to separate as in the powder pattern of a low symmetry material, the least squares procedure is performed against the whole powder pattern. This *Rietveld* procedure will be further discussed in Sec. 4.13.8. The most common application of our ability to compute a powder pattern, in the materials laboratory, is to check any features of the pattern (like preferred orientation of the crystallites, solid solution effects, etc.) and any structural modifications which may have taken place. The most commonly used program for this purpose is POWD10 by D. K. Smith (1982).

4.7.8.1 Systematic Extinctions

A centered Bravais lattice permits us to divide the atoms in a unit cell into groups associated with each lattice point. Since a lattice point represents some grouping of atoms, each lattice point in a centered cell must represent exactly the same group. Thus, in a body centered cell, for each atom located at x, y, z there will be an identical atom located at $x + \frac{1}{2}, y + \frac{1}{2}, z + \frac{1}{2}$. Thus the structure factor can be factored into a sum over the atoms represented by each lattice point,

$$F_{hkl} = \left\{ \sum_{j=1}^{(\text{number of atoms in cell})/2} f_j \cdot \right.$$

$$\left. \cdot \exp\left[2\pi i\left(h x_j + k y_j + l z_j\right)\right] \right\} \cdot$$

$$\cdot \left\{ \sum_{n=1}^{(\text{number of atoms in cell})/2} f_n \cdot \right. \qquad (4\text{-}36)$$

$$\left. \cdot \exp\left[2\pi i\left(h x_n + \frac{h}{2} + k y_n + \frac{k}{2} + l z_n + \frac{l}{2}\right)\right] \right\}$$

If the sum of $h + k + l$ is even, the second term in Eq. (4-36) will contain an exponent with an integer in it. An integral number of 2π's will have no effect on the value of this term and in this case the equation reduces to,

$$F_{hkl} = \sum_{j=1}^{(\text{number of atoms in cell})/2} 2 f_j \cdot$$

$$\cdot \exp\left[2\pi i\left(h x_j + k y_j + l z_j\right)\right] \qquad (4\text{-}37)$$

However, if the sum of $h + k + l$ is odd then the second term in Eq. (4-36) will contain an exponent with a $2\pi(0.5)$ term. This causes each term in the second sum to be negative. Thus, for each atomic scattering vector in the first sum there is an equal and opposite scattering vector in the second sum. The result is that the structure factor (and hence the intensity for all reflections with the sum of $h + k + l =$ odd) is exactly equal to zero. This condition is called a *systematic extinction*. A similar analysis will show different extinction conditions for the other Bravais lattices. Screw axis and glide planes will also introduce systematic extinctions into various classes of reflections. Reflections which happen to have zero intensity due to the locations of the atoms in the unit cell and not due to a systematic symmetry condition are called accidently absent.

4.7.8.2 Primary and Secondary Extinction and Microabsorption

All of our theoretical development until now has referred to what is called the ideally imperfect crystal. This means a crystal with slight misalignments between the small mosaic blocks. These slight misalignments are usually due the relief of strain energy resulting from crystal imperfections. In some cases very perfect crystallites approach an ideally perfect crystal where the Bragg planes extend long distances with no variation in *d* value. In these very perfect crystals a dynamical effect reduces the diffracted intensity of a beam as it passes through the crystal. Each successively deeper Bragg plane in a crystal will reflect some portion of the intensity of the incident beam, through the Bragg angle back out of the crystal. If these planes are perfectly parallel, as they will be in a perfect crystal, then the beam heading out of the crystal will encounter more Bragg planes above it, which must be exactly in position to again reflect the diffracted beam back into the crystal. This doubly reflected wave will be exactly out of phase with the incident beam (because each reflection causes a phase shift of $\pi/2$) causing a net reduction in the observed intensity of a diffraction peak. This effect is called *primary extinction* and was first described by Darwin (1922). Zachariasen (1945) has developed a quantitative description of this effect.

There is another dynamical effect in perfect crystallites called *secondary extinction*. This results when the strongest reflections diffract most of the intensity in the incident beam, out of the crystal before the beam can penetrate to any significant depth. The lower lying planes thus never have a chance to diffract the amount of intensity they are capable of reflecting. This causes the most intense reflections to have a lower relative intensity than weaker reflections.

Both primary and secondary extinction effects cause the calculated intensities to differ from those observed in crystals with a high degree of perfection. The common advice to control this otherwise hard to quantitatively model phenomenon, is to introduce strains and imperfections into the crystallites by grinding. The traditional thinking is that crystallites in the tens of micron size range would not show appreciable extinction effects. However, recent studies by Cline and Snyder (1983, 1987) have shown that extinction will have significant effects on intensities even from crystallites as small as one micron.

There is one other effect which can disturb the relative intensities of diffraction peaks in a poly-phase mixture. Microabsorption occurs when crystallites of phase α lie above or below crystallites from phase β. Since the incident beam will spend some percent of its time inside an α crystallite, it will not be absorbed as if it passed through a medium having the average mass absorption coefficient of the mixture. Instead, if it spends more time in an α crystallite it will act as if it had been absorbed by a material with a mass absorption closer to that of pure α. This phenomenon was first described by Brindley (1945) and is of most concern in quantitative phase analysis applications. This, like extinction, will be reduced in importance as crystallite size decreases.

4.7.9 Vision, Diffraction and the Scattering Process

We have seen that one of the things that happens when electromagnetic radiation falls on matter is the process of coherent scattering. The incident radiation is scattered in all directions with no change in phase. Consider this process when visible light falls on a clock and a person's eye intercepts some of the scattered radiation. We of course, know that an image results from this process, projected on the retina. What is not so evident is that the image of the clock has been encrypted onto the scattered radiation and that the lens of the eye decrypts this image.

One must think about this process to be convinced that it has to be true. For example, placing a sheet of paper between the clock and the eye causes the image to disappear from the retina. Another person looking at a right angle between the clock and the first person (i.e., looking at the line of view to the clock rather than at the clock itself), sees nothing of the clock's image. Thus, the lens in the eye does not see the encoded information but instead transforms it to an image. We can describe this process mathematically: the encrypted image is the *Fourier transform* of the original image. The lens of the eye performs another Fourier transformation recreating the image to project on the retina. The process of holography captures this Fourier transform on film, so that when one looks through the film, the eye sees the encrypted information and transforms it to the image of the object whose Fourier transform had been captured. The result is the appearance of a three dimensional image behind the film.

When X-rays fall onto the atoms making up a crystal, the scattered waves carry the Fourier transform of the image of the crystal structure. If we had a lens that could bend x-rays we could use it to image the atoms in the crystal. In fact, if we had an X-ray laser we could use a visible light laser to create a hologram of the crystal structure which could be projected on a screen! However, we still await a method for producing practical coherent X-rays. While waiting, we continue to use the techniques which have evolved since 1913.

Since we do not have a material that can act as a lens to X-rays we must simulate this process with a mathematical lens on a computer. The Fourier transform of the electron density at the point in the unit cell, x, y, z (ϱ_{xyz}) is

$$\varrho_{(xyz)} = \frac{1}{V} \sum_h \sum_k \sum_l F_{hkl} \cdot$$
$$\cdot \exp[-2\pi(hx + ky + lz)] \quad (4\text{-}38)$$

where V = Volume of the unit cell. Note: The term needed in this Fourier series is F_{hkl}, not F_{hkl}^2. Since the phase of each Bragg reflection is not measurable, the intensity only yields $|F_{hkl}^2|$.

Because F_{hkl} is a complex number, we can only measure $|F_{hkl}|$. This means that the phase angle of the structure factor $\{\exp[2\pi i(hx_j + ky_j + lz_j)]\}$ is lost during the measuring process and must be inferred from other techniques. The obtaining of this term is called the phase problem. Today, due to the early work of Patterson (1938) and the more recent work of Karle and Hauptmann (recognized in 1985 with the Nobel prize), the problem is solvable by direct computer calculations for about 90% of all crystal structures (when single crystal diffraction data are available). When F_{hkl} is known from solving the phase problem, the above Fouier series can be computed for all x, y, z positions in a unit cell to give a picture of the electron density (or nuclear spin density in the case of neutron diffraction – see Chap. 19 in Volume 2 B by von Dreele for further information on neutron diffraction).

4.7.10 X-Ray Amorphography

The observed intensities, without their respective phases, may be used as the coefficients in the Fourier synthesis shown in Eq. (4-38), in place of the F_{hkl} terms. This "$|F^2|$ synthesis" was first interpreted by Patterson (1935) to be a bond distance map of the unit cell rather than the atom map one obtains from the conventional Fourier. The *Patterson map* is directly computed from the observed intensities and contains peaks which correspond to all of the interatomic bond distance vectors between the atoms in the unit cell. Each of the bond distance vectors has its origin at the origin of the Patterson function. Thus, the Patterson map looks like a pin-cushion with different length vectors protruding in all directions from a common origin. The heights of the peaks in the Patterson function correspond to the product of the electron density of the two atoms comprising each vector. This three dimensional bond distance map has been widely used to establish the phases of the F_{hkl} terms, so that Eq. (4-38) can be used to produce an image of the atoms in the unit cell.

If we plot all of the vectors in the three dimensional bond distance map on a one dimensional axis as a function only of distance r, we will have

$$g(r)\,dr = 4\pi^2 \varrho(r)\,dr \quad (4\text{-}39)$$

in which $g(r)$ is known as the *radial distribution function* (RDF). This function has maxima corresponding to each shell of neighboring atoms as one moves out in distance, r, from each atom in the material. Debye (1915) was the first to point out that the Fourier transformation of the scattering from an amorphous array of atoms would yield the RDF for a glass. This theory was brought to maturity by Warren and Gingrich (1934) and today represents one of the principal tools in the analysis of glass and liquid structure. For an excellent tutorial review of X-ray and neutron scattering from amorphous materials, see Wright (1992). See Chap. 4 by Gaskell in

Volume 9 of this series for a complete description of amorphous scattering techniques.

4.8 X-Ray Absorption Spectroscopy (XAS)

4.8.1 Extended X-Ray Absorption Fine Structure (EXAFS)

In Sec. 4.4.6 the phenomenon of the absorption edge was described and shown in Fig. 4-8. The fine structure of this absorption spectrum, which can be seen with high resolution experiments directly at the absorption edge, is related to the slight energy shifts in the ground state of the K electron induced by the bonding molecular orbitals. For many years it had been noted that the absorption edge had an extended fine structure. This is illustrated in Fig. 4-34 which shows the absorption of copper atoms in a glass with the composition of the superconducting phase $Bi_2Sr_2CaCu_2O_8$ (Bayya et al., 1991). During the 1970's an explanation of this extended X-ray absorption fine structure (EXAFS) emerged. If we think of the ionized K electron emerging from an atom as a spherical wave, then we can picture how this wave will be reflected back onto itself by the atoms in the coordination sphere around the ionizing atom. Thus, the EXAFS will depend on the number and type of atoms in the coordination sphere.

The linear absorption coefficient μ, introduced in Sec. 4.4.6, must now be thought of as having two components. The first is μ_0, the absorption of an atom in the absence of any neighbors. The other term is χ, the oscillatory component of μ, arising from self-interference of the photo-electron reflected back onto itself by neighboring atoms. χ is related to μ and μ_0 by (Lengeler, 1990 a),

$$\chi = \frac{\mu - \mu_0}{\mu} \qquad (4\text{-}40)$$

It is clear that the amplitude of the back-reflected wave will be related to the repulsive field of (i.e., number of electrons on) the neighboring atoms and the number of neighboring atoms in each successive shell away from the central atom. The amount of interference will further depend on the phase of the back reflected wave which, in turn, will depend on the distance of the scattering neighbor atom from the central atom being ionized and the wavelength of the photo-electron. Scattering theory shows that the EXAFS signal χ is,

$$\chi(k) = \frac{1}{k} \sum_i \frac{N_i}{R_i^2} F_i(k) \cdot \qquad (4\text{-}41)$$

$$\cdot \exp\{(-2\sigma_i^2 k^2) \sin[2kR_i + \phi_i(k)]\}$$

where N_i is the number of atoms in ith coordination shell at a distance R from the absorbing atom. F_i is the backscattering amplitude from neighboring atoms and ϕ is the phase shift experienced by the ejected photoelectron. σ_i is the mean relative displacement between the absorbing and scattering atoms (a Debye-Waller type factor).

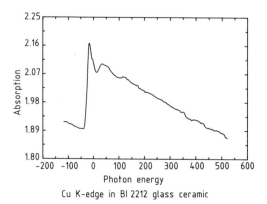

Cu K-edge in Bl 2212 glass ceramic

Figure 4-34. The copper K-edge absorption spectrum of a glass with composition $Bi_2Sr_2CaCu_2O_8$.

The radial distribution function Eq. (4-39) was described in Sec. 4.7.10 as the one dimensional projection of all of the bond distances in a structure. This function is obtained by taking the Fourier transformation of the X-ray scattering from either a crystal or an amorphous material. The Fourier transformation of the background corrected EXAFS signal, produces the atomic radial distribution function for the absorbing atom in a material. This function, which shows the probability of encountering other atoms as you move, in distance, away from the central atom, offers a powerful characterization tool for amorphous, disordered and quasi-crystalline materials. It also is considerably easier to interpret, in that it contains only the shells around the absorbing atom rather than all of the neighboring shells from all of the atoms, superimposed. Later we will discuss methods which use X-ray line broadening to determine the size of very small crystallites however, when these atom clusters get as small as a few hundred, the diffraction signal fades into background. For these cases the essentially amorphous clusters may be analyzed by EXAFS as has been done by Apai et al. (1979) for Cu and Ni clusters and Zhao et al. (1989) for Ag, Mn, Fe and Ge clusters.

As an example of amorphous analysis we will look at the processing of the Cu atom EXAFS signal from a glass with the composition of the superconducting compound $Bi_2Sr_2CaCu_2O_8$, shown in Fig. 4-34 (Bayya et al., 1991). This spectrum was measured at Brookhaven National Laboratory using the National Synchrotron Light Source Facility. Copper K-edge spectra were collected on the glass, the glass-ceramic which resulted from heating the glass, and CuO as a reference material.

EXAFS oscillations, measured by the interference function χ, were extracted from the absorption spectra shown in Fig. 4-34 by fitting a background function and normalizing using a spline fit (Teo, 1986). A Fourier transform of these oscillations was taken to obtain the atomic RDF which is shown in Fig. 4-35. The distances in this transformed data are displaced from the actual distances by a phase shift. The peak corresponding to the cation-oxygen coordination shell was back-transformed. Numerical single shell fitting was applied to the Fourier filtered data using a theoretically calculated back scattering amplitude factor for oxygen (Teo et al., 1977).

The first major peak in the spectra shown in Fig. 4-35 corresponds to the Cu−O distance. In order to determine the structural parameters of the glass and glass-ceramic, CuO was used as a model compound. Using the known values of the Cu−O bond distance and coordination number in CuO, the phase function for a Cu−O pair was determined. Theoretically calculated values of the back scattering amplitudes were used. The phase function determined from CuO and the theoretical amplitude function were then transferred to the Fourier spectra in the glass and glass-ceramic, for curve fitting, in order to

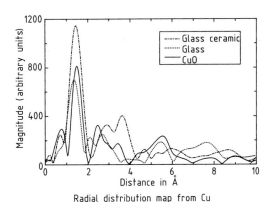

Figure 4-35. The Fourier-transform of the Cu K-edge EXAFS $k^3 \chi(k)$ in CuO, $Bi_2Sr_2CaCu_2O_8$ glass and glass-ceramic.

determine actual Cu–O distances and the coordination numbers. Figure 4-36 shows the curve fitting of $k^3 \chi(k)$ in these samples where $k = 2\pi/\lambda$ and is the magnitude of the wave vector.

The best fit least-squares refined bond-distances and coordination numbers are listed in Table 4-6. The Cu–O bond distance in the $Bi_2Sr_2CaCu_2O_8$ glass-ceramic was found to be 1.91 Å with 6 oxygens surrounding each copper atom. These values are in good agreement with those reported for the crystalline material via conventional X-ray structure analysis. However, a shorter Cu–O bond distance, 1.83 Å, was observed in the glass. This reduction of Cu–O bond distances in glass may be attributed to the possible changes in oxida-

tion state of the Cu ion and to the absence of the crystal field energy which, if present, would pull the atoms apart from their minimum energy positions. For an excellent review of all of the techniques of X-ray Absorption Spectroscopy see Lengeler (1990b).

4.8.2 Reflectometry

In recent years the materials industries have turned more and more to the use of thin films. The required characterization of films, often less than 1 μm in thickness, involves chemical content, phase and crystal structure, thickness, surface roughness and density. All of these things may be learned by lowering the angle of the primary X-ray beam to a *grazing incidence* where the phenomenon of total reflection occurs. At angles, on the order of 0.5°, the incident beam sets up a standing wave in the surface of the specimen. It penetrates into the material to a depth ranging from 20 Å for highly absorbing material to 70 Å for water. A number of exciting applications have recently developed based on this phenomenon (Lengeler, 1990b).

Since the incident X-ray beam penetrates so shallowly into the material, all X-ray analytical techniques, like fluorescence, diffraction and absorption become inherently surface sensitive under the condition of total reflection. As the angle of incidence is raised from the critical angle to a few times the critical value, the incident beam increases its penetration into the specimen, up to about 1 μm. This fact has recently led a number of researchers to examine the fluorescence and diffraction signals from thin films, as a function of depth, almost Å by Å!

At grazing incidence *Fresnel reflection* will also occur from a specimen. This basic reflection phenomena is dependent on the

Table 4-6. Cu K-edge results for Cu–O bond distance and number of coordinating oxygens.

Compound	Cu–O length R (Å)	Coordination number, N
CuO (model compound)	1.96	4.0
$Bi_2Sr_2CaCu_2O_8$ glass	1.83	3.9
$Bi_2Sr_2CaCu_2O_8$ glass-ceramic	1.91	6.0

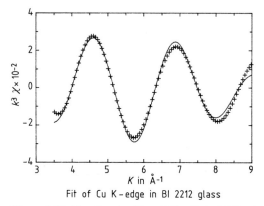

Fit of Cu K-edge in Bl 2212 glass

Figure 4-36. The Fourier-filtered $k^3 \chi(k)$ (solid lines) and simulated curve (dashed lines) of Cu K-edge EXAFS oscillations in $Bi_2Sr_2CaCu_2O_8$ glass.

density and surface roughness of the material and both of these parameters may easily be determined. If the specimen is a thin film deposited on a substrate of different composition, then the X-ray beam at grazing incidence will reflect off of both the film and substrate interfaces producing an interference pattern like that shown in Fig. 4-37. The interference pattern immediately yields the thickness of the film. Grazing incidence X-ray reflectometry is one of the most powerful new tools in the materials characterization repertoire. A recent summary of these techniques can be found in Huang et al. (1991).

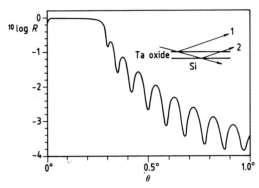

Figure 4-37. Reflectivity of 9800 eV photons from a Ta oxide layer on Si deposited at 410 °C by CVD (from Lengeler, 1990).

4.8.3 X-Ray Tomography

The use of tomographic imaging developed in the world of medicine from the need to do non-destructive testing on humans. The most common form of tomography passes radio frequencies through a brain and monitors their absorption due to hydrogen nucelar magnetic resonance. The absorption experiment is, in principal, exactly the same as that pictured in Fig. 4-4. The difference is that the brain (or the source and detector) is rotated through many orientations to collect three dimensional absorption data. The creative part of tomography developments was the writing of the computer algorithm which can reconstruct a complete three dimensional image of the brain or other organ of interest.

Recently X-ray crystallographers (Kinney et al., 1990) have taken advantage of the high intensity and wavelength tunability of synchrotron radiation to collect three dimensional absorption data on solid materials. The medical tomographic algorithms have been adapted to give a full three dimensional picture of the interior of solid objects, nondestructively. At the mo-

ment, the resolution of this technique approaches one micrometer, opening a wide variety of ceramic problems to scrutiny. There are only a few examples of full three dimensional quantitative microstructural examinations of ceramic bodies, due to the extreme difficulty of sectioning, polishing and characterizing an entire piece, so tomography offers a new window for complete microstructural analysis.

4.9 X-Ray Powder Diffraction

The preceding analysis has allowed us to predict the angle at which X-rays will diffract from a crystal. The intensity of a diffracted beam is another easily measured parameter and will be a function of the types and locations of atoms in the unit cell. Together the angle of diffraction and the intensity of the diffracted peak will be characteristic of a particular crystal structure. Since no two atoms have exactly the same size and X-ray scattering ability (i.e., number of electrons), the d_{hkl}'s between all of the possible sets of atomic planes and the intensities of diffracted beams will be unique for every material. This uniqueness allows us to identify any material, just as unique fingerprints allow the identification of any person.

We have seen that a single crystal with a particular set of atomic planes oriented toward the X-ray beam will diffract X-rays at an angle θ determined by the distance between the planes. However, most materials are not single crystals but are composed of billions of tiny crystallites. This type of material is referred to as a powder or a polycrystalline aggregate. Most materials in the world around us, ceramics, polymers and metals, are polycrystalline because they were fabricated from powders. In any polycrystalline material there will be a great number of crystallites in all possible orientations. Thus, when a powder is placed in an X-ray beam all possible interatomic planes will be seen by the beam but diffraction from each different type of plane will only occur at its characteristic diffraction angle θ. Thus, if we change the angle, 2θ, an X-ray detector makes with a specimen we will see all of the possible diffraction peaks which can be produced from the differently oriented crystallites in the powder. Figure 4-38 shows how a d^*_{hkl} from each crystallite becomes a cone of vectors which intercept the Ewald sphere forming a cone of diffraction. Thus, instead of a dot pattern, a powder pattern is a series of concentric rings. Since each point in the whole three dimensional reciprocal lattice produces a ring, a powder pattern is much more complex than its single crystal equivalent. In general, single crystal diffraction is more simple and preferred. However, since most of the high-interest phases of materials science are difficult to impossible, to grow as single crystals, we are forced use powder diffraction to study them.

4.9.1 Recording Powder Diffraction Patterns

The first powder diffraction patterns were recorded by directing an X-ray beam at a powdered specimen in a capillary tube or glued to the end of a thin glass rod, and recording the diffracted X-ray beams on a strip of film surrounding the sample. The cylindrical film wraps around the Ewald sphere so that it intercepts the cones, recording arcs. The camera which carries out this diffraction measurement is called a Debye-Scherrer camera. This was the standard way of recording powder diffraction patterns until the 1960's. The inherent poor resolution of this technique and the cumbersomeness of film techniques has caused it to fall into disuse. However, a form of this camera called the Gandolfi camera has found some amount of modern use. The Gandolfi camera is designed to obtain a powder pattern from a small single crystal. The crystal is mounted on the end of a glass rod and attached to a goniometer head. This goniometer is then turned through all of the motions needed

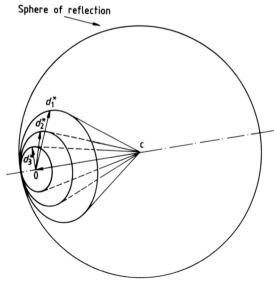

Figure 4-38. The Ewald construction for a powder.

to bring each reciprocal lattice vector onto the sphere of reflection, in all orientations. Thus, the crystallite is moved through all of the orientations that would be found in a powder and the Debye-Scherrer film arrangement records the diffraction pattern. The usefulness of this procedure is in obtaining full powder patterns from very small samples.

Beginning in 1945 and continuing through the 1950's William Parrish and his colleagues (Parrish and Gordon, 1945) developed the powder diffractometer for recording powder patterns electronically. The diffractometer is shown schematically in Fig. 4-39 and a modern realization of it in Fig. 4-40. The para-focusing geometry is called Bragg-Brentano but should more appropriately be called Parrish geometry. In order to maintain the para-focusing geometry, the detector has to move twice as fast as the sample in scanning through the diffraction angle. In the original, non-automated diffractometers, the detector is moved at a constant rate of speed, with a synchronous motor. The sample stage is coupled to the detector through a 2:1 reducing gear producing the required $\theta : 2\theta$ motion. The signal from the detector is coupled to the pen of a chart recorder, whose paper is also moving at a constant speed, producing an X-ray diffraction recording like that shown in Fig. 4-41.

Today, most patterns are obtained from automated powder diffractometers with a few from high resolution focusing cameras. The para-focusing principle in the diffractometer comes from keeping the X-ray source (i.e., the X-ray tube target), the specimen and the receiving slit on a common circle, called the focusing circle. To meet the condition of Bragg's law we must move the sample through an angle of θ while the detector is scanned through an angle of 2θ. This $\theta - 2\theta$ motion implies that the radius of the focusing circle is continually changing throughout a diffraction pattern scan. The systematic errors for a powder diffractometer fall into two categories. The first is related to the instrument itself and include such things as the error in the setting of a true zero angle and the eccentricity in the gears which drive θ and 2θ. The second error category is related to the sample. By far the most serious of these is due to a displacement of the sample from the focusing circle. This, of course, can never be avoided in that even if the sample surface is perfectly tangent to the focusing circle, the X-rays penetrate into the sample to an av-

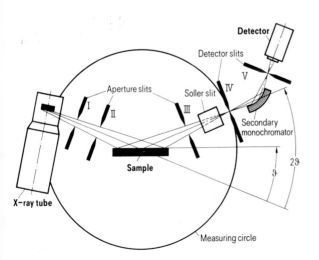

Figure 4-39. Schematic of an X-ray diffractometer of a Bragg-Brentano parafocusing diffractometer.

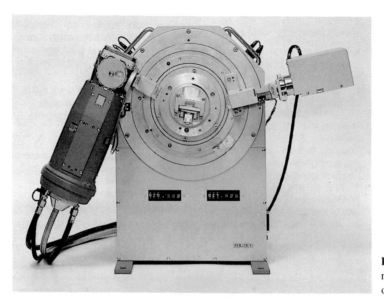

Figure 4-40. A modern automated diffractometer (Courtesy of Siemens AG).

erage depth dependent on the sample's absorption coefficient. Thus, there is always a "sample displacement" (or "sample transparency") error present. To correct the errors caused by this effect see Sec. 4.9.2.5.

Each peak, in the pattern shown in Fig. 4-41, corresponds to diffraction from a particular set of interatomic planes whose spacing d_{hkl}, may be calculated from the Bragg equation since we use X-rays of known wavelength (1.54060 Å for the $K_{\alpha 1}$ line from a Cu target X-ray tube). Although the pattern shown is characteristic of sodium chloride or table salt it would look different if we had used X-rays of a different wavelength. To remove this wave-

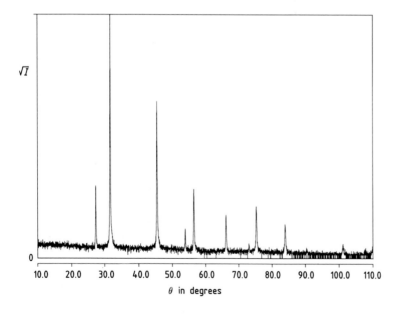

Figure 4-41. The diffraction pattern of NaCl.

length effect when the pattern is published we use only the wavelength independent d_{hkl} values, as shown in Fig. 4-58. Likewise, the intensities we observe are a function of the incident beam intensity which will vary from one laboratory and instrument to another, thus, the published values are normalized so that the highest peak is equal to 100.

4.9.1.1 Diffractometer Optics and Monochromators

The need to make the beam monochromatic follows directly from Bragg's law. If more than one wavelength is present in the beam, each set of planes will form a diffraction cone for each wavelength. This would hopelessly complicate the diffraction pattern. We have seen that the wavelength spectrum from a sealed X-ray tube contains the intense K_α and K_β lines and the continuous bremsstrahlung. The high energy continuous radiation and the beta peak can be reduced in intensity by the use of a beta filter, however, a crystal monochromator is much preferred. A crystal monochromator is a single crystal placed in the X-ray beam and oriented so that one set of its planes meet the Bragg condition. At the angle dictated by Bragg's law for the oriented d_{hkl}, of the monochromator crystal, only one wavelength will diffract. Thus, we use diffraction itself to get a monochromatic beam for our powder diffraction experiment. The most common arrangement for a diffractometer is to use a quasi-single crystal graphite monochromator which, due to its high *mosaicity*, allows the closely spaced $K_{\alpha 1}$ and $K_{\alpha 2}$ wavelengths through. Mosaicity is a measure of the misalignment of the small perfect crystalline domains which make up a single crystal. The elimination of all spectral wavelengths from the X-ray tube except the $K_{\alpha 1}$ and $K_{\alpha 2}$ greatly improves the quality of the powder pattern. This, most common type of monochromator, is placed in the diffracted beam, in front of the detector, as illustrated in Figs. 4-39 and 4-40. Placing the monochromator in the diffracted beam not only removes spectral impurities from the X-ray tube but also any fluoresced X-rays from the specimen.

Monochromators may also be placed on the incident X-ray beam. Figure 4-42a illustrates the conventional Parrish geometry diffractometer, which places the specimen in a reflecting position, with such a monochromator in place. A second diffractometer geometry, which places the sample in transmission, uses what is called Seemann-Bohlin optics and is illustrated in Fig. 4-42b. For this truly focusing geometry the powder sample is spread on a thin film with grease. Incident beam monochromators are usually made from single crystal of quartz, silicon or germanium, cut along a particular crystallographic direction. These very perfect crystals (i.e., with very low mosaicity) will separate the $K_{\alpha 1}$ and $K_{\alpha 2}$ components of the incident beam and allow complete rejection of the $K_{\alpha 2}$. These types of monochromators sacrifice intensity for wavelength resolution. To compensate for the loss in intensity, a position sensitive detector may be used in place of one that detects radiation at only one angle. The most common device employing such a monochromator is the Guinier camera, where the sample is mounted in transmission geometry and a position sensitive, piece of film is placed around the focusing circle. The true focusing condition produced by Seemann-Bohlin geometry gives a resolution slightly better than the parafocusing Parrish optics. However, the removal of the $K_{\alpha 2}$ spectral component is a much larger factor in improving the resolution over any inherent design feature.

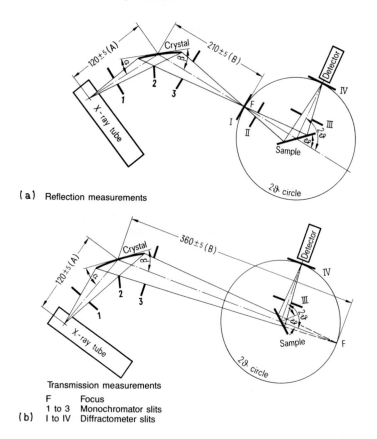

(a) Reflection measurements

Transmission measurements

F   Focus
1 to 3   Monochromator slits
(b) I to IV Diffractometer slits

Figure 4-42. Focusing incident beam monochromators used with (a) Bragg-Brentano optics and (b) Seemann-Bohlin transmission geometry.

4.9.1.2 The Use of Fast Position Sensitive Detectors (PSD)

The use of electronic PSD's in scanning mode has been developed by Göbel (1982a). In this mode the active window of perhaps 10° 2θ, is passed over the full 2θ range to be measured, the full pattern being summed into a multi channel analyzer (MCA). The attachment of such a scanning PSD to a Seemann-Bohlin transmission diffractometer, illustrated in Fig. 4-43, can produce a digital powder pattern, equivalent in resolution to a Guinier camera, in less than five minutes.

The development of fast PSD's has also led to the ability to record the diffraction peaks in a region of 2θ in as little as a few

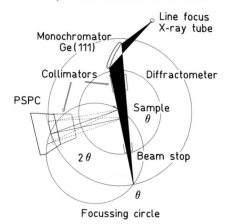

Focussing transmission diffractometer rapid data collection with PSPC

Figure 4-43. Diffractometer using Seemann-Bohlin transmission geometry and a position sensitive proportional counter (PSPC) (Göbel, 1982b).

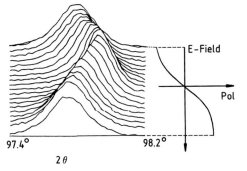

Figure 4-44. Dynamic XRD of the 400 peak of a lead zirconium titanate material as a function of electric field (from Zorn et al., 1985).

ature profile, collecting diffraction patterns every few minutes. High-temperature reactions can be mapped over hundreds of different temperatures in an overnight run. Figure 4-45 shows such a series of patterns for the peritectic melting of the high temperature superconductor $YBa_2Cu_3O_{7-\delta}$. The dynamic observation of the peritectic melting of the $YBa_2Cu_3O_{7-\delta}$ superconducting phase shows the presence of $BaCuO_2$ beginning at 950 °C. Conventional quench studies missed this important observation (Snyder et al., 1992).

4.9.1.3 Very High Resolution Diffractometers

Angular resolution can only be achieved by sacrificing intensity. For this reason the highest resolution diffractometers have been built at high intensity synchrotron facilities. The simplest of these has been built at the National Synchrotron Light Source (NSLS) at Brookhaven National Laboratory by D. Cox and colleagues (1986). It is nothing more than two germanium crystals, one placed in the incident beam in

μs, when dealing with oscillatory phenomena. An elegant application of this speed is shown in Fig. 4-44 where a peak of a PZT piezoelectric was recorded dynamically, into 20 different banks of MCA channels, which were addressed according to the instantaneous value of the electric field (Zorn et al., 1985). The full electrostriction tensor was derived from measurements like these.

Another important application of PSD's is their use in scanning mode with a high-temperature stage (Zorn et al., 1987). This permits the following of a complex temper-

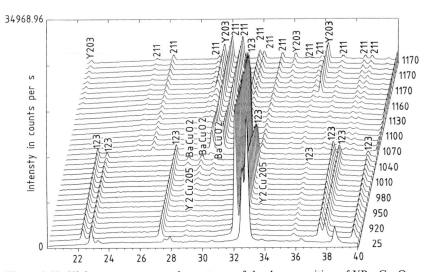

Figure 4-45. High temperature powder patterns of the decomposition of $YBa_2Cu_3O_{7-\delta}$.

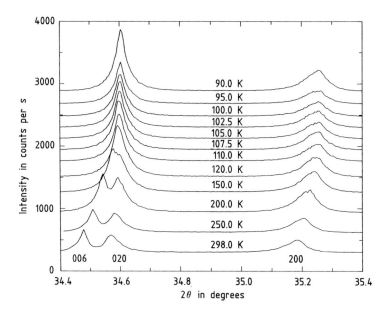

Figure 4-46. High-resolution synchrotron pattern of 123 showing complementary peak asymmetries in the 020 and 200 peaks due to martensitic strain distortion (Rodriguez et al., 1990).

front of the sample, the other in the diffracted beam after the sample. This "two monochromator" system, uses a parallel beam geometry, rather than focusing, and eliminates most of the instrumental effects which cause diffraction profiles to broaden, allowing measurement of the profile shape due only to the sample. Figure 4-46 (Rodriguez et al., 1990) shows the powder pattern of $YBa_2Cu_3O_{7-\delta}$ obtained at various temperatures on this extremely high-resolution diffractometer. The observed patterns clearly show the opposite shaped asymmetric broadening of the (020) and (200) reflections. This asymmetry is caused by the compression of **b**, and expansion of **a**, as the oxygens located at $0, 1/2, 0$, in a tetragonal cell, order on to the **a** axis of an orthorhombic cell. This observation permits the quantitative assessment of the strain energy which until now could only be qualitatively inferred from the martensitic twinned nature of $YBa_2Cu_3O_{7-\delta}$ crystals. A second design for a high-resolution diffractometer has been developed by Hart and Parrish (1989) and is shown in

Fig. 4-47. This instrument, like the previous two crystal instrument, uses a parallel beam geometry rather than focusing or para-focusing. In addition, it uses a type of monochromator where two diffracting surfaces, cut exactly at the Bragg angle, have been carved into a single crystal of silicon. This *channel-cut* monochromator introduces no instrumental aberrations into the diffraction experiment.

Another use of a parallel beam diffractometer is illustrated in Fig. 4-48. Here the incident beam is directed at the sample at a very shallow glancing angle to meet the

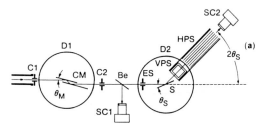

Figure 4-47. The Hart-Parrish (1989) parallel beam high-resolution diffractometer. CM is a channel-cut monochromator, S is the specimen, HPS a long soller slit assembly and SC2 the detector.

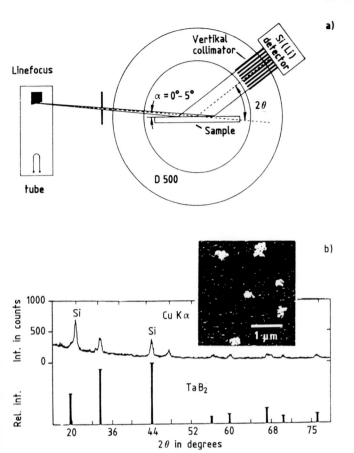

Figure 4-48. Diffraction geometry and pattern from 20 ng of TaB_2 on B-implanted, etched, $TaSi_2$ via glancing angle parallel beam experiment, from Göbel (1990).

condition of total reflection. The enhanced diffraction intensity, resulting from this geometry, allowed a full diffraction pattern to be obtained from 20 ng of sample (Göbel, 1990). A man-made multilayer monochromator (see Chap. 8 of Volume 15 of this series for more information on multilayers) has been used to monochromate the incident beam. These procedures are particularly suitable for thin film analysis. Grazing angle experiments have also been used to examine structural variations via diffraction from films as a function of depths of as little as 50 Å (Lim, 1987).

The development of high-resolution neutron diffractometers have also proved invaluable in materials characterization. These instruments have been particularly important for the case of high-temperature superconducting ceramic systems. The first significant insight into the cause and nature of the Cooper pair condensation temperature T_c, has come from ILL experiments at Grenoble (France) (Cava et al., 1990), using a high-resolution neutron diffractometer. Highly accurate bond distance measurements allowed the computation of the electron population on the ions, which showed that T_c directly relates to the charge transferred from the chain to the plane copper atoms of the $YBa_2Cu_3O_{7-\delta}$ superconductor.

4.9.2 The Automated Diffractometer

The automation of the X-ray powder diffractometer during the 1970's, led to a renaissance of this method. Many of the hardware advances already described are the direct result of the stimulation of this field brought about by automation. The other two broad areas of advances are in computer procedures and method development. We will survey each of these areas beginning with the automation of the diffractometer, which naturally leads to the evolution of software and finally to the development of new applications techniques.

The microelectronics revolution of the 1970's brought inexpensive computers and interfaces to the X-ray laboratory, although the principal thrust in the early 1970's was to develop the hardware interfaces needed to allow a computer to control a diffractometer. This work rapidly gave way to the much more serious problem of devising algorithms for the control of the instrument, and the processing of the digital data, which has occupied the subsequent decade and remains an active area of research.

There are four aspects to the automation of a diffractometer which are illustrated in Fig. 4-49:

(1) The replacement of the synchronous $\theta-2\theta$ motor with a stepping motor and its associated electronics.
(2) The replacement of a conventional analog scalar/timer with one which can be remotely set and read.
(3) The conversion of the various alarms, limit switches and shutter controls to computer readable signals.
(4) The creation of a computer interface which will allow a computer to control items 1 through 3.

These four items, illustrated in Fig. 4-50, are easily obtained today by modules which often plug directly into the bus of a modern minicomputer.

The major impact of computer automation on improving the accuracy of the measurement of diffraction angles has come from the algorithms which bring much more "intelligence" to the process, than has been conventionally used in manual measurements. There are two areas here which need to be considered, the first is the algorithms which control the collection of data and the second, those which reduce the data to d values and intensities. The first generation of control algorithms were principally non-optimizing, move and count methods. Only limited progress has been made on the development of algorithms which bring full optimization to the various aspects of data collection. The techniques used to process the digitized step scan data produced by an automated diffractometer, or an automated film reading densitometer, represent a fundamentally new approach to materials characterization. The manual methods for most characterization techniques have very rapidly fallen into disuse due to the increased power brought by numerical analysis procedures. To appreciate this major development of the last twenty years, we will examine the procedures which have been developed for processing automated diffraction scans. A number of researchers and companies have taken somewhat different approaches to this problem. We will look at the first of the published algorithms (Mallory and Snyder, 1979) as representative of the problems and solutions to them.

4.9.2.1 Background Determination

The operation of differentiating peaks from background noise may be performed in two discrete steps (Snyder, 1983a). The first is to linearize the pattern in order to

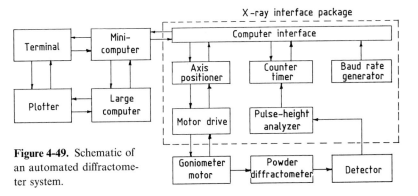

Figure 4-49. Schematic of an automated diffractometer system.

remove the low angle curvature due to the divergence of the X-ray beam and the broad maximum resulting from amorphous scattering. This is performed by connecting and then smoothing the minimum intensity points in each 0.25° 2θ segment of the pattern. A polynomial passed trough these minima is subtracted from the observed pattern, producing the linearized pattern shown in Fig. 4-51. The second step is to determine the threshold of statistically significant data (Mallory and Snyder, 1980). This is performed in a manner similar to the linearization step, but here the

maximum intensity in each 0.25° pattern segment are collected and, after some values which clearly belong to peaks are rejected, they are fit to a polynomial which represents the threshold, above which points are significantly different from background. Both of these procedures are illustrated in Fig. 4-51. This pattern was obtained from a five milligram sample placed on a glass slide with a small depression etched in it to minimize sample displacement error. Due to the very small sample size the pattern was counted at 20 seconds per 0.04° step. The raw data shown in the

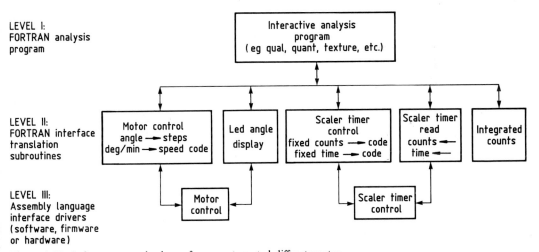

Figure 4-50. Software control scheme for an automated diffractometer.

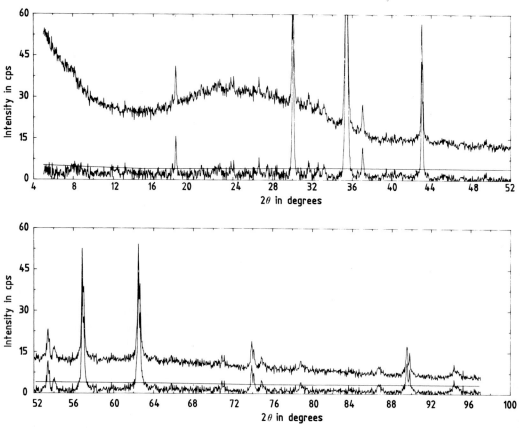

Figure 4-51. Upper trace is the observed pattern extended over both plots. Lower trace is after linearization and solid line is the threshold level.

upper curve exhibit both the effect of the amorphous sample holder and the increased low angle intensity due to the 1° divergence slit. The lower tracing shows the pattern after linearization. Note that all traces of the distortions have been removed. The smooth line on top of the background noise of the linearized pattern is the computer established threshold level. The peaks rising above this line are statistically significant.

4.9.2.2 Data Smoothing

Statistical fluctuations and the possible presence of noise spikes in the intensity measurements can lead to the detection of false peaks in the regions above the threshold. In order to avoid this, a quadratic or cubic polynomial is fit to an odd number of adjacent raw data points using a least squares regression. The point in the middle of the interval is replaced with the point computed from the interpolating polynomial. This process is illustrated in Fig. 4-52 for the linear case. Subsequent smoothed data points are produced in a similar manner by selecting the odd number of raw data points to start at each successive raw data point in the peak or peak group. As this "digital filter" slides over the data, statistical fluctuations are greatly reduced.

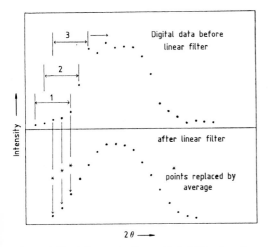

Figure 4-52. A five-point linear digital filter to reduce noise. The upper plot shows observed data, the lower shows the averaged data points.

However, there is also a corresponding loss in peak resolution. This loss increases with the 2θ step width and the number of points used in the filter. Savitzky and Golay (1964) have produced an extremely efficient computational method for applying this filter to digital data.

4.9.2.3 Spectral Stripping

For most work the $K_{\alpha 1}$ diffraction peaks can be readily recognized by a computer algorithm based on their location and height. However, when it is desired to completely remove the $K_{\alpha 2}$ peaks from the raw data, a modified Rachinger procedure is used (Ladel et al., 1970). The *Rachinger method* uses our knowledge of the exact wavelengths of the $K_{\alpha 1}$ and $K_{\alpha 2}$ lines and their intensity ratio of 2:1. As the computer scans through the raw data from low angle to high, half of the intensity at each 2θ is subtracted from the forward position calculated for the $K_{\alpha 2}$ peak. Figure 4-53 shows the results of this procedure for a portion of the quartz diffraction pattern.

4.9.2.4 Peak Location

A second derivative method is used to locate peaks because the first derivative is insensitive to shoulders which indicate overlap. The Savitzky Golay (1964) polynomial smoothing procedure automatically produces a curve whose second derivative may be evaluated. The minima in the second derivative, indicate peaks in the raw data. The number of peaks found in a procedure will depend on the amount of data smoothing and the signal to background ratio, which is a simple function of the count time. The number of false peaks found can be minimized by correlating the smoothing parameters with the count time.

4.9.2.5 External and Internal Standard Methods for Accuracy

The accuracy of the peak locations will depend in some degree on each of the above steps (Snyder, 1984b). However, to achieve absolute accuracy it is essential that we know the X-ray wavelength exactly and that the data be corrected for the aberrations introduced by the instrumental measurement technique. The sample displacement error described above is by far

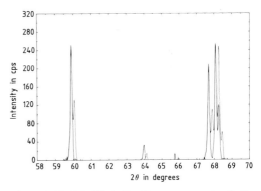

Figure 4-53. Modified Rachinger removal of $K_{\alpha 2}$ peaks. The outer enevelopes are the raw data. The inner peaks are the result after removal of the $K_{\alpha 2}$ peak.

the most serious. Displacements in the micron range can cause peak shifts of hundredths of a degree.

Wavelength Accuracy

In the early years of X-ray analysis the wavelength of the radiation was determined from Bragg's law by calculating the d_{hkl} values from the known NaCl lattice parameter and using the observed diffraction angles from the NaCl powder diffraction pattern. The lattice parameter is obtained from the density of NaCl (ϱ), the atomic weights of the Na and Cl atoms in the unit cell (ΣW) and Avagadro's number (L) via,

$$a = V^{1/3} = \left(\frac{\dfrac{\Sigma W}{L}}{\varrho} \right)^{1/3} \qquad (4\text{-}42)$$

The computed value of d_{200} was used in the Bragg equation with the observed θ for this peak, to calculate the wavelength. This method has evolved over the years through the use of ruled gratings (Bearden, 1967), to modern calibration in terms of the wavelength of the ^{3}He–^{20}Ne laser (Deslattes and Henins, 1973). The current value for the wavelength of the Cu $K_{\alpha 1}$ line is 1.54060 Å.

Internal Standard Method

The internal standard method uses a substance of very well known lattice parameters like the NBS Standard Reference Material Si 640 (Hubbard et al., 1987; Dragoo, 1986). Since the lattice parameter of the cubic Si standard has been very accurately determined (5.4060 Å), we can compute exact values for all of the d_{hkl} values and hence, the expected 2θ's of the diffraction peaks. To measure the pattern of a material accurately, some of this Si is mixed with the specimen and the combined pattern is measured. Since the average μ/ϱ of the mixture will determine the average depth from which diffraction will occur, then the sample displacement error will be the same for all lines. Thus, if we calculate the $\Delta 2\theta$ values between the observed and calculated positions for the Si lines, we can plot a correction curve of $\Delta 2\theta$ vs. 2θ. This curve is then used to correct the observed 2θ's of the material mixed with the internal standard.

External Standard Method

The use of computer automated procedures can affect accuracy in two ways: the first is to remove the computational drudgery from making internal standard corrections and the second is to allow the routine application of an external standard calibration curve (Mallory and Snyder, 1979). Approximately half of the error in peak location is due to instrumental error (Snyder et al., 1982). All instrumental error can be removed running a series of known standards and plotting $\Delta 2\theta$ (i.e., $2\theta_{obs} - 2\theta_{calc}$) vs. 2θ, as shown in Fig. 4-54. The least squares fitting of a polynomial to this curve allows the polynomial coefficients to be stored on a disk file. These may be used to correct all patterns for instrumentally induced error. However,

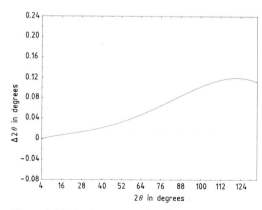

Figure 4-54. An instrumental calibration curve.

it should be emphasized that the error in peak location can be reduced by another factor of two, if an internal standard is used. This is illustrated in Table 4-7 using the F_N powder pattern quality figure of merit described below in Sec. 4.12.2.

Table 4-7. The effect of calibration on the figure of merit F_N. Notation: $F_N(|\overline{\Delta 2\theta}|, N_{poss})$.

Method	Arsenic trioxide F_{29}	Quartz F_{30}
No correction	9.9 (0.049, 59)	16.4 (0.052, 35)
External standard	15.4 (0.026, 59)	30.0 (0.028, 35)
Internal standard	42.0 (0.012, 59)	66.1 (0.013, 35)

Zero Background Holder Method

In recent years "zero background holders" (ZBH) have come into wide use. These are diffraction specimen holders made from a single crystal of quartz or silicon, which has been cut along a reciprocal lattice direction in which no diffraction will occur. The technique thus makes use of Bragg extinction to remove all X-ray scattering except that which is due to the specimen. If the specimen is carefully ground to less than 10 µm particle size, and spread on a thin (1 µm) layer of grease on the ZBH, the total specimen displacement can be held to less than 10 µm. Even the small 2θ error which would be introduced from such a displacement can be eliminated by preparing and using, an external standard calibration curve measured in exactly the same manner on a ZBH. It should also be noted that if only a monolayer of powder is exposed to the X-ray beam, all matrix absorption effects are eliminated.

4.9.3 Software Developments

During the 1960's as mainframe computers evolved (e.g., IBM 7094, CDC 6600), crystallographic computing was performed in a non-interactive environment, where cards containing coded data were input to a program operating in a batch stream. Due to the very high costs of computers, laboratory automation was nearly non-existent, so we may refer to these early developments as the zeroth generation of laboratory computational software. The 1970's saw the evolution of the laboratory computer (e.g., PDP 8 and PDP 11/10) and the first generation of process control software. Input, to control programs, was from cards and paper-tape in a non-interactive environment. The second generation of laboratory software evolved on the next generation of laboratory computers (e.g., PDP 11/34) and conducted an interactive dialogue with the user to setup and conduct an automated experiment and to analyze the data (Mallory and Snyder, 1979). This rapidly evolved into a third generation of software (ca. 1985) which employed forms and help screens on video terminals, rather than interactive dialog (Snyder et al., 1981 b). This generation saw the evolution of complex file structures containing all of the information required to process the data from an automated experiment.

A fourth generation of software is currently evolving where all interaction with the user is done through a "point and click" video interface with a mouse. The trend is to eliminate all keyboard activity and convert crystallographic methods, which used to rely on the user evaluating numbers, to visual evaluation of graphical data. For example Fig. 4-62 shows a screen from one of the first of these programs (Zorn, 1991). Most interaction is done with the mouse and all evaluation is done in the visual display window. The figure illustrates the overlaying of PDF reference patterns onto an experimental pattern. The program display is, of course, in color. The elimination of numerical d-I methods of phase identification have allowed a much larger volume of data to be processed per experiment. The problem here is that software development continues to be a scientific orphan done mostly on bootleg time. The trend to visual analysis, with mouse driven interfaces, requires substantial development efforts which both companies and universities have traditionally found difficult to muster.

4.9.3.1 The Accuracy of Diffraction Intensities

One of the traditional limitations of X-ray powder diffraction is the large uncertainty in observed intensity values (Snyder, 1983 b), due primarily to the preferred orientation of crystallites in powders. This effect arises from the tendency of crystallites to lie on the faces determined by their growth habits. This causes the powder to expose a non-random set of orientations to the X-ray beam thereby distorting the relative intensities of the diffraction lines. This problem has strongly limited the application of powder diffraction techniques to qualitative, quantitative and structural analysis. This problem can be seen in the very first published powder patterns and has remained in need of a general solution. The spray drying of powders has been shown to remove preferred orientation and produce intensity values which are accurate to a few percent (S. T. Smith et al., 1979; Cline and Snyder, 1985, 1987).

Spray drying is a technique in which a powder is suspended in a non-dissolving liquid. A small amount of an amorphous binder like polyvinyl alcohol is added to the suspension along with a defloculant to keep the particles in a fifty weight percent suspension. The slurry is then atomized through a spray nozzle into a heated chamber where the droplets dry before falling back onto a collector tray. The dried droplets, which are small spheres with a typical average agglomerate size of 50 μm, are shown in Fig. 4-55. Since a sphere is isotropic, the crystallites composing it show a random distribution of orientations to the X-ray beam. The powder pattern from a spray dried sample shows no intensity distortion from preferred orientation.

4.9.3.2 The Limit of Phase Detectability

The limit of phase detectability, by manual X-ray powder diffraction, is usually stated as being in the one to two percent concentration range. However, with automated systems, which can count for longer periods of time, the limit depends only on the number of counts collected and the signal to background ratio. Thus, to increase the sensitivity of the method we simply need to increase the number of counts collected. This may be done by increasing the count time an automated instrument

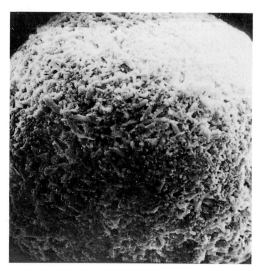

Figure 4-55. Spray dried agglomerates of acyclic wollastonite (200 ×).

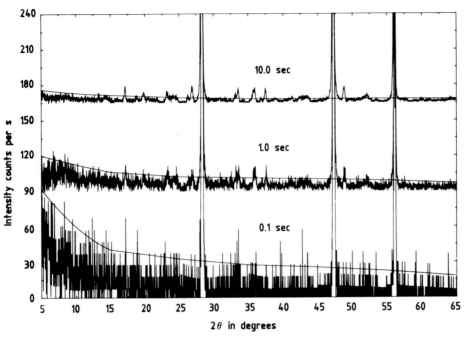

Figure 4-56. Scans of an impure Si sample at 0.1, 1.0 and 10.0 s/step.

spends at each angular step in a pattern. This is illustrated in Fig. 4-56 which shows the powder pattern of a hot pressed sample of silicon counted for 0.1, 1.0 and 10.0 seconds per point. These patterns were passed through the linearization and threshold finding procedures previously described. The increase in the number of peaks breaking through the threshold line, with increasing count time, is due to the decreasing relative error. All of the peaks other than the three lines due to silicon are below 1% relative intensity. These peaks have been identified as being due to minor impurity concentrations of SiC, Si_3N_4 and SiO_2. Figure 4-57 shows the theoretical limit of detectability as a function of count time for three different mixtures. For many materials detectability of 10 ppm is obtainable in a overnight run.

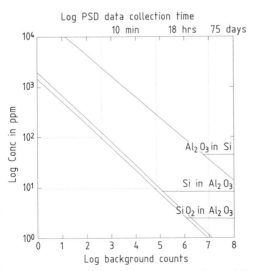

Figure 4-57. Theoretical limit of phase detectability for three mixtures.

4.10 Phase Identification by X-Ray Diffraction

We have seen that the set of diffracted d's and I's for any material is as characteristic as a fingerprint is for a human. We must now face the same problem the crime solvers have had to face – an observed set of fingerprints is useless unless you have an extensive collection of reference patterns for comparison. Our problem is even more complex in that unknown materials are commonly composed of more than one phase. When this occurs, the observed pattern contains all the diffraction peaks of each phase present. This is analogous to trying to unravel multiple finger prints superimposed on one another. We are fortunate in that usually powder diffraction patterns, particularly from high symmetry materials, are less complicated than a finger print.

4.10.1 The Powder Diffraction Data Base

Scientists have answered the need for a powder diffraction pattern database by creating an organization called the International Centre for Diffraction Data[1] (ICDD), which publishes the *Powder Diffraction File* (PDF) in annual installments. The ICDD is a nonprofit organization established by a number of international scientific organizations. The organization has historically sponsored the determination of powder patterns, through its association with the U.S. National Bureau of Standards (now called N.I.S.T.) and a number of other laboratories. However, most of the patterns published in the *Powder Diffraction File* are obtained from literature articles. They are reviewed by editors and published in book, and computer formats (e.g., CD-ROM). An example of such a pattern is shown in Fig. 4-58. The data base is published in sets with about 2000 new patterns released each year. Currently there are about 60 000 patterns in the file.

---

[1] 1601 Park Lane, Swarthmore, PA 19081, U.S.A.

A number of manual and computer searching methods have been developed over the years to accomplish phase identification. These methods place an emphasis on the accuracy of the intensities (Hanawalt method) or the d_{hkl} values (Fink method) or on the elements present (alphabetical search). Each year an alphabetical listing and a Hanawalt search manual are published for the updated PDF (by the ICDD). At non-regular intervals *Fink Search Manuals* and *Common Phase Manuals* are published. The *Common Phases* are a list of about 2500 of the most frequently encountered materials, its use often speeds phase identifications. In any search method, the ability to recognize a reference pattern in an unknown depends on the quality of the d's and I's in both the reference and unknown. The problem of phase identification is shown picturally in Fig. 4-59. The error in the d's and I's depend on a great number of factors. The experimental technique used to measure the pattern is one of the first quality indications to a user of the data base.

4.10.2 Phase Identification Strategies

The Hanawalt method involves sorting the patterns in the PDF according to the d-value of the 100% intensity line. This list is then broken into small intervals of d (with a ± 0.01 overlap in d between intervals) and each interval is sorted on the d-value of the second most intense line. Subsequent lines are listed in order of decreasing intensity.

Each phase is entered multiple times under the category of each of its most intense lines. The reason for the multiple entries is to help avoid the greatest weakness of this search method: distortion of the intensities due to preferred orientation. A sample of the Hanawalt index is shown in Fig. 4-60

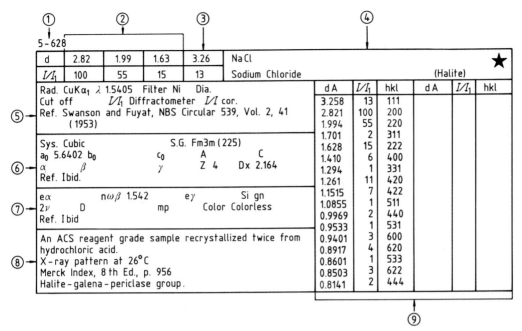

Figure 4-58. Example PDF card format: Standard 3×5 in. ICDD diffraction data card (card 628 from Set 5) for sodium chloride. Appearing on the card are: (1) set and file number, (2) three strongest lines, (3) lowest-angle line, (4) chemical formula and name of substance, (5) data on diffraction method used, (6) crystallographic data, (7) optical and other data, (8) data on specimen, and (9) diffraction pattern. Intensities are expressed as percentages of I_1, the intensity of the strongest line on the pattern. Most cards have a symbol in the upper right corner indicating the quality of the data: ★ (high quality), i (lines indexed, intensities fairly reliable), C (calculated pattern), and O (low reliability). A blank indicates undetermined quality.

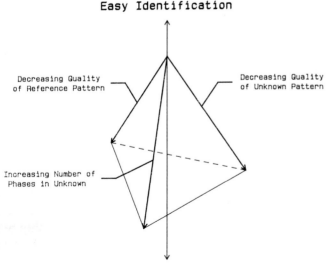

Figure 4-59. The parameters affecting phase identification.

2.84 − 2.80 (± 0.01)									File No.	I/Ic
i 2.83_x	2.00_3	1.64_3	1.42_1	1.27_1	1.07_1	2.32_1	1.16_1	$EuNbO_3$	26−1417	4.10
c 2.83_x	2.00_6	1.64_2	1.27_1	1.16_1	1.42_1	0.94_1	0.90_1	$Li_{0.4}Ag_{0.6}Br$	26− 858	
2.83_x	2.00_6	1.64_6	1.07_6	4.62_4	2.42_4	1.42_4	1.27_4	K_2CaSiO_4	19− 943	
2.83_x	2.00_9	1.63_9	1.42_7	1.27_7	2.31_6	4.90_5	4.01_5	$Sr(Mg_{0.33}Nb_{0.67})O_3$	17− 181	
i 2.83_x	2.00_4	1.62_4	2.79_3	1.63_3	1.77_3	1.27_3	4.00_2	$NdScO_3$	26−1275	2.90
2.83_x	2.00_8	1.62_8	1.77_7	1.95_6	1.65_6	3.99_5	1.50_4	$PmScO_3$	33−1091	
2.83_x	2.00_7	1.27_3	1.64_2	1.16_2	0.95_1	0.90_1	1.42_1	$AgSnSe_2$	33−1194	
2.83_x	2.00_7	1.26_3	1.63_3	1.15_3	1.41_2	3.27_2	1.70_1	$ErSe$	18− 490	
2.82_x	2.00_x	6.24_5	1.63_5	3.85_3	3.35_3	5.93_2	5.07_2	Yb_2Se_3	19−1434	
2.82_x	2.00_8	4.67_6	4.52_6	3.98_6	2.38_4	1.82_4	3.78_2	Ca_3ReO_6	34−1328	
2.82_x	2.00_5	2.23_3	1.82_2	1.41_1	1.26_1	1.15_1	1.28_1	$SiCl_4$	10− 220	
2.82_x	2.00_9	1.74_9	7.12_8	3.31_8	2.13_8	3.51_7	3.10_7	NbS_2I_2	20− 811	
i 2.82_x	2.00_4	1.63_4	1.41_8	1.07_1	1.26_1	2.30_1	0.82_1	$Li_{0.25}SrNb_{0.75}O_3$	25−1383	
2.82_x	2.00_8	1.63_8	1.41_8	1.27_8	1.07_8	2.30_5	1.98_5	Pb_2CoWO_6	17− 491	
2.82_x	2.00_8	1.63_8	1.41_8	1.07_8	2.30_5	1.16_5	4.62_3	Pb_2CoWO_6	17− 494	
o 2.81_x	2.00_x	1.62_x	1.27_x	1.65_7	1.64_7	1.78_6	3.95_5	Ca_3TeO_6	22− 156	
o 2.81_x	2.00_x	1.54_x	6.32_8	4.48_8	2.24_8	2.11_8	1.88_8	$Pt(NH_3)_4PtCl_4$	2− 817	
2.79_x	2.00_x	1.91_x	1.23_x	1.19_x	2.47_8	1.96_8	1.85_8	MnP	7− 384	
2.84_x	1.99_6	3.39_5	3.45_5	2.25_4	3.61_3	3.27_3	2.93_3	$LiBi_3S_5$	33− 791	
2.84_8	1.99_x	1.68_8	3.24_6	1.13_6	1.07_6	0.94_6	3.04_4	$CuS_{0.5}CN$	28− 402	
i 2.83_x	1.99_x	2.89_x	2.75_x	1.60_8	2.01_6	1.62_5	1.66_4	$YInO_3$	21−1449	
* 2.82_9	1.99_x	2.30_8	1.41_4	1.63_2	0.89_1	1.20_1	1.15_1	$KMgF_3$	18−1033	0.90
* 2.82_x	1.99_6	1.63_2	3.26_1	1.26_1	1.15_1	1.41_1	0.89_1	NaCl/Halite, syn	5− 628	4.40
c 2.82_x	1.99_8	1.26_3	1.63_2	1.15_2	0.94_1	0.89_1	1.41_1	$BePd$	18− 225	
i 2.81_x	1.99_4	3.98_3	2.79_3	1.62_3	2.30_3	1.63_2	4.02_2	$La_4Mn_4O_{11}$	35−1354	

Figure 4-60. Sample of the *Hanawalt Search Manual*.

and a flow chart of the search strategy is given in Fig. 4-61.

The Fink method is an alternative to the Hanawalt method when preferred orientation is suspected of distorting the relative intensities. The Fink method selects the eight strongest lines of each pattern in the PDF and rotates them to make entries in the search manual. The four most intense lines are used as entry points. The remaining seven lines are then entered in order of decreasing d-value. The deemphasis on the intensity allows materials with strong preferred orientation or with intensity distortions due to the use of a radiation other than Cu K_α, to be identified. In fact, this is the method of choice for identifying elec-

tron diffraction patterns or patterns obtained by energy dispersive techniques where the published X-ray intensities are of little value.

One of the first algorithms developed to take advantage of the large and rapid data handling capacity of the digital computer was developed at Pennsylvania by Johnson Jr. and Vand (1967). This approach compares each reference pattern in the PDF to the unknown pattern and computes a figure of merit for each match. The figure of merit is based on such quantities as the average error in matching the d's and I's. The absolute value of the merit figure is not important, since it will be computed for all reference patterns and the best

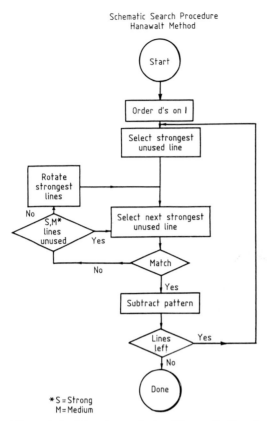

Schematic Search Procedure
Hanawalt Method

Start

Order d's on I

Select strongest
unused line

Rotate
strongest
lines

No

S,M*
lines
unused

Yes

Select next strongest
unused line

No

Match

Yes

Subtract pattern

Lines
left

Yes

No

Done

*S = Strong
M = Medium

Figure 4-61. Flowchart of the Hanawalt Search Method.

criterion. However, when a good match with the wrong chemistry occurs, it most likely means that the unknown is isostructural with the reference and thus, still provides useful information.

The probability of a computer search producing correct candidate phases for matching is greatly enhanced by allowing the program to reject compounds based on their chemistry. True unknowns should always be examined by X-ray fluorescence first, to establish the chemical constraints to be used in the search. This advice applies equally well to the second generation type of computer search algorithms as well as the first.

The development of the full digital version of the *Powder Diffraction File* and its production on a CD-ROM has dramatically improved our abilities in phase identification and characterization. The latest version of the PDF contains nearly 60 000 phases and due to the high-density format of the CD-ROM, is readily accessible on a PC. The new digital version called PDF-2 contains all of the unit cell, indices, experimental conditions, etc. contained in the full database. A second generation of computer search algorithms (Snyder, 1981) have evolved into current search/match software which uses the CD-ROM database and performs identifications in about 60 seconds. Access to this database is essential to speed up data analysis, in that the instrumental developments mentioned earlier, have led to the ability to produce data much faster than it can be analyzed using second or third generation software or the even slower manual methods. The evaluation of the results of a search have moved to the fully visual in recent years. Figure 4-62 shows a screen from the phase identification section of Zorn's (1991) program SHOW.

fifty matches listed for evaluation. This first approach to computer phase identification had a number of shortcomings (Cherukuri et al., 1983). In addition, to the obvious problems, which will be caused by systematic d_{hkl} and I errors in the reference or unknown patterns, a number of other points can cause difficulties. If the unknown pattern contains a large number of lines then it becomes quite likely that some of the reference patterns with only a few lines will match well. Thus, these patterns will tend to rise to the top of the list. Even when a reference matches a large number of lines it still may contain chemical components known not to be present, so chemistry should always be used as a search

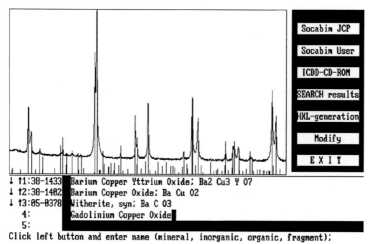

↓ 11:30-1433 Barium Copper Yttrium Oxide; Ba2 Cu3 Y O7
↓ 12:30-1402 Barium Copper Oxide; Ba Cu O2
↓ 13:05-0378 Witherite, syn; Ba C O3
  4:           Gadolinium Copper Oxide
  5:
Click left button and enter name (mineral, inorganic, organic, fragment);

Figure 4-62. A screen from the fourth generation program, SHOW by G. Zorn.

4.10.3 The Crystal Data Database

The ICDD also publishes the *Crystal Data* database which has been built up over the years at the U.S. National Bureau of Standards (now called the National Institute of Science and Technology). This database contains the unit cell information on every material for which a report has been published. At the moment, it contains references to about 150 000 unit cells. Mighell and Himes (1986) have developed a computer algorithm which will take a unit cell derived, for example, from the powder pattern of an unknown, and compute all possibly related super- and sub-cells and then search the crystal data base for the phase or possible isomorphous phases.

As an example of the use of this database, we can look at a recent analysis of the crystal structure of BeH_2 (G. S. Smith et al., 1988). Due to the low symmetry and extremely low X-ray scattering ability of BeH_2, this structure defied analysis for many years. Using the very high-resolution

diffractometer on beamline X7A at the BNL/NSLS synchrotron source mentioned previously, a very good quality pattern for this material was obtained. Computer indexing (G. S. Smith, 1987), of this pattern indicated an orthorhombic unit cell. A search of the *Crystal Data* database produced a reference to a high-pressure form of ice (i.e., OH_2). The very well resolved peaks allowed isolation of integrated intensities and the solution of this crystal structure based on this model.

4.10.4 The Elemental and Interplanar Spacings Index (EISI)

As mentioned above, the *Crystal Data* database contains about 150 000 references to published unit cells. As just described, this can be used with the modern computer indexing procedures to do phase identification and propose crystal structure models. Recently, using Eq. (4-20), powder patterns were computed from the unit cell information for all of the materials in *Crystal Data*. These calculated patterns have had the space group systematic extinct reflections removed, leaving a few *d*-values which might not occur in the observed pattern. These, occasional accidentally absent reflections are due to a particular structure factor having a low value, causing the intensity to be below the observational threshold. Since crystal structure information is not available for most of these phases, the intensities can not be computed. The new computed "*d*-value" patterns were added to the existing observed patterns in the PDF, producing a database with over 200 000 patterns. Because this database, based only on high *d*-values, is particularly suited to identifying patterns from electron diffraction, it has been called the electron diffraction database (EDD) or Max-*d* index. When the entries in the EDD are arranged by the elements present in each phase, as they would be determined by X-ray fluorescence, the resulting search manual is called the elemental and interplanar spacings index (EISI) and is also available from the International Centre for Diffraction Data.

As an example of the use of this new index, and an illustration that it is also useful for X-ray and neutron data, as well as electron diffraction, we can consider a recent application from the author's laboratory. We have been substituting various ions, including indium, into the high temperature superconducting material, $Bi_2Sr_2CaCu_2O_8$. One of the samples produced a phase whose powder pattern is not in the PDF. On carrying out the phase identifications using the EISI, a perfect match for the phase $In_{0.5}Sr_{0.5}O_3$ was observed, except that the database pattern contains two extra lines which are accidentally absent in the actual observed pattern. Thus, this index permits access to a much wider database for phase identification, even for XRD data.

4.11 Quantitative Phase Analysis

Quantitative phase analysis by X-ray powder diffraction dates back to 1925 when Navias (1925) quantitatively determined the amount of mullite in fired ceramics, and to 1936 when Clark and Reynolds (1936) reported an internal-standard method for mine dust analysis by film techniques. In 1948 Alexander and Klug (1948) presented the theoretical background for the absorption effects on diffraction intensities from a flat brickette of powder. Since then there have been numerous methods developed, based on their basic equations. Methods applicable to a wide range of phases and samples include the *method of standard additions* (also called the addition of analyte or *spiking method*) (Lennox, 1957), the *absorption diffraction method* (Alexander and Klug, 1948; S. T. Smith et al., 1979 b) and the *internal-standard method* (Klug and Alexander, 1974). Formalisms have been established to permit inclusion of overlapping lines and chemical constraints in the analysis (Copeland and Bragg, 1958). For an up-to-date summary of modern quantitative analysis procedures see Snyder and Bish (1989) and Snyder and Hubbard (1984).

Quantitative analysis using any method is a difficult undertaking. It requires careful calibration of the instrument using carefully prepared standards and many repetitions during the set up-phase to establish the required technique. In general, quantitative analysis of a new phase system will require a minimum of a few days and often a week of setup time. After the technique and standards have been established, a well-designed computer algorithm, such as the NBS*QUANT84 system (Snyder et al., 1981 b; Snyder and Hubbard, 1984) can produce routine analyses in as little as a few minutes to an hour per sample, depending on the precision desired.

Perhaps it is the difficult nature of quantitative analysis that has limited the number of carefully done literature examples. This, in turn, has resulted in a concentration of effort on a specific analysis technique, rather than on understanding the sample and instrument-dependent limitations of quantitative phase analysis by X-ray powder diffraction.

4.11.1 The Internal-Standard Method of Quantitative Analysis

The internal-standard method is the most general of any of the methods for quantitative phase analysis. In addition, this method lends itself most easily to generalization into the *RIR* (reference intensity ratio) method, and even further can be generalized into a system of linear equations which allow for the use of overlapped lines and chemical analysis constraints.

The intensity of a diffraction line *i* from a phase α, in the form of a flat plate, can be rewritten from Eq. (4-34) as

$$I_{i\alpha} = \frac{K_{i\alpha}}{\mu} \tag{4-43}$$

where μ is the linear absorption coefficient of phase α and

$$K_{i\alpha} = \frac{I_0 \lambda^3 e^4}{32 \pi r m_e^2 c^4} \frac{M_i}{2 V_\alpha^2} |F_{i\alpha}|^2 \cdot$$
$$\cdot \left(\frac{1 + \cos^2 2\theta_i \cos^2 2\theta_m}{\sin^2 \theta_i \cos \theta_i} \right) \tag{4-44}$$

The intensity of a diffraction line *i* from phase α in a mixture of phases is given by

$$I_{i\alpha} = \frac{K_{i\alpha} X_\alpha}{\varrho_\alpha (\mu/\varrho)_m} \tag{4-45}$$

where: X_α = weight fraction of phase α, ϱ_α = density of phase α, $(\mu/\varrho)_m$ = the mass

absorption coefficient of the mixture, cf. Eq. (4-13).

The fundamental problem in quantitative analysis lies in the $(\mu/\varrho)_m$ term in Eq. (4-45). To solve for the weight fraction of phase α, we must be able to compute $(\mu/\varrho)_m$ and this, as can be seen from Eq. (4-13), requires a knowledge of the weight fractions of each phase. The internal-standard method is based on elimination of the absorption factor $(\mu/\varrho)_m$, by dividing two equations of type (4-45), giving

$$\frac{I_{i\alpha}}{I_{js}} = k \frac{X_\alpha}{X_s} \tag{4-46}$$

which is linear in the weight fraction of phase α. The subscripts *i* and *j* refer to different *hkl* reflections. The *k* is the slope of the internal standard calibration curve: a plot of $(I_{i\alpha}/I_{js})$ vs. (X_α/X_s).

Equation (4-46) is linear and forms the basis of the internal-standard method of analysis. The addition of a known amount, (X_s), of an internal standard to a mixture of phases, which may include amorphous material, permits quantitative analysis of each of the components of the mixture by first establishing the values of *k* for each phase [i.e., the slope of the internal standard plot of $(I_{i\alpha}/I_{js}) X_s$ vs. X_α], from standards of known concentration.

Use of a pre-established calibration curve, defining the slope *k*, permits the weight fraction of any phase α in the original mixture to be computed. Note that *k* is a function of *i,j,α*, and *s* for a constant weight fraction of standard.

4.11.1.1 $I/I_{corundum}$

The slope *k* of the plot of $(I_{i\alpha}/I_{js}) X_s$ vs. X_a is a measure of the inherent diffracted intensities of the two phases. Visser and DeWolff (1964) were the first to propose that *k* values could be published as materials constants, if the concentration and the diffraction lines of the phases α and s were specified and s itself were agreed to by all researchers. Their proposal was to use corundum, as the universal internal standard, and to use the maximum or 100% lines of phase α and corundum, mixed in a 50% by weight mixture. This I/I_{corundum} or I/I_c value has been widely accepted and now is published with patterns for standards in the *Powder Diffraction File*.

The simple 2-line procedure for measuring I/I_c is quick but suffers from several drawbacks. Preferred orientation commonly affects the observed intensities unless careful sample preparation methods (Cline and Snyder, 1985) are employed. Other problems include extinction (Cline and Snyder, 1987), inhomogeneity of mixing, and variable crystallinity of the sample due to its synthesis and history (Cline and Snyder, 1983; Gehringer et al., 1983). All of these effects conspire to make the published values of I/I_c subject to substantial error. For greater accuracy, multiple lines from both the sample (α) and the reference phase (corundum, in this case) should be used (Hubbard et al., 1976). Such an approach often reveals when preferred orientation is present and provides realistic measurement of the reproducibility in the measurement of I/I_c. In general, spray drying the samples is the best way to eliminate this serious source of error (S. T. Smith et al., 1979 a, b; Cline and Snyder, 1985).

4.11.1.2 The Generalized Reference Intensity Ratio

The concept of the I/I_c value as a materials constant leads naturally to a broader definition, which permits the use of reference phases other than corundum, lines other than just the 100% relative intensity line, and arbitrary concentrations of the two phases α and s. The most general definition of the *RIR* has been given by Hubbard and Snyder (1988) as

$$RIR_{\alpha,\,s} = \frac{I_{i\alpha}}{I_{js}} \frac{I_{js}^{\text{rel}}}{I_{i\alpha}^{\text{rel}}} \frac{X_s}{X_\alpha} \tag{4-47}$$

In Eq. (4-47) we see that the *RIR* is a function of α and s only, not of i, j, or X_s. It, of course, remains the slope of the calibration curve for phase α with internal standard s but now has been normalized so that it may be computed from any pair of diffraction lines in a calibration mixture. We also see that I/I_c values are simply *RIR* values where s is corundum. The *RIR*, when accurately measured, is a true constant and allows comparison of the absolute diffraction line intensities of one material to another (Chung, 1974 a, b; Hubbard et al., 1976; Hubbard and Snyder, 1988). It also enables quantitative phase analysis in a number of convenient and useful ways.

4.11.1.3 Quantitative Analysis with *RIR*s

The internal-standard method of quantitative analysis when stated in terms of an *RIR* comes from rearranging Eq. (4-47)

$$X_\alpha = \frac{I_{i\alpha}}{I_{js}} \frac{I_{js}^{\text{rel}}}{I_{i\alpha}^{\text{rel}}} \frac{X_s}{RIR_{\alpha,\,s}} \tag{4-48}$$

The *RIR* value in Eq. (4-48) may be obtained by careful calibration, by determining the slope of the internal-standard plot, or it may be derived from other *RIR* values via

$$RIR_{\alpha,\,s} = \frac{RIR_{\alpha,\,s'}}{RIR_{s,\,s'}} \tag{4-49}$$

where s' is any common reference phase. When s' is corundum, the *RIR* values are simply I/I_c values. Hence, we can combine I/I_c values for phases α and s to obtain the reference intensity ratio for phase α relative to phase s. Taking s to be an internal standard and substituting Eq. (4-49) into Eq. (4-48) and rearranging we have

$$X_\alpha = \frac{I_{i\alpha}}{I_{js}} \frac{I_{js}^{rel}}{I_{i\alpha}^{rel}} \frac{RIR_{s,c}}{RIR_{\alpha,c}} X_s \qquad (4\text{-}50)$$

This equation is quite general and allows for the analysis of any crystalline phase in an unknown mixture, as long as the RIR is known, by the addition of an internal standard. However, if all four of the required constants ($I_{i\alpha}^{rel}$, I_{js}^{rel}, $RIR_{s,c}$ and $RIR_{\alpha,c}$) are taken from the literature the results should be considered as only semi-quantitative, since each of them may contain significant error.

Equation (4-50) is valid even for complex mixtures which contain unidentified phases, amorphous phases, or identified phases with unknown RIR^s.

4.11.1.4 Standardless Quantitative Analysis

The ratio of the weight fractions of any two phases whose RIR^s are known may always be computed by choosing one as X_α and the other as X_s. That is

$$\frac{X_\alpha}{X_\beta} = \frac{I_{i\alpha}}{I_{j\beta}} \frac{I_{j\beta}^{rel}}{I_{i\alpha}^{rel}} \frac{RIR_{\beta,s}}{RIR_{\alpha,s}} \qquad (4\text{-}51)$$

If the RIR values for all n phases in a mixture are known, then $n-1$ equations of type (4-51) may be written. Karlak and Burnett (1966) and later Chung (1974a, b; 1975) pointed out that if no amorphous phases are present an additional equation holds

$$\sum_{m=1}^{n} X_m = 1. \qquad (4\text{-}52)$$

Equation (4-52) permits analysis without adding any standard to the unknown specimen by allowing us to write a system of n equations to solve for the n weight fractions via

$$X_\alpha = \frac{I_{i\alpha}}{RIR_\alpha I_{i\alpha}^{rel}} \cdot$$
$$\cdot \left[1 \Big/ \left(\sum_{k=1}^{\text{number of phases}} \frac{I_{j,k}}{RIR_k I_{j,k}^{rel}} \right) \right] \qquad (4\text{-}53)$$

Chung referred to the use of Eq. (4-52) as the matrix flushing or the adiabatic principle, but it should be called the normalized RIR method. It is important to note that the presence of any amorphous or unidentified crystalline phase invalidates the use of Eq. (4-53). Of course, any sample containing unidentified phases cannot be analyzed using the normalized RIR method in that the required RIR^s, will not be known.

When the RIR values are known from another source, for example, published I/I_c values, it may be tempting to use them along with the I^{rel} values on the PDF card to perform what some call a completely "standardless" quantitative analysis using Eq. (4-52). These I^{rel} and I/I_c values are seldom accurate enough to be used directly in quantitative analysis. The analyst should accurately determine the relative intensities and RIR values for an analysis by careful calibration measurements. In fact the word "standardless" is a misnomer. Standards are always required however, the RIR method allows you to use them from the literature for semi-quantitative analysis.

4.11.1.5 Constrained X-Ray Diffraction (XRD) Phase Analysis – Generalized Internal-Standard Method

A quantitative analysis combining both X-ray diffraction and chemical or thermal results, with a knowledge of the composition of the individual phases, can yield results of higher precision and accuracy than is generally possible with only one kind of observation. The analysis becomes more complex when several phases in a mixture have similar compositions and/or potential compositional variability, but it is possible with appropriate constraints during analysis to place limits on the actual compositions of the constituent phases.

The most general formulation of these ideas has been described by Copeland and Bragg (1958). Their internal-standard equations for multicomponent quantitative analysis, with possible line superposition and chemical constraints, take the form of a system of simultaneous linear equations. In terms of RIR values each equation is of the form

$$\frac{I_i}{I_{js}} = \left(\frac{I_{i1}^{rel}}{I_{js}^{rel}} RIR_{1,s}\right)\frac{X_1}{X_s} + \left(\frac{I_{i2}^{rel}}{I_{js}^{rel}} RIR_{2,s}\right) +$$

$$+ \ldots + \left(\frac{I_{ik}^{rkl}}{I_{js}^{rel}} RIR_{k,s}\right)\frac{X_k}{X_s} + \varepsilon \quad (4\text{-}54)$$

where

- I_i is the intensity of the ith line from the mixture of k phases which may contain contributions from one or more lines from one or more phases in the mixture,
- I_{js} is the intensity of a resolved line j of the internal standard s,
- X_k is the weight fraction of phase k in the mixture of sample plus internal standard,
- ε is a least-squares error term.

Line overlap is allowed for by having each intensity ratio (I_i/I_{js}) have contributions from multiple lines from multiple phases. As many terms, involving contributions to the intensity ratio, as needed are included in each linear equation. The Copeland-Bragg analysis results in n equations in n unknowns which may be solved via least squares.

Quantitative elemental data, obtained from X-ray fluorescence analysis for example, can be added to the system of equations, without increasing the number of unknowns, making it overdetermined. Coupling both the elemental and diffraction data will result in more accurate quantitative analysis of multicomponent mixtures. More importantly, estimates of the standard deviation of the results will be more accurate. Both the line overlap and chemical analysis features of the Copeland-Bragg formalism have been included in the AUTO and NBS*QUANT84 systems of Snyder and Hubbard (1981, 1982, 1984).

4.11.1.6 Full-Pattern Fitting

With the availability of automated X-ray powder diffractometers, digital diffraction data are now routinely available on computers and can be analyzed by a variety of numerical techniques. Complete digital diffraction patterns provide the opportunity to perform quantitative phase analysis using all data in a given pattern, rather than considering only a few of the strongest reflections. As the name implies, full-pattern methods involve fitting the entire diffraction pattern, often including the background, with a synthetic diffraction pattern. This synthetic diffraction pattern can either be calculated and fit dynamically from crystal structure data or can be produced from a combination of observed or calculated standard diffraction patterns.

The use of the whole pattern by D. K. Smith and his colleagues (D. K. Smith et al., 1987, 1988), involves collection of reference or standard data on pure materials using fixed instrumental conditions. These patterns of standards are processed to remove background and any artifacts, and the data may be smoothed. For standard materials that are unavailable in appropriate form, diffraction patterns can either be simulated from Powder Diffraction File data or calculated powder patterns. The next step in standardization is analogous to procedures used in conventional reference intensity ratio quantitative analyses. Data are collected on each of the samples of the standard that have been mixed with a known amount of corundum, α-Al_2O_3, in order to determine the RIR. These patterns yield the basic calibration data used in the quantitative analysis. The final step involves collection of data for the unknown samples to be analyzed, using conditions identical to those used in obtaining the standard data. Background and artifacts are also removed from unknown sample data.

Reference intensity ratios are obtained from the whole patterns by initially assigning the standard material a reference intensity ratio of 1.0. Then the values of "apparent" X_α and X_s are computed from Eq. (4-50). The ratio of the "apparent" weight fractions of α to s gives the correct RIR value. For example (D. K. Smith et al., 1987), a standard mixture of 50% ZnO and 50% corundum yields results of 57.8% ZnO and 42.2% corundum. The reference intensity ratio for ZnO is therefore 57.8/42.2 = 1.37.

The digital patterns of the unknown mixture are fit to the standards with a least-squares procedure minimizing the expression

$$\delta(2\theta) = I_{unk}(2\theta) - \sum_k X_k RIR_k I_k(2\theta) \quad (4\text{-}55)$$

where $I_{unk}(2\theta)$ and $I_k(2\theta)$ are the diffraction intensities at each 2θ interval for the unknown and each of the standard phases. Intensity ratios derived using this procedure are equivalent to peak-height RIR^s, rather than integrated-intensity RIR^s because the minimization is conducted on a step-by-step basis.

This whole pattern fitting procedure is simply a method for measuring RIR^s from whole pattern standards. Once this is accomplished the method becomes a conventional RIR quantitative analysis procedure, which requires standards and calibration. To make it "standardless" one must invoke the normalization assumption, i.e., Eq. (4-52).

4.11.1.7 Quantitative Phase Analysis Using Crystal Structure Constraints

Quantitative phase analysis using calculated patterns is a natural outgrowth of the *Rietveld method* (1969), originally conceived as a method of refining crystal structures using neutron powder diffraction data. Refinement is conducted by minimizing the sum of the weighted, squared differences between observed and calculated intensities at every step in a digital powder pattern. The Rietveld method requires a knowledge of the approximate crystal structure of all phases of interest (not necessarily all phases present) in a mixture (Hill and Howard, 1987; Bish and Howard, 1988).

The input data to a refinement are similar to those required to calculate a diffraction pattern i.e., space group symmetry, atomic positions, site occupancies, and lattice parameters. In a typical refinement, individual scale factors (related to the weight percents of each phase) and profile, background, and lattice parameters are varied. In favorable cases, the atomic positions and site occupancies can also be successfully varied. The method consists of fitting the complete experimental diffraction pattern with calculated profiles and backgrounds, and obtaining quantitative phase information from the scale factors for each phase in a mixture.

The $K_{i\alpha}$ of Eq. (4-44) can be divided into two terms. The first is

$$k = \left[\left(\frac{I_0 \lambda^3}{32 \pi r} \right) \left(\frac{e^4}{2 m_e^2 c^4} \right) \right] \quad (4\text{-}56)$$

which depends only on the experimental conditions and is independent of angle and sample effects. The second part

$$T_{i\alpha} = \frac{M_i}{V_\alpha^2}|F_{i\alpha}|^2 \left(\frac{1 + \cos^2 2\theta_i \cos^2 2\theta_m}{\sin^2 \theta_i \cos \theta_i}\right) \quad (4\text{-}57)$$

depends on the crystal structure and the specific reflection in question. Equation (4-43), for a pure phase, can now be written in terms of Eqs. (4-56) and (4-57) as

$$I_{i\alpha} = \frac{k T_{i\alpha}}{\mu_\alpha} \quad (4\text{-}58)$$

In a mixture, the intensity of reflection i from phase α is obtained from Eq. (4-45) as

$$I_{i\alpha} = k T_{i\alpha} \frac{x_\alpha}{\varrho_\alpha(\mu/\varrho)_m} \quad (4\text{-}59)$$

Here all of the structure sensitive parameters, which the Rietveld procedure optimizes, are in the T_α term.

In a multiple-phase mixture, other terms must be introduced to scale the computed powder pattern for each phase, before summing them together to fit the pattern of the mixture. In addition, terms to aid in the pattern fitting will be needed. The quantity minimized in Rietveld refinements is the conventional least squares residual

$$R = \sum_j w_j |I_j(\text{o}) - I_j(\text{c})|^2 \quad (4\text{-}60)$$

where $I_j(\text{o})$ and $I_j(\text{c})$ are the intensity observed and calculated, respectively, at the jth step in the data and w_j is the weight. Thus, as is the case in whole pattern analysis, it is more appropriate to consider the intensity at a given 2θ step rather than for a given reflection. In a single phase powder pattern of phase α, the intensity at each step j, is determined by summing the contributions from background and all neighboring Bragg reflections as

$$I_j(\text{c}) = S \sum_i T_{i\alpha} G(\Delta\theta_{j,i\alpha}) P_i + I_{jb}(\text{c}) \quad (4\text{-}61)$$

where S is the conventional Rietveld scale factor, which puts the computed intensities on the same scale as those observed, P_i is a sometimes used preferred orientation function for the ith Bragg reflection, $G(\Delta\theta_{j,i\alpha})$ is the profile shape function, and $I_{jb}(\text{c})$ is the background (Wiles and Young, 1981). The Rietveld scale factor, S, includes all of the constant terms in Eq. (4-56) along with the absorption coefficient from Eq. (4-58)

$$S = \frac{k}{\mu} \quad (4\text{-}62)$$

where S is the constant required to scale the observed and calculated patterns, and μ is the linear absorption coefficient for the sample. For a multi-phase mixture, Eq. (4-62) can be rewritten summing over the n phases in a mixture (e.g., Hill and Howard, 1987) as

$$I_{ic} = I_{ib} + \sum_n S_n \sum_j T_{jn} G_{ijn} \quad (4\text{-}63)$$

The scale factor for each phase can now be written as

$$S_\alpha = k \frac{X_\alpha}{\varrho_\alpha(\mu/\varrho)_m} \quad (4\text{-}64)$$

Therefore, in a Rietveld analysis of a multicomponent mixture, the scale factors contain the desired weight fraction information as

$$X_\alpha = \frac{(\mu/\varrho)_m}{k} S_\alpha \varrho_\alpha \quad (4\text{-}65)$$

However, the sample mass absorption coefficient is not known and we are forced to apply the usual internal-standard analysis, calibrating the RIR^s of the phases to be analyzed, or applying the normalization assumption, constraining the sum of the weight fractions of the phases considered to unity. In fact, Hill and Howard (1987) recently showed the Rietveld scale factor acts in the role of a reference intensity ratio permitting conventional RIR analysis.

In order to apply the normalized RIR approach in the Rietveld method, we can think of Eq. (4-52) as

$$X_\alpha = \frac{X_\alpha}{X_\alpha + X_\beta + \dots} \qquad (4\text{-}66)$$

Equation (4-66) can be solved for the weight fractions of each of the diffracting phases by substituting Eq. (4-65) into it, giving

$$X_\alpha = \frac{S_\alpha \varrho_\alpha}{\sum_n S_n \varrho_n} \qquad (4\text{-}67)$$

We see that this procedure is exactly analogous to the normalization assumption in which reference intensity ratios are measured prior to analysis. Instead of measuring reference intensity ratios to put all intensities on an absolute scale, the Rietveld method calculates what we should call normalized RIR_N values, in that they refer to the relative scale between the observed and calculated patterns

$$RIR_{N,\alpha} = S_\alpha \varrho_\alpha \qquad (4\text{-}68)$$

The conventional internal-standard method of analysis, applied to Rietveld quantitative analysis, requires that a known weight fraction of a crystalline internal standard be added to the unknown mixture or that one of the components be of known concentration. Thus, if X_s is known, then the normalized RIR can be used directly, to determine the weight fractions for other phases in the sample. For example, the weight fraction for the α phase is determined by

$$X_\alpha = \frac{S_\alpha \varrho_\alpha}{\left(\dfrac{RIR_{N,s}}{X_s}\right)} \qquad (4\text{-}69)$$

S_α is a refined parameter, and ϱ_α can be calculated from the refined composition and cell parameters of each phase. There-

fore, the weight fraction of phase α, (X_α) can be easily determined. This internal-standard method does not constrain the sum of the weight fractions, as does the normalized RIR method.

The total weight fraction of any amorphous components, can also be determined with this method, by adjusting the Rietveld background polynomial to fit the broad amorphous scattering profile. The difference between the sum of the weight fractions of the crystalline components and 1.0 is the total weight fraction of the amorphous components. O'Connor and Raven (1988) used this method to conclude that their quartz contained an 18% amorphous component.

4.11.2 The Absolute Reference Intensity Ratio: RIR_A

Equations (4-64) shows that the Rietveld scale factors for all phases, including any amorphous material, contain the constant term $(\mu/\varrho)_m$. If this term could be factored out and refined independently, we could put the RIR^s onto an absolute scale by referring all of them to the scale of the calculated pattern. The problem is that a multiplicative factor cannot be independently refined using least squares. Only two other approaches remain. The first is to compute the overall $(\mu/\varrho)_m$ at the end of each cycle of least squares and then use it in computing the weight fractions to initiate the next cycle. The other is to break between cycles, of a nearly converged system, and run the simplex algorithm to determine the value of $(\mu/\varrho)_m$. The latter method suffers from removing the variable from the statistical environment of the least squares and if any correlations develop between this and any other parameter the procedure becomes invalid.

The absolute reference intensity ratio (Snyder, 1991 a) can be defined as

$$RIR_{A, \alpha} = \frac{(\mu/\varrho)_m}{k} S_\alpha \varrho_\alpha \qquad (4\text{-}70)$$

4.11.3 Absorption-Diffraction Method

In the internal-standard method, Eq. (4-45) was written twice, once for the unknown and again for the internal standard. The ratio of these gave Eq. (4-46) in which the absorption coefficient of the unknown sample cancelled out. In the absorption-diffraction method, we again write Eq. (4-45) twice: once for line i of phase α in the unknown and again for line i of a pure sample of phase α. The ratio here gives

$$\frac{I_{i\alpha}}{I_{i\alpha}^0} = \frac{(\mu/\varrho)_\alpha}{(\mu/\varrho)_m} X_\alpha \qquad (4\text{-}71)$$

where $I_{i\alpha}^0$ is the intensity of line i in pure phase α. Equation (4-71) is the basis of the absorption-diffraction method for quantitative analysis. There are several methods for implementing Eq. (4-71). A few cases will be discussed in some detail. Each of these cases is dependent only on the validity of Eq. (4-71), and since we have made no assumptions in the derivation, this equation is rigorous.

It is more common than one might imagine that quantitative X-ray diffraction phase analysis is carried out on samples whose chemical composition is known. When this is the case, the mass absorption coefficients of the pure phase, to be analyzed, and of the mixture can be computed from Eq. (4-13). Thus, if $I_{i\alpha}$ and $I_{i\alpha}^0$ are measured from the unknown and a pure sample of α respectively, the weight fraction of the analyte (i.e., X_α) may be computed from Eq. (4-71).

Leroux et al. (1953) were the first to propose that quantitative analysis may be performed on unknowns by measuring the density and determining the linear absorption coefficients of the unknowns experimentally. This may be done using absorption experiments on specimens of different thickness and employing the mass absorption law given in Eq. (4-10). The difficulty with this approach is that the measurement of μ is extremely error prone. The low precision of this measurement dramatically limits the accuracy of a quantitative analysis.

It happens now and then that the absorption coefficient of the mixture and the phase to be analyzed are the same. In this case Eq. (4-71) reduces to

$$\frac{I_{i\alpha}}{I_{i\alpha}^0} = X_\alpha \qquad (4\text{-}72)$$

since $(\mu/\varrho)_\alpha$ is exactly the same as $(\mu/\varrho)_m$. Common examples of this case are the analysis of cristobalite in a quartz and amorphous SiO_2 matrix or the analysis of the cubic, tetragonal, and monoclinic forms of ZrO_2 in pure ZrO_2 bodies. In cases where we have mixtures of polymorphs, the absorption-diffraction method of analysis becomes very attractive.

Another special case is that of a lightly loaded filter where the sample approximates a monolayer and presents a special case of Eq. (4-71). As long as all the particles lie alongside each other and do not shade each other from the X-ray beam, then all matrix absorption effects are eliminated. There can, of course, be no preferential absorption. Each crystallite of each phase diffracts as if it were a pure phase. Thus, the absorption coefficient for each phase is the same as for the pure phase. This causes Eq. (4-72) in case 3 to become operable. A plot of $I_{i\alpha}/I_{i\alpha}^0$ vs. X_α will be linear and immediately allows for quantitative analysis. This procedure is valid up to the limit of concentration of particles where crowding invalidates the assumption that $(\mu/\varrho)_\alpha = (\mu/\varrho)_m$. This method is commonly used, for example, in the analysis of respirable silica in air-filtered samples collected on silver membranes.

The last example of the use of the absorption-diffraction method is that of a binary mixture. In this case, Eq. (4-45) becomes

$$I_{i\alpha} = \frac{K_{i\alpha} X_\alpha}{\varrho_\alpha[X_\alpha(\mu/\varrho)_\alpha - (\mu/\varrho)_\beta] + (\mu/\varrho)_\beta} \quad (4\text{-}73)$$

Equation (4-45) for the pure phase α may be written as

$$I_{i\alpha}^0 = \frac{K_{i\alpha}}{\varrho_\alpha(\mu/\varrho)_\alpha} \quad (4\text{-}74)$$

Dividing Eq. (4-73) by Eq. (4-74) gives

$$\frac{I_{i\alpha}}{I_{i\alpha}^0} = \frac{X_\alpha(\mu/\varrho)_\alpha}{X_\alpha(\mu/\varrho)_\alpha + X_\beta(\mu/\varrho)_\beta} \quad (4\text{-}75)$$

Equation (4-75) can easily be used to compute the standard plot of $I_{i\alpha}/I_{i\alpha}^0$ vs. X_α for all possible compositions of mixtures of α and β.

4.11.4 Method of Standard Additions or Spiking Method

Lennox (1957) was the first to develop the spiking method, which has been widely adopted in X-ray fluorescence spectroscopy as the method of standard additions. This method, like the internal-standard method, is perfectly general applying to any phase α in a mixture. The only requirement is that the mixture also contains, among other phases, a phase β, which has a diffraction line unoverlapped by any line from α, which may be used as a reference. Phase β is not analyzed and does not even need to be identified.

In a sample containing α and β, the ratio of the intensities from a line from each phase can be obtained from Eq. (4-45) as

$$\frac{I_{i\alpha}}{I_{j\beta}} = \frac{K_{i\alpha} \varrho_\beta X_\alpha}{K_{j\beta} \varrho_\alpha X_\beta} \quad (4\text{-}76)$$

Using the method of standard additions, some of the pure phase α is added to the mixture containing the unknown concentration of α. After adding Y_α grams of the α phase, per gram of the unknown, the ratio $I_{i\alpha}/I_{j\beta}$ becomes

$$\frac{I_{i\alpha}}{I_{j\beta}} = \frac{K_{i\alpha} \varrho_\beta (X_\alpha + Y_\alpha)}{K_{j\beta} \varrho_\alpha X_\beta} \quad (4\text{-}77)$$

where

- X_α = the initial weight fraction of phase α,
- X_β = the initial weight fraction of phase β,
- Y_α = the number of grams of pure phase α added per gram of the original sample.

In general, the intensity ratio is given by

$$\frac{I_{i\alpha}}{I_{j\beta}} = K(X_\alpha + Y_\alpha) \quad (4\text{-}78)$$

where K is the slope of a plot of $I_{i\alpha}/I_{j\beta}$ versus Y_α, with Y_α in units of grams of α per gram of sample. Multiple additions are made to prepare a plot in which the negative x intercept is X_α, the desired concentration of the phase α in the original sample.

4.11.4.1 Amorphous Phase Analysis

The area under the amorphous scattering hump, in the approximate range between $15°$ and $30°$ 2θ, is related to the concentration of the amorphous phase present, in exactly the same manner as the intensity of a diffraction peak is related to concentration. If an unobscured section of the amorphous scattering region is integrated, the intensity may be treated according to any of the above methods. The only consideration which must be emphasized, is the points at which background is measured. Usually only a small error will result if a low-angle background in the vicinity of $10°$ 2θ, and a high-angle background in the range of $40°$ 2θ, are chosen. This technique was used by Hubbard to estimate the amount of amorphous silica in an NBS standard reference sample of quartz.

4.12 Indexing and Lattice Parameter Determination

Many of the patterns in the powder diffraction file give the hkl associated with each of the d-values. In this section we will discuss how these values may be obtained from powder diffraction patterns. In fact, many of the PDF cards have had their hkl's computed from lattice parameters obtained from single crystal X-ray diffraction measurements. However, current computer methods can routinely determine the indices of lines in patterns of unknowns if, and only if, the patterns are measured very accurately and corrected for systematic errors.

Indexing is the procedure used in assigning hkl values to the set of planes which give rise to each diffraction line. There are various procedures for indexing but in all procedures more difficulty is encountered as the crystal symmetry is decreased, i.e., cubic crystals are easy while triclinic are the most difficult. The procedure which is followed is dependent upon the information that is known about the crystal, that is, whether the unit cell dimensions and/or the crystal system is known.

4.12.1 Accuracy and Indexing

If the cell dimensions are known, then the pattern can be indexed by calculation,

regardless of the crystal system. This is a result of the fact that d_{hkl} values are a geometric function of the size and shape of the unit cell. The relationship between d_{hkl} and the real unit cell is cumbersome and usually stated in a different form for each crystal system. However, the functional relation between the square of the reciprocal lattice vectors $(d_{hkl}^* = 1/d_{hkl})$ and the size and shape of the reciprocal unit cell, previously derived as Eq. (4-20), applies easily to all crystal systems.

For any indexing procedure to work, the data must be very accurate. The three most common errors result from:

– Uncorrected instrumental and sample related errors (e.g., diffractometer specimen displacement error). These can cause $\Delta 2\theta$ errors greater than $0.1°$.
– Debye-Scherrer data containing absorption errors, in the low angle reflections. This causes shifts in d values which follow a $\cos\theta$ relation.
– The presence of one or more extraneous lines due to a second phase or the presence of a $K_{\alpha 2}$ or K_β peak. A single incorrect line will cause most indexing methods to fail, in that they all attempt to index all lines.

These problems may be eliminated by using diffractometer or focusing Guinier camera data, and correcting the observed 2θ's using the internal standard method. It can not be emphasized too strongly that only diffraction patterns with 2θ values of the highest possible accuracy should be used in the modern indexing procedures. The numerical procedures are particularly vulnerable to experimental error.

4.12.2 Figures of Merit

A figure of merit is a mathematical function which rates the quality of a match between any two things. In indexing powder patterns, we need some type of objective criterion to rate the different matches. Since d_{hkl} spacings are related to the intuitive, measured, 2θ values through a nonlinear sin function, we cannot rely on the users qualitative estimate of the correctness of a proposed cell. A number of authors prefer to use Q values, for both indexing algorithms and for figures of merit, rather than d_{hkl}^{*2} because the values are much larger. Q is defined as

$$Q_{hkl} = 10\,000\, d_{hkl}^{*2} = \frac{10^4}{d_{hkl}} \qquad (4\text{-}79)$$

There are two figures of merit in common use today. Most of the popular computer indexing and lattice parameter refinement procedures use either one or both of them. The M_{20} figure of merit, proposed by de Wolff (1969), is defined as

$$M_{20} = \frac{Q_{20}}{2\,|\Delta Q|\, N_{20}} \qquad (4\text{-}80)$$

where $|\Delta Q|$ is the average discrepancy between Q_{obs} and Q_{calc}, Q_{20} is the Q value of the 20th line and N_{20} is the number of possible, space group allowed, lines out to the value of Q_{20}. This figure of merit works well in determining the correctness of an indexing but has the disadvantage of having a different absolute value for each different crystal system.

Smith and Snyder (1979) have proposed another figure of merit called F_N which is defined as

$$F_N = \frac{N}{|\Delta 2\theta|\, N_{poss}} \qquad (4\text{-}81)$$

where N is the number of observed lines used in the calculation, N_{poss} is the number of possible, space group allowed, lines out to the 2θ of line N, and $|\Delta 2\theta|$ is the average error between the observed and calculated 2θ values. This figure of merit is useful both in evaluating the quality of powder patterns and in establishing the correctness of an indexing. The advantage for indexing is that its value is independent of the crystal system, so a common scale can be applied to all crystal systems. Another advantage is that the reciprocal of F_N establishes an upper limit on the average $\Delta 2\theta$, which has some intuitive meaning.

4.12.3 Indexing Patterns with Unknown Lattice Parameters

The indexing of a unit cell with unknown lattice parameters is a formidable problem. The problem becomes more difficult as we go to lower symmetry because the number of unknowns we must determine increases. The d_{hkl}^{*2} equation can be rewritten as

$$Q_{hkl} = h^2 A + k^2 B + l^2 C + kl D + \\ + hl E + hk F \qquad (4\text{-}82)$$

The problem of indexing can then be stated as having to find the six unknowns A through E. For example, we previously saw that in the cubic crystal class the d_{hkl}^{*2} equation reduces to

$$d_{hkl}^{*2} = (h^2 + k^2 + l^2)\, a^{*2} \qquad (4\text{-}83)$$

With only one unknown lattice parameter the problem of indexing a cubic powder pattern is always solvable. However, it still requires solving one equation for two unknowns: $(h^2 + k^2 + l^2)$ and a^{*2}. Therefore, the simple mathematical procedures for solving, for example, two equations for two unknowns do not apply, in that we have only one equation. To supply the extra information needed to avoid this difficulty, we must take an indirect approach and apply a relationship between the lattice parameter and d, which can not be written in the form of an explicit equation. Thus, the method will be an algorithm rather than a simple mathematical formula evaluation.

The algorithm for cubic is based on our being able to further reduce the d_{hkl}^{*2} equation even further to

$$d_{hkl}^{*2} = n^2 a^{*2} \qquad (4\text{-}84)$$

where $n^2 = h^2 + k^2 + l^2$. Notice that when one d_{hkl}^{*2} value is subtracted from any another

$$\Delta d^{*2} = d^{*2} - d^{*2'} = (n^2 - n^{2'}) a^{*2} \quad (4\text{-}85)$$

the result is an integral multiple of a^{*2}. Thus, if two reflections differ in n^2 by only 1, the result of subtracting their d^{*2} values is a^{*2}. Once a^{*2} is known we can calculate all possible d_{hkl} values from the d_{hkl}^{*2} equation.

Once the value of a^{*2} has been established, we need to consider how to obtain the "best" value. This is the subject of lattice parameter refinement described in the next section.

There are three fundamental approaches to indexing lower symmetry patterns. Since the problem gets progressively more difficult numerically as we decrease symmetry all three approaches employ computers. The three approaches are (1) intelligent algorithms, (2) searches of solution space and (3) searches of index ($h\,k\,l$) space.

The intelligent procedures try to determine the lattice parameters by applying known relationships from both geometry and crystallography and even previous experience with indexing problems. They are "intelligent" only in the sense of computer artificial intelligence and that is the context in which the word will be used. Two quite successful methods of this type have been developed by Visser (1969) and Werner (1964).

Search methods do a systematic search of solution space (i.e., look at all possible solutions within some boundaries). There are two types of index searching procedures: those that work in parameter space and those that examine index or $h\,k\,l$ space. Some of the search methods make assumptions about the regions of solution space that should be examined while others are exhaustive.

We should pause here to discuss the concept of a solution space. This space for an exhaustive parameter search, is envisioned as having up to six dimensions in a plane (corresponding to functions relating to a, b, c, α, β and γ) and the F_N figure of merit plotted perpendicular to this plane. For cubic problems this means a two dimensional problem: vary all possible values of a along the x axis with F_N plotted vertically. As described above, there are an infinite number of unit cells that can describe a lattice. In fact, there are also a great number of incorrect cells that will almost describe the lattice (within reasonable error limits). Each of these solutions will correspond to a maximum of F_N in solution space. Our assumption is that the highest maximum corresponds to the correct unit cell.

An exhaustive indexing procedure working in parameter space examines all possibe values of a, b, c, α, β and γ within some reasonable limits, usually adjustable by the user. The lattice parameters are divided into small increments and each possible set is used in an attempt to index the unknown pattern. For the triclinic case this means using a six dimensional space with an extremely large number of attempted indexings. The usual limit on exhaustive parameter space procedures is to only look at possible cells with less than a few thousand cubic angstroms in volume. Louër and Louër (1972) have established an extremely successful program based on this approach.

Exhaustive procedures in index or $h\,k\,l$ space examine all possible combinations of h, k and l which can be used to index each of the lines in the unknown pattern. Here

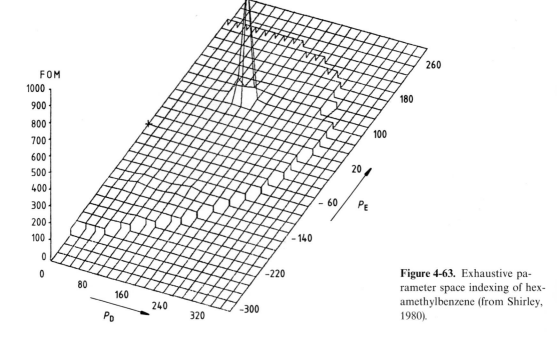

Figure 4-63. Exhaustive parameter space indexing of hexamethylbenzene (from Shirley, 1980).

our solution space is four dimensional for all symmetries: axis in integer units of h, k and l in a plane, F_N perpendicular to that plane. The advantage here is that hkl space is "naturally" quantized into integers, while in parameter space a reasonable quantum, or division factor, must be determined. If the spatial increment is too large the correct solution may be stepped over. A very powerful algorithm based on hkl space searching has been developed by Taupin (1973b).

Shirley (1980) has shown two extreme cases for an exhaustive parameter search. Figure 4-63 shows a very smooth solution space for the indexing of hexamethylbenzene. The cosine terms, D and E from Eq. (4-82), are in the plane with a sharpened figure of merit (FOM) vertical. Figure 4-64, on the other hand, shows the extremely lumpy or solution prone space for the indexing of Cu-phthalocyanine · HCl.

Before the days of powerful, fast PC's, exhaustive search procedures were only practical in relatively high symmetry systems and in a few exceptional cases (see Warren, 1969). For example the classic case of a manual orthorhombic indexing was performed by Jacob and Warren (1937) in the determination of the crystal structure of α-uranium. As the symmetry decreases to triclinic, the amount of computation required became impractical. However, today monoclinic and triclinic problems are routinely solved on researchers desk-top workstations. Usually the Niggli reduced cell, described in Sec. 4.6.5, is determined and then an analysis is performed to look for higher implied symmetry.

4.12.4 The Refinement of Lattice Parameters

Once a powder pattern is indexed we may routinely write from one to six d_{hkl}^{*2}

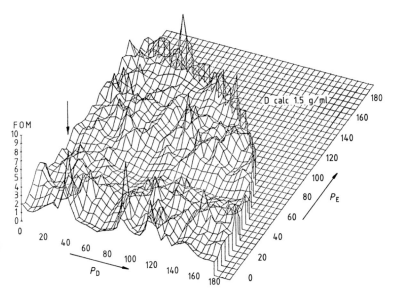

FOM

D calc 1.5 g/ml

P_E

P_D

Figure 4-64. Exhaustive parameter space indexing of Cu-phthalocyanine · HCl (from Shirley, 1980).

equations with all terms known except the one to six lattice parameters. These equations may be solved simultaneously to give values for the lattice parameters, but these will not be the most precise values.

The precise and accurate measurement of lattice parameters, or unit cell dimensions, have been used for various purposes such as

- Determining thermal-expansion coefficients.
- Finding the true density of a material.
- Providing a direct measure of interatomic distances in the case of simple crystal structures.
- Studying interstitial and substitutional solid solution.
- Developing more satisfactory concepts of bonding energies.
- Developing phase-equilibrium diagrams.
- Determining stresses in materials such as steels.

In order to obtain accurate lattice parameters, the minimization and elimination of a number of errors is necessary. All sources of error, except the very accurately known

wavelength, affect the accuracy of the lattice parameters through their effect on 2θ. Those errors that affect 2θ may be classified into two categories: (1) random errors and (2) systematic errors. Random errors are those made in locating and recording the center of the diffraction peak. These errors vary randomly with the angle 2θ. Random errors can be minimized by using back reflection ($2\theta > 90°$) lines, since precision in d-spacing or lattice constant is dependent upon $\sin\theta$ in the Bragg equation and not upon 2θ, the measured quantity. $\sin\theta$ changes very slow with 2θ near $180°$ and, therefore, a peak location error, $\Delta 2\theta$ will have a minimum effect upon the d-spacing. This is illustrated in Fig. 4-65. Systematic errors are those whose magnitude depends upon the position of the line, that is, upon the angle 2θ. In the cylindrical camera method, these systematic errors decrease as θ increases, vanishing completely when $\theta = 90°$. Therefore, in order to minimize these errors, high angle lines can be used. In the case of the diffractometer method, the sources of error cause the error in the line positions to vary in a compli-

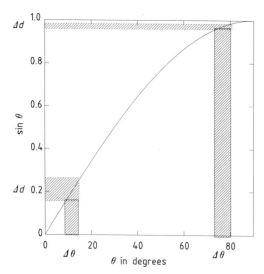

Figure 4-65. Variation of $\Delta \sin \theta$ with $\Delta \theta$.

cated manner with θ but must also minimize at high angle.

Systematic error is best eliminated, for any experimental techniques, by using an internal standard. With systematic error handled in this manner, the best approach for obtaining the most precise parameters from observed data, is to use the method of least squares, minimizing the differences between the observed d_{hkl} values and d_{hkl} values calculated from the d_{hkl}^{*2} equation, by varying the lattice parameters. This very powerful method will simultaneously produce estimates of the standard deviations of the lattice parameters. A general approach to this analysis has been incorporated into the powder pattern evaluation program NBS*LSQ by C. R. Hubbard.

4.13 Analytical Profile Fitting of X-Ray Powder Diffraction Patterns

The inherent asymmetry of many powder diffraction profiles has been a principal hindrance in extending the application of diffraction techniques beyond simple phase identification. The resurgence of developments in X-ray analytical procedures in recent years is primarily the result of the availability of computer automated powder diffractometers (Mallory and Snyder, 1979). The ability to collect digitized representations of the line profiles and apply numerical methods to their analysis, has led to a number of new and exciting applications (Snyder, 1991 b). The specimen shape contribution to the observed profile, when deconvolved from it, allows insights into the anisotropic strain and size of the crystallites as well as things like the ordering lengths of dopants or ions on lattice sites.

The most exciting of the profile fitting developments has been the continuing development of a whole pattern X-ray refinement method to obtain crystal structure information, begun by Rietveld (1969), who applied it to neutron diffraction. This method goes beyond the simple computation of intensities from Eq. (4-34) by distributing the calculated intensity over the kind of profile found using a particular instrument. The normally symmetric-Gaussian nature of neutron diffraction profiles has aided this application of this method to develop to its current, rather mature state. The two most common profile shape functions used in profile fitting are the Gaussian,

$$G(z) = I_0 \exp\left[-(\ln 2)\, z^2\right] \qquad (4\text{-}86)$$

and Lorentzian,

$$L(z) = \frac{I_0}{(1 + z^2)^m} \qquad (4\text{-}87)$$

where

$$z = \frac{2\theta - 2\theta_0}{(\beta/2)}$$

$2\theta_0$ is the peak position, β is the full width at half maximum, I_0 is the peak intensity and $m = 1$. If the m term in the Lorentzian expression is made 1.5 (and z is multiplied by 0.5874), the profile shape function is called an intermediate Lorentzian. If m is made 2.0 (and z is multiplied by 0.4142), the profile function is called a modified Lorentzian. If the m term is used as an adjustable parameter and z is multiplied by $4(2^{1/m} - 1)$ then the function is called a Pearson VII. Figure 4-66 shows the relative shapes of these profile shape functions.

The application of the Rietveld method to X-ray patterns has been slower to develop, primarily because of the asymmetric and non-Gaussian nature of, and multiple spectral components in, most X-ray diffraction profiles (Snyder, 1992). These asymmetric profiles are not easily modeled and thus it is difficult to distribute the calculated intensity of the profile shape. Malmros and Thomas (1977) and Young et al. (1977) gave one of the first applications of this technique to X-ray data and the work of Wiles and Young (1981) marks the beginning of the much wider development of this method. However, the various successes fitting asymmetric profiles with symmetric functions, reported to date, have often been the result of defect disorder which tends to obliterate the inherent peak asymmetry and produce a more symmetric shape.

4.13.1 The Origin of the Profile Shape

There once was a time when a discussion of the shape of X-ray powder diffraction profiles could have been happily limited to the effects seen using a Bragg-Brentano geometry diffractometer with a sealed tube source. In fact, this has been done very well in the text by Klug and Alexander (1974). Today, a broader discussion is required to cover the common use of both pulsed and steady state neutron and of synchrotron and rotating anode X-ray sources, with incident and/or diffracted beam monochromators. Even the advances in X-ray detectors have an impact on this topic.

A diffraction line profile is the result of the *convolution* of a number of independent contributing shapes, some symmetric and some asymmetric. The process of convolution is one in which the product of two functions is integrated over all space. If can be represented as

$$h_{2\theta} = g_{2\theta} * f_{2\theta} = \int_{-\infty}^{+\infty} g_{2\theta'} f_{2\theta - 2\theta'} \, d(2\theta') \quad (4\text{-}88)$$

where $h_{2\theta}$ is the final observed profile and $g_{2\theta'}$ and $f_{2\theta}$ are shape functions contribut-

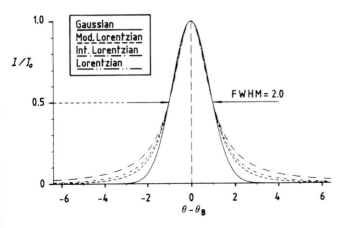

Figure 4-66. The Gaussian, modified-, intermediate- and Lorentzian profile shape functions.

ing to the resulting profile. Each point in the convolution is the result of summing the product of g and f over all possible values for $f_{2\theta}$. It is clear that, if this operation cannot be performed analytically, a lot of computer time will be used to evaluate it numerically. However, this may be reduced by the proper choice of functions, or by carrying it out in Fourier space, where convolution is mathematically more simple.

The components contributing to a diffraction profile can be divided into three categories.

4.13.1.1 Intrinsic Profile: S

The dynamical effects of an X-ray beam in a perfect crystal produce a reflection whose inherent width is called the Darwin width, after the author of the first dynamical treatment of diffraction. This inherent width is simply the result of the uncertainty principal ($\Delta p \Delta x = h$), in that the absorption coefficient of the specimen requires that the location of the photon in a crystal be restricted to a rather small volume. This means that Δp and in turn $\Delta \lambda$ ($\Delta p = h/\Delta \lambda$ by the de Broglie relation), must be finite, producing a finite width to a diffraction peak. The Darwin profile can usually be represented by a Lorentzian profile shape function (Parrish et al., 1976).

In addition to the inherent width, there are two physical sample effects which will broaden the profile shape function S, which the specimen contributes to the observed profile. The Scherrer crystallite size broadening is described by Eq. (4-21) and strain broadening of the specimen profile S, follows Eq. (4-22). Both of these specimen broadening effects generally effect the peak shape symmetrically.

4.13.1.2 Spectral Distribution: W

The most common X-ray source continues to be the sealed target X-ray tube. The inherent spectral profile, from the K transition in a Cu target X-ray tube, has a width of 1.18×10^{-4} Å and has been shown to be Lorentzian and not completely symmetric (see Frevel, 1987). The inherent width and asymmetry is usually overwhelmed by the fact that the various components of radiation in a polychromatic beam will each spread out as 2θ increases (i.e., with $\tan\theta$). This spectral dispersion is so great that this term dominates the convolution of diffraction lines at high angle, making them relatively symmetric and quite broad.

Monochromatization of an X-ray or neutron beam, using an incident beam monochromator, limits the breadth of the wavelength profile function W, to the Darwin width of the monochromator crystal and its intrinsic mosaicity. The use of both an incident and a diffracted beam monochromator in a parallel beam configuration, produces a broadening effect, in the diffraction profile, which depends on the distance, in 2θ, away from the point of optimum focus. (The point of optimum focus depends on the orientation of the two monochromators.) This distance in turn, depends on the crystal and Bragg reflection chosen for the monochromator and analyzer crystals (Cox et al., 1983, 1988). At the point of optimum focus, using perfect germanium crystals on a synchrotron source, the spectral profile is so narrow that the observed breadth seen is due primarily to the collimator divergence.

4.13.1.3 Instrumental Contributions: G

There are five possible profile contributions depending on the instrumental arrangement.

1. *The X-ray source image.* In a closed tube system this can be approximated with a symmetric Gaussian curve with a full width at half maximum (β) of 0.02° using a

take off angle of 3°. The effects of incident and diffracted beam monochromators are described by Cox et al. (1988) and do not introduce any asymmetry to the Gaussian shape. The long beam lengths on neutron and synchrotron ports allow for nearly perfect parallel optics. This, when coupled with both an incident and diffracted beam monochromator leaves the instrumental profile, at optimum focus, nearly a delta function (Cox et al., 1983). The use of focusing optics with an incident beam slit following the monochromator, while greatly increasing the intensity, introduces significant but symmetric broadening (Parrish et al., 1986).

2. *Flat specimen.* To maintain the Bragg-Brentano focusing condition the sample should be curved so that it follows the focusing circle. Since the focusing circle continuously changes radius with 2θ most experimental arrangements use a flat specimen, tangent to the focusing circle. This "out of focus" condition introduces a $\cot\theta$ dependence and produces a small asymmetry in the profile. It is particularly noticeable at low angle where the irradiated length of the sample is large. This term is not present on those neutron and synchrotron devices that use a cylindrical sample bathed in the beam.

3. *Axial divergence of the incident beam.* This also follows a $\cot\theta$ dependence and causes a substantial asymmetry in the profile particularly at low angles.

4. *Specimen transparency.* As the absorption coefficient of the sample decreases, the X-ray beam penetrates ever deeper, making the effective diffracting surface farther and farther off of the focusing circle. This produces a substantial asymmetric convolution term for low μ materials.

5. *Receiving slit.* For instruments using a receiving slit, another symmetric term contributes to the observed profile.

4.13.1.4 Observed Profile: *P*

Each of the terms described above conspire, via convolution, to produce a final diffraction profile that will range from the very asymmetric, in the case of sealed tube Bragg-Brentano systems, to quite symmetric, nearly Gaussian profiles, on neutron and some synchrotron instruments.

The observed diffraction profile, as stated by Jones (1938) and applied to powder diffraction systems by Taupin (1973a) and Parrish et al. (1976), is the result of the convolution of a specimen profile (S) and a combined function modeling the aberrations introduced by the diffractometer and wavelength dispersion. Taupin and Parrish grouped these terms together as $W * G$. The overall line profile can be expressed as

$$Profile = (W * G) * S + background \quad (4\text{-}89)$$

where the ($*$) represents the convolution operation. Since both W and G are fixed for a particular instrument/target system, ($W * G$) may be regarded as a single entity which we will refer to as the instrumental profile P. The specimen function S for a sample with no defect broadening has only the Darwin width, which can be approximated with a delta function (i.e., a line with height but no width). Using a delta function for S in Eq. (4-89) yields

$$Profile = P + background \quad (4\text{-}90)$$

Hence, for a pattern of an ideal sample, with background appropriately accounted for, the profiles are identical to the profiles of P. However, Parrish et al. (1976) and Howard and Snyder (1983) have indicated that the intensity ratio of components is affected by the setting of the monochromator, when one is present and hence, the W component of P must also be evaluated.

4.13.2 Modeling of Profiles

Khattak and Cox (1977) have shown fundamental problems in representing X-ray diffraction lines with either the Gaussian or simple Lorentzian functions. Of more recent interest are the Voigt (Langford, 1978; Cox et al., 1988), the pseudo-Voigt (see Young and Wiles, 1982) and the split-Pearson VII of Brown and Edmonds (1980). The Voigt function is the result of an analytical convolution of a Gaussian and a Lorentzian. It therefore ranges from pure Lorentzian to Gaussian type, depending on the ratio of both half widths. The function normalized to 1 has the form

$$V(x, \beta_l, \beta_g) = L(x/\beta_l) * G(x/\beta_g) \qquad (4\text{-}91)$$

where L is the Lorentzian function with a full width at half maximum of β_l and G is the Gaussian function with half width of β_g and $x = \Delta 2\theta$. The Voigt-function can be calculated numerically using the complex error function.

The pseudo-Voigt conveniently allows the refinement of a mixing parameter determining the fraction of Lorentzian and Gaussian components needed to fit an observed profile. See Hastings et al. (1984) and Cox et al. (1988) for applications of this function. The pseudo-Voigt, although symmetric, allows for a flexible variation of the two most common profiles, ranging from the broad β Gaussian to the narrow β Lorentzian. However, Smith et al. (1987) have found an example of synchrotron profiles that are narrower than Lorentzian's and cannot be fit by the Voigt function.

The Pearson VII function also allows for the variation in shape from Lorentzian to Gaussian (Howard and Snyder, 1983) and is also symmetric. A split Pearson function, combined with a Gauss-Newton or a Marquardt least-squares optimization algorithm, gives excellent fits to asymmetric X-ray diffraction lines, obtained under a wide variety of conditions (Howard and Snyder, 1983, 1985, 1989). The approach here is to split each diffraction profile into a low angle and high angle part, by dividing at the profile maximum, and fitting a separate Pearson VII to each side. The Pearson VII function has the form

$$I = I_0/(1 + a x^2)^m \qquad (4\text{-}92)$$

where $a = (2^{1/m} - 1)/(\beta/2)^2$ and m is the shape factor whose value determines the rate at which the tails fall and β is the full width at half maximum. The split-Pearson VII is illustrated in Fig. 4-67. The principal problem with this function is that its clumsy form prevents the analytical determination of convolutions, thus forcing extensive numerical approximations.

There are several demands on a profile shape function.

- The function must fit non-symmetric peaks.
- It should be mathematically as simple as possible, to make possible the calculation of all derivatives with respect to the variables.
- It must allow simple computation of the integral intensity.
- The convolution with a Lorentzian or a Gaussian function modeling S should be possible analytically.

Tomandl (Snyder, 1992), has proposed an asymmetric function which can be directly convoluted with the instrumental profile. Recently Hepp and Baerlocher (1988) have proposed a similar treatment for an asymmetry function in their "learned shape function" approach.

4.13.3 Description of Background

The description of the background in a powder diffraction pattern is critical to

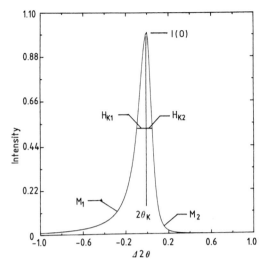

Figure 4-67. The Split-Pearson VII function uses two halves of the Pearson VII function with a common peak angle and intensity. The integral breadth β and the exponents m for each side are varied independently.

4.13.4 Unconstrained Profile Fitting

The SHADOW algorithm, of Howard and Snyder (1985 b), was developed as a general profile or pattern fitting program, which permits the use of a wide variety of shape functions and either a Gauss-Newton or Marquardt optimization algorithm. This was used in Fig. 4-68 to fit two peaks measured with an incident beam monochromator, to a simple unconstrained split Pearson function. The fit is quite good with the individual profiles looking like realistic diffraction lines. Figure 4-69 shows the same two peaks measured on a similar diffractometer but with different resolving power. Here we see that the low angle side of the split Pearson fitting the right peak has lifted in an unrealistic manner. In fact, there are a number of ways to place a group of profiles under an envelope, if the only criterion is to minimize the differences between the summed profiles and the observed envelope. All that any optimization

profile refinement because any background function must correlate strongly with the profile function. Two commonly used methods for describing background involve selecting points between peaks and interpolating between them (Rietveld, 1969), or refining the coefficients of a polynomial along with the profile parameters (Wiles and Young, 1981). While Sabine (1977) has described an analytical method for neutron powder diffraction by which the background scattering is analytically described and included in the refinement, there is no simple technique available for X-ray powder diffraction. In addition, no method proposed to date (Snyder, 1983 a) will routinely produce an accurate description of X-ray background in complex patterns. The most serious problem in finding the true background occurs in low symmetry materials, where the powder pattern presents a continuum of peaks at high diffraction angle.

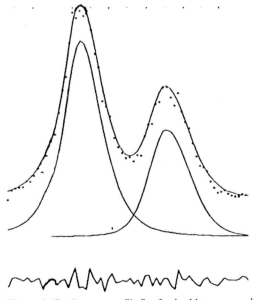

Figure 4-68. Correct profile fit of a doublet measured with only $K_{\alpha 1}$ radiation, using a split Pearson function.

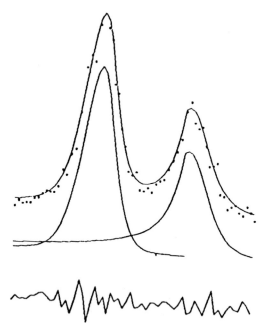

Figure 4-69. Incorrect profile fit of a doublet measured with only $K_{\alpha 1}$ radiation, using a split Pearson function.

procedure can do is to minimize this difference. Thus, it is clear that unconstrained profile fitting can always lead to failure to give a correct fit, unless we constrain the profiles used to be of the shape that are allowed for a particular experimental configuration. Any general procedure for profile fitting must then allow for the application of experimental constraints on the profile shape.

4.13.5 Establishing Profile Constraints: The *P* Calibration Curve

In order to establish constraints on the allowed shapes of the profiles from a particular instrument, Parrish and co-workers (1976) proposed a calibration procedure. The profiles of a standard, which shows no sample broadening, are fit with the split-Pearson VII function. Ideally, one would like a uniform particle size of about 5 μm,

for an unstrained, high-symmetry material to use as a standard. In the Howard and Snyder approach, the 11 profiles of a standard sample of Si, observed with Cu K_α radiation are refined using the constrained split-Pearson function described above. Each profile is refined separately in a region with enough points to allow a precise description of the profile. The β values for each Pearson VII component, obtained from the refinement, are used to determine the value of the coefficients in the polynomial expression derived for neutron diffraction by Caglioti, Paoletti, Ricci (1958) as

$$(\beta_k)^2\, 2\theta = U \tan^2(\theta_k) + V \tan(\theta_k) + W \tag{4-93}$$

Two equations of this type were established by using a least-squares regression, analyzing for both the low angle and the high angle sides of the split-Pearson VII coefficients, from the fit to the standard's profiles. Similarly, a polynomial was used for the shape factors, m where

$$m = a'\,(2\theta_k)^2 + b'\,(2\theta_k) + c' \tag{4-94}$$

Least squares regression of this function versus the two sets of shape factors completes the establishment of an analytical expression for evaluating P. Example polynomials for the high and low angle split Pearson parameters (m and β) from the calibration of a conventional diffractometer are illustrated in Fig. 4-70. This calibration procedure allows the evaluation of the instrumental P profile at any 2θ angle where a random specimen may have profiles that require fitting.

4.13.6 Modeling the Specimen Broadening Contribution to an X-Ray Profile

A powder diffraction pattern, for a material to be analyzed for size and strain, is determined. The observed peak positions

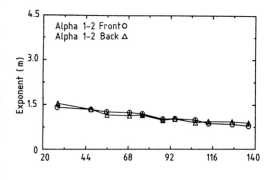

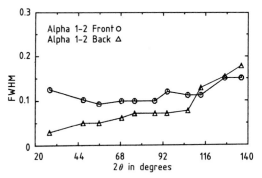

Figure 4-70. Variation of the split Pearson coefficients as a function of 2θ.

The convolution gathers intensity contributions from all points in both P and S. The only approximation in the numerical convolution comes from the fact that the n limits are finite. The broader the full width of S, the greater the smearing of the profiles and loss of apparent resolution.

Since all non-specimen related broadening terms, like the $\tan\theta$ spectral broadening, have been accounted for in P, pure crystallite size and strain effects may be modeled by Eqs. (4-21) and (4-22). The slope of the curves β vs. $\lambda/\cos\theta_k$ and β vs. $4\tan\theta_k$ give $1/\tau$ and ε respectively or, more simply, a Williamson-Hall plot (1953) of $B_k\cos\theta_k$ vs. $4\varepsilon\sin\theta_k$ gives a line of slope ε and y-intercept λ/τ. Program SHADOW uses these simple models to allow for the simultaneous determination and refinement of τ and/or ε. Figure 4-71 shows a measured region of two size broadened samples of Al_2O_3 along with their generated complete profiles and deconvoluted S profiles.

are used to generate P profiles from the instrumental function, which is expressed as the four polynomials in 2θ shown in Fig. 4-70. Lorentzian or Gaussian profile functions are generated to model the specimen broadening function S. It is generally accepted that S may be represented by a Lorentz function when profile broadening is caused by small crystallite size. When strain is responsible, S has often been assumed to be represented by a Gaussian function but sometimes a Lorentz profile shape is required.

In the absence of an analytical convolution function for the profile models, numerical techniques must be employed.

The convoluted profile is obtained from

$$(P) * S_{(i)} = \sum_{j=-n}^{+n} (P)_{(i-j)} S_{(j)} \qquad (4\text{-}95)$$

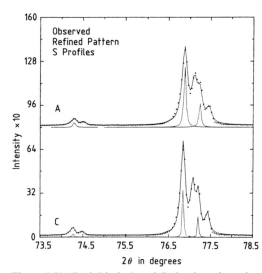

Figure 4-71. Both Linde A and C aluminas show significant line broadening. The inner curves are the deconvoluted S functions. The integral breadth of these S values show that the X-ray crystallite size of linde A is about half that of Linde C.

4.13.7 Fourier Methods for Size and Strain Analysis

We have explained both the origin and a method for analyzing size and strain in terms of the intuitive observed profiles. However, these easily understood procedures are relatively new. Before the days of inexpensive, powerful computers, the only way to carry out such an analysis was to simplify the computation by carrying it out in Fourier space. The traditional way of analyzing crystallite size and strain effects, was developed by Warren and Averbach (1969). In this method, the Fourier transformation of the diffraction peaks of a broadened sample and a standard material exhibiting no sample broadening, are computed. The deconvolution procedure is carried out in Fourier space. Since all intuition is lost on performing a Fourier transformation, this process can only be followed mathematically. The procedure presented above has the advantage of being understandable in terms of easily pictured profile shapes. If either procedure is carried out correctly the same information about size and strain will result.

4.13.8 Rietveld Analysis

With the P and S functions properly modeled for X-ray diffraction profiles, this method can now be applied to the pattern-fitting-structure-refinement procedure of Rietveld. The formalism developed above allows for the discarding of a number of empirical parameters in the Rietveld method, replacing them with τ and ε. The removal of all empirical parameters from the refinement, leaves in their place only parameters associated with known physical effects. All empirical parameters associated with the peak asymmetry are incorporated into the P instrumental function and are fixed during refinement.

When both strain and crystallite size effects are present, the integral breadth of the S profile is assumed to be a linear addition of the two components. That is, $\beta = \beta_\varepsilon + \beta_\tau$. The basis for this assumption is that the convolution of two Lorentzian functions yields another Lorentzian function. Thus,

$$L(\text{strain}) * L(\text{size}) * P = L(\text{combined}) * P \tag{4-96}$$

The value for the full width of the new convoluted profile is simply a sum of the component β's. It is assumed that the order of convolution does not matter.

The X-ray version of the Rietveld procedure, as developed by Young, Mackie and von Dreele (1977) and later by Wiles and Young (1981), was modified by Howard and Snyder (1985) to incorporate the P deconvolution and modeling of S described above. The introduction of an analytical expression describing S broadening simplifies profile refinement. This constrained broadening reduces the number of parameters undergoing refinement while characterizing and quantifying the source of the broadening. However, the cost is a considerable increase in the execution time of the program. To eliminate the need for numerical convolution an analytically convolutable function is required like the Tomandl function.

4.14 Crystal Structure Analysis

All of the developments in hardware, software and application methods described in this article have produced a number of recent results which have a direct impact on the advancement of materials science. To conclude this article, we will survey some of the most exciting applications and how they have converged to produce exciting new results in the area of crystal structure analysis. These examples will also serve as illustrations of how all of the principles and techniques previously described can be applied to materials systems.

For example, the X-ray structure determination of the modulated structure of the $Bi_2Sr_2CaCu_2O_8$ superconductor was recently achieved by taking advantage of anomalous dispersion and obtaining diffraction patterns on each side of the absorption edge (Gao et al., 1989). Atomic resolution electron microscopy has made this structure dramatically clear. Figure 4-72 from Eibl (1990) dramatically shows the modulated wave misaligning the unit cells with respect to each other. If the wavelength of this modulation were a multiple of one of the lattice parameters, a superlattice would result. However, in this case the wavelength is not such a multiple so the modulation is said to be incommensurate.

Beyond the scanning transmission electron microscope, a wealth of new atomic resolution probes have developed in the last five years like, the scanning tunneling microscope which has spawned the atomic force microscope, friction force microscope, magnetic force microscope, electrostatic force microscope, chemical bonding force microscope, scanning thermal microscope, optical absorption microscope, scanning ion-conductance microscope, scanning acoustic microscope .... All have become tools of modern applied crystallography.

4.14.1 Structure of $YBa_2Cu_3O_{7-\delta}$

The crystal structure of the first superconducting material to break the liquid nitrogen barrier has probably been determined by more independent studies than any other. In our laboratory we isolated the diffraction peaks common to those

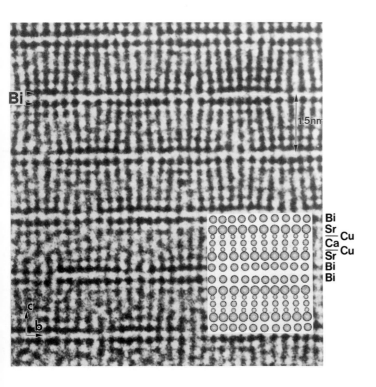

Figure 4-72. Lattice image of the modulated structure of the high temperature superconducting phase $Bi_2Sr_2CaCu_2O_8$ (from Eibl, 1990).

specimens (out of 40 compositions) which showed superconductivity. Microprobe analysis showed that the cations were in the approximate ratio of 1:2:3 for Y:Ba:Cu. Calibration of the XRD lines with a Si internal standard allowed each of the modern indexing programs, described in Sec. 4.12 to propose an orthorhombic unit cell of approximately $4 \times 4 \times 12 \, \text{Å}^3$. Use of the NBS*LATTICE program (Mighell and Himes, 1986), which computes super and sub cells related to the orthorhombic cell, and searches the *Crystal Data* database for matches, produced suggested matches for $BaTiO_3$ and a number of other cubic perovskites with unit cells near $4 \times 4 \times 4 \, \text{Å}^3$. In addition, it proposed a number of double celled perovskites and even one triple celled material. On placing the Y, Ba and Cu atoms into a triple celled perovskite structure, a Rietveld refinement led to convergence. This was far from the first structure analysis of this material but it occurred before the published reports of the structure arrived on our campus!

4.14.2 The Structures of the γ and η Transition Aluminas

As an example of the kind of analysis permitted by the new diffraction tools described in this article and currently available in the materials laboratory, we will conclude this article by describing a classical structural problem in an extremely important ceramic system (Zhou and Snyder, 1991). The transition aluminas refer to the group of partially dehydrated aluminium hydroxides, other than the anhydrous α-alumina (corundum). The diffuse character of their powder patterns reflects a high degree of structural disorder, but the similarity of the patterns indicates certain structural aspects that pervade all these phases.

The most disordered of these materials, γ and η, are catalytically active, and of high economic value, and have been studied for many years, but have resisted structure analysis.

4.14.2.1 Profile Analysis

The evolution of the profile fitting and Rietveld refinement procedures mentioned above allows new approaches to these previously intractable patterns. The deconvolution of the specimen broadening profile S from each of the profiles in the γ and η patterns immediately showed an unusual phenomenon (Zhou and Snyder, 1991). The patterns exhibit anomalous reflection broadening in two aspects: (1) irregular broadening among peaks of different $h k l$'s [even from the same crystal zone – see the (111), (222), and (333) reflections in Fig. 4-73 and (2) the formation of an odd peak shape with broadened base and sharp top that can not be fit well by a single profile function. The differences between the γ and η patterns are in the degree of these anomalies.

Attributing the anomalous broadening to be indicative of the "crystallite size" or ordering lengths of the various sublattices that dominate the scattering for the different peaks, the following structural information can be extracted from profile fitting and pattern calculations, based on a spinel structure.

- The oxygen sublattice in both η and γ-alumina is fairly well ordered because the sharp (222) reflection is dominated by the scattering from the oxygen sublattice. The ordering domain of the oxygen sublattice estimated from the β value of the specimen convolute S, of the (222) reflection is 9.8 (2) nm for η-alumina and 20 (1) nm for γ-alumina. These numbers are comparable with the average crystal-

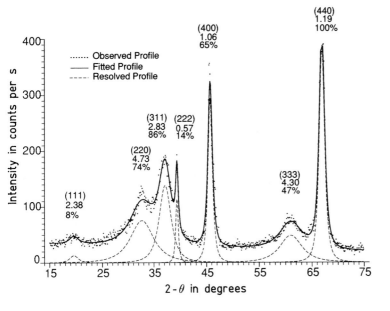

Figure 4-73. Profile fit of XRD pattern of γ-alumina. On top of peaks are hkl, β and relative intensity.

lite sizes of the precursor bayerite and boehmite. In fact, the η value agrees well with the average long dimension of TEM observed lamellae in the dehydroxylated tablets. Thus, it is the framework of close packed oxygens which supports the original crystallite shape and accounts for the pseudomorphosis of dehydroxylation.

- Both η and γ-alumina have octahedral coordinated Al as well as tetrahedral coordinated Al ions, since the (220) reflection is only due to scattering from the tetrahedral Al sublattice.
- The tetrahedral Al sublattices in both η and γ-alumina are very disordered because the (220) reflection is the most diffuse in the pattern. The ordering domain of the tetrahedral Al sublattice estimated from the β value of S, for the (220) reflection is 1.6(2) nm for η-alumina and 1.9(2) nm for γ-alumina. These dimensions also correspond to the TEM observed striations on the pseudomorphic tablets which result on heating.

The disparity in the size of the ordering domains of the different sublattices within a single spinel lattice, manifests itself in the formation of odd-shaped peaks (i.e., sharp top with broadened base) routinely observed on the XRD powder patterns of the transition aluminas. This disparity can be understood by considering the fact that the starting hydroxides have only octahedral Al ions and quasi cubic close-packed oxygen ions. Since the tetrahedral Al ions are the newcomers, from the dehydroxylation, they most reflect the diffusion induced disorder.

4.14.2.2 Rietveld Analysis

In the Rietveld refinement of powder diffraction patterns, the profiles are assumed to vary isotopically with diffraction angle. When this assumption is very poor the "two-step" method is used where the integrated intensities of the individual S profiles are treated as single crystal intensities and the traditional single crystal structure solving and refinement procedures can be employed. By contrast, this notation would refer to the Rietveld as the "one step" method. However, in the case of the transition aluminas, the extra statistical precision which comes from using all of the digital points in a pattern is essential due to the small number of reflections. The compromise here was to use a "one and one half step" procedure: namely, the anomalous β's from profile fitting were specified for each peak in the pattern and not refined during the Rietveld refinement. The Rietveld procedure was applied to neutron patterns collected on deuterated samples of γ and η-alumina, as well as on X-ray patterns. Structures were refined using traditional Fourier techniques by starting with the well determined oxygen atoms and iteratively refining the model and computing difference Fourier patterns to locate the other atoms in the structure.

The refinement results show that Al ions in η-alumina occupy only one octahedral site (16d), in $Fd3m$, one tetrahedral site (48f), and one *quasi-trihedral* site (32e). The quasi-trihedral Al ions are slightly displaced 0.0162 nm away from the oxygen plane toward the second octahedral site (16c) 0.0958 nm away from site (16c). In γ-alumina however, Al ions occupy only one octahedral site (16d), one tetrahedral site (8a), and one *quasi-octahedral* site (32e). The quasi-octahedral Al ions are 0.0373 nm away from the octahedral site (16c) and 0.0748 nm away from the nearest oxygen plane. The extraordinary presence of nearly pure three coordinated Al ions in η-alumina and of the quasi-octahedral Al ions in γ-alumina can be accounted for by considering them to exist in the surface of the crystallites. The size of these peaks as found in the difference Fourier map indicated a concentration of 13%. Using the crystallite size as determined in this study

and confirmed by transmission electron microscopy, a simple computation predicts that 13% of the Al ions must lie in the surface of such small crystallites. These surface Al ions are clearly responsible for the strong Lewis-acidity (electron-accepter property) observed on the surfaces of η and γ-alumina and must result in the catalytic activity of these phases. Thus the recent advances in applied crystallography, presented in this article, have converged in this case to produce the first structural model for catalytic activity in the most widely used industrial catalyst system.

The ability to determine the ordering lengths associated with the various sublattices in the many important disordered materials systems, offers the exciting prospect of finally understanding the structures of, and structure-property relationships in, these materials.

4.15 References

Alexander, L., Klug, H. P. (1948), *Anal. Chem. 20*, 886–894.

Apai, G., Hamilton, J. F., Stohr, J., Thompson, A. (1979), *Phys. Rev. Lett. 43*, 165–169.

Bayya, S. S., Kudesia, R., Snyder, R. L. (1991), *Superconductivity and Its Application*, Vol. 3: Kao, Y. H., Coppens, P., Kwok, H. S. (Eds.). Washington: Am. Phys. Soc., pp. 306–314.

Bearden, J. A. (1967), *Rev. Mod. Phys. 39*, 78.

Bellier, S. P., Doherty, R. D. (1977), *Acta Metallurgica 25*, 521.

Bish, D. L., Howard, S. A. (1988), *J. Appl. Cryst. 21*, 86–91.

Bragg, W. L. (1913), *Proc. Cambridge Phil. Soc. 17*, 43.

Bragg, W. H., Bragg, W. L. (1913), *Proc. Roy. Soc. London 88 A*, 428.

Brindley, G. W. (1945), *Phil. Mag. 36 (7)*, 347.

Brown, A., Edmonds, J. W. (1980), *Adv. in X-ray Anal. 23*, 361.

Buerger, M. (1964), *The Precession Method*. New York: Wiley.

Bunge, H.-J. (1982), *Texture Analysis in Materials Science*. London: Butterworths.

Caglioti, G., Paoletti, A., Ricci, F. P. (1958), *Nuclear Instruments and Methods 3*, 223–226.

Cava, R. J., Hewat, A. W., Hewat, E. A., Batlogg, B., Marezio, M., Rabe, K. M., Krajewski, J. J., Peck, W. F., Rupp, L. W. (1990), *Physica C 165*, 419–433.

Cherukuri, S., Snyder, R. L., Beard, D. (1983), *Adv. X-ray Anal. 26*, 99–105.

Chung, F. H. (1974a), *J. Appl. Cryst. 7*, 519–525.

Chung, F. H. (1974b), *J. Appl. Cryst. 7*, 526–531.

Chung, F. H. (1975), *J. Appl. Cryst. 8*, 17–19.

Clark, G. L., Reynolds, D. H. (1936), *Ind. Eng. Chem., Anal. Ed. 8*, 36–42.

Cline, J. P., Snyder, R. L. (1983), *Adv. in X-ray Anal. 26*, 111–118.

Cline, J. P., Snyder, R. L. (1985), *Advances in Material Characterization II:* Snyder, R. L., Condrate, R. A., Johnson, P. F. (Eds.). New York: Plenum Press, pp. 131–144.

Cline, J. P., Snyder, R. L. (1987), *Adv. in X-ray Anal. 30*, 447–456.

Coolidge, W. D. (1913), *Phys. Rev. 2*, 409.

Copeland, L. E., Bragg, R. H. (1958), *Anal. Chem. 30*, 196–206.

Cox, D. E., Hastings, J. B., Thomlinson, W., Prewitt, C. T. (1983), *Nucl. Instrum. Methods 208*, 273–278.

Cox, D. E., Hastings, J. B., Cardoso, L. P., Finger, L. W. (1986), *Mat. Sci. Forum 9*, 1.

Cox, D. E., Toby, B. H., Eddy, M. M. (1988), *Aust. J. Phys. 41*, 117–131.

Cromer, D. T., Waber, J. T. (1965), *Acta Cryst. 18*, 104.

Darwin, C. G. (1914), *Phil. Mag. 27*, 315.

Darwin, C. G. (1922), *Phil. Mag. 43*, 800.

Debye, P. (1913), *Ann. der Physik 43*, 49.

Debye, P. (1915), *Ann. der Physik 46*, 809.

Deslattes, R. D., Henins, A. (1973), *Phys. Rev. Lett. 31*, 972.

Delhez, R., de Keijser, Th. H., Mittemeijer, E. J., Langford, J. I. (1988), *Aust. J. Phys. 41*, 213–227.

de Wolff, P. M. (1968), *J. Appl. Cryst. 1*, 108–113.

Dragoo, A. L. (1986), *Powder Diffraction 1*, 294.

Eibl, O. (1990), *Physica C 168*, 215–238.

Ewald, P. P. (1921), *Z. Kristallogr. 56*, 129.

Frevel, L. K. (1987), *Powder Diffraction 2 (4)*, 237–241.

Friedrich, W., Knipping, P., M. v. Laue (1913), *Ann. der Physik 41*, 971.

Gao, Y., Sheu, H. S., Petricek, V., Restori, R., Coppens, P., Darovskikh, A., Phillips, J. C., Sleight, A. W., Subramanian, M. A. (1989), *Science 244*, 62.

Gehringer, R. C., McCarthy, G. J., Garvey, R. G., Smith, D. K. (1983), *Adv. in X-ray Anal. 26*, 119–128.

Geiger, H., Müller W. (1928), *Phys. Z. 29*, 839.

Göbel, H. E. (1982a), *Adv. in X-ray Anal. 24*, 123–138.

Göbel, H. E. (1982b), *Adv. in X-ray Anal. 24*, 315–324.

Göbel, H. E. (1990), *Analytical Techiques for Semiconductor Materials and Characterization*, Vol. 90-11: Kolbesen, B. O., McCaughan, D. V., Vandervorst, W. (Eds.). Pennington, NJ: The Electrochemical Society Inc. pp. 238–260.

Hargittai, I. (1990), *Quasicrystals, Networks, and Molecules of Fivefold Symmetry*. New York: VCH.

Hastings, J. B., Thomlinson, W., Cox, D. E. (1984), *J. Appl. Cryst. 17*, 85–95.

Hart, M. Parrish, W. (1989), *Mater. Res. Soc. Symp. Proc. 143*, 185–195.

Hepp, A., Baerlocher, Ch. (1988), *Aust. J. Phys. 41*, 229–236.

Hermann, C. (1931), *Z. Kristallogr. 76*, 559.

Hill, R. J., Howard, C. J. (1987), *J. Appl. Cryst. 20*, 467–474.

Howard, S. A., Snyder, R. L. (1983), *Adv. in X-ray Anal. 26*, 73–81.

Howard, S. A., Snyder, R. L. (1985), *Advances in Material Characterization II:* Snyder, R. L., Condrate, R. A., Johnson, P. F. (Eds.). New York: Plenum Press, p. 43–56.

Howard, S. A., Snyder, R. L. (1989), *J. Appl. Cryst. 22*, 238–243.

Huang, T. C., Cohen, P. C., Eaglesham, D. J. (Eds.) (1991), *Advances in Surface and Thin Film Diffraction*. Warrendale: Materials Research Society.

Hubbard, C. R., Snyder, R. L. (1988), *J. Powder Diffraction 3*, 74.

Hubbard, C. R., Evans, E. H., Smith, D. K. (1976), *J. Appl. Cryst. 9*, 169–174.

Hubbard, C. R., Robbins, C. R., Snyder, R. L. (1983), *Adv. X-ray Anal. 26*, 149–157.

Hubbard, C. R., Robbins, C. R., Wong-ng, W. (1987), *Standard Reference Material Silicon 640 b*. Available from National Institute of Science and Technology Office of Standard Reference Materials, Gaithersburg, MD 20899, USA.

Inokuti, Y., Maeda, C., Ito, Y. (1987), *Trans. Iron and Steel Inst. of Japan 27*, 139, 302.

International Centre for Diffraction Data – JCPDS, 1601 Park Lane, Swarthmore, PA 19081.

International Tables for Crystallography (1983), Vol. A: *Space-Group Symmetry:* Hahn, T. (Ed.). Dordrecht: D. Reidel.

Jacob, C. W., Warren, B. E. (1937), *J. Chem. Soc. 59*, 2588.

Johnson, G. G., Vand, V. (1967), *Ind. Eng. Chem. 59*, 19–31.

Jones, F. W. (1938), *Proc. R. Soc. London Ser. A 166*, 16–43.

Karlak, F., Burnett, D. S. (1966), *Anal. Chem. 38*, 1741–1745.

Khattak, C. P., Cox, D. E. (1977), *J. Appl. Cryst. 10*, 405–411.

Kinney, J. H., Stock, S. R., Nichols, M. C., Bonse, U., Breunig, T. M., Saroyan, R. A., Nusshardt, R., Johnson, Q. C., Busch, F., Antolovich, S. D. (1990), *J. Mat. Res. 5 [5]*, 1123–1129.

Klug, H. P., Alexander, L. E. (1974), *X-ray Diffraction Procedures*, 2nd ed. New York: John Wiley and Sons.

Kossel, W. (1936), *Ann. der Physik 27*, 694.

Ladell, J., Zagofsky, A., Pearlman, S. (1970), *J. Appl. Cryst. 8*, 499–506.

Langford, J. I. (1978), *J. Appl. Cryst. 11*, 10–14.

Langford, J. I., Delhez, R., de Keijser, Th. H., Mittemeijer, E. J. (1988), *Aust. J. Phys. 41*, 173–187.

Lawton, S. L., Jacobson, R. A. (1965), *The Reduced Cell and Its Crystallographic Applications*, U.S.A.E.C., *Ames Laboratory Report IS-1141*. Ames, IA: Iowa State University.

LeGalley, D. P. (1935), *Rev. Sci. Instr. 6*, 279.

Lengeler, B. (1990a), Adv. Mater. *2*, 123–131.

Lengeler, B. (1990b), *Photoemission and Absorption Spectroscopy:* Campana, M., Rosci, R. (Eds.). New York: North Holland, p. 157–202.

Lennox, D. H. (1957), *Anal. Chem. 29*, 767–772.

Leroux, J., Lennox, D. H., Kay, K. (1953), *Anal. Chem. 25*, 740–748.

Lim, G., Parrish, W., Ortiz, C., Bellotto, M., Hart, M. (1987), *J. Mater. Res. 2*, 471.

Louër, D. Louër, M. (1972), *J. Appl. Cryst. 5*, 271–275.

Mallory, C. L., Snyder, R. L. (1979), *Adv. X-Ray Anal. 22*, 121–132.

Mallory, C. L., Snyder, R. L. (1980), in: *Accuracy in Powder Diffraction, National Bureau of Standards Special Publication 567*. Gaithersburg: Nat. Bur. Stand., p. 93.

Malmros, G., Thomas, J. O. (1977), *J. Appl. Cryst. 10*, 7–11.

Mauguin, C. (1931), *Z. Kristallogr. 76*, 542.

Mighell, A. D. (1976), *J. Appl. Cryst. 9*, 491–498.

Mighell, A. D., Himes, V. L. (1986), *Acta Cryst. A 42*, 101–105.

Navias, A. L. (1925), *J. Amer. Ceram. Soc. 8*, 296–302.

Newsam, J. M., Yang, C. Z., King, H. E. J., Jones, R. H., Xie, D. (1991), *J. Phys. Chem. Solids*, in press.

O'Connor, B. H., Raven, M. D. (1988), *J. Powder Diffraction 3*, 2–6.

Parrish, W., Gordo, S. (1945), *Am. Minerol. 30 (5-6)*, 326.

Parrish, W., Huang, T. C., Ayers, G. L. (1976), *Am. Cryst. Assoc. Monograph 12*, 55–73.

Parrish, W., Hart, M., Huang, T. C. (1986), *J. Appl. Cryst. 20*, 79–83.

Patterson, A. L. (1935), *Z. Kristallogr. A 90*, 517.

Rietveld, H. M. (1969), *J. Appl. Cryst. 2*, 65–71.

Rodriguez, M. A., Matheis, D. P., Bayya, S. S., Simmins, J. J., Snyder, R. L., Cox, D. E. (1990), *J. Mater. Res. 5 (9)*, 1799–1801.

Sabine, T. M. (1977), *J. Appl. Cryst., 10*, 277–280.

Savitzky, A., Golay, M. (1964), *Anal. Chem. 36 (8)*, 1627–1639.

Scherrer, P. (1918), *Nachr. Ges. Wiss. Göttingen*, 98–100.

Schönflies, A. (1891), *Krystallsysteme und Krystallstruktur*. Leipzig.

Shechtman, D., Blech, I., Gratias, D., Cahn, J. W. (1984), *Phys. Rev. Let. 53*, 1951.

Shirley, R. (1980), in: *Accuracy in Powder Diffraction, National Bureau of Standards Special Publication 567*. Gaithersburg: Nat. Bur. Stand., pp. 361–382.

Smith, D. K., Nichols, M. C., Zolensky, M. E. (1982), *POWD10. A FORTRAN IV program for calculating X-ray powder diffraction patterns – version 10*. University Park, Pa: The Pennsylvania State University.

Smith, D. K., Johnson, Jr., G. G., Scheible, A., Wims, A. M., Johnson, J. L., Ullmann, G. (1987), *Powder Diffraction 2*, 73–77.

Smith, D. K., Johnson, Jr., G. G., Wims, A. M. (1988), *Aust. J. Phys. 41*, 311–321.

Smith, G. S., Snyder, R. L. (1979), *J. Appl. Cryst. 12*, 60–65.

Smith, G. S., Johnson, Q. C., Cox, D. E., Snyder, R. L., Smith, D. K., Zalkin, A. (1987), *Adv. in X-Ray Anal. 30*, 383–388.

Smith, G. S., Johnson, Q. C., Smith, D. K., Cox, D. E., Snyder, R. L., Zhou, R. S., Zalkin, A. (1988), *Solid State Commun. 67*, 491–494.

Smith, S. T., Snyder, R. L., Brownell, W. E. (1979a), *Adv. X-ray Anal. 22*, 77–88.

Smith, S. T., Snyder, R. L., Brownell, W. E. (1979b), *Adv. X-ray Anal. 22*, 181–191.

Snyder, R. L. (1981), *Adv. X-ray Anal. 24*, 83–90.

Snyder, R. L. (1983a), *Advances in Material Characterization*, Rossington, D. R., Condrate, R. A., Snyder, R. L. (Eds.). New York: Plenum Press, pp. 449–464.

Snyder, R. L. (1983b), *Adv. X-ray Anal. 26*, 1–11.

Snyder, R. L. (1991a), *The Use of Reference Intensity Ratios in Quantitative Analysis, X-ray and Neutron Structure Analysis in Materials Science II*, in press.

Snyder, R. L. (1991b), *Applied Crystallography in Advanced Ceramics, EPDIC1 Proceedings of the First European Powder Diffraction Conference:* Delhez, R., Mittermeijer, E. J. (Eds.), Zürich: Trans Tech Publications.

Snyder, R. L. (1992), *Profile Analysis*, in: *Rietveld Analysis:* Young, R. A. (Ed.), in press.

Snyder, R. L., Bish, D. (1989), *Modern Powder Diffraction: Reviews in Mineralogy*, Vol. 20: Bish, D. L., Post, J. E. (Eds.). Washington, DC: Mineralogical Society of America, pp. 101–145.

Snyder, R. L., Carr, W. L. (1974), *Interfaces of Glass and Ceramics:* Frechette, V. D. (Ed.). New York: Plenum, pp. 85–99.

Snyder, R. L., Hubbard, C. R. (1984), *NBS*QUANT84: A System for Quantitative Analysis by Automated X-ray Powder Diffraction, NBS Special Publication*.

Snyder, R. L., Hubbard, C. R., Panagiotopoulos, N. C. (1981a), *AUTO: A Real Time Diffractometer Control System*. Gaithersburg: U.S. National Bureau of Standards, NBSIR 81-2229.

Snyder, R. L., Hubbard, C. R., Panagiotopoulos, N. C. (1981b), *Adv. X-ray Anal. 25*, 245–260.

Snyder, R. L., Rodriguez, M. A., Chen, B. J., Göbel, H. E., Zorn, G., Seebacher, F. B. (1992), *Adv. X-ray Anal.*, in press.

Steurer, W. (1990), *Z. Kristallogr. 190*, 179.

Tanner, B. K. (1976), *X-ray Diffraction Topography*. Oxford: Pergamon Press.

Taupin, D. (1973a), *J. Appl. Cryst. 6*, 266–273.

Taupin, D. (1973b), *J. Appl. Cryst. 6*, 380–385.

Teo, B. K. (1986), *EXAFS: Basic Principles and Data Analysis*. Heidelberg: Springer-Verlag.

Teo, B. K., Lee, P. A., Simons, A. L. Eisenberger, P., Kineaid, B. M. (1977), *J. Am. Chem. Soc. 99*, 3854–3857.

Thomson, J. J. (1906), *Conduction of Electricity through Gases*, 2nd ed. London: Cambridge University Press.

Visser, J. W. (1969), *J. Appl. Cryst. 2*, 89–95.

Visser, J. W., de Wolff, P. M. (1964), "Absolute Intensities". *Report 641.109*. Delft, Netherlands: Technisch Physischer Dienst.

Waller, I. (1928), *Z. Phys. 51*, 213.

Warren, B. E. (1969), *X-ray Diffraction*. Reading, MA: Addison-Wesley.

Warren, B. E., Averbach, B. L. (1950), *J. Appl. Phys. 21*, 595.

Warren, B. E., Gingrich, N. S. (1934), *Phys. Rev. 46*, 368.

Werner, P.-E. (1964), *Z. Kristallogr. 120*, 375–387.

Wiles, D. B., Young, R. A. (1981), *J. Appl. Cryst. 14*, 149–151.

Williamson, G. K., Hall, W. H. (1953), *Acta Metall 1*, 22–31.

Wright, A. (1992), "Neutron and X-ray Amorphology": *Analytical Techniques in Glass Science*, El-Bayoumi, O. H., Simmons, C. J. (Eds.). Columbus, OH: Amer. Ceram. Soc.

Young, R. A., Wiles, D. B. (1982), *J. Appl. Cryst. 15*, 430–438.

Young, R. A., Mackie, P. E., VonDreele, R. B. (1977), *J. Appl. Cryst. 10*, 262–269.

Zachariasen, W. H. (1945), *Theory of X-ray Diffraction in Crystals*. New York: Dover Publications, Inc.

Zhao, J., Ramanathan, M., Montano, P. A., Shenoy, G. K., Schulze, W. (1989), *Mat. Res. Soc. Symp. Proc. 143*, 151–156.

Zhang, Z., Ye, H. Q., Kuo, K. H. (1985), *Philos. Mag. A52*, L49.

Zhou, R. S., Snyder, R. L. (1991), *Acta Cryst. B47*, 617–636.

Zorn, G. (1991), *Programs DIFF and SHOW*. München: Siemens, Private communication.

Zorn, G., Wersing, W., Göbel, H. E. (1985), *Japanese J. of Appl. Phys. 24 (2)*, 721–723.

Zorn, G., Hellstern, W., Göbel, H., Schultz, L. (1987), *Adv. X-ray Anal. 30*, 483–491.

General Reading

For more tutorial information on X-ray diffraction and applications:

Cullity, B. D. (1978), *Elements of X-ray Diffraction*, 4th ed. Reading, MA: Addison-Wesley.

This is an excellent introductory treatment fully covering the fundamentals of X-ray diffraction.

For more depth on any particular topic and a good historical literature review:

Klug, H. P., Alexander, L. E. (1974), *X-ray Diffraction Procedures*, 2nd ed. New York: Wiley and Sons.

This book gives a comprehensive development of all of the methods of powder diffraction with a complete literature survey up through the early 1970's.

For a modern treatment of the techniques of applied crystallography:

Bish, D. L., Post, J. E. (Eds.) (1989), *Reviews in Mineralogy*, Vol. 20, *Modern Powder Diffraction*. Washington, DC: Mineralogical Society of America.

This is an excellent, fully up-to-date treatment of current applications of powder diffraction. It compliments the first two books listed in that they do not cover the modern applications of X-ray diffraction which this article concentrated on.

For an introduction to single crystal techniques:

Stout, G. H., Jensen, L. H. (1968), *X-ray Structure Determination*. New York: The Macmillan Co.

This is an excellent introduction to single crystal structure analysis techniques.

The "bible" of space group theory is:

Hahn, T. (Ed.) (1983), *International Tables for Crystallography*, Vol. A: *Space-Group Symmetry*. Dordrecht, Holland: D. Reidel.

An excellent introduction to classical crystallography is:

Phillips, F. C. (1963), *An Introduction to Crystallography*, 3rd ed. New York: Wiley.

A good review of the new area of quasi-crystals may be found in:

Hargittai, I. (Ed.) (1990), *Quasicrystals, Networks, and Molecules of Fivefold Symmetry*.

A good presentation of EXAFS techniques may be found in:

Teo, B. K. (1986), *EXAFS: Basic Principles and Data Analysis*. Heidelberg: Springer-Verlag.

5 Light Microscopy

Rainer Telle and Günter Petzow

Max-Planck-Institut für Metallforschung, Institut für Werkstoffwissenschaft,
Pulvermetallurgisches Laboratorium, Stuttgart, Federal Republic of Germany

List of Symbols and Abbreviations

a, b, c	axes of a crystal system
A	amplitude
A_i	amplitude of wave i
A_n	numerical aperture
A_0	amplitude of incident beam in a reference medium
c	vacuum light velocity
d	thickness, linear distance
$\boldsymbol{D}$	dielectric displacement
D	coefficient of quasi-elastic force
e	charge of an electron
e_g	d-electron orbitals
$\boldsymbol{E}$	electric field strength
$\boldsymbol{E}_0$	electric field amplitude
$\boldsymbol{E}_y$	y-component of the electric field amplitude
$\boldsymbol{H}$	magnetic field strength
$\boldsymbol{H}_0$	magnetic field amplitude
I	intensity
I^A	intensity in analyzer plane
I_i, I_r	intensities of incident and reflected beams
I_0	intensity of beam in air
$I_{1,2}$	intensity of component 1, 2
I_{res}	resulting intensity
I_{max}	maximum intensity
I_{min}	minimum intensity
k	constant in Abbe's equation of resolution
k	absorption coefficient
k_0	absorption coefficient of a reference medium
$\boldsymbol{k}, \boldsymbol{k}_1, \boldsymbol{k}_2$	wave vector
$\boldsymbol{k}^A, \boldsymbol{k}_1^A, \boldsymbol{k}_2^A$	wave vector in analyzer plane
K	refractive energy
K_i	refractive energy of element i
l	character of elongation
m	counting parameter $= 0, 1, 2, 3, \ldots$
m	mass
m_i	mass fractions of particular elements
M	Mallard's constant
n	refractive index
n', n''	refractive index of reference media 1 and 2
n_0	refractive index of a reference medium (air)
n_i	refractive index of medium in which the incident beam propagates
$n_{i,1,2,3}$	principal refractive indices of the indicatrix
n_1	smallest principal refractive index of the indicatrix
n_2	intermediate principal refractive index of the indicatrix

n_3	largest principle refractive index of the indicatrix
n_c	refractive index of compensator
n_e	refractive index of extraordinary beam
n_o	refractive index of ordinary beam
n_s	refractive index of sample
n_{refr}	refractive index of refracting medium
n_X	smallest principal refractive index of the indicatrix
n_Y	intermediate principal refractive index of the indicatrix, *optical normal*
n_Z	largest principal refractive index of the indicatrix
n_α	smallest principal refractive index of the indicatrix
$n_{\alpha'}$	smallest refractive index of an arbitrary section of the indicatrix
n_β	intermediate principal refractive index of the indicatrix, *optical normal*
$n_{\beta'}$	intermediate refractive index of an arbitrary section of the indicatrix
n_γ	largest principal refractive index of the indicatrix
$n_{\gamma'}$	largest refractive index of an arbitrary section of the indicatrix
n_ϑ	aperture angle-dependent number of isochromates
$\mathfrak{n}$	complex refractive index of opaque substances
$\mathfrak{n}_{ij}$	components of the complex indicatrix
N	number of atoms per unit volume
N_i	number of atoms of a particular element
$\boldsymbol{P}$	dielectric polarization tensor
r	stands for $2V_{red}$
R	reflectivity
R', R''	reflectivity of sample in contact with reference medium of n', n''
R_i	ionic refractivity in Lorentz-Lorenz equation for a particular element
s_{refl}	direction of reflected beam
s_i	direction of incident beam
$s_{1,2}$	directions
t, t_i	time
t_{2g}	d-electron orbitals
T	vibration velocity
v	phase velocity, propagation velocity
v	stands for $2V_{violet}$
v_e	velocity of extraordinary beam
v_o	velocity of ordinary beam
v_{refr}	velocity of refracted beam
V	unit cell volume
V	half the angle of bisectrix
V_{obs}	observed half angle of bisectrix
V_X, V_Z	half the angle of bisectrix along the principal X-axis, e.g., Z-axis of the indicatrix
x	distance coordinate
$x_{i, 1, 2, 3}$	principal axes of a coordinate system
$\mathfrak{x}, \mathfrak{y}, \mathfrak{z}$	principal axes of a complex coordinate system
X, Y, Z	principal axes of the indicatrix

Δz	depth of focus
Z	wave resistivity
$2V$	angle of bisectrix, optical angle, angle between the optical axes
$2V_{red}, 2V_{violet}$	angle of acute bisectrix for color red, e.g. violet
$2V_X, 2V_Z$	angle of bisectrix along the principal X-axis and Z-axis of the indicatrix
α	smallest refractive index of the indicatrix
β	intermediate refractive index of the indicatrix
γ	largest refractive index of the indicatrix
Γ	path difference
Γ_i	path difference of incident beam at differential interference-contrast illumination
Γ_n	path difference of reflected beams due to different refractive indices
Γ_{refl}	path difference of reflected beams due to re-entering Nomarsky prism
Γ_{top}	path difference of reflected beams due to topography effects
Γ_{total}	total path difference
Γ_ϑ	aperture-angle dependent path difference
δ	phase difference
Δ	principal double refraction, birefringence
Δ'	partial birefringence of arbitrary indicatrix section
Δ_ϑ	angle-dependent birefringence
ε	dielectric permittivity
$\boldsymbol{\varepsilon}$	dielectric permittivity tensor
ε_{ij}	components of $\boldsymbol{\varepsilon}$
ε_0	specific dielectric permittivity
ϑ	aperture angle
$\varkappa$	index of absorption
λ	wavelength
λ_0	wavelength of incident beam in a reference medium
μ	magnetic permeability
μ_0	specific magnetic permeability
ν	frequency
ξ	frequency-dependent factor in Lorentz-Lorenz equation
ϱ	density
τ	transmittance
φ	angle
φ_i	incident angle
φ_{ic}	critical angle of incidence, Brewster's angle
φ_{refl}	angle of reflection
φ_{refr}	angle of refraction
φ_{tot}	angle of total reflection
ω	angular frequency
ω_0	eigenfrequency
$\wedge$	symbol for angle

EDX	energy dispersive X-ray analysis
EELS	electron energy loss spectroscopy
SEM	scanning electron microscopy
STEM	scanning transmission electron microscopy
TEM	transmission electron microscopy
WDX	wavelength dispersive X-ray analysis

5.1 Introduction

Light microscopy is usually divided into two separate domains of application: microscopy using a *reflected beam* and using a *transmitted beam*. The predominant equipment of metallographic laboratories in material sciences is still the incident-light microscope which is also used for the characterization of transparent materials such as advanced ceramics or ceramic-metal composites. On the other hand, in the geosciences the examination of minerals and rocks is routinely performed using transmission light microscopy, and only highly specialized ore geologists are really experienced in microscopy using a reflected beam. Furthermore, transmission microscopy is widely used for glasses, traditional ceramics and polymers.

The scope of this chapter is to point out that light microscopy using reflected and transmitted beams are *both* useful methods for the quantitative characterization of physical, chemical and microstructural properties of matter. In comparison to electron microscopy, light microscopes can easily be handled in routine operations and do not usually require any specially trained personnel. Moreover, they are of low cost and very variable in equipment and applicability. Table 5-1 shows some typical properties, advantages and disadvantages of light microscopes compared to electron microscopes.

The fundamental ideas of light microscopy are based upon the more geometrical relationships of light propagation introduced by Snellius (1591–1662), Huygens (1629–1695), Newton (1643–1727), Fresnel (1788–1827), Foucault (1819–1868) and others. A more physical understanding of the interaction of electromagnetic waves with matter was achieved by the theory of wave mechanics (Maxwell, 1831–1879),

whereas quantum mechanics (1915–1930: Planck, Bohr, Schrödinger, Heisenberg) yield an atomistic and statistical model of light and its interaction with matter which is valuable for the description of certain optical effects such as light absorption and emission.

Abbe, Schott and Zeiss accomplished the fabrication of special glasses of high surface performance and various refractive indices enabling the production of the first polarization microscopes for transmitted light equipped with condenser lens and optical accessories. In 1878 Martens can be considered the first who recognized the importance of microstructural analysis in metals sciences, and LeChatelier constructed the first inverted reflected-beam microscope in the same year. Tammann established metallography as a descriptive science in 1905, but the important breakthrough in the determination of opaque substances and in the understanding of the physical phenomena of reflected light was achieved by mineralogists in 1910–1930 (Schneiderhöhn, Murdock, Ramdohr) after the quantitative theory of reflected light was developed by Berek in 1910 and his "ore microscope" was available from Zeiss.

5.2 Interaction of Light with Solid Matter

In spite of the fact that solid matter consists of a three-dimensional arrangement of distinct atoms and thus represents a microscopically discontinuous structure, the interaction with light can be successfully treated as a continuum effect. This is due to the fact that the spacings between the atoms are three orders of magnitude smaller than the wavelength of visible light, which ranges between approximately 400 nm (violet) and 760 nm (red) (corre-

Table 5-1. Comparison of microscopic methods.

Criterion	Optical microscopy		Electron microscopy	
	Reflected beam	Transmitted beam	SEM	TEM
Maximum magnification	$2000 \times$	$2000 \times$	$50\,000 \times$	$> 300\,000 \times$
Sample size	$0.01 - 10 \text{ cm}^2$	$0.01 - 10 \text{ cm}^2$	$10^{-4} - 10 \text{ cm}^2$	$2 - 3 \times 10^{-3} \text{ cm}^2$
Yields information on:				
Texture	limited	quantitative	limited	little e.g. coherency
Phase identification	limited	excellent: optical constants	limited: atomic number contrast	excellent with diffraction
Phase composition	none	limited: optical constants	many with EDX and WDX	many with STEM, EELS
Crystal structures	little	crystal system	none	space groups
Crystal growth	limited	many: nucleation, changes in composition etc.	little	many: nucleation, precipitates, etc.
Grain boundaries	little	limited	limited	atomic structure
Internal stresses	none	quantitative stress analysis	none	quantitative stress analysis
Sample preparation	easy	medium	easy	difficult
Routinely operational	yes	yes	yes	no
Permanent personnel demand for operation	none	none	technician	scientist, technician
Costs [US $]	10 000.— − 50 000.—	3 500.— − 50 000.—	25 000.— − 200 000.—	≫ 150 000.—

sponding to a frequency between 7.5×10^{14} and $3.7 \times 10^{14} \text{ s}^{-1}$), whereas the lattice constants of crystals vary between about 0.15 and 2 nm. The transition to a discontinuous description of matter therefore occurs upon shifting to probes of smaller wavelength such as electron irradiation or X-rays, which can then be used for the characterization of crystal structures. These methods are described elsewhere in this Volume.

The optical properties of either amorphous or crystallized solid matter are due to the modification of interacting light. The propagation velocity and mode of vibration may be influenced by volume effects if light penetrates transparent matter, by sur-

face effects if light is reflected from opaque matter or by interfacial effects if light is refracted at the transition between media of different optical behavior. All these interactions depend upon the wavelength and are based on the dielectric polarization of the medium by the electric term of the electromagnetic waves.

5.2.1 Physical Properties of Light

Light is a form of radiant energy absorbed or emitted by spontaneous energy changes of bonding electrons initiating transitions between energy levels in the outer shell of an atom. In the *electromagnetic theory* by Maxwell, light is regarded as superimposed oscillating electric and magnetic fields carrying energy through space in the form of continuous waves. Its behavior is adequately described by Maxwell's equations. According to the *quantum theory*, energy is transported discontinuously in individual bundles called *photons*. The effects of interaction of light with matter observed in optical microscopy are wave-like in character and will thus be explained by means of wave mechanics.

Electromagnetic waves are transverse waves because the electric vector E, also referred to as the electric field strength, oscillates perpendicular to the magnetic vector H (magnetic field strength), and also perpendicular to the direction of propagation x. The magnitudes of both vectors vary such that maxima and minima are reached simultaneously. For "normal" light, e.g., light emitted by the sun, a candle or a bulb, the plane of oscillation is not fixed, as the azimuths of vibrations are arbitrary. Devices that make the vectors vibrate in defined azimuths are called *polarizers*. Light exhibiting one constant plane of oscillation for each vector is *linearly polarized* or *plane polarized* (Fig. 5-1) and possesses a wavelength λ of

$$\lambda = 2\pi v/\omega \qquad (5\text{-}1)$$

with v being the phase velocity and ω the angular frequency. The oscillation plane of H is known as the *plane of polarization*, and the plane of vibration of E, the *plane of vibration*. At any time, both planes are defined by the vectors H_0 and k or E_0 and k, respectively, with H_0 and E_0 being complex vectors (field *amplitudes*) and k being the wave vector. The source of light can thus be described as a harmonic oscillator causing a time and space dependent periodic (sinusoidal) change of the electric and

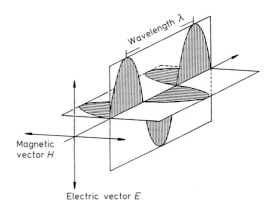

Figure 5-1. Vibration and propagation of plane polarized light.

magnetic vector

$$E = E_0\, e^{i(kx-\omega t)} = E_0\, e^{i\omega(x/v - t)};$$
$$H = H_0\, e^{i(kx-\omega t)} \qquad (5\text{-}2)$$

where x represents the actual coordinates and v the phase propagation velocity. The term $(x/v - t)$ then defines the status of oscillation at a given point x at a given time t. This corresponds to a *periodic dielectric displacement*

$$D_{(\omega, t)} = \varepsilon_{(\omega, t)}\, \varepsilon_0\, E_{(\omega, t)} \qquad (5\text{-}3)$$

where ε is the dielectric permittivity of a medium and ε_0 that of a vacuum.
With

$$H = \varepsilon\varepsilon_0 v E \quad \text{and} \quad E = \mu\mu_0 v H \qquad (5\text{-}4)$$

where μ represents the magnetic permeability and the propagation velocity of an electromagnetic wave – i.e., the rate of phase displacement – is given by the equation

$$v = (\varepsilon\varepsilon_0\mu\mu_0)^{-1/2} \qquad (5\text{-}5)$$

Since for a vacuum ε and μ are both equal to 1, the light velocity in vacuum c can be calculated by

$$c = (\varepsilon_0\mu_0)^{-1/2} = \qquad (5\text{-}6)$$
$$= 2.997925 \pm 0.0000011 \times 10^8 \text{ m/s}$$

Conversely, Eq. (5-5) indicates that the velocity of electromagnetic waves in matter is smaller than in a vacuum since with $\mu \approx 1$ the following equation is obtained as a good approximation for all non-magnetic media:

$$v = c/\varepsilon^{-1/2} \qquad (5-7)$$

This so-called Maxwell relationship between the pure optical measures v and c and the electrical measure ε proves that optical effects can be described by the electromagnetic theory. A generalization of that relationship is, however, not possible, since the phase velocity of light waves depends upon their frequency v, which means that ε is also a function of v. This behavior is called *dispersion,* and its explanation can only be a compromise between Maxwell's continuum theory and an atomistic approach. The c/v ratio is an important characteristic optical constant for matter and is defined as the *refractive index n*:

$$\qquad (5-8)$$
$$n = \frac{c}{v} \text{ or, more accurately, } n_{(v)} = \frac{c}{v_{(v)}}$$

Eqs. (5-5) and (5-8) infer that the refractive index of a material interacting with an electromagnetic wave is a function of its electric and magnetic properties and chemical composition as well as a function of the frequency of that particular wave. Methods for the measurement of n, some approaches for its calculation from the chemical composition and its symmetry-dependence are presented in Sec. 5.3.

5.2.2 Geometry of Light Propagation and Interference

"Normal" light waves are not sinusoidal, do not consist of a single wavelength (but rather a mixture of wavelengths, i.e., "white" light), and are not plane polarized. Yet the derivation of coherency, interference and absorption is based on ideally sinusoidal, plane polarized waves of a distinct wavelength (*monochromatic* light) (Fig. 5-1). In discussing wave propagation geometry, the *wave front* concept is more likely and is therefore mostly used to explain reflection and refraction. Since waves vibrate systematically and repetitively, particular points on waves which are in a comparable position in both space and time are said to be *in the same phase.* Sinusoidal waves are in phase, e.g., if at an instant in time the crests and troughs are in the same actual or relative positions. A *wave front* is defined as a surface passing through all points of equal state, i.e., of equal phase. Consequently wave fronts enclose the center of wave initiation (spherical wave fronts). If the source of waves is point-like and infinitely far away or, alternatively, if it consists of a linear array of single point sources, which is assumed to be the case for the effects observed in microscopy, then the wave fronts are *planes.* The radius perpendicular to the tangent plane of spherical wave fronts or the line perpendicular to planar wave fronts are called *wave-front normals* or *wave normals.* Hence a wave front advances in the direction of the wave normal whereas a *ray* is the propagation direction of a single wave, i.e., the *direction of energy flow.* Measurement of the light energy yields the intensity I which is the energy transmitted with time across a unit area perpendicular to the direction of wave propagation. The *amplitude A* of a simple sinusoidal electromagnetic wave equals the maximum displacement of the electric and magnetic vector from their equilibrium positions and cannot be measured directly because of the high frequencies. However, from the energy of a simple harmonic oscillator it can be shown that I is proportional to A^2 (i.e., E^2, H^2).

The *phase difference* is defined as the difference between phases of two points on a wave at a given time, or as the difference between the phases of two waves at different times at a given distance from the origin or a fixed reference point (Fig. 5-2). Since the term $(\boldsymbol{k}\,\boldsymbol{x} - \omega\,t)$ in Eq. (5-2) defines the *phase angle* as a function of space and time, considering only the y-components $E_y^{(1)}$ and $E_y^{(2)}$ of two waves, the equation can also be written as follows:

$$E_y^{(1)} = E_0^{(1)} \cos 2\pi\left(\frac{t}{T} - \frac{x_1}{\lambda} - \delta_1\right) \qquad (5\text{-}9)$$

$$E_y^{(2)} = E_0^{(2)} \cos 2\pi\left(\frac{t}{T} - \frac{x_2}{\lambda} - \delta_2\right) \qquad (5\text{-}10)$$

where $E_0^{(1)}$ and $E_0^{(2)}$ are the amplitudes, $T = v^{-1}$ is the vibration velocity, λ is the wavelength, x_1 and x_2 represent the distance from the origin, and δ_1 and δ_2 are phase factors:

$$(5\text{-}11)$$

$$\delta_1 = (k\,x_1 - \omega\,t) \text{ and } \delta_2 = (k\,x_2 - \omega\,t)$$

The phase difference δ is then

$$\delta = \delta_1 - \delta_2 = (k\,x_1 - \omega\,t) - (k\,x_2 - \omega\,t) =$$
$$= k\,x_1 - k\,x_2 \qquad (5\text{-}12)$$

or

$$\delta = \delta_1 - \delta_2 = (k\,x - \omega\,t_1) - (k\,x - \omega\,t_2) =$$
$$= \omega\,t_1 - \omega\,t_2 \qquad (5\text{-}13)$$

The *path difference* Γ between two waves moving in the same direction along the x-axis is the distance of movement in the x-direction between comparable points on the waves. The relation of Γ to δ is as follows:

$$\Gamma = \frac{\delta}{2\,\pi}\lambda \quad \delta \text{ in radians} \qquad (5\text{-}14)$$

The interaction between two waves propagating in parallel can be described by the rule of undisturbed superposition which involves a simple addition of field vectors of the particular waves for all common points in space x_i or in time t_i, respectively. The sum of Eqs. (5-9) and (5-10) results in electric field vector E_y:

$$E_y = E_y^{(1)} + E_y^{(2)} \qquad (5\text{-}15)$$

For waves with the same v or λ the resulting intensity I_{res} is then

$$I_{res} = \frac{Z}{2}\left[E_0^{(1)2} + E_0^{(2)2} + 2\,E_0^{(1)}\,E_0^{(2)} \cdot \qquad (5\text{-}16)\right.$$
$$\left. \cdot \cos 2\pi\left(\frac{x_2 - x_1}{\lambda} + \delta_2 - \delta_1\right)\right]$$

where Z is the wave resistivity of the medium and is given by $Z = \sqrt{\mu/\varepsilon}$. Consequently, the intensities I_1 and I_2 of the initial waves cannot be simply added except for

$$(5\text{-}17)$$

$$2\,E_0^{(1)}\,E_0^{(2)} \cos 2\pi\left(\frac{x_2 - x_1}{\lambda} + \delta_2 - \delta_1\right) = 0$$

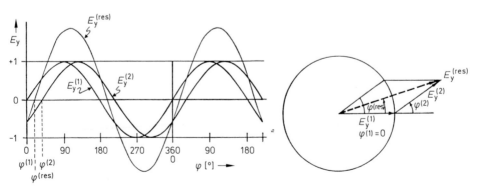

Figure 5-2. Phase difference of two coherent waves (after Galopin and Henry, 1972).

All other cases, i.e., all deviations from the additivity of intensities, are called *interference*. The term of Eq. (5-17) is named *interference term*. A maximum intensity results at

$$\left(\frac{x_2 - x_1}{\lambda} + \delta_2 - \delta_1\right) = n$$

with $n = \ldots -2, -1, 0, 1, 2 \ldots$ (5-18)

because with $\cos 2\pi n = 1$, Eq. (5-16) gives

$$I_{max} = I_1 + I_2 + 2\sqrt{I_1 I_2}$$ (5-19)

in which particular crests or troughs coincide to give *constructive interference*, $I_{max} > I_1 + I_2$.
For

(5-20)

$$\left(\frac{x_2 - x_1}{\lambda} + \delta_2 - \delta_1\right) = n + \frac{1}{2}; n = 0, 1, 2 \ldots$$

when $\cos \pi (m + \frac{1}{2}) = -1$, a crest and a trough coincide with the resulting intensity

$$I_{min} = I_1 + I_2 - 2\sqrt{I_1 I_2}$$ (5-21)

(*destructive interference,* $I_{min} < I_1 + I_2$).
  In a particular case, when $I_1 = I_2$ (i.e., $E_0^{(1)} = E_0^{(2)}$), Eqs. (5-20) and (5-21) yield $I_{max} = 4 I_1$ and $I_{min} = 0$, respectively. Since according to the rule of energy conservation intensity cannot simply be extinguished, every point in space and time with $I_{max} > I_1 + I_2$ corresponds to another point with $I_{min} < I_1 + I_2$. Consequently *interference* can be defined as a spatial redistribution of energy.
  Of course, local intensity extinctions can only be observed if they remain at a constant position during the observation time. This is equivalent to the condition that $\delta_2 - \delta_1$ is constant with time, i.e., that light which oscillates along individual rays is in phase or has a constant phase difference yielding a systematic, periodic interference. Waves behaving this way are said to be *coherent*. Only coherent waves can cause

interference. For all other conditions – which are present in the majority of cases under normal conditions – Eq. (5-16) has to be averaged over all values of $\delta_2 - \delta_1$, which gives Eq. (5-17) and thus the additivity of the particular intensities. Quasi-coherent light can be produced, for example, by diffraction gratings, by reflection and refraction from thin films, by masers and lasers. For instance, light that passes through a polarizing microscope is rendered quasi-coherent in an upper polarizer after propagating through a crystal which has produced definite phase differences among the various components. Thus interference effects such as extinction and color fringes become visible. White light is generally incoherent since it oscillates completely randomly and aperiodically even if it is plane polarized. It can result in interference, however, if it is resolved into nearly monochromatic components propagating in the same direction.

5.2.3 Optical Properties of Solid Matter

  Matter can be divided into two classes of fundamentally different optical behavior: The first class is said to be *optically isotropic* and exhibits no direction-dependent variation of optical properties. This means that light waves are transmitted with equal velocity in all directions. The property of isotropy is due to a statistically homogeneous or three-dimensional array of atoms or molecules with the same periodicity in the principal directions. Materials behaving in this way are gases (including vacuum), homogeneous liquids, amorphous matter (glass), and crystals of cubic structure. The second class is said to be *optically anisotropic*. Due to the systematic variation of physical and chemical properties in preferential lattice directions, the light velocity and mode of polarization also change as a function of the transmission direction. An example of this is all material which crystallizes differently from the cubic structure. Optically anisotropic behavior can also be introduced to amorphous or cubic matter by applying external stresses: Since an elastic response to compression or tension results in a unidirectional displacement of atoms (strain) the optical properties also become direction-dependent (*stress-induced double reflection*). Similar effects are observed under the influence of electric and magnetic fields (*Kerr effect, Cotton effect*).

In the following chapters, the optical effects are discussed in relation to isotropy and anisotropy. The geometric explanations of light propagation in both media are based on *Huygens' principle,* which states that any point or particle excited by the impact of wave energy becomes a new point source of energy. This is mathematically equivalent to *Fermat's principle,* which states that light propagates along a path requiring a minimum amount of time. The optical effects addressed below are *reflection,* i.e., the interaction of light with an optical interface neglecting the penetration of the media forming the interface; *refraction,* which deals with the bending of the propagation direction of light on passing an optical interface and penetrating a medium; and *absorption,* which involves

the conversion of light energy into other types of energy.

5.2.3.1 Reflection

According to the fundamental law of reflection, the angles of incidence and reflection, measured relative to the normal of the reflecting surface, are equal and in the same plane as the *plane of incidence* (Fig. 5-3). Incident light waves with a common wave front generate point sources of reflected waves which also possess a common wave front.

In the case of flat, accurately polished surfaces a *regular reflection* is obtained (surface roughness ≪ wavelength), whereas rough surfaces exhibit a *diffuse reflection,* as all hills and ridges act as distinct mirror planes (surface roughness ≥ wavelength). The intensity and wavelength of the reflected beam depends on the nature of the surface and the optical properties of the reflecting medium. Complications arise if a material is more or less opaque to certain wavelengths (selective absorption). Moreover, all gradations exist between transparent and opaque matter of metallic and nonmetallic luster. Generally, substances with delocalized electrons (electrically conducting or semiconducting) are opaque and exhibit excellent reflectivity, whereas materials with predominantly ionic or co-

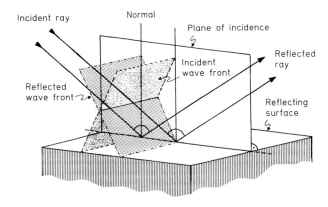

Figure 5-3. Geometry of light reflection at a plane interface.

valent character are non-opaque but are sometimes colored. This is due to the fact that mobile electrons may absorb any wavelength, whereas bonded electrons interact only with discrete wavelengths.

Since for microscopic examination of materials in reflected light a vertically incident beam is usually used, the following discussion is limited to surface orientations normal to the incident beam. The theory and the methods have been described in detail by Berek (1931a, 1931b and 1937) and Cameron (1961).

The ratio of the intensity of the reflected beam I_r to the incident beam I_i is defined as the *reflectivity* R which is given by

$$R = I_r/I_i = \frac{(n - n_0)^2 + k^2}{(n + n_0)^2 + k^2} =$$
$$= \frac{(n - n_0)^2 + n^2 \varkappa^2}{(n + n_0)^2 + n^2 \varkappa^2} \qquad (5\text{-}22)$$

where n is the refractive index, k the absorption coefficient, $\varkappa$ the index of absorption in the tested medium, and n_0 the corresponding refractive index of the medium in which the light propagates before entering the absorbant (see Sec. 5.2.3.3). In air n_0 is equal to 1.00029 (Na$_D$ line, 1 atm, 20 °C), and in a vacuum n_0 is exactly equal to 1, which gives the following relation for both media without major error:

$$R = I_r/I_0 = \frac{(n - 1)^2 + k^2}{(n + 1)^2 + k^2} =$$
$$= \frac{(n - 1)^2 + n^2 \varkappa^2}{(n + 1)^2 + n^2 \varkappa^2} \qquad (5\text{-}23)$$

For transparent materials k is close to zero which simplifies Eq. (5-23) to:

$$R = \frac{(n - 1)^2}{(n + 1)^2} \qquad (5\text{-}24)$$

In a similar way, a *coefficient of transmission* τ is attributed to the intensity ratio of an incident ray and a refracted ray. For coupled reflection and refraction at interfaces of transparent media, $R + \tau = 1$.

Like n or k, R is also a function of λ (*dispersion of the reflectivity*) and depends on the electrical conductivity. In anisotropic substances R varies with the orientation of the polarization plane relative to the symmetry elements (*reflection anisotropy, bireflectance*). Examples of materials which exhibit strongly anisotropic behavior are graphite, molybdenum sulfide and copper sulfide. Reflectivity is measured on carefully polished samples by a photometer. Monochromatic light of at least three different wavelengths (e.g., $\lambda_0 = 450$, 560 and 640 nm) must be used. For calibration, a standard material of known R and k is measured, e.g., a single crystal of tungsten carbide or silicon carbide.

n and k can be obtained from Eq. (5-23) if the reflectivity is measured as a function of two immersion media for a given wavelength. If R' of a sample is determined in a medium of known refractive index n_0', a second examination with another medium of known n_0'' is carried out to give R''. n is then calculated according to

$$n = \frac{n_0''^2 - n_0'^2}{2\,[n_0''\,(1 + R'')/(1 - R'') - n_0'\,(1 + R')/(1 - R')]} \qquad (5\text{-}25)$$

For air ($n_0' \approx 1$), Eq. (5-25) gives:

$$n = \frac{n_0''^2 - 1}{2\,[n_0''\,(1 + R'')/(1 - R'') - (1 + R')/(1 - R')]} \qquad (5\text{-}26)$$

The absorption coefficient k is obtained from:

$$k = [(n+n_0'')^2 R'' - (n-n_0'')^2]^{1/2} (1-R'')^{-1/2} \tag{5-27}$$
$$= [(n+n_0')^2 R' - (n-n_0')^2]^{1/2} (1-R')^{-1/2}$$

Some absorption and reflectivity data for metals and other opaque substances are listed in Table 5-2. The high reflectivity of metals is due to either high coefficients of absorption or to high refractive indices. Differentiation of opaque phases by measuring the optical constants is, however, difficult and not always clear, but many phases can be visually distinguished by their reflectivity and colors if they are in contact with a known compound. As a rule, silver is the brightest phase in polished sections followed by other noble metals. Other metals exhibit 60–70% reflectivity, arsenides between 50 and 60%, followed by complex sulfides (40–50%) and antimonides (30–40%). Oxides and some sulfides such as ZnS, MoS$_2$, Ag$_2$S exhibit 10–30% reflectivity. Extensive tables exist for ore minerals, but only a small amount of data is available for intermetallic phases and alloys.

The general relation between n and R according to Eq. (5-24) (Fig. 5-4) exhibits a minimum for $n = 1$ and approaches $R = 1$ (100% reflectivity) for $n = 0$ and $n = \infty$. If absorption and effects due to surface roughness are neglected, a certain *luster* can be attributed to a particular portion of this curve. Materials with $n = 1.3$–1.9 and $R < 0.09$ (e.g., halogenides, carbonates, sulfates, phosphates, arsenates, silicates and some oxides) show *glassy luster*. *Diamond luster* is attributed to materials with $n = 1.9$–2.6 and R between 0.09 and 0.20 (diamond, SnO$_2$, TiO$_2$, ZrO$_2$, ZnS, sulfur). The luster at n ranging between 2.6 and 3.0 and $R = 0.20$–0.25 is considered *semimetallic* (Mo$_2$S, Ag$_2$S, HgS, Cu$_2$O, Fe$_2$O$_3$). Materials with $n > 3$ and $R > 0.25$ possess

Table 5-2. Optical properties of some materials in reflected light (measured for Na$_D$-line, $\lambda = 589$ nm).

Phase	Crystal system	Opt. character	n	$n\varkappa$	R [%]
Ag	cubic	isotropic	0.181	3.67	95
Au	cubic	isotropic	0.366	1.82	85
Cu	cubic	isotropic	0.64	2.62	80
Hg	liquid	isotropic	1.73	4.96	78
Pt	cubic	isotropic	2.06	4.26	70
Sb	cubic	isotropic	3.0	5.01	74
Sb$_2$S$_3$	orthorhombic	anisotropic	4.44–5.17	0.62–0.37	35
ZnS sphalerite	cubic	isotropic	2.38	0.01	17.5
α-Fe	cubic	isotropic	2.36	3.40	56
Diamond	cubic	isotropic	2.43		45
Graphite	hexagonal	strongly anisotropic	1.1		5–23!
Fe$_2$O$_3$ hematite	rhombohedral	anisotropic	2.87–3.15	0.32–0.42	28
Al$_2$O$_3$	trigonal	anisotropic	1.765–1.770		5.7–6.7
SrTiO$_3$	cubic	isotropic	2.35		16.7
α-SiC	hexagonal	anisotropic	2.65–2.69		19–21
α-Si$_3$N$_4$	hexagonal	anisotropic	2.05–2.11		11.8–12.7
β-Si$_3$N$_4$	hexagonal	anisotropic	2.05–2.11		11.8–12.7
AlN	hexagonal	anisotropic	2.11–2.18		13–14
h-BN	hexagonal	anisotropic	1.71–1.72		7.5
c-BN	cubic	isotropic	2.5		?

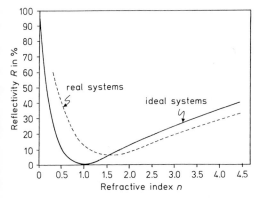

Figure 5-4. Dependence of reflectivity upon refractive index; note the difference between ideal systems ($k = 0$, observation in vacuum) and real systems (dashed line, $k \neq 0$, observation in oil).

metallic luster (MoS_2 and other sulfides and arsenides, metal carbides, metal borides, most of the metallic elements, alloys and intermetallic phases). Gold and silver, however, are examples of elements with a metallic appearance at $n < 1$ (Table 5-2). Platinum is an exception from Eq. (5-24), since for $n = 2.06$ a reflectivity of 12% instead of 70% is expected. This is due to the fact that absorption has to be taken into account as an additive term according to Eq. (5-23).

Randomly polarized light reflected from polished flat surfaces is partially linearly polarized. The amount and kind of polarization depends on the angle of incidence, the refractive indices and the crystal symmetry of the reflecting material and the quality of the reflecting surface. The correlation between the crystal symmetry and the optical constants of absorbing and reflecting matter is rather complex and will be discussed in Sec. 5.2.3.3. Generally, the incident light impacting the surface at angle φ_i is partially refracted at an angle of φ_{refr} and partially reflected by $\varphi_{refl} = -\varphi_i$ (Fig. 5-5). In *non-absorbing*, isotropic matter there is a critical angle of incidence φ_{ic}

at which the reflected beam is nearly totally linearly polarized. φ_{ic} is known as the *Brewster angle* or the *polarization angle*. At the Brewster angle, the following relationship holds:

$$(5\text{-}28)$$

$$n = \frac{\sin \varphi_i}{\sin \varphi_{refr}} = \frac{\sin \varphi_i}{\sin (90° - \varphi_i)} = \tan \varphi_{ic}$$

Moreover, the refracted beam is also partly linearly polarized and oscillates perpendicularly to the reflected ray. Consequently almost no reflection occurs for linearly polarized incident light with an electric vector vibrating in the plane of incidence because the vibrations induced in the dielectric medium would then oscillate parallel to the virtually reflected ray. This fact would violate the rule that energy cannot be transmitted in the direction of oscillation.

In *non-absorbing anisotropic* crystals, a plane polarized incident beam oscillating parallel to s_i is split into two perpendicularly vibrating components s_1 and s_2 which are usually reflected to different extents (*anisotropy of reflection*, Fig. 5-6). Both

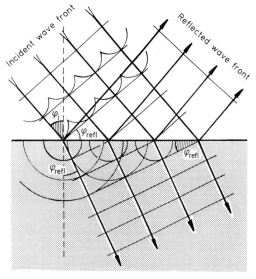

Figure 5-5. Geometry of combined reflection and refraction.

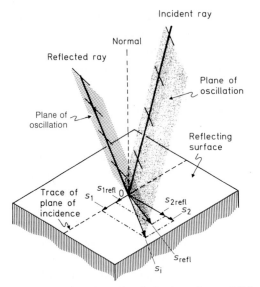

Incident ray

Normal

Reflected ray

Plane of oscillation

Plane of oscillation

Reflecting surface

Trace of plane of incidence

s_1 s_{1refl} O s_{2refl} s_2

s_{refl}

s_i

Figure 5-6. Rotation of polarization plane of light upon reflection.

beams superimpose again forming a linear oscillation with a resulting vibration direction s_{refl} which appears slightly rotated towards the direction of the stronger reflected beam. The reflected beam then exhibits a different azimuth of oscillation relative to the incident beam and a smaller intensity. In special orientations s_i may be positioned parallel to s_1 or s_2 which results in no variation in the polarization direction. This case is known as *uniradial reflection* and gives maximum reflectivity. Unpolarized incident light consequently becomes partially polarized light due to the influence of the reflected polarized light with the highest intensity. On *metallic* surfaces perpendicular incident plane polarized light is reflected with a phase displacement π and thus forms a stationary wave with a node at the reflecting surface (Wiener's experiment) since E steadily approaches zero at the metallic surface, whereas H is reflected without any displacement.

5.2.3.2 Refraction

Light waves are bent or *refracted* on passing from one transparent medium to another when the density or optical properties of the media differ. The capability of a material to refract light is called *refringence*. The *angle of refraction* φ_{refr} is defined as the angle between the refracted ray and the normal to the interface between the media (Fig. 5-5). φ_{refr} is a function of the angle of incidence φ_i as well as the wavelength λ, or in other words, a function of the velocity v in the particular dielectric medium. At the interface between isotropic media the geometric situation is determined by *Snellius' law*:

$$\frac{v_i}{v_{refr}} = \frac{\lambda_i}{\lambda_{refr}} = \frac{\sin \varphi_i}{\sin \varphi_{refr}} = \frac{n_{refr}}{n_i} \qquad (5\text{-}29)$$

where n_{refr} is the *refractive index* of the refracting medium, and n_i is the refractive index of the medium in which the beam propagates prior to incidence (e.g., air).

According to Snellius' law there is no refraction at an incidence normal to the interface. If light passes from a transparent medium of high n_i into one of lower n_{refr} a critical angle of incidence φ_{tot} exists at which the light is totally internally reflected at the interface (*total reflection*):

$$\text{for} \quad n_i \gg n_{refr} \rightarrow \frac{n_i}{n_{refr}} = \sin \varphi_{tot} \qquad (5\text{-}30)$$

The effect of total reflection is of particular importance for the determination of refractive indices, for the construction of optical instrumentation as well as for the identification of pores and inclusions in transparent materials (see Sec. 5.3.2).

Since in *isotropic* media the dielectric properties at optical frequencies are given by

$$D = \varepsilon_0 \varepsilon E \qquad (5\text{-}31)$$

the light velocity v_{refr} is equal in all directions and thus Eq. (5-31) holds for all φ_i. The numerical operation of v_{refr} into the directions of light transmission, known as the *ray velocity tensor surface,* then results in a sphere (Fig. 5-7). The bending of the wave front and ray direction can be derived using *Huygens' construction,* i.e., by drawing the tangential line from E, which is the origin of the refracted wave (b), to C, which is the arrival point of the advancing refracted wave (a) at that time.

In an *anisotropic medium,* however, Eq. (5-31) has to be replaced by inserting the second rank dielectric constant tensor ε:

$$D_i = \varepsilon_0\, \varepsilon_{ij}\, E_j \qquad (5\text{-}32)$$

This means that the light velocity now depends on the transmission direction and that D has a different direction to E. As a result of Maxwell's equations, two waves of *different velocity* and *perpendicular polarization plane* propagate through the anisotropic crystal. On of these waves behaves as in the *isotropic* case and is therefore referred to as an *ordinary wave* or *ordinary beam* o whereas the other is called

extraordinary e. Beam splitting is known as *double refraction* or *birefringence.* The geometric representation of the velocity of the extraordinary beam v_e as a function of the transmission direction gives an ellipsoid known as the *extraordinary ray velocity surface,* which touches the sphere of the ordinary beam velocity v_o at a circular cross-section. The effect of wave splitting can also be illustrated by Huygens' construction. In Fig. 5-7, beam e propagating from B to C_2 is faster than o propagating from B to C_1 and hence creates a wave front with a different angle of refraction. Hence, Snellius' law is valid for the ordinary wave but not for the extraordinary beam. In addition, the *wave front* e is not perpendicular to its propagation direction (*beam* e) which means that the wave normal and the ray direction are no longer parallel. The deviation is, nevertheless, small but results in an ovaloid for the wave normal instead of an ellipsoid for the ray direction.

Since the representation of v_o and v_e by using two interconnected index surfaces is not very comprehensive and does not contain any information about the polariza-

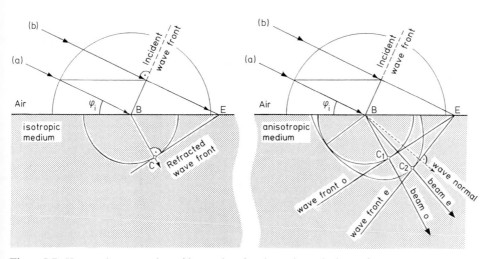

Figure 5-7. Huygens' construction of isotropic refraction using velocity surfaces.

tion directions another kind of figure was introduced by Fletcher (1892, based on the ideas of Fresnel) called the *indicatrix,* which shows simultaneously the direction dependence of both refractive indices as well as that of the particular polarization mode. The indicatrix is defined as the tensor surface of the *reciprocal dielectric tensor* or the *tensor of dielectric impermeability.* Its relationship to the crystal system is given in a system of principal axes x_i $(i = 1, 2, 3)$ by

$$\frac{x_1^2}{\varepsilon_{11}} + \frac{x_2^2}{\varepsilon_{22}} + \frac{x_3^2}{\varepsilon_{33}} = 1 \quad \text{and for } n_i^2 = \varepsilon_{ii}:$$

$$\frac{x_1^2}{n_1^2} + \frac{x_2^2}{n_2^2} + \frac{x_3^2}{n_3^2} = 1 \tag{5-33}$$

with ε_{ij} being the principal dielectric constants. The indicatrix can thus form an ellipsoid with the principal axes $n_1 < n_2 < n_3$.

The basic advantage of the geometric construction of the indicatrix is that the vectors representing the refractive indices are shown perpendicular to the wave normal, i.e., they are located in the plane of the wave front and plotted with n on the trace for the particular polarization direction (Fig. 5-8). In other words, the indicatrix has the following important properties. If a straight line is drawn through the origin O in an arbitrary direction P, the central section of the indicatrix perpendicular to that line will be an ellipse. Then the two wave fronts normal to the beam OP that may propagate through the crystal have *by definition* refractive indices *equal to the largest and the smallest elongations* OA and OB of that ellipse which are always the *semi-axes.* The displacement vector D in the plane polarized wave with a refractive index n_i equal to OA vibrates parallel to OA. Similarly, the displacement vector in the wave with refractive index equal to OB vibrates parallel to OB. From this it follows that the two possible waves with the wave normal to x_1 have the refractive indices n_2 and n_3; D in the two waves is then parallel to x_2 and x_3, respectively. Similarly, the wave normal to x_2 corresponds to two waves of refractive indices n_3 and n_1 with D parallel to x_3 and x_1, respectively. A similar statement is valid for the wave normal to x_3. Thus, n_1, n_2, n_3 are called the *principle refractive indices* of waves oscillating in the direction of the principal axes x_1, x_2, and x_3, respectively.

Generally n_1, also known as n_α, α or n_x is the smallest and n_3, also known as n_y, γ or n_z, is the largest index of refraction whereas the intermediate index n_2 is called n_β, β or n_y. Consequently, there are an infinite number of random directions of

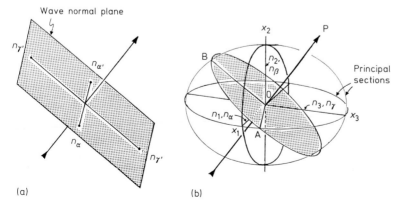

(a)   (b)

Figure 5-8. Construction of the indicatrix: (a) projection of vibration directions onto the wave front, (b) resulting indicatrix for three-dimensionally continued procedure.

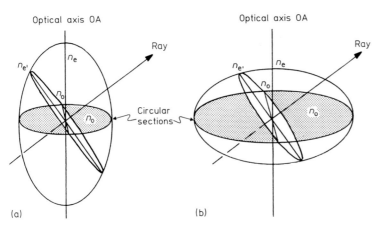

Optical axis OA

Ray

Circular sections

(a)

Optical axis OA

Ray

(b)

Figure 5-9. Uniaxial indicatrix: (a) optically positive, (b) optically negative.

wave propagation possible in the crystal with intermediate refractive indices $n_{\gamma'}$ being $n_\gamma < n_{\gamma'} < n_\beta$ and $n_{\alpha'}$ being $n_\beta < n_{\alpha'} < n_\alpha$. The difference $n_{\gamma'} - n_{\alpha'}$, i.e., the ratio of the semi-axes of the particular cross-section of the indicatrix, gives the *partial birefringence* Δ' of that wave normal direction and ranges between zero (normal direction parallel to optical axes) and the maximum *birefringence* Δ which is attributed to $n_\gamma - n_\alpha$ and is a material constant.

Symmetry, shape and orientation of the indicatrix relative to the crystal are defined by the crystal symmetry, the chemical composition, the wavelength and the temperature. Generally, for a *cubic*, i.e., *isotropic* crystal, the indicatrix becomes a *sphere*. Since all sections are circles, there is no beam splitting, no direction of preferential polarization and hence no birefringence. The size of the sphere and thus the principal refractive index n is solely defined by the chemical composition and the character of the atomic bonds. Since the symmetry of the indicatrix must be compatible with that of the crystallographic system in *trigonal, tetragonal and hexagonal* systems, the indicatrix is necessarily an *ellipsoid of revolution* about the principal symmetry axis c (Fig. 5-9). Hence, there are only two principal refractive indices n_1

and n_3. The central section perpendicular to the principal axis c, and only this section, is a regular circle with radius n_0 where o refers to an "ordinary beam". Hence, for a wave propagating along the principal axis c there is no double refraction which means that this direction behaves isotropically. This preferential axis of isotropy is called the *optical axis* (OA), and the crystal is said to be *uniaxial*. The propagation direction of the *extraordinary beam* with refraction index n_e is thus perpendicular to the optical axis and vibrates in parallel to it. The crystal is said to exhibit an *optically positive character* "(+)" for $n_e - n_0 > 0$ (i.e., if $n_e = n_3, n_0 = n_1$) and an *optically negative character* "(−)" for $n_e - n_0 < 0$ ($n_0 = n_3, n_e = n_1$). An optically positive uniaxial indicatrix is thus elongated in the direction of the optical axis, whereas a negative indicatrix appears compressed (Fig. 5-9).

In the *orthorhombic, monoclinic and triclinic* systems the indicatrix is a triaxial ellipsoid. There are two circular sections and hence two privileged wave normal directions of quasi-isotropy in which there is no double refraction. These two directions are called the *optical axes* and the crystal is said to be *biaxial* (Fig. 5-10). In *positive biaxial* crystals n_β approaches n_α, the indi-

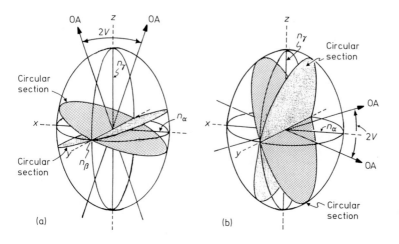

Figure 5-10. Biaxial indicatrix: (a) optically positive, (b) optically negative.

catrix therefore approaches the shape of a prolate ellipsoid of revolution, i.e. the form of the uniaxial positive indicatrix. In the *negative biaxial* case, as n_β approaches n_γ, the indicatrix assumes the shape of a negative uniaxial indicatrix with a horizontal optical axis (Fig. 5-10). The circular sections always pass through the *Y*-axis called the *optical normal direction*, which means that n_β is always the refractive index of the quasi-isotropic direction. The relative positions of the optical axes are hence a function of n_α and n_γ since they are located in the *X-Z*-plane, also known as the *plane of optical axes* or *optical plane* with n_β normal to it. Consequently the optical character of the biaxial indicatrix can also be derived from the angle between the optical axes. The smaller (acute) one is known as the *optical angle 2 V*. In *positive crystals* 2 V is bisected by the vibration direction of n_γ, i.e. the *Z*-axis of the indicatrix, which is then called *acute bisectrix 2 V*. In *negative* crystals the vibration direction of n_α, i.e. the *X*-axis, divides 2 V which means that the *Z*-axis is now the *obtuse bisectrix*. The optical angles are usually specified as $2 V_X$ and $2 V_Z$ and can be measured directly by means of *conoscopy* (Sec. 5.3.2.2) or calculated by

$$\cos^2 V_X = \frac{n_\gamma^2}{n_\beta^2}\left(\frac{n_\beta^2 - n_\alpha^2}{n_\gamma^2 - n_\alpha^2}\right) \quad \text{and}$$

$$\tan^2 V_Z = \frac{n_\gamma^2}{n_\alpha^2}\left(\frac{n_\beta^2 - n_\alpha^2}{n_\gamma^2 - n_\beta^2}\right) \tag{5-34}$$

or, for material of small double refraction, by *Mallard's approximation*:

$$\tan V_Z \approx \left(\frac{n_\beta - n_\alpha}{n_\gamma - n_\beta}\right)^{-1/2} \tag{5-35}$$

Generally, 2 V is a very sensitive measure for small variations in symmetry, chemical composition or mechanical stresses.

The *crystallographic position of the indicatrix* relative to the crystal axes depends on the symmetry system to which the particular crystal belongs. In the *orthorhombic* system the principal axes of the indicatrix are always parallel to the principal crystallographic axes which also fit into the symmetry elements of both figures. In the *monoclinic* system *one* principal axis of the indicatrix must be parallel to [010] of the crystal whereas the inclination of the other axes relative to [100] or [001] is free. This means that both indicatrix and crystal share a common mirror plane. In the *triclinic* system there is no correlation between indicatrix and crystal axes at all. In this case the position of the indicatrix is determined only by the type, concentration and position of the atoms in the crystal structure interacting with the light, which means it is a characteristic material constant.

The indicatrix only exists, of course, as a well-defined description of the optical behavior of material for a particular wavelength. The dependence of n, $2V$ and the orientation upon λ is called *dispersion*. In colorless materials the refractive indices decrease with increasing wavelength (*normal dispersion*). The slopes of the functions $n_\alpha, n_\beta, n_\gamma$ versus λ may be equal, convergent or divergent. This means that for uniaxial systems the shape of the indicatrix varies with λ, whereas the direction of the optical axis must not. In the orthorhombic system $2V$ varies in relation to λ and may even approach uniaxiality for a particular λ with a subsequent change of the optical character (i.e., rotating the optical plane by $90°$). An example of this behavior is a modification of TiO_2 known as Brookite. In monoclinic crystals the indicatrix may also rotate about the Y-axis (optial normal), being as it is parallel to the crystallographic [010] direction, and this creates a change in the inclination angle between the other principal and crystallographic axes (*inclined dispersion*). *Crossed dispersion* occurs if one of the principal axes of the indicatrix is the common acute bisectrix for all wavelengths and if it is parallel to [010] while the other axes are freely oriented. The case of the obtuse bisectrix X of all colors being parallel to [010], and the acute bisectrix Z as well as the optic normals having different orientations for different wavelengths, is known as *horizontal or parallel dispersion*. In the triclinic system the behavior of the indicatrix in relation to λ is irregular.

For a complete description of the optical properties of a material a *set of constants* has to be known: $n_\alpha, n_\beta, n_\gamma, \Delta, 2V_Z$ and the optical character given in $(+)$ or $(-)$ which determines the shape and size of the indicatrix. The relative orientation of the indicatrix is usually given in terms of the angle between the principal axes of the indicatrix and the crystallographic axes e.g.: $X = c$, $Y = b$, $Z = a$ (orthorhombic system) or $X \wedge c = 20°$, $Y = b$ (monoclinic system). The dispersion is described by the relation between $2V_{red}$ and $2V_{violet}$ in terms of $2V_{red} < 2V_{violet}$, $2V_{red} \gg 2V_{violet}$ etc. or in abbreviation $r < v$, $r \gg v$, respectively. If necessary, information on colors may also be added e.g., $X =$ pale green, $Z =$ dark green (see Sec. 5.2.3.3).

5.2.3.3 Absorption

Although all rules of light propagation can be readily derived by Maxwell's theory only the *Quantum theory* gives explanations for the effects of emission and absorption. Absorption results from the transformation of wave energy into other types of energy, mostly heat. Particular wavelengths usually excite electrons of the outer shells of an atom (*optical electrons*) and are thus eliminated from the set of frequencies of the incident beam. As a result the absorbing material appears colored and the intensity of the incident beam is weakened.

In terms of an atomistic description, absorption is a resonance effect. An oscillating electric field $E = E_0 \cos \omega t$ applies a force $e E_0 \cos \omega t$ on every charge unit of an atom creating an oscillating charge displacement and hence a dielectric polarization:

$$P = D - \varepsilon_0 E = (\varepsilon - 1)\varepsilon_0 E_0 \cos \omega t \quad (5\text{-}36)$$

Eq. (5-36) is, however, a function of v since the dipole formation does not always occur

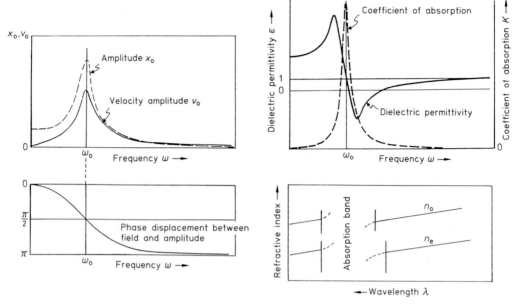

Figure 5-11. Frequency-dependence of ε and wavelength-dependence of n in the vicinity of an absorption band.

in phase with the applied field **E**. If it is assumed that the oscillating dipole reacts quasi-elastically, vibrating with an eigenfrequency of $\omega_0 \approx (D/m)^{1/2}$ (D being the quasi-elastic force and m the mass of the oscillator) then the dispersion curve resembles a resonance curve. On increasing the applied frequency ω the phase displacement of the amplitude increases and approaches $\pi/2$ at $\omega = \omega_0$. At the point of maximum amplitude the induced current is in phase with the oscillating field (resonance) which results in a maximum of energy absorption (Fig. 5-11). Consequently the dispersion curve for both ε and n undergoes a sharp reversal in slope (*abnormal dispersion*), and for portions of the spectrum the refractive index increases with increasing wavelength. Since particular transmission directions in a crystal may absorb light to a different extent – i.e. the dispersion curve for a particular n may exhibit a different position for the reversal point compared to the others – a direction de-

pendence of coloring may be observed in polarized light known as *pleochroism* and described in terms of e.g.: X = weak, Y = strong, Z = very strong or equivalently $X < Y < Z$.

The color depends in a complex way, not only on the interaction of light with selectively absorbing chemical constituents, but also on their structural arrangement, type of chemical bonding, impurities and defects. So an originally colorless transparent material may contain colored inclusions (*pigments*) and hence appear colored. Then inhomogeneities such as precipitates or caverns with dimensions of several hundreds of nanometers may generate internal scattering leading to a colored luster (*opalescence, labradorescence*). Furthermore, crystal imperfections such as interstitial atoms or vacancies may cause local charge accumulations or lattice distortions which interact with light (*color centers*). The simplest type of point defect creating colors is the F-center which is a negative ion vacancy containing captured electrons. The most important origin for selective absorption is, however, *ion coloring* which is, contrary to aforementioned cases, an intrinsic property of the crystal structure. The chemical elements which are most effective in selective absorption are cations of the transition metals, e.g., Ti, V, Cr, Mn, Fe, Co, Cu, and Zn, since they occur in several oxidation states. The effect of absorption by the excitement of electrons in incompletely occupied d- or f-orbitals (d-f-transitions) is called *central ion coloring*. The position of the absorption bands is determined by the type and symmetrical arrangement of the next-neighbor anions or charged complexes (coordination shell) which influence the splitting of the energy levels of the orbitals, and hence the transition energy, from one level to another. Small variations in the interatomic distances, symmetry or charges may thus strongly affect the colors of the crystal. An example is alumina, Al_2O_3, generally colorless, which becomes ruby-red if Cr^{3+} ($3d^3$-configuration) is added (*allochromatic color*) whereas the corresponding pure compound chromia, Cr_2O_3, exhibits a deep green color (*idiochromatic color*). Heavy variations in color due to the distortion of the co-ordination polyhedra around the central ion (*Jahn-Teller effect*) have been found e.g. in Cu^{2+}-compounds ($3d^9$-configuration) where the colors may change from green to deep blue, or in Mn^{3+}-compounds ($3d^4$-configuration, pink and red colors).

Compared to central ion coloring which covers electron transitions within the shell of a single ion, electron hopping between overlapping orbitals of adjacent ions may create an absorption of a thousandfold intensity and result in nearly opaque materials. This effect is known as *charge-transfer coloration* and is observed between two cations, two anions or between cation and anion. The absorption bands are rather broad and spread from the ultraviolet to the visible light. The basic mechanism is a repetitive oxidation-reduction process between differently charged ions with overlapping d-orbitals. Examples the neighboring pairs Fe^{2+}-Fe^{3+}, Mn^{2+}-Mn^{3+} or Ti^{3+}-Ti^{4+}. Opportunities for such an arrangement occur preferentially by coupled substitution of the constituents in the solid solution series thus preserving the charge neutrality of the crystal structure. For example in a silicate structure exhibiting Si^{4+} in an tetrahedral coordination [IV], i.e. surrounded by four oxygen ions, and Fe^{2+} and Mg^{2+} in an octahedral configuration [VI], i.e. surrounded by six oxygen ions, substitution of Si^{4+} by Al^{3+} must be accompanied by substitution of the bivalent cations in the octahedral position by trivalent cations:

$$^{[VI]}(\underset{\text{octahedron}}{Fe^{2+}, Mg^{2+}}) + ^{[IV]}\underset{\text{tetrahedron}}{Si^{4+}} \leftrightarrows$$

$$\leftrightarrows ^{[VI]}(\underset{\text{octahedron}}{Fe^{3+}, Al^{3+}}) + ^{[IV]}\underset{\text{tetrahedron}}{Al^{3+}} \quad (5\text{-}37)$$

If octahedra of differently charged central Fe-ions now possess a common face, d-orbital overlapping enables oscillating electron transitions and thus charge transfer coloration. Generally, charge transfer is possible between octahedral coordination polyhedra with a common face or common edges (t_{2g} orbitals overlapping) and between tetrahedrally coordinated ions with common edges (e_g orbitals overlapping, common faces are generally impossible). The phenomenon of *pleochroism*, i.e. direction dependent selective absorption, is caused by a preferential unidirectional periodic arrangement of absorbing atoms as interconnected chains or clusters of coordination polyhedra. Thus the particular orbitals enabling electron transitions between certain energy levels or between overlapping orbitals are in parallel and thus are simultaneously excited by light waves of a certain oscillation direction. For example, in cordierite (also called *dichroite*), orthorhombic $^{VI}Mg_2{}^{IV}Al_3[AlSi_5O_{18}]$, iron may be incorporated as $^{VI}Fe^{2+}$ for Mg^{2+} and $^{IV}Fe^{3+}$ for Al^{3+} which results in adjacent octahedra of $^{VI}Fe^{2+}$ and $^{IV}Fe^{3+}$-tetrahedra with common edges. As t_{2g}-orbitals and e_g-orbitals overlap parallel to the crystallographic a-axis, the following optical behavior is observed: n_α vibrates in parallel to [001] and exhibits weak central ion coloring absorption at wave numbers of 8750 and 10 200 cm^{-1}; n_β vibrates in parallel to [100] and absorbs strongly at 17 700 cm^{-1} and weakly at 10 750 cm^{-1} giving a violet color whereas n_γ vibrates in parallel to [010] and exhibits an intermediate absorption at 17 400 cm^{-1} (light blue). This corresponds to energy level splitting in t_{2g}-orbitals and e_g-orbitals with an energy difference of approximately 50 kcal/mol which matches the energy for chemical bonding.

An excellent tool for the calculation of energy level splitting and the width of absorption bands affected by charges, interatomic spacings or symmetrical arrangements of next-neighbour ions and hence for the prediction of color effects is provided by the *ligand field* or *crystal field theory* introduced by Bethe in 1929.

The decrease in intensity of radiation passing through a medium is known as *extinction* and is a composite effect of scattering and absorption. If scattering is neglected, the resulting amplitude A of monochromatic light of wavelength λ_0 and initial amplitude A_0 after penetrating an absorbing medium of thickness d equals:

$$A = A_0 e^{-2\pi k d/\lambda_0} \quad \text{(Beer's Rule)} \quad (5\text{-}38)$$

where k represents a characteristic material constant known as the *absorption coefficient*. It depends on the refractive index n according to

$$k = n\varkappa \quad \text{yielding} \quad n = \lambda_0/\lambda = k/\varkappa \quad (5\text{-}39)$$

where $\varkappa$ is the index of absorption and λ is the wavelength in the absorbing medium. With $I = A^2$ we obtain for the intensity:

$$I = A_0^2 e^{-4\pi k d/\lambda_0} \quad (5\text{-}40)$$

The term $4\pi k/\lambda_0$ is called the *modulus of extinction*. In transparent materials, k is usually $\ll 1$ but it may be $3-5$ for opaque substances. Like n, k is a rather complex function of λ. Equation (5-40) gives for the light vector E [see Eqs. (5-2 and 5-9)]:

$$E = E_0 e^{2i\pi[t/T - d/\lambda_0(n-ik)]} \quad (5\text{-}41)$$

which means that for absorbing media a *complex index of refraction* $\mathfrak{n}$ has to be inserted with

$$\mathfrak{n} = n - ik = n(1 - i\varkappa) \quad (5\text{-}42)$$

Consequently, in the presence of absorption, the symmetry-dependent optical properties of matter can no longer be described using the relatively comprehensive indicatrix, except in the case of the cubic system where $\mathfrak{n}$ equals nk. The generalization of the indicatrix for anisotropic systems gives a second-order surface with complex refractive indices as radial vectors (Berek, 1937). Using gothic characters for complex units we obtain for the absorbing indicatrix:

$$\frac{\mathfrak{x}^2}{\mathfrak{n}_1^2} + \frac{\mathfrak{y}^2}{\mathfrak{n}_2^2} + \frac{\mathfrak{z}^2}{\mathfrak{n}_3^2} = 1 \quad (5\text{-}43)$$

which becomes, after a coordinate transformation, a real system with the principal axes x, y, and z:

$$(5\text{-}44)$$

$$\frac{x^2}{\mathfrak{n}_{11}^2} + \frac{y^2}{\mathfrak{n}_{22}^2} + \frac{z^2}{\mathfrak{n}_{33}^2} + \frac{2xy}{\mathfrak{n}_{12}^2} + \frac{2yz}{\mathfrak{n}_{23}^2} + \frac{2xz}{\mathfrak{n}_{13}^2} = 1$$

In similarity to non-absorbing *optically anisotropic* matter, two wave fronts propagate through the absorbing crystal but, contrary to the non-absorbing material, they possess not only two different refractive indices but also two different absorption coefficients. Both waves are usually elliptically polarized and oscillate perpendicularly to one another with the same direction of rotation. The special condition of linearly polarized light exists only in optical symmetry planes.

In optically uniaxial systems – i.e. in *trigonal, tetragonal and hexagonal* crystals – the complex indicatrix consists of two surfaces for n and nk with a common revolution axis of different lengths. The surfaces are, however, not ellipsoids, as in the case of non-absorbing matter, but ovaloids. However, an ordinary and an extraordinary ray of linearly polarized light also exists.

Optical axes in the sense of quasi-isotropically behaving directions do, however, not exist any more in the *orthorhombic, monoclinic and triclinic* systems. With increasing absorption the optical axes split, in perpendicular to the particular optical symmetry plane, into two axes of preferential polarization status. These four axes are called the *winding axes*, because light propagating in these directions does not oscillate arbitrarily but is *circularly polarized* with an opposite direction of rotation (Fig. 5-12). In addition, there are two principal directions in which only one refractive index is effective but two absorption indices resulting in linearly polarized light and alternatively, another two directions of equal absorption coefficients but different refractive indices.

For a complete description of the optical properties of an absorbing triclinic crystal in monochromatic light twelve independent constants have to be known: three

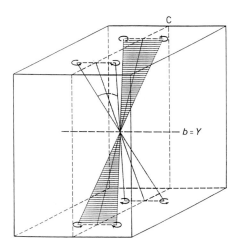

Figure 5-12. Position of winding axes in an orthorhombic crystal (after Berek, 1931 a and 1931 b). In the original publication the polarization along the winding axes was attributed to the motion of ether particles as rays moving parallel to each axis.

refractive indices, three absorption coefficients and six trigonometric functions which correlate the symmetry elements of the crystal to n_i and k_i. With increasing symmetry of the crystal, the relations become less difficult, so in uniaxial systems the properties can be derived from the set of n_o, k_o, n_e, and k_e. Examples for the variation of R, k, and n are shown in Fig. 5-13. In the monoclinic case, the principal axes x and $\mathfrak{z}$ are complex whereas in orthorhombic crystals the principal axes of both the non-absorbing and the absorbing indicatrix are real and parallel to the axes of the crystallographic system. The non-absorbing indicatrix is drawn in dashed lines. In triclinic systems there are no correlations between the principal axes of n, $\mathfrak{n}$, k, R, and the crystallographic system. Finally, it should be noted again that size and mutual orientation of the tensor surfaces of n, $\mathfrak{n}$, k, and R depend on the wavelength λ (dispersion).

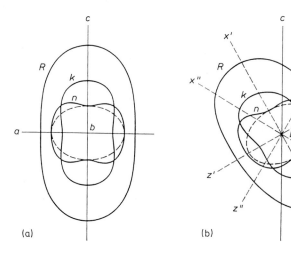

(a)   (b)

Figure 5-13. Sections of optical representation surfaces of opaque matter (after Berek, 1937): (a) orthorhombic system, (b) monoclinic system. Symbols: a, b, c axes systems; n complex refractive index; R reflectivity, k index of absorption, x, x', x'', z, z', z'' are complex principal axes of the surfaces, whereas y and b are simple.

5.3 Instrumentation for Light Microscopy

The intention of this section is to give an overview on the opportunities for materials characterization by light microscopy rather than to describe or to judge the quality of the numerous technical equipment in detail. This can readily be obtained from handbooks on metallography, optical physics (see reference list) and from brochures provided by the producers of optical instrumentation. However, the set of microscope components will be schematically addressed and commented on where necessary for the understanding of the examination procedures. Generally, this section focuses on the methods used to get most information from a sample by making use of the principles of crystal optics and crystal physics as described in Sec. 5.2.

5.3.1 Reflected Light Microscopy

5.3.1.1 Microscope Equipment

Incident-light microscopes for the inspection of opaque materials are available in two types of construction, namely *upright* and *inverted instruments* which refers to the position of the sample surface either being on top or upside-down. Both systems are optically equivalent. For serious applications the microscope should be equipped with a fully rotatable and adjustable stage with graduation and x-y-micrometer. The *illumination system* is usually placed in an external housing to prevent heat transfer to the optical system. It may consist of a low voltage tungsten or carbon filament lamp for routine observations and – ideally – an additional xenon-arc or tungsten filament lamp for photography or examinations at crossed polarizers. An adjustable *condenser lens* in front of the light source combined with a *field diaphragm* focuses the light on the optical path. *Filters* such as monochromators or neutral-density filters which equalize the wavelength-dependent light intensity may be inserted after that point. A yellow-green filter is used to reduce the effect of lens defects for black-and-white films. Other filters may be applied for the adaptation of the light temperature to that of the film material. A separate insert is provided as a polarizing filter called the *polarizer* which produces plane polarized light and should be rotatable about 360°. In less expensive instruments the polarizer consists of a foil covered with parallel acicular crystals of a strongly pleochroitic iodine chinine sulfate (*herapathite*) which allows light transmission only in a plane polarized state. High-quality polarizers are fabricated from a calcite crystal cut in a special orientation and glued together again in such a way that the ordinary beam passing through one part of the crystal undergoes total reflection at the interface to the other section of the crystal, whereas the extraordinary beam is refracted and hence leaves that assembly completely plane polarized. This polarizer is known as *Nicol's prism* and hence all polarizing filters are usually called *Nicols*.

The *aperture diaphragm* determines the *aperture angle* and is placed before light entering the vertical illuminator. The optimum setting for the aperture angle depends on the particular objective lens and influences intensity, contrast, sharpness and depth of field. With increasing magnification the aperture diaphragm should be closed more. The *vertical illuminator* also referred to as the *opaque illuminator* consists of a semi-transparent mirror or prism which bends the light from the lamp towards the objective lens and allows the light reflected from the sample to pass straight through to the ocular (Fig. 5-14).

The *objective lens* is the most important component of the microscope and produces the image of the sample. The *numerical aperture A_n* which is a measure of the capability of collecting light is a function of the refractive index n of the medium between the specimen and the lens, i.e. air or oil, and the half-angle ϑ of the most oblique light rays that enter the lens:

$$A_n = n \sin \vartheta \qquad (5\text{-}45)$$

The objective lenses are divided according to their chromatic correction in achromates, fluorites, and apochromates. The most common lenses are *achromates* which are corrected spherically for a specific color, e.g. yellow-green, and for longitudinal chromatic aberration for two colors, e.g. red and green. Therefore, achromates can only be used for black-and-white photography, the use of yellow-green filters hence yields optimum results. In *fluorites* the color-specific spherical correction is obtained by the use of calcium fluoride CaF_2 instead of glasses. *Apochromatic objectives* consist of a combination of several lenses that are also fabricated from CaF_2 and special glasses. They provide both spherical and chromatic correction to a higher extent but yield lower light intensity due to internal reflection losses. *Plano objectives* are adapted best to flat samples and hence are widely used in metallography. Generally, all objectives used in polarized light

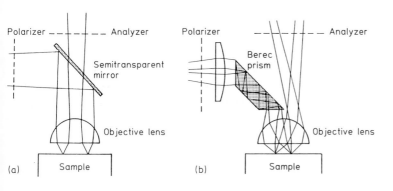

Polarizer ——$|$—— Analyzer   Polarizer   ——$\backslash$—$|$— Analyzer

Semitransparent mirror

Berec prism

Objective lens   Objective lens

(a)   Sample   (b)   Sample

Figure 5-14. Opaque illuminator (a) and Berek compensation prism (b) for generating polarized light illumination at almost perpendicular incidence.

must be free of strain. For thousandfold magnification oil immersion is recommended, since by increasing n in Eq. (5-45) a higher numerical aperture is achieved so giving improved resolution. The minimum resolvable lateral distance d between two points can be estimated by Abbe's equation of resolution:

$$d = \frac{k\lambda}{n\sin\vartheta} = \frac{k\lambda}{A_n} \qquad (5\text{-}46)$$

where k is a constant which is 0.61 for luminescent points and 0.5 for opaque points if they are completely resolved and equal 1 for an intensity decrease of 20% between the two maxima of emitted or scattered light. Other formulas for the determination of resolution power are reported by Zieler (1969) giving different suggestions for the maximum reasonable magnification. The use of oil as an immersion medium also enhances sharpness and color contrast.

The *depth of focus* Δz which is the distance between the positions of focal points at which the image appears at acceptable sharpness can be estimated by the Rayleigh Equation

$$\Delta z = \frac{\lambda}{n\sin^2\vartheta} \qquad (5\text{-}47)$$

which may be combined with Eq. (5-46) to give a relationship between longitudinal and lateral aberration:

$$\Delta z = \frac{4n}{\lambda}\left(\frac{\lambda}{2n\sin\vartheta}\right)^2 = \frac{n\lambda}{A_n^2} \qquad (5\text{-}48)$$

Continuing along the optical path, behind the objective lens system it is possible to insert additional optical accessories such as *filters* and *sensitive tints* ($\lambda/4$ plate, first-order-red-plate, quartz-wedge, Berek compensator, Nomarsky prism) and others which produce a particular path difference of separately propagating and differently polarized light waves and hence are only

useful in combination with the *analyzer*, which is another polarization filter or Nicol prism yielding plane polarized light vibrating perpendicularly to that passing through the polarizer. By definition, the polarizer produces light which vibrates vertical to the eye-field and the analyzer allows transmission of horizontally polarized light. Nevertheless, the analyzer should be able to be rotated by at least 90° and it should possess a graduation scale.

The virtual image produced by the objective lens is magnified and adapted, for inspection by the eyes, by the *oculars*. Moreover, as some lens defects may be corrected by the oculars they must match with the type of objective used. For documentation, a separate adapter for photo or video cameras may be provided.

5.3.1.2 Principles and Methods

Bright-field illumination is the most common method of sample inspection. The incident beam passing through the vertical illuminator and the objective lens strikes the sample perpendicular to its surface and is reflected back through the objective lens. Surface features in an oblique position relative to the incident beam yield a darker contrast due to the intensity losses of the rays reflected with angles larger than the aperture angle. Thus perfectly flat surfaces act as a mirror whereas pores, pulled-out particles, inclusions, cracks and bonding defects can easily be recognized by their very strong contrast. In general, cavities and ridges can be distinguished by focusing the center point as cavities show internal reflections whereas ridges do not. If the surface is not perfectly flat, e.g. if phases with a strong difference in hardness face each other, the step formation around the harder particles creates a thin bright contrast known as the *Schneiderhöhn line* which always moves into the softer area if the distance between sample and objective is enlarged. The Schneiderhöhn line can thus be used for the identification of phases.

Other microstructural features such as grain boundaries and twinning planes may be visible in the case of metals with a strong *amplitude contrast* relative to other compounds in the sample, or due to an anisotropic behavior. Thus even small differences in reflectivity or color of adjacent crystals of different composition or orientation may become visible. Although the human eye is very sensitive to recognizing *differences* in contrast and color, statements based solely on that impression may, however, be erroneous. If different phases are in contact at least one of them has to be identified definitely in order to be able to distinguish the others by the comparison of their reflectivity and color relative to the known one. Absolute methods for the measurement of reflectivity and other optical constants are presented in Secs. 5.2.2.1 (reflection) and 5.2.2.3 (absorption).

Ceramic materials exhibit rather poor reflectivity which makes phase determination by optical impression alone nearly impossible. Another disadvantage is that light is both reflected and partially transmitted to deeper regions. *Internal reflections* inside crystallites are thus typical for non-opaque substances and are created by multiple reflections on particle interfaces, pores or cracks beneath the surface of the sample. Occasionally, transparent particles at the surface are cut wedge-like and then exhibit *Newtonian fringes* which are rainbow-colored interference effects due to the path difference between both rays reflected at the surface and the interface. Internal reflections are avoided by operating with polarized light or – for a better result

– by coating the samples with strongly reflecting metallic layers. The thickness of the coating should not exceed 20–30 nm. Suitable materials are gold or aluminum.

In order to make microstructural features such as grain boundaries and twinning planes more visible etching techniques have to be applied which create an artificial surface roughness (see Sec. 5.4.4) and thus enhance the contrast.

Some microscopes are equipped with a condenser lens, a vertical illuminator mirror or an aperture diaphragm which can be decentralized so that the incident beam strikes the sample at an inclined angle. This method is called *oblique illumination* and produces a pronounced three-dimensional appearance of surface roughness. Since deviation from the normal incident beam is limited and the interpretation of the image is difficult, it is only recommended for etched specimens if no differential interference contrast is available.

Dark-field illumination is preferentially used for the observation of powder specimens or certain surface roughness effects such as pores, etched grain boundaries or twinning planes or composites with phases of different hardness or polishing behavior. Also artefacts from the grinding and polishing treatment can readily be identified. In this technique only the obliquely reflected light is collected by the objective lens whilst perpendicular reflected light is blocked. As a result, the contrast of the bright-field illumination appears inverted, dark features become bright and bright contrasts become dark.

Phase contrast is an effect based on the phase displacement of light reflected by a certain crystallite relative to the environment. The particular extent of phase displacement is a function of the refractive indices, the dielectric constant, the incident angle and the direction of polarization

(Beyer, 1974). It may also be generated from optical path differences, i.e. by surface steps of the order of nanometers in height. This technique requires a special vertical illuminator, a different set of objectives and a $\lambda/4$-plate to introduce an additional phase shift of $\pi/2$ which converts the invisible phase contrast into a recognizable amplitude contrast. Since the observed effects cannot be interpreted easily the use of phase contrast illumination is limited and in metallography almost completely replaced by interference-contrast techniques.

Reflected beam microscopy with *polarized light* is applied in metallography for the observation of strongly anisotropic metals such as beryllium, α-titanium, zirconium or uranium and certain intermetallic phases of hexagonal structure, but is extensively used for ore microscopy because beyond all doubt it is an important tool for the optical identification of phases. Contrary to transmission optical microscopy one of the most important conditions, namely that the light is completely linearly polarized and vibrating in the same plane all over the entire field of view, is not accomplished in reflected beam. Deviations from this requirement occur if the incident angle differs from $90°$, which is always the case in the margins of the observed area. Moreover, on placing the mirror in the vertical illuminator the ray undergoes a phase displacement differing from $0°$ or $180°$ which also yields elliptically polarized light to a certain extent. These aberrations can be avoided by the use of the *Berek compensation prism* instead of a plane mirror in the vertical illuminator (Fig. 5-14).

If the incident beam is unpolarized but the reflected beam is monitored with the inserted analyzer its polarization state can be observed by rotating the stage and monitoring the reflectivity. Although the orientation-dependent variations in reflected intensity are only a few percent, isotropic and anisotropic materials can be distinguished. If rotated under plane polarized light, anisotropic substances exhibit a variation in the intensity and color of the reflected light. Their reflectivity changes between two extremes which are called *uniradial reflectivities*. The difference between these extremes is referred as *bireflectance* or *double reflectivity*. The bireflectivity approaches zero if the section of the crystal is normal to the optical axes or the winding axes or to another special cut. Reflectivity, bireflectivity and their dependence on wavelengths are significant optical constants and can be measured by means of microphotometers. In the same way the color may vary due to anisotropic absorption (see Sec. 5.2.2.3). This effect is called *reflection pleochroism* and can be a very typical property of certain phases.

Under *crossed polarizers,* isotropic and anisotropic crystals can be distinguished in the same way as with only one polarizer but the anisotropy effects appear much clearer. Isotropic material generally stays dark whereas anisotropic matter shows azimuth-dependent variations in reflectivity. Since reflection of absorbing matter usually yields slightly elliptically polarized light, even if the incident ray is completely plane polarized, rotating a crystal about $360°$ results in four positions of minimum reflectivity but not of complete extinction. As already stated above for bireflectance specially oriented sections may also yield quasi-isotropic behavior. Spectacular effects of reflection pleochroism may be obtained in some colored substances which can even be intensified by using oil immersion. A famous example is covellite, CuS, which shows a deep blue color when exposed in air and a variety of orientation depending colors ranging from dark red to purple when observed in oil.

If polished sections are only evaluated qualitatively by a reflected beam, the polarizers are not usually crossed completely, but to approximately 87° since the intensity yield is too small to allow observation by eye and the exposure time for photographs may exceed several minutes. In particular ceramic materials which possess less than 10% reflectivity cannot be readily examined with crossed polarizers. Anisotropic properties of metals with low reflectivity and ceramics can, however, be made visible by *interference layer contrasting*. Contrast enhancing layers can be deposited on the surface of polished sections by *reactive sputtering* from a metallic cathode (Pt, Au, Pb, Fe, Al...) in an oxygen atmosphere or by evaporation (ZnS, ZnSe, ZnTe). Figure 5-15 shows a contrasting chamber mounted on a microscope stage. The sample holder can be rotated in such a way that the sample faces either the glass lid, for inspection by the microscope, or the cathode. After evacuation the chamber is flushed with oxygen at approximately 20 Pa partial pressure. A voltage of 2000–2500 V is applied between the sample and the cathode as a function of the distance. When the plasma discharge is ignited the current is adjusted to about 1–2 mA so

generating the epitaxial deposition of a film of oxidic cathode materials (Bartz, 1973; Hofmann and Exner, 1974; Blumenkamp et al., 1980). The optimum thickness for optically effective layers is about 10–20 nm.

Other possibilities for the deposition of coloring coatings are *heat tinting,* i.e. exposing the sample to air at elevated temperatures in order to grow epitaxially thin oxide layers on certain preferential metallic phases which are sensitive to corrosion (Cathcart et al., 1967), *color etching* which involves a reactive solvent assisted deposition of precipitation layers and *anodizing* meaning an electrolytic process for reactive film deposition during electro-polishing (see Sec. 5.4.4).

The optical effect of contrasting is caused by multiple reflections of the incident beam at the interfaces of sample/coating and coating/air which introduces path differences as a function of layer thickness, refractive indices and absorption coefficients of both the sample and the coating material. Certain wavelengths are then extinguished by interference and the resulting complement colors are observed. Since the polarization mode of the reflected beam, as well as the optical constants which are effective on the particular crystal surface, depend upon the crystal orienta-

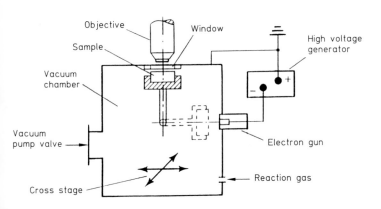

Figure 5-15. Schematic sketch of a contrasting chamber (after Mills et al., 1985).

tion (see Sec. 5.2.2.1) strong contrasts may be observed even in single phase materials. Because of their poor reflectivity and comparatively low refractive indices the effects for ceramics are not as good as for metals but satisfactory results can be obtained if the following conditions are met: The ceramic phases should exhibit a high absorption, opaque materials are preferred. Platinum, iron, lead or cryolithe (Na_3AlF_6) are the most suitable coating materials. The thickness should not exceed 5 nm.

Differential interference-contrast illumination is widely used to accomplish a three-dimensional impression of surface topography similar to oblique illumination but requires the use of polarized light. After having passed through the vertical illuminator, the plane polarized beam enters a *Nomarski prism* which is a modified *Wollaston prism* and is split into two perpendicular plane-polarized waves with a path difference Γ_i (Fig. 5-16). The two rays are then reflected at two separate points on the surface of the sample. Due to the different heights of these points, an additional path difference Γ_{top} is superimposed. Moreover, phase displacements are also introduced if the points of reflection are situated on compounds of different refractive indices, or particles of different orientation (Γ_n). On re-entering the prism another path difference Γ_{refl} is added to both rays, but only waves with a total path difference $\Gamma_{total} = \Gamma_i + \Gamma_{top} + \Gamma_n + \Gamma_{refl}$ of odd multiplicity of $\lambda/2$ pass through the analyser after recombination. In the symmetrical case, the path difference generated by the Nomarski prism is equal for the incident and the reflected beam and the intensity of light passing through the analyser is a function of the path difference created by the topography and the optical constants. If the Nomarski prism can be decentralized Γ_i and Γ_{refl} can be made unequal which increases the total path difference and thus enhances the color contrast. In practice, differential interference contrast microscopy should be carried out on etched samples.

In Color Table 5-1 some microstructures prepared by various contrasting techniques are presented. Figure 1 shows an aged NbAl40 alloy, which is treated by reactive Pt/O_2 sputtering and photographed with crossed polarizers. Note the orientation-dependent change in colors. In Fig. 2, the microstructure consists of an as-cast Al-Fe alloy with Al_2Fe-platelets (yellow, brown, red-brown) in an Fe_2Al_5 matrix (large violet and grey-blue areas), Pt/O_2 coated, parallel polarizers. Note the tremendous change in color due to anisotropy effects, both phases are tetragonal. Figure 3 presents a Pd16.5In16.5Sn (at.%) alloy which was etch-polished with ammonia persulphate and ferricyanide solution and alumina. The later contrasting procedure with Fe in O_2-atmosphere revealed clearly the martensitic transformation of the $Pd_2(In,Sn)$ phase. The as-cast microstructure of a Pd27In alloy is shown in Fig. 4. Contrasting with In-cathode in O_2-atmo-

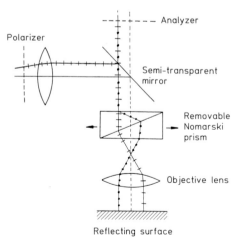

Figure 5-16. Polarization mode of light passing through a Normarski prism.

sphere develops the primary yellow-brown Pd_3In crystals with martensite ribbons inside. Note the orientation contrast. The matrix consists of a Pd_3In-Pd_2In eutectic which has also undergone martensitic tranformation (green lamellae). Preparation was performed by vibration polishing in sodium-thiosulfate-ferricyanide solution. Figure 5 represents a bronze etch-polished with Klemm III solution. In Fig. 6, the grain growth of a molybdenum-nickel solid solution in the vicinity of a large pore is monitored by a repeated cooling and heating treatment. The intermediate stages of coarsening are made visible by shock-polishing and subsequent Murakami etching due to slight variations in composition. A ceramic material – YBaCuO superconductor – is shown in Fig. 7. The twinning is due to the orthorhombic-to-tetragonal transformation upon cooling from the sintering temperature. The material was carefully ground and polished in oil because of its degradation in water. The section was photographed under crossed polarizers without any contrasting procedure. Ni-coated agglomerates of Fe-powder have been solid-state sintered, ground, polished and etched with Klemm I solution yielding the microstructure in Fig. 8. Note that white Ni-phase has diffused along grain boundaries of colored Fe-particles and hence has disintegrated the agglomerate.

5.3.2 Transmission Light Microscopy

Light microscopy of transparent materials using a transmitted beam is a well established domain of the geosciences and became the most important tool for the geological and mineralogical characterization of rocks and minerals.

The application of transmission light microscopy is, of course, not limited to natural matter but also provides valuable information on artificially produced or modified transparent materials such as advanced ceramics, glasses and polymers. In contrast to light microscopy using a reflected beam the images obtained by transmitted polarized light can be evaluated more quantitatively since the observed effects are much easier to measure and to understand.

5.3.2.1 Microscopic Equipment

In general, the construction of a transmitted light microscope is similar to that for use with reflected light for the basic components such as the light source, illumination system, diaphragms, polarizers and other filters, objective lenses and eye pieces. Since these parts have been described extensively in Sec. 5.3.1.1 the following presentation will focus on the specific components of microscopes for transmitted light and how they operate.

First of all, since the samples are always subject to transmission and hence are placed in a straight beam between the polarizer, optically active accessories and the analyser, only upright instruments are reasonable. The *sample* usually consists of a thinned or compressed glass-covered object glued onto a glass plate, or of particles disseminated in a planar array between two glasses and immersed in a liquid of known refractive index. Consequently only strain free *plano objectives* are useful. It is essential for examination in polarized light that the *stage* is fully rotatable about 360°, exhibits a precise graduation and can be locked at any position. Modern microscopes provide automatic locking upon rotating about 90°. For certain applications, such as direction-dependent measurement of refractive indices on one single crystal or for quantitative texture analysis, multiaxially rotatable *universal stages* are available which permit the rotation of the sam-

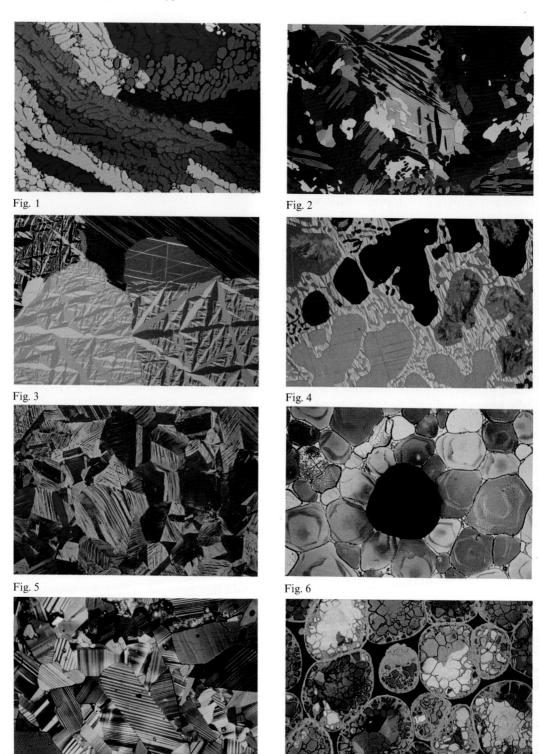

Fig. 1

Fig. 2

Fig. 3

Fig. 4

Fig. 5

Fig. 6

Fig. 7

Fig. 8

Color Table 5-1. Microscopy in reflected light.

Figure 1. NbAl40 alloy, annealed, Pt/O_2-contrasted. (Courtesy K. Kaltenbach, U. Täffner.)

Figure 2. As-cast Al-Fe alloy: Pt/O_2-contrasted; yellow equi-axed grains and reddish brown plates: $FeAl_2$; blue and gray matrix phase: Fe_2Al_5; (different coloration due to anisotropy effects). (Courtesy R. Laag, L. Täffner.)

Figure 3. PdIn16.5 alloy: with martensite lamellae; Fe/O_2-contrasted. (Courtesy E. Schmidt, U. Schäfer.)

Figure 4. As-cast PbIn27 eutectic alloy with martensite in matrix and dendrites; Pt/O_2-contrasted, polarized light. (Courtesy G. Müller, V. Carle.)

Figure 5. Bronze with twins and martensite, color-etched with Klemm III, polarized light. (Courtesy U. Täffner.)

Figure 6. Mo-powder liquid phase sintered with Ni, repeatedly sintered and cooled in three cycles, etched in Murakami agent and shock-polished; note the three distinct steps in grain growth leading to shape accommodation and densification of the pore in the center, residual Ni at the grain boundaries. (Courtesy K. Kang, U. Täffner.)

Figure 7. YBaCuO-superconductor material with multiple twinning due to the tetragonal-orthorhombic transition upon cooling from fabrication temperature, polarized light. (Courtesy M. Aslan, U. Schäfer.)

Figure 8. Ni-coated polycrystalline Fe-spheres; color-etched with Klemm I; note the disintegration of Fe-particles due to the penetration of grain boundaries by Ni. (Courtesy W. A. Kaysser, U. Täffner.)

ple into any desired orientation. For the observation of temperature-dependent reactions stages have been produced which can be heated or cooled.

Unlike the illumination system of the reflected beam microscope, the transmission microscope is equipped with an additional swing-out auxiliary array of *condenser lenses* and an *iris diaphragm* placed between the polarizer and the sample. This component can be centered and focused and provides a convergent illumination of the specimen. In contrast to reflected light microscopy this facility is not only used to increase the light intensity yield of the objective lens at high magnifications (*Köhler illumination*) but is also of particular physical significance: Convergent beam examination enables observation of the sample simultaneously at different angles whereas normal illumination operates with almost parallel beams. For so-called *conoscopic observation* a removable *Amici-Bertrand lens* is placed between the analyzer and the ocular which converts the microscope into a low-power telescope focused at infinity. This means that the observer looks through the sample rather than at the sample. If an anisotropic object is examined between crossed polarizers at high magnification (large numerical aperture) an interference figure can be seen which provides information about the optical character and the crystal symmetry. Figure 5-17 illustrates the light path in direct (*orthoscopic*) and conoscopic view.

Finally, optical accessories such as *compensators* ($\lambda/4$-plates, gypsum plates, Berek compensators, quartz wedges, etc.) are used more widely than for reflected light microscopy. One or more slots are provided between the objective lens and the analyser for the insertion of these components at an angle of 45° relative to the vibration plane of polarizer and analyzer.

5.3.2.2 Principles and Methods

Bright field illumination is used to recognize colors and interfaces of translucent materials as well as to distinguish between translucent and opaque matter. As discussed in Sec. 5.2.2.3, *colors* result from selective absorption of certain wavelengths and may vary with the orientation of the absorbing crystal. Since the polarizer is never usually removed in any kind of illumination technique, orientation-dependent effects such as pleochroism can readily be observed. *Interfaces* become visible due to refraction and reflection effects caused by the mismatch in refractive indices between two adjacent materials. As an example, cracks in glasses and ceramics can easily be recognized because of the transition of light from matter to the penetrating air but also material of the same composition may reveal clearly visible grain boundaries due to orientation effects. Another example is a blend of immiscible liquids such as water and oil.

The fact that light is scattered by interfaces due to refraction, ordinary reflection and even total reflection resulting in a strong contrast is used for the determination of the refractive indices n. Since light is always scattered and refracted towards the material with the higher refractive index a thin bright line known as the *Becke line* is observed in the vicinity of optical inhomogeneities such as pores or inclusions, as shown in Fig. 5-18, which may be considered to act as lenses creating convergent rays above the object with the higher refractive index. If, however, the image is taken out of focus by raising the tube slightly from the object the Becke line always moves towards the medium with the higher n. This means that the *relative* refractive index can be distinguished visually. The absolute value of n is obtained if

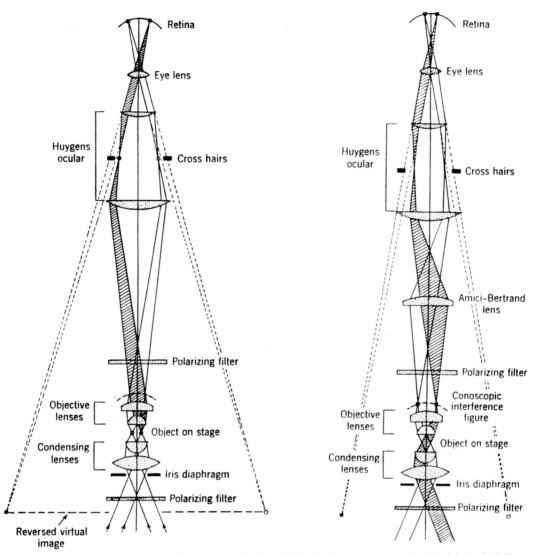

Figure 5-17. Light propagation in the orthoscopic view (left) and the conoscopic view (right); dashed lines indicate refracted and reflected rays (after Wahlstrom, 1979).

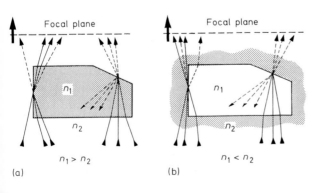

Figure 5-18. Monochromatic light propagation producing a Becke line in the vicinity of an optical inhomogeneity [(a): $n_1 > n_2$, (b): $n_1 < n_2$]; redrawn after Wahlstrom (1979).

a fragment of the material to be examined is immersed in liquids of known refractive index and repeatedly subjected to the comparative treatment. If the refractive indices of both immersion medium and crystal approach each other the interface contrast gets weaker and weaker. If monochromatic light is used the contour disappears completely as both refractive indices match each other, and the fragment becomes invisible. In white light equality of refractive indices can only be obtained for a particular wavelength because of dispersion. Thus slightly colored contours appear as the complement wavelengths still undergo refraction and reflection. Since they vary rather characteristically with very small changes of Δn, they can be utilized for a precise determination of n with an accuracy of the order of ± 0.001. The refractive indices have to be related to the wavelength in which they have been measured.

In the case of anisotropic matter two or three refractive indices have to be determined depending on the crystallographic system. This can readily be done if there are clear criteria for the orientation of the indicatrix relative to the morphology of the crystal fragment such as facets, cleavage planes or twin boundaries which have to be related to the extinction angles or the orientation of the optical axes observed under crossed polarizers. The amount of *double refraction (birefringence)* is obtained by the difference between the largest and the smallest refractive index measured.

Suitable immersion media are listed in Table 5-3, additional materials have been reported by West (1936). In particular the liquids or melts of high refractive index are rather toxic and should be handled with extreme care. Generally, the actual temperature of the immersion medium has to be taken into account since n varies by 0.0007 per Kelvin.

Since the immersion method for the determination of refractive indices requires small particles of the order of $20-500$ μm it is not applicable for large crystals or gemstones. For this case *refractometers* have been constructed by which the critical angle for total reflection, i.e., Brewster's angle is measured. These and other methods are extensively discussed by Wahlstrom (1979). Refractive indices can also be estimated from the chemical composition of matter. Gladstone and Dale have empirically stated that n is a function of the refractive energy K of a substance and its density ϱ, where K is composed of the mass fractions m_i of the particular elements in the compound multiplied by the specific refractive energy K_i of these elements (given in oxide formula):

$$n = \varrho K + 1 \quad \text{with} \quad K = \sum K_i m_i \quad (5\text{-}49)$$

By measuring the refractive indices of simple metal oxides the refractive energies of particular ions are calculated, which can then be composed to more complex compounds. Thus, the mass fraction of certain elements in a solid solution series can be determined. For example, the refractive energies of Fe^{2+} and Fe^{3+} are 0.169 and 0.308, respectively (Jaffe, 1956).

Lorentz and Lorenz have established a formula based on the electromagnetic interaction of light with atoms introducing a dipole moment. If absorption is negligible Eq. (5-50) holds:

$$\frac{n^2 - 1}{(n^2 - 2)\varrho} = \frac{4\pi}{3} e^2 \xi N = K \quad (5\text{-}50)$$

where e represents the charge of an electron, ξ is a frequency-dependent factor and N is the number of atoms per unit volume. Inserting the unit cell volume V, the number of atoms N_i of a particular element in the compound and the ionic refractivities R_i of that element the Lorentz-Lorenz

Table 5-3. Refractive indices of selected immersion media [589 nm (Na_D)].

Substance	n at 20 °C	Dispersion	Remarks
Vacuum	1.00000		
Air (1 atm.)	1.000294		
Water (deionized)	1.333	slight	dissolves chlorides and some sulfates
Acetone	1.357	slight	dissolves many minerals
Ethanol	1.362	slight	
Methyl butyrate	1.386	slight	
Ethyl valerate	1.393	slight	
Amyl alcohol	1.409	slight	
Kerosene	1.448	slight	
Carbon tetrachloride	1.460	slight	
Russian alboline	1.470	slight	
American alboline	1.477	slight	
Benzene	1.502	slight	toxic
Chlorobenzene	1.525	slight	very toxic
Eugenol	1.541	moderate	toxic
Monobromobenzene	1.560	moderate	toxic
Bromoform	1.590	moderate	toxic
Cinnamon oil	1.606	moderate	
Monoiodobenzene	1.620	moderate	toxic
Monochloronaphthalene	1.626	moderate	toxic
Monobromonaphthalene	1.658	moderate	toxic
Methylene iodide	1.737	strong	toxic
Sulfur saturated methylene iodide	1.778	strong	toxic
Antimony iodide saturated methylene iodide	1.868	strong	toxic
Sulfur	1.998	very strong	
Selenium	2.72_{Li}	very strong	very toxic

equation gives:

$$\frac{V}{1.6602} \frac{n^2 - 1}{n^2 + 2} = \sum N_i R_i \qquad (5\text{-}51)$$

The values for atomic or ionic refractivities are listed in the particular handbooks of physics and chemistry (e.g., Batsanov, 1961; Allen, 1956; McConnell, 1967). As predicted by the crystal field theory, the interaction of ions with light depends, however, not only on the particular species but also on the crystallographic environment i.e. the ligands in the coordination shell and the interatomic distances. Nevertheless both equations agree more or less with the average refractive indices of ionic substances measured.

The most useful information on transparent matter concerning the crystal system and other characteristics for materials identification, twinning, orientation, texture and stresses is gained by *microscopy with crossed polarizers*. According to Sec. 5.2.3.2, isotropic, i.e. amorphous or unconstrained cubic matter, can easily be distinguished from anisotropic materials by staying dark at all orientations relative to the transmitted beam whereas anisotropic crystals possess only distinct directions of isotropy or quasi-isotropy, namely if the transmitted beam propagates parallel to an optical axis. In all other cases of incidence the cross-section of the indicatrix where the beam hits yields an ellipse which results in splitting of the propagating waves into different velocities and polarization modes (double refraction, birefringence) now vibrating in perpendicular polarization planes. Passing the analyzer, both rays recombine by superimposing the path difference Γ as well as the vibration directions of the wave vectors. Since only the horizontally vibrating component of the resulting ray is transmitted (*interference*), plane polarized light with a wavelength and amplitude different to that entering the crystal is generated. The principles of these effects are illustrated in Fig. 5-19.

It is now obvious that the observed wavelength λ and intensity I are functions of (i) the wavelength of the incident beam (usually white light), (ii) the optical properties of the crystal such as size and symmetry of the indicatrix (extent of double refraction) and coefficients of absorption, (iii) the relative orientation of the crystal, i.e. of the indicatrix to the transmitted beam, (iv) the orientation of the indicatrix

in respect to polarizer and analyzer, and (v) the thickness of the sample. These dependences will be described in the following paragraphs.

The interaction of plane polarized light coming from the polarizer with the crystal depends on the particular cross-section of the indicatrix which is normal to the transmitted beam. Neglecting absorption and dispersion and assuming parallel incident light this arbitrary cross-section is an ellipse with principal axes $n_{\alpha'}$ and $n_{\gamma'}$. This means that this cross-cut of the crystal exhibits a partial birefringence Δ' of

$$\Delta' = n_{\gamma'} - n_{\alpha'} \qquad (5\text{-}52)$$

resulting in two separately propagating rays of velocity $v_1 = c/n_{\alpha'}$ and $v_2 = c/n_{\gamma'}$. The time for crossing a crystal plate of thickness d is given by

$$t_1 = d\frac{n_{\alpha'}}{c} \qquad t_2 = d\frac{n_{\gamma'}}{c} \qquad (5\text{-}53)$$

The path difference Γ between both rays is thus a function of the thickness d:

$$\Gamma = (t_2 - t_1)c = (n_{\gamma'} - n_{\alpha'})d = d\,\Delta'$$
$$[\Gamma \text{ in nm, } d \text{ in mm}] \qquad (5\text{-}54)$$

At an arbitrary orientation of the principal axes of the indicatrix section relative to

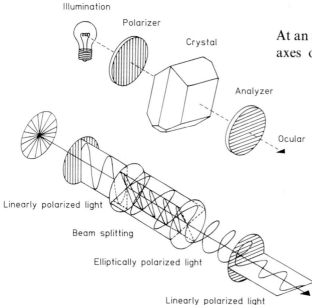

Illumination

Polarizer

Crystal

Analyzer

Ocular

Linearly polarized light

Beam splitting

Elliptically polarized light

Linearly polarized light

Figure 5-19. Resolution and interference of plane polarized light passing through an anisotropic crystal.

polarizer and analyzer the particular polarization mode of the waves leaving the crystal, as well as the intensity of the observed beam behind the analyzer, can be obtained by a vector resolution of the incident wave vector k into the vectors k_1 and k_2 within the crystal vibrating along the principal axes:

$$k_1 = k \cos \varphi \quad \text{and} \quad k_2 = k \sin \varphi \qquad (5\text{-}55)$$

with φ being the angle between k and k_1. This means that the light leaving the anisotropic crystal consists of two waves with a path difference due to the birefringence and a difference in amplitudes due to the rotation of the vibration plane (Fig. 5-20). Taking this fact into account both waves

may be composed again by vectorial addition. It has to be considered, however, that both waves originally vibrate perpendicular to one another and propagate separately but can be described physically as elliptically, circularly or plane polarized light suitable for vector analysis of wave motion and propagation. The polarization modes of the resulting wave as a function of the path difference are listed in Table 5-4 for $\varphi = 45°$. A more detailed study on the vector analysis of the polarization modes for arbitrary φ is presented by Wahlstrom (1979).

At the analyzer, the waves coming from the crystal may interfere since they can be considered to be coherent and in the same plane. The projection of the components

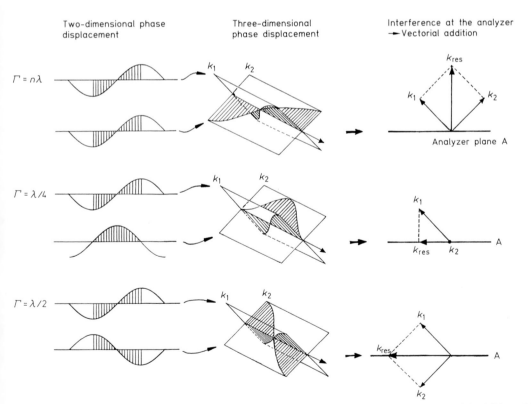

Figure 5-20. Linear displacement of oscillating perpendicular plane polarized waves and vectorial addition of k at the analyzer plane: Unlike linear interference, extinction is obtained at $\Gamma = n\lambda$ and maximum intensity at $\Gamma = \lambda/2$.

Table 5-4. Polarization modes of waves leaving an anisotropic crystal. (Assumptions: Incident beam is vertically plane polarized, principal axes of indicatrix cross-section are inclined 45° relative to polarizer.)

Γ	Polarization mode
$m\,\lambda,$ $m = 0, 1, 2, 3 \ldots$	vertically plane polarized
$(m + \frac{1}{8})\lambda$	elliptically polarized, clockwise rotating
$(m + \frac{1}{4})\lambda$	circulary polarized, clockwise rotating
$(m + \frac{3}{8})\lambda$	elliptically polarized, clockwise rotating
$(m + \frac{1}{2})\lambda$	horizontally plane polarized
$(m + \frac{5}{8})\lambda$	elliptically polarized, counterclockwise rotating
$(m + \frac{3}{4})\lambda$	circularly polarized, counterclockwise rotating
$(m + \frac{7}{8})\lambda$	elliptically polarized, counterclockwise rotating

onto the analyzer plane gives:

$$k_1^A = k_1 \sin \varphi = k \sin \varphi \cos \varphi \quad \text{and}$$

$$k_2^A = k_2 \cos \varphi = -k \sin \varphi \cos \varphi \qquad (5\text{-}56)$$

Since the vibration planes of polarizer and analyzer are mutual perpendicular, k_1^A and k_2^A are of the same length but of opposite direction, complete extinction would occur at the analyzer. But considering a phase difference δ between k_1 and k_2 behind the crystal as being

$$\delta = \frac{2\,\pi}{\lambda}\Gamma = \frac{2\,\pi}{\lambda}(n_{\gamma'} - n_{\alpha'})\,d \qquad (5\text{-}57)$$

(λ being the vacuum wavelength) and inserting for the intensity $I_i = |k_i|_{\max}^2$ gives

$$I^A = I_1 + I_2 + 2\sqrt{I_1 I_2}\cos \delta \qquad (5\text{-}58)$$

Combination of Eqs. (5-57) and (5-58) yields for the observed intensity behind the analyser the so-called Fresnel formula:

$$I^A = A^2 \sin^2 2\,\varphi \sin^2 \frac{\delta}{2} \qquad (5\text{-}59)$$

The intensity thus depends on φ and δ resulting in extinction for $\sin(\delta/2) = 0$ and for $\varphi = 0$ or $\varphi = \pi/2$, i.e. if the vibration planes of the resulting waves behind the crystal are parallel to polarizer and analyzer. A maximum intensity is obtained at $\sin^2 2\,\varphi = 1$ and $\varphi = \pi/4$, i.e. in a 45° position (Fig. 5-21). Geometrically speaking, at the extinction position one of the principal axes of the indicatrix section is parallel to the polarizer and, thus, there is no beam splitting but the unaffected transmitted beam is hence completely blocked by the analyzer.

Allowing the observation of an intensity at $\varphi \neq 45°$ and varying Γ or δ, respectively, we obtain extinction of particular wavelengths λ at $\Gamma = m\,\lambda$ (i.e., $\delta = 2\,m\,\pi$, $m = 0, 1, 2 \ldots$) and maximum intensity of that wavelength for $\Gamma = (2\,m + 1)\,\lambda/2$ [i.e., $\delta = (2\,m + 1)\,\pi$] due to destructive and constructive interference.

The above-mentioned effects are very useful for the characterization and identification of translucent crystals. Since the four positions of extinction upon rotating the crystal on the stage about the incident beam yields the exact position of the prin-

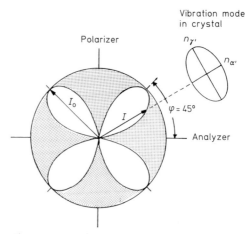

Figure 5-21. Pointer diagram of azimuth-dependent intensity of transmitted light upon rotation of a birefringent crystal between crossed polarizers.

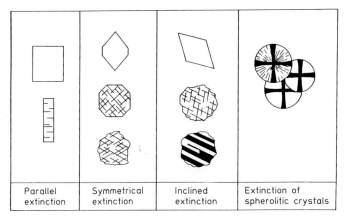

| Parallel extinction | Symmetrical extinction | Inclined extinction | Extinction of spherolitic crystals |

Figure 5-22. Rules of extinction yield evidence for crystal symmetry.

cipal axes of the indicatrix its orientation relative to the crystallographic axes can be determined if the crystal exhibits clear morphological features such as facets, edges, twins, cleavage planes, epitaxial inclusions or coatings. For systems of high symmetry such as the trigonal, tetragonal, hexagonal, and orthorhombic system, at least one characteristic mark should be parallel to the principal axes of the indicatrix (*straight extinction*). The case where only pyramidal facets are visible in these systems is called *symmetric extinction* (Fig. 5-22). In monoclinic and triclinic systems the indicatrix may be inclined relative to the crystal axes (*oblique extinction,* see also Sec. 5.2.3.2). The angle between crystal axes, e.g. *a*, and a principal axis, e.g. *X*, is known as the *angle of extinction* and listed as e.g. "$X \wedge a$". Since the orientation of the indicatrix is a very sensitive measure of the chemical composition of crystals the angle of extinction is widely used for the identification of members of a solid solution series e.g. in the mineral families of feldspars and pyroxenes. On growing eutectically from melts upon cooling, precipitates may change their chemical composition as a function of temperature and liquidus composition and hence consist of several zones of varying extinction angle. For example,

Fig. 1 in Color Table 5-2 shows a micrograph of a Ti-bearing augite crystal of volcanic origin under crossed polars with a zone structure clearly visible. Crystals with mosaic structure show diffuse extinction of particular segments called *undulating extinction*. If the crystal exhibits dispersion of the optical axes there is not complete extinction at one crystal position but only for a certain wavelength. Consequently, the complement colors become visible upon rotating the crystal giving a bluish-gray or dirty-brown impression known as *abnormal extinction*.

A cross of *extinction* becomes visible if botryoidal aggregates or clusters of acicular crystals are present e.g. as in crystallized glasses and glass ceramics. As several crystals of these aggregates are always in an extinction position, the dark cross with the branches parallel to the polarizer and analyzer does not disappear on rotating the specimen. Aggregates which never show extinction at any position may consist of many very small particles which overlap and thus cause an accumulated birefringence called *aggregate polarization*.

The fact that the path difference Γ is directly proportional to both crystal thickness *d* and birefringence Δ can be utilized for measuring thicknesses if the optical

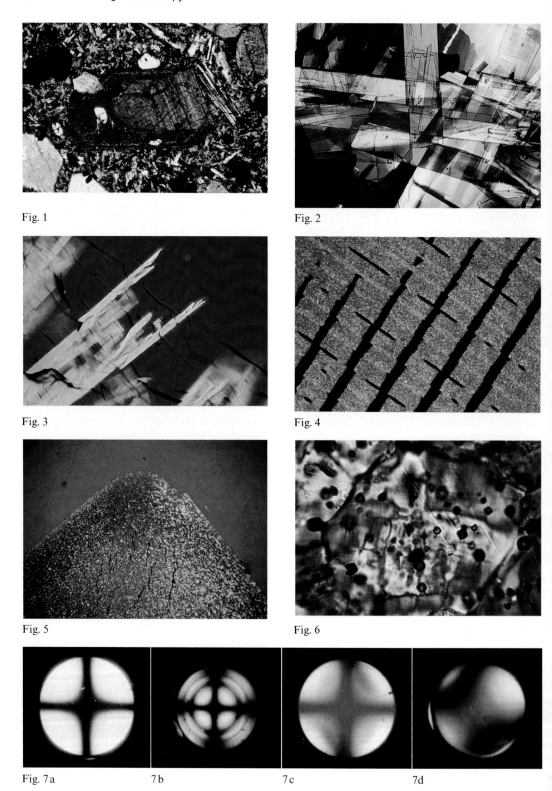

Fig. 1

Fig. 2

Fig. 3

Fig. 4

Fig. 5

Fig. 6

Fig. 7a   7b   7c   7d

Color Table 5-2. Microscopy in transmitted light.

Figure 1. Titano-augite crystal in volcanic rock with zone structure due to inclined extinction of varying angle (changes in chemical composition during grain growth); + polarizers.

Figure 2. Overlapping crystals of MoO_2; note the changes in interference colors due to orientation and thickness; + polarizers.

Figure 3. Cordierite crystals growing into a glassy phase and causing cracks due to changes in specific volume; note negative character of elongation; + polarizers, compensator 551 nm.

Figure 4. Layered texture of tape-casted, alumina-mullite, multi-layer substrate with opaque Mo-conductor lines; + polarizers, compensator 551 nm.

Figure 5. Edge of an injection-molded alumina bar; note change in interference colors due to fluidal texture; + polarizers, compensator 551 nm.

Figure 7. Conoscopic figures: (a) α-SiC (hexagonal, uniaxial), crystal from thin section, (b) α-SiC, thick single crystal, note isochromate fringes, (c) α-SiC as in (a) with compensator superimposed (optically positive), (d) titano-augite (biaxial) with dispersion of optical axes ($r < v$).

Figure 6. Strain-induced birefringence in cubic Mg-partially stabilized zirconia due to undissolved MgO-particles and pores (dark spots). Note that colors of second order can be generated; diagonal bright lines are deformation patterns; the fine perpendicular network of striations is due to epitaxially intergrown tetragonal ZrO_2 precipitates; + polarizers.

constants of the crystal and its orientation are known, i.e. if Δ' is given or, on the other hand, for the recognition of unknown phases, if the thickness is known. As Γ results in extinction at a particular wavelength λ at $\Gamma = m\lambda$ and maximum intensity at $\Gamma = (2m+1)\lambda/2$, at a given thickness d certain wavelengths from the incident set of white light are extinguished or weakened. The resulting complement colors known as *interference colors* are then a very sensitive function of the thickness and birefringence, which are listed in the so-called *Michel-Lévy Chart* which is usually printed as a colored table correlating path differences in nanometers, thicknesses in millimeters or micrometers and extents of birefringence (Fig. 5-23). This is also the reason why the thickness of a sample in geoscience is normalized to $20-30\ \mu m$,

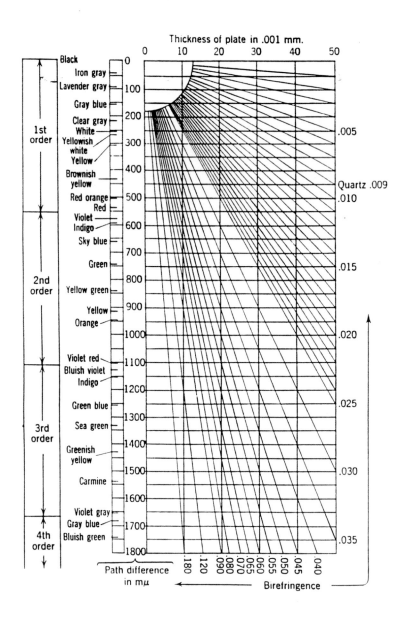

Figure 5-23. Michel-Lévy Chart of interference colors.

which makes the identification of minerals much easier.

Since the superposition of sinusoidal intensity fluctuations for distinct wavelengths yields a periodic change of interference colors with a repeatedly appearing purple-red coloration for whole numbers of 551 nm path difference, the succession of interference colors may be arranged in several *orders*. The first order starts with $\Gamma = 0$, i.e. $d = 0$ or $\Delta' = 0$, respectively, i.e. black, then dark gray (40 nm), lavender gray (97 nm), greenish white (234 nm), yellow (306 nm), orange (470 nm), red (536 nm), and purple-red (551 nm). The second order series which consists of violet, indigo, blue, green, yellow, orange and red has basically the same sequence but is clearer and brighter, whereas at the higher orders the color separation becomes weaker and weaker due to the irregular mixing of wavelengths, path differences and luminosities. Finally, a so-called *white of the higher order* is visible.

Since with some experience the interference colors may be used for the identification of chemical compounds and their orientation in multiphase materials the thickness of sample material or thin sections has to be defined. In geosciences, this definition is based on the most common mineral, quartz, which has to be thinned to a grayish appearance yielding, more or less, exactly 20 µm thickness. The color impression may vary very strongly if crystallites of smaller diameter than the sample thickness overlap and hence may create additive or subtractive interference. This implies that with for mechanical reasons decreasing grain size an examination of distinct particles in ceramic materials gets more and more difficult. Hence, the thickness of sections has to be adapted for the information required, recognizing that with decreasing thickness the intensity and luminosity of interference colors decreases too.

Many possibilities have been derived for the determination of optical constants and crystal orientations from the correlation between path difference and birefringence, and many optical accessories have been developed for quantitative measurements. The following paragraph deals with the most important types of *compensators* and points out some practical applications.

Generally, all optical accessories are inserted in the tube at a slot provided in a 45°-position relative to polarizer and analyzer (i.e., usually "northwest-southeast" relative to the eye-field) in order to gain maximum intensity, Eq. (5-59). In principle, compensators add or subtract a known path difference to that caused by the birefringence of the sample and hence generate variations of the interference colors. An increasing path difference (addition) is accomplished if the sample crystal and the compensator crystal are oriented in such a way that the vibration planes of both fast and slow rays of the sample (index s) and the compensator (index c) are parallel ($n_{\alpha's} \parallel n_{\alpha c}$ and $n_{\gamma's} \parallel n_{\gamma c}$) and just work like a thicker sample, whereas the opposite orientation ($n_{\gamma's} \parallel n_{\alpha c}$ and $n_{\alpha's} \parallel n_{\gamma c}$) diminishes the path difference (subtraction, *compensation, retardation*). Figure 5-24 illustrates geometrically how retardation is obtained by retarding the faster ray and accelerating the slower ray of the sample within the compensator. To understand the observed phenomena quantitatively the vibration direction of n_γ and the path difference are marked on the compensator. Figure 2 in Color Table 5-2 represents an array of molybdenum oxide crystals showing additive interference colors with thickness increasing towards the center of the particles, and subtractive colors at the overlapping areas of perpendicular columns.

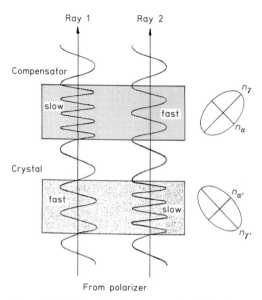

Figure 5-24. Principle of compensation: If the fast ray in the crystal travels along the slow direction of the compensator, and, alternatively, if the slow ray of the crystal passes along the fast direction of the compensator, retardation occurs.

The most common compensator is a special cut of a gypsum crystal creating a path difference of 551 nm (*"red of 1st order"*). This color is extremely sensitive to slight variations in wavelength changing immediately to blue and other colors of higher orders if a path difference is added, or to yellow or gray if a path difference is retarded. That is why it is also known as *sensitive tint plate.* The gypsum plate can be used for the determination of the *optical character* by examination of the *character of elongation l* or by means of conoscopy, as explained later. The character of elongation is attributed to crystallites which possess a preferred growth direction such as platelets or needles, columns and fibers. Rotating the significant growth direction, or usually a section thereof, to a 45°-position, i.e. aligned "northeast-southwest" relative to the eye-field, and inserting the compensator *blue* interference colors indi-

cate a *positive* character (addition) and *yellowish colors* a *negative* character of elongation (subtraction). In the case of uniaxial crystal systems (trigonal-rhombohedral, tetragonal, hexagonal) growing in the direction of the c-axis (acicular crystallites) the character of elongation is equal to the optical character, whereas it is of opposite sign for platelet particulates. In other systems the character of elongation can only be correlated to the optical character if the orientation of the indicatrix relative to the crystallographic axes is known. Optical constant tables usually give data on this property such as $l = (+)$ or $(-)$, and, as an example for a monoclinic or triclinic crystal, $l \wedge [001] = 7°$. Figure 3 in Color Table 5-2 shows an example of a cordierite crystal growing into a glassy phase of similar-composition. The character of elongation is negative.

Another accessory for determining the optical character and fast or slow transmission directions in a sample is the quarter-wave plate (λ-*quarter plate*), which is usually made from mica, quartz or birefringent plastics which gives a path difference of $\lambda/4$ for a certain wavelength, usually for sodium light. Monochromatic elliptically polarized light is thus converted into plane polarized light. It can thus be used for the very sensitive compensation of all wavelengths, except for sodium light, as well as for the quantitative measurement of very small path differences if it is combined with a precisely rotatable analyzer (*Senarmont Method,* see Wahlstrom, 1979). Furthermore, the polarization mode of optically active crystals can be determined. In a similar way λ-*plates* are crystal plates which are able to extinguish a particular λ from the white light according to Eq. (5-57) and are used as very sensitive filters. For example the two Na_D-lines can be separated by a 30 mm quartz crystal.

If the preferred growth direction of the particular phase is known compensation with a sensitive tint plate may be easily used for *texture analysis* of ceramic parts. Parallel crystallites forming areas of the same interference colours reveal characteristic microstructural features introduced by the fabrication technique. As an example, axial hot pressing of fibers (whiskers) or platelets causes a layered structure, but inhomogeneities due to friction between the powder filling and the die wall on moving the plungers are made visible. Textures formed during slip casting, tape casting, injection moulding and ram or auger extrusion are likely to be seen (Clinton et al., 1986). Figure 4 in Color Table 5-2 shows a sintered tape-cast, alumina-mullite, multilayer chip carrier with punched and molybdenum-paste filled conductor lines. Figure 5 in Color Table 5-2 presents a cross-section of an injection molded alumina bar in the vicinity of an edge; note the change in color due to the variation in flux directions. Counting the number of yellow or blue particles or measuring the covered areas, respectively, the extent of texture formation is obtained quantitatively and orientation maps can be designed. For a uniform random distribution the number of particles of both blue and yellow colors should be equal. The use of a universal stage makes three-dimensional analysis possible which can be plotted as a pole diagram.

Other types of compensators create a variable but exactly calibratable path difference. A widely used accessory of this kind is the quartz wedge (*Soleil compensator*) which yields a thickness-dependent change in Γ. The disadvantage is that very low path differences cannot readily be observed because of the difficulty of preparing a thin edge. The problem was solved by the *Soleil-Babinet compensator* which uses two displaceable quartz wedges of different crystal orientation which yield a zero line due to subtractive interference. The *Berek compensator* generates a variable path difference by being tiltable about a principal axis of the indicatrix. Thus, path differences may be adjusted due to variations in thickness relative to the transmitted beam, as well as due to the change of the indicatrix orientation.

Observation of virtual images of anisotropic crystals using a convergent beam at crossed polars, also known as interference figure microscopy (*conoscopy*), reveals information on the number of optical axes, the optical axes angle and the optical character of a crystal. Inserting the condenser, focusing the object at a high magnification ($1000 \times$, large aperture angle) and inserting the Amici-Bertrand lens produces an interference figure which results from the incident light forming a cone above the condenser lens. The clearest and most informative figures are obtained in uniaxial systems if the crystal is observed in parallel, or almost in parallel, to the optical axis, or in biaxial systems if observed in parallel to the acute bisectrix. These *conoscopic figures* or *optical-axis figures* consist of two systems of extinction patterns called *isogyres* and *isochromates*. In monochromatic light both systems are black whereas in white light the isochromates show the same color sequences as a quartz wedge. The isogyres refer to areas of equal vibration directions being parallel to polarizer and analyzer and hence being extinguished. Since the crystal is watched simultaneously from many directions, depending on the aperture angle ϑ the transmitted beams travel along different distances causing a particular path difference Γ_ϑ for all ϑ of

$$\Gamma_\vartheta = n_\vartheta \lambda = d \Delta_\vartheta / \cos \vartheta \qquad (5\text{-}60)$$

where n_ϑ is the number of observed isochromates and $\varDelta_\vartheta$ is the particular birefringence of the transmission direction ϑ. The isochromates thus represent figures of constant path differences $\varGamma_\vartheta$ which appear bright in monochromatic light at $\varGamma_\vartheta = \lambda/2$, $3\lambda/2$, $5\lambda/2$... and black at $\varGamma_\vartheta = 0$, λ, 2λ, 3λ.... For uniaxial crystallographic systems a view parallel to the optical axis shows a cross-shaped isogyre with the branches parallel to polarizer and analyzer. This can be explained by considering the vibration directions of an ordinary beam oscillating tangentially and an extraordinary beam oscillating radially with reference to the optical-axis figure (Fig. 5-25; Fig. 7a, b in Color Table 5-2). The isochromates form concentric rings of narrower interspacings and weaker intensities with increasing distance from the center. The center of the figure also referred to as the *melatope* can be interpreted as the position of the optical axis. If the optical axis is exactly adjusted to the axis of the microscope the interference figure does not change upon rotating the stage since the uniaxial indicatrix is an ellipsoid of revolution (Fig. 5-9). A slight mismatch of the orientation makes the cross precesses until at coarser misorientations only single branches propagate through the eye-field.

Biaxial systems exhibit hyperbolic curves as isogyres if the sample is observed parallel to the acute bisectrix. The vertices of the hyperbolas (melatopes) again represent the intersection of the optical axes and their distance corresponds to the optical axes angle $2V_Z$. Starting from a uniaxial conoscopic figure the formation of distinct curves in the biaxial system can be explained by a separation of pairs of adjacent branches of the cross due to the splitting of the axis. Biaxial crystals with a very small optical axis angle $2V_Z$ hence exhibit a conoscopic figure similar to that of a uniaxial crystal. The isochromates form so-called *Cassinian curves* with mirror symmetry in respect to the optical normal plane. Upon rotating a well-centered acute bisectrix the hyperbolas join forming a cross with branches parallel to the polarizer and analyzer if the optical axes plane is also parallel to the polars, and they swing out to the largest separation distance if the optical axes plane is oriented 45° (Fig. 5-26). With increasingly decentralized orientation only single hyperbolic curves are visible with curvatures dependent on the optical axis angle. If the transmitted light propagates parallel to the obtuse bisectrix diffuse isogyre branches may propagate through the eye-field upon rotating the

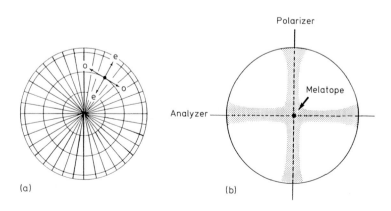

(a)

(b)

Polarizer

Analyzer

Melatope

Figure 5-25. Optical axes figures, unaxial case: (a) vibration directions of n_o and n_e, (b) conoscopic figure.

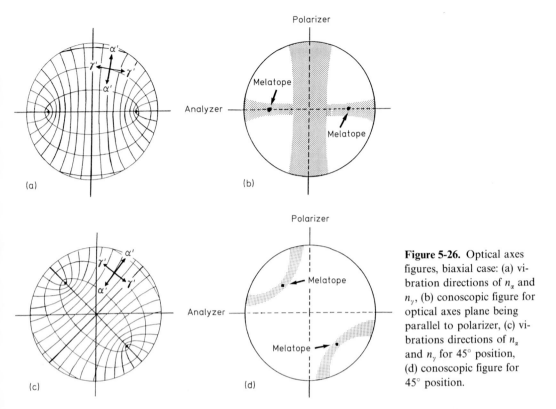

Figure 5-26. Optical axes figures, biaxial case: (a) vibration directions of n_α and n_γ, (b) conoscopic figure for optical axes plane being parallel to polarizer, (c) vibrations directions of n_α and n_γ for 45° position, (d) conoscopic figure for 45° position.

stage, but derivations from these figures can only be made after considerable experience.

Generally, for the quantitative explanation of conoscopic figures a model was developed by Becke (1905 and 1909) called the *Skiodrome method* which considers the three-dimensional orientation of the vibration planes of the rays propagating through a crystal as a function of the angle of incidence and in respect to the position of crossed polars (see also Rosenbusch and Wülfling, 1921).

The optical axes figures are very useful for the identification of transparent matter by (i) determination of the symmetry of a crystal, (ii) defining the orientation of the indicatrix in respect to the crystal axes, (iii) measuring the optical axes angle $2 V_Z$ of the acute bisectrix, (iv) determination of dispersion and (v) the optical character, as described in the following paragraphs.

Since all trigonal-rhombohedral, tetragonal and hexagonal systems show optical uniaxiality the orientation of the c-axis, being the optical axis, relative to the crystal morphology can easily be characterized. In monoclinic crystals, the optical normal, i.e. the Y-direction perpendicular to the optical axes plane yields the direction of the crystallographic b-axis. Searching large areas using conoscopy gives information of texture which can be quantified if the orientations of the optical axis or the acute bisectrix are transferred to a pole diagram. A universal stage, rotatable in all space directions, may be a useful accessory for this task.

Considering the refraction of light emerging from the crystal, the optical axes angle

$2V$ can be calculated from the scalar distance d between the melatopes if the optical axes plane is oriented in a 45° position relative to analyzer and polarizer. This is of particular importance for the determination of members of a solid solution series because $2V$ is a sensitive measure of chemical concentration differences. Applying Snellius' law (Eq. 5-29) to the geometrical situation presented in Fig. 5-27 we obtain:

$$\varphi_i = V_{obs}; \quad \varphi_{refr} = V \quad \text{and thus}$$

$$\sin V = \frac{\sin V_{obs}}{n_\gamma} \qquad (5\text{-}61)$$

The limit for an observed $2V_{obs}$ is given by the numerical aperture of the objective lens which should thus be large. Then, measuring the scalar distance d between the melatopes using a ruler ocular, V_{obs} in air is calculated by:

$$\sin V_{obs} = d/(2M) \quad \text{or} \quad \sin V = \frac{d}{2M n_\gamma} \qquad (5\text{-}62)$$

with M being Mallard's constant which has to be ascertained for a particular microscope by measuring V_{obs} on a crystal with known V and n_γ according to Eq. (5-62). Formulas for the construction of tables or charts containing correlations between $2V$ and $2V_{obs}$ have been published by Winchell (1946) and Tobi (1946). An-

other method for the determination of $2V$ makes use of the curvature of the isogyres in the 45°-position. If the isogyre remains a straight line in all positions on rotation, then $2V$ equals 90°. As the curvature angle of the isogyre approaches 90° the optical angle approaches uniaxiality. The angles inbetween may be calculated for the particular refractive indices and numerical apertures. Optical axes figures can also be used for the determination of *strain-induced deformation of crystals structures* if the optical constants of the unconstrained crystal are known. The method may be preferentially applied to uniaxial systems since any distortion of the structure at an oblique angle to the *c*-axis results in biaxiality. The amount of axis splitting may be used to calculate the applied stress. For this procedure, however, the elasto-optical tensor must be known. An example for strain-induced double refraction is shown in Fig. 6 in Color Table 5-2. Cubic and thus isotropic Mg-stabilized zirconia is constrained by MgO-inclusions producing interference color of a higher order in the vicinity of those particles (King, 1971). Note also the deformation lamellae (light diagonal lines) and the fine perpendicular network of striations due to epitaxial tetragonal zirconia precipitates.

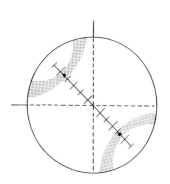

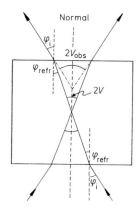

Figure 5-27. Determination of the acute bisectrix angle $2V_\gamma$ by measuring the apparent distance of the melatopes $2V_{obs}$ (after Wahlstrom, 1979).

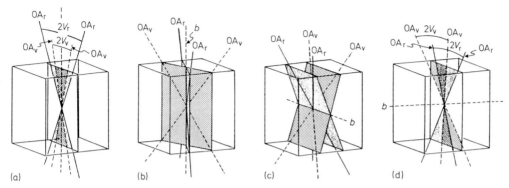

Figure 5-28. Dispersion of the optical axes: (a) symmetrical dispersion, (b) crossed dispersion, (c) horizontal dispersion, (d) inclined dispersion.

As already described in Sec. 5.2.3.2 some materials exhibit pronounced *dispersion* i.e. wavelength-dependent deformation of the indicatrix. This means that the above-mentioned statements are valid only for monochromatic light. In the case where $2V$, the position of the acute bisectrix or even the position of the optical axes plane varies, so-called *abnormal* optical axes figures are observed in biaxial systems, whereas no change occurs in uniaxial systems since the indicatrices for all λ possess the same axis of revolution. Remembering the fact that isogyres are areas of extinction of certain waves vibrating in parallel to polarizer and analyzer and taking into account that the melatopes may change their position as a function of λ, it is obvious that in white light illumination dispersion in biaxial systems must reveal isogyre fringes of complement colors. For *symmetrical dispersion* in orthorhombic crystals i.e. for a variation of $2V$ combined with a fixed position of the acute bisectrix, the isogyre curves exhibit reddish-brown and bluish-gray margins if rotated to a 45°-position. If the red coloration appears on the margin close to the center of the eye-field, extinction of violet light occurs at that position, whereas red light is extinguished on the other margins, apart from the center,

producing bluish impressions as the complement color. This fact means that the optical angle for red light $2V_{\mathrm{red}}$ is larger than that for violet light $2V_{\mathrm{violet}}$, i.e., $r > v$. On the other hand, the opposite effect is observed for $r < v$ (Fig. 5-28). The so-called *crossed dispersion, horizontal dispersion* and *inclined dispersion* are special features of monoclinic crystals and are described in Sec. 5.2.3.2. The conoscopic figures are shown schematically in Fig. 5-27 for the parallel and 45°-positions, respectively. Figure 7d in Color Table 5-2 presents an example of a conoscopic figure of a titano-augite crystal with $r < v$ (same sample as in Fig. 1).

Recognition of the *optical character* by conoscopy makes use of a sensitive tint plate (gypsum 1st order red plate) inserted in a 45°-position. In principle, the method is based on the retardation and enhancement of optical path differences producing a very sensitive change in coloration, as described above for the determination of the character of elongation. In *uniaxial systems*, optical character means the relation between the wave velocity and the vibration planes of ordinary and extraordinary rays, that is, to know whether the extraordinary beam is the *faster* one (*positive* character) or the *slower* one (*negative* character). Light emerging at any point of the crystal in conoscopic view now consists of two components, the extraordinary beam e containing waves vibrating in a plane having a radial line as the trace, and the ordinary component o containing waves vibrating in a plane the trace of which is tangential (perpendicular to e). Note that in the 45°-position the traces of the extraordinary (radial) components in the 2nd and 4th quadrants are parallel to the vibration direction of the (tangential) ordinary beam in the 1st and 3rd quadrants, and vice versa (Figs. 5-25 and 5-29). Thus in optically positive crystals the vibration direction of the faster extraordinary ray in the 1st and 3rd quadrants is parallel to the faster direction of the compensator plate, whereas the slower ordinary ray oscillates in parallel to the slower direction of the compensator which produces additive interference colors (blue) in the 1st and 3rd quadrants close to the isogyre cross. Figure 7c in Color Table 5-2 shows an optically positive conoscopic figure of an α-SiC-crystal with compensator superimposed. Likewise, in the 2nd and 4th quadrants, faster and slower beams of the crystal vibrate perpendicular to the faster and slower directions of the sensitive tint yielding subtractive interference colors (orange-red) close to the isogyre center. In negative crystals the effects are opposite to

those obtained in positive crystals. In both cases the isogyre becomes purple since a path difference of 1st order red is introduced. Consequently, it is only important to know which quadrant is actually visible even if the conoscopic figure is drastically off center. If crystals of high birefringence are observed it is more suitable to use a quartz wedge as a compensator. On pushing the quartz wedge into the slot a positive crystal makes the concentric isochromates move closer to the center of the 1st and 3rd quadrants whereas they move outside in the 2nd and 4th quadrant. On removing the quartz wedge the opposite behavior is observed.

In *biaxial crystals* an acute bisectrix figure is required for the determination of the optical character. A crystal is called *positive* if the Z-axis of the indicatrix, being the vibration direction of n_γ, is the *acute bisectrix*. In a negative crystal, X, being the vibration direction of the n_α, is the acute bisectrix. Thus in positive crystals of 45°-position, i.e. with the optical normal running northeast-southwest or, on the other hand, the optical axes plane running northwest-southeast, an enhancement of the path difference at the convex margin of the isogyre hyperbolas yielding a spot of blue interference colors, is observed, whereas the concave margin exhibits retardation and hence an orange-red spot

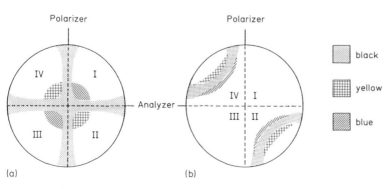

Figure 5-29. Recognition of optical character: (a) positive uniaxial crystal, (b) positive biaxial crystal.

(Fig. 5-29). In the same orientation, a negative crystal behaves in an opposite way showing orange-red colors at the convex side and blue spots on the concave side of both hyperbolas.

In conclusion, the *identification* of a transparent material by polarization microscopy in transmitted light requires the following *procedure:*

1. Check for clearly visible morphological features such as facets, cleavage, aligned inclusions (without analyzer) and twins (with crossed polarizers).
2. Check birefringence under crossed polarizers:
   - no birefringence: crystal may either be *isotropic* or oriented perpendicular to an optical axis, continue at 4.
   - birefringence: estimate extent of birefringence, continue at 3.
3. Check character of extinction:
   - symmetric extinction in three space directions: crystal is *orthorhombic*.
   - symmetric extinction in one direction: crystal is *monoclinic*; measure angle of extinction.
   - only oblique extinction: crystal is *triclinic;* measure angle of extinction continue at 5.
4. Check conoscopic figure:
   - no optical axis figure visible: material is isotropic (*cubic or amorphous*).
   - if optical axis figure available continue at 5.
5. Check optical axes:
   - uniaxial: crystal is either *trigonal, tetragonal or hexagonal*.
   - biaxial: check optical angle $2V$ and dispersion.
   - check optical character.
6. If possible: Measure refractive indices accurately. Check color and pleochroism with one polarizer.
7. Look up in tables for identification.

Since advanced ceramics exhibit small grain sizes, usually in the range $0.5-10$ µm, a complete optical analysis may be difficult because of overlapping crystallites. But generally, thin sections of most of the commercially available materials provide valuable information on grain growth via liquid phases, residual sintering additives (e.g., opaque carbon in SiC-ceramics), pores, cracks, textures, and other microstructural features which cannot easily be recognized by reflected beam techniques. Table 5-5 gives some optical data for transparent ceramic materials as far as is yet available.

5.4 Sample Preparation Techniques

Sampling, sectioning and surface preparation are the most important steps for the physical and microscopic characterization of materials. They strongly influence the value of the information obtained and the conclusions are usually transferred from a small specimen to a large part. First of all, it is critical that the collected specimen represents the average features of the larger sample (*average sampling*) or contains all of the information necessary for observation of a specific detail of interest (*defined sampling*). This is particularly important for the characterization of single defects such as cracks and origins of fracture, inhomogeneities of all kinds e.g. inclusions, pores, chemical and physical gradients, areas of mechanical deformation etc. For averaging, either a large number of specimens has to be analyzed, which were selected from well-distributed sites in the large part, or large samples have to be examined. In the case of defined sampling it has to be ascertained that the specimen also represents, in part, the average microstructure in order to clarify the rela-

Table 5-5. Optical properties of ceramic materials.

Phase Formula name	Angle of optical axes	Optical character	Indices of refraction			Reflectivity (%)	Symmetry	Cleavage
			n_α	n_γ	Δ			
Oxides and silicates								
Al_2O_3								
Corundum	0	(−)	1.765	1.770	0.008	5.7−6.7	trigonal	(0001)
$Al_6Si_2O_{13}$			1.642−	1.654−	0.012−	6.0−6.2	ortho-rhombic	(010)
Mullite	20−50	(+)	1.653	1.679	0.031			
MgO								
Periclase	isotropic		1.73		0	7.2	cubic	?
$2MgO \cdot 2Al_2O_3 \cdot 5SiO_2$								
Cordierite	0−90	(+) to (−)	1.524	1.574	0.05− 0.018	4.2	ortho-rhombic	(010)
Indialithe (high-temperature cordierite)	0	(−)	1.528	1.539	0.004	4.2	hexagonal	none
TiO_2								
Anatase	0−5	(−)	2.56	2.63	0.07	18−19	tetragonal	(001)
TiO_2								
Brookite	var.	(−) to (+)	2.59	2.71	0.12	20−21.5	ortho-rhombic	none
TiO_2								
Rutile	0	(+)	2.61	2.90	0.12	21−25	tetragonal	(110)
m-ZrO_2	30	(−)	2.19− 2.24	2.20− 2.24	0.07	13−14	monoclinic	(001)
t-ZrO_2	0	(−)	*	*	*	13−14	tetragonal	(001)
c-ZrO_2	isotroipic		*		0	13−14	cubic	{111}
Y-TZP	0	*	<2.20	<2.24	*	*	tetragonal	?
Mg-PSZ	isotropic		≪2.20		0	*	cubic	?
$ZrSiO_4$								
Zircon	0−2	(+)	1.92	1.96	0.04− 0.08	10−11	tetragonal	none
$CaTiO_3$								
Perovskite	isotropic		2.38		0	16−17	cubic	none
$SrTiO_3$	isotropic		2.35		0	16.7	cubic	none
$BaTiO_3$	isotropic		2.34	2.48	0	16.5−17.2	cubic	none
Carbides								
C								
Diamond	isotropic		2.40−2.43		0	17	cubic	{111}
Graphite	0	?	1.1	?	?	5−25	hexagonal	(0001)
α-SiC	0	(+)	2.65	2.69	0.04	19−21	hexagonal	none
β-SiC	isotropic		2.65		0	19−21	cubic	none
B_4C	0	(+)	>2.1	>2.2	0.1	15−20	trigonal	none
Nitrides								
α-Si_3N_4	0	(+)	2.05	2.11	0.06	11.8−12.7	hexagonal	none
β-Si_3N_4	0	(+)	2.05	2.11	0.06	11.8−12.7	hexagonal	none

Table 5-5 (Continued).

Phase Formula name	Angle of optical axes	Optical char- acter	Indices of refraction			Reflec- tivity (%)	Symmetry	Cleav- age
			n_α	n_γ	Δ			
Si_2N_2O								
Sinoite	2–10	(−)	1.74	1.85	0.11	8.2– 8.8	ortho- rhombic	none
AlN	0	(+)	2.11	2.18	0.07	13–14	hexagonal	none
h-BN	0	?	1.71	1.72	0.02	7.5	hexagonal	(0001)
c-BN	isotropic		2.5?		0	?	cubic	{110}

* Property depends upon dopants.

tionship between the particular point of interest and the unchanged environment. Furthermore, the exact position and orientation of the specimen relative to the part should be reconstructable.

Another problem is the influence of the preparation method on the microstructure. Incorrect techniques may introduce damage by hot and cold working. Material transfer from the cutting or grinding wheel to the treated surface may also occur and be accompanied by tribochemical reactions. Damage by surface fatigue, e.g. initiation or opening of pre-existing cracks, particle pullout or stress-induced multiple

twinning and phase transformations, are an even more severe problem for brittle materials, such as ceramics, and hence may result in erroneous conclusions.

5.4.1 Sectioning

The kind and depth of damage introduced to a specimen depends on the mechanical and chemical properties of the treated material and on the cutting, grinding, and polishing processes. Figure 5-30 shows typical depths of deformation for several materials, and various cutting techniques (after Vander Voort, 1984). It is ob-

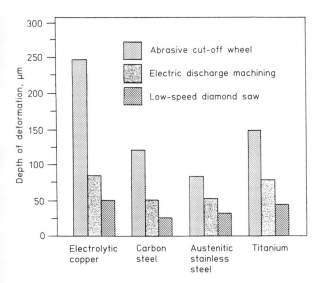

Figure 5-30. Method-dependent deformation depth in metal cutting (after Vander Voort, 1984).

vious that cutting with abrasives is the most severe process, followed by wire sawing and electric discharge machining because it introduces mechanical defects. Critical parameters of *abrasive wheel cutting* are normal and tangential forces affecting the sample which are functions of wheel speed and feed rate, size and concentration of abrasives, and amount and chemical composition of coolants. *Wire sawing* with bonded or slurry-dispersed diamond, boron carbide, silicon carbide, corundum or garnet grit is mostly applied for the sectioning of single crystals, e.g. of semiconductors, but is also suitable for cutting parts of complex geometry or composition e.g. catalyst carriers, electronic devices, fiber-reinforced materials such as ceramics, glasses and polymers (McLaughlin, 1974).

The applicability of *electric discharge machining* is limited to electrically conductive materials or composites having a minor amount of non-conducting materials. It is widely used for shaping hard metals, metal single crystals and is also applied to ceramics (Barrett, 1974). The sample, as the anode, is submerged in a dielectric liquid (kerosene or oil), and a continuously wound copper wire, as the cathode, is moved close enough to the workpiece to initiate a glow discharge. The removal process is due to particle erosion by partial melting, local thermal shock and impact by shock wave fronts and collapsing cavitation bubbles moving in the liquid (Petrofes and Gadalla, 1988). This operation avoids the direct introduction of mechanical stresses to the material but severe thermal shock damages and chemical alterations at the surface may occur due to the locally high temperatures. The depth of damage can extend to several hundred micrometers. Similar considerations are valid for *laser cutting* which is also based

on local evaporation and partial melting. In the case of materials with low thermal conductivity and low fracture toughness (e.g., glasses and ceramics) the process is accompanied by considerable thermal shock. *Ultrasonic erosion* of parts is based on particle pullout by continuous multiple impacts due to sonotrode-induced vibrations of abrasive particles of boron carbide or silicon carbide. This method is suitable for work hardened metals, gray cast iron or glasses and ceramics. The operation is mostly used for the preparation of complex shapes but is also applicable instead of diamond core drilling or cutting. For brittle and hard materials it is one of the softer methods since the depth of damage depends upon the abrasive grit size but does not usually exceed the average grain size.

Probably the most careful method for cutting metals or ionically bonded non-metals with almost no surface damage is *chemical* and *electrochemical sectioning*. Sawing and milling is carried out with acid infiltrated threads or ropes or with acid jets of 150 μm width. The procedure is usually electrically assisted by using the thread or nozzle as a cathode and the sample as an anode. These treatments take rather a long time and are thus limited to the cutting of single crystals or thin sections.

Microtome cutting can be successfully used for resin imbedded powder specimens for optical control of agglomerate size.

5.4.2 Mounting

Mounting of samples is usually necessary to facilitate the handling of the specimen during the subsequent operation as well as to guarantee a defined positioning during examination. Furthermore, damage to both the sample and the metallographic equipment is avoided if the material is conveniently shaped and sized.

The mounting material should not influence the sample as a result of chemical reaction or mechanical stresses. It should exhibit excellent adhesion to the specimen, satisfactory strength, hardness and chemical inertness against dissolving and etching agents. If X-ray or scanning electron microscopy investigations are additionally planned, the mounting material should be amorphous and electrically conducting. Electric conductivity is also required for electrical polishing and etching.

Favoured mounting materials are duroplastics such as phenolic resin, epoxy resin, polyester resins and acrylates which are either compacted hot under axial pressure or are mixed with a hardener for polymerization. Special polymer mixes with copper, iron or graphite powder are available for the preparation of conductive specimens. Materials with a high porosity or powder specimens are infiltrated by vacuum impregnation with epoxy resin.

5.4.3 Grinding, Lapping, Polishing

Surface layers which have been altered or damaged by cutting and sawing have to be removed carefully by subsequent grinding, lapping and polishing with decreasing grit size. *Mechanical grinding* is a microscale, chip-forming machining process in which every abrasive grain removes material from the sample by acting as a discrete cutting tool i.e. by scratching. The specific mechanisms of material removal depend on the mechanical properties of the sample (Fig. 5-31). In the grinding process, the abrasive material is cemented onto a tool – e.g., a grinding wheel or grinding paper – exposing a defined array of sharp cutting edges. On the other hand, in the *lapping* process abrasive particles are not

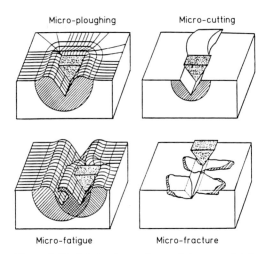

Figure 5-31. Mechanisms of material removal (after Zum Gahr, 1987).

firmly fixed but loosely attached to a soft-metal disk or a net-covered disk. Hence surface roughness from grinding steps is removed by the micro-impact of rolling abrasive particles. In a similar way, a final surface finish is achieved by plastic deformation and adhesive material pull-out during the *polishing* treament using rolling abrasive particles of decreasing grit size. Figure 5-32 shows a comparison of surface damage in metals and ceramics. In metallic materials the damage is mostly due to plastic deformation (formation and movement of dislocations) whereas ceramics react in a brittle manner with particle pull-out, i.e. transgranular or intergranular fracture due to the release of elastic stresses after impact by an abrasive particle. The formation of dislocations beneath the surface and superplastic deformation, probably due to sliding of grain boundaries and twin boundaries, or due to the activation of slip planes at high shear rates, has, however, also been observed in ceramic materials (Rice, 1974). Figure 5-33 shows grinding grooves in alumina indicating plastic deformation, combined with areas of pre-

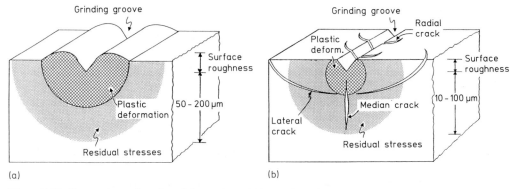

Figure 5-32. Comparison of surface deformation of (a) metals and (b) ceramics due to grinding.

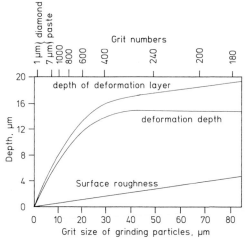

Figure 5-33. Plastic deformation and particle pull-out in diamond-ground alumina ceramic (coolant: water-oil suspension; courtesy Park).

Figure 5-34. Surface roughness and deformation depth of steel as a function of grit size of grinding medium (redrawn after Freund, 1968).

dominant particle-pullout. Generally, both roughness and depth of deformation should be decreased by a step-by-step lapping and polishing treatment with decreasing particle size. Figure 5-34 presents the dependence of the deformation depth upon the abrasive grit size for a steel sample (0.05 wt.% C, 0.15 wt.% Mn, 0.20 wt.% Si, after Lihl and Mayer, 1960). It should be noticed, however, that even careful hand-polishing of hard SiC with 0.25 μm diamond paste yields a totally amorphous surface layer of 5–20 nm thickness. Favourite grinding and polishing grits are listed in Table 5-6.

Mechanical damage of near-surface areas can be avoided by *electrolytic and chemical polishing*. Smoothing is accomplished by selective dissolution of hills and ridges causing roughness due to higher electric and chemical potentials in the vicinity of small radii of curvature. Brightening of surfaces is related to the elimination of microstructural irregularities of the order of 0.01 μm and the local suppression of etching due to the deposition of a thin passivation layer. Suitable electrolytes for anodic electro-polishing are diluted phosphoric, sulfuric or perchloric acid, acetic acid or alcohol. Since electrolytic polishing is limited to conducting materials it is only

Table 5-6. Grinding, lapping and polishing grits.

Material	Hardness in Mohs' scale	Application, Remarks
Pumice	<5	Mild grinding material for soft metals, e.g. for silver or gold
Diatomaceous earth	<5	Paste for pre-polishing treatment of brass and noble metals
Magnesia (Periclase, MgO)	5.5	Polishing of Mg, Al and related alloys
Cassiterite (SnO_2)	6–7	Polishing of sulfidic ores
Hematite, limonite (Polish red, Fe_2O_3-FeOOH)	5-6	Polishing of soft metals and alloys, final polishing of ceramics
Quartz (SiO_2)	7	Fine polishing of metals and ceramics
Chromia (Polish green, Cr_2O_3)	7–8	Polishing of hard metals, e.g. steels, ferrochrome
Smirgel (emery) (50–70% Al_2O_3, balance SiO_2, Fe_2O_3, Fe_3O_4, TiO_2)	7–8	Grinding and polishing
Alumina (γ- and α-Al_2O_3)	8	Universal polishing material, also applicable for softer ceramics, hard metals and composites
Natural corundum (90–95% Al_2O_3)	8	Coarse grinding of steel and brass
Electro-corundum Smelter corundum Noble corundum	8	Most important grinding material for steel, cast iron, sinter metals and hardened steel; used in disks, papers, as grits; various qualities available
Silicon carbide (Carborundum, SiC)	9–9.2	Important grinding, lapping and polishing material for universal use as disks, papers and grits
Boron carbide (Tetrabor, B_4C)	9.5	Grinding of hard metals, lapping of metals, ceramics and diamonds
Borazon (cubic boron nitride, CBN)	9.8–10	Grinding, honing and lapping of Al-Si alloys (piston alloys)
Diamond (cubic carbon, C)	10	Important grinding, lapping and polishing material for universal use as disks, papers and grits; especially applied for glass, ceramics and gemstones

applicable to certain ceramics such as transition metal borides, carbides and silicides. For example, boron carbide can be satisfactorily treated with a 1 wt.% aqueous solution of KOH or NaOH. *Chemical polishing* is also a very mild method for smoothing the surface. The agents are usually strongly oxidizing like nitric acid, sulfuric acid, chromic acid or hydrogen peroxide.

For some metals, high-quality surfaces can be obtained together with a significant time saving compared to mechanical methods. However, corners and edges are attacked more than for the average sample, and surface steps are smoothed but not flattened. Moreover, the treatment of composite materials is very difficult because the cathodic or anodic behavior of the different phases in respect to the matrix may vary and thus cause selective dissolution of certain components.

Combined techniques, such as alternating mechanical and electrolytic polishing or etch-polishing, give improved results in the case of soft or tough metals since deformation and reaction layers at the surface of the sample are easily removed. Moreover, refractory metals, multiphase alloys and coated materials can be satisfactorily treated. Etch-polishing is also applicable to some ceramics, e.g. grain boundary phases in AlN or Al_2O_3 become visible if NaOH is added to the abrasive suspension.

5.4.4 Etching

After the final polishing treatment examination of the sample in reflected light reveals the most information on microstructural flaws and inhomogeneities such as pores, pits, cracks, inclusions, impurities, and corrosion attack. Furthermore, differences in hardness of multiphase materials may be observed due to relief formation. Other microstructural features, such as grain boundaries, can sometimes only be exposed by etching. The term of etching in a metallographic sense is not usually limited to destructive methods based on chemical, electrochemical or physical at-

tack but also covers optical methods for making the true microstructure visible, such as special illumination and compensation techniques, phase contrasting or interference contrasting (Fig. 5-35). These techniques have been described in Sec. 5.3. The *destructive methods* of etching can be divided into electrochemical and physical techniques. Electrochemical etching, which may or may not be assisted by an external voltage source, is widely applied to metallic materials, whereas ceramics are reasonably etched by physical methods such as thermal etching, plasma etching etc.

Electrochemical etching results from reduction (cathodic reaction) and oxidation (anodic reaction) of metals in contact with chemical solvents. The extent of these reactions depends on the electrochemical potential of the metals relative to the standard potential of a reference electrode. Therefore, microstructural components of different composition are attacked at different rates. Furthermore, different chemical potentials also originate from crystal imperfections, such as grain boundaries, deformation zones and dislocations, or from local passivation or oxidation layers (*precipitation etching, color etching*) and concentration fluctuations in the electrolyte. Thus grain boundaries and etch pits become visible, and different phases can be distinguished by interference colors (Picklesheimer, 1967). This process may be assisted by applying an external voltage

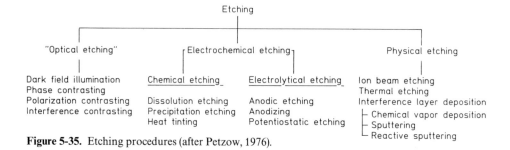

Figure 5-35. Etching procedures (after Petzow, 1976).

(*anodizing*, *potentiostatic etching*). Compositions of suitable etchants may be taken from Beraha and Shpigler, 1977; Vander Voort, 1984; Mills et al., 1985.

Oxide layers can also be generated by *heat tinting* (annealing of the sample in an oxidizing atmosphere) yielding a specific coloration of oxygen-sensitive metals and alloys (Beraha and Shpigler, 1977; Weck and Leistner, 1982 and 1983).

The application of *plasma etching* is currently limited to silicon nitride ceramics (O'Meara et al., 1986; Täffner et al., 1990). An induction-coupled plasma of O_2 and CF_4 containing fluor radicals is generated and reacts with Si_3N_4 forming SiF_4. Since glassy grain boundary phases are not attacked a clear relief is obtained, see Fig. 5-36.

Thermal etching makes use of thermally induced effects such as surface and grain boundary diffusion, coarsening, recrystallization, evaporation and recondensation which change the morphology of the surface. Since metallographic sections are randomly oriented relative to the truncated grain boundaries and the crystallographic orientation of the particles, surface tensions are generated between adjacent crystallites and the contacting vapor phase (air) which tend to establish equilibrium conditions, i.e. to minimize the surface energies. Consequently, grooves are formed at the grain boundaries by surface diffusion and selective evaporation and recondensation of matter towards areas of lower total energy. On the other hand, small defects such as pits or scratches heal. Hence, smoothed surfaces are obtained (Fig. 5-37). The required temperature for the initiation of thermal etching within a reasonable time is approximately half of the homologous temperature. In order to avoid microstructural changes during the process use of temperatures of $100-250\,^\circ C$

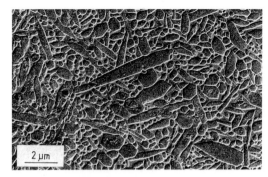

Figure 5-36. Microstructure of plasma-etched sialon ceramic (15% Al_2O_3/Y_2O_3 additive): note that the β-silicon nitride particles are attacked whereas the glassy phase resists (courtesy Krämer).

below the fabrication temperature is recommended.

Ion beam etching makes use of selective material dissociation of the specimen under ion bombardment. Electrically conductive samples are required. For ceramics, satisfactory results are reported for porcelain, Si-infiltrated SiC, and zirconia (Bierlein et al., 1958; Schlüter et al., 1980). The method is widely applied for surface cleaning and thinning of TEM samples.

5.4.5 Preparation of Thin Sections

Generally, all non-metallic phases are suitable for examination by transmission

Figure 5-37. Microstructure of thermally etched Al_2O_3-15 vol.% ZrO_2 ceramic. The white particles are zirconia.

light microscopy if they are prepared as transparent foils. Even boron carbide was successfully thinned to a transparent appearance although it exhibits a semi-metallic luster. Limitations result only from the particle size and the mechanical behavior of the material. Because of the wavelength-dependent resolution in light microscopy, the smallest particle size to be readily observed is of the order of 1 μm. For mechanical stability, however, a thickness of approximately 5–30 μm is more favorable for a thin section. Consequently, in very fine grained ceramics particles may overlap which creates difficulties in discrimination of certain optical effects. Hence a particle size greater than 5 μm is most suitable. In geosciences, the thickness of the thin section is normalized to 20–30 μm according to the interference colors of quartz. For the characterization of ceramics such a rule does not yet exist, thus the thickness may be chosen according to the particle size and the effects desired to be observed. In some cases, the thickness of section has to be adapted to the transparency of the material, e.g. boron carbide has to be sliced down to a thickness of approximately 5 μm.

The preparation of thin sections generally requires a stiff and very precise cut-off and polishing machine. It has to be considered that several square-centimeters of sample must be sliced, ground and polished to a film of a uniform overall thickness of 10–30 μm and a final surface roughness of less than 0.1 μm on both surfaces. In the first step, the samples are treated like polished sections. The ceramic is cut, embedded in epoxy resin, ground, lapped and polished. The finished surface is glued onto a glass plate (Fig. 5-38). Instead of an epoxy resin, Canadian Balsam is often used, especially if the index of refraction of the resin has to be known. Ca-

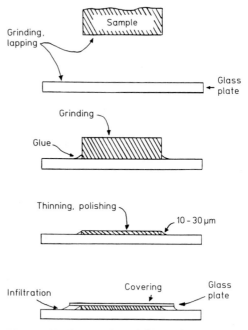

Figure 5-38. Preparation of thin sections.

nadian Balsam is, however, very brittle, must be handled hot and turns yellow after several years. Epoxy resins, on the other hand, tend to crystallize after several years of storage which can make working with polarized light difficult or even impossible. In order to avoid pores and bubbles between the sample and the sample carrier, gluing should be carried out in a vacuum infiltration chamber. If the glue gets hard the sample is cut to a residual thickness of approximately 500 μm, and ground to 80–100 μm which seems to be the appropriate thickness to avoid particle pull-out. The samples can then be lapped and polished automatically or by hand. Since the section is now approaching its final thickness it has to be treated very carefully and examined repeatedly under the microscope using polarized light. A uniform thickness is accomplished if all areas of the section show similar interference colors. In general, the final thickness is reached if all particles show gray, yellow-brown or blue-green color depending upon the birefringence. Wedge-shaped sections exhibit a dark-gray color at one corner and heavily lustering colors on the other one. As a last step, the sample is covered with a thin glass plate and infiltrated with diluted epoxy resin or Canadian Balsam. Without the glass cover the sample can also be utilized for studies in reflected beam as well as for microprobe analysis.

5.5 References

Allen, R. D. (1956), *American Mineralogist 41*, 245.

Barrett, J. (1974), in: *Metallographic Specimen Preparation – Optical and Electron Microscopy:* McCall, J. L., Mueller, W. M. (Eds.). New York: Plenum Press, pp. 69–76.

Bartz, G. (1973), *Praktische Metallographie 10*, 311.

Batsanov, S. S. (1961), *Refractometry and crystal structure.* New York: Consultants Bureau.

Becke, F. (1905), in: *Tschermaks Mineralogische und Petrographische Mitteilungen 24*, 1; (1909), *28*, 290.

Beraha, E., Shpigler, B. (1977), *Color Metallography:* Metals Park, OH: ASM.

Berek, M. (1931a), in: *Centralblatt für Mineralogie, Abt. A No. 6*, pp. 198–209.

Berek, M. (1931b), in: *Neues Jahrbuch für Mineralogie, Abhandlungen Beilage 64, Abteilung A*, pp. 123–136.

Berek, M. (1937), in: *Fortschritte der Mineralogie 22*, pp. 1–104.

Beyer, H. (1974), *Theorie und Praxis der Interferenzmikroskopie.* Leipzig (GDR).

Bierlein, T. K., Newkirk, H. W., Mastel, B. (1958), in: *J. Am. Ceram. Soc. 41*, p. 196.

Blumenkamp, H. J., Wallura, E., Hoven, H., Koizlik, K., Nickel, H. (1980), in: *Berichte der Kernforschungsanlage Jülich, Jül-1673.*

Cameron, E. N. (1961), *Ore microscopy.* New York: John Wiley and Sons.

Cathcart, J. V., Peterson, G. F., Sparks, C. J. (1967), in: *Surfaces and Interior Chemical and Physical Characteristics:* Burke, C. J., Reed, T. A., Weiss, G. A. (Eds.). Syracuse: University Press.

Clinton, D. J., Morell, R., McNamee, M. (1986), *Brit. Ceram. Trans. J. 85*, 175.

Freund, H. (Ed.) (1968), *Handbuch der Mikroskopie in der Technik*, 8 Volumes. Frankfurt/Main: Umschau-Verlag.

Galopin, R., Henry, N. F. M. (1972), *Microscopic Study of Opaque Minerals.* Cambridge: W. Heffer and Sons, Ltd.

Hofmann, S., Exner, H. E. (1974), *Zeitschrift für Metallkunde 65 (12)*, 778.

Jaffe, H. W. (1956), *American Mineralogist 41*, 757.

King, A. G. (1971), *J. Am. Ceram. Soc. 53 (7)*, 424.

Lihl, F., Mayer, H. (1960), *Zeitschrift für Metallkunde 51*, 186.

McConnell, D. (1967), *Geochim. et Cosmochim. Acta 39*, 1479.

McLaughlin, H. B. (1974), in: *Metallographic Specimen Preparation – Optical and Electron Microscopy:* McCall, J. L., Mueller, W. M. (Eds.). New York: Plenum Press, pp. 55–68.

Mills, K., Davis, J. R., Destefani, J. D., Dieterich, D. A., Crankovic, G. M., Frissell, H. J. (Eds.) (1985), *Metals Handbook 9th. ed., Vol. 9: Metallography and Microstructures.* Metals Park, OH: ASM International.

O'Meara, C., Nilsson, P., Dunlop, G. L. (1986), *Journal de Physique, Colloque C1, Supplement au n° 2, Tome 47*, C297.

Petrofes, N. F., Gadalla, A. M. (1988), *Ceramic Bulletin 67*, 1048.

Petzow, G. (1976), *Metallographisches Ätzen, Metallkundlich-Technische Reihe 1.* Stuttgart, Berlin: Gebr. Borntraeger, new edition in 1991/92.

Picklesimer, M. L. (1967), *Microscope 53*.

Rice, R. W. (1974), in: *Ceramics for High-Performance Applications:* Burke, J. J., Gorum, A. E., Katz, R. N. (Eds.). Chestnut Hill (MA): Brook Hill Publ. Co., pp. 287–343.

Rosenbusch, H., Wülfling, E. A. (1921/1924), *Mikroscopische Physiographie I, Part 1.* Stuttgart: Nägele und Obermiller.

Schlüter, P., Wallace, J. S., Elssner, G. (1980), *Sonderbände der Praktischen Metallographie 11.* Stuttgart: Dr. Riederer-Verlag, p. 291.

Täffner, U., Hoffmann, M. J., Krämer, M. (1990), *Z. Praktische Metallographie 27*, 385.

Tobi, A. C. (1946), *Am. Mineralogist 41*, 516.

Vander Voort, G. F. (1984), *Metallography: Principle and Practice.* New York: McGraw-Hill.

Wahlstrom, E. E. (1979), *Optical Crystallography, 5th. ed.* New York: John Wiley and Sons.

Weck, E., Leistner, E. (1982), *Metallographic Instructions for Colour Etching by Immersion – Part I: Klemm Colour Etching.* Düsseldorf: Deutscher Verlag für Schweißtechnik.

Weck, E., Leistner, E. (1983), *Metallographic Instructions for Colour Etching by Immersion – Part II: Beraha Colour Etchants and Their Different Variants.* Düsseldorf: Deutscher Verlag für Schweißtechnik.

West, C. D. (1936), *American Mineralogist 21*, 245.

Winchell, H. (1946), *American Mineralogist 31*, 43.

Zieler, H. W. (1969), *Microscope 17*, 249.

Zum Gahr, K. H. (1987), *Microstructure and Wear of Materials, Tribology Series 10.* Amsterdam: Elsevier.

General Reading

Bell, E. E. (1967), in: *Encyclopedia of Physics XXV/2a:* Flügge, S. (Ed.). Berlin, Göttingen, Heidelberg: Springer, pp. 1–58.

Blum, M. E., French, P. M., Vander Voort, G. F., (Eds.) (1987), *Field Metallography, Failure Analysis and Metallography, Microstructural Science 15.* Columbus (OH): The Int. Metallogr. Soc.; and Metals Park (OH): ASM International.

Cameron, E. N. (1957), *Economic Geology 45,* 719.

Cherkasov, Y. U. (1960), *International Geology Review 2,* 218.

Drude, P. (1929), *Theory of Optics.* London: Longman & Green.

Perryman, E. C. W. (1952), in: *Polarized Light in Metallography:* Conn, G. K., Bradshaw, F. J. (Eds.). London: Butterworths, pp. 70–89.

Ramachandran, G. N., Ramaseshan, S. (1961), in: *Encyclopedia of Physics XXV/1:* Flügge, S. (Ed.). Berlin, Göttingen, Heidelberg: Springer, pp. 1–217.

Schneiderhöhn, H. (1952), *Erzmikroskopisches Praktikum.* Stuttgart: E. Schweizerbart'sche Verlagsbuchhandlung.

Saenz, N. T. (1989), *Proc. ISM/ASM Annual Meeting.* Charlotte (SC).

Samuels, L. E. (1962), *Metallurgia 66 (396),* pp. 187.

Wilcox, R. E. (1956), *American Mineralogist 41,* 683.

Wright, F. E. (1919), *American Philosophical Soc., Transactions 58,* 401.

Zieler, H. W. (1974), *The Optical Performance of the Light Microscope, Microscope Publications Ltd.,* Part 1 (1972), Part 2 (1974).

Recommended Periodicals

Metallography, An International Journal on Materials Structure and Behaviour, Bagnall, Ch. (Ed.), Journal of the International Metallographical Society. New York: Elsevier.

Praktische Metallographie/Practical Metallography, Petzow, G. (Ed.). Stuttgart: Dr. Riederer-Verlag.

6 Atomic Spectrometry

Eileen M. Skelly Frame

GE Corporate Research and Development, Materials Characterization Laboratory,
Schenectady, NY, U.S.A.

Peter N. Keliher †

Chemistry Department, Villanova University, Villanova, PA, U.S.A.

List of Symbols and Abbreviations

a	absorptivity
A	absorbance
b	absorption path length
c	concentration of the analyte
I	energy after passing through the atomic reservoir
I_0	incident energy or light
k	confidence factor
m	slope
M_j	total magnetic quantum number
s_{bk}	standard deviation of the blank measurement
S	analytical signal
T	transmittance

AAS	atomic absorption spectrometry
AES	atomic emission spectrometry
AFS	atomic fluorescence spectrometry
ASTM	American Society for Testing and Materials
CEWM-AA	continuum echelle wavelength modulated atomic absorption spectrometer
DCP	direct current plasma
DL	detection limit
EDL	electrodeless discharge lamp
EDTA	ethylenediamine tetraacetic acid
ESCA	electron spectroscopy for chemical analysis
ETA	electrothermal atomization
FANES	furnace atomization nonthermal emission spectrometry
FIA	flow injection analysis
GC	gas chromatography
GD	glow discharge
GF	graphite furnace
HCL	hollow cathode lamp
HPLC	high-performance liquid chromatography
ICP	inductively coupled plasma
LEAFS	laser excited atomic fluorescence
LOD	limit of detection
LOQ	limit of quantitation
MIBK	4-methyl-2-pentanone
MIP	microwave induced plasma
MS	mass spectrometry
MX	sample solution of metal plus anion
NAA	neutron activation analysis
NIST	National Institute of Standards and Technology
PMT	photomultiplier tube
RSD	relative standard deviation

SAS	Society for Applied Spectroscopy
SRM	standard reference materials
SSMS	spark source-MS
VDL	vapour discharge lamp
VIS-NIR	visible near infrared
XRF	X-ray fluorescence spectroscopy

6.1 General Introduction

This chapter will consider atomic spectrometry (also called atomic spectroscopy) from an analytical chemistry point of view. There are three major branches (with many minor branches within this) of atomic spectrometry. These are atomic emission spectrometry (AES), atomic absorption spectrometry (AAS), and atomic fluorescence spectrometry (AFS). In almost all cases, these techniques require very high temperature *atomic reservoirs* which provide a means to generate atoms. These atom reservoirs include chemical combustion flames such as air-acetylene flames and nitrous oxide-acetylene flames, arc and spark discharge devices, and plasmas. A *plasma* is a self-sustaining discharge of an inert gas, normally argon. Various types of plasmas are currently being used with the most popular being the inductively coupled plasma (ICP). Direct current plasmas (DCP) and various types of microwave plasmas are also used. It should be noted that analytical atomic spectrometric techniques are *normally* used for the determination of metals and metalloids although they can also be used, in some cases with appropriate instrumental modifications, for the determination of non-metals.

In AES, electrons in the ground state of the atom are excited thermally by the atom reservoir to excited states. Atomic emission occurs when these electrons return to the ground state. In AAS, an external light source is used to excite electrons so that in AAS the signal is dependent on the population of atoms in the ground state. In AAS, the atoms raised to excited states by the external light source are not further monitored. AFS can be considered as an excitation of the electrons of an atom followed by monitoring of the subsequent emission of the atom.

In all of these techniques, it is the outermost (valence) electrons of atoms that are monitored and the radiation (emitted or absorbed) normally consists of sharp, well defined lines in the ultraviolet (200–400 nanometers (nm) wavelength), visible (400–800 nm), or near-infrared (800–900 nm) regions of the spectrum. By monitoring an appropriate spectral region, it is possible to do *qualitative* analysis whereas the measurement of the relative emission or absorption of a particular line is used for *quantitative* analysis.

6.2 Historical Background

The first spectra observed by man were the continuum spectra of the sun, the moon, and the stars. The first spectrum observed by man was, of course, the rainbow. Unable to explain this effect, however, the early cavemen ascribed this to a supernatural power. Indeed, the "pot of gold" at the end of the rainbow goes back into ancient times. It was not until 1704 that Sir Isaac Newton stated "Do not all fix'd Bodies, when heated beyond a certain degree, emit light and shine; and is not this Emission perform'd by the vibrating motion of their parts?" Newton used a glass prism to study the spectrum of the sun to observe a *rainbow* spectrum of seven colors – red, orange, yellow, green, blue, indigo, and violet. (The acronym Roy G. Biv is useful to remember the visible spectrum order going from longest to shortest visible wavelength.)

6.2.1 19th Century Spectroscopy

The development of a modern knowledge of atomic spectrometry began, in a primitive way, in the early part of the 19th century. An English physician, William H.

Wollasten, observed *dark lines* in the spectrum of the sun. These were considered to be boundaries between colors. Wollaston noted:

"The line A that bounds the red side of the spectrum is somewhat confused . . . The line B between the red and green, in a certain position of the prism, is perfectly distinct; so also are D and E, two limits of the violet. But C, the limit of the green and blue, is not so clearly marked as the rest; and there are also, on each side of this limit, other distinct dark lines, F and G, either of which, in an imperfect experiment, might be mistaken for the boundaries of these colours." Wollaston (1802).

It must be emphasized that Wollaston did not understand the nature of the dark lines. We now know that these are absorption lines but it was not until many years after Wollaston's experiments that the true nature of the dark lines was discovered. Wollaston's major interest was in the spectral colors that he observed. He did repeat his experiment using both candlelight and static electric light and so was one of the first to observe emission spectra. As noted by McGucken (1969) in his excellent book "Nineteenth Century Spectroscopy", Wollaston observed that the blue light of the lower part of a candle flame gave five images. The first was broad and red and terminated by a bright yellow line, the second and third were green, and the fourth and fifth blue. As for the blue electric light, it also gave several, although somewhat different images. At this point Wollaston stopped his experiments, thinking it not necessary to describe appearances which seemed to vary according to the intensity of the light and which he was not able to explain. It was to be many years before the nature of the bright yellow light would become clear to 19th century spectroscopists.

In 1814, Fraunhofer allowed sunlight to pass through a narrow slit into a darkened room and then into a prism placed on a theodolite. A theodolite is an instrument for measuring horizontal and vertical angles by means of a small telescope turning on a horizontal and a vertical axis. He made a surprising discovery.

"I wished to see if in the colour-image from sunlight there was a bright band similar to that observed in the colour image of lamplight. But instead of this I saw with the telescope an almost countless number of strong and weak vertical lines, which are, however, darker than the rest of the colour-image; some appearing to be almost perfectly black." Fraunhofer (1814).

Fraunhofer had independently re-discovered the dark lines that had first been observed by Wollaston in 1802. Ironically, today these absorption lines are referred to as Fraunhofer lines rather than Wollaston lines. Fraunhofer found a total of 574 dark lines, the strongest of which he mapped. He did not, however, have any understanding as to the nature of the lines. He saw them as borders between colors.

As noted by McGucken (1969), Fraunhofer's studies appeared at a time of intense interest and rapid progress in optics. One of the most interesting phenomena of contemporary optics was polarization, discovered in the interval between the publication of Wollaston's and Fraunhofer's papers. This subject was being studied by Brewster. In 1822, Brewster published a paper on a lamp designed to produce monochromatic yellow light. This paper was the first of many papers by Brewster and others that tried to provide an explanation of the Fraunhofer lines. Various theories were "proven" and then rejected. During this time, primitive attempts at developing chemical analysis based on the spectra of flames were developed. Also in

1823, Herschel described flame emission spectra using an alcohol lamp as a flame excitation source.

In 1825, Talbot (see Talbot, 1826) impregnated the cotton wick of a spirit lamp with common salt in order to produce a brighter homogeneous light than that given by Brewster's monochromatic lamp. Talbot found it difficult, however, to determine precisely the source of the homogeneous yellow light.

"I have found that the same effect takes place whether the wick of the lamp is stepped in the muriate, sulphate, or carbonate of soda, while the nitrate, chlorate, sulphate, and carbonate of potash agree in giving a bluish tinge to the flame. Hence the yellow rays may indicate the presence of soda, but they, nevertheless, frequently appear where no soda can be supposed to be present." Talbot (1826).

According to McGucken (1969), Talbot also found that candles, and platinum touched by the hand or rubbed with soap, also gave the yellow light. We know, of course, today that this light is caused by the excitation of sodium atoms. Talbot (1826) noted that common salt sprinkled on platinum gave this light while the salt decrepitated, and this effect could be renewed at will by wetting the platinum. This latter circumstance led Talbot to think that the light was due to water of crystallization, rather than to sodium, the more so as it was also given by wood, ivory, and paper, whose only common constituent with sodium salts is water. But then Talbot was faced with the problem of explaining why potassium salts should not also produce the yellow light. Talbot finally concluded that water could not produce the yellow light as it was also produced by sulfur "which was supposed to have no 'analogy' with water". McGucken (1969) has noted that Talbot's paper provides an excellent illustration of the problem of the ubiquitous yellow line, which was to perplex spectroscopists for a considerable time to come and constitute a major obstacle to progress.

In 1834, Talbot wrote a very important paper entitled "Facts Relating to Optical Science". In this paper, he observed that it was extremely easy to distinguish strontium emission in a flame from lithium emission.

"The strontia flame exhibits a great number of red rays well separated from each other by dark intervals, not to mention an orange, a very definite bright blue ray. The lithia exhibits one single red ray. Hence I hesitate not to say optical analysis can distinguish the minutest portions of these two substances from each other with as much certainty, if not more, than any other known method." Talbot (1834).

Today, most high school chemistry students use a Bunsen burner and a platinum wire to perform simple chemical determinations for various metals. They learn that sodium gives a yellow color, lithium gives a red color, calcium gives an orange color, strontium gives a red color, etc. They may not, however, recognize that atomic line emission is responsible for the *colors* of the Group I metals such as lithium and sodium while molecular band emission gives the colors for the Group II elements such as calcium and strontium. Talbot was not clear on this point, either, but he did recognize the difference between a single red line (lithium) and the broad band emission of the strontium. We recognize today that the bright blue ray that Talbot observed with strontium is the strontium atomic line at 460.73 nm. Although this line is very intense, it cannot be seen visually by the eye since it is obscured by the very intense red strontium oxide band emission.

In his important 1834 paper, Talbot also made the following observation.

"An extensive course of experiments should be made on the spectra of chemical flames, accompanied with accurate measurements of the relative position of the bright and dark lines, or maxima and minima of light which are generally seen in them. The definitive ray emitted by certain substances as, for example, the yellow rays of the salts of soda, possess a fixed and inviolable character, which is analogous in some measures to the fixed proportion in which all bodies combine according to the atomic theory. It may be expected, therefore, that optical researches, carefully conducted, may throw some additional light upon chemistry." Talbot (1834).

Talbot's statement regarding optical research throwing additional light upon chemistry is certainly as true today as it was in 1834. Another important paper appeared in 1834: this paper was written by Charles Wheatstone. As noted by McGucken (1969), Wheatstone's most important discovery was that the spectrum of electric light varied according to the metallic electrodes employed. The number, positions, and colors of the bright lines varied in each case, and each spectrum was so different from the others that by this means, the metals that he studied (bismuth, cadmium, lead, mercury, tin and zinc) could readily be distinguished from one another. Wheatstone noted that if the electric spark was passed between two different metals, the bright lines of each were simultaneously seen.

In 1848, John William Draper showed that the dark lines of flame spectra were spurious, being due to the presence of some incombustible substance in the flame. He did not elaborate on this. McGucken (1969) has noted that while Draper reduced the apparent complexity of flame spectra in this respect, he nevertheless added to it in another. In testing the assertion of others that certain flames give monochromatic light, he examined the flames of many substances and found that while they were of different colors, prismatic analysis showed them to contain many colors. For example, the supposedly monochromatic alcohol flame burned on a wick impregnated with common salt was found to give other, although less strong, colors in addition to the yellow. Draper concluded that many flame spectra were less simple than had been previously thought. The first published sketches of flame spectra were made in 1845 by William Allen Miller. His paper was entitled "Experiments and Observations on Some Cases of Lines in the Prismatic Spectrum Produced by the Passage of Light Through Coloured Vapours and Gases, and from Certain Coloured Flames". Miller drew the spectra of calcium, copper, and barium chlorides, boric acid, and strontium nitrate, each of which displayed several bright lines and bands, and all having the intense yellow line in common. This yellow line was the only one visible in the "brilliant" spectra of iron, platinum, steel, and zinc ignited in a oxyhydrogen jet. It was also given by sodium chloride but Miller noted that the latter, like the first five substances showed "a marked tendency to the occurrence of bands in other parts". According to McGucken (1969), the consequence of results like Draper's and Miller's was to obstruct the extension of chemical spectrum analysis, so that it became impossible to detect further lines characteristic of only one element.

In 1851, Antoine Masson studied electric spectra and confirmed Wheatstone's finding that metals have characteristic spectra. He also observed that all metallic spectra shared "common lines" with respect to position in the spectrum but differing in intensity with each metal. Masson gave the first detailed sketches of the electric spectra of antimony, bismuth, cadmium, carbon, copper, iron, lead, tin, and zinc. In 1853, Anders Ångström noted that the so-called electric spectrum of a metal was actually an overlapping of two spectra, one corresponding to the metal of the electrode and the other to the particular gas enveloping it. McGucken (1969) has observed that this was a significant discovery, for until that time, only absorption spectra of gases had been known and it had not been suspected that gases could also give emission spectra.

In 1854, Leon Focault passed sunlight through a carbon electric arc spectrum. He noted that the double yellow line of the arc corresponded exactly with the double yellow line of the sunlight. To his surprise, however, he found that the arc absorbed the yellow lines from the sunlight. Thus, Focault concluded that "the arc offers a medium which emits, on its own account, the yellow D rays, and which at the same time absorbs them when these rays come from somewhere else".

In 1857, William Swan attempted to explain the phenomena of artificial (non-sunlight) light by examining the flames of burning hydrocarbons. In order to do this, he had to use a colorless flame and, fortunately, he had become aware of a new burner devised by Bunsen and Roscoe (today known, of course, as "the Bunsen burner") and so was able to use this new burner for his studies. As noted by McGucken (1969), Swan's use of Bunsen's flame is an instance of a curious interaction, for the flame allowed Swan to make his discovery, which in turn provided Bunsen and Kirchhoff with one of the clues to an extended spectrum analysis. Swan described the flame of the Bunsen burner as

consisting of at least two distinct portions, an envelope of a pale lavender tint enclosing a luminous hollow cone of a strong bluish-green color. He noticed that because of its inherent luminosity, this outer envelope was highly susceptible to having its color influenced by "foreign matter". Swan observed that the flame displayed "an abundance of yellow scintillations" in dusty air. Swan then determined "how small a portion of matter would in this way render its presence sensible". This was the first recorded instance of a reported "detection limit" in the literature.

"One-tenth of a grain of common salt, carefully weighed in a balance indicating 1/100 of a grain, was dissolved in 5000 grains of distilled water. Two perfectly similar slips of platinum foil were then carefully ignited by the Bunsen lamp, until they nearly ceased to tinge the flame with yellow light; for to obtain the total absence of yellow light is apparently impossible. One of the slips was dipped into the solution of salt, and the other into distilled water, the quantity of the solution of salt adhering to the slip, being considerably less than 1/20 grain, and both slips were held over the lamp until the water had evaporated. They were then simultaneously introduced into opposite sides of the flame; when the slip which had been dipped into the solution of salt, invariably communicated to a considerable portion of the flame a bright yellow light, easily distinguishable from that caused by the slip which had been dipped into pure water. It is thus proved that a portion of chloride of sodium, weighing less than 1/1 000 000 of a grain is able to tinge a flame with bright yellow light; and as the equivalent weights of sodium and chlorine are 23 and 35.5 it follows, that a quantity of sodium not exceeding 1/2 500 000 of a troy grain renders its presence in a flame sensible. If it were possible to obtain a flame free of yellow light, independently of that caused by the salt introduced in the experiment, it is obvious that a greatly more minute portion of sodium could be shown to alter appreciably the colour of the flame. It therefore follows, that much caution is necessàry in referring the phenomena of the spectrum of a flame to the chemical constitution of the body undergoing combustion. For the brightest line in the spectrum of the flame of a candle, – the yellow line R of Fraunhofer, – can be produced in great brilliancy, by placing an excessively small portion of salt in a flame, in whose spectrum that line is faint or altogether absent. The question then arises, whether this line in the candle flame is due to the combustion of the carbon and hydrogen of which tallow is chiefly composed, or is caused by the minute traces of chloride of sodium contained in most animal matter. When indeed we consider the almost universal diffusion of the salts of sodium, and the remarkable energy with which they produce yellow light, it seems highly probable that the yellow line R which appears in the spectra of almost all flames, is in every case due to the presence of minute quantities of sodium." Swan (1857).

The importance of Swan's 1857 paper cannot be overemphasized. There were two important discoveries that came about as a result of Swan's work. It led immediately to a simplification of many spectra, including all electric spectra of metals. Perhaps more important, however, as noted by McGucken (1969), is the fact that it underlined the unprecedented sensitivity of atomic spectroscopy. If any certain knowledge was to be gained, one was required to operate in conditions of the highest purity. One could no longer afford to be careless of impurities as some previous workers had been.

In 1859, van der Willigen used metallic wires as pole pieces in an investigation of electric spectra. When these poles were moistened with water, he observed a momentary but very intense yellow emission which he attributed, as a result of Swan's paper, to sodium impurities. Using hydrochloric acid, van der Willigen observed an increase of intensity of the various metallic spectral lines. He explained this by noting that the metallic lines were similarly intensified when the metals were burned in chlorine. He reasoned that the metallic chlorides produced, being volatile, allowed the metallic parts to become widely dispersed. In extending the experiments with water and acids, van der Willigen moistened platinum (as platinum did not show any "noticeable" characteristic bands) pole pieces with weak solutions of barium chloride, calcium chloride, calcium nitrate, strontium chloride, etc. and observed metallic bands which were characteristic of the particular metal. He found the chlorides particularly bright and calcium nitrate was found "*serviceable*" but he did not observe the effect with sulfates. He concluded that he could bring all metals that he observed "into the spectrum".

An important point should be noted here. Van der Willigen had noted that metal chlorides give spectra identical to those of the component metals. This gave rise to what McGucken (1969) has called "the triumphant achievement of Bunsen and Kirchoff". In their classic work entitled "Chemical Analysis by Spectrum Observations", Bunsen and Kirchoff (1860), extended the work of van der Willigen to include the unresponsive sulfates and other salts, so that by placing practically any salt in a flame its component metal could be immediately determined. The apparatus developed by Bunsen and Kirchoff was much more refined than any previously used systems. It had a collimator as previously devised by Swan and the prism was housed in a blackened box, in order to cut out extraneous light. As noted by McGucken (1969), Bunsen's skill and knowledge as a chemist ensured that Swan's admonition concerning pure sources was scrupulously observed. The purification of the salts was carried out as much as possible in platinum vessels and as many as fourteen precipitations or crystallizations were performed. The salt to be examined was melted on the eye of a suitably bent platinum (following van der Willigen, 1859) wire and held in a colorless flame before the collimator. By this means Bunsen and Kirchoff systematically examined the bromides, carbonates, chlorides, hydrated oxides, iodides, and sulfates of barium, calcium, lithium, sodium, and strontium. As a result of their experiments, they concluded that:

"The different bodies with which the metals employed were combined, the variety of the chemical processes occurring in several, and the wide differences of temperature which these flames exhibit, produce no effect upon the position of the bright lines in the spectrum which are characteristic of each metal." Bunsen and Kirchoff (1860).

This was a most important discovery. Instead of there being a different spectrum for each salt, Bunsen and Kirchoff discovered that there are only as many spectra as there are metals. Bunsen und Kirchoff noted that a metal gives the same characteristic spectrum no matter whether, or how, it is combined chemically. An accompanying colored plate showed the spectra of the chlorides. In order to demonstrate still more conclusively that each of the metals studied always produced the same intense spectral lines, Bunsen and Kirchoff compared the flame spectra of the chlorides

with their electronic spectra. They produced the latter by sending an electrical discharge through a glass tube containing two platinum electrodes, to which small pieces of the chosen metal were fastened.

McGucken (1969) has observed that the spectra described by Bunsen and Kirchoff were oversimplified. The limitations of the apparatus employed allowed Bunsen and Kirchoff to see only the more conspicuous characteristics of a spectrum. However, for the purpose of chemical analysis, that was all that was necessary. The analyst only needed to know which line was given by a particular element in order to distinguish the various salts.

It was after Bunsen and Kirchoff had begun their collaboration that Kirchoff came upon the explanation of the Fraunhofer lines by himself. Kirchoff viewed the spectrum of sunlight passed through a salted flame placed before the slit of his apparatus. This was the method that he and Bunsen had used in comparing flame spectra with the solar spectrum. If the sunlight was sufficiently weak, two bright lines appeared in place of the two dark D lines. On the other hand, if the sun's intensity exceeded a certain limit, the two D lines appeared more distinct than if the salt flame had not been employed. As noted by McGucken (1969), this latter fact surprised Kirchoff as being totally unexpected. Even more striking, Kirchoff found that the lithium flame, whose bright line had no corresponding dark line in the solar spectrum (there isn't much lithium in the sun), could produce a dark line in the latter when sunlight was passed through a lithium-containing flame. Only gradually did Kirchoff come to an explanation of this effect. In an 1860 paper entitled "On The Relations Between the Radiating and Absorbing Powers of Different Bodies for

Light and Heat", he stated "for waves of this length the radiating power of the flame is very considerable, while for waves of lengths corresponding to the other visible colours it is imperceptible. Accordingly, the power of absorption of the lithium flame must be great for waves of this length, but very small for those constituting the other visible rays. If, therefore, a continuous spectrum be formed by suitable means and a lithium-flame be placed between the source of light and the slit of the apparatus, the spectrum is only affected in the place of the lithium line, its brightness being increased in that part of the radiation of the flame, while on the other hand it is diminished by its power of absorption for waves of that particular length". Kirchoff (1860).

McGucken (1969) has observed that when Kirchoff next attempted to explain "reversals", he was much more lucid although his explanation was still incomplete. Kirchoff took the sodium lines as an example in his "Researches on the Solar Spectrum" (1862–1863) and without referring to his 1860 theoretical paper, he stated that the reversal may be easily explained upon the "supposition that the sodium flame absorbs rays of the same degree of refrangibility as those it emits, whilst it is perfectly transparent for all other rays". Kirchoff then stated:

"It is plain that if the platinum wire emits a sufficient amount of light, the loss of light occasioned by absorption in the flame must be greater than the gain of light from the luminosity of the flame; the sodium lines must then appear darker than the surrounding parts, and by contrast with the neighbouring parts they may seem to be quite black, although their degree of luminosity is necessarily greater than that which the sodium flame would have produced." Kirchoff (1862–1863).

McGucken (1969) has observed that while Kirchoff did not fully explain the origin of the Fraunhofer lines, he was able in several instances to demonstrate experimentally their physical connection with the spectra of metals. However, the lack of a complete explanation was no barrier to the task of determining the composition of the sun's atmosphere, which was at this point begun by Kirchoff and others. Kirchoff's work was an extremely important experimental achievement since it laid the groundwork for the development of practical atomic spectrometry. What Bunsen and Kirchoff did was to lay the experimental foundations of spectrum analysis. McGucken (1969) concludes his excellent chapter on the origins of spectrum analysis by stating "A multitude of theoretical questions remained for posterity's consideration".

After the pioneering work of Bunsen and Kirchoff, there was not too much activity during the latter part of the 19th century. However, in 1870, Janssen described a quantitative method for the determination of sodium. He used a Bunsen burner as the excitation source and introduced the sample into the flame on a platinum wire. The sodium concentration was estimated by a *visual* comparison of the intensity of the emitted light with that emitted by a series of known standard concentrations. Refining this approach, Champion, Pellet, and Grenier (1873) used two flames for the determination of sodium, one for the unknown and the other for a series of standards. The spectra were observed simultaneously and an accuracy of two to five percent for sodium in plant ash was reported. In 1877, Gouy showed conclusively that radiation intensity from a flame was a function of flame size and the quantity of substance introduced into a flame. To control these factors, Gouy (1879) designed a pneumatic atomizer to inject a controlled amount of sample into a flame. Increased precision and accuracy was clearly demonstrated.

Schrenk (1988) has recently published an in-depth historical account of the development of high-energy excitation sources for analytical emission spectrometry. He notes that towards the end of the 19th century, spectroscopic excitation sources utilizing spark and dc arc excitation had become very useful tools in atomic spectrometry. Spark excitation progressed from friction machines to induction coils while dc arcs went from battery operated to the dc generator. Schrenk (1988) also notes that photographic emulsions of sufficient sensitivity to record spectra were developed and immediately used to record spectra for future study. Thus, the way was opened for a rapid expansion of emission spectroscopy.

6.2.2 20th Century Spectroscopy

As noted by Schrenk (1986), there was apparently little further research in flame emission methods of analysis from the time of Gouy until the pioneering work of Lundegardh in Sweden starting about 1928. Until that time, flames used for excitation were Bunsen type natural gas or air burners, or were alcohol lamps. These flames operate at relatively low temperatures, so only easily excited elements were observed. Lundegardh (1929) used an air-acetylene flame to produce a higher temperature chemical combustion flame than had been used by previous workers. Lundegardh also improved the pneumatic atomizer developed by Gouy. Lundegardh determined spectral line intensities using his own designed photoelectric densitometer and calibration curves were construct-

ed with standard solutions. Lundegardh reported analytical data for twenty four elements. Ells and Marshall (1939) were the first workers in the United States to use the *Lundegardh technique* for the determination of various metals in soil samples. This was followed by papers by Griggs (1939) and Cholak and Hubbard (1944).

Schrenk (1986) considers the "modern period" of flame excitation atomic spectroscopy to have begun with the introduction of a commercial flame photometer (Perkin-Elmer Corporation) in the mid 1940s. This unit was developed by Barnes and co-workers (1945) and included a modified Meeker air/natural-gas burner as the "atom reservoir". In 1948, the Beckman Corporation introduced the flame-aspirator attachment for their Model DU spectrophotometer. This was a direct-injection burner; sometimes called (erroneously) a *total consumption* burner. This burner has become universally known as the *Beckman burner*. Further developments in the development of burners of chemical combustion flames are discussed in Schrenk's 1986 review while his 1988 article reports on the development of high energy excitation sources (arc, spark, plasmas) for analytical emission spectrometry.

The development of practical analytical AAS and AFS did not occur until the middle of the 20th century. Atomic fluorescence was first reported by Wood (1905) and later by Nichols and Howes (1923a, 1923b) but these reports were not concerned at all with potential analytical applications. In 1939, Woodson described an analytical technique for the determination of mercury vapor in air but did not extend the method any further. It should be noted that mercury is the only element (other than the inert gases) that has a reasonably high monoatomic vapor pressure at room temperature. This allowed Woodson to develop an analytical method for mercury but it was not generally used and was, in fact, "rediscovered" many years later by Hatch and Ott (1968).

In 1955, independent papers by Walsh and by Alkemade and Milatz clearly showed the tremendous potential of analytical AAS. Walsh's paper provided an in depth theoretical treatment and he is now considered the "father" of analytical AAS. Hollow cathode lamps (HCLs) were used as excitation sources and the HCL signal was modulated so that atomic absorption measurements could be made without interference from the atomic emission from the flame. In a very significant and important paper, Willis (1965) proposed the use of the pre-mixed nitrous oxide-acetylene flame for refractory elements. This increased the number of elements that could be determined by flame AAS from about 35 to 70. In the 1960s, various workers used a variety of "non-flame" devices in order to reduce detection limits and, in some cases, to conserve sample. Important papers include those by L'vov (1961), and Woodruff et al. (1968).

Robinson (1961) demonstrated that atomic excitation by radiation directed at a flame was possible, bringing AFS from the realm of "academic interest" to practical use. Winefordner and Vickers (1964) used metal vapor discharge lamps as excitation sources for AFS and reported excellent detection limits for a variety of elements. In related work, Veillon et al. (1966) used a continuum xenon arc source for analytical AFS. About this time, West et al. (1966a, 1966b) at Imperial College, London were also studying AFS. These workers used laboratory constructed microwave (2450 MHz) electrodeless discharge lamps (EDLs) for AFS studies and were able to make useful EDLs for over thirty elements.

Commercially available AAS systems were introduced in the early 1960s. The first unit was developed by Hilger and Watts in England and this was soon followed by instrumentation from the Perkin-Elmer Corporation. Today, at least fifteen manufacturers offer AAS complete systems of components. In 1970, Technicon Instruments introduced a six channel AFS system; this was, however, withdrawn after about two years. Today, one company (Baird Corporation) offers a multielement ICP-AFS instrument.

6.3 Fundamental Concepts of Atomic Spectrometry

The purpose of this section is to provide information on the fundamental concepts of atomic spectrometry/spectroscopy with an emphasis on the similarities as well as the differences between atomic emission spectrometry (AES), atomic absorption spectrometry (AAS), and atomic fluorescence spectrometry (AFS). As the term atomic implies, all three techniques deal with atomic spectra. An atom, of course, is the smallest amount of an element that retains its characteristic properties. In all atomic spectrometric methods, atoms are created by manipulation of gaseous, liquid, or solid samples. This is accomplished in what can be called an *atom reservoir*. As noted previously, this is simply a device, normally a high temperature device, that is used to generate atoms. Various atom reservoirs including chemical combination flames, arc and spark discharges, and high temperature plasmas are discussed in Sec. 6.4.

An atom consists of a positively charged nucleus containing protons (positively charged) and neutrons (no charge) as well as planetary electrons (negatively charged) in a series of shells (or orbits) at discrete distances from the nucleus. The 1st shell is referred to as the K shell, the 2nd shell is referred to as the L shell, etc., and each shell contains sub-shells which, for historical reasons, are labeled s, p, d, f, g, etc. The number of sub-shells within a main shell is controlled by the number of the shell. Thus, for example, the 1st (K) shell contains only one sub-shell, the 2nd (L) shell contains two sub-shells (an s and a p), and the 3rd (M) shell contains three sub-shells (an s, p, and a d). The number of positively charged protons is equal to the atomic number of the element, and in any neutral atom that would exist or be generated in an atom reservoir, the number of protons is equal to the number of electrons. Neutrons add mass or weight to an atom but the electrons essentially control the atomic spectrum for any element. A sodium atom, for example, has an atomic number of eleven and, therefore, would have eleven protons in its nucleus and eleven electrons. Two electrons would be in the K shell, eight in the L shell (two in the L_s and six in the L_p), and one electron would be in the M_s shell. This is referred to as the ground state configuration. This may also be designated as follows: $Na = 1s^2, 2s^2, 2p^6, 3s^1$. The electrons in the outermost shell of an atom are called valence electrons; these are the electrons which are most loosely held and most easily excited or removed. The electron in the M_s shell $(3s^1)$ would be referred to as a valence electron. Magnesium has two valence electrons and aluminum has three valence electrons. Designations are $Mg = 1s^2, 2s^2, 2p^6, 3s^2$ and $Al = 1s^2, 2s^2, 2p^6, 3s^2, 3p^1$.

Thermal excitation within the atom reservoir can cause an electron (or electrons) to become excited; that is, raised to higher energy levels. Energy levels increase in the order $K < L < M$ and $s < p < d < f$ and so on. In the case of sodium, the M_s p valence electron can be easily raised to the M_p level. It will, almost immediately, return to the ground state configuration with the emission of yellow light at 589.0 and 589.6 nm. This behavior is observed yearly by millions of students who place sodium salts in Bunsen burners. Each element will emit energy of *particular* wavelengths which will be characteristic of that element. This is the fundamental principle behind AES. In other words, this type of spectrum is associated with quantized changes in atomic energy produced by variations in the orbital motions of valence electrons. The resulting wavelengths are (generally speaking) quite distinct from each other. The term *line spectra* is also used to describe these wavelengths. As noted previously, the intensity of the radiated emission is directly related to the concentration of the element in a sample.

If an external radiation source is directed into the atom reservoir the absorption of the radiated source may be monitored, this is the fundamental principle behind AFS. The three techniques are shown, schematically, in Fig. 6.1.

Atom reservoirs are, in general, very high temperature sources that emit, in addition to atomic spectra, other forms of light. Molecular (sometimes called "band") spectra are associated with energy changes of molecules and these give rise to spectral band groups. These are generally more complex than atomic spectra. Another type of light emission is true continuous emission spectra. The main interest in these spectra (from an analytical atomic spectrometric point of view) is the potential interferences that they can cause in chemical analysis. These interferences are discussed in Sec. 6.4.5.

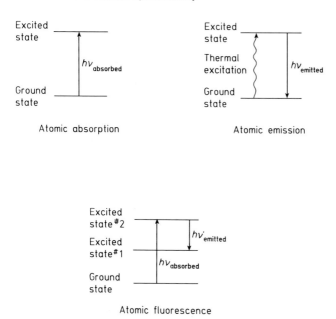

Figure 6-1. Schematic diagram of the three basic atomic spectrometric techniques. In atomic absorption spectrometry, photons of a specific wavelength from an external source are absorbed by ground state electrons. The amount of light absorbed is related to the number of ground state atoms present. In atomic emission, atoms are excited thermally (e.g., by collisions) to excited states. Photons are emitted as electrons return to the ground state. The intensity of emission is related to the number of excited atoms present. In atomic fluorescence spectrometry, atoms are excited by the absorption of photons from an external source. Emission of photons occurs as electrons return to lower energy levels.

6.3.1 Basic Principles of Atomic Emission Spectrometry

Conceptually, AES is the simplest and the oldest of the three related analytical techniques. Sample is introduced into the atom reservoir and the emitted light is monitored through a wavelength isolation device called a *monochromator*. This is normally accomplished by a prism or a grating. A light sensitive detector is used to record the desired emission wavelength(s); this may be done using photographic film (the term spectrography is used in this case) or with a light sensitive device; almost always using a photomultiplier tube (PMT). This is discussed further in Sec. 6.4.

6.3.2 Basic Principles of Atomic Absorption Spectrometry

In AAS, ground state atoms in the "atom reservoir" absorb energy in the form of light of a particular wavelength.

In addition to the components discussed above, AAS requires a separate external light source that is directed into the atom reservoir. This is almost always a *hollow cathode lamp (HCL)*. The basic arrangement of the components is shown in Fig. 6.2. In AAS, the atom reservoir is usually either a chemical combustion flame or a graphite furnace. This is discussed further in Sec. 6.4.

The amount of light absorbed from the HCL (or other source) is given by the transmittance, T, where

$$T = I/I_0 \tag{6-1}$$

I_0 refers to the incident energy (before passing through the atom reservoir) and I refers to the energy after passing through the atom reservoir. The term A refers to the absorbance which is a logarithmic term.

$$A = \log(I_0/I) = -\log T \quad \text{or}$$
$$A = abc \tag{6-2}$$

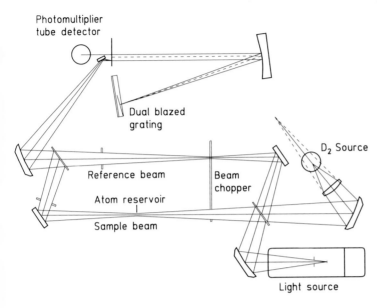

Photomultiplier
tube detector

Dual blazed
grating

Reference beam

Atom reservoir

Sample beam

Beam
chopper

D₂ Source

Light source

Figure 6-2. Double-beam atomic absorption spectrometer. Radiation from hollow cathode or electrodeless discharge lamp and radiation from broadband source, such as a deuterium lamp (D_2 source, used for background correction), is split, chopped, and focused onto the atomizer (sample beam) and into a reference beam. Beams are recombined, sent through a monochromator to a PMT detector (courtesy of The Perkin-Elmer Corporation).

In this equation, $A = abc$ (the Beer-Lambert Law), a is a constant (for a particular element at a particular wavelength) called the absorptivity, b refers to the absorption path length, and c is the concentration of the element. Thus, absorbance is directly proportional to concentration. In practice, it is best to work at concentrations that give signals in the middle of the absorbance range. Most modern day atomic absorption spectrometers give concentrations directly, comparing known standards to unknown samples. In order to distinguish the atomic absorption signal from emission from the atom reservoir, the HCL (or other source) signal is either mechanically chopped or electronically modulated to provide an alternating current on the source. The electronics of the instrument distinguish the alternating current from the direct current of the atom reservoir. This same arrangement is also used in AFS.

6.3.3 Basic Principles of Atomic Fluorescence Spectrometry

In AFS, an appropriate light source is used to excite the atoms in the atom reservoir and the resulting emission (fluorescence) signal is measured. Thus, AFS can be considered as a combination of absorption and emission. Although the fluorescence signal is emitted in all directions, it is most usual to measure it at an angle of 90 degrees from the light source. A typical arrangement is shown in Fig. 6-3. The most common AFS technique is to measure the emitted radiation at the same wavelength as that used for excitation. This is referred to as *resonance fluorescence*, and is shown in Fig. 6-4. In *direct line fluorescence*, the electron does not return to the ground state but to some intermediate state, with the emission of radiation. It then returns to the ground state via radiationless transfer. Thus, the observed fluorescence (emitted radiation) will al-

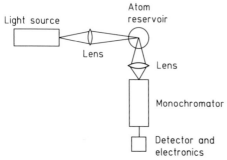

Figure 6-3. Typical single beam atomic fluorescence spectrometer. Radiation from source (lamp or laser) is focused onto the atom reservoir. Fluorescence photons are collected at a 90° angle to incident radiation, sent to a wavelength selector and PMT detector.

cence, partial thermal activation can also contribute to the fluorescence intensity when a high temperature flame is used as the atom reservoir. As noted earlier, however, most analytical studies have involved resonance AFS.

In AAS, the intensity of the incident light (I_0) does not change the absorbance signal for a particular concentration. This is because the measured ratio I_0/I does not change by increasing the signal intensity. By contrast, the intensity of the light source is directly proportional (up to certain limits) to the fluorescence intensity and, hence, to the concentration in AFS. Thus, the AFS technique is *potentially* capable of providing much lower detection limits than the corresponding AAS technique. Unfortunately, despite much research work with various spectral sources (most commonly microwave excited electrodeless discharge lamps), this potential has not been totally realized. However, the only commercially available AFS instrument, the AFS 2000 (Baird Corporation,

ways be of longer wavelength (lower energy) than the excitation wavelength. This is also true for *stepwise fluorescence* where an electron is initially excited to an excited state and undergoes deactivation (often by a collision or other radiationless process) to a lower excited state. It then undergoes fluorescence from that state to the ground state. A reverse process to stepwise fluores-

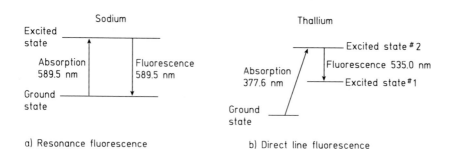

a) Resonance fluorescence

b) Direct line fluorescence

c) Stepwise fluorescence

Figure 6-4. Types of atomic fluorescence. (a) In resonance fluorescence, the absorption and fluorescence transitions have the same upper and lower electronic states. (b) In direct line fluorescence, the absorption and fluorescence transitions have the same upper state, but different lower states. (c) In stepwise fluorescence, the absorption and fluorescence transitions have different upper states (reprinted from Robinson, J. W., 1990, by courtesy of Marcel Dekker, Inc. and the author).

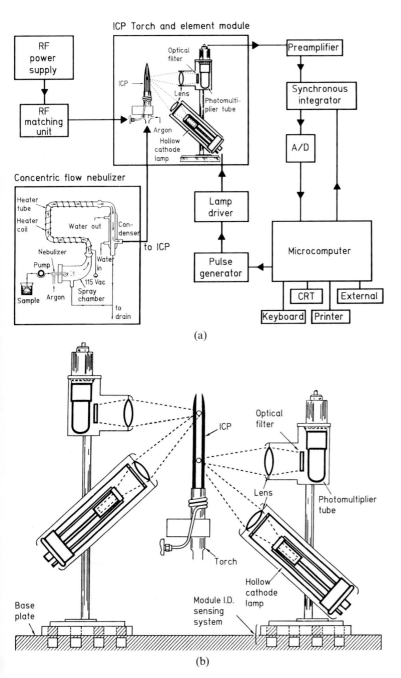

ICP Torch and element module

Concentric flow nebulizer

(a)

(b)

Figure 6-5. Commercial atomic fluorescence spectrometer. (a) Schematic diagram of the Baird AFS 2000, showing the ICP atom reservoir and one external light source, a specially designed hollow cathode lamp. Sample is introduced into the ICP via a peristaltic pump and concentric flow nebulizer, and fluorescence intensity is measured with a PMT detector. (b) A close-up of the atom reservoir and external light sources. Each hollow cathode lamp has its own PMT detector; up to 12 lamps and detectors can be placed around the atom reservoir, allowing simultaneous detection of up to 12 elements (courtesy of the Baird Corporation).

Bedford, MA) does use 12 specially designed HCLs as excitation sources into an inductively coupled plasma (Fig. 6-5). Recent results show superior detection limits with an added advantage of being virtually free from any spectral interferences.

6.4 Instrumental Requirements

As noted previously, instrumentation for AES involves an atom reservoir, a wavelength dispersion device (monochromator), electronics, and a light detector or detectors, normally a photomultiplier tube or tubes in the case of multi-element systems. AAS and AFS instrumentation requires, in addition to the above, an appropriate radiation source that is directed into the atom reservoir. Each of these components will be discussed in detail.

6.4.1 Sources for AAS and AFS

Two general types of light sources are used in AAS and AFS. Continuum sources emit light over a broad wavelength range and spectral line sources emit radiation mainly as discrete lines. Sources may also be categorized in terms of the source emitting radiation continuously or in short bursts of energy at regular intervals, that is, in pulses. For both AAS and AFS, it is important that the source chosen be stable and relatively intense in order to obtain a useful analytical signal. This allows good measurement precision which, in turn, provides good detection limits. Spectra produced by line sources should be relatively pure although they will normally emit, in addition to the analytical atomic lines, filler gas, ionic, and some impurity lines in the spectrum. A further desirable characteristic is that the sources have long operational and storage lives. They should be relatively robust although dropping a source will, in almost all cases, destroy it.

6.4.1.1 Hollow Cathode Lamps (HCL)

The HCL is, by far, the most commonly used spectral source in AAS and it also finds some use in AFS although conventional HCLs *do not provide a great deal of intensity*. As noted in Sec. 6.3.4, this is a limitation in AFS but is not a limitation in AAS. A typical HCL is shown in Fig. 6-6. The HCL consists of a cathode made from the metal (or an alloy of the metal) of interest enclosed in a sealed glass tube with a

metal-to-glass seal at the base of the lamp. A plastic cover is cemented to the glass and electrical connections are made with a special lamp socket built into the base of the HCL. The anode is simply an electrical wire designed to complete the circuit. The HCL window is usually constructed of quartz rather than glass to allow the low ultraviolet radiation wavelengths of most elements to pass through. At these low wavelengths (below roughly 320 nm), glass is opaque. The HCL itself is filled with an appropriate inert gas, such as neon or argon. The inert gas pressure is normally about 5 torr. The lamp is initiated by striking a potential of several hundred volts between the electrodes. This is controlled by the HCL geometry.

The filler gas is ionized at the anode to produce positive ions. The filler gas ions strike the oppositely-charged cathode, and sputter a cloud of excited metal atoms into the space inside the cathode. The excited metals atoms relax back to the ground state with emission of characteristic wavelengths of light. It is this light which is directed into the atom reservoir. The sputtered atoms then diffuse back to the cathode or to the glass lamp envelope. The operator controls the power into the lamp, normally 5 to 12 milliamps. The HCL should be run at a minimum power (in AAS) to obtain a good signal since running it at a higher power does not provide any gains and might result in some self absorption or self reversal effects. This is caused by absorption of some of the emitted radiation by the cloud of sputtered atoms from the cathode in the window region of the lamp and will be more pronounced with more volatile elements.

The problem of self-reversal (Fig. 6-6) can be minimized through use of a modified demountable hollow cathode lamp, as described by Robinson (1990). Filler gas

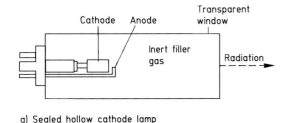

a) Sealed hollow cathode lamp

Emission profile

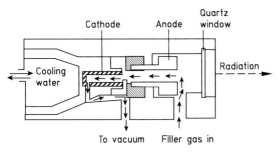

b) Demountable hollow cathode lamp

Emission profile

Figure 6-6. (a) Diagram of typical commercial sealed hollow cathode lamp. The emission profile from sealed lamps is often distorted by self-absorption, the absorption of the center of the emission line by atoms sputtered from the cathode. (b) Diagram of modified demountable hollow cathode lamp. Self-absorption is eliminated, even for very volatile elements, by flowing filter gas (Fig. 6-6b reprinted from Skelly, 1982).

flows constantly through this lamp, sweeping out the cloud of sputtered atoms from the cathode and greatly decreasing self-absorption. Robinson and Skelly (1981) demonstrated the utility of a modified demountable HCL for the very volatile element mercury.

Although multi-element HCLs are commercially available, there is usually some slight degradation for a particular metal and most HCLs are single-element.

All commercially available AAS instruments are primarily designed for use with HCLs. The factors responsible for the tremendous popularity of HCLs are: Lines may be produced over the entire optical region from the infrared to the low (vacuum) ultraviolet; intense spectra may be excited using only low currents; extremely sharp lines are quite readily obtained; the lamps and their associated power supplies are relatively simple and inexpensive to construct; they give a stable, noise-free output over long periods of time. It is cer-

tainly safe to say that HCLs *dominate* AAS and probably about 98% (or higher) of AAS measurements are done with HCLs.

As discussed in Section 6.3.3, the fluorescence intensity is directly proportional to the source intensity. A boosted-output hollow cathode lamp, designed by Lowe (1971), is used on the Baird AFS-2000, to provide improved detection limits compared to conventional HCLs. The cathode of a boosted-output HCL is a hollow cylinder, with a flat, circular anode located below the cylinder. During operation, a discharge occurs between the cathode and anode, as in a normal HCL, which produces sputtering and relatively inefficient excitation. In addition, a second discharge is produced between the anode and a filament cathode designed to emit electrons on passage of a current. Through collisions, the emitted electrons from the second cathode produce additional high energy ions of filler gas, which then excite additional sputtered atoms. The result is more intense emission from the lamp due to more efficient excitation and reduction of self-reversal. Fluorescence signals for a variety of elements show relative increases in intensity of 4–20 times those obtained with conventional HCLs run at the same current (Demers, 1987).

6.4.1.2 Electrodeless Discharge Lamps (EDL)

EDLs (also called in older literature electrodeless discharge tubes – EDTs) have been used for many years as spectral sources in atomic spectrometry. As the name implies, there are no electrodes associated with these devices. They are simply low pressure gas discharges in quartz bulbs (tubes or lamps) that contain the metal (either as the metal itself or a salt of the metal) of interest as well as a small amount of an inert gas, usually argon. They are surrounded by a device which provides radiofrequency or microwave frequency radiation into the EDL. Although a variety of exciting frequencies have been used, the usual radiofrequency is 27 MHz and the most common microwave frequency is 2450 MHz. The actual discharge is initiated by means of a Tesla coil which causes some of the inert gas atoms to become ionized. The electrons that are produced gain energy by oscillating in the field and, in a short period of time, there is sufficient energy to cause intense atomic line emission of the chosen metal.

Radiofrequency excited EDLs are currently commercially available and have been designed primarily as *alternative* spectral sources in AAS for those elements for which it is difficult to manufacture high quality HCLs with good intensity. The EDL itself is surrounded by a coil connected to a radiofrequency generator. They are available only for a few relatively volatile metals with resonance lines in the low ultraviolet region of the spectrum (e.g., arsenic 193.7 nm; selenium 197.3 nm). The higher intensity and improved signal-to-noise ratios for these EDLs can lead to better detection limits in AAS as compared with HCL devices. However, this will not only depend on the EDL/HCL characteristics but also on the optical design of the particular AAS instrument used.

Although microwave excited EDLs are commercially available, most of these have been laboratory constructed on simple vacuum lines. They have been utilized almost exclusively for AFS studies and a large number of papers on these devices appeared in the literature from about the mid 1960s through the mid 1970s. The EDL itself was placed in a microwave cavity to contain the microwave discharge. Originally, medical diathermy units were used to generate the 2450 MHz radiation. Later, generators designed specifically to power EDLs were developed. However, concern about the reproducibility of laboratory methods of preparation, the subsequent performance of the EDLs, and developing interest in other areas of atomic spectrometry (most particularly ICP-AES) has greatly reduced the number of publications on microwave excited EDLs. Very recently, Sneddon et al. (1989) have published an extensive in depth review of EDLs and the reader is directed to this paper for further details on these devices.

6.4.1.3 Vapor Discharge Lamps (VDL)

VDLs have been available for many years for volatile elements such as cadmium, cesium, potassium, mercury, sodium, rubidium, thallium, and zinc. The familiar sodium street lamp is, in fact, a VDL. Although these sources have been used for both AAS and AFS, their limitations, fairly broad emitted lines, and the clear superiority of HCLs render their use in analytical atomic spectrometry as essentially historical. They are, however, still useful in some instances for wavelength calibration. Mercury vapor arc lamps are used in apparatus designed specifically for the atomic fluorescence determination of mercury (Muscat et al., 1972).

6.4.1.4 Continuum Sources

One of the limitations of AAS is the requirement to have a separate (or almost separate if multi-element sources are used) spectral source for each element. The ability to use one continuum source for all elements would seem, therefore, to have an important advantage. Unfortunately, however, analytical results comparable to spectral line results will only occur if a very high resolution monochromator is employed. Keliher and Wohlers (1976) used a high resolution echelle monochromator with a continuum (xenon lamp) source and were able to obtain characteristic concentrations that were comparable with conventional HCL-AAS. In an extension of this work, Zander et al. (1976, 1977) developed a system that incorporated wavelength modulation in conjunction with a continuum source (Eimac Lamp) and an echelle monochromator. This was referred to as a continuum echelle wavelength modulated-atomic absorption spectrometer and called CEWM-AA. More recently, Harnly et al. (1979) developed a multi-element version of the CEWM-AA system which they referred to as SIMAAC. (A review of continuum-source AAS was published in 1984 by O'Haver). Despite the interest in continuum source (particularly multi-element) AAS, no commercial manufacturer has developed a system and it is unlikely that this will occur in the future.

Continuum sources have also been used in many academic AFS studies (e.g., Chuang and Winefordner, 1975). However, the intensity *per unit wavelength* of most continuum sources is quite low compared with that of a line source at a strong emission line. Their use today in AFS must be considered historical. A comparison of continuous and line sources has been made by Perkins and Long (1988).

6.4.1.5 Lasers

Lasers are distinguished from other optical sources by their extremely coherent and unidirectional beam of radiation, the large amount of energy available over a small spectral region, and the extremely small area into which the light beam may be concentrated. Despite these potential advantages, there has been almost no work using lasers in AAS and only limited academic projects using lasers for exciting AFS. This is due in great part to the very high expenses associated with variable wavelength lasers. Small, air-cooled argon ion lasers, which have recently become available, have addressed the cost, as well as complexity and reliability issues, and the use of lasers in commercial instruments is expected to grow. AFS studies with tunable lasers have been carried out by Winefordner and coworkers (Omenetto et al., 1973) and by Michel and coworkers (Pereli et al., 1987).

6.4.2 Atom Reservoirs

The function of all of the atom reservoirs described below is to produce atoms and it is appropriate here, therefore, to describe the processes which occur when a sample solution of metal plus anion, MX, is placed or transported into an atom reservoir. This can be represented as follows:

$$[M^+X^-]_{sol} \xrightarrow{\text{nebulization}} [M^+X^-]_{mist}$$

$$[M^+X^-]_{mist} \xrightarrow{\text{evaporation}} [M^+X^-]_{solid}$$

$$[M^+X^-]_{solid} \xrightarrow{\text{fusion}} [M^+X^-]_{liquid}$$

$$[M^+X^-]_{liquid} \xrightarrow{\text{vaporization}} [M^+X^-]_{gas}$$

$$[M^+X^-]_{gas} \xrightarrow{\text{dissociation}} M_{gas} + X_{gas}$$

The atomic processes which the metallic atom might undergo can be represented as follows:

Absorption: $\quad M^0 + h\nu \rightarrow M^{0*}$

Ionization: $\quad M^0 \rightarrow M^+ + e^-$

Oxidation: $\quad M^0 + O^0 \rightarrow MO$

Emission: $\quad M^{0*} \rightarrow M^0 + h\nu$

Combination: $\quad M^0 + Y^0 \rightarrow MY$

where M^0 represents a ground state atom, $h\nu$ represents light absorption or emission, M^{0*} represents an atom in an excited state, O represents oxygen, and Y represents any other foreign atom. Clearly, ionization, oxidation, and combination are undesirable since they reduce the atom population in the *atom reservoir*. Various atom reservoirs enhance or reduce these effects and these will be discussed, where appropriate, in the individual descriptions of these atom reservoirs.

6.4.2.1 Chemical Combustion Flames

Two different types of burner systems have been used in analytical flame spectrometry, the direct injection burner, shown in Fig. 6-7, and the pre-mixed burner, an example of which is shown in Fig. 6-8. In the direct injection burner (also called the total consumption burner), the fuel, support gas, and sample meet at the burner head. With this sort of burner, however, there is a limit in physical size to the flame (a very definite disadvantage in

AAS) and a flame flicker fluctuation that leads to high noise levels. Furthermore, the flame itself is quite audible, generating an extremely unpleasant high pitched, whining type of sound. The only important advantage of the direct injection burner is that it can be used with oxygen as a support gas. Although much of the past work in flame emission spectrometry (and some limited work in AAS) was done with direct injection burners, they are of very limited use today.

In the conventional pre-mixed burners, the support gas and the fuel are intimately mixed in a spray chamber. The support gas is normally used to aspirate the liquid sample and a bead of some sort, usually a metal or a glass bead, is used to disperse the large drops from the nebulizer into smaller drops. Only about 5 to 7% of the sample actually reaches the burner head, the rest goes down the drain. Burner heads may be cylindrical (normally used for flame AES) or slot type for AAS. It is absolutely essential that the flow velocity of the gases be greater than the burning velocity, otherwise a mini-explosion (called a flashback) in the premix chamber will occur. For that reason, it is necessary to choose the correct burner head for the particular fuel and

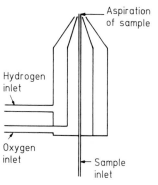

Figure 6-7. Schematic of a direct injection burner (reprinted from Robinson, J. W., 1990, by courtesy of Marcel Dekker, Inc. and the author).

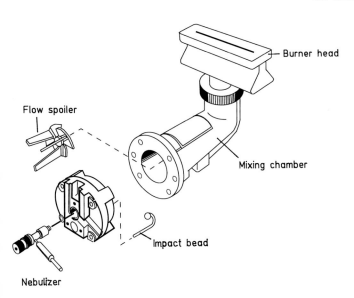

Figure 6-8. Premixed burner: The Perkin-Elmer dual option burner system. The burner consists of an adjustable nebulizer for sample aspiration, a choice of impact bead or flow spoiler to disperse the nebulized sample, a corrosion-resistant mixing chamber and corrosion-resistant burner head. The flow spoiler gives optimum precision and sensitivity with nitrous oxide/acetylene flames and gives extended analytical range; the impact bead gives maximum sensitivity using air/acetylene flames (courtesy of The Perkin-Elmer Corporation).

support gas chosen. For example, the AAS burner head chosen for the air-propane flame would have a width of 1.5 mm and a slot length of 10 cm. The air-acetylene burner head would have the same length but a width of 0.5 mm. A nitrous oxide-acetylene burner head would have a width of 0.45 mm and a length of 6 cm. A nitrous oxide-acetylene burner head could support an air-acetylene flame although the path length reduction would cause a decrease in analytical signal. However, an air-acetylene burner head could not be used to support a nitrous oxide-acetylene flame. A flashback would certainly occur. In practice, virtually all modern commercial AAS systems have safeguard systems that do not allow operation if the wrong burner head for a particular flame is installed. Also, automatic shutdown of gases occurs if the flame, for whatever reason, goes out.

Air-Acetylene

This is the workhorse flame of AAS. Approximately 35 elements are routinely and conveniently determined using this flame. Any flame temperature is somewhat dependent upon the stoichiometry (relative composition) of the fuel and support gas as well as the flow rates of the gases but, bearing this in mind, the temperature of this flame is about 2100 °C. Elements can be divided into two broad categories, flame intensive and flame sensitive. Atomization for elements in the first category is not particularly dependent upon flame stoichiometry (i.e., the ratio of oxidant to fuel) whereas atomization for elements in the second category can be very dependent upon flame stoichiometry. A stoichiometric flame is one in which the amounts of oxidant and fuel used are chemically equivalent to one another. An oxidizing flame is one in which there is an excess of oxidant; a reducing flame is one in which there is an excess of fuel. In the air-acetylene flame, copper and iron are examples of elements in the flame insensitive category and chromium and molybdenum are examples of elements in the flame sensitive category. In other words, for some elements, the ratio of fuel to oxidant in the gas mixture greatly influences the extent of free atom formation in the flame. Furthermore, the atoms may only be in a limited part of the flame. In comparison with AAS, the air-acetylene flame has found only limited use in AFS. This is partially due to the much lower background found with hydrogen flames which leads, in AFS, to much lower detection limits.

Air-Propane

The air-propane flame has a lower temperature than the air-acetylene flame, about 1800 °C, and also a smaller burning velocity. For that reason, a wide slot width can be used with the air-propane head. This flame provides slightly better detection limits for the alkali metals such as lithium, sodium, and potassium. This is because these elements are easily ionized and there is less ionization in the lower temperature air-propane flame as compared to the air-acetylene flame. These elements are commonly determined not only by AAS but also by flame emission spectrometers that can be dedicated, quite simple instruments using only filters for wavelength selection. These are popularly called flame photometers. The use of the air-propane flame is restricted to these few elements.

Nitrous Oxide-Acetylene

The air-acetylene flame has been described as the workhorse flame of AAS; the nitrous oxide-acetylene flame can be described as the salvation flame. Until 1965, only about 35 metals could be determined by AAS *using a pre-mixed burner.* However, approximately 32 metals could not be determined since these elements did not form free atoms in low temperature flames. In other words, oxidation and combination processes were predominant. In a significant paper, Willis (1965) showed that nitrous oxide could be used as an oxidant gas in conjunction with acetylene to provide an extremely convenient atom reservoir to allow the determination of many refractory elements such as, for example, aluminum, beryllium, silicon, titanium, and vanadium. The nitrous oxide-acetylene flame had the advantage of being a very high temperature source, about 3000 °C, but it also had a burning velocity low enough to allow its use with conventional pre-mixed burner systems. In sharp contrast, the oxygen-acetylene flame has only a slightly hotter temperature, about 3200 °C, but an extremely high burning velocity. This precludes its safe use in pre-mixed burner systems.

Particular care, however, must be taken with the nitrous oxide-acetylene flame. In almost all cases, the flame is not ignited directly. Instead, an air-acetylene flame is ignited and the flame made very fuel rich. At that point, the air in the oxidant line is replaced, via a valve switch, with the nitrous oxide. When the flame is extinguished, the reverse procedure is followed, to avoid flashback. Use of the nitrous oxide-acetylene flame almost doubled the number of elements that could be determined by AAS. It also allowed the same number of new elements to be determined by flame-AES. In fact, Pickett and Koirtyohann (1968) observed that the pre-mixed nitrous oxide-acetylene flame burner system designed for AAS was equally suitable for high temperature flame AES. With respect to relative detection limits, Pickett and Koirtyohann reported that 24 elements gave better (i.e., lower) detection limits by emission measurement than by AAS, and 17 elements were about the same. As a generality, elements having wavelengths above 320 nm were usually better determined by flame AES whereas lower wavelengths elements were better determined by AAS.

Entrained Air Flames

An entrained air flame is a flame where the oxidant (support gas) is replaced with an inert gas such as nitrogen or argon. The term inert is used in this case to describe a gas that does not burn or support combustion. A fuel is introduced into the burner in the usual way and combustion occurs when the fuel reacts with the oxygen in the atmosphere. In the case of a pre-mixed burner system, no flashback is possible since there is no explosive mixture in the premix chamber. If a hydrocarbon is introduced as a fuel, solid carbon particles are formed and continue to burn as they move upwards from the burner. The unburned carbon leaves the flame as soot. This makes these flames totally unsuitable for analytical purposes. If hydrogen is introduced as the fuel, however, an extremely low background flame is obtained. This is, in fact, the major advantage for using this entrained air (either N_2-H_2-EA or Ar-H_2-EA) flame for AAS and AFS. This is particularly important at the low wavelengths associated with some elements such as arsenic and selenium. It must be emphasized that the temperatures of these flames, particularly when aspirating aqueous solutions, can be as low as 300 °C. Due to the low temperature of the flame there is considerable chemical interference observed when aspirating actual samples into entrained air hydrogen flames. For that reason, these flames have been most often used with hydride generation techniques. These flames have also been used to monitor chemiluminescence molecular band emission for species such as HPO for phosphorus and S_2 for sulfur. This is the basis behind the popular flame photometric detector used in gas chromatography.

Other Flames

Although the oxygen-acetylene flame cannot be safely used with conventional pre-mixed burner systems, it can be used with direct injection burners. Oxygen-hydrogen flames, which have even higher burning velocities than oxygen-acetylene flames can also be used with direct injection burners.

The nitrous oxide-hydrogen flame, which has a temperature of about 3000 °C, can safely be used with a nitrous oxide-acetylene burner head but, although it is at roughly the same temperature as the nitrous oxide-acetylene flame, it does not provide the same reducing atmosphere to allow the efficient generation of atoms for refractory elements such as aluminum, vanadium, etc. and is, therefore, not used today.

Natural gas (normally mostly methane) can be used in inexpensive portable AAS units where low detection limits are not required. Air-natural gas has a temperature somewhat lower than air-propane with a roughly comparable burning velocity. Likewise, liquid fuels such as octane, pentane, and even kerosene have been used in portable AAS units. They offer no particular advantages except that gas cylinders do not have to be carried into the field. Balanced against that, however, a special second aspirator must be used to introduce the liquid fuel into the spray chamber. Liquid hydrocarbon fuels can be used with air as oxidant but not with nitrous oxide.

Separated Flames

A separated flame is created when a conventional flame is surrounded with a wall of an inert gas such as argon or nitrogen. The effect is to extend the inner core of the flame, where the majority of free atoms exist. This device is particularly useful with the high temperature nitrous oxide-acetylene flame. In flame AES, improved signal-to-noise ratios can be obtained, leading to improved detection limits and in AAS, separated flames can be used with purged (removal of oxygen) systems to allow the determination of some non-metals such as iodine, phosphorus, and sulfur at vacuum ultraviolet wavelengths. In AFS, separated flames remove intense radiation (from the flame) from the optical system, permitting the use of solar blind detectors and increasing detection limits (Robinson, 1990).

6.4.2.2 Vapor Generation and Electrothermal Atomization

In addition to the use of various flames for the generation of atoms in AAS and AFS, other atomization devices have also been used. These have been developed primarily for AAS but find applicability in AFS. In some cases they can be used to excite atoms for AES.

The Cold Vapor Method for Mercury

Mercury is a unique element as it is the only element (except for the noble gases) that shows an appreciable vapor pressure at room temperature. It is thus not necessary to provide a high temperature atom reservoir to determine mercury by AAS. As noted in Sec. 6.2.3, Woodson (1939) first took advantage of this phenomenon to determine mercury vapor in air. The technique was popularized, however, by Hatch and Ott (1968) who used the technique to determine mercury in a variety of geological samples. Today, the technique is sometimes known as the Hatch and Ott method, or more commonly, the *cold vapor method*. A reducing agent such as stannous chloride is added to the sample solution

which causes free elemental mercury vapor to be released. The mercury vapor is swept by a flow of air or nitrogen into a long path absorption cell in the light path of an atomic absorption spectrometer. A mercury HCL or EDL is used as the light source, and the absorption by the room temperature vapor is measured. Water vapor is normally removed through a drying tube. Conventional air-acetylene AAS gives a detection limit of about one ppm; the cold-vapor mercury AAS technique gives detection limits of about 0.05 ppb. In other words, more than a thousandfold improvement in detection limits can be obtained using the cold-vapor technique. Many variants of the originally popularized technique have been used including some involving cold-vapor AFS. Dedicated AAS instruments for mercury are commercially available; they have the advantage of being relatively portable and inexpensive.

Hydride Generation Techniques

A vapor generation method, using an apparatus somewhat similar to that used for mercury, can be used for elements which form volatile hydrides. These elements are antimony, arsenic, bismuth, germanium, lead, selenium, tellurium, and tin. In practice, hydride generation techniques are used mostly for arsenic and selenium. In 1969, Holak reported on the determination of arsenic by AAS following arsine (AsH_3) evolution. The arsine, generated by a zinc-hydrochloric acid reaction, was collected in a liquid nitrogen trap and then warmed and passed into an air-acetylene flame. This very simple introduction technique overcame many of the problems associated with conventional flame AAS and drastically reduced detection limits. The principal advantages of using hydride generation in conjunction with

Table 6-1. Atomic spectroscopy detection limits (microgram/liter) (courtesy of Perkin-Elmer Corporation).

Element	Flame AS	Hg/hydride	GFAA	ICP emission	ICP-MS
Ag	0.9		0.005	1	0.04
Al	30		0.04	4	0.1
As	100	0.02	0.2	20	0.05
Au	6		0.1	4	0.1
B	700		20	2	0.1
Ba	8		0.1	0.1	0.02
Be	1		0.01	0.06	0.1
Bi	20	0.02	0.1	20	0.04
Br					1
C				50	50
Ca	1		0.05	0.08	5
Cd	0.5		0.003	1	0.02
Ce				10	0.01
Cl					10
Co	6		0.01	2	0.02
Cr	2		0.01	2	0.02
Cs	8		0.05		0.02
Cu	1		0.02	0.9	0.03
Dy	50				0.04
Er	40				0.02
Eu	20				0.02
F					100
Fe	3		0.02	1	1
Ga	50		0.1	10	0.08
Gd	1200				0.04
Ge	200		0.2	10	0.08
Hf	200				0.03
Hg	200	0.008	1	20	0.03
Ho	40				0.01
I					0.02
In	20		0.05	30	0.02
Ir	600		2	20	0.06
K	2		0.02	50	10
La	2000			1	0.01
Li	0.5		0.05	0.9	0.1
Lu	700				0.01
Mg	0.1		0.004	0.08	0.1
Mn	1		0.01	0.4	0.04
Mo	30		0.04	5	0.08
Na	0.2		0.05	4	0.06
Nb	1000				0.02
Nd	1000				0.02
Ni	4		0.1	4	0.03
Os	80				0.02
P	50000		30	30	20
Pb	10		0.05	20	0.02
Pd	20		0.25	1	0.06

Table 6-1. (continued)

Element	Flame AS	Hg/hydride	GFAA	ICP emission	ICP-MS
Pr	5000				0.01
Pt	40		0.5	20	0.08
Rb	2		0.05		0.02
Re	500			20	0.06
Rh	4			20	0.02
Ru	70			4	0.05
S				50	500
Sb	30	0.1	0.2	60	0.02
Sc	20			0.2	0.08
Se	70	0.02	0.2	60	0.5
Si	60		0.4	3	10
Sm	2000				0.04
Sn	100		0.2	40	0.03
Sr	2		0.02	0.05	0.02
Ta	1000				0.1
Tb	600				0.01
Te	20	0.02	0.1	50	0.04
Th					0.02
Ti	50		1	0.5	0.06
Tl	9		0.1	40	0.02
Tm	10				0.01
U	10000			10	0.01
V	40		0.2	2	0.03
W	1000			20	0.06
Y	50			0.2	0.02
Yb	5				0.03
Zn	0.8		0.01	1	0.08
Zr	300			0.8	0.03

AAS include preconcentration, separation of the analyte from potential matrix interferences, and highly efficient sample introduction. Table 6-1 compares the advantage of flame, electrothermal atomization, and hydride generation AAS for the detection and quantification of the hydride forming elements, as well as other elements by AAS and AES.

Two different, generally useful reduction techniques have been used to generate hydrides. The Zn/HCl reduction method was the first system to gain acceptance. Sulfuric acid has occasionally been used in combination with or instead of hydrochloric acid. In this method, acidic samples in solution are treated with reducing agents to generate the covalent hydrides. However, a better method of hydride generation is to use a sodium borohydride acid reduction. The hydride formed by either method is swept from the reaction vessel using an inert gas into a suitable atom reservoir. The element may be measured by AAS, AFS, or AES. In the latter case, the atom reservoir is almost always an ICP (inductively coupled plasma). It should be noted that the sodium borohydride reaction can also be used to release elemental mercury vapor, but not as the hydride. Hershey and Keliher (1989) have recently published an in depth review article on the current status of arsenic and selenium determination using hydride generation AAS and AES. Nakahara (1990) has reviewed hydride generation as a sample introduction technique.

Electrothermal Atomization Techniques

Electrothermal atomization (ETA) techniques (also referred to as furnace techniques) for atomic spectrometry, most often AAS, provide an extremely useful alternative to chemical combustion flames for the generation of atoms. Commercial ETAs are small, electrically heated tubular furnaces, usually made of graphite (Fig. 6-9). The first report of an electrically heated carbon tube furnace came from Boris L'vov in 1961. Although his work showed the very high sensitivity which has popularized the use of carbon furnaces, L'vov's design was very complex. In 1968, Massman described a much simpler furnace. Modifications of the Massman furnace were introduced commercially in 1969 (Koirtyohann, 1980). Although there have been a great number of designs of these devices, the basic principle underlying the

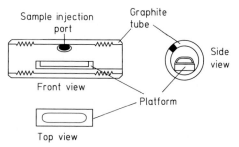

Figure 6-9. Graphite furnace tube with L'vov platform. The graphite furnace allows determination of over 40 elements in microliter sample volumes with detection limits 100–1000 times better than flames. The tubes and platforms are usually made of pyrolytic graphite. The graphite is resistively heated to atomize the sample placed on the platform. The role of the L'vov platform is to delay atomization until the furnace atmosphere has reached equilibrium conditions (courtesy of The Perkin-Elmer Corporation).

technique is extremely simple. A small sample (about 50 to 100 µg) is placed in the furnace and a flow of an inert gas, usually argon or nitrogen, is directed around the furnace. The furnace is then heated, by resistance heating, until a drying temperature of just over 100 °C is obtained. This releases the solvent vapor. This is referred to as the drying step. At the end of the drying step, the sample will be present as a layer of dried salts. In the second step, called the ash, char or pyrolysis step, the sample and furnace are heated to between 800 and 1500 °C. The actual temperature will depend upon the particular element and sample to be determined. This causes the decomposition and removal of organic materials as well as the removal of relatively volatile inorganic compounds. Finally, the sample and furnace are heated to a temperature in excess of 2000 °C (again, the actual temperature depends upon the element and sample) and the metal atoms are released into the optical path of the AAS unit. This is called the atomization step. Modern commercially available ETA systems for AAS include power supplies where the operator chooses the time and temperature for each step. The furnace may be heated gradually or ramped. In some cases, the solvent can boil too vigorously and the ramp procedure can prevent losses by spattering.

Several important points must be made about ETA-AAS techniques particularly as they relate to flame-AAS techniques. The use of ETA devices as atom reservoirs results in detection limits that are orders of magnitude below those that can be obtained with flames. This due to the fact that the dilution factor is about 1000 times less for a sample in a furnace than in a flame atomizer. Furthermore, ETA devices require only very small amounts of sample which can be of great importance when the analyst is sample limited, as, for example, with biological samples such as blood. ETA signals are, however, *transient*, as opposed to *steady state* for flame-AAS, and it is necessary for the AAS system to respond rapidly to the dry, ash, and atomization steps.

Background and matrix interference effects can be significantly larger in furnace AAS than in flames, due to the same concentration factor that gives furnace AAS its sensitivity. Background correction is critical in furnace AAS and will be discussed in Sec. 6.4.5.

In order to provide accurate and precise analysis, the sample must be atomized into a thermally stable environment. This can be accomplished through the use of a pyrolytic graphite platform, known as a L'vov platform, in the furnace. The function of the platform is to delay the vaporization and atomization of the sample until the furnace atmosphere has reached equilibrium conditions.

ETA methods often require the use of a matrix modifier, a reagent added to the sample in the furnace to permit as high a temperature as possible during the ashing step. The matrix modifier acts either to retain the analyte in a non-volatile form while allowing removal of the matrix, or to selectively volatilize matrix components, resulting in lower background signals. The use of matrix modification and the elaborate temperature programming needed make furnace AAS a slow analytical method compared to flame AAS or plasma emission, but recent advances in background correction and slurry sampling have dramatically increased sample throughput (Slavin et al., 1990).

While furnaces, like chemical flames and hydride techniques, are usually used for the analysis of solutions, a significant advantage is that they can be used for the direct analysis of solid samples. A weighed amount of solid may be placed in the furnace and the heating program adjusted to completely atomize the material. Solid sampling accessories are available for commercial furnaces. Slurries of solids may be analyzed, as discussed in papers by McCurdy et al. (1990), Carnrick et al. (1989) and Bradshaw and Slavin (1989). Robinson and coworkers developed a radiofrequency-heated carbon bed atomizer, the quartz "T", shown in Fig. 6-10, which has proven effective and highly sensitive for the direct analysis of solids, gases, and liquids (Robinson and Skelly, 1981, 1982, 1983 a, b).

Quite clearly, ETA devices should be used when the analyst is sample limited or when the metal present in the sample is at very low concentrations. Generally speaking, flame-AAS is useful at concentration levels down to about 0.1 ppm (100 ppb) whereas ETA-AAS can be used to determine many metals at sub-ppb levels.

Although ETA devices are used primarily in AAS, they have found some applica-

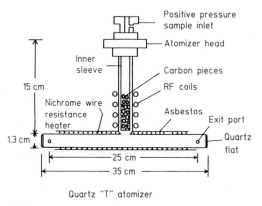

Quartz "T" atomizer

Figure 6-10. The quartz "T" atomizer. The carbon bed in the stem of the "T" is heated by radiofrequency induction and maintained at a temperature of 1450 °C. The atomizer, with its large carbon bed and slow flow rate, permits the complete decomposition and atomization of solid, liquid and gaseous samples with no sample preparation. The flow through design minimizes background absorption and eliminates volatilization losses (reprinted from Skelly, 1982).

tions in AES. Michel and Irwin (1988, 1990) have pioneered a technique known as ETA-LEAFS (for electrothermal atomization-laser excited atomic fluorescence). Also, *direct* emission from ETAs has been used for chemical analysis but this approach must be considered very secondary to their primary usage in AAS. Furnace emission spectrometry has largely been supplanted by the glow-discharge furnace emission approach known as furnace atomization nonthermal emission spectrometry (FANES), discussed in Sec. 6.4.2.3. ETAs have been used in AES where the ETA is used to dissociate the sample and the resulting gas flow is then introduced into an ICP to provide the extremely low detection limits associated with an ETA as well as the convenience, particularly for multi-element analysis, of the ICP. The introduction at the 1991 Pittsburgh Conference of a new commercial instrument combining the advantages of GF-AAS and ICP-AES may lead to more routine use of

furnace emission for analysis. Aurora Instruments (Vancouver, British Columbia, Canada) unveiled a graphite furnace-capacitivity coupled plasma (defined in Sec. 6.4.2.3), which atomizes samples using conventional graphite furnace techniques, but excited the atoms in a plasma. The combination reduces interference, provides large linear dynamic range and excellent detection limits, according to a recent report (Ciurczak and Miller-Ihli, 1991).

6.4.2.3 Plasmas

A plasma is an electrically conductive ionized gas. Fassel (1973) has defined a plasma as "any luminous gas in which a significant fraction (more than one percent) of its atoms or molecules are ionized". Although this definition would certainly include chemical combustion flames such as the air-acetylene and nitrous oxide-acetylene flames, it is conventional to restrict the term *plasma* to those supported by electrical means. Also, popular usage of the term plasma further restricts the term to flamelike plasmas, referred to by Fassel (1973) as electrical flames, so that non-flamelike plasmas such as the ac arc, dc arc, and ac spark would not be included. An important point must be noted here. Plasmas are generally thousands of degrees hotter than chemical combustion flames and generally exhibit the spectrum of the gas discharge. Although a variety of gases (including gas combinations) have been used for plasmas, the most common gas, by far, is argon. Plasmas may be divided into two general categories: static or low frequency plasmas, and high frequency plasmas. High frequency plasmas, in turn, may be further divided into single-electrode high frequency plasmas and electrodeless high frequency plasmas. The inductively coupled radiofrequency plasma

(ICP) and the microwave induced plasma (MIP) belong in this latter category. The direct current plasma (DCP) is an example of a static plasma. Keliher (1978) has given a complete in depth review of the historical development of various plasmas.

Energy must be applied to ionize the gas used to create the plasma. The energy may be applied (coupled) in different ways: resistive, in which the plasma is generated with an electric arc; capacitive, in which the plasma is produced by introduction of a gas into a capacitance; and inductive, in which the plasma is produced by introduction of a gas into an electromagnetic field created by an inductive coil.

The Direct Current Plasma (DCP)

DCPs are generated by a DC arc discharge where a primary arc is struck in a chamber between anode and cathode electrodes. The resulting plasma is a well defined temperature gradient column. A variety of electrode geometries have been used with DCPs and the electrodes themselves can be graphite, tungsten, or doped tungsten. DCPs have been called plasma jets, plasmatrons, or transferred plasmas. The first flamelike DCP was developed by Weiss (1954). Margoshes and Scribner (1959) were the first to realize the potential of the device for analytical chemistry. The *plasma jet* developed by Margoshes and Scribner used graphite disk electrodes for both the anode (lower electrode) and cathode (upper electrode). A standard DC arc power supply was used with very high plasma currents, about 15 to 20 amps. Margoshes and Scribner (1959) applied their plasma to the spectrographic analysis of stainless steel for chromium, iron, and nickel. Developments on various types of DCPs continued through the 1960s and 1970s, as described by Keliher (1978).

Elliott (1971) described a right-angle two electrode plasma which he called a *SpectraJet*. This was a very important development since the original SpectraJet is the ancestor of the only present day commercially available DCP. The original SpectraJet was manufactured by Spectra-Metrics, Inc. (Danvers, MA) as part of a high resolution echelle grating spectrometer. It is important to note that the Elliott DCP was made specifically to be used with an echelle grating spectrometer. As of this writing, the SpectraSpan 7 from Fisons/ARL Instruments is the only commercially available DCP. Its three electrode DCP resembles an inverted "Y" and a schematic is shown in Figure 6-11. There are two bottom anodes set at an angle of about 60° and a top cathode. Argon is introduced through sheaths surrounding the electrodes. The aerosol is introduced at a rate of about 1.4 mL/min below the intersection of the two lower arc columns and the spectral observation region is a small area just below this intersection. Due to the optics of the echelle spectrometer, only this

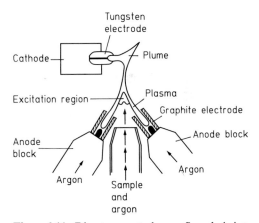

Figure 6-11. Direct current plasma. Sample is introduced as an aerosol between the graphite electrodes. Observation of emission in the region below the inverted "Y" junction minimizes plasma background (courtesy of FISONS Instruments/ARL).

small region is observed. Thus, this device is extremely well suited to be used with echelle optics and would be much less favorable on conventional grating spectrometers. Today, this DCP/echelle system is used in thousands of laboratories all over the world. Ebdon and Sparkes (1985) have published a comprehensive review of DC-P-AES and discussed the many diverse applications of the DCP, including the analysis of agricultural, clinical, geological, and metallurgical samples.

The Inductively Coupled Plasma (ICP)

An inductively coupled plasma is sustained by an induction coil connected to a radio frequency generator. The generator is usually operated between 4–50 MHz.

Reed (1961 a, 1961 b, 1962) successfully generated a stable argon plasma at 1 atm pressure using a commercial radio frequency heating unit. His torch consisted of a quartz tube with a brass base having a tangential gas entry, placed within the work coil of the generator. Reed's pioneering efforts in the design of ICPs led to their use for analysis by Greenfield and coworkers (Greenfield, 1965; Greenfield et al., 1964 and 1968) and Fassel and coworkers (Wendt and Fassel, 1965; Dickinson and Fassel, 1969).

Due to early observations of significant interelement interferences and the rapid growth of flame atomic absorption spectrometry during this time period, ICP spectrometry did not gain rapid acceptance as an analytical technique (Ingle and Crouch, 1988). The first commercial ICP spectrometer appeared more than 10 years after the first description of ICP sources (Ingle and Crouch, 1988).

A typical ICP source is shown in Fig. 6-12. The torch consists of two concentric quartz tubes surrounded by an induction

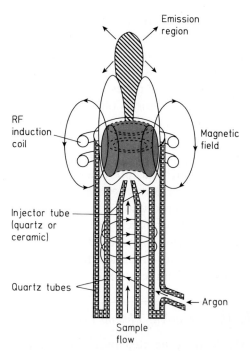

Figure 6-12. Inductively coupled plasma. The plasma is formed by a tangential stream of argon flowing between the quartz tubes and through the RF induction coil (reprinted from Robinson, J. W., 1990, by courtesy of Marcel Dekker, Inc. and the author).

coil. The plasma is formed by a tangential stream of argon flowing between the quartz tubes. Radiofrequency power is applied through the coil and an oscillating magnetic field is formed. A spark from an electrical discharge, such as a Tesla coil, creates *seed* electrons and ions in the argon. The gas becomes conductive; electrons and ions flow in a closed annular path, constrained by the induced magnetic field. This induced current heats the argon gas to temperatures near 10 000 K and causes additional ionization needed to sustain the plasma. The process of plasma formation is so rapid as to be instantaneous.

Commercial inductively coupled Ar plasmas have an annular or toroidal (*doughnut*) shape, which allows injection of sample aerosol through the center of the plasma. This ensures efficient atomization, an optically thin emission source (i.e., one with minimal self-absorption) and a chemically inert atmosphere. The high temperatures, relatively long residence times, and high electron densities found in the plasmas lead to minimization of matrix and ionization interferences, and a wide dynamic range.

Three separate regions can be identified in the argon plasma. The primary reaction zone, or plasma core is the region of highest gas temperature. It extends from inside the induction coil to a few millimeters above the coil and is characterized by intense continuum radiation. Desolvation, vaporization, atomization, ionization and excitation occur in the primary reaction zone but the region is of little analytical utility due to the high background radiation. The interconal or observation zone is a diamond shaped zone extending 1–3 cm above the induction coil. This region is optically thin and relatively free from continuum radiation; some background emission from Ar lines and OH bands can be observed. This is the most analytically useful portion of the plasma; emission from elements with a wide range of excitation energies can be measured with high signal/noise (S/N) ratios. The third region of the plasma is called the secondary reaction zone, tail flame or plume. Temperatures in this region are significantly lower, on the order of chemical combustion flames. Chemical recombination occurs, as evidenced by the appearance of oxide bands for elements such as yttrium, but the region is useful for analysis of easily excited elements.

The ICP is in many respects an ideal emission source (Ingle and Crouch, 1988), although spectral interference (overlap of emission lines) is common. These can be minimized by the use of a high resolution monochromator and judicious choice of analytical wavelength. The high temperature plasma provides numerous emission lines to choose from for a given element. Multielement analysis is possible with rapid scanning monochromators or by coupling the ICP atom reservoir to a polychromator.

ICP emission spectroscopy has grown rapidly since 1975, and has become a widely accepted technique for trace single and multielement analysis.

The Microwave Induced Plasma (MIP)

A microwave induced plasma is a discharge operating at a frequency greater than 300 MHz. Energy to an MIP is provided by a standing electro-magnetic wave resonating in a confined cavity or by surface microwave propagation along a plasma column, in a type of atmospheric pressure microwave plasma cavity known as a *Surfatron* (Goode and Baughman, 1984; Abdellah et al., 1982).

The MIP has become widely accepted as an element-specific detector for gas chromatography (GC). The GC-MIP, using a helium plasma, has been successfully developed as a selective detector for phosphorus, sulfur, the halogens, mercury, arsenic and antimony (Keliher, 1978). Relatively little emission from non-metals occurs in an argon plasma; the higher energy states in helium make He plasmas capable of producing both atomic and ionic emission from non-metals (Caruso et al., 1987).

Microwave induced plasmas have been generated with argon, nitrogen and helium. Because they operate at much lower power levels than ICPs, MIPs do not tolerate large amounts of sample or solvent. Direct injection of solution seriously affects plasma stability, but preliminary desolvation, ultrasonic nebulization and injection of gaseous analytes have proven successful (Ingle and Crouch, 1988).

Hewlett-Packard offers a commercial gas chromatography system with an atomic emission detector using an MIP as the atom reservoir. The system uses an innovative movable photodiode array detector to provide qualitative and quantitative multi-element analysis. Elements which can be measured and detection limits are shown in Table 6-2.

Arc and Spark Discharges

DC arc and high voltage spark discharges have been used as atomic emission atom reservoirs for over 100 years, although commercial instrumentation was only introduced in the 1940s.

Schrenk (1988) gives an excellent detailed history of high energy excitation sources in his review paper.

The DC arc is the most common arc used in atomic spectroscopic analysis. In 1874, Lockyer and Roberts used a battery-powered DC arc and a spectroscope to study the composition of alloys. They photographed the spectrum produced, becoming the first spectroscopists to successfully photograph a spectrum (Schrenk, 1988). Later studies by Lockyer used a DC generator developed by Siemens to power the arc.

The DC arc consists of a continuous discharge of $1-30$ A between a pair of electrodes. Arc temperatures fall in the $3000-8000$ K range; graphite electrodes are generally used, since metal electrodes melt or vaporize at these temperatures. The sample is placed on the anode (lower) electrode. Powders, chips or filings can be placed in a cup-shaped lower electrode. Solutions can be dried on electrodes and the residues analyzed. The sample is va-

Table 6-2. Hewlett Packard 5921A atomic emission detector for gas chromatography (courtesy of Hewlett Packard).

Element [a]	Wavelength (nm)	Minimum detectable level (pg/s)	Selectivity over carbon	Range	Compound
Carbon	193	1	–	2×10^4	nitrobenzene
Carbon	248	4	–	2×10^4	nitrobenzene
Carbon	248 (2nd order)	15	–	2×10^4	nitrobenzene
Hydrogen	486	4	–	5×10^3	nitrobenzene
Hydrogen	656	2	–	–	nitrobenzene
Deuterium	656	8	50 (over H)	1×10^4	n-decane (perdeuterated)
Chlorine	479	40	8 000	1×10^4	trichlorobenzene
Bromine	478	60	2 000	1×10^3	bromohexane
Fluorine	690	80	20 000	2×10^3	fluoroanisole
Sulfur	181	2	8 000	1×10^4	tertiarybutyl-disulfide
Phosphorus	178	1	5 000	1×10^3	triethylphosphate
Nitrogen	174	50	2 000	2×10^4	nitrobenzene
Oxygen	777	120	10 000	5×10^3	fluoroanisole
Silicon	252	85	1 000	2×10^3	tetraethyl-orthosilicate

[a] Other elements which have no specifications, but have been measured in the 1 to 100 pg/s range are: tin (303 nm), lead (406 nm), mercury (185 nm), iodine (184 nm).

porized into the arc and emission intensities are integrated, either photographically or electronically, over the entire burn time, a period of a few minutes duration.

More than 70 elements have been excited in a DC arc; detection limits are excellent, but the precision is poor. The arc does not remain on one position between the electrodes, but wanders from side to side, decreasing the accuracy and precision of data. The arc temperature depends on the sample matrix, which can result in serious matrix interference in quantitative work. The outer portion of the arc is cooler than the center, resulting in self-absorption of emitted radiation. Selective volatilization, i.e. the atomization of more volatile elements first, spectral overlap and fluctuating emission intensities with time all contribute to difficulties in quantitative analysis and to the use of the DC arc primarily for qualitative and semiquantitative work.

Arcs have also been used as atomizers for other atomic emission sources. Hieftje and coworkers used a microarc discharge, consisting of a tungsten cathode and stainless steel anode, to introduce microliter volumes of solution into MIPs and ICPs (Keilsohn et al., 1983).

The high voltage spark discharge is an intermittent discharge, lasting only a few microseconds. The electrode material is sampled many times from different places on the surface, which improves the precision of the analysis. Samples for spark excitation can be machined into proper shape to serve as an electrode or mixed with graphite and pressed into an electrode. Solutions can be analyzed with a rotating disk electrode. Less than microgram quantities of material are vaporized by an individual spark; this minimizes matrix and interelement effects but contributes to poorer detection limits in comparison to the DC arc. Selective volatiliza-

tion is also minimized by the short, high energy sampling.

The process that occurs during the buildup and decay of a spark is complex and not completely understood. A qualitative description of spark formation and decay is given by Ingle and Crouch (1988).

Spark discharges are most often used in the analysis of solid samples and they are used for the introduction of solid samples into other atom reservoirs, such as plasmas. The Jobin-Yvon Division of Instruments S.A. makes a commercial spark ablation sampler for its ICP emission spectrometer, in which the spark samples and vaporizes solid samples into the gas stream of the ICP torch.

Spectra produced with a spark discharge are complex and suffer from high background, but the spark discharge plays an important role in the analysis of solid samples (Ingle and Crouch, 1988).

The Glow Discharge (GD)

The glow discharge is a reduced pressure gas discharge, generated between two planar electrodes in a tube filled with a gas such as argon. The gas is present at a pressure of a few torr; the discharge is produced by bombardment of the cathode with filler gas ions and excitation of the ejected (sputtered) atoms of cathode (sample) material by collision. The emission spectrum of the metallic cathode is produced. Grimm (1968) designed a glow discharge lamp in which conductive solid samples served as part of the cathode (Fig. 6-13). Samples could be readily inserted for analysis. A Grimm discharge source-optical emission spectrometer is commercially available from Jobin Yvon Division, Instruments S.A. (Fig. 6-14).

Spectra produced by a glow discharge source exhibit lower background and nar-

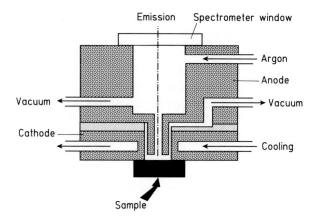

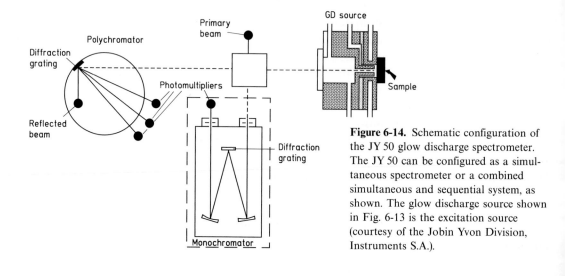

Figure 6-13. Glow discharge source. An emission spectrum of the sample, which is part of the cathode, is produced by a low pressure gas discharge between the two planar electrodes (courtesy of Jobin Yvon Division, Instruments S.A.).

rower emission lines than spark sources. Linear calibration curves are achieved over a wide range of concentrations. The GD-emission technique exhibits much lower levels of interelement and matrix interferences than arc and spark emission. This allows the use of the same calibration standards for different families of materials. The Jobin Yvon instrument is capable of analyzing both conducting and nonconductive samples. Non-conductive compacted powders can be analyzed without coating or blending a conductive material with the sample. Detection limits are on

the order of 0.1–10 ppm for most elements, with precisions of 0.5–2% RSD. Since the mechanism of sampling in the glow discharge is sputtering of atoms from the sample, the GD source can be used to provide both surface and depth profile analysis of samples.

A graphite tube is used both for atomization of a sample and as the hollow cathode of a glow discharge for excitation of the sample in the technique known as furnace atomization nonthermal excitation spectrometry, or FANES (Falk et al., 1981). The sample is pipetted onto the

Figure 6-14. Schematic configuration of the JY 50 glow discharge spectrometer. The JY 50 can be configured as a simultaneous spectrometer or a combined simultaneous and sequential system, as shown. The glow discharge source shown in Fig. 6-13 is the excitation source (courtesy of the Jobin Yvon Division, Instruments S.A.).

cathode and dried and ashed as in a conventional graphite furnace. A high current is applied to atomize the sample under low pressure, and an emission pulse signal is produced as the atoms are excited in the GD (Falk et al., 1984; Harrison et al., 1990).

The use of glow discharge techniques in analytical chemistry, with particular attention to atomic spectroscopy has been reviewed recently by Harrison et al. (1990).

6.4.3 Optical Systems

Instrumentation for atomic spectrometry is designed to disperse the light from the atom reservoir, to isolate the desired wavelengths from the rest of the spectrum emitted and to detect the desired wavelengths of light.

Instruments generally consist of an entrance slit to limit the area of the atom reservoir being viewed, a dispersing device such as a prism or diffraction grating, and some combination of exit slit(s) or aperture(s) and detector(s).

An instrument with one entrance and one exit slit of approximately the same size is called a *monochromator* – one wavelength is viewed at a time. Most commercial AAS systems and many commercial ICP systems (sequential systems) use monochromators. An instrument with multiple exit slits, allowing the simultaneous viewing of multiple wavelengths, is called a *polychromator*. Commercial AFS and simultaneous ICP systems employ polychromators. An example of a combination sequential and simultaneous ICP system is the JY 70 from Jobin Yvon Division, Instruments S.A., which uses both a monochromator and polychromator to provide rapid simultaneous analysis, but the flexibility of a sequential system (Fig. 6-15).

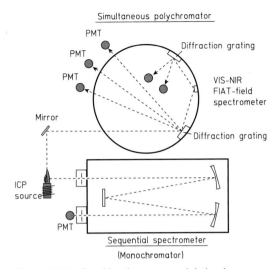

Figure 6-15. Combination sequential-simultaneous ICP emission spectrometer. Such a combination permits rapid multielement analysis using the polychromator and preselected wavelengths. The monochromator adds the flexibility to monitor additional elements or alternate wavelengths in case of spectral interferences (courtesy of Jobin Yvon Division, Instruments S.A.).

An instrument which combines a monochromator or polychromator with a photoelectric detector is called a *spectrometer*; an instrument which uses a large exit aperture and a spatially sensitive detector such as a photographic film is called a *spectrograph*.

Other components of optical systems focus, collimate, direct and modulate the light from the atom reservoir. An excellent review of basic optics and spectrometer components can be found in Ingle and Crouch (1988).

6.4.4 Detectors

Two types of detectors are generally used for atomic spectrometry, photon detectors and multichannel detectors. Detectors are characterized by their sensitivity, response time and noise.

6.4.4.1 Photon Detectors

Although a variety of photon detectors exist, most commercial atomic spectrometers rely on the photomultiplier tube (PMT). A photomultiplier tube contains a photosensitive cathode and a collection anode separated by a series of emissive electrodes (dynodes). Each dynode has a positive potential relative to the previous one. A photon striking the cathode ejects a photoelectron, which is accelerated to the first dynode, where several electrons are ejected; these electrons are in turn accelerated to the next dynode, where multiple electrons are ejected. This multiplication process continues to the anode, where a large number of electrons are collected for each photon striking the cathode.

The multiplication factor (the gain) is on the order of $10^4 - 10^7$ electrons per photon. Because of the high gain, PMTs are very sensitive. For reliable analytical data, the PMT power supply must be stable and vibrations minimized, and the PMT operated within its linear current range.

Window material and photoemissive surfaces vary, allowing optimization of the wavelength range detected. Selection of the correct PMT for the wavelength range to be measured is critical.

PMTs, like other photon detectors, exhibit a dark current, that is, an electrical output in the absence of radiation. In PMTs, the dark current is due primarily to thermal emission of electrons. At very low light levels, the dark current noise is the limiting source of noise in the detector (in other words, it is the factor in the detector which limits the precision of the measurement). At higher light levels, shot or flicker noise is more important than dark current noise.

6.4.4.2 Multichannel Detectors

Multichannel detectors provide simultaneous detection of the dispersed spectrum from an atom reservoir.

Photographic Film

Photographic detectors have been used historically in arc-spark emission instruments. Photographic films are sensitive detectors, and can be optimized for different regions of the spectrum. The major drawback is the time required for both development and quantitative measurement of the density of the exposed portions of the film.

Photodiode Arrays

Photodiodes are extremely fast, simple, highly linear detectors. A photodiode is a p-n junction diode which produces current upon absorption of a photon. The limiting current produced is proportional to the amount of incident radiation.

Arrays of photodiodes are now being used as spectroscopic detectors on commercial equipment, replacing PMTs. Leco Corporation currently markets a simultaneous ICP system using a photodiode array detector, as does Hewlett-Packard in its previously discussed GC-atomic emission system.

6.4.5 Background Emission and Absorption

Chemical flames, arcs, plasmas, and other atom reservoirs often exhibit emission, absorption and fluorescence in the absence of any sample. This emission, absorption, and fluorescence is referred to as *background*. Samples placed in atom reservoirs can also exhibit emission or absorption due to species other than the analyte; such emission, absorption, or fluorescence is also referred to as background. The nature of the background and its effect on the analysis of a sample differs for each technique and will be discussed in turn.

6.4.5.1 Causes of Background Emission, Absorption and Fluorescence

In flames, background emission is of two types: continuous emission and band emission. Continuous emission can be caused by blackbody radiation from hot solid particles in the flame, by reactions such as those of CO and NO with oxygen atoms to produce CO_2 and NO_2, and by radiative recombination of ions and electrons (Ingle and Crouch, 1988). Band spectra are emission from molecular species, most often OH radicals, hydrocarbon radicals and the like.

Absorption for all types of flames is severe below 220 nm. Molecular species in the sample can give rise to broad-band absorption. Background fluorescence can arise from molecular species as well.

Plasmas are relatively inert chemically and operate at much higher temperatures than flames; consequently, background is significantly reduced compared to flames, but is not eliminated entirely. Continuum radiation is produced from ion-electron recombination and bremsstrahlung. Atomic emission lines (from argon and carbon) and band emission from OH, NO and other molecular species are present. Fluorescence may be observed from the molecular species present.

With electrothermal atomizers such as graphite furnaces, blackbody radiation in the visible region of the spectrum occurs and can pose problems for the analysis of elements such as barium. Intense band absorption and emission from molecular species in the sample are often observed.

Arcs exhibit severe background emission unless controlled-atmosphere excita-tion is used (Ingle and Crouch, 1988). During the early stage of spark formation, emission from ionized atmospheric species and continuum background are observed. During the sampling phase, intense continuum emission is observed near the cathode (Ingle and Crouch, 1988).

6.4.5.2 Background Correction

Background emission or absorption, if not measured and corrected, can result in serious analytical errors.

In emission methods, the preferred correction technique is to measure the background at a point or points away from the analyte emission line and to subtract the background reading from the analyte signal. Most modern instruments provide software with a variety of baseline subtraction routines, so that this may be done automatically. The analyst merely chooses the points at which the background should be measured. If the background for a given analyte emission line is very intense or irregular, choosing an alternate analytical line with less background may be the best course of action.

In fluorescence methods, modulation of the light source is normally used to discriminate against background emission from the atom reservoir and against thermal emission from analyte in the atom reservoir. Thermal emission would otherwise result in a direct interference with the fluorescence measurement. Modulation may be accomplished electronically or with a mechanical chopper. Modulation does not eliminate background fluorescence or light scattering at the source wavelength. With continuum sources, wavelength modulation with a quartz refractor plate, in conjunction with source modulation, can distinguish background fluorescence and light scattering from the analytical signal.

Non-resonance fluorescence can be used to completely eliminate scattering and to optimize the signal to background ratios.

Several methods are available for correction of broad-band (molecular) absorption in atomic absorption spectroscopy using line source (HCL, EDL) excitation: continuum source, Zeeman effect, and Smith-Hieftje methods are most common.

Continuum source background correction is available on most commercial AAS instruments. Light from the line source (HCL, EDL) is passed through the atom reservoir alternately with light from a continuum source, such as a deuterium lamp, tungsten lamp, or arc lamp. With the line source, absorption from the analyte and the background is measured. With the continuum source, only background absorption is measured due to the very narrow line width of atomic absorption lines compared to the spectral slit width. The different between the two signals is that due to the analyte. The method assumes that the background is uniform over the spectral band pass. Structured or non-uniform background will lead to positive or negative errors.

Zeeman background correction takes advantage of the fact that a magnetic field will split normally degenerate spectral lines into multiple lines with different polarization characteristics. Analyte and background absorption can be discriminated by the polarization of the lines. Usually only atoms are affected by the magnetic field; molecules and solid particles which cause the background absorption are not affected.

In the absence of a magnetic field, quantum mechanical states which differ only in the resultant total magnetic quantum number, M_j, are degenerate, i.e., they are all of equal energy. Only one spectral line would arise from these degenerate states.

In a magnetic field, there are slight energy differences between these states, giving rise to multiple absorption lines of slightly different energies. In normal Zeeman splitting, the magnitude of the splitting is directly proportional to the magnetic field strength. The normal Zeeman effect for the calcium absorption line at 732.6 nm is shown in Fig. 6-16. In the absence of a magnetic field, one transition at 732.6 nm occurs; in a magnetic field, three transitions occur, two of which are shifted away from the 732.6 nm line. The unshifted line is polarized parallel to the magnetic field (π), and the shifted lines are polarized perpendicular to the magnetic field (σ).

The magnetic field can be placed around the atomizer or around the source. For electrothermal atomizers, it is convenient to place the magnet around the atomizers and apply a transverse magnetic field. For an AC magnetic field, a polarizer which transmits only the σ component from the source is placed in the optical system. When the magnetic field is off, the analyte

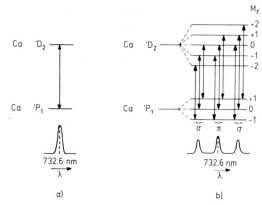

Figure 6-16. The normal Zeeman effect. (a) In the absence of a magnetic field, calcium exhibits a single absorption line at 732.6 nm. (b) In a magnetic field, the degeneracy of the electronic states is removed, and calcium exhibits three absorption lines, the unshifted 732.6 nm line, and two lines shifted to higher and lower wavelengths, respectively.

plus background signal is measured. When the magnetic field is on, the analyte σ components are shifted away from the monitored wavelength. Only the background absorption is measured. Subtracting the two signals results in the analyte absorbance.

The need to incorporate a polarizer in the optical system significantly reduces light throughput. If a longitudinal magnetic field is used around the atomizer, a polarizer is not needed, because the absorption π component does not absorb the source π component, which is out of plane. Perkin-Elmer Corporation (Nowalk, Connecticut) has just introduced a commercial graphite furnace atomizer with longitudinal Zeeman background correction, with a very high signal to noise ratio and maximum light throughput. The realization of the longitudinal Zeeman effect was made possible by use of a new high-density magnet material, which has the added advantage of reducing stray magnetic fields significantly.

Much larger and more powerful magnets would be needed if a flame atomizer were used instead of an electrothermal atomizer. As an alternative, the magnet can be placed around the source, and the emission line split as described above. This kind of Zeeman correction can be used with any kind of atomizer. However, special stabilized lamps (EDLs or glow discharge) are required as sources.

The advantages of the Zeeman correction technique are

(1) only one source is required, eliminating problems of intensity differences, alignment and drift which occur with two sources, and permitting the correction of very high background signals;

(2) background correction takes place at the analytical wavelength or very close to it, resulting in more accurate correction of

structured, changing or otherwise non-uniform background.

The major disadvantage, aside from cost and complexity, is that calibration curves exhibit greater non-linearity than uncorrected or continuum-corrected curves, and tend to *roll-over*, i.e. exhibit decreasing signal with increasing concentration at absorbances of approximately 0.88 or greater.

The Smith-Hieftje background correction system employs a pulsed hollow cathode lamp (Smith and Hieftje, 1983). The hollow cathode lamp is operated at normal current for part of a modulation cycle, and then is pulsed at high current (100–500 mA) for part of the cycle [Fig. 6-17(a)]. At high current, the normal narrow emission line profile becomes broad and exhibits self-reversal [Fig. 6-17(b)]; in other words, it looks more like a broad-band

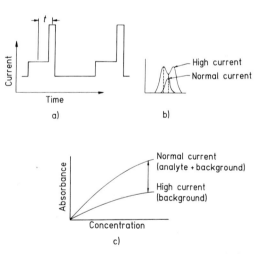

Figure 6-17. (a) Schematic of the current modulation applied to an HCL using Smith-Hieftje background correction. Absorbance is measured at two points separated by time t. (b) Emission line profiles at normal and high current showing the line-broadening and self-reversal at high current. (c) Calibration curves during normal and high current operation. Subtraction of the two signals results in the corrected analyte absorbance [modified from Smith and Hieftje (1983) with the kind permission of the Society for Applied Spectroscopy].

source than an atomic source. The amount of analyte absorbance during the high current pulse is effectively reduced to nothing, and only the background is measured. At normal current, both analyte and background absorbance are measured as usual; subtraction of the high current absorbance from the normal current absorbance results in the analyte absorbance [Fig. 6-17(c)].

This method of background correction requires that HCLs be stable when pulsed, but has the advantage of using only one source and does not require the addition of polarizers to the optical system. The Smith-Hieftje technique is simple and inexpensive compared with Zeeman systems and can be used with any atomizer. Nonlinearity and roll-over do occur, but are less severe than with Zeeman correction. Thermo-Jarrell Ash atomic spectrometers incorporate the Smith-Hieftje background correction system.

Emission from flames is eliminated as an interference in atomic absorption spectroscopy by modulating the light source, as described above for atomic fluorescence. Background correction techniques for AAS have been reviewed by Carnrick and Slavin (1989).

6.5 Sample Introduction

Quantitation in atomic spectroscopy is achieved by comparing the response of a known standard containing the analyte of interest with the response of an unknown sample. It is critical to quantitation that the sample and standard be introduced into the atom reservoir in a uniform and reproducible manner.

Samples to be analyzed by atomic spectroscopy will be solid, liquid or gaseous. The ideal sample introduction system will efficiently transfer the sample to the atom reservoir, be free from interferences, reproducible over time, independent of the sample type, useful for all atomic spectroscopic techniques and have no memory effects (Sneddon, 1990). As might be expected, no one sample introduction system possesses all of the above characteristics. Sample introduction systems will be discussed according to the physical form of the sample to be analyzed.

6.5.1 Liquids

Liquids are the samples most commonly (and most easily) analyzed by commercial atomic spectroscopy equipment. Many samples, such as water, urine, plating solutions, oil and similar fluids, can be introduced into the atom reservoir as is or following dilution with an appropriate solvent. Solid materials can be dissolved and handled as liquid samples (the usual procedure for atomic absorption spectrometry and plasma emission spectrometry). Liquids are generally introduced into flames and plasmas via a nebulizer and spray chamber arrangement.

The choice of nebulizer is governed by the type of solution to be analyzed, the elements to be determined and the detection limits desired, as well as by cost and availability.

Nebulizers designed with fine capillaries tend to clog easily, making the analysis of solutions containing large amounts of dissolved solids difficult. Very acidic solutions can corrode metal nebulizers, and can result in analytical errors, such as erroneously high iron results from a stainless steel nebulizer. Solutions containing hydrofluoric acid or high concentrations of fluoride ion can etch glass nebulizers, resulting in contamination of samples and damage to the nebulizer.

Detection limits are related to the efficiency of the nebulizer. Detections limits generally decreased in the following order, irrespective of the analyte: V-type > pneumatic > thermospray > grid > ultrasonic.

6.5.1.1 Nebulizers

Pneumatic Nebulizers

Pneumatic nebulizers convert a liquid solution to a fine mist, which optimizes both transport into the flame or plasma and the processes of desolvation and atomization in the flame or plasma. There are several different types of pneumatic nebulizers in common use: concentric, cross-flow, Babington type, and grid or frit.

In concentric nebulizers, the channel for the nebulizer (aspiration) gas surrounds the capillary channel for sample solution. Concentric nebulizers are used in all commercial AAS instruments. These nebulizers are made of a variety of metal alloys, a fact which should be remembered if very concentrated acids or corrosive solutions are to be analyzed. They are sturdy, reliable, low cost and do not require a pump to deliver solution to the nebulizer. Concentric glass nebulizers, like the Meinhard nebulizer, are used extensively in ICP spectrometers, generally in conjunction with a peristaltic pump for sample delivery. The Meinhard nebulizer works well for dilute solutions, but tends to clog with solutions containing high concentrations of dissolved solids. While a clogged metal concentric nebulizer can be cleared by use of a fine, burr-free wire (available from the instrument manufacturer) or by cleaning in an ultrasonic bath, such rigorous treatment cannot be used on the fine glass capillary in Meinhard nebulizers. If the clogged nebulizer cannot be cleaned by

soaking, a fused silica needle, such as those used for sample injection onto capillary GC columns, can be used to clear the nebulizer. An additional drawback to the glass Meinhard nebulizer is that it cannot be used for solutions containing HF since HF dissolves glass. Concentric nebulizers work best with dilute (low solid/salt) solutions. Both metal and glass concentric nebulizers will handle organic solvent solutions as well as aqueous solutions.

Cross-flow nebulizers are so-called because the nebulizer gas and sample solution meet at right angles. Both fixed and adjustable cross-flow nebulizers are available, although Cresser (1990) notes that adjustable nebulizers of this type are difficult to align and prone to drift. Cross-flow nebulizers are similar to concentric nebulizers in performance, and are most useful for dilute aqueous solutions.

Babington-type nebulizers operate by passing a thin film of liquid sample over the exit port of the nebulizer gas. Babington-type nebulizers, such as the plastic V-groove nebulizer, are excellent for solutions with high concentrations of dissolved solids and have been used for analysis of slurries. The plastic V-groove nebulizer is also ideal for use with hydrofluoric acid solutions.

Grid and frit nebulizers, which operate by pumping solution through or over a porous frit or grid, have the advantages of using much less sample than other types of nebulizers and of being particularly good for organic solvent introduction into ICPs. Frits tend to clog and are easily contaminating, resulting in memory effects (the measurement of an erroneously high analyte concentration in a sample caused by "leftover" analyte from previous samples run through the nebulizer).

Grid nebulizers, on the other hand, work well with high concentrations of dissolved salts/solids and do not exhibit memory effects.

Thermospray Nebulizer

The thermospray nebulizer, developed by Vestal and coworkers (Blakley and Vestal, 1983; Vestal and Fergusson, 1985) consists of a heated stainless steel capillary through which the sample solution is pumped. The solution is heated to boiling and forms a fine mist of droplets or salt crystals. Thermospray nebulizers are more efficient than conventional nebulizers, but precision and corrosion problems have been reported (Elgersma et al., 1986).

Thermospray nebulizers have been coupled to HPLC and flow injection systems, since this nebulizer is capable of handling a wide variety of solution flow rates.

Ultrasonic Nebulizers

Ultrasonic nebulizers use high frequency sound waves to form an aerosol from a bulk solution. The aerosol is created on excitation of a solution with high frequency sound when droplets of liquid are broken off from the peaks of the waves generated in the solution. The droplet size is a function of the frequency of the sound waves used (Lang, 1962). The design and use of ultrasonic nebulizers has been reviewed by Denton et al. (1990).

Ultrasonic nebulizers are highly efficient and form a very fine homogeneous aerosol; this results in exceptionally low detection limits. Detection limits are often 2 orders of magnitude lower for ultrasonic nebulizers than for pneumatic nebulizers. Their disadvantages include complexity of operation, poor stability, memory effects, and, with some designs, the need for very large sample volumes (20–50 ml versus 1–2 ml for pneumatic nebulizer). The memory effects require long rinse times be-

Table 6-3. Comparison of nebulizer detection limits (microgram/liter).

	ICP/ ultrasonic nebulizer[a]	ICP/ pneumatic nebulizer[b]	ICP/MS[b]	GF-AAS[b]
Fe	0.06	1	1	0.02
Cu	0.07	0.9	0.03	0.02
Ca	0.009	0.08	5	0.05
Pb	2	20	0.02	0.05
As	2	20	0.05	0.2
Mg	0.01	0.08	0.1	0.004

[a] Denton et al. (1990); [b] data courtesy of The Perkin-Elmer Corporation.

tween samples, making these nebulizers slower than pneumatic nebulizers.

As can be seen from Table 6-3, an ultrasonic nebulizer coupled to an ICP provides detection limits comparable to graphite furnace AAS, with the multielement advantage of ICP, and limits comparable to ICP-MS at a much lower cost.

6.5.1.2 Liquid Sample Introduction into Graphite Furnaces

The sensitivity of the graphite furnace (for atomic spectroscopy) is so high that only very small quantities of sample are analyzed. Where a pneumatic nebulizer uses 3–10 mL of solution per minute, a typical graphite furnace aliquot for analysis is only 10–50 µl. It is obvious that the accuracy and precision of the graphite furnace analysis depends upon the ability to accurately and precisely deliver microliter quantities of solution to the furnace.

Solutions can be injected manually or automatically. Manual injection can be accomplished with a syringe or micropipette. Micropipettes with plastic tips are most commonly used, to avoid contamination problems from metals leached from syringes. It is advisable to use only clear plastic tips, and to acid-leach the tips before use, to minimize contamination problems. If manual micropipettes are used for sample delivery, fixed volume pipettes are generally more accurate than variable volume pipettes. Manual pipetting is inherently less precise than automated pipetting, especially when one needs to pipette sample, matrix modifier and addition of standards, as is normally required for furnace AAS.

Autosamplers take the tedium as well as the imprecision out of pipetting, and allow the unattended analysis of samples and standards. Two types of commercial autosamplers are available for furnace work-discrete samplers and aerosol deposition samplers. Discrete samplers, which inject the required volume into the furnace in the form of a large droplet (similar to the way an eye dropper dispenses solution), are offered by most AAS instrument manufacturers. An aerosol-deposition autosampler, which sprays the sample aliquot through a capillary tube and deposits it in a fine mist on the furnace, is available on Thermo-Jarrell Ash spectrometers.

Autosamplers combined with computer-controlled spectrometers can automatically dilute samples, add matrix modifiers, prepare working standards from a stock standard, and perform analyses by the method of standard additions.

Optimum precision and accurary for both manual pipetting and autosamplers is obtained with dilute aqueous solutions. The delivery of viscous samples and organic solvents is less precise, although addition of surfactants to viscous samples can improve the precision of delivery.

6.5.2 Solids

Solid materials often form the major part of an atomic spectroscopist's analytical workload, but as is clear from Sec. 6.4, with the exception of arc/spark and GD spectrometers, most commercial instruments have been designed for the analysis of solutions. This requires that the sample be converted into a soluble form with no loss of analyte or contamination during the process. Sample preparation techniques will be discussed in Sec. 6.6, but suffice it to say that the process of converting solid samples to solutions capable of being aspirated into an atom reservoir is the most time-consuming and often the most difficult part of an atomic spectrometric analysis. The direct analysis of solids is a very attractive idea to the practicing atomic spectroscopist.

The advantages of direct solid sampling are obvious: elimination of time-consuming decomposition techniques, elimination of analyte loss due to volatilization, insolubility, precipitation, adsorption onto glassware, elimination of dilution, which results in loss of sensitivity, and elimination of contamination of the sample during the decomposition process, to mention a few.

The disadvantages of direct solid sampling are perhaps not so obvious. Solids generally tend to be heterogeneous materials. In atomic spectrometric techniques only a small amount of sample ($<$ mg quanties) is analyzed, giving rise to the question of how representative is the portion sampled of the bulk composition. Direct solid sampling results in a high concentration of the matrix, often leading to strong matrix interferences. (The *matrix* is everything in the sample other than the analyte; matrix interferences are discussed in Sec. 6.6.) Solid powders generally exhibit a range of particle sizes, which can lead to variable atomization efficiencies. Quantitative analysis is difficult since standard reference materials are not available for many solid samples and the preparation of accurate solid standards, especially for trace elements, is not an easy task.

Direct solid sampling can be broken down into two main categories: methods which present solids and powders directly into atom reservoirs and methods which generate aerosols from solids external to the atom reservoir (Ng, 1990). Examples of the first category (direct injection) are arc, spark, and glow discharge emission systems, powder injectors for furnaces and plasmas, fluidized bed injectors and slurry injectors for furnaces. Examples of the second category (indirect injection) include laser and spark ablation samplers, electrothermal vaporizers and slurry nebulizers for flames and plasmas. An up-to-date review of many of these techniques may be found in Sneddon, 1990.

6.5.2.1 Direct Solid Sampling

Arc, Spark and GD Emission Systems

Arc, spark and glow discharge emission spectrometers have been used for many years for direct analysis of solids in the metals industry. As already discussed, samples are limited to those that conduct electricity. Non-conducting solids can be analyzed by mixing with high purity carbon or copper powder to provide conductance. Mixing, of course, dilutes the sample, introduces the possibility of contamination and may give rise to spectral interferences. For example, copper emission lines may overlap and obscure analyte emission lines. Quantitation is difficult without matrix-matched standards, and the precision and detection limits are poor compared to solution techniques. However, the ability to perform rapid semiquantitative analysis of solids with minimal sample preparation makes arc, spark and GD-AES attractive techniques for a variety of materials.

Electrothermal Solid Sampling Systems

Walter Slavin, a noted leader in the development of graphite furnace AAS, believes "that the graphite furnace is the best adapted of all spectroscopic methods for solid sampling" (Slavin, 1990). Residence time in the furnace is long compared to flame or plasma methods, resulting in more complete vaporization of the sample. The basic design of furnaces and other electrothermal atomizers makes the introduction of solids relatively easy; one doesn't have to worry about destabilizing a plasma or plugging a nebulizer orifice.

Graphite cup furnaces and modified tube furnaces have been developed for easy introduction of solid samples, especially powders. Cup furnaces generally exhibit lower sensitivity than tube furnaces due to shorter path lengths and the ease with which free atoms can escape into the atmosphere. Carry-over (contamination) from sample residues can require extensive heating of the furnace prior to analysis of another sample: disposable graphite cups can minimize this problem, although the cups are expensive. A variety of crucibles, boats, spoons and specially designed micropipets have been used to introduce solids and powders into furnaces (Ng, 1990). Atomic emission spectroscopy of solids inserted into furnaces has been used successfully, as in the LEAFS technique of Michel and coworkers already discussed.

Other types of electrothermal atomizers have been used for direct analysis of solids. The Robinson Quartz "T" atomizer, discussed in Sec. 6.5, was designed with a wide ($\sim \frac{1}{4}''$) orifice, allowing solid samples such as pieces of hair, biological tissue, paper, etc., to be dropped onto the carbon bed via a pair of tweezers. Samples could be introduced at a rate of approximately 1 per minute, with no need to cool or disassemble the atomizer. The flow-through design eliminated volatilization losses, and the sensitivity was increased in comparison to commercial tube furnaces due to the long path length and ability to handle relatively large samples.

The major drawback to quantitative analysis using the solid insertion techniques described above is that very small samples have to be weighed accurately and transferred completely to the furnace. A small error in transfer or weight will result in a large error in concentration of the analyte. (Static electricity can make weighing small amounts of powdered samples a very frustrating job.)

Many workers attempted to eliminate problems with handling powders by mixing them in liquids to form slurries. The work of Miller-Ihli in 1988 clearly demonstrated that the major cause of imprecision in the analysis of slurries was the rapid settling of sample particles. She obtained excellent recoveries for a variety of elements in NIST (National Institute of Standards and Technology) Standard Reference Materials (SRMs) by using a small ultrasonic agitator to stir the slurry as it was being sampled for graphite furnace-AAS. A relatively large amount of sample (10 mg or so) can be weighed and slurried for analysis, minimizing the effect of weighing errors on the results and increasing the probability that the sample taken truly represents the bulk composition. A commercial automated ultrasonic mixer for furnace autosamplers is available from Perkin-Elmer Corporation.

It is possible in many cases to optimize furnace parameters to achieve very high atomization efficiencies. This permits calibration with aqueous standards, either directly or by the methods of standard additions, discussed in Sec. 6.6.4.3, and avoids the need to obtain or prepare solid calibration standards.

Flame Solid Sampling Systems

Two early efforts to analyze solids directly by AES are described by Ng (1990): the rolled paper method of Ramage (1929) and the rotating disk method of Lundegardh (1934). In the method of Ramage, powdered samples were deposited on ashless filter paper, which was rolled up and burned in a natural gas flame. Lundegardh placed powdered samples on an asbestos disk and rotated the disk into an acetylene-oxygen flame.

Delves (1970) described an analytical technique to allow the rapid determination of lead in blood samples by flame AAS. The method has since been extended to allow the determination of other volatile metals in various samples, including solids. In the method, now known after its originator, an aliquot of untreated sample is placed into a small nickel cup. The nickel cup is placed into an air-acetylene flame and vaporized into a long, open-ended silica tube mounted in the flame. The elongated tube serves to keep the atoms in the light path long enough to be measured. The advantage of the method is that only very small samples are required and no pretreatment is needed, allowing rapid screening of complex matrices for volatile metals.

Modifications of the Delves cup include tantalum boats and metal rods for direct insertion of solid samples into flames.

The limitation to these approaches is that only volatile elements, such as lead, cadmium and mercury, can be determined due to the low atomization temperature attained. Precision is surprisingly good, generally 5–20% RSD is reported compared to 1–2% for solution work.

Plasma Solid Sampling Systems

Plasma sampling systems fall into two main groups, those that use cups, boats or rods similar to those described for flames and those that entrain powdered samples in the plasma gas. Pettit and Horlick (1986) developed an automated 24-cup carousel for direct introduction of solids into an ICP. The sample cups were made of graphite and computer control allowed optimization of position in the plasma. They also demonstrated the feasibility of using a variety of NIST SRMs as solid calibration standards.

Gravity-feed, mechanical agitation, high velocity gas jets, tangential gas flows and fluidized-bed chambers have all been used to suspend solid powders in a gas stream which carries them into ICPs for analysis (Ng, 1990). The samples have to be in the form of fine powders of uniform particle size for good results. Analytical data have been reported for a variety of materials, including NIST SRMs. Problems encountered included high background emission, intense atomic emission from major constituents, and poor precision due to powder aggregation and non-uniform uptake into the plasma (surges). The preparation of standards and samples is an area requiring more research before direct solid injection into plasmas is a routine analytical tool.

6.5.2.2 Indirect Solid Sampling

A number of external sources for aerosol or atomic vapor generation have already been discussed. These included glow discharge sources and spark ablation. Electrothermal vaporization as a sample introduction technique is summarized in Sec. 6.5.3.2. This section will focus on two techniques, slurry nebulization and laser ablation. Both techniques result in formation of an aerosol which is then introduced into an atom reservoir for atomization and excitation. The atom reservoir is usually a plasma or flame.

Slurry nebulization may gave some advantage over laser ablation in bulk analysis of extremely heterogeneous solids, such as coal, rocks and biological tissue, because the ability to homogenize a large sample (several grams) increases the likelihood of analyzing a representative sample (McCurdy et al., 1990). The amount of material actually sampled by the laser may be too small to permit bulk analysis of such materials without multiple sampling of the surface. On the other hand, laser ablation can provide spatial distribution and depth profiling information not available from slurry samples.

Slurry Nebulization

The aim of slurry nebulization, as stated by McCurdy et al. (1990) is to maintain the best features of flame, furnace and plasma analysis while minimizing the time, effort, cost and safety hazards associated with conventional preparation of heterogeneous solid samples. They have reviewed the topic in great detail, so only a summary of the more important points will be presented here.

Factors which affect the accuracy of a slurry analysis include particle size, aerosol transport efficiency, atomization efficiency, magnitude of matrix effects, and partial or complete solubilization of analytes. The ideal slurry method would not require sieving of the sample, would allow rapid preparation of large numbers of samples and would permit accurate results using aqueous calibration standards.

Traditional ball milling or shatterbox (swing) milling, used to prepare NIST SRMs, for example, results in powders with particle diameters too large for effective aerosol transport in conventional plasmas and flames. Conventional plasmas and flames have spray chambers of complex geometrics optimized for the transport of solutions, with cutoff diameters of <12 μm, and in some cases, <3 μm. These designs will exhibit severe mass transport losses for slurries.

A significant amount of research has been focused on preparation of small particle size slurries and on design and optimization of nebulizer/spray chamber assemblies. Three slurry preparation techniques appear to offer promise: use of an efficient particle size reduction mill, the McCrone Micronising Mill (McCrone Associates, Chicago, Illinois); use of the *bottle and bead* method developed by Ebdon and Lechotycki (1986) and Ebdon and Evans (1987), and for soft tissues, use of a sonic/cavitational tissue homogenizer, such as those from Brinkman and Tekmar. These techniques result in particles with median diameters <7 μm.

Conventional capillary nebulizers, ICP injector tips, AAS burner heads and narrow-bore peristaltic pump tubing are all prone to clogging with the aspiration of slurries. For ICP and DCP analysis, successful slurry nebulization can be achieved with Babington-type nebulizers and wide-bore pump tubing. The use of wider injector tips and burner slots (high solids burner heads) have been reported.

McCurdy, Fry and coworkers have demonstrated that use of an unobstructed, impactorless spray chamber, efficient particle size reduction to insure that all of the material is within the transport range of the spray chamber, and sharply reduced nebulizer flow rates to increase residence time in the plasma can result in 100% recovery of sulfur in a NIST coal SRM using only aqueous standards. Other researchers have reported variable success with slurry analysis, probably due to a combination of aerosol transport losses and incomplete atomization.

Slurry techniques generally result in a five-fold loss of sensitivity compared to solution analysis, due mainly to the reduced flow rates employed, and are subject to contamination from the milling equipment used. Additional improvements in time required for slurry preparation and commercially-available spray chambers and reliable slurry samplers for plasmas and flames are needed before slurry methods become routine for sample introduction into flames and plasmas.

Laser Ablation

When a high-powered laser is focused on a solid surface, light energy is converted into thermal energy, resulting in the vaporization and removal of small amounts of material from the surface. This technique of sampling solids, known as laser ablation or laser vaporization was developed by Brech and Cross in 1962. Although the laser can be used both to vaporize and excite emission from a solid sample, this approach often suffered from poor precision, poor sensitivity and matrix effects (Denoyer et al., 1991). Significant improvements were made by separating the sampling and atomization steps: the laser is used only to produce a sample aerosol, which is then transported to an atom reservoir, such as an ICP. This permits independent optimization of each step.

When laser light is absorbed by a solid, a variety of phenomena occur, including surface heating, melting, vaporization, dissociation and phase changes. The interactions of laser and target are very complex and not well understood, but obviously need to be controlled if quantitative data are to be obtained (Dittrich and Wennrich, 1990).

The amount of material evaporated from a solid surface depends on the laser characteristics and on the properties of the sample: reflectivity, density, specific heat, thermal conductivity and boiling point, to mention a few. Quantitative analysis of bulk composition depends on reproducible evaporation of sample and standards and efficient transport of the ablated sample to the atomizer. As is the case with other high energy sampling sources like arcs and sparks, laser ablation can result in selective volatilization of elements out of the solid, resulting in vapor phase concentrations which do not reflect bulk composition. The most accurate results are obtained with matrix-matched standards, although good semi-quantitative data can be obtained without them (Denoyer et al., 1991). Use of an internal standard when possible is highly recommended for quantitative work.

The laser spot size can vary from ~ 30 μm to 6 mm, so laser ablation has the ability to provide information on the spatial distribution of elements (e.g., inclusions) in a sample. Spatial resolution falls between that of ESCA (electron spectroscopy for chemical analysis) and Auger spectroscopy (Denoyer et al., 1991). Preparing standards for such analyses is virtually impossible, and the elemental information obtained is semi-quantitative.

Laser ablation can be used on both electrically conductive and non-conductive materials. The sample preparation techniques used for laser ablation are the same as those used for microanalysis. Large solid samples are polished or etched, powders are compacted with or without binder or are fused with lithium metaborate into disks. Samples can be embedded in epoxy and cut for microanalysis.

Laser ablation has been coupled with ICPs, DCPs, MIPs, arcs, and FANES systems, providing multielement detection by atomic emission from the ablated sample.

Lasers have been used to sample solids for flame and furnace atomic absorption spectrometry, with quantitative, single element detection. Commercial AA spectrometers can be used in conjunction with a solid state laser.

Laser ablation has been used with atomic fluorescence to provide absolute detection limits of $\sim 10^{-13}$ g (Kwong and Measures, 1979), but there are problems with stray light in AFS systems.

Laser ablation has been used as a solid sampling tool for the atomic spectrometric analysis of rocks, glasses, ceramics, metal oxides, pigments, alloys, high purity metals, plastics and biological tissues, among other materials. A good list of survey and review papers, and other references can be found in Dittrick and Wennrick (1990).

6.5.3 Gases

The introduction of a gaseous sample into an atom reservoir has significant analytical advantages over the introduction of solutions or solids. Transport efficiency is much higher (potentially 100%) than that of pneumatic nebulizers (usually 2–15%), and the sample is homogeneous (Nakahara, 1990).

6.5.3.1 Hydride Generation/Cold Vapor Introduction

The hydride generation technique and the cold vapor mercury technique are the most common gas phase sample introduction methods used. The hydride generation technique is applicable to As, Bi, Ge, Pb, Sb, Se, Sn and Te. The volatile hydrides are generated by treatment of a sample solution with acid and a reducing agent such as sodium borohydride, and the volatile hydride is swept into the atom reservoir. The cold vapor technique for mercury uses acid and a reducing agent to form elemental mercury vapor which is swept into the atom reservoir. Commercial systems, both manual and automated, are available from major atomic spectroscopy instrument manufacturers. Flow injection systems for automated hydride/cold vapor analyses are available (discussed in Sec. 6.5.4).

6.5.3.2 Electrothermal Introduction

Electrothermal devices, both modified commercial graphite atomizers and research designs, have been used to efficiently vaporize solutions and solids for introduction of gaseous samples into ICPs, DCPs, MIPs, and other atom reservoirs. The major advantage of electrothermal vaporization devices is that the vaporization step is separated from the atomization/excitation step, allowing both processes to be independently optimized. Electrothermal vaporization as a sample introduction technique has been reviewed by Ng and Caruso (1990).

6.5.3.3 Other Methods of Gas Sample Introduction

Gas chromatographs have been interfaced with atomic spectroscopy systems to provide element-specific detection. The sample introduced into the atom reservoir is, of course, a gas. GCs have been combined with AAS, AFS, and AES systems. The advantages of such a combined system are the ability to detect specific elements in eluted species (e.g., the ability to speciate organometallic compounds), and the ability to detect molecular functional groups by derivative element tagging, as well as the

efficient delivery of vaporized sample to the atomizer (Uden, 1990).

6.5.4 Flow Injection Analysis

The term flow injection analysis (FIA) describes an automated continuous analytical method in which an aliquot of sample is injected into a continuously flowing carrier stream, which may contain a reagent. A transient signal is monitored as the analyte or its reaction product flows past a detector (Ruzicka and Hansen, 1975; Tyson, 1985).

A FIA system consists of four basic parts, a pump for regulation of flow, an injection valve to insert sample volumes accurately and reproducibly, a manifold, the term used for tubing, fittings, mixing coils, and other parts used to carry out the desired reactions and a flow-through detector. In the case of FIA systems for atomic spectroscopy, the AAS, AFS or ICP spectrometer is the detector.

FIA techniques for atomic spectroscopy have been reviewed by Tyson (1985) and Valcarcel (1990), and a monograph on the topic has been published (Burguera, 1988). FIA methods for atomic spectroscopy include automated on-line dilution, automated cold-vapor mercury/hydride determination, on-line matrix-matching, on-line preconcentration and extraction for electrothermal AAS (Fang and Welz, 1989). Flow injection methods can handle samples with very high dissolved solids content (Mindel and Karlberg, 1981), and permit use of organic solvents that would otherwise extinguish the flame in flame AAS (Tyson, 1984).

FIA methods are advantageous in flame AAS by automating complicated sample pretreatment and preconcentration, dilution and preparation of standard additions. FIA permits the analysis of much smaller volumes (e.g., 200 μL) than used in conventional flame AAS. The sensitivity is less and the precision is generally poorer than that obtained with conventional nebulization (Tyson, 1985).

Modern AA spectrometers are equipped with electronics capable of handling the transient responses from FIA, which are similar to those generated by graphite furnaces. Commercial FIA systems are already available, such as the FIAS-200 from Perkin Elmer.

6.6 Practical Materials Characterization by Atomic Spectrometry

Characterization of a material by atomic spectrometry can be as simple (relatively-speaking) as identifying the presence of a specific element in a sample (qualitative analysis) or as complex as accurately determining the concentration of every metallic or metalloid element in a sample (quantitative analysis). One should keep in mind that atomic spectrometry provides information about the *elemental* composition of a sample; no information about oxidation state, chemical form or molecular structure is available directly from the atom reservoir, although such information may be provided from coupled techniques such as GC-AAS, ion-chromatography-ICP-AES, and so on. It should also be remembered that, in general, non-metals cannot be determined by atomic spectrometry.

Materials characterization is a process that includes collection of the sample to be brought to the laboratory, subsampling, sample pretreatment, acquisition of data, data processing and evaluation. While it is not possible in a chapter such as this to cover all aspects of the process in detail, the principles governing a practical analytical method will be considered. In all cases, the first step in the materials characterization process should be discussion of the problem to be solved with the analytical chemist. Information about the source of the sample, approximate composition, stability, reactivity, and so on will help the analyst (in this case, the atomic spectroscopist) to choose the appropriate technique for preparing and analyzing the sample. Such information improves the turnaround time, reduces costs and improves the chances of accurate and precise results being obtained.

6.6.1 Sample Collection, Storage and Handling

A great deal of time and money spent on analysis of samples will be wasted unless the samples have been collected, stored and prepared in such a way that they are representative of the original bulk material. It can be seen from Table 6-1 that most atomic spectrometric techniques have detection limits well below the ppm concentration range, so samples must be collected and stored to prevent loss of analytes by precipitation, adsorption or volatilization, contamination from external sources and concentration changes due to evaporation and other physical processes.

The sample brought to the laboratory must be representative of the bulk material to be characterized, i.e., the laboratory sample must reflect the true amount and distribution of the analyte in the bulk material. The application of statistics plays a major role in deciding how to obtain representative samples from large lots of material. Sampling of particulate materials has been studied by Gy (1979), and others. Procedures for sampling can be found in basic analytical chemistry texts, specialized texts on sampling, and in guidelines published by agencies which issue standard analytical procedures, such as the Environmental Protection Agency, and the American Society for Testing and Materials (ASTM). Sampling and sample handling are covered in the 2 volume NBS (now NIST) publication edited by Lafleur (1976). A basic tutorial on sampling is the text by Woodget and Cooper (1987). Collection of samples for specific analyses is often covered in specialized texts, for example, sampling for the determination of precious metals is discussed by Van Loon and Barefoot (1991), with specific procedures, sources of error and the like for these particular elements.

Solids are often heterogeneous and difficult to sample. Segregation due to particle size differences must be avoided. Melting a solid to reduce heterogeneity and sampling the molten material is a technique used for metals and alloys where practical. Samples from large pieces of metal are obtained by drilling, shaving or chipping. Large solid samples, or samples with a wide range of particles sizes, such as soil, must be crushed, ground and sieved to produce a homogeneous sample. Samples can be contaminated during these procedures by the metals in the drilling and grinding equipment. Van Loon and Barefoot (1991) note that malleable metals like gold can be lost from the sample by adhesion to grinding surfaces.

Pure liquids and gases are homogeneous, but mixtures can settle into layers due to differences in density. Laminar flow in liquid mixtures can result in inhomogeneity.

The container used for the sample can be a significant source of error. Most solid samples, such as alloys, polymers, glass, ceramics, and the like, may be placed in a clean glass or plastic container. Solids which are sensitive to light, air or moisture need to be packaged accordingly, in amber or foil-wrapped jars, under dry nitrogen, over desiccant. Solids such as biological tissues may need to be frozen or freeze-dried if not analyzed immediately after collection.

Liquids, especially aqueous solutions of trace metals, can easily be contaminated by their containers. Glass should not be used for aqueous solutions of trace metals. Problems encountered in the use of glass containers are loss of trace metals from solution by adsorption onto the container walls, and leaching of elements such as sodium from the glass into solution. High density linear polyethylene, polypropylene and Teflon bottles, with caps of the same material, are suitable for collection and storage of aqueous samples, after the bottles have been properly cleaned. All containers (and caps) must be acid-washed and rinsed in deionized reagent grade water. A cleaning procedure developing by Moody and Lindstrom (1977) is recommended, especially for cleaning of new containers. Containers must be dried and stored in a dust-free environment.

Dilute (<100 ppm trace metals) aqueous solutions should be acidified to $pH < 2$ with high purity nitric acid immediately upon collection. If silver is to be determined, a separate sample in an amber bottle should be collected. Special preservation is also needed for gold (Van Loon and Barefoot, 1991). Natural water samples may need to be filtered upon collection and prior to storage.

It is often beneficial for the analyst to provide cleaned containers, with the appropriate preservative, to the customer. Alternatively, pre-cleaned bottles and jars for sample collection are commercially available from several vendors.

The sample brought to the laboratory should be sufficient to allow replicate determinations, when possible. The analyst must insure that subdivision of the initial sample results in a representative sample for analysis and that homogenizing and drying procedures do not change the composition of the sample. Generally, solid samples should be dried to constant weight before being subdivided. Metal and alloy samples should be cleaned to remove cutting fluids before being weighed.

Liquid samples should be well-mixed prior to taking aliquots for analysis. If a liquid has separated into 2 or more phases, separate analysis of each phase is recommended. Liquid samples should be tightly capped to minimize evaporation.

Care must be taken in the physical manipulation of the sample, to avoid contamination. Samples must be protected from contaminants in the laboratory air, so containers should remain tightly closed as much as possible. Solid samples should be handled with talc-free vinyl gloves or with plastic tweezers. Analysts and persons collecting samples should be aware that hand lotions, cosmetics and similar personal care products often contain large amounts of metals and can be a significant source of contamination.

6.6.2 Sample Preparation

As discussed previously, some atomic spectrometers are capable of analyzing solid samples directly. Often, the solid sample must be ground to a fine powder, machined into a specific shape or polished to give a flat surface. These operations must be undertaken so as to avoid contamination of the sample with analytes of interest and to avoid loss of analytes during processing.

The most accurate and precise results are obtained from the analysis of dilute aqueous solutions. Although some samples are water-soluble and require only dissolution, most samples must undergo some chemical treatment to render the analytes of interest soluble in water. The art and science of sample decomposition for trace element analysis is best learned from the analytical literature. Comprehensive summaries of sample decomposition techniques have been compiled, such as the text by Bock (1979).

A brief overview of sample decomposition techniques is presented, with the understanding that detailed procedures may be found in textbooks on quantitative analysis, sample preparation and pretreatment, and in the analytical chemistry literature.

6.6.2.1 Inorganic Samples

Inorganic solid samples may be converted into soluble salts by digestion with hot concentrated mineral acids or by fusion with suitable salts such as sodium carbonate or lithium tetraborate. Mineral acids used to digest samples for trace metals determinations must be of very high purity. There are several commercial suppliers of high purity acids; such acids are packaged in Teflon bottles and are supplied with certificates of analysis. High purity sub-boiling distilled acids are also available from the National Institute of Standards and Technology.

Nitric and hydrochloric acids are most frequently used for atomic spectroscopic work, although hydrochloric acid may result in loss of volatile chloride species in graphite furnace AAS work if care is not taken in developing the furnace heating program. Hydrofluoric acid is needed to dissolve silica-based matrices and some metals and alloys, such as tungsten and niobium. Loss of volatile fluorides (e.g., SiF_4, BF_3) and precipitation of the insoluble rare earth and alkaline earth fluorides must be considered when HF is used. Sulfuric acid is used when required but the purity, from the standpoint of trace metals content, is usually less than that of nitric or hydrochloric acids and several important metal analytes (e.g., Ba, Pb) form insoluble or sparingly soluble sulfates. Obviously, sulfur cannot be determined in sample solutions containing sulfuric acid. Perchloric acid, phosphoric acid and fluoboric acid, as well as hydrogen peroxide solutions are also used to digest inorganic samples. Many materials require mixtures of acids to effect dissolution; the best known example is the use of a mixture of hydrochloric and nitric acids (aqua regia) to dissolve some of the platinum group metals and gold.

Fusions with alkaline or oxidizing compounds at high heat may be necessary to solubilize refractory materials, ores, minerals and similar matrices. The flux material must be of high purity, cannot contain an analyte element (Li, Na, K salts are commonly used for fusions), and must be carried out in a crucible that will not contaminate the sample or alloy with analyte elements. Fusions result in solutions with high concentrations of dissolved solids, so care in choosing a nebulizer is needed, and matrix-matching of standard may be necessary for accurate analyses (see Sec. 6.6.4.3).

Inorganic solutions may require digestion to convert analytes to the same chemical form or oxidation state, as is needed for cold vapor mercury and hydride techniques.

6.6.2.2 Organic Samples

Organic samples, such as biological material and polymers, are usually prepared for atomic spectrometric analysis by removing the organic matrix through application of high temperatures or digestion with hot concentrated mineral acids. Decomposition of organic samples in a furnace at high temperatures is called ashing or *dry ashing*, while mineral acid digestion is often referred to as *wet ashing*. Dry ashing is often done in a muffle furnace, with the temperature slowly raised from room temperature to 600–800 °C, although low temperature *plasma ashing* units using excited oxygen are also available. The organic material burns away to leave an inorganic residue which is dissolved in water or dilute acid for analysis. Loss of volatile elements from the sample can result in inaccurate analysis.

Wet ashing of organics generally requires a mixture of hot oxidizing acids (nitric, perchloric, sulfuric) to convert the matrix to CO_2 and water while solubilizing the inorganic constituents. Nitric and perchloric acids will oxidize most organics; however, extreme care should be exercised to avoid the hazards associated with perchloric acid. Samples should be heated first with nitric acid to destroy easily oxidized materials, and cooled prior to addition of the perchloric acid. Perchloric acid digestions should be carried out in a special fume hood (see Sec. 6.6.3). Sulfuric acid and hydrogen peroxide are an effective digestion mix, particularly for samples in which antimony is to be determined. Use of perchloric acid can cause volatilization and loss of antimony.

In some cases, organic solids and liquids can be dissolved or diluted in solvents such as xylene or 4-methyl-2-pentanone (also called methylisobutylketone or MIBK) and analyzed without further treatment. Metals in engine oil are often measured by atomic emission after dilution of the oil with kerosene or xylene, for example. Organic solvents can be used to extract and concentrate metal complexes from aqueous matrices; lead can be complexed with EDTA and extracted into MIBK for determination by atomic absorption spectrometry.

Sample digestions and ashing are carried out in some type of vessel or container. Flasks, beakers, evaporating dishes and crucibles are available in a variety of materials. The material chosen depends on the analytical requirements. Plastic labware (made of polyethylene, polypropylene, polystyrene, Teflon®, Nalgene®) will resist attack by hydrofluoric acid, and is useful if low levels of sodium (a common contaminant in glass) are to be determined. Plastic labware is limited to low temperature dissolutions, although heatable beakers made of Nalgene® with graphite bottoms are available and can be used directly on hot plates at temperatures up to 260 °C. Most common laboratory glassware is borosilicate glass (e.g., Pyrex®) and cannot be used for digestion when boron is to be determined; quartz and Vycor® are boron-free alternatives. Quartz vessels can also be used to advantage if low levels of sodium are to be measured in the sample. Quartz, Vycor®, Pyrex® and other glasses are not resistant to hydrofluoric acid. Crucibles of platinum, nickel, zirconium, porcelain and Vycor® may be used for fusions; crucible material depends on the sample, the elements to be determined and on the flux to be employed.

6.6.2.3 Microwave Digestion

Conventional wet acid digestions carried out on hot plates or over flames in open vessels are subject to contamination and loss of volatile elements, such as mercury, arsenic, selenium, tin and boron. These digestions often require hours of heating to effect decomposition. To eliminate loss of volatiles, and to increase the rate of decomposition by digesting samples at elevated pressures and temperatures, sealed vessels such as the Carius sealed quartz tube and Teflon-lined steel-jacketed bombs (e.g., Parr bombs, Parr Instruments Co., Moline, Illinois) have been used. The Carius tubes or bombs are heated in a conventional oven or furnace. These sealed vessel methods are time-consuming and can be hazardous; Carius tubes frequently explode unexpectedly.

Heating with microwave energy in acid digestions was first demonstrated by Koirtyohann et al. in 1975 (Abu-Samra et al., 1975). For ten years, use of microwave heating for sample digestion remained a curiosity, with studies carried out in home-appliance microwave ovens and open beakers or closed plastic bottles (Kingston and Jassie, 1988). Beginning in 1984, renewed interest in microwave-based sample preparation, sparked by the development of commercial microwave ovens designed for laboratory use, has resulted in a proliferation of microwave dissolution techniques for a wide variety of materials. Sample digestion can take place in 3–5 minutes instead of hours, in inert, sealed Teflon vessels with safety valves to prevent over-pressurization and explosion. Studies have shown that matrices like soil, rock and biological tissue can be digested in minutes with no loss of volatile metals and significantly lowered blank levels (Kingston and Jassie, 1988; Skelly and DiStefano, 1988). Advantages include the improvement in accuracy and significant decrease in sample preparation time; in addition, microwave digestion methods can be easily automated and robot-controlled, allowing rapid processing of large number of samples. Applications, methodology and safety precautions are covered in the text by Kingston and Jassie, 1988. Microwave heating is also being used for drying of samples, moisture determinations and dry ashing. CEM Corporation (Matthews, NC) makes a microwave muffle furnace to fit inside their laboratory microwave digestion system. Microwave sample preparation for atomic spectrometry is now well-established and in daily use in laboratories throughout the world.

6.6.3 The Laboratory

Optimally, sample preparation and analysis for trace metals by atomic spectrometry should be carried out in a specially designed analytical clean laboratory (Moody, 1982; Skelly and DiStefano, 1988) in which all incoming air is filtered through high efficiency filters and exposed metal surfaces are eliminated or minimized. However, most atomic spectrometry laboratories are located in ordinary laboratories. Regular cleaning and good housekeeping practices are needed if analyses are to be accurate. Common sense must be employed both in setting up the laboratory and in daily operation. For example, grinding and sieving of samples should be performed in an area separate from the analytical area.

Most sample decompositions and extractions require the use of corrosive or volatile chemicals; therefore, the laboratory must be equipped with fume hoods. Special fume hoods are needed if perchloric acid is used to decompose samples. The fume hood must be on a separate ventilation system, must be constructed of materials which will not form explosive perchlorates on exposure to perchloric acid vapors, and must be equipped with a water wash-down system to remove condensed acid.

A source of high purity (deionized) water must be available in the laboratory. Commercial reagent grade water systems, composed of ion exchange resins, activated charcoal and filters will produce deionized water, free from trace elements, organics and bacteria.

Laboratory equipment needs will depend in part on the type of samples to be analyzed, but the laboratory should be equipped with one or more of the following: analytical balance, grinding equipment, drying oven, muffle furnace or other ashing system, hot plate, steam bath, pH meter, ultrasonic bath, drying lamps, microwave digestion system, automatic fusion fluxer.

A variety of volumetric labware, both glass and plastic, will be required. Different sizes of pipettes and micropipettes should be available for taking aliquots of samples and standards.

6.6.4 Calibration

Most atomic spectrometric methods are *comparative* methods. A comparative method is one which requires calibration against external standards of known analyte concentration in order to obtain quantitative results. The atomic spectrometer is calibrated by measuring the analytical signal for known standards. The mathematical function relating the two parameters, signal and concentration (the calibration curve) is determined by plotting the data on graph paper or by fitting them to a suitable equation. An analytical signal from an unknown and the calibration curve are used to determine analyte concentration in the unknown. Quantitative analytical work is best performed in the portion of the calibration curve where the relationship between signal and concentration is a straight line.

6.6.4.1 Standards

The calibration standards are usually prepared from highly purified elements or compounds containing a known amount of analyte, dissolved in acid and deionized water. Standard solutions used for instrument calibration contain low concentrations of the analyte; usually less than 100 ppm. Dilute standards (<100 ppm analyte) are not stable for long periods of time and should be freshly prepared each day. Standard solutions of 1000 ppm analyte or greater are termed stock solutions, and are stable for longer periods of time if properly acidified and kept tightly closed. Stock solutions should be replaced each year. Aqueous stock solutions of most elements are available commercially from a variety of suppliers. These solutions are generally reliable, but should be checked periodically for stability, and for lot-to-lot variations.

Many commercial standards are prepared from salts rather than from pure metals, and unexpected problems or interferences may arise if the analyst is not aware of the exact composition of the standard. For example, a commercial aqueous silicon standard made from a salt containing both silicon and fluorine was reported to have etched glass vessels, resulting in erroneously high silicon concentrations in the diluted standards. Similarly, a commercial tungsten standard, described as being dissolved in water, has a pH of 12, which could lead to etched glassware and the precipitation of alkali-insoluble elements when used in mixed element standards or standard addition calibrations (described in Sec. 6.6.4.3). Standard Reference Material (SRM) solutions containing 10 000 ppm analyte have been prepared for a large number of elements by the National Institute of Standards and Technology and may be used as stock standards in their own right, or to verify the concentration of other commercial standards. Stock metal standards in oil for preparation of calibration curves in organic solvents are available under the name Conostan from Conoco, Inc., Ponca City, Oklahoma, as well as from NIST.

High purity metals and organometallic compounds are available from many commercial vendors for in-house preparation of standards.

The Office of Standard Reference Materials at NIST produces a wide variety of materials with certified compositions which can be used as standards, or as part of method development and quality control activities. These Standard Reference Materials (SRMs) can be used to check analyte recovery, accuracy, precision and interferences in atomic spectrometric measurements. Materials available include water, alloys, ores, cements, biological tissues, soils, and oil.

All of the reagents used in sample preparation should be added to the calibration standards. A reagent blank, which contains all of the reagents used and the solvent, but no analyte, is required to determine the analytical signal. The signal from the blank is subtracted from the signals of the standards and samples to provide a correction for any signal due to components other than the sample.

The accuracy of the analysis can be no better than the accuracy with which the concentration of analyte in the standards is known. The purity and stability of the materials used to prepare the standards must be confirmed (Ingle and Crouch, 1988). Ideally, the chemical form of the standards should be identical to that of the analyte.

6.6.4.2 Accuracy and Precision

The *accuracy* of a determination of analyte concentration in a sample indicates how close the measured concentration is to the true concentration of the analyte in the sample. The *precision* indicates the reproducibility in the measured concentration. Precision can also indicate the variability (homogeneity) of the sampled population. If one thinks in terms of shooting at a target, accuracy is hitting the bullseye; precision is having a tightly-clustered shot group. It is possible to have an analysis which is precise but not accurate (a tightly-clustered shot group in the upper right corner of the target, for example). It is difficult to have an accurate analysis which is not precise; the chance that widely-scattered numbers will average to give the correct number is very small. The ideal analysis is, of course, both accurate and precise.

Accuracy is affected by both *systematic* (or *determinate*) and *random* (*indeterminate*) errors. Systematic errors cause the measured concentration to be either too high or too low by either a constant amount or a fixed percentage. Theoretically, a systematic error can be corrected if the cause of the error is known. Random error is caused by uncontrolled fluctuations in the analysis that affect the magnitude of the measured concentration. Random error results in both positive and negative deviations from the true concentration, unlike a given systematic error which is either positive or negative. Random error can be reduced by controlling the source of the fluctuation (by use of temperature controllers or line conditioners to eliminate fluctuations in temperature and voltage, for example). Random error can

also be reduced by increasing the number of samples analyzed from a given population.

The accuracy and precision of an analysis are evaluated by the application of statics. A basic understanding of standard deviation, probability distributions, hypothesis testing, confidence intervals and propagation of error is needed to manipulate and understand data generated by atomic spectrometric analysis. Basic texts on probability and statistics as well as specialized texts on statistics for analytical chemistry should be consulted for detailed examples.

The *detection limit* (DL), also known as the limit of detection (LOD) is defined as the smallest concentration that can be reported as being present in a sample with a specified level of confidence (Ingle and Crouch, 1988). The detection limit is defined operationally as that concentration giving rise to an analytical signal, S, equal to a confidence factor, k, times the standard deviation of the blank measurement s_{bk}. If the calibration curve is linear near the DL, with slope m, the DL can be calculated directly (Ingle and Crouch, 1988):

$$DL = \frac{k\,s_{bk}}{m} \qquad (6\text{-}3)$$

The factor k is usually chosen to be 2, which corresponds to a 95% confidence limit, or 3, which corresponds to a 99% confidence limit. To determine DL, standards are used to construct a calibration curve, from which m is determined, and the blank is measured repeatedly to determine s_{bk}. Detection limits reported by instrument manufacturers are obtained with pure standards (usually just the analyte in deionized water) and completely optimized instruments. Detection limits obtained in real samples are often not as good as reported DLs.

A related quantity is the *limit of quantitation* (LOQ). It is the lowest concentration which can be determined with a specified precision: usually, the LOQ is calculated as the concentration for which $S = 10 s_{bk}$.

6.6.4.3 Errors

For an accurate determination, the calibration curve established from the standards must be valid for the samples, i.e. sample and standard of the same analyte concentration must yield the same analytical signal. If this assumption is not true, an error in the determination will occur. Errors can be classified as systematic or random. A systematic error results in a concentration which is either higher or lower than the true concentration, regardless of the number of measurements made. Random errors, caused by fluctuations in experimental variables result in concentrations which are distributed around the mean concentration. The mean concentration measured is the true concentration, if there are no systematic errors in the analysis. Systematic errors affect the accuracy of the analysis; random errors affect both the accuracy and precision of the analysis.

Matrix interferences or errors can occur when differences exist between the chemical composition of the samples and standards. Accurate determinations by ICP emission spectrometry, for example, require the same uptake rate for samples and standards. If the acid concentration in the samples is significantly different than that in the standards, the difference in viscosity may affect the uptake rate and result in an error. Light scattering from sample solutions containing high concentrations of dissolved solids can result in erroneously high absorbances compared to dilute aqueous standards. Large amounts of non-analyte species (called concomitants) can enhance or suppress the signal from an analyte by changing the bulk properties of the sample or by chemical reactions with the analyte which affect the efficiency of atomization. An example of enhancement by changing a bulk property, the electron density in a flame, is the addition of cesium to solutions for flame AAS determination of sodium. In a flame, sodium is easily ionized, resulting in depopulation of the ground state, low absorbance signals and a non-linear calibration curve. Addition of a large concentration of the more easily ionized element cesium increases the electron density in the flame and suppresses the ionization of sodium, which enhances the sodium absorbance and results in a linear calibration curve. A well-known chemical interference in flame AAS is the determination of calcium in samples containing large amounts of phosphorus, such as biological tissue and dairy products. Calcium phosphate does not atomize in chemical flames as efficiently as, say, calcium nitrate, which is the usual form of the calibration standard. Calcium concentrations in the sample would be erroneously low, because equal concentrations of calcium were producing unequal numbers of ground state calcium atoms and, hence, unequal analytical signals. Addition of large amounts of lanthanum to the solutions ties up the phosphate as the lanthanum salt and eliminates the interference.

Flame and furnace AAS suffer from these chemical interferences more than ICP and DCP emission, since the higher temperatures encountered in plasmas are more efficient at atomizing samples. Plasma emission is not completely free from matrix effects, which should be kept in mind when developing methods.

Emission techniques are more prone to errors from spectral interference, the overlap of the analyte spectral profile with that of a concomitant, than absorption techniques. Spectral interference errors can often be avoided by selection of alternate emission lines, or by application of experimentally-determined correction factors. Direct spectral interference in AAS is rare, but a significant example is the interference of iron at the selenium 196.1 nm resonance absorption line, which poses a problem for the analysis of biological and environmental samples.

Matrix interferences can be minimized or corrected with sample treatment or instrumental techniques. Analyte can be separated from the matrix, or the interferent can be separated from the sample using ion exchange, solvent extraction, precipitation and other techniques. Dilution of the sample can minimize matrix interference. In the matrix-matching approach, the matrix is duplicated synthetically in the blank and standards. The matrix-matching approach can be used when the major constituents of the matrix are known and high purity materials are available. For example, trace metals in nickel alloys can be determined by preparing the calibration standards in the appropriate amount of high purity nickel.

The use of an internal standard can correct for several random and systematic errors. The internal standard is another element, not present in the sample, which is added to all samples, standards, and blanks. A calibration curve is constructed with the analyte concentration and the ratio of the analyte to reference analytical signal. This technique can compensate for instrument drift, random fluctuations during analysis, and, in some cases, enhancement or suppression effects. Use of an internal standard does not compensate for spectral interferences.

The method of standard additions is used to compensate for matrix effects when it is not possible to duplicate the sample matrix. In this method, several aliquots of the sample are taken, and different, known quantities of analyte are added to each aliquot. The analytical signal for the original sample solution is measured; then the analytical signals from the standard addition solutions (i.e., the aliquots with added analyte) are measured. A plot of analytical signal versus added analyte is made and the concentration of analyte in the sample is determined from the absolute value of the x-axis intercept. The method of standard additions does not correct for background or spectral interference. The method is useful for identifying analyte interferences. If the slope of the standard addition calibration is different than the slope of a calibration curve prepared from conventional standards, either analyte was not fully recovered in sample preparation or a matrix interference exists.

6.6.5 Atomic Spectrometry and Other Elemental Analysis Techniques for Materials Characterization

It has been shown that atomic spectrometric techniques are capable of determining metals and metalloids in a wide variety of samples. The detection limits for most elements (Table 6-1) are in the sub-ppm range. Good qualitative and semiquantitative data can be obtained on solid samples by techniques such as DC arc emission spectrometry; excellent quantitative data can be obtained on solutions of samples by ICP, DCP, AAS, and AFS techniques. But there are other instrumental methods for obtaining elemental composition, such as X-ray fluorescence spectroscopy (XRF), neutron activation analysis (NAA), and plasma source mass spectrometry, to include ICP-MS, GDMS, and spark source-

MS (SSMS). The choice of techniques to be used to characterize materials depends on the chemical nature of the material, the concentration range of the analyte, the availability of standards, the level of precision and accuracy needed and the availability of instrumentation.

XRF is an excellent method for obtaining rapid, qualitative analysis of most solid and liquid samples. It is capable of detecting non-metals, such as the halogens, which cannot be measured by atomic spectrometry, has a wide dynamic range and detection limits on the order of 10–100 ppm for most elements. Obtaining quantitative information from XRF measurements requires standards; the preparation of homogeneous solid phase standards can be a time-consuming process and difficult to accomplish without contamination. In the characterization of an unknown, a good strategy is to screen the sample by XRF to identify what elements are present, and whether they are major, minor or trace components, and then to quantitate the elements by atomic spectrometry.

Neutron activation analysis can provide quantitative elemental analysis for a wide variety of solid materials, with detection limits in the sub-ppm to sub-ppb range for most elements. There are some materials which are unsuitable for NAA because they contain strong neutron absorbers; samples high in boron or cobalt, for example, cannot be analyzed by NAA. NAA can detect most metals and non-metals, and has excellent sensitivity for the rare earth elements, which can be difficult to determine by atomic spectrometry due to spectral overlap and poor sensitivity. Both XRF and NAA are non-destructive techniques, so the sample can usually be recovered for other analysis if needed.

Plasma source mass spectrometry is becoming a popular characterization technique since the appearance of commercial ICP-MS and GDMS equipment in the last few years. Detection limits are often at the part-per-trillion level. GDMS and SSMS are designed to analyze solid samples; ICP-MS handles solutions in the same way an ICP emission system does, but solids can be analyzed using laser ablation for sampling. As is the case with most solid sample techniques, qualitative and semiquantitative data are easily obtained; quantitative analysis is difficult without closely matrix-matched standards. Complex matrices can give rise to mass overlaps, especially in low-resolution mass spectrometers, and in ICP-MS, argon and argon-containing ions can obscure analyte masses, as is the case with the major calcium isotopes.

Atomic spectrometry has in many cases replaced time-consuming classical elemental analysis (gravimetry, titrimetry, colorimetry) while enabling analysis of very small, complex samples for trace elements, which could not have been analyzed using classical methods. Atomic spectrometry is a vital tool in modern materials characterization.

6.6.6 Atomic Spectrometry Literature and Professional Societies

Specific methods for atomic spectrometric determination of almost any element in almost any matrix can be found in the literature. In addition to the general analytical chemistry journals, such as Analytical Chemistry, Analytica Chimica Acta, and The Analyst, there are journals and publications which focus on spectroscopy and atomic spectrometry. The ICP Information Newsletter, edited by R. M. Barnes, University of Massachusetts, Amherst, is an excellent source of information, references and articles for the practicing atomic spectroscopist. Spectrochimica Acta, Part B, the Journal of Analytical Atomic Spectroscopy, and Spectroscopy Magazine provide articles of general interest to spectroscopists as well as specific methods of analysis. Applied Spectroscopy is the journal of the Society for Applied Spectroscopy (SAS), a professional society for spectroscopists. Membership in SAS includes a subscription to the journal. Other societies with interest groups in atomic spectrometry and materials characterization are the American Chemical Society and the American Society for Testing and Materials. Similar professional societies exist in many other countries.

Journals in specialized fields, e.g. ceramics, polymers, and metals, are also a source of analytical methodology for specific materials.

6.7 Acknowledgements

E. M. Skelly Frame wishes to thank Professors Ramon Barnes, Gary Hieftje, Roy Koirtyohann, Thomas O'Haver, James Robinson, and James Winefordner, Dr. Frank Fernandez of The Perkin-Elmer Corporation, Dr. Don Demers of Baird Corporation, and Marc Nouri, J-Y Division, Instruments SA, Inc. for their assistance in the completion of this chapter following the untimely death of Dr. Peter Keliher.

6.8 References

Abdellah, M. H., Coulombe, S., Mermet, J. M., Hubert, J. (1982), *Spectrochim. Acta 37B*, 583.
Abu-Samra, A., Morris, J. S., Koirtyohann, S. R. (1975), *Anal. Chem. 47*, 1475.
Alkemade, C. T. J., Milatz, J. W. M. (1955), *J. Opt. Soc. Am. 45*, 583.

Bock, R. (1979) *A Handbook of Decomposition Methods In Analytical Chemistry* (translated by I. L. Marr). London: International Textbook Company Limited.
Barnes, R. B., Richardson, D., Berry, J. W., Hood, R. L. (1945), *Ind. Eng. Chem. Anal. Ed. 17*, 605.
Blakley, C. R., Vestal, M. L. (1983), *Anal. Chem. 55*, 750.
Bradshaw, D., Slavin, W. (1989), *Spectrochim. Acta 44B*, 1245.
Brech, F., Cross, L. (1962). *Appl. Spectrosc. 16*, 59.
Bunsen, R., Kirchoff, G. (1860), *Ann. der Physik 110*. 160.
Burguera, J. L. (1988), *Flow Injection Atomic Spectroscopy*. New York: Marcel Dekker, Inc.
Carnrick, G. R., Slavin, W. (1989), *Am. Lab (Feb.)* 90.
Carnrick, G. R., Daley, G., Fotinopoulos, A. (1989). *At. Spectrosc. 10*, 170.
Caruso, J. A., Wolnik, K., Fricke, F. L. (1987), in: *Inductively Coupled Plasmas in Analytical Atomic Spectrometry:* Montaser, A., Golightly, G. W. (Eds.). New York: VCH Publishers, Inc.
Champion, P., Pellet, H., Grenier, M. (1873), *C.R. Acad. Sci. Paris 76*, 707.
Cholak, J., Hubbard, D. M. (1944), *Ind. Eng. Chem., Anal. Ed. 16*, 333.
Chuang, F. S., Winefordner, J. D. (1975), *Appl. Spectrosc. 29*, 412.
Ciurczak, E. W., Miller-Ihli, N. (1991), *Spectroscopy, 6(4)*, 23.
Cresser, M. (1990), in: *Sample Introduction in Atomic Spectroscopy:* Sneddon, J. (Ed.). New York: Elsevier, pp. 13–35.
Delves, H. T. (1970), *Analyst 95*, 431.
Demers, D. (1987), *Am. Lab. (Aug.)*, 30.
Denoyer, E. R., Fredeen, K. J., Hager, J. W. (1991), *Anal. Chem. 63(8)*, 445A.
Denton, M. B., Freelin, J. M., Smith, T. R. (1990), in: *Sample Introduction in Atomic Spectroscopy:* Sneddon, J. (Ed.). New York: Elsevier, pp. 73–106.
Dickinson, G. W., Fassel, V. A. (1969), *Anal. Chem. 41*, 1021.
Dittrich, K., Weinnrich, R. (1990), in: *Sample Introduction in Atomic Spectroscopy:* Sneddon, J. (Ed.). New York: Elsevier.
Ebdon, L., Evans, E. H. (1987), *J. Anal. At. Spectrom. 2*, 317.
Ebdon, L., Lechotycki, A. (1986), *Microchem. J. 34*, 340.
Ebdon, L., Sparkes, S. (1985), *ICP Information Newsletter 10(10)*, 797.
Elgersma, J. W., Maessen, F. J. M. J., Niessen, W. M. A. (1986), *Spectrochim Acta 41B*, 1217.
Elliott, W. G. (1971), *Am. Lab. 3(8)*, 45.
Ells, V. R., Marshall, C. E. (1939), *Soil Sci. Soc. Am. Proc. 4*, 131.
Falk, H., Hoffman, E., Ludke, Ch. (1981), *Spectrochim. Acta 36B*, 767.

Falk, H., Hoffman, E., Ludke, Ch. (1984), *Spectrochim. Acta 39B,* 283.

Fang, Z., Welz, B. (1989), *J. Anal. At. Spectrom. 4,* 543.

Fassel, V. A. (1973), *Electrical Plasma Spectroscopy,* Colloq. Spectrosc. Int. 16th Heidelberg 1971. London: Adam Hilger.

Fraunhofer, J. von (1814), *Denkschriften der Königlichen Akademie der Wissenschaften zu München 5,* pp. 193–226

Goode, S. R., Baughman, K. N. (1984), *Appl. Spectrosc. 38,* 775.

Gouy, A. (1879), *Ann. Chim. Phys. 18,* 5.

Greenfield, S. (1965), *Proc. Soc. Anal. Chem. 2,* 111.

Greenfield, S., Jones, I. L., Berry, C. T. (1964), *Analyst 89,* 713.

Greenfield, S., Smith, P. B., Breeze, A. E., Chilton, N. M. D. (1968), *Anal. Chim. Acta 41,* 385.

Griggs, M. A. (1939), *Science 89,* 134.

Grimm, W. (1968), *Spectrochim Acta 23B,* 7, 443.

Gy, P. M. (1979), *Sampling of Particulate Materials: Theory and Practice.* Amsterdam: Elsevier.

Harnly, J. M., O'Haver, T. C., Golden, B., Wolf, W. R. (1979), *Anal. Chem. 51,* 2007.

Harrison, W. W., Barshick, C. M., Klingler, J. A., Ratliff, P. H., Mei, Y. (1990), *Anal. Chem. 62,* 18, 943A.

Hatch, W. R., Ott, W. V. (1968), *Anal. Chem. 40,* 14, 2085.

Hershey, J. W., Keliher, P. N. (1989), *Appl. Spectrosc. Rev. 25,* 3–4, 213.

Holak, W. (1969), *Anal. Chem. 41,* 1712.

Ingle, J. D., Crouch, S. R. (1988), *Spectrochemical Analysis.* New Jersey: Prentice Hall, Inc.

Janssen, M. J. (1870), *C.R. Acad. Sci. Paris 71,* 626.

Keilshon, J. P., Deutsch, R. D., Hieftje, G. M. (1983), *Appl. Spectrosc. 37,* 101.

Keliher, P. N., Wohlers, C. C. (1974), *Anal. Chem. 46(6),* 682.

Keliher, P. N., Wohlers, C. C. (1976), *Anal. Chem. 48(1),* 140.

Keliher, P. N. (1978), in: *Phys. Methods in Modern Chem. Anal. Vol. 1:* Kuwana, J. (Ed.), New York: Academic Press.

Kingston, H. M., Jassie, L. B. (Eds.) (1988), *Introduction to Microwave Sample Preparation.* Washington, D.C.: American Chemical Society.

Kirchoff, G. (1860), *Phil. Mag. 20,* 1.

Kirchoff, G. (1862–1863), *Researches on the Solar Spectrum and the Spectra of the Chemical Elements* (translated by H. E. Roscoe). London: Part 1 (1862), Part 2 (1863).

Koirtyohann, S. R. (1980), *Spectrochim. Acta 35B,* 663.

Kwong, H. S., Measures, R. M. (1979), *Anal. Chem. 51,* 428.

L'vov, B. V. (1961), *Spectrochim. Acta 17,* 761.

Lafleur, P. D. (Ed.) (1976), *Accuracy in Trace Analysis: Sampling, Sample Handling and Analysis,* Vols.

1 and 2, NBS Spec. Publ. No. 422. Washington, D.C.: U.S. Government Printing Office.

Lang, R. J. (1962), *J. Acoust. Soc. Am. 34,* 6.

Lockyer, J. N., Roberts, W. C. (1874), *Phil. Trans. 164,* 495.

Lowe, R. M. (1971), *Spectrochim. Acta 26B,* 201.

Lundegardh, H. (1929), *Die Quantitative Spectralanalyse der Elemente,* Part I. Jena: G. Fischer.

Lundegardh, H. (1934), *Die Quantitative Spectralanalyse der Elemente,* Part II. Jena: G. Fischer.

Margoshes, M., Scribner, B. F. (1959), *Spectrochim. Acta 15,* 138.

Massman, H. (1968), *Spectrochim. Acta 23B,* 215.

McCurdy, D. L., Weber, A. E., Huges, S. K., Fry, R. C. (1990), in: *Sample Introduction in Atomic Spectroscopy:* Sneddon, J. (Ed.). New York: Elsevier, pp. 37–72.

McGucken, W. (1969), *Nineteenth Century Spectroscopy.* Baltimore: The Johns Hopkins Press.

Michel, R. G., Irwin, R. (1988), *J. Anal. At. Spectrom. 3(8),* 1059.

Michel, R. G., Irwin, R. (1990), *J. Anal. At. Spectrom. 5(7),* 603.

Miller-Ihli, N. (1988), *J. Anal. At. Spectrom. 3,* 73.

Mindel, B. D., Karlberg, B. (1981), *Lab. Pract. 721.*

Moody, J. M., Lindstrom, R. M. (1977), *Anal. Chem. 49,* 2264.

Moody, J. R. (1982), *Anal. Chem. 54,* 1358A.

Muscat, V. I., Vickers, T. J., Andren, A. (1972), *Anal. Chem. 44,* 218.

Nakahara, T. (1990), in: *Sample Introduction in Atomic Spectroscopy:* Sneddon, J. (Ed.). New York: Elsevier, pp. 255–288.

Newton, I. (1704), *Optics or Retreaties of the Refractions, Inflections and Colors of Light.* Revised by Dover Publications, New York, 1952.

Ng, K. C. (1990), in: *Sample Introduction in Atomic Spectroscopy:* Sneddon, J. (Ed.). New York: Elsevier, pp. 147–163.

Ng, K. C., Caruso, J. A. (1990), in: *Sample Introduction in Atomic Spectroscopy:* Sneddon, J. (Ed.). New York: Elsevier, pp. 165–193.

Nichols, E. L., Howes, H. L. (1923a), *Phys. Rev. 22,* 425.

Nichols, E. L., Howes, H. L. (1923b), *Phys. Rev. 23,* 472.

O'Haver, T. C. (1984), *Analyst 109,* 211.

Omenetto, N., Fraser, L. M., Winefordner, J. D. (1973), *Appl. Spectrosc. Rev. 7,* 147.

Pereli, F. R., Jr., Dougherty, J. P., Michel, R. G. (1987), *Anal. Chem. 59,* 1784.

Perkins, L. D., Long, G. L. (1988), *Appl. Spectrosc. 42 (7),* 1285.

Pettit, W. E., Horlick, G. (1986), *Spectrochim. Acta 41B,* 699.

Pickett, E. E., Koirtyohann, S. R. (1968), *Spectrochim. Acta 23B,* 235.

Ramage, H. (1929), *Nature 123,* 601.

Reed, T. B. (1961a), *J. Appl. Phys. 32,* 821.

Reed, T. B. (1961 b), *J. App. Phys. 32*, 2534.

Reed, T. B. (1962), *Int. Sci. Technol. 6*, 42.

Robinson, J. W. (1961), *Anal. Chem. Acta 22*, 254.

Robinson, J. W. (1990), *Atomic Spectroscopy*. New York: Marcel Dekker, Inc.

Robinson, J. W., Skelly, E. M. (1981), *Spectrosc. Letters 14(7)*, 519.

Robinson, J. W., Skelly, E. M. (1982), *Spectrosc. Lett. 15(8)*, 631.

Robinson, J. W., Skelly, E. M. (1983a), *Spectrosc. Lett. 16(1)*, 59.

Robinson, J. W., Skelly, E. M. (1983b), *Spectrosc. Lett. 16(2)*, 117.

Ruzicka, J., Hansen, E. H. (1975), *Anal. Chem. Acta 78*, 145.

Schrenk, W. G. (1986), *Appl. Spectrosc. 40(1)*, 19.

Schrenk, W. G. (1988), *Appl. Spectrosc. 42(1)*, 4.

Sharp, B. L. (1988), *J. Anal. At. Spectrom. 3*, 613.

Skelly, E. M. (1982), *The Direct Determination and Speciation of Mercury Compounds in Environmental and Biological Samples by Carbon Bed Atomic Absorption Spectroscopy*. Ph.D. Dissertation, Louisiana State University, Baton Rouge, Louisiana.

Skelly, E. M., DiStefano, F. T. (1988), *Appl. Spectrosc. 42(7)*, 1302.

Slavin, W. (1990), *Spectroscopy 5, 9*, 24.

Slavin, W., Miller-Ihli, N. J., Carnrick, G. R. (1990), *American Lab. 10*, 80.

Smith, S. B., Jr., Hieftje, G. M. (1983), *Appl. Spectrosc. 37*, 419.

Sneddon, J. (1990), *Sample Introduction in Atomic Spectroscopy*. New York: Elsevier.

Sneddon, J., Browner, R. F., Keliher, P. N., Winefordner, J. D., Butcher, D. J., Michel, R. G. (1989), *Prog. Anal. Spectros. 12(4)*, 369.

Swan, W. (1957), *Trans. Roy. Soc. Edinburgh 21*, 411.

Talbot, W. H. F. (1826), *Edinburgh J. Sc. 5*, 77.

Talbot, W. H. F. (1934), *Phil. Mag. 4*, 112.

Tyson, J. F. (1984), *Anal. Proc. 21*, 377.

Tyson, J. F. (1985), *Analyst 110*, 419.

Uden, P. C. (1990), in: *Sample Introduction in Atomic Spectroscopy:* Sneddon, J. (Ed.). New York: Elsevier, pp. 195–223.

Valcarcel, M. (1990), in: *Sample Introduction in Atomic Spectroscopy*: Sneddon, J. (Ed.). New York: Elsevier, pp. 289–327.

Van Loon, J. C., Barefoot, R. R. (1991), *Determination of the Precious Metals*. Chichester (England): John Wiley and Sons, Ltd.

van der Willigen, V. S. M. (1859), *Ann. der Physik 106*, 610.

Veillon, C., Mansfield, J. M., Parsons, M. L., Winefordner, J. D. (1966), *Anal. Chem. 38*, 204.

Vestal, M. L., Fergusson, G. J. (1985), *Anal. Chem. 57*, 2373.

Walsh, A. (1955), *Spectrochim. Acta 7*, 108.

Weiss, R. (1954), *Z. Phys. 138*, 170.

Wendt, R. H., Fassel, V. A. (1965), *Anal. Chem. 37*, 920.

West, T. S. et al. (1966a), *Anal. Chim Acta 36(3)*, 269.

West, T. S. et al. (1966b), *Talanta 13*, 805.

Willis, J. B. (1965), *Nature (London) 207*, 715.

Winefordner, J. D., Vickers, T. J. (1964), *Anal. Chem. 36*, 165.

Wollaston, W. H. (1802), *Phil. Trans. of the Royal Society, London Series A 92*, 365.

Wood, R. W. (1905), *Phil. Trans. 10*, 513.

Woodget, B. W., Cooper, D. (1987) *Samples and Standards*. Chichester (Great Britian): John Wiley and Sons.

Woodruff, R., Stone, R. W., Held, A. M. (1968), *Appl. Spectrosc. 22(5)*, Part 1, 408.

Woodson, T. T. (1939), *Rev. Sci. Instrum. 10*, 308.

Zander, A. T., O'Haver, T. C., Keliher, P. N. (1976), *Anal. Chem. 48(8)*, 1166.

Zander, A. T., O'Haver, T. C., Keliher, P. N. (1977), *Anal. Chem. 49(6)*, 838.

General Reading

Alkemade, C. T. J., Herrmann, R. (1979), *Fundamentals of Analytical Flame Spectroscopy*. New York: Wiley.

Barnes, R. M. (Ed.) (1976), *Emission Spectroscopy*. Stroudsburg (PA): Dowden, Hutchinson and Ross.

Boumans, P. W. J. M. (1987), *Inductively Coupled Plasma Emission Spectroscopy*, Part I and II. New York: Wiley-Interscience.

Dean, J. A., Rains, T. C. (Eds.) (1969, 1971, 1975), *Flame Emission and Atomic Absorption Spectrometry*, Vol. I–III. New York: Marcel Dekker.

Ingle, J. D., Crouch, S. R. (1988), *Spectrochemical Analysis*. New Jersey: Prentice Hall, Inc.

Montaser, A., Golightly, D. W. (Eds.) (1987), *Inductively Coupled Plasmas in Analytical Atomic Spectrometry*. New York, Weinheim: VCH Publishers.

Robinson, J. W. (1990), *Atomic Spectroscopy*. New York: Marcel Dekker.

Sneddon, J. (1990), *Sample Introduction in Atomic Spectroscopy*. New York: Elsevier.

Winefordner, J. D., Schulman, S. G., O'Haven, T. C. (1972), *Luminescence Spectrometry in Analytical Chemistry*. New York: Wiley-Interscience.

7 Thermoanalytical Methods

Patrick K. Gallagher

Departments of Chemistry and Materials Science and Engineering,
The Ohio State University, Columbus, OH, U.S.A.

List of Symbols and Abbreviations

C_p	heat capacity
H	entropy
L	length
L_T	dimension change at temperature T
L_0	dimension change at standard temperature
m/e	mass-charge ratio
T	temperature
T_c	crystallization temperature
T_g	glass transition temperature
T_m	melting temperature
V	volume
α	thermal expansion coefficient
ASTM	American Society for Testing Materials
CGA	consumed gas analysis
CVD	chemical vapor deposition
DMA	dynamic mechanical analysis
DSC	differential scanning calorimetry
DTA	differential thermal analysis
DTG	differential thermogravimetry
EGA	evolved gas analysis
EGD	evolved gas detection
ESR	electron spin resonance
ETA	emanation thermal analysis
FTIR	Fourier transform infrared spectroscopy
ICTA	International Confederation for Thermal Analysis
IUPAC	International Union of Pure and Applied Chemistry
LVDT	linear voltage differential transformer
MS	mass spectroscopy
MS–EGA	mass spectroscopy – evolved gas analysis
N.I.S.T.	National Institute of Standards and Technologies
NMR	nuclear magnetic resonance
TA	thermal analysis
TD	thermodilatometry
TIC	total ion current
TG	thermogravimetry
TM	thermomagnetometry
TMA	thermomechanical analysis
TS	thermosonimetry

7.1 Introduction

The definition of thermal analysis put forth by the Nomenclature Committee of the International Confederation for Thermal Analysis (ICTA) and subsequently accepted by IUPAC and ASTM is "a group of techniques in which a physical property of a substance, or its reaction products, is measured as a function of temperature whilst the substance is subjected to a controlled temperature program". Table 7-1 summarizes most of the common techniques generally ascribed to thermal analysis.

This definition presents two major problems in writing this chapter. It encompasses virtually every possible method and measurement provided it is performed under a controlled temperature environment. Secondly, the value of the methods extends well beyond that which is conventionally referred to as "analytical chemisty". It is, therefore, necessary to excercise considerable judgement in what constitutes the appropriate scope and the selection of applications for a chapter primarily oriented toward chemical analysis.

The approach taken for this specific chapter is revealed in the table of contents. The four major, and most generally accepted techniques will be considered in detail. These are (1) thermogravimetry (TG), where the mass of the sample is monitored;

Table 7-1. Principal thermoanalytical methods.

Property	Technique	Acronym
Mass	Thermogravimetry	TG
Apparent mass[a]	Thermomagnetometry	TM
Volatiles	Evolved gas detection	EGD
	Evolved gas analysis	EGA
	Thermal desorption	
Radioactive decay	Emanation thermal analysis	ETA
Temperature	Differential thermal analysis	TA
Heat[b] or heat flux[c]	Differential scanning calorimetry	DSC
Dimensions	Thermodilatometry	TD
Mechanical properties	Thermomechanical analysis	TMA
	Dynamic mechanical analysis	DMA
Acoustical properties	Thermosonimetry (emission)	TS
	Thermoacoustimetry (velocity)	
Electrical properties	Thermoelectrometry (resistance)	
	(voltage)	
	(current)	
	(dielectric)	
Optical properties	Thermooptometry (spectroscopy)[d]	
	Thermoluminescence (emission)	
	Thermomicroscopy (structure)	
	Thermoparticulate analysis	TPA

[a] Change induced by an imposed magnetic field gradient; [b] power compensated DSC; [c] heat flux DSC; [d] absorption fluorescence, Raman, etc. Non-optical forms of spectroscopy, e.g., NMR, ESR, Mößbauer, etc. are also applicable. From Gallagher (1992a): reproduced by permission of JAI, Inc.

(2) differential thermal analysis (DTA) and its offspring differential scanning calorimetry (DSC), where the difference in temperature between the sample and an inert reference material or the heat evolved or absorbed by the sample is measured; (3) evolved gas analysis (EGA) or evolved gas detection (EGD), where the specific nature of the evolved products is followed either qualitatively or quantitatively (EGA) or just the fact that gas evolved is determined (EGD) as a function of temperature; and (4) thermomechanical methods, where the dimensions are measured under a specific load or some mechanical modules is measured under a specified stress. Strictly speaking thermodilatometry, the measurement of dimensional change, is not thermomechanical analysis (TMA), however, it is generally considered part of that class of methods because the general apparatus used can frequently perform both techniques.

The remaining techniques are gathered into a section on miscellaneous methods. Here, only the measurements made under conditions of scanning temperature, i.e., dynamic measurements as opposed to isothermal measurements, will be considered.

It will become increasingly obvious to the reader that invariably the use of two or more techniques is better than one. Frequently non-thermoanalytical techniques are also necessary to unequivocally solve a particular problem. Confining the discussion to thermoanalytical methods, it is clear, for example, that merely knowing the extent of mass loss (TG) is far less valuable than also knowing what volatile species are evolved (EGA).

Consequently, unless the nature of the products evolved is obvious, e.g., the thermal decomposition of $CaCO_3$, it is also advisable to perform an EGA experiment.

There are several modes of coupling these experiments. The most common approach is to perform separate experiments using different equipment and separate samples. While this may be most convenient, it has several drawbacks. These are: (1) the sample may not be equivalent, (2) the thermal environment and sample temperature may not correlate identically for both temperature sensors, and (3) the ambient atmosphere and flow rate may vary.

In order to minimize or even eliminate these uncertainties in comparing the measurements two approaches to coupled methods are used. The "combined techniques" involve the use of separate matched samples but subject to a nearly identical thermal environment, i.e., in the same furnace undergoing exactly the same temperature program. As an example consider a sample suspended from above on a balance beam for TG while a sample and reference are supported from below for DTA. Combined methods are rarely used today. The more common and clearly superior approach is to utilize a single sample but following several aspects such as mass change (TG) and gas evolution (EGA or EGD). These are referred to as "simultaneous techniques".

The general mode of operation in thermal analysis is to heat or cool the sample under some predetermined temperature program. Typically this involves a series of constant heating or cooling rates, perhaps incorporating some isothermal periods. A growing tendency (Rouquerol, 1989), however, is to pre-establish the desired changes in the experimental variable and accept whatever temperature program is required. The power to the sample chamber is then controlled by feedback from the particular sensor associated with the property being measured. The distinction between these modes of operation is shown

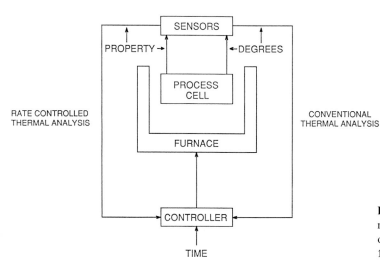

RATE CONTROLLED
THERMAL ANALYSIS

CONVENTIONAL
THERMAL ANALYSIS

Figure 7-1. Comparison of normal and "controlled rate" modes of thermal analysis (Gallagher, 1991 a).

in Fig. 7-1 (Gallagher, 1991 a). The advantages of this "controlled rate" processing will be discussed subsequently.

Virtually all the analytical techniques have benefitted immensely from the advent of the computer and microprocessor. This is particularly true for thermal methods. They are generally similar to chromatographic measurements in that both methods measure relatively slowly changing and frequently small signals over long periods of time. In addition their mathematical analyses have much in common, e.g., integration, differentiation, base line subtraction, smoothing, etc. Such methods are ideally suited for computer control with digital data acquisition and analysis.

The preceding paragraphs have addressed the first problems described and have established the scope of methods to be covered from this broad topic of thermal analysis. The second problem, which of these many techniques are applicable to chemical analysis and how, will now be discussed briefly. None of these techniques, with the exception of EGA and certain spectroscopic methods, directly identify the species evolved or their concentrations. The usefulness of most ther-

moanalytical techniques for chemical analysis involves indirect correlations. Fortunately, there are many such applications and they are often simple, accurate, relatively quick, and require only small amounts of sample. Consequently, they are frequently the characterization method of choice.

These indirect methods, however, do require calibration to relate the measurement, e.g., melting point, heat evolved, or temperature of a phase transformation to the concentration. Knowledge of the specific process involved in the change of property can make the mass loss or area under a peak in DTA/DSC a highly sensitive, quantitative analytical measurement. Because of these indirect correlations between the thermal analytical measurement and the chemical analysis, several representative applications will be described for each technique.

One of the most common applications of thermal methods is to study the kinetics of various reactions and processes, particularly heterogeneous kinetics. While this is frequently a controversial subject, these reservations usually are concerned with the mechanistic inferences used in the derivation of numerical rates and the resulting Arrhenius parameters rather than with the actual determination of the fraction reacted as a function of time. It is this latter information that constitutes the analytical chemical aspect. Again, however, one must have outside knowledge to connect the thermoanalytical measurement with the reaction of process. Are there overlapping events? Can you establish clearly the start and end of the reaction or process? Such information may be relatively simple, e.g., the decomposition of pure $CaCO_3$, or it may be hopelessly complex as for some multicomponent mineral. With these general thoughts in mind it is time to consider the individual techniques in more detail.

7.2 Thermogravimetry (TG)

7.2.1 The Basic Instrument

The schematic for a typical thermobalance is presented in Fig. 7-2. The sample can be linked to the balance in several ways. It can be suspended directly from the balance beam to hang down into a furnace or controlled temperature environment. Since heated air rises, this means of suspension generally requires careful baffling or cooling to prevent the heat from affecting the balance.

Alternatively the sample and furnace arrangement can be placed above the furnace so that the heat is less likely to affect the balance. The greater mass of the counterbalance used in this method of suspension, however, must be subtracted from the ultimate capacity of the instrument. A third approach is to fix the sample as a horizontal extension of the beam. This also minimizes the heating effect and frequently appears less influenced by the flow patterns set up within the balance and furnace chambers.

The modern electromagnetic balances are remarkable instruments showing relatively little dependence on vibration, high sensitivity, and temperature stability. This

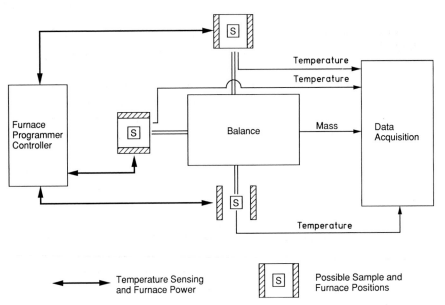

Figure 7-2. General schematic representation of a TG instrument. From Gallagher (1992a): reproduced by permission of JAI, Inc.

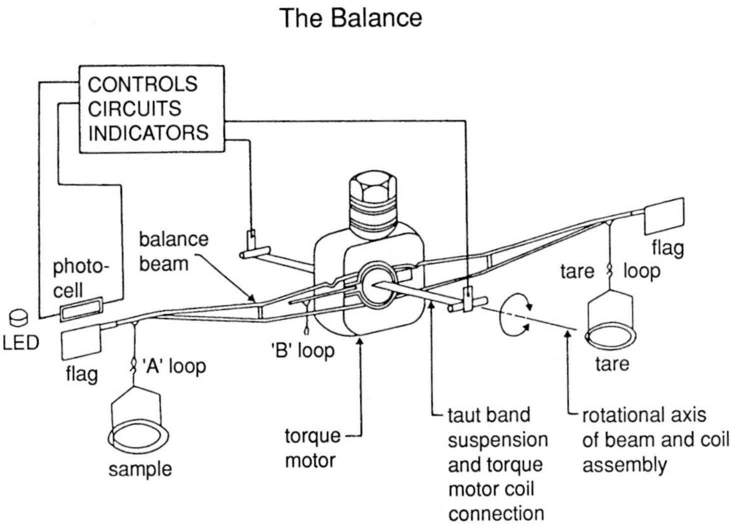

The Balance

Figure 7-3. Schematic of the Cahn Electrobalance. From Brown (1988): reproduced by permission of Chapman and Hall.

class of balances has evolved from the original Cahn electrobalance. A schematic of this type of balance is shown in Fig. 7-3.

The beam position is monitored by a photodetection scheme. Assuming that the sample suspension has been tared and the balance is in equilibrium, addition of a sample to the left side of the beam will cause the right side of the beam to be displaced upward and sufficient current is supplied to the electric torque motor to restore it to the original beam position. The restoring force, and hence the current, is proportional to the change in weight. A typical sensitivity of 0.1 µg is possible. Any portion of the sample's mass can be tared within the capacity range of the particular balance.

The overall capacity of the system is determined by the strength of the balance beam and the linear range of the torque motor. Alternatively, the effective capacity of the balance can be increased, at the expense of sensitivity, by coupling the sample suspension closer to the fulcrum of the balance. The mass of the sample holders and suspension system will, of course, subtract from the actual range of sample mass that

can be utilized. Many manufacturers produce several models having different capacities and sensitivities.

Nearly all of the modern, commercially available TG equipment is based on similar electrobalances, however, there are other special purpose mass sensors. The resonance frequency of a piezoelectric crystal, generally quartz (SiO_2), varies with the mass deposited on the crystal surface. As little as 1 pg cm^{-2} can be detected in this manner (Plant, 1971). This sensitivity allows the vaporization or deposition rates of sub-monolayers to be measured. These detectors have a limited range of temperature and require careful compensation for the many factors which lead to a drift of the signal with time and small changes in temperature. This has been partially achieved through the use of a matched crystal for compensation (Boersma and Van Empel, 1975).

Thermal methods depend upon uniform precisely controlled heating or cooling of the sample. Several physical arrangements of the balance compartment and furnace are depicted in Fig. 7-2. These show the furnace as external to the sample compartment. There are, however, some instruments, designed primarily for small samples, which have the furnace inside the sample chamber and hence in the same gaseous environment as the sample.

This allows for miniature furnaces capable of high heating and cooling rates. Such conditions conserve sample and allow for a more rapid turn around time for the instrument. There are some situations, however, that require a large sample size. Examples are when using a relatively heterogeneous sample like a mineral, coal, etc.; when a specific sample geometry is required such as foils for corrosion studies; or when the highest precision is necessary for very small changes in mass such as for thin films or minute changes in oxygen stoichiometry.

The heating elements of the furnaces are most often based upon resistive heating. Nichrome and Kanthal are alloys commonly used in the range of temperature up to 1000 to 1200 °C. Fused quartz tubes and accessories are frequently used to contain the sample and atmosphere. Aluminum is often used as a sample container up to 600 °C because of its thermal conductivity, ease of fabrication and low cost.

Molybdenum disilicide or silicon carbide heating elements and platinum or platinum alloys are used in the range up to 1500 to 1700 °C. Ceramic refractories such as alumina or mullite are used to contain controlled atmospheres and platinum or alumina to hold the sample. Only a few manufactures, e.g., Linseis, Netzch, and SETARAM, make instruments for use above these temperatures. The heating elements are molybdenum, tungsten, or graphite and require protective atmospheres free of oxygen. The other materials of construction and temperature sensors at $>1700\,°C$ also are very limited and expensive.

Radiant heating, such as shown in Fig. 7-4, is capable of even faster controlled heating rates approaching $500\,°C$ min^{-1} for temperatures up to $1200\,°C$. Various focal arrangements for the infrared radiation will heat a point, plane, surface or volume. The walls containing the atmosphere can be essentially transparent to the radiation and therefore remain at much lower temperatures. But any deposition on these walls will drastically alter the thermal characteristics of the system. It is not possible, however, to cool samples at anything approaching those rates while simultaneously following their mass.

A typical TG curve and its differential, DTG, curve are presented in Fig. 7-5 (Gallagher, 1992a). These are for $CaC_2O_4 \cdot H_2O$ which is frequently used as an example because of the three well resolved steps in the thermal decomposition. The ordinate is usually wt.% rather than actual mass so that multiple curves can be

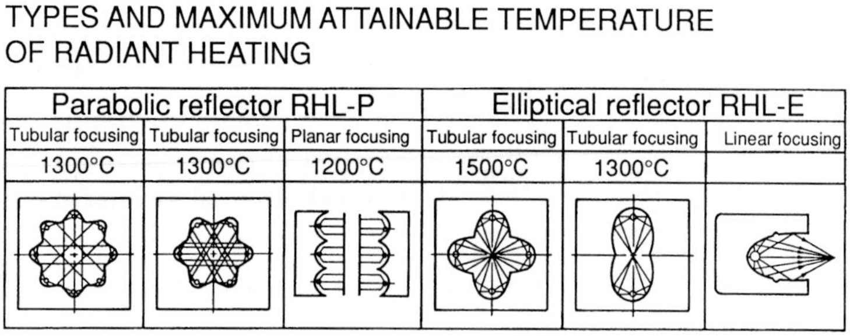

TYPES AND MAXIMUM ATTAINABLE TEMPERATURE OF RADIANT HEATING

Parabolic reflector RHL-P			Elliptical reflector RHL-E		
Tubular focusing	Tubular focusing	Planar focusing	Tubular focusing	Tubular focusing	Linear focusing
1300°C	1300°C	1200°C	1500°C	1300°C	

Figure 7-4. Examples of radiant furnaces available for thermal analysis. Reproduced by permission of Ulvac Sinko Riko Co.

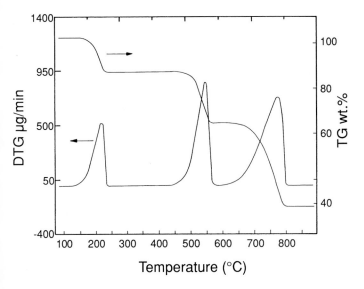

Figure 7-5. TG and DTG curves for the thermal decomposition of $CaC_2O_4 \cdot H_2O$, 10.22 mg heated in Ar at $20\,°C\,min^{-1}$. From Gallagher (1992a): reproduced by permission of JAI, Inc.

easily compared on a normalized basis. Occasionally the weight data and temperature data will be plotted as a function of time. This allows verification of the heating rate but is generally less convenient for purposes of discussion and comparison with other curves.

The DTG curves enhance resolution and are more easily compared to other differential measurements discussed in later sections. Differentiation, however, is a noise amplifier and data are frequently smoothed during the software manipulation to generate and plot the derivative. Such curves are also of interest for kinetic purposes, since they represent the actual rate of the reaction. The stoichiometry, however, is more readily evaluated from the original integral presentation.

The three steps obvious in Fig. 7-5 correspond to the loss of H_2O to form the anhydrous oxalate, the loss of CO to form the carbonate, and finally the loss of CO_2 to form CaO. The observed weight losses may be compared with the theoretical values to establish the stoichiometry of the process and the starting material.

7.2.2 Atmosphere, Sample Considerations, and Primary Sources of Error

The influences of atmosphere and sample are best considered in conjunction with the sources of error for TG. These are summarized in Table 7-2. Factors in the left hand portion of Table 7-2 are associated with the potential errors in measuring the mass while those on the right hand side address the main errors encountered in determining the temperature of the sample. The electronic drift problems are relatively

Table 7-2. Major factors affecting thermogravimetry.

Mass	Temperature
Buoyancy	Heating rate
Atmospheric turbulence	Thermal conductivity
Condensation and reaction	Enthalpy of the process
Electrostatic and magnetic forces	Sample, furnace, and sensor arrangement
Electronic drift	Electronic drift

From Gallagher (1992a): reproduced by permission of JAI, Inc.

minor and only become factors for experiments of duration beyond several hours. The other factors mentioned in Table 7-2 have a dependence on the atmosphere, flow, and sample properties.

Buoyancy corrections arise from the change in density of the gas phase with temperature. The mass of the atmosphere displaced with increasing temperature decreases, giving a slight apparent weight gain as the sample along with its container and suspension system are heated. One cm^3 of dry air weighs 1.3 mg at 25 °C but only 0.3 mg at 1000 °C. It is obvious that corrections are necessary for the most accurate work. There are compensation methods based on instrumental modifications or a software based subtraction method for making the necessary corrections.

The hardware approach involves exposing both the active and a nearly identical tare side of the balance beam to the same temperature program. Under these conditions the effects cancel to an extent depending on the degree of match between the two sides. The atmosphere should be identical and any changes in sample volume matched. Such compensation can only be approximated for samples which undergo decomposition. Both the sample size and the surrounding atmosphere will change during a decomposition.

Manufacturers have developed two methods of instrumental compensation. One approach is to use a simple, symmetrical balance and sample arrangement with dual furnaces controlled by the same or identical programmer controllers. The second approach is to have an identical electrobalance with a matching sample suspension system in the same thermal environment as the active sample. The output of the second balance is constantly subtracted from that of the active one in real time. This method essentially duplicates the temperature system while the other method duplicates the weighing system. Each approach has its set of advantages and disadvantages.

The software approach is based on storage of a blank run, performed under conditions as identical as possible with an inert dummy sample, and subsequent subtraction from the actual experiment. Again this process is only an approximation and requires a selection of blank experiments to cover each set of experimental conditions. For this reason the hardware mode is preferable.

The atmospheric turbulence is determined by several factors, e.g. the, flow, pressure, and geometrical considerations. Wendlandt (1988) discusses these factors in some detail. Clearly compromises are necessary to maximize atmospheric interactions with the sample and yet minimize the noise level introduced by the interaction. These factors will be temperature, flow, and pressure dependent; so further compromise is required to arrive at an optimum configuration for a suitable operating range of experimental parameters. More restrictive operating conditions lead to better optimization. Baffling and shaping of the sample compartment and suspension are important. Flow and temperature profiles determine the pattern of convection currents. Operation in vacuum reduces the aerodynamic noise and the buoyancy corrections. High pressures exacerbate these same problems.

The prior discussion in this section has dealt with the general aspects of atmospheric effects. There are, however, specific effects associated with the frequent reversibility of the chemical equilibria involved. The partial pressures of CO_2 and H_2O will determine the extent of carbonate, hydroxide, and hydrate decompositions at a particular temperature. Consequently, the ability of these decomposition products to be swept away will control the degree of decomposition. Similarly, if a reactant is in the gas phase, then the ability of the flow system to deliver that component at a sufficient rate will also influence the change in mass at any time.

This dependence is demonstrated in Fig. 7-6 which shows the influence of sample holders on the decomposition of $CaCO_3$ (Paulik and Paulik, 1986). The holder that presents the sample as a thin layer having good exposure to the flowing gas has the lowest decomposition temperature. The labyrinth sample holder fills with CO_2 after only a minute part of the total decomposition. Because of the tortuous diffusion path, the remainder of the decomposition occurs in effectively pure CO_2 at a much higher temperature. The normal range of sample holders falls somewhere between these extremes.

The packing of the sample is another variable. Is it a loose powder; a pressed or pelletized piece; or a dense amorphous chunk, single crystal, or sintered polycrystalline specimen? These factors will affect the escape of volatile products and the access of any gaseous reactants. For these reasons each reversible reaction can take place over a wide range of temperature.

The deliberate manipulation of such processes can be used to enhance the resolution between events that might otherwise overlap.

From the above discussion on the effects of variations in the atmosphere on experiments involving the large and important class of reversible reactions, it is obvious why difficulties often arise when studying the kinetics and mechanisms of such processes (Gallagher and Johnson, 1973).

Changes in mass not directly associated with the sample may be unintentionally detected. Condensation of volatile decomposition products on cooler portions of the sample suspension system will alter the observed mass loss. Thought must be given to the direction of flow and the possibility of reactive gaseous species reaching the vulnerable parts of the balance mechanism, temperature sensor, or any internal furnace windings.

Portions of the TG system can become contaminated with products from the reaction. This may lead to the introduction of impurities during subsequent experiments in the contaminated system. Contamination is particularly a problem when working with relatively volatile oxides such as Pb, Zn, Hg, Cd, etc. Most condensates of organic material, however, can be easily

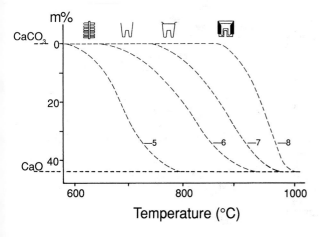

Figure 7-6. Influence of sample holder on the decomposition of $CaCO_3$ (Paulik and Paulik, 1986).

eliminated by subsequent firing to temperatures above 600 °C in an oxidizing atmosphere.

Similarly, any reaction of the atmosphere with the sample holder can influence the apparent change of mass. Vaporization or oxidation at higher temperatures of parts of the sample system can be a problem. Reactions between the sample and its holder must be considered. Materials like alkali and alkaline earth carbonates can be reactive with ceramic or metal parts or metals products may alloy with Pt sample pans, particularly under reducing conditions.

Electrostatic forces can be particularly annoying. They can cause the sample holder to cling to the nearest wall or otherwise perturb the mass signal. Dry atmospheres will increase the likelihood of the problem. Facilitating the equalization of charges between the sample suspension and chamber walls will decrease the effect. Wiping glass surfaces with conducting liquids, providing thin metallic films connected to electrical ground, or weakly ionizing the atmosphere are some of the common methods of minimizing electrostatic influences.

Establishing the actual temperature of the sample, particularly when it is undergoing reaction, is a more difficult matter than following the change in mass. The possibilities of electronic drift are somewhat greater because there are the cold junction compensating circuits in addition to the usual amplifier and recording electronics. Nevertheless, these are trivial compared to the other aspects of the problem.

The other four factors related to temperature in Table 7-2 are highly interactive. Changes in the heating or cooling rate, thermal conductivity of the atmosphere and materials of construction, and the heat absorbed or evolved by the reaction process will tend to influence those differences that arise due to the geometrical relationships established by the placement of the sample, furnace, and temperature sensor.

Assuming that the temperature sensor and sample are in reasonable proximity to each other and that the furnace control is good with a homogeneous and sufficiently large hot zone, then the major sources of error in the recorded sample temperature are associated with the enthalpy of the reaction processes.

The magnitude of the enthalpy of reaction in relation to the heat capacity of the products and reactants is frequently overlooked. The simple decomposition of $CaCO_3$ to form CaO has an enthalpy of approximately 168 kJ mol^{-1}. In comparison with the heat capacity of $CaCO_3$ this is sufficient to lower the temperature over 1000 °C in a hypothetical adiabatic environment (Gallagher and Johnson, 1973). Exothermic reactions are even more of a problem and thermal runaway reactions are common during heating.

The question of the actual temperature at a reacting interface is one of the major problems in thermal analysis. Add to this the problems described earlier concerning the effects of atmosphere on reversible reactions, and the demand for exercising great care in the interpretation of kinetic and mechanistic results becomes obvious. The Arrhenius parameters can vary dramatically with the experimental parameters, see for example Huang and Gallagher (1992) and texts on this topic, e.g., Brown et al. (1980).

The intermediate step in the thermal decomposition of $CaC_2O_4 \cdot H_2O$ depicted in Fig. 7-5 involves the evolution of CO. Had the atmosphere been air or O_2 instead of inert Ar there would have been the highly exothermic oxidation of CO to CO_2 occurring at the surface and in the interstices of the powder. This heat is not effectively dissipated and the temperature of the sample is temporarily raised hundreds of degrees. If the sample was visible it would have been seen to glow a bright red in spite of what the nearby thermocouple might imply.

Thermobalances vary widely in their ability to detect such events. When the system does measure these processes the plot of mass or wt.% versus temperature shows anomalous behavior as the curve folds back on itself. Under these conditions the use of time as the abscissa gives a more presentable plot. Since the thermocouple can not accurately reflect the true sample temperature during that period, there is no advantage to the plot versus temperature.

The thermal conductivity of the atmosphere influences the thermal transport properties between the furnace, sample and thermocouple. Consequently, higher thermal conductivity enhances the response time for temperature control and permits better heat dissipation or absorption by the sample (Caldwell et al., 1977). Depending on the flow rate, it also increases the heat loss from the furnace and raises the power level necessary to obtain the same temperature.

7.2.3 Controlled Rate Techniques

The controlled rate principle was introduced in Sec. 7.1 and Fig. 7-1. The rate of change in temperature is controlled by a preset rate of weight loss or change in partial pressure. This has also been referred to as quasi-isothermal or quasi-isobaric TG (Paulik and Paulik, 1986). Figure 7-7 illustrates the schematic of an apparatus designed for this purpose (Rouquerol, 1989). Feedback from either the balance mechanism or the total pressure gauge can be used to control the change in temperature. Controlled rate plots for the thermal de-

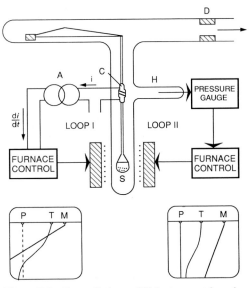

Figure 7-7. Controlled rate TG instrument based on either constant rate of change in weight or pressure (Rouquerol, 1989).

composition of $CaCO_3$ are presented in Fig. 7-8 (Paulik and Paulik, 1986). The reversible decomposition has now been constrained to occur over a much narrower range of temperature compared with that observed in Fig. 7-6 for the conventional approach. Resolution is greatly enhanced under these circumstances.

The Hungarian Optical Works Company has produced "The Derivatograph – Q" since 1972 which is capable of controlled rate experiments. More recently, Thermal Analysis (TA) Instruments has announced a model capable of such TG measurements. The newer instrument does not actually establish a rate of weight loss for the experiment but does slow the process during the regions of change in mass according to an operator controlled setting. This leads to enhanced resolution and allows the operator to set a fast rate initially knowing that the process will be slowed during regions of interest. The elapsed experimental time may be shortened under these operating conditions.

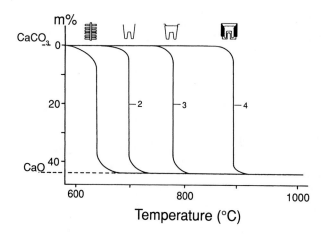

Figure 7-8. Examples of controlled rate TG using $CaCO_3$ (Paulik and Paulik, 1986).

7.2.4 High Precision Isothermal Methods

Liberal interpretation of a controlled temperature program allows consideration of isothermal as well as dynamic studies. It is reasonable since they utilize the same instruments. Isothermal studies of equilibria are used for detailed studies of variable stoichiometry and to determine phase diagrams. Kinetic uses are to measure rates of reactions involving changes of mass. Both of these classes of experiments can make severe demands on the technique.

As an example of phase studies consider the use of isothermal TG to determine the oxygen nonstoichiometry in such technologically important materials as ferrites. Small changes in weight must be accurately measured at high temperatures. The nickel zinc ferrite system has been studied by Bracconi and Gallagher (1979). Careful compensation must be made for buoyancy, loss of zinc from the sample, and loss of platinum from the sample holder and suspension system in order to obtain accurate values for the small changes in oxygen content.

Figure 7-9 shows the results after the corrections have been made. Equilibrium is approached from higher and lower temperatures in order to reveal regions of metastability. As is often found, the phase boundary between the single and two phase region shows no metastability upon passing from the two phase to the single phase region but a strong tendency toward metastability when approached from the other direction. This is due to the need to nucleate a second phase precipitate in one direction but not in the other. The difficulty in nucleation leads to metastability.

Another interesting and particularly important use of TG to establish nonstoichiometry is concerned with the High-T_c superconductors (Gallagher, 1991 b). In this instance the changes are relatively great and the temperatures lower. Consequently, the experimental difficulties are much less.

Isothermal TG is a common method for studying corrosion and oxidation of metals (Evans, 1960). High precision is necessary to investigate the early stages of such processes. Similarly, high precision is also required to study the reactions of thin films. The large amount of substrate present represents the overwhelming part of the total weight. Consequently, small changes in only a very minor part of the

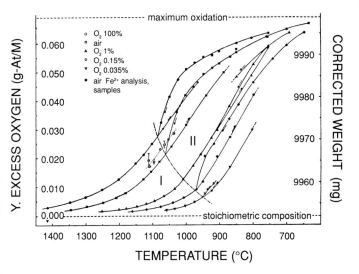

Figure 7-9. Changes of mass for $Ni_{0.685}Zn_{0.177}Fe_{2.138}O_{4-x}$ as a function of temperature and partial pressure of oxygen (Bracconi and Gallagher, 1979).

total sample are being measured (Gallagher et al., 1986).

7.2.5 Thermomagnetometry (TM)

Superimposing a magnetic field gradient in the vicinity of the sample converts TG to TM. Depending upon the direction of the field gradient, an apparent weight gain or loss will occur, if the sample is paramagnetic. This, of course, is the principle of the Faraday balance for measuring magnetic susceptibility. The term TM, however, is not applied to this application. It is applied to the method of using variations in this apparent weight change in the presence of a magnetic field gradient to determine the magnetic transformation temperature and the formation or demise of magnetic materials. Warne and Gallagher (1987) have briefly reviewed applications to that time.

The technique is also capable of greatly increasing the sensitivity towards reactions involving magnetic materials. The kinetics associated with the oxidation and reduction of thin cobalt films was followed by TM (Gallagher et al., 1987). Figure 7-10 shows TM curves for the reactions of 90 nm of Co on Al_2O_3. Although the tra-

ditional TG change in mass accompanying the oxidation of so little Co would not be detected on the scale shown in Fig. 7-10, the oxidation of the Co by O_2 and the subsequent reduction of the oxide produced by H_2 are clearly evident.

These apparent changes in mass are due to the loss of the magnetic attraction from the ferromagnetic Co as it is oxidized to paramagnetic Co_3O_4 at those temperatures. Similarly, the magnetic attraction returns as the oxide film is reduced back to the metal in H_2. The magnetic field gradient in Fig. 7-10 was generated by a small

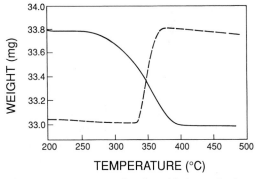

Figure 7-10. TM curves for the oxidation and subsequent reduction of thin films (90 nm) of Co on sapphire (Gallagher et al., 1987).

permanent magnet. The use of strong electromagnets enables one to follow the reactions at levels approaching a monolayer.

The reaction of Co with Si to form a silicide was also studied by TM even though there is no change in mass associated with the reaction. It must be stressed, however, that the technique is following the disappearance or appearance of the magnetic phase, Co. This may have mechanistic implications in the kinetics, e.g., there may be several intermediates in the formation of the Co_3O_4 so that the decrease in Co would not correspond directly with the formation of the product. Had one been able to follow the process by conventional TG, it is possible that entirely different rates and rate laws might have been determined.

The assumption that the magnetic attraction is directly proportional to the mass of Co is certainly wrong for very small particles of Co and might be wrong if the microstructure changes the crystalline alignment with the field gradient. The fact that the mass returned to essentially the original value after having gone through the oxidation–reduction cycle would suggest that these effects were not major factors in this case.

7.2.6 Standards

Standards for mass are readily available in a vast array of values for calibration at or near room temperature and require no elaboration. Operation over wide range of temperature, however, does add uncertainties due to considerations discussed earlier such as buoyancy. Consequently, the calibration for mass is made at room temperature and effort is made to keep the sensor mechanism at a reasonably constant temperature. Changes are ascribed to other causes and incorporated into any blank or background correction.

The sample temperature has traditionally been a problem because contact between the temperature sensor and sample tend to disturb the weighing process. Therefore the sensor, usually a thermocouple, was nearby as opposed to in direct contact with the sample. The interplay between flow and thermal transport problems are, therefore, important.

Two methods commonly used to establish the relation between the sensor's temperature and that of the actual sample temperature are the use of magnetic standards and the "fused link". In a comparison study they have been found to work equally well (Gallagher and Gyorgy, 1986). The advantage of these methods is that there is a definite temperature which is clearly associated with a well defined event that can be measured without contact, i.e., Curie or Néel temperature or the melting point.

In the "fusible link" approach a weight is suspended using a link made of a metal whose melting point is well known. The link is essentially in the normal sample position so that the observed melting temperature would be representative of the actual sample. When the link metls there is a sharp indication in the TG curve (McGhie, 1983; McGhie et al., 1983).

The magnetic technique uses TM of a metal or alloy whose transition temperature has been accurately determined by other reliable techniques (Norem et al., 1970). By definition the Curie or Neel temperature is the temperature at which the magnetic effect disappears. Consequently, the extrapolated end point is used as the calibration point. Because there is no interaction between the samples, it is convenient to include several magnetic standards in the same experiment.

The National Institute of Standards and Technology, N.I.S.T., supplies a series of magnetic metals for this purpose. The calibration procedures and recommended temperatures were established by the Committee for Standardization of ICTA in an extensive round robin study (Garn, 1980).

Since these metals are subject to oxidation, it is necessary to use an inert gas whose thermal conductivity best approximates the gas to be used in the subsequent studies. The magnetic field strength should be no more than necessary to obtain a reasonably measurable effect. Higher fields will broaden the transition and slightly shift the observed temperature. Figure 7-11 presents an example for the heating and cooling of Ni (Gallagher and Gyorgy, 1986). The extrapolated endpoints are illustrated on both the integral and differential curves.

The thermal transport properties of the system vary with the heating rate. Thus the apparent temperature at which the transition occurs is a function of the heating or cooling rate. The extrapolation to zero rate should converge as shown in Fig. 7-12 (Gallagher and Gyorgy, 1986). It is necessary to perform the calibration using the same conditions of heating and flow rates as those to be used in the subsequent studies.

The advent of simultaneous techniques and equipment (see Sec. 7.7) allows the use of DTA standards to calibrate TG in a simultaneous TG/DTA apparatus. Several studies have been made using such equipment (Hongtu and Laye, 1989; Zhong and Gallagher, 1991). The latter study included magnetic TG standards as well as the DTA standards for comparison.

7.2.7 Selected Applications

Space limitations restrict the discussion of applications to a representative few. The origins of TG were associated with the de-

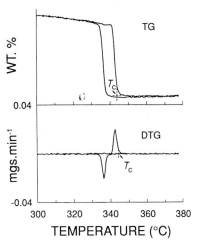

Figure 7-11. TM curves for Ni at $10\,°C\,min^{-1}$ (Gallagher and Gyorgy, 1986).

termination of the temperature and composition at constant weight for analytical gravimetric precipitates. It is also used extensively to assay hydrates for their water content or alkaline earth and rare earth oxides for the absorbed water and carbon dioxide. These materials are notorious for their non-conformance to the formula on the label, particularly after having been opened and exposed.

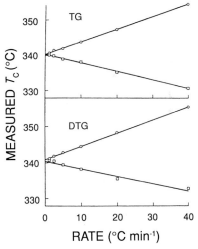

Figure 7-12. Example of the variation in calibration temperature for TG as a function of heating or cooling rate (Gallagher and Gyorgy, 1986).

Naturally occurring mixed carbonate minerals such as dolomite ($Ca_{1-x}Mg_xCO_3$) can be analyzed for the Ca and Mg contents based upon their decomposition in CO_2 (Wiedemann and Bayer, 1985). While the decompositions of the two carbonates tend to overlap in air making resolution difficult, the temperatures of decomposition are shifted in CO_2, see Fig. 7-6, sufficiently that the decompositions are clearly separated permitting analysis. In addition, upon cooling only the CaO recarbonates providing a check on the analysis.

Work with thin films is complicated by the added mass due to the substrate. Occasionally the circumstances are fortuitous so that this background can be precisely subtracted to yield an accurate analysis of the film. Such an example is the determination of the thickness of a carbon film deposited as a hermetic barrier on a fused silica optical fiber. The TG curve shown in Fig. 7-13 was produced by heating mg of coated fiber in flowing oxygen to burn away the carbon and then reheating the remaining uncoated fiber in the same manner. The latter curve was subtracted to produce an accurate representation of the weight of the original film and its combustion (Gallagher, 1992b).

Another way of obtaining data for films is separating them from the substrate where possible. This assumes that the process does not alter the material and that the substrate and film do not have significant interactions. Polyimide is a relatively stable organic material often used as a dielectric film in electronic applications. The temperature that this material can tolerate without degradation during subsequent processing and use is therefore important information.

Portions of a several μm thick film were mechanically striped from a silicon wafer and subjected to TG in various atmospheres to determine their thermal stability. Results are presented in Fig. 7-14 (Gallagher, 1992b). Clearly there is an oxidative degradation which destroys the film at lower temperatures with increasing partial pressures of oxygen. There is a slight weight gain initially as a result of the early stages of oxidation. Working in inert atmospheres obviously enhances the stability of the polyimide.

There is currently great interest in the newer forms of carbon, "fullerenes". These are extracted from carbon soot by benzene, toluene, etc. The fractions of fullerenes can be estimated by TG (Gallagher et al., 1991). The fullerenes are more volatile than the normal graphite and evolve in several stages as indicated in Fig. 7-15. The relative contents of ful-

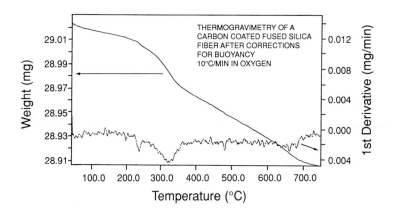

Figure 7-13. Corrected TG and DTG curves for a carbon coated fused quartz fiber heated in oxygen at $10\,°C\,min^{-1}$, see text (Gallagher, 1992b).

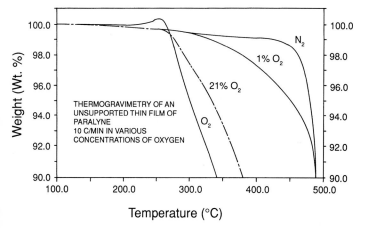

Figure 7-14. TG curves for stripped polyimide film in various partial pressures of oxygen at $10\,°C\,min^{-1}$ (Gallagher, 1992 b).

lerenes can be readily compared for different sources of soot. Supplementary information such as mass spectrographic EGA (Sec. 7.4) is necessary to establish the specific nature of the volatiles.

The ability to vary the atmosphere at different stages greatly enhances the usefulness of TG. Alymer and Rowe (1984) have demonstrated that proximate analysis of coal by TG and TM gives comparable results to that obtained by the much more involved and time consuming methods previously used. Figure 7-16 indicates the basic scheme of the analysis.

The coal specimen is heated in an inert atmosphere to determine the moisture content and then to a higher temperature, 750 to 950 °C, to measure the volatile content. The flowing atmosphere is then changed to oxygen once constant weight had been achieved. At this stage the fixed carbon is burned to leave the ash.

Upon cooling the iron present in the ash is hematite and its exact amount is determined by TM. The magnetic field gradient is applied and the sample reheated in a reducing atmosphere to form iron metal. The amount of iron is determined from the magnetic saturation as evidenced by the apparent weight gain.

Figure 7-17 shows a comparison of the results obtained by this TG-TM method with that from the conventional ASTM method. The comparison is favorable and this new method is being adopted.

Many other mixtures of organic and inorganic materials are amenable to this type of analysis. The proportions of oil, polymer, carbon black, mineral filler, and ash in elastomers has been thoroughly investigated (Maurer, 1969). Similarly, the loading of various materials, e.g., fiberglass, $CaCO_3$, $Al(OH)_3$, etc., into thermosetting

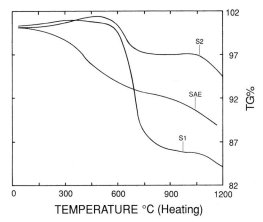

Figure 7-15. TG curves of two sources of soot indicating their content of fullerenes, $20\,°C\,min^{-1}$ (Gallagher et al., 1991).

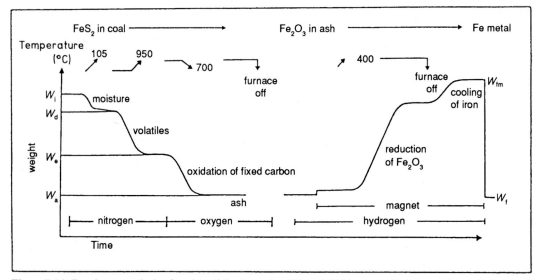

Figure 7-16. Proximate analysis of coals and lignites using TG and TM in controlled atmospheres (Aylmer and Rowe, 1984).

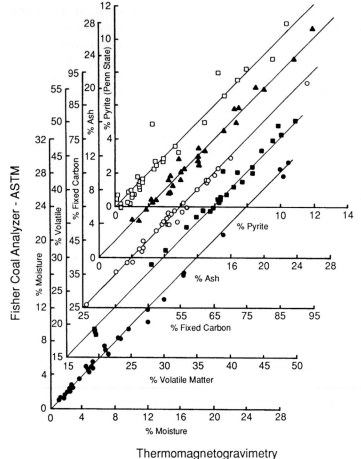

Thermomagnetogravimetry

Figure 7-17. Comparison of the results between the TG-TM proximate analysis and the ASTM Fisher coal analyzer (Aylmer and Rowe, 1984).

resins has been analyzed by TG based methods (Prime, 1980).

Haglund (1982) describes the usefulness of TM to characterize several alloy systems. One of his examples is presented in Fig. 7-18 which indicates how the magnetic transformation temperature of the Co phase changes with WC content in Co cemented carbide tools (Fukatsu, 1961). The shape of the TM curves also show variations with changes in microstructure resulting from different heat treatments.

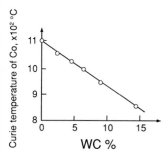

Figure 7-18. Magnetic transformation temperatures of the cobalt binder phase in WC-Co alloys (Fukatsu, 1961).

The aqueous corrosion of metals in sealed systems was cleverly studied by TM (Charles, 1982). The magnetic attraction decreased as the carbon steel was consumed by aqueous 5% ammoniated EDTA solution in sealed glass vessels. Another use of TM is to detect the appearance of magnetic intermediates during reactions. Magnetite was shown to form during the decomposition of siderite, $FeCO_3$, in oxidizing atmospheres but was not detected in inert atmospheres (Gallagher and Warne, 1981). The incorporation of various naturally occurring impurities into the spinel phase was also demonstrated by the concomitant changes in the magnetic transition temperatures.

7.3 Differential Thermal Analysis (DTA) and Differential Scanning Calorimetry (DSG)

7.3.1 Basic Principles and Apparatus for DTA and DSC

Following the temperature of a substance as a function of time, thermal analysis, has been a scientific tool since the beginning. Le Chatelier (1887a, b), however, added the feature of considering the rate of change of temperature, differential thermal analysis. His very clever experimental method displayed marks at equal increments of time and the spacing between marks corresponded to the change in temperature. Marks were close together during slow rates and further apart during rapid changes.

One could still use this fundamental approach today, but a better method than simply differentiating the heating or cooling curve has evolved. Placing an inert reference material, having a similar heat capacity, along side the specimen of interest and following the difference in temperature between the reference and sample has several advantages (Roberts-Austen, 1899a, b). Unintended fluctuations in the rate of heating and cooling are much less likely to cause a significant disturbance in the signal. Both the sample and reference will react similarly canceling the potential effect leaving the baseline unperturbed.

The principle and a schematic of a DTA apparatus are illustrated in Fig. 7-19. As the specimen is heated or cooled in a controlled manner its temperature will depart from the normal rate as it undergoes a reaction or transformation. Consider the heating curve, if the event is endothermic, then the sample will slow its rate of heating while it is undergoing the particular process. Similarly, if the event was exother-

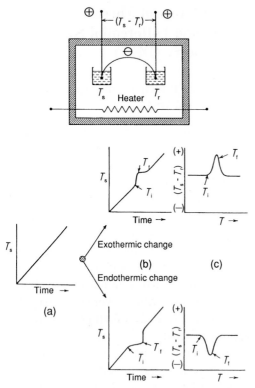

Figure 7-19. General principles and schematic for a DTA apparatus. The subscript s denotes "sample", r "reference", i "initial", and f "final". From Brown (1988): reproduced by permission of Chapman and Hall.

data acquisition system. The ΔT is generally plotted as a function of either the actual sample temperature, reference temperature, or time. Each of these methods has advantages which will become apparent later.

The sample and reference systems are closely matched thermally and arranged within the furnace or oven in a symmetrical fashion so that they are both heated or cooled in the identical manner. Under these circumstances the ΔT signal will be essentially zero until the sample undergoes one of the events described earlier.

The main use of DTA is to detect thermal processes and qualitatively characterize them as endothermic or exothermic, reversible or irreversible, first order transition or higher order transition, etc. The temperature of the event is usually determined accurately. This type of information along with the dependence upon the specific atmosphere make the method particularly valuable for the determination of phase diagrams.

Ideally the area under the DTA peak should be proportional to the enthalpy of the process that gave rise to the peak. There are many factors, however, which influence the curve and which are not compensated in the traditional simple DTA plot. The changes in thermal transport properties of the system, detector sensitivity, etc. with temperature will generally diminish the response of the DTA with increasing temperature.

The DSC was very cleverly developed to either avoid these difficulties or to quantitatively compensate for their effects. There are two types of DSC instruments. The initial type, from which the technique derives its name, is called the "power compensating" version and was developed by the Perkin-Elmer Co (O'Neill, 1964; Watson et al., 1964).

mic, then the sample's temperature would increase at a more rapid rate during that period. A change in the thermal conductivity or heat capacity of the sample would give a change in slope during thermal analysis and an offset or step in the baseline of a DTA curve.

The typical DTA apparatus, shown in Fig. 7-19, utilizes a pair of matched temperature sensors, generally thermocouples, one of which is in contact with the sample or its container and the other in contact with the reference material or its container. The output of the differential thermocouple is amplified and fed to the recorder or

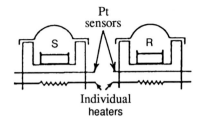

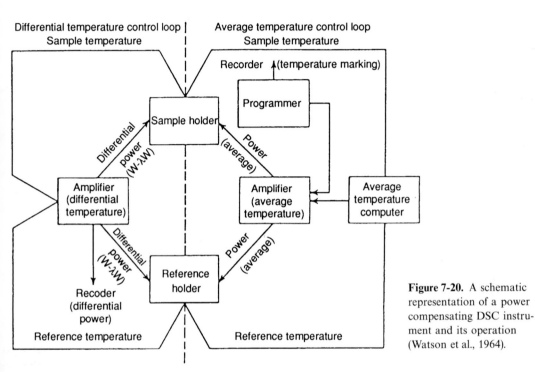

Figure 7-20. A schematic representation of a power compensating DSC instrument and its operation (Watson et al., 1964).

A schematic representation of this instrument is shown in Fig. 7-20. The inventive concept keeps the value of ΔT equal to zero by placing the temperature sensors, Pt resistance thermometers, into a bridge circuit. Any imbalance is used to drive a heater in the appropriate sample or reference portion of the cell. The power needed to keep the bridge circuit in balance is proportional to the change in heat capacity or enthalpy occurring. The integral of the power over the time of the event gives the energy difference between the sample and reference.

If power is supplied to the sample, the process is endothermic. If it is supplied to the reference side, then the process is exothermic. Power applied to the sample has a positive sign so that the integral yields a positive value of ΔH consistent with the endothermic process. This means, however that the sign convention is different between DSC and DTA. Positive power correlates with a negative value of ΔT so that an endothermic peak points upward in DSC and downward in DTA.

The second type of DSC is referred to as operating in a "heat flux" mode. It is similar in operation to a conventional DTA except that the quantitative compensation for the problem areas, such as the temperature dependence of thermal transport and sensor sensitivity, are carefully built into the associated hardware and software. Both instruments are capable of giving satisfactory data. Some instruments have a built in switch to operate in either the DTA or heat flux DSC modes.

Table 7-3 lists many of the processes that are commonly investigated by DTA and DSC. The nature of the effect, i.e., whether endothermic, exothermic, or shift in baseline, is also indicated. Metastable to stable, most decompositions, and polymerization reactions are not reversible. Hence, comparing the heating curve with the cooling curve or a one taken during a reheat is often informative as to the nature of the phenomena observed.

Several of these processes can be illustrated in Fig. 7-21 depicting the typical DTA pattern of a glass. The first obvious event is the positive departure of the base line associated with the ΔC_p at the glass transition. Following that are two overlapping exothermic processes due to the metastable to stable, amorphous to crystalline transformations. These ordered phases melt at higher temperatures giving rise to the two endothermic events.

Reasonable cooling rates would only reveal the two exotherms associated with the

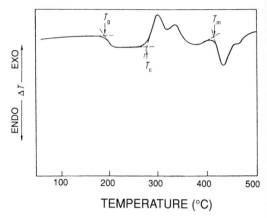

Figure 7-21. DTA curve for a glass (0.5 BeF$_2$, 0.3 KF, 0.1 CaF$_2$, 0.1 AlF$_3$), 25 mg at 20 °C min^{-1} in N$_2$ (Vogel and Gallagher, 1985).

crystallization of the melt. There may be significant supercooling accompanying the process so that the onset of the melting and freezing curves do not correspond. There would not be reversible processes analogous to the original crystallization peaks and the glass transition. A second heating would only show the endothermic peaks due to melting.

DSC can also be used to determine the heat capacity, C_p, of a substance by comparison with a known standard, usually sapphire (single crystal Al$_2$O$_3$). The method is summarized in Fig. 7-22. The basis is the comparison of the differences in the power level for the empty pan, the pan and sample, and the pan with the reference ma-

Table 7-3. Events detected by DTA and DSC and their effects during heating.

Transformation	Observation	Reaction	Observation
1st order	Endothermic	Decomposition	Either
Higher order	Step in base	Liquid-solid	Either
Vaporization	Endothermic	Solid-solid [a]	Either
Fusion	Endothermic	Polymerization	Exothermic
Metastable to stable	Exothermic	Chemisorption and catalysis	Exothermic

[a] Also liquid-liquid or gas-gas reactions. From Gallagher (1992a): reproduced by permission of JAI, Inc.

terial. The comparative aspect negates the need to determine the instrumental constant. The relative displacements in Fig. 7-22b are proportional to the products of the mass and C_p for the sample and the reference materials.

Figure 7-22c indicates the effect a change in heat capacity, ΔC_p, accompanying some reactions can have on the baseline. There are many other factors which can affect the nature of the baseline, some of which are discussed in the next section.

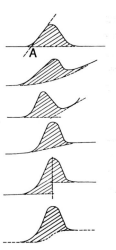

Figure 7-23. Typical DTA and DSC curves illustrating several methods of defining the baseline. From Gallagher (1992a): reproduced by permission of JAI, Inc.

(a)

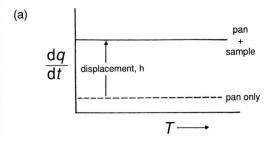

(b)

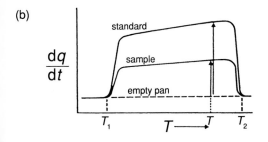

(c)

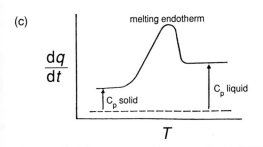

Figure 7-22. Measurement of heat capacity by DSC. From Brown (1988): reproduced by permission of Chapman and Hall. (a) Displacement of the baseline. (b) Scans with the sample and with the reference. (c) Effect of ΔC_p for a reaction on the DSC curve.

It is important to have well defined points for comparisons of curves and association with the cause of the event. Construction of a proper baseline is essential to determine the area of the peak accurately.

Some of these points are considered in Fig. 7-23. The beginning of the event is associated with the initial departure of the curve from the baseline. This is a highly subjective evaluation and easily influenced by the degree of amplification. The extrapolated onset, point A in Fig. 7-23a, is generally used to alleviate this variability. The other curves illustrate ways in which the baseline has been determined. For further information on the topic see some of the recent texts in the field, e.g., Wendlandt (1988), Brown (1988), and Wunderlich (1990).

Some instruments have the ability to adjust the baseline through adjustments in hardware. All computerized instruments have the opportunity to subtract a stored baseline determined under nearly identical circumstances to the actual experiment.

7.3.2 Experimental Parameters and Sample and Atmosphere Considerations

The decomposition of $CaC_2O_4 \cdot H_2O$ was used as a representative decomposition to illustrate TG. The DTA curves for this same decomposition are shown in Fig. 7-24. The three steps can be clearly associated with the changes in mass shown in Fig. 7-5. The initial process is the endothermic loss of water from the monohydrate to form the anhydrous material.

The second step corresponding to the loss of CO and formation of the carbonate is particularly informative. The intrinsic endothermic peak is evident for the DTA curve measured in Ar. The curve in O_2, however, exhibits an entirely different character. It is very highly exothermic because the CO is oxidized on the surface and in the pores and interstices of the powder. This very exothermic oxidation completely overwhelms in mildly endothermic decomposition of the oxalate.

The ΔT signal does not adequately represent the real increase in temperature within the sample. Much of the heat is dissipated in the gas phase and radiated away, particularly in an open container. The true ΔT is several hundred degrees as indicated by the red glow of the sample. There is essentially an uncontrolled thermal runaway situation. It is impossible to predict how much of the oxidation process will occur within the detectability of the DTA sensors. This will depend greatly on the flow patterns and thermal characteristics of the particular DTA instrument.

The final step in the decomposition process is the loss of CO_2 from the carbonate to form the oxide. This, like the hydrate decomposition, does not involve oxidation or reduction and is an endothermic process in either atmosphere. It must be remembered that both the loss of H_2O and CO_2 are reversible processes so that the temperature at which they occur is dependent upon the partial pressure of the product in the atmosphere surrounding the sample. Consequently, the choice of atmosphere and its flow rate, pattern, and purity are factors influencing the temperature of the reaction.

Table 7-4 lists factors which influence the nature of the DTA or DSC patterns observed. These effects are highly interactive. For example, variations of the parameters described in the first three lines of the Table will determine how changes in the heating rate, the last line of the Table, influence the results.

The critical thermal transport between the sample, reference, and furnace is af-

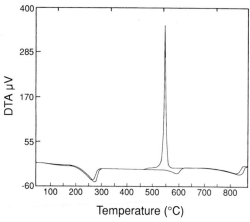

Figure 7-24. DTA curves for the thermal decomposition of $CaC_2O_4 \cdot H_2O$ at $20\,°C\ min^{-1}$ in O_2 and Ar. From Gallagher (1992a): reproduced by permission of JAI, Inc.

Table 7-4. Classes of variables in DTA and DSC.

Construction – thermal linkages – sensor response
Atmosphere – reactions – thermal conductivity
Sample size – packing – particle size
Heating rate – hysteresis

From Gallagher (1992a): reproduced by permission of JAI, Inc.

fected by (1) the physical arrangement between them; (2) the choice of sensor, its size, and position; and the materials of construction are all crucial factors in determining these important thermal linkages. Along with the sensor sensitivity, these factors determine the shape and size of the peak corresponding to the thermal event. The stronger the thermal coupling between the sample and reference the smaller the observed ΔT but the faster the return to the base line with a concomitant improvement in resolution of events. Space limitations preclude an extensive discussion of these factors and the reader is again referred to the standard texts.

The nature of reactions between the sample and the atmosphere have already been described in some detail. The pattern and rate of the flow around and over the sample plays a major role in controlling the processes. The thermal conductivity of the atmosphere is an important link in the thermal connection between the source of heat, the sample, and the reference (Berg et al., 1975; Caldwell et al., 1977). The thermal properties of vacuum or low molecular weight gases such as hydrogen and helium are much different from the normal gases at atmospheric pressure. Similarly, work at high pressures substantially alters the characteristics of the system.

The relative humidity of the gas stream is important in determining the decomposition temperatures of hydrates and hydroxides as well as the oxidizing power of an otherwise inert gas such as nitrogen. Avoiding traces of reactant gases such as oxygen, moisture and carbon dioxide demands attention to the plumbing details.

The affect of sample size and heating rate have been discussed earlier when describing TG. Not only does the flow pattern influence the ability of the gas phase

to react with the sample, but also the sample packing determines how well the two phases can interact. The sample packing also affects the thermal conductivity and, hence, the critical thermal linkages. Dilution of the sample with an inert material has been used to control the size of the effect and aid in achieving a linear baseline, particularly if the diluent corresponds closely to the reference material.

As in TG, the faster the heating rate the greater the opportunity for a lag in temperature between the sample and the sensor. There is, however, another aspect to be considered when performing DTA or DSC. The amplitude of the ΔT signal is increased with increasing heating rate because the thermal event takes place in a shorter period of time. Consequently, the rate of change in the process and hence the signal is greater.

Optimization of a system requires numerous compromises in the operating parameters and the instrumental design. Various adjustments are necessary to achieve the maximum sensitivity, accuracy, and resolution. Table 7-5 indicates some of these considerations. A careful approach with ample sample is to use a fast heating rate and large sample for the initial experiment in order to maximize the ability to detect the minor events. Then a sec-

Table 7-5. Operational compromises for DTA and DSC.

Parameter	Maximum resolution	Maximum sensitivity
Sample size	Small	Large
Heating rate	Slow	Fast
Sample – reference	Linked	Isolated
Particle size	Small	Large
Atmosphere	High conductivity	Low conductivity

From Wendlandt (1988): reproduced by permission of Academic Press.

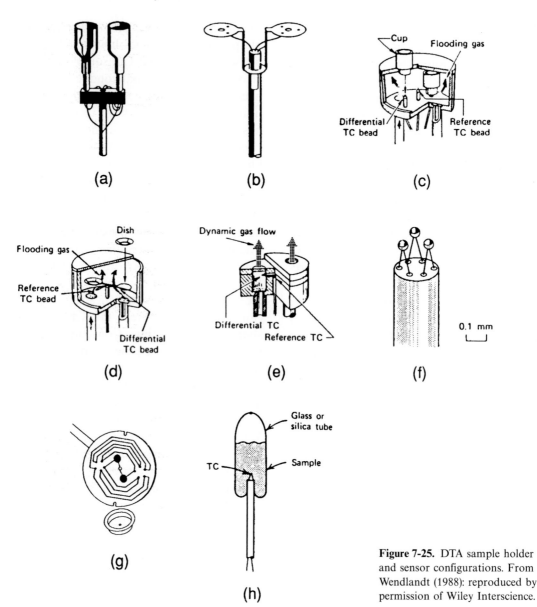

Figure 7-25. DTA sample holder and sensor configurations. From Wendlandt (1988): reproduced by permission of Wiley Interscience.

ond experiment with smaller sample and slower heating rate in order to best resolve the events and determine the associated temperatures more accurately.

Many of the above factors are reflected in the wide range of commercially available sample and sensor configurations. Figure 7-25 displays some of the various types. These accommodate a range in sam-

ple size and sample containers as well as distinctly different flow patterns. The pre-heated gas flows around the sample for most of these arrangements.

In Fig. 7-25e, however the gas flow passes directly through the sample powder analogous to a fluidized bed. This achieves the maximum interaction between sample and gas, either providing a ready supply of reactant gas or sweeping a product gas away. The disadvantages, however, are that fine particles may be swept away and the flow rate will change as the sample reacts, sinters, etc. The latter effect can substantially alter the baseline.

Since the escape of the volatiles is not the overwhelming consideration as it is in TG, it is possible to extend the self-generated atmosphere to a sealed sample approach. This enables the simulation of a closed environment or the protection of the sample from reaction with the flowing atmosphere in the instrument. Manufacturers provide stainless steel containers which seal and contain substantial pressures or simple presses and sample pans which will cold weld and hold up to several atmospheres of pressure. Many varieties of sealed sample containers have been developed by individual investigators.

The common temperature sensor is a thermocouple selected to have an adequate sensitivity and be chemically inert to the gas and other materials in direct contact through the temperature range of interest. This sensor can be placed in direct contact with the sample but is more often in con-

tact with the sample holder or platform. Some imaginative examples are the ring thermocouple in Fig. 7-25d, the micro DTA in Fig. 7-25f, and the thin film thermopile shown in Fig. 7-25g.

7.3.3 Standards

Calibration of both the temperature and the ΔT or energy axis for DTA and DSC under the specific set of experimental parameters in use is clearly necessary. The Standardization Committee of the International Confederation for Thermal Analysis (ICTA) has developed a series of materials for this purpose in conjunction with the National Institute for Standards and Technology (N.I.S.T.). These specific standards are based on solid$_1$ to solid$_2$ phase transitions (Garn, 1975). These and other generally accepted standards are listed in Table 7-6. The melting points are those used to establish the current International Temperature Scale [see Wunderlich (1990) or Quinn (1990)].

There appear to be some differences in the calibration factors determined at temperatures above about 700 °C for the two

Table 7-6. Some standard materials for DTA and DSC.

	Melting points [a]			Solid state polymorphic transformations		
Material	Temperature °C	Enthalpy J/g		Material	Temperature °C	Enthalpy J/g
In	156.5985	28.42		KNO_3	127.7	50.48
Sn	231.928	59.2		$KClO_4$	299.5	99.32
Pb	327.502	23.16		Ag_2SO_4	412	59.85
Zn	419.527	112.0		SiO_2	573	12.1
Al	660.323	400.1		K_2SO_4	583	48.49
Ag	961.78	104.7		K_2CrO_4	665	51.71
Au	1064.18	63.7		$BaCO_3$	810	89.00
Cu	1084.62	205.4		$SrCO_3$	925	133.23

[a] International Temperature Scale 1990 for temperatures and Hultgren et al. (1973) for the enthalpies; [b] NBS–ICTA Certificates GM-758, GM-759, GM-760 for the temperatures and Barin (1989) for the enthalpies. From Gallagher (1992a): reproduced by permission of JAI, Inc.

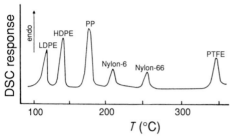

Figure 7-26. DSC of sample plastic waste. From Brown (1988): reproduced by permission of Chapman and Hall. LDPE: Low density polyethylene; PP: Polypropylene; HDPE: High density polyethylene; PTFE: Teflon.

categories of standards (Zhong and Gallagher, 1991). The metals melt at lower temperatures than predicted. The specific causes of the difference between the two classes of standards is not clear at this time. Purity is a greater factor for melting points and emissivities become more significant at higher temperatures. It suggests that the type of standard closest to the samples of interest should be used for calibration at elevated temperatures.

The distinctions between DSC and DTA are such that DSC is much preferred for measuring values of enthalpy. DTA, however, can be used to estimate the value of ΔH when necessary. A calibration curve prepared using a standardized set of conditions is used to provide a factor to convert the area of a DTA peak to energy. An example of such a calibration curve is shown in Fig. 7-26 (Zhong and Gallagher, 1991). The reduction in sensitivity with increasing temperature is dramatic. The points represent both types of standards given in Table 7-6. There does not appear to be any reason to distinguish between the two classes of standards for this purpose.

7.3.4 Selected Applications

DTA and DSC have found frequent and wide ranging use from qualitative analysis,

to quantitative analysis, to many process related applications. DSC has been run in a simultaneous mode with X-ray diffraction to observe the crystallization melting and fusion in poleyethylene (Crowder et al., 1984). It has been used extensively in the pharmaceutical and organic chemical industries for purity analysis (Morros and Stewart, 1976). An example of the use of DSC to provide qualitative essentially fingerprint type information is illustrated in Fig. 7-26. The melting points of each component in the collection of plastic scrap can be used to indicate the materials present (Brown, 1988).

In a more quantitative fashion the temperature of crystallization or the glass transition temperature for polyblends of terpolymer and polyvinyl acetate can be used to determine the relative amounts of each component in the mixture (Anderson et al., 1979). Figure 7-27 shows the linear decrease in the temperature of crystallization as a function of the PVC content. The

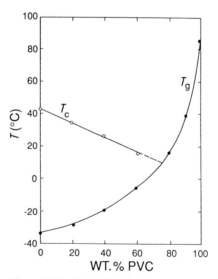

Figure 7-27. Crystallization temperature, T_c, and glass transition temperature, T_g, as a function of PVC content for blends of terpolymer and PVC (Anderson et al., 1979).

glass transition temperature increases in a smooth non-linear fashion from $-34\,°C$ for terpolymer to $86\,°C$ for PVC.

The relative areas under the decomposition peaks have been used to determine the proportion of iron present as the carbonate minerals, ankerite or siderite, in feroan dolomites (Warne et al., 1981). The sharpness and separation of the individual peaks is enhanced by the use of CO_2 as the atmosphere because of the reversible nature of the decompositions. The DTA curves in Fig. 7-28 for a series of prepared mixes show that as little as 1 wt.% of either mineral can be detected in the presence of the other.

The second order ferroelectric phase transformation in lithium niobate, $Li_{1-x}NbO_{3.0-0.5x}$, has been shown to be a function of the cation vacancy concentration and used to establish the congruent composition for crystal growth (O'Bryan et al., 1985; Gallagher and O'Bryan, 1988; Gallagher et al. 1988). The cation vacancy

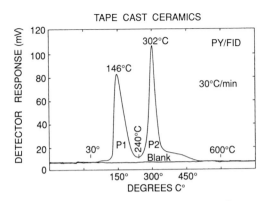

TAPE CAST CERAMICS

Figure 7-29. EGD curve for a tape cast ceramic material using a flame ionization detector. (Unpublished work, Gallagher and the Ruska Co.)

content can be directly related to the extent of Li deficiency, X, or the extent of aliovalent ion substitution, e.g., Ti^{4+}, Mg^{2+}, etc.

Figure 7-29 is a plot of the transition temperature as a function of cation vacancy concentration. The dependence is linear and independent of the cause of the vacancies. There is a critical vacancy content beyond which the structure can no longer tolerate them and further aliovalent substitution or Li deficiency leads to second phase formation. An unchanging transition temperature of $963\,°C$ indicates the attainment of that critical concentration.

DTA is used extensively to determine phase diagrams. In many alloy systems it is possible to use the liquidus temperature as an analytical tool to determine the alloy's composition. The lead tin system is of special interest because of its use as soft solder. In circuit board production the solder is frequently deposited by electroplating. Considerable variations in the composition can result from process variables such as current density, temperature, bath depletion, etc. DSC was used to determine the composition in small sections of the deposit to follow the variations and help optimize the process (Kuck, 1986).

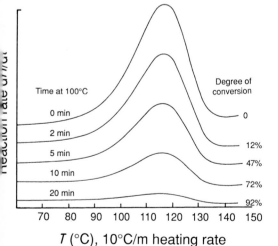

T (°C), 10°C/m heating rate

Figure 7-28. DSC curves for the evaluation of degree of cure for resin held at $100\,°C$ for the indicated times (Swarim and Wims, 1976).

The above examples have dealt with the use of the transition temperature to obtain analytical information. It is also possible to use the quantity of heat absorbed or evolved for this purpose. The evaluation of catalysts for high pressure hydrogenation is such an example. Kosak (1976) used the area under the exotherm in a high pressure DSC cell to estimate the extent of the hydrogenation reaction. Under a specified set of conditions, the area was used to establish the optimum catalyst – support system and as a quality control tool thereafter.

The exotherm associated with the crosslinking of the uncured resin remaining has been used successfully as a sensitive indicator of the degree of initial cure in thermosets (Swarim and Wims, 1976). Figure 7-28 shows how the amount of uncured resin remaining can be determined from the residual exotherm in comparison with the exotherm for the complete reaction. The resin was held isothermally at $100\,^{\circ}C$ for various times prior to determining the DSC curve. The area of the residual peak, compared to that for complete reaction, is indicative of the fraction uncured. It is claimed that accurate values are attainable in the range of 0 to 98 % cure.

The advent of photocuring has led to the modification of DSC instruments to allow for the exposure of both the sample and reference compartments to a controlled source of radiation. By matching the radiant energy to both cells, it is possible to follow the photocuring process accurately. The wavelength of the light can be scanned to determine the spectral dependence of the photochemistry. Isothermal studies can be used to evaluate the kinetics of the process.

A DSC or DTA instrument can be used to sense the onset of an exothermic event with considerable accuracy and sensitivity. This has been exploited to determined re-actions of the gas phase with the sample surface. The oxidative degradation of many materials, particularly electrical insulation, has been studied along with the effects of various antioxidants (Bair, 1980). Similarly, the onset of catalytic activity can be detected and used efficiently to screen potential catalysts (Gallagher et al., 1976). The poisoning effects of various additives to either the catalyst or the gas stream can also be effectively studied by these techniques.

7.4 Evolved Gas Analysis and Detection

7.4.1 Sampling Considerations

The distinction was made in the introduction, Sec. 7.1, that EGD refers to the qualitative detection that volatile species are being evolved while EGA pertains to the determination of the specific species being evolved and, where possible, the quantitative determination of the amount. It should also be noted that EGA or EGD can be equally effective by following the consumption of a reactive gas such as O_2, H_2, CO_2, H_2O, etc. Perhaps a term like "consumed gas analysis", CGA, might be more appropriate under those circumstances.

There are many instances where the EGA or EGD apparatus is designed to function as a stand alone device providing the temperature control and all other functions. More often it is used in a simultaneous mode accepting the output carrier gas stream from a TG, DTA, DSC, or even simultaneous TG/DTA.

The particular method of sampling the gas is determined by the nature of the specific sensor and/or the sensitivity that is required. When sensitivity is an issue, it is frequently necessary to accumulate products by trapping them over a period of time and analyzing them periodically. Trapping is also used when the analytical instrument must be physically removed or is too valuable to be dedicated to an EGA role. The stored samples can then be brought to the instrument for analysis at a more convenient time.

If the analytical instrument is a dedicated on line instrument, the rate of sampling is dictated by the nature of the instrument and the rate of change in the gas phase. Some analytical methods, such as gas chromatography, are inherently discontinuous in operation. Other methods, such as FTIR, can be sampled continuously or discontinuously depending upon the frequency of sampling appropriate for the specific problem under investigation.

Factors influencing this choice are (1) the rate of change in the composition of the gas stream, (2) the storage capacity and speed of the data acquisition system, (3) the time needed to scan a range of wavelength or mass numbers, etc. The latter consideration is highly variable. If the species evolved are not known, then a wide range is generally scanned in order to qualitatively determine the evolved products. Once this has been accomplished, another experiment can be performed in which only the relevant wavelengths, mass numbers, etc. are measured in a much faster manner.

Compatibility of the carrier gas and the product gases must be considered, both in terms of direct reactions and analytical interferences. Similarly materials of construction should not alter the composition through absorption, catalysis, etc. Many of the gaseous products evolved have a significant vapor pressure only at the temperature of the reaction. Consequently, they will tend to condense in cooler portions of the exhaust train prior to reaching the sensor. As a result, heated transfer lines are often necessary.

The delay between the "simultaneous" signals is controlled by the space velocity of the gas stream. Reducing the volume between the sample and detector or increasing the flow rate of the gas will shorten the delay and enhance the direct correlation. The carrier gas rate can not be increased indefinitely, however, as it will generally impact on the noise level or sensitivity of the TG, DTA, DSC, etc. and dilute the concentration of the evolved gases.

Thermoanalytical measurements are performed over a wide range of atmospheric pressure. The previous discussion has concentrated on conditions near atmospheric pressure. There are situations when the gas analyzer and the other thermoanalytical devices require different conditions of pressure. The usual example involves the use of mass spectrometry which requires a vacuum of 10^{-5} torr or less. A variety of interfaces have been designed for this purpose and some will be described when discussing the mass spectrometer in the next section. When the conditions defy optimization it becomes advisable to perform the measurements separately.

7.4.2 Detectors and Analyzers

There are a number of commonly used detectors which are based upon physical properties of the gas phase and are not specific to the chemical nature of the gases. If the system is static, i.e., not a flowing atmosphere, then either pressure or volume can be used as a detector for EGD. The apparatus depicted earlier in Fig. 7-7 is an example of this method. Gas density can be used in a flowing system. A carrier gas such as helium would tend to maximize the signal. Thermal conductivity is an excellent choice because of its sensitivity and the ready availability of detectors. If the specific nature of the reactions involved are known, these methods can also become quantitative analytical tools.

Other methods used for EGD are somewhat more selective. This can be a benefit or disadvantage depending on the nature of the problem. Broad band optical absorption or emission has been successfully used. Filters can narrow the range and improve the selectivity. Aerosols can also be detected in this manner.

Flame ionization detectors have been adapted from gas chromatography and are very sensitive for organic matter. Figure 7-29 shows an EGD curve of a dried ceramic slurry used for tape casting. The organic solvents, binders, plastisizers, lubricants, surfactants, etc. must be carefully removed before the final sintering process. The curve clearly indicates the two stage removal of the residual organic materials after drying.

Another selective approach to EGD involves passing the exhaust gases through liquids or over solids which will absorb certain types of gases, e.g., acidic or basic. The absorbents could be specific to a particular gas, or to a specific class of gases only one of which is known to be in the gas stream. The method can be quantitative EGA under these conditions.

There are many gas analyzers which are commercially available for process stream analysis or environmental monitoring that can be easily adapted for EGA. Dew point detectors can be used to determine the water loss from a sample or that formed by reactions such as the hydrogen reduction of oxides. The EGA curve for the thermal decomposition of hydrated barium hydroxide is compared with the simultaneous DTG curve in Fig. 7-30 (Gallagher and Gyorgy, 1980). The agreement is excellent

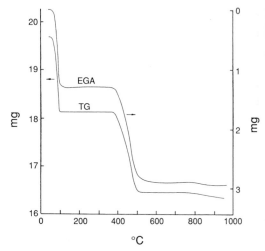

Figure 7-30. Comparison of the TG and EGA curves for water loss from hydrated barium hydroxide (Gallagher and Gyorgy, 1980).

and the dew point and flow rate can be combined to give quantitative information.

The evolution of oxygen or its consumption from a gas mixture can be followed using standard oxygen analyzers based upon the emf generated in special ceramic cells operated at high temperatures. The flow of the experimental exhaust through such a cell will detect the changes taking place, provided that the initial oxygen partial pressure is low enough to exhibit a significant change during the time of the reaction.

When oxidation-reduction phenomena are being studied, the use of an oxygen analyzer on the exit stream is a good practice. Low partial pressures of oxygen are difficult to maintain because of outgassing of the system during heating and the aspirating effect of a flowing gas past even a slight leak. The purity of the source gases is also checked in this manner.

Emanation thermal analysis, ETA, is a unique form of EGA based on the evolution of a radioactive gas from the sample as it is heated (Balek, 1987). The radioactive parent, ^{228}Th or ^{224}Ra, is incorporated into the sample during its synthesis. Radioactive ^{220}Rn is formed by α decay and trapped within the sample. Alternatively the sample can be labeled by ion implantation.

During heating the radioactive gas is released at a rate that is influenced by reactions, transformations, changes in surface area, porosity, etc. The radioactivity of the gas stream is monitored with high sensitivity using a flow through counter. Figure 7-31 shows a comparison of the DTA, EGA, and ETA curves for the decomposition of $Th(C_2O_4)_2 \cdot 6H_2O$ (Balek, 1969). The first peaks correspond to stepwise loss of water. The intermediate peak around 350 °C corresponds to the decomposition of the oxalate to an amorphous or microcrystalline oxide. The final peak is due to

the amorphous to crystalline transition in the oxide.

Commercial instrumentation is available for ETA and interest is slowly developing in the method. Differences in the rate of emanation from different crystallographic orientations have been observed in single crystal tungsten (Balek, 1987). It has also been extensively used for studies of the reactivity of many oxides by Ishii et al. (1981).

There are two methods of analysis widely used for EGA because of their general ability to detect and analyze most gases. FTIR is less sensitive and versatile than mass spectroscopy (MS), however, it is simpler and less expensive to use. Gas chromatography is less widely used, at least in real time analysis, because of the time required for analysis. There have been examples, however, where it has been used successfully. As an example it has been coupled with DSC to determine the extent of CO oxidation for catalyst screening (Gallagher et al., 1976).

Figure 7-31 shows a heated cell arrangement that is suitable for EGA. Heated fused silica tubing is used to connect the heated cell with the thermoanalytical module. The cell windows and walls are heated above 200 °C and the tubing heated to a slightly lower temperature to prevent materials from depositing in the FTIR cell.

An example of the resulting curves is presented for the decomposition of $CaC_2O_4 \cdot H_2O$ in Fig. 7-32. These curves were part of the same simultaneous study depicted in Figs. 7-5 and 7-24. The overall integrated intensity of the IR absorption shown in Fig. 7-32a is dependent upon the absorbency indices of the particular species evolved. Hence, the relative amounts of volatiles can not be directly inferred as they can in DTG curves. In this instance the water absorption is weaker than the

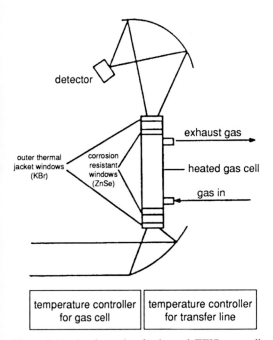

Figure 7-31. A schematic of a heated FTIR gas cell used for EGA. Reproduced by permission of Bio-Rad Co., Digilab Div., Cambridge, MA.

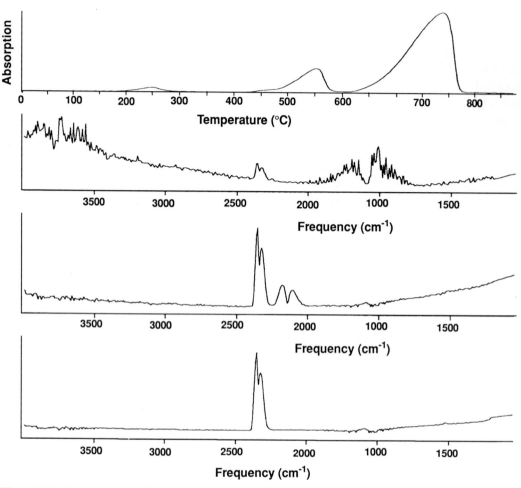

Figure 7-32. EGA curves for FTIR analysis of the thermal decomposition of $CaC_2O_4 \cdot H_2$ at 20 °C min^{-1} in Ar. From Gallagher (1992a): reproduced by permission of JAI, Inc.

others and the first peak is under-represented compared to the DTG curve in Fig. 7-5.

The curves shown in Figs. 7-32 b, c and d are the FTIR scans taken at the peak positions corresponding to the three major evolutions indicated in Fig. 7-32 a. The spectrum for the first peak is that of water vapor. The particularly informative aspect of EGA is illustrated by the spectrum corresponding to the second peak, the decomposition of the anhydrous oxalate by the evolution of CO. A large amount of CO_2 is

indicated in addition to the CO. This reveals the extent of the disproportionation which has taken place.

Thermodynamically, CO is unstable below about 800 °C with respect to the formation of CO_2 and C. The extent to which this reaction occurs depends upon the catalytic nature of the system and is unpredictable. The EGA indicates that in this particular instance the disproportionation goes to a significant extent. Again, it must be remembered that the apparent relative amount is distorted by the differences in the strengths of the respective IR bands. The powder at this point is either grey, black, or brown depending upon the amount of carbon deposited. In an oxidizing atmosphere this disproportionation does not occur because the CO released is immediately oxidized by the oxygen in the atmosphere. The final step in the decomposition shows the evolution of CO_2 in Fig. 7-32d from the thermal decomposition of the carbonate.

Advantages of FTIR over MS are that the interpretation is generally simpler and the measurement does not require a vacuum. FTIR has been particularly valuable in studies of organic materials and polymers.

The interface between the mass spectrometer and the thermoanalytical module is generally complicated by the difference in operating pressures of the two modules. Wendlandt (1988) and Langer (1980) describe some of the techniques used to accomplish the connection for a variety of circumstances. However, when the simultaneous measurement is also conducted in vacuum the interface is simplified. The major problem under those circumstances is the same as that faced by FTIR, the prevention of condensation between the sample and the detector. Heated transmission lines remain the most viable approach.

This problem can be alleviated by separate MS-EGA measurements where the sample can be heated directly in the vacuum system with a short direct line of sight to the ionizing elements of the mass spectrometer (Gallagher, 1978, 1984).

Figure 7-33 shows the results for this system again using the thermal decomposition of $CaC_2O_4 \cdot H_2O$ as an example. The nature of Fig. 7-33 is similar to that for Fig. 7-32. The first curve is the total ion count representing the integral over the entire range of mass scanned. Once more three major peaks are evident.

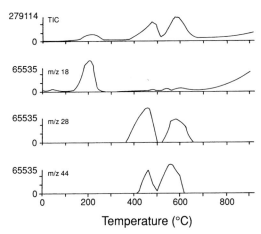

Figure 7-33. EGA curves for mass spectroscopic analysis of the thermal decomposition of $CaC_2O_4 \cdot H_2O$ at $20\,°C\,min^{-1}$ in vacuum (Gallagher, unpublished work).

The remaining three portions of the figure are curves formed by continuously plotting the intensity of the indicated peaks as a function of temperature. Alternatively, the scans taken at the maximum of each peak in Fig. 7-33a could have been plotted as was done in Fig. 7-32. A complicating aspect of mass spectroscopy is that the peaks in the mass spectra are not solely those of the parent species but involve many daughter species created during the ionization step. Consequently the "cracking patterns" of the evolved species must be considered. For example, the curve for CO^+ shown in Fig. 7-33c includes that amount arising from the cracking pattern of CO_2 as well as that originating from CO and N_2.

In both Figs. 7-32 and 7-33 specific wavelengths or amu could have been selected and the scan restricted to those values so that many more points are accumulated as a function of time and a smoother more detailed plot achieved. This would normally be done on a follow up experiment once the species evolved had been

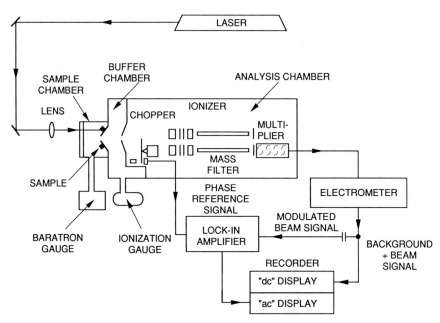

Figure 7-34. A laser heated differentially pumped MS-EGA instrument (Lum, 1977).

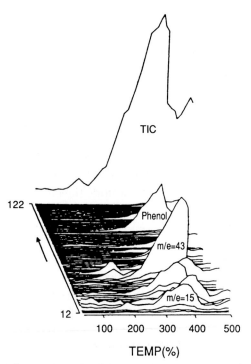

Figure 7-35. MS-EGA curves for the decomposition of Novolac epoxy (Lum and Feinstein, 1980). *m/e* is mass to charge ratio of the ionized species.

clearly established through the use of a broad scan.

Another variation on this method is to heat the sample in a separate chamber, which can be at atmospheric pressure, using a laser and allowing the vapor to pass through differentially pumped orifices into the vacuum of the mass spectrometer (Lum, 1977). The beam can be chopped and phase lock amplification used to facilitate the interpretation of parent–daughter fragmentation and to make correction for the background. Figure 7-34 shows a schematic of such an instrument.

An example of the output is shown in Fig. 7-35 for the decomposition of a Novolac epoxy, a common semiconductor encapsulant (Lum and Feinstein, 1980). TIC represents total ion current as in Fig. 7-33 a. The particular software devised for this instrument allows for the presentation of data in a convenient 3-D manner. Several of the curves are labeled with respect to their parent species.

Figure 7-36 shows a similar differentially pumped interface to a simultaneous TG-DTA instrument (Kaisersberger, 1980). This results in simultaneous TG-DTA-EGA similar to that described earlier for the FTIR. In this instance the walls of the interface are directly in the furnace and, therefore, are at a relatively high temperature, approaching that of the decomposition temperature. Any atmosphere can be flowed over the sample providing considerable experimental lattitude. The sensitivity is reduced, however, compared to that of a separate system incorporated directly in the vacuum. The alignment of the orifices in these differentially pumped systems is critical in order to produce a molecular beam with minimum interactions and contacts with the walls before reaching the mass spectrometer.

7.4.3 Selected Applications

The complex cyanides represent an instructive set of examples on the applications of EGA in inorganic chemistry. The thermal decomposition of rare earth hexacyanoferrate(III) or ammonium rare earth hexacyanoferrate(II) compounds in air or oxygen is an excellent way of preparing the stoichiometric rare earth orthoferrites (Gallagher, 1968). Extensive effort using many techniques was made to understand the details of the decomposition process (Gallagher and Schrey, 1968).

It remained, however, for EGA to provide the major clue that the prominent early step in the decomposition was the hydrolytic reaction between the waters of crystallization and the complex cyanide ion to form the rare earth oxyhydroxide and hydrogen cyanide (Gallagher and Prescott, 1970). The EGA also indicated the loss of cyanogen as the iron(III) was reduced to iron(II). The cyanogen was not

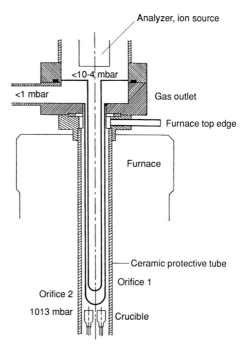

Figure 7-36. A simultaneous TG-DTA-MS-EGA apparatus (Kaisersberger, 1980).

evolved during the decomposition of the hexacyanoferrate(II) materials.

Again the loss of cyanogen was the indication of the reduction process taking place during the decomposition of complex platinum cyanides (Gallagher and Luongo, 1975). The detection of nitrogen in the EGA for the ammonium dihydrogen arsenate(V) was the evidence for the reduction of arsenic(V) by the ammonium ion (Gallagher, 1976). The specific changes associated with such redox reactions are frequently difficult to establish without the benefit of EGA.

There have been numerous examples of the application of EGA to mineralogy and geochemistry. Morgan (1976) demonstrated the use of several different EGA techniques simultaneously to follow the decomposition Cambrian shale. Different detectors measured the evolution of H_2O, CO_2, and SO_2 as a function of temperature. Other examples are described in a brief review by Morgan et al. (1988).

Detailed studies of the reaction of nickel oxide with various iron oxides and sands containing iron and titanium were conducted by MacKenzie and Cardile (1990). The formation of nickel ferrite and magnetite are followed by oxygen evolution, among other techniques.

Asbestos is a mineral of considerable interest and technological importance. It had been proposed that TG and DTA could be used to analyze the mixtures of asbestos with the six minerals that naturally occur in conjunction with deposits of asbestos (Cossette et al., 1980). Templates had been prepared which could overlay TG and DTA curves to separate the various temperature ranges associated with the decomposition of the individual minerals. From the weight losses it would then be possible to determine the specific amounts of each mineral. During the course of a round robin on the procedure, however, EGA revealed that there was considerable overlap in several areas between the decomposition of carbonate and hydroxide minerals (Khorami et al., 1984). Conse-

quently, the method never became widely accepted.

The patina on copper metal formed by corrosion products has been characterized by EGA (Nassau et al., 1987). Comparison of copper taken from roofs and that taken from the Statue of Liberty during its recent refurbishing was made in this manner. An example of a MS scan taken at 530 °C is shown in Fig. 7-37.

Obviously the patina or corrosion layer is composed of several components. There is evidence of, at least, hydroxides, sulfates, carbonates, and chlorides. Specific identifications were made by comparison of the EGA and X-ray diffraction results with those of assorted pure minerals and compounds.

In addition to corrosion studies, another area of metallurgy that has benefitted from EGA is the reduction of oxides to form metals. There have been a number of studies suggesting that an external magnetic field would affect the rate of reduction of metal oxides by hydrogen, see for example Rowe et al. (1979) and references therein. These studies were generally conducted by TG and, hence, the problems associated with sample movement and variation in the magnetic properties of very fine particles were of concern. Following the course of the reaction using EGA of the water vapor formed, however, does not suffer from such difficulties.

Figure 7-38 shows a comparison of the simultaneous DTG and EGA curves for the reduction based on weight loss and the dew point of the gas stream (Gallagher et al., 1982). EGA curves are also shown for data obtained using a separate EGA cell which could subsequently be used between the pole faces of an electromagnet to study the potential influence of a magnetic field on the rate of the reduction. The curves in all three sets are qualitatively

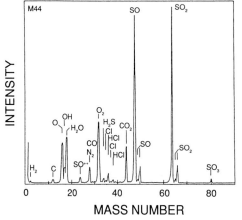

Figure 7-37. A MS-EGA scan taken on corroded copper metal at about 530 °C (Nassau et al., 1987).

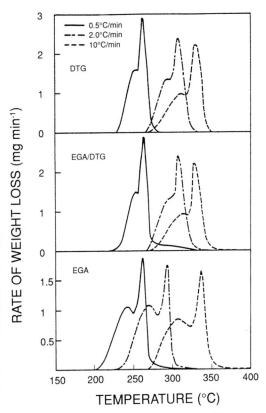

Figure 7-38. Comparison of DTG and EGA curves for the reduction of NiO by H_2 (Gallagher et al., 1982).

important in VLSI semiconductor technology. Films are made by several methods. Sputtering is a common method that frequently leaves a small amount of the ions used to vaporize the target trapped in the final film. Argon is the most common gas used for this purpose and the amount of trapped or occluded argon is of concern. It varies markedly depending on the experimental conditions used during the sputtering process. MS-EGA and Rutherford backscattering have been successfully used to determine the amounts of Ar in amorphous magnetic films of GdCo and correlate the properties and Ar content with the sputtering conditions (Hong et al., 1986; Bacon et al., 1986).

During high temperature encapsulation of VLSI circuits in borophosphosilicate glass, the Ar has been shown to evolve and cause blistering and loss of adhesion of underlying $TaSi_2$ films. The temperature

very similar. Shifts in the apparent reaction temperature as a result of the heating rate are clearly evident.

Experiments performed using several oxides in the presence of a strong magnetic field did not indicate any influence of the field upon the rate of reduction (Gallagher et al., 1981; Rowe et al., 1983). Figure 7-39 shows the results of a series of experiments for Fe_2O_3. The curve appears to be unaffected by the presence of the magnetic field.

The sensitivity of MS is exploited to do EGA of thin films where the amount of sample is particularly small and therefore difficult to measure by other thermoanalytical techniques. Films are particularly

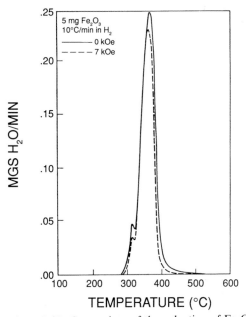

Figure 7-39. Comparison of the reduction of Fe_2O_3 by H_2 in the presence and absence of an external magnetic field (Gallagher et al., 1981).

profile of this evolution and the relative amounts of Ar have been studied by EGA (Levy and Gallagher, 1986). The loss of Ar shown in Fig. 7-39 at about 1000°C is responsible for the problems. This is the temperature at which the encapsulating glass is flowed. The peak in Fig. 7-39 near 300°C is due to gas evolution during the amorphous to crystalline transition of the TaSi$_2$ film. The sputtering conditions were varied and EGA used to optimize conditions which minimized the incorporation of Ar.

Hydrogen is another gas that is commonly trapped or bound in films. The source of hydrogen is the starting compounds in the chemical vapor deposition (CVD) or plasma deposition processes. Hydrogen in polysilicon has dramatic effects upon the electrical properties. The amount of hydrogen incorporated into the film has been shown by EGA to be primarily determined by the temperature of the deposition process.

Films of BN are of interest as photomasks in the X-ray lithography process. There usefulness, however, has been shown to be limited because of the strains present in the films prepared by CVD. Duncan et al. (1988) have shown that the strain is created by unbound but incorporated hydrogen. Figure 7-40 shows an EGA curve of hydrogen in such a film. The shape of the curve indicates several types of binding are evolved. Thermal anneals at 500°C are effective in removing the lightly held material and relieving the stresses in the film. The exact annealing conditions were determined by EGA. NMR and IR spectroscopy of the films were used to determine the various bonding sites for the hydrogen.

The MS-EGA technique in which the sample is heated directly in the vacuum is most practical for studying the degradation of III–V and II–VI semiconductors.

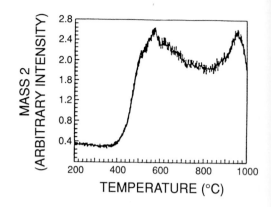

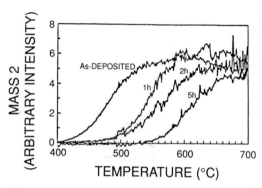

Figure 7-40. EGA curves for hydrogen in BN films at 20°C min^{-1} (Duncan et al., 1988).

Kinsborn et al. (1979) have utilized the method to show how the gold electrical contact on GaAs will degrade at temperatures much below the onset of degradation for pure GaAs. This was shown to be the result of Ga dissolving in the Au film leaving free As. The volatilization of the As was observed by MS-EGA at temperatures achieved during subsequent processing. Barrier films were ultimately developed to prevent the Ga diffusion.

The very early stages of the degradation of InP were studied by Gallagher and Chu (1982) utilizing MS-EGA. The very high sensitivity enabled them to derive meaningful kinetic data from as little as 0.03% decomposition of the InP.

Polymers are the primary field of application for thermal analysis. The thermal degradation of polyvinyl chloride pipe is shown in Fig. 7-41. The nature of the volatile products is important from an environmental standpoint when fire occurs, since this polymer is used so much as an electrical insulator and material of construction. Both TG and FTIR-EGA curves are shown which enables identification of the products coming off at different stages in the decomposition process. The first weight loss peak can be ascribed to benzene, the second to hydrogen chloride, and the third peak to various hydrocarbons. Similar studies were made by Wieboldt et al. (1988).

The volatile components from combustion are important. The addition of materials to a polymer in order to reduce its flammability is a major application, particularly pertaining to textiles. Flame retardants act in several different ways to adsorb the heat or reduce the propagation of the flame front. Pearce et al. (1980) discuss these aspects and describe the usefulness of EGA for such investigations.

7.5 Thermomechanical Methods

Thermomechanical methods consist of three classes of measurements. Thermodilalatometry is concerned with dimensional

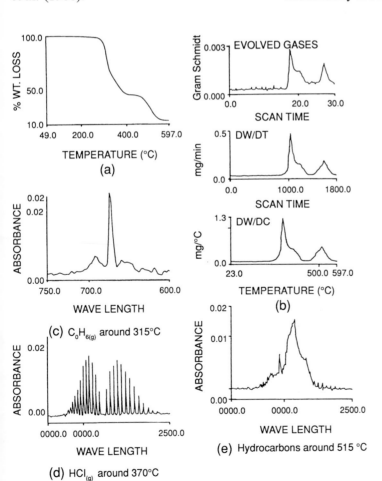

Figure 7-41. Comparison of TG and EGA curves for polyvinyl chloride pipe. Reproduced by permission of Bio-Rad Co., DigiLab Div., Cambridge, MA.

changes as a function of temperature under conditions of essentially zero external loading. Thermomechanical analysis also pertains to dimensional changes as a function of temperature but under the application of a static stress. Dynamic mechanical analysis denotes the measurement of various mechanical parameters under a dynamic or oscillatory stress as the temperature varies in a controlled manner.

7.5.1 Thermodilatometry (TD)

The thermal expansion of a substance is generally measured by following the change of length in a particular direction as a function of temperature without any stress imposed other than the sample's mass. This is experimentally simpler than following the volume and offers the ability to determine the extent of anisotropy for that material.

The change in dimension is proportional to the starting dimension and is generally expressed as a dimensionless quantity $(L_T-L_0)/L_0$ where L_T is the length at temperature T and L_0 is the length at some standard temperature, generally 298 K. The quantity L_T-L_0 is frequently abbreviated ΔL_T. The coefficient of thermal expansion, α, at any temperature is the derivative of $\Delta L/L_0$ with respect to temperature at the desired temperature and has the units of reciprocal temperature. The volume expansion can be similarly expressed by substituting volume, V, for length.

Thermal expansion data is frequently expressed and tabulated as a third order polynomial in temperature. The constants are evaluated by a curve fitting procedure over ranges which do not include phase transformations or other disturbances. The polynomial can be differentiated easily to give a second degree polynomial expressing the value of α as a function of temperature.

Many techniques can be used to follow the change in length, see for example the review by Murat (1983). If the sample is large and high accuracy is not required, it is adequate to use a simple scale or micrometer. For accurate measurement of smaller samples a linear voltage differential transformer (LVDT) is used most often commercially. This device is shown schematicly in Fig. 7-42. The sign of the output voltage changes with the direction of motion simplifying the distinction between expansion and contraction. Changes on the order of 1 µm or less can be readily detected by this device.

Still greater sensitivity can be attained using optical methods. Laser interferometry is well suited for such measurements. Commercial instruments are available with an accuracy of 0.02 µm or 1/32 of the laser wavelength used. The temperature range of these instruments is restricted by loss of reflection from the mirrored surfaces. An upper temperature of around 700 °C is typical while the LVDT based instruments are used to 2000 °C. The sample and reference samples require parallel polished surfaces for the optical measurements.

Figure 7-43 shows representations of both types of instruments. The optical dilatometer is shown in a differential

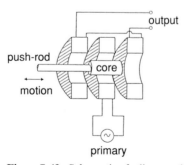

Figure 7-42. Schematic of a linear voltage differential transformer. From Brown (1988): reproduced by permission of Chapman and Hall.

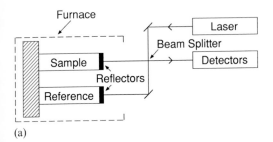

(a)

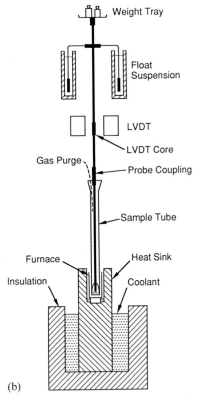

(b)

Figure 7-43. Representative thermodilatometers. (a) Laser interferometer thermodilatometer (Linseis, 1975). (b) LVDT based dilatometer. Reproduced by permission of Perkin-Elmer Co., Norwalk, CT.

metals having well defined thermal expansion. The differential mode helps reduce the correction necessary for the expansion of the sample holder and transmission rod, provided that the sample and reference length are approximately equal. These corrections are essential for the most accurate work.

The material of construction is usually fused quartz because of its low thermal expansion and ease of working. At higher temperatures high density alumina is used. The sample temperature is determined by a thermocouple in good contact with the sample. The LVDT is maintained in a thermostated environment outside of the furnace. The output voltage of the LVDT (mV μm^{-1}) is frequently standardized using mechanical gauge blocks of sheet metal.

Besides the measurement of simple thermal expansion data, TD is useful to detect phase transformations. Figure 7-44 from Wunderlich (1990) shows the variation of molar volume with temperature as a function of temperature. Two types of curves are indicated on cooling from the liquid. The first curve shows an abrupt discontinuity in the volume at the first order phase

mode. The LVDT can also be used in a differential fashion by connecting the reference push rod to the coil and the sample rod to the core. When used differentially, the thermal expansion is measured relative to a standard material. Such a standard material is an oriented single crystal of sapphire, fused quartz, or various pure

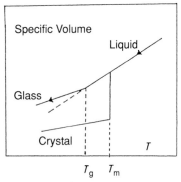

Figure 7-44. Typical changes in molar volume with temperature for a liquid to solid transformation. From Murat (1983): reproduced by permission of John Wiley and Sons.

transformation corresponding to freezing at temperature, T_m. The second curve shows supercooling of the liquid with the formation of a glassy phase. This transition is higher order and therefore does not show a discontinuity. Instead there is a change in slope of the thermal expansion curve at the glass transition temperature, T_g.

7.5.2 Thermomechanical Analysis (TMA)

The instrument shown in Fig. 7-43 b is capable of both TD and TMA. The stress applied in TMA is in the form of a weight added to the platform at the top of the instrument and a variety of probes are used to transmit that stress to the sample.

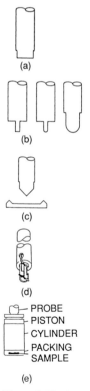

(a)

(b)

(c)

(d)

PROBE
PISTON
CYLINDER
PACKING
SAMPLE

(e)

Figure 7-45. Assorted probes used in TMA. From Wendlandt (1988): reproduced by permission of Wiley Interscience.

Some of these are depicted in Fig. 7-45. The broad flat faced probe in Fig. 7-45 a is used for simple thermal expansion, TD. A variety of penetration probes are shown in Fig. 7-45 b. The probes illustrated in Figs. 7-45 c and d are for bending and tensile modes respectively. Volume expansion can be measured using a transmitting fluid to surround the sample in a cylinder and piston arrangement such as that indicated in Fig. 7-45 e.

Three simple examples are selected from Wendlandt (1988) to illustrate the nature of such measurements. Figure 7-46 a shows the penetration mode as a function of temperature for a polymer heated at $20\,°C$ min^{-1} under a 5 g load. The mixed polymer exhibits two glass transformations which are clearly revealed in this fast simple TMA measurement.

Creep measurements can be made by varying the load under isothermal conditions as shown in Fig. 7-46 b for a plastic material. Finally Fig. 7-46 c shows the extension or tensile measurement of a nylon fiber. Jaffe (1980) discusses the use of thermal methods to study fibers.

7.5.3 Dynamic Mechanical Analysis (DMA)

DMA is an extension of TMA based on the periodic application of stress and strain to the material as the temperature is varied. Through this method it is possible to determine the mechanical moduli and loss mechanisms. There are many commercially available instruments which are extensively utilized, primarily in the area of polymers. There are two general modes of operation based upon the measurement of the mechanical resonance frequency, or the effects of an imposed predetermined frequency. For details concerning TMA

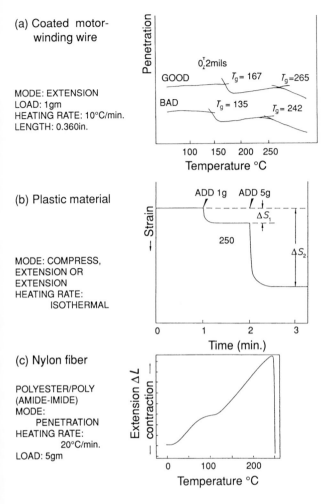

(a) Coated motor-winding wire

MODE: EXTENSION
LOAD: 1gm
HEATING RATE: 10°C/min.
LENGTH: 0.360in.

(b) Plastic material

MODE: COMPRESS,
EXTENSION OR
EXTENSION
HEATING RATE:
    ISOTHERMAL

(c) Nylon fiber

POLYESTER/POLY
(AMIDE-IMIDE)
MODE:
    PENETRATION
HEATING RATE:
    20°C/min.
LOAD: 5gm

Figure 7-46. Selected examples of TMA. From Wendlandt (1988): reproduced by permission of Wiley Interscience.

and DMA the reader is referred to the text of Wunderlich (1990) and references therein.

The TA Instruments model is shown in Fig. 7-47. It is capable of operating in both the resonant and imposed frequency modes. The sample environs are controlled to the desired temperature and the stress and strain applied as a predetermined function of time, frequently sinusoidal. It is also possible to perform relaxation and creep studies with this instrument.

Feedback from the LVDT that follows the amplitude and frequency of the oscillation is used to control the motion of the complex pendulum holding the sample.

The driving force of the electromechanical transducer necessary to maintain a constant amplitude of vibration is related to the damping of the sample.

The stress strain response curves reveal the mechanical nature of the material. Elastic solids show a direct relation of the resultant strain to the applied stress, i.e., they obey Hooke's laws. Viscous materials, on the other hand, obey Newton's laws whereby the applied stress is proportional to the rate of strain. Many materials exist in the intermediate range and are referred to as viscoelastic materials. These require a combination of both laws to accurately describe their mechanical behavior.

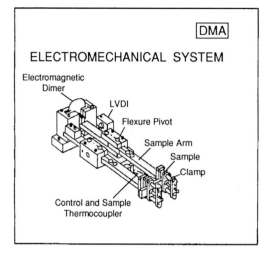

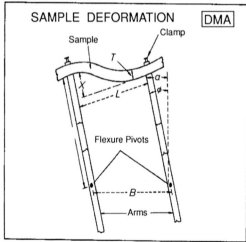

Figure 7-47. The TA Instruments DMA module. Reproduced by permission of TA Instruments, Inc., New Castle, DE.

7.5.4 Selected Applications

Matching the thermal expansion of mating materials is important to minimize stress and preserve the alignment of various elements in a device. Lithium niobate has been used in electrooptic devices because of its ability to change refractive index with electrical field. When utilizing this ability to switch light transmission between optical fibers, it is essential that the optical alignment be preserved over a use-

ful temperature range. Consequently, the thermal expansion of the material becomes a critical quantity.

Single crystals are grown using the lithium deficient congruent composition but can be made stoichiometric through the in-diffusion of lithium oxide. The TD curves for both compositions and crystallographic orientations are shown in Fig. 7-48 (Gallagher and O'Bryan, 1985). The expansion along the a-axis is much greater than along the c-axis. The maximum in the c-axis expansion is unusual and related to the electrostrictive strain. The data can be fitted well to a third degree polynomial as indicated in Fig. 7-49. The anisotropic changes in the thermal expansion between the stoichiometric and congruent compositions offset each other so that the volume

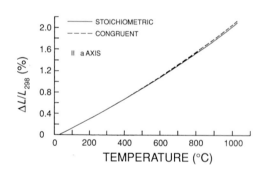

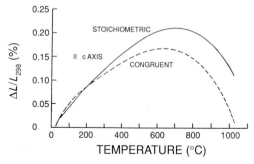

Figure 7-48. Thermal expansion (TD) curves of single crystal lithium niobate having both the stoichiometric and the congruent composition (Gallagher and O'Bryan, 1985).

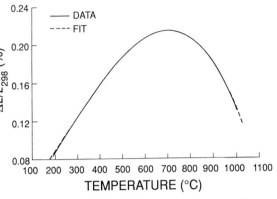

Figure 7-49. Degree of polynomial fit to the TD data for the congruent composition parallel to the *c*-axis from Fig. 7-48. $\% (\Delta L/L_{298}) = 8.705 \times 10^{-2} - 7.966 \times 10^{-4} T + 3.108 \times 10^{-6} T^2 - 1.008 \times 10^{-9}$.

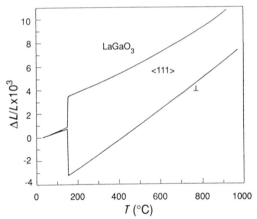

Figure 7-50. Thermal expansion (TD) curves at for LaGaO$_3$ along and orthogonal to the $\langle 111 \rangle$ direction (O'Bryan et al., 1990).

expansion of the two materials are virtually identical.

The change in thermal expansion through a first order crystallographic phase transformation from orthorhombic to rhombohedral can be clearly seen for LaGaO$_3$ in Fig. 7-50 (O'Bryan et al., 1990). This material is of interest as a potential substrate for thin films, however, the stresses induced between the film and substrate would be substantial at the phase transformation temperature.

There is an interesting contrast in the thermal expansion curves for two of the high-T_c superconducting materials. Ba$_2$YCu$_3$-O$_{7-x}$ will contract during the uptake of oxygen (decrease in x) from the structure while the Pb$_2$Sr$_2$YCU$_3$O$_{8+\delta}$ system expands under the same circumstances, see Fig. 7-51 (O'Bryan and Gallagher, 1989). These reactions are most evident in the temperature range of about $400 - 550\,°C$.

The accompanying orthorhombic to tetragonal phase transformation occurs simultaneously. This transformation is much more a function of oxygen stoichiometry than temperature. Hence, it is possible to make the materials transform isothermally by varying the partial pres-

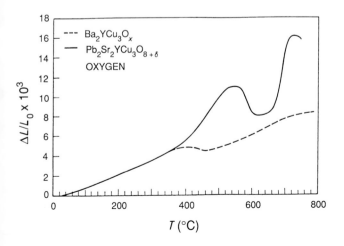

Figure 7-51. Thermal expansion (TD) curves for some sintered high-temperature superconductors as a function of temperature (O'Bryan and Gallagher, 1989).

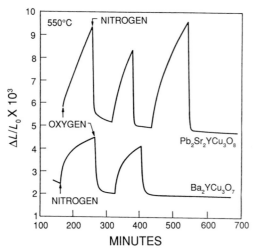

Figure 7-52. Isothermal expansion curves for selected high-temperature superconductors for different partial pressures of oxygen (O'Bryan and Gallagher, 1989).

sure of oxygen. This is illustrated in Fig. 7-52. The contrasting behavior of the two materials is evident by their opposite reactions to the change of atmosphere.

Shrinkage is an important property of a material or process whether it involves a fabric, a decomposition or sintering. The shrinkage and rate of shrinkage are shown in Fig. 7-53 for a drawn partially crystal-

lized fiber (Jaffe, 1980). The shrinkage behavior is characteristic for each of the four typical regions. Solvent loss and simple thermal expansion occur in the first region. Relaxations in the amorphous region control the second region surrounding the glass transition. Rearrangements associated with perfecting the polymer's structure and dependent upon the thermal history of the sample dominate the third region. Finally, the catastrophic shrinkage associated with melting occurs in region four.

Shrinkage is widely used to study sintering behavior of compacted powders, particularly metals and ceramics. The sintering of UO_2 is an important aspect of the preparation of nuclear fuels. Figure 7-54 shows the TD curve for a compact heated in flowing hydrogen (Backmann et al., 1968). The early changes below 600 °C are due to the loss of moisture and some oxygen. At higher temperatures and in the isothermal region the shrinkage is due to densification of the powder compact. The kinetics of shrinkage are often followed in this manner.

The use of feedback from the dilatometer to establish the optimum temperature program for sintering was proposed by

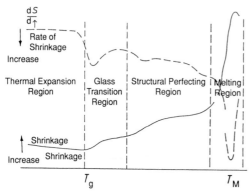

Figure 7-53. Representative changes in shrinkage occurring for a drawn partially crystallized polymer fiber (Jaffe, 1980).

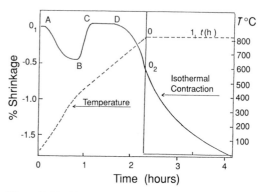

Figure 7-54. Shrinkage of compacted UO_2 powder heated in hydrogen (Backman et al., 1968).

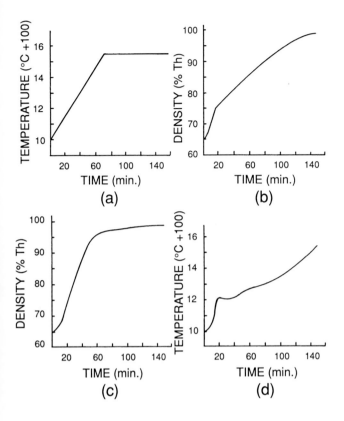

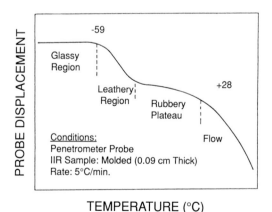

Figure 7-55. An example of rate controlled sintering for alumina (Huckabee and Palmour, 1972).

Palmour and Johnson (1967). An example of its use is shown in Fig. 7-55 for the sintering of an alumina compact (Huckabee and Palmour, 1972). The conventional sintering program and the resulting density versus time are shown. In addition the predetermined shrinkage profile designed to optimize the sintering process is indicated along with the resulting densification profile. The rate controlled process requires less energy and yields a denser product having a finer grain size.

TMA and DMA are widely used to characterize polymeric materials. The penetration reveals different regions in the behavior of an elastomer as indicated in Fig. 7-56 from a review by Maurer (1978). Information can be inferred regarding such factors as the glass transition temperature and variations in cross-linking and molecular weight (Maurer, 1980).

The high damping of elastomers is important for many acoustic and vibrational applications. DMA is valuable for monitoring quality and heat build up of potential materials. Figure 7-57 shows these

Figure 7-56. TMA of isobutene-isoprene rubber. From Maurer (1978): reproduced by permission of Franklin Institute Press.

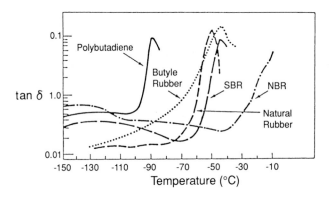

Figure 7-57. Comparative damping of various elastomers. Reproduced by permission of the DuPont Co.

characteristics for several natural and synthetic rubbers (DuPont Co).

The final example shows the use of DMA to characterize the degree of cure for some phenolic resins. The comparative modulus and damping are presented in Fig. 7-58. A large peak occurs for the uncured sample in the damping at the glass transition and a smaller one related to the cure process. The completely cured sample shows no damping and the partially cured specimen exhibits an intermediate response.

7.6 Miscellaneous Methods

There are a wide range of analytical methods and measurements of various properties that can be determined as a function of temperature. Many of the analytical techniques are described elsewhere in this series. Most of these techniques have found limited use in thermal analysis in comparison to the methods that have been discussed earlier. Due to the restrictions in space it is appropriate to describe very briefly a limited number of them. For expanded coverage see Wendlandt (1988).

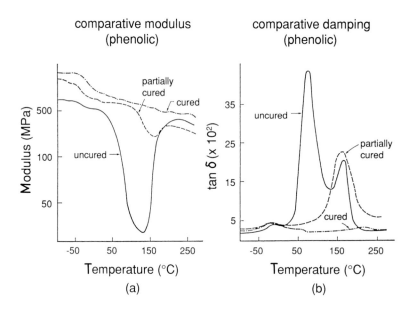

Figure 7-58. TMA curves of phenolic resins at several stages of cure. Reproduced by permission of the DuPont Co.

7.6.1 Electrical Properties

There are a number of electrical properties that can be observed. The most commonly used techniques follow changes in ac or dc conductivity, capacitance or dielectric properties, thermally stimulated discharge currents, and the emf developed between dissimilar electrodes in contact with the sample.

Changes in resistance can be measured at constant current or voltage. There are many occurrences that will give rise to a change in resistance for a powder. These can result from the intrinsic changes in the material itself due to cyrstallographic transitions or changes in defect concentrations. They can also result from variations in the contact resistance between particles arising from surface adsorption, sintering, external pressure, melting, etc. Particularly large and abrupt changes can result from the formation of a liquid phase (Wendlandt, 1970). Perturbations in reactivity or transitions may arise from the effects of the imposed electric field gradient (MacKenzie and Hadipour, 1980).

Dielectric measurements frequently relate to DMA measurements where mechanical loss peaks can be correlated with the phase change and attenuation of the ac electrical signal. The dielectic measurements can be performed by scanning the frequency range, e.g., 50 Hz to 1 MHz, isothermally or by varying the temperature at a fixed frequency. Changes in the dielectric constant can be large compared to the enthalpy of a phase transition for ferroelectric materials and, hence, may be a more sensitive means of detecting the transformation.

7.6.2 Optical Properties

A vast array of spectroscopic methods can be adapted to provide information

while the temperature of the sample is changing in a predetermined manner. Several examples in the Sec. 7.4 described FTIR and mass spectrometric examination of the gas phase. The techniques in this section pertain to observations of the sample itself. They share the similar ability to scan a range of energy as the sample is heated or the option to fix on a specific energy and obtain a continuous curve for that particular species.

Again, the reader is referred to Wendlandt (1988) for a more extensive description of this class of techniques. Reflectance spectroscopy has been particularly useful to study transitions and decompositions of inorganic coordination complexes. Transmission IR spectroscopy has been used to follow changes as a function of temperature for a material supported in a KBr pellet (Hisatsune et al., 1970).

Emission spectroscopy has been used to follow the fluorescence of europium ion during decomposition of the oxalate (Gallagher et al., 1970). Even Mößbauer spectroscopy has been used to follow the changes in oxidation state and coordination during the thermal decomposition of appropriate oxalates (Gallagher et al., 1970; Gallagher and Kurkjian, 1966).

Other optical methods involve thermomicroscopy, particularly using polarized light; thermoluminescence and oxyluminescence. Recently a scanning electron microscope capable of operating at pressures approaching atmospheric has been coupled with a hot stage. This suggests many interesting possibilities.

7.6.3 X-Ray Diffraction

There are several manufacturer's of hot stages that are suitable for X-Ray diffractometers. These will operate from cryogenic temperatures to beyond 2000°C.

These devices are capable of determining lattice constants as a function of temperature, detecting phase transformations, and following the course of solid state reactions.

There have been several clever adaptations which allow the simultaneous measurement of DSC or TG with the X-ray diffraction pattern. These have been mentioned earlier in this chapter. High energy radiation from a synchrotron has been ingeniously utilized to obtain simultaneous X-ray diffraction patterns and DSC measurements for a variety of biological systems (Caffrey, 1991).

7.6.4 Acoustic Techniques

Thermoacoustimetry is associated with the velocity and absorption of deliberately imposed sonic waves. The technique has found use primarily in the characterization of polymers. It is extensively reviewed by Chatterjee (1980). Thermosonimetry (TS) is concerned with the acoustic emissions arising from mechanical forces operating at phase transformations. Both methods have been known for a long time. The quality of a bell or the "tin cry" associated with the sub-ambient phase transition in metallic tin are evidence of the effects. The thermoanalytical exploitation of TS has been extensively discussed in a recent workshop (Lonvick, 1987).

A typical apparatus for studying TS is shown schematically in Fig. 7-59 (Lonvick, 1987). A fused quartz rod is used as a stethoscope to convey the acoustic emission to the piezoelectric detector. TS can be used in two modes. The acoustic signals, within the range of the detector, may

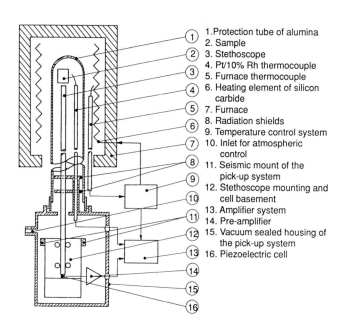

1. Protection tube of alumina
2. Sample
3. Stethoscope
4. Pt/10% Rh thermocouple
5. Furnace thermocouple
6. Heating element of silicon carbide
7. Furnace
8. Radiation shields
9. Temperature control system
10. Inlet for atmospheric control
11. Seismic mount of the pick-up system
12. Stethoscope mounting and cell basement
13. Amplifier system
14. Pre-amplifier
15. Vacuum sealed housing of the pick-up system
16. Piezoelectric cell

Different stethoscope sample holders

Figure 7-59. A typical apparatus for TS (Lonvick, 1987).

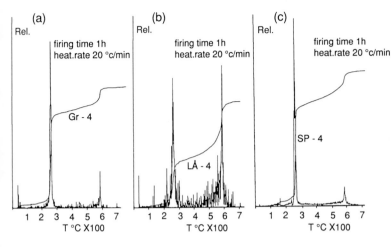

Figure 7-60. Differential and integral TS curves for several specimens of quartz (Lonvick, 1987). (a) Sample from Greece. (b) Sample from Sweden. (c) Sample from Spain.

be summed as the temperature is scanned to present a curve similar to the differential curve shown in Fig. 7-60 (Lonvick, 1987). Alternatively the frequency spectrum of the emission can also be determined, however, the complexity of the interpretation is great and this mode of operation has not been used extensively.

The acoustic emissions can be integrated to give a smoother curve as seen in Fig. 7-60. These curves clearly show the ability to distinguish the origan of the quartz specimens. The low temperature activity is due to the collapse of pores and the higher thermosonic activity is attributed to the phase transformation occurring in pure quartz around 573 °C. Thermosonic emissions are frequently observed to anticipate the transition slightly.

Decrepetation caused by the escape of trapped gases also lead to strong sonic signals. Cracking resulting from stresses also gives rise to strong thermoacoustic signals. The volume and phase changes associated with the exchange of oxygen between the high-T_c superconductors and the atmosphere are significant, see Figs. 7-51 and 7-52.

These effects place the ceramic in tension as it cools in an oxidizing atmosphere

and leads to the formation of microcracking. The microcracking is clearly evident in Fig. 7-61 (Richardson and De Jonge, 1990). This microcracking has a devestating affect on the superconductivity of the material by interfering with the connecting path through the sample. TS was used to follow conditions that minimized the microcracking.

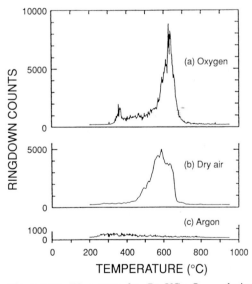

Figure 7-61. TS curves for $Ba_2YCu_3O_{7-x}$ during cooling at 50°C min^{-1} (Richardson and De Jonge, 1990).

7.7 Concluding Remarks

7.7.1 Summary

This chapter has served as only an sketchy introduction to the topic of thermal analysis. The major techniques have been described in sufficient detail so that the readers can recognize their potential applications to their problems and research and make an intelligent selection of equipment. The techniques are conceptually very simple; however, their strength and versatility have been amply demonstrated, even within the confines of this brief chapter.

The class of techniques is not limited to those listed in Table 7-1. The limitations are the bounds of our imaginations and the extent of our ingenuity. The manufacturers do not lead in the development of new applications or techniques. They follow their perception of the market place and its response to the innovation of individual investigators. This is not a criticism of them, but a simple corollary of economics.

The subject of heterogeneous kinetics was only alluded to on several occasions. It is a major and controversial topic that is strongly related to thermal analysis. Thermoanalytical methods are the primary tools used for such studies. The interplay between the sample temperature, the surroundings, and related experimental measurements has been introduced. It is a complex topic but technology demands that a best effort be made.

Hopefully the education in thermal analysis is not concluded. Besides the conventional texts in the field there are proceedings of the many meetings devoted to thermal analysis. Virtually every country that is active in science has a national society devoted to thermal analysis and calorimetry. The literature is scattered because of the diversity of applications and the materials investigated. There are, however, several journals and abstracting documents devoted specifically to the topic.

7.7.2 Future Directions

Because everyone recognizes the pitfalls of predictions and extrapolations, it is not necessary to elaborate on them. There are, however, obvious areas where progress will be made. The continuing advance in computer power and the accompanying decline in cost bodes well for the marriage between the measuring instrument; its data acquisition system; and the sophistication and convenience in data analysis, storage, and presentation.

Miniaturization of sensors and electronic circuitry will allow thermal analysis to move into the body, on board space craft, and wherever the need may arise. The smaller sensors require less power and disturb the system less so that more accurate and rapid measurements can be made. Optical sensors and information transfer will allow accurate and sensitive remote sensing of parameters such as temperature. All of these developments enable smaller samples and enable one to investigate inhomogenieties on an ever shrinking scale.

The tendencies toward simultaneous measurements and the use of feedback control will increase. This will provide a better basis for understanding the subject under investigation. More meaningful data is collected faster.

The immediate contributions of thermal analysis to the characterization of the high profile new materials such as superconductors, carbon clusters, etc., indicates that the methods have achieved a degree of respectability and visability. Such trends will certainly continue.

7.8 References

Anderson, E. W., Bair, H. E., Johnson, G. E., Kwei, T. K., Padden, R. J., Williams, D. (1979), *Adv. Chem. Ser. 176*, 413–419.

Aylmer, D. M., Rowe, M. W. (1984), *Thermochim. Acta 78*, 81–92.

Backman, J. J., Cahour, P., Cizeron, G. (1968), *Mem. Scient. Rev. Metallurgie 65*, 481–495.

Bacon, D. D., Hong, M., Gyorgy, E. M., Gallagher, P. K., Nakahara, S., Feldman, L. C. (1986), *Appl. Phys. Lett. 48*, 730–732.

Bair, H. E. (1980), in: *Thermal Characterization of Polymeric Materials*. New York: Academic Press, Chap. 9.

Balek, V. (1969), *J. Mater. Sci. 4*, 919–924.

Balek, V. (1987), *Thermochim. Acta 110*, 222–235.

Barin, I. (1989), *Thermochemical Data of Pure Substances*. Weinheim: VCH publishers.

Berg, L. G., Equanov, V. P., Kiyaev, A. D. (1975), *J. Therm. Anal. 7*, 11–19.

Boersma, F., Van Empel, F. J. (1975), in: *Progress in Vacuum Microbalance Techniques*, Vol. 3. London: Heyden and Sons, pp. 9–17.

Bracconi, P., Gallagher, P. K. (1979), *J. Am. Ceram. Soc. 62*, 171–176.

Brown, M. E. (1988), *Introduction to Thermal Analysis*. London: Chapman and Hall.

Brown, M. E., Dollimore, D., Galwey, A. K. (1980), *Reactions in the Solid State*. Amsterdam: Elsevier Scientific Publishers.

Caffrey, M. (1991), *Trends in Anal. Chem. 10*, 156–162.

Caldwell, K. M., Gallagher, P. K., Johnson, D. W. (1977), *Thermochim. Acta 18*, 15–19.

Charles, R. G. (1982), in: *Thermal Analysis Vol. 1*. New York: J. Wiley and Sons; pp. 264–271.

Chatterjee, P. K. (1980), in: *Treatise on Analytical Chemistry*, Vol. 12, Part I. New York: John Wiley and Sons, Chap. 9.

Cossette, M., Winer, A. A., Steele, R. (1980), in: *Proc. of the Fourth Int. Conf. on Asbestos*. Torino, Italy, pp. 201–212.

Crowder, C. E., Fawcett, T. G., Harris, W. C., Newman, R. A., Whiting, L. F. (1984), *Procedings 13th NATAS Conf*. Colonia, NJ: NATAS, pp. 447–448.

Duncan, T. M., Levy, R. A., Gallagher, P. K., Walsh, M. W. (1988), *J. Appl. Phys. 64*, 2990–2994.

Evans, U. R. (1960), *The Corrosion and Oxidation of Metals*. New York: St. Martin's Press.

Fukatsu, T. (1961), *J. Japan Soc. Powder Metallurgy 8*, 183–190.

Gallagher, P. K. (1968), *Mat. Res. Bull. 3*, 225–232.

Gallagher, P. K. (1976), *Thermochim. Acta 14*, 131–139.

Gallagher, P. K. (1978), *Thermochim. Acta 26*, 175–183.

Gallagher, P. K. (1984), *Thermochim. Acta 82*, 325–334.

Gallagher, P. K. (1991a), *Mikrochim. Acta II*, 391–399.

Gallagher, P. K. (1991b), *Thermochim. Acta 174*, 85–98.

Gallagher, P. K. (1992a), in: *Advances in Analytical Geochemistry*. Greenwitch, CT: JAI Press Inc., in press.

Gallagher, P. K. (1992b), *J. Thermal Anal.*, in press.

Gallagher, P. G., Chu, S. N. G. (1982), *J. Phys. Chem. 86*, 3246–3250.

Gallagher, P. K., Gyorgy, E. M. (1980), in: *Thermal Analysis Vol. 1*. Basel, CH: Birkhäuser Verlag, pp. 113–118.

Gallagher, P. K., Gyorgy, E. M. (1986), *Thermochim. Acta 109*, 193–206.

Gallagher, P. K., Johnson, D. W. (1973), *Thermochim. Acta 6*, 67–83.

Gallagher, P. K., Kurkjian, C. R. (1966), *Inor. Chem. 5*, 214–219.

Gallagher, P. K., Luongo, J. P. (1975), *Thermochim. Acta 12*, 159–164.

Gallagher, P. K., O'Bryan, H. M. (1985), *J. Am. Ceram. Soc. 68*, 147–150.

Gallagher, P. K., O'Bryan, H. M. (1988), *J. Am. Ceram. Soc. 71*, C56–C59.

Gallagher, P. K., Prescott, B. (1970), *Inor. Chem. 9*, 2510–2512.

Gallagher, P. K., Schrey, F. (1968), in: *Thermal Analysis*, Vol. 2. New York: Academic Press, pp. 929–952.

Gallagher, P. K., Warne, S. St. J. (1981), *Thermochim. Acta 43*, 253–267.

Gallagher, P. K., Schrey, F., Prescott, B. (1970), *Inor. Chem. 9*, 215–219.

Gallagher, P. K., Johnson, D. W., Vogel, E. M. (1976), in: *1976 Catalysis in Organic Synthesis*. New York: Academic Press, pp. 113–136.

Gallagher, P. K., Gyorgy, E. M., Jones, W. R. (1981). *J. Chem. Phys. 75* (52), 3847–3849.

Gallagher, P. K., Gyorgy, E. M., Jones, W. R. (1982), *J. Therm. Anal. 23*, 185–192.

Gallagher, P. K., Sinclair, W. R., Bacon, D. D., Kammlott, G. W. (1986), *J. Electrochem. Soc. 130*, 2054–2056.

Gallagher, P. K., Gyorgy, E. M., Schrey, F., Hellman, F. (1987), *Thermochim. Acta 121*, 231–240.

Gallagher, P. K., O'Bryan, H. M., Brandle, C. D. (1988), *Thermochim. Acta 133*, 1–10.

Gallagher, P. K., Zhong, Z., Kane, S. (1991), *Trends Anal. Chem. 10*, 279–282.

Garn, P. D. (1975), *J. Therm. Anal. 7*, 593–601.

Garn, P. D. (1980), in: *Thermal Analysis*, Vol. 1. Basel, CH: Birkhäuser Verlag, pp. 593–594.

Haglund, B. (1982), *J. Thermal Anal. 25*, 21–43.

Hisatsune, I. C., Adl, T., Beahm, E. C., Kempf, R. J. (1970), *J. Phys. Chem. 74*, 3225–3235.

Hong, M., Gyorgy, E. M., van Dover, R. B., Nakahara, S., Bacon, D. D., Gallagher, P. K. (1986), *J. Appl. Phys. 59*, 551–556.

Hongtu, F., Laye, P. G. (1989), *Thermochim. Acta 153*, 311–319.

Huang, J., Gallagher, P. K. (1992), *Thermochim. Acta 192*, 35–46.

Huckabee, M. L., Palmour, H. (1972), *Am. Ceram. Soc. Bull. 51*, 574–578.

Hultgren, R. (1973), *Selected Values of the Thermodynamic Properties of the Elements.*

Ishii, T., Tashiro, N., Takemura, T. (1991), *Thermochim. Acta 180*, 289–301.

Jaffe, M. (1980), in: *Thermal Characterization of Polymeric Materials.* New York: Academic Press, pp. 709–792.

Kaisersberger, E. (1980), in: *Thermal Analysis,* Vol. 1. Basel, CH: Birkhäuser Verlag, pp. 251–257.

Kinsbron, E., Gallagher, P. K., English, A. T. (1979), *Solid State Electronics 22*, 517–524.

Khorami, J., Choquette, D., Kimmerle, F. M., Gallagher, P. K. (1984), *Thermochim. Acta 76*, 87–96.

Kosak, J. (1976), in: *Catalysis in Organic Synthesis.* New York: Academic Press, pp. 137–148.

Kuck, V. (1986), *Thermochim. Acta 99*, 233–249.

Langer, H. G. (1980), in: *Treatise on Analytical Chemistry,* Vol. 12, Part I. New York: John Wiley and Sons, Chap. 6.

Le Châtelier, H. (1887a), *Compt. Rend. Hebd. Seanc. Acad. Sci. Paris* 104, 1143 and 1517.

Le Châtelier, H. (1887b), *Bull. Soc. Fr. Miner. 10*, 204–210.

Levy, R. A., Gallagher, P. K. (1986), *J. Electrochem. Soc. 132*, 1986–1991.

Linseis, M. (1975), *Thermal Analysis*, Vol. 3. Chichester: John Wiley and Sons, pp. 913–920.

Lonvick, K. (1987), *Thermochim. Acta 110*, 253–264.

Lum, R. M. (1977), *Thermochim. Acta 18*, 73–94.

Lum, R. M., Feinstein, L. G. (1980), *Proc. Electron. Components Conf. 30*, 113–120.

MacKenzie, K. J. D., Hadipour, N. (1980), *Thermochim. Acta 35*, 227–238.

MacKenzie, K. J. D., Cardile, C. M. (1990), *Thermochim. Acta 165*, 207–217.

Maurer, J. J. (1969), *Rubber Chem Technol. 42*, 110–158.

Maurer, J. J. (1978), in: *Thermal Methods in Polymer Analysis.* Philadelphia: Franklin Institute Press, pp. 129–161.

Maurer, J. J. (1980), in: *Thermal Characterization of Polymeric Materials.* New York: Academic Press, Chap. 6.

McGhie, A. R. (1983), *Anal. Chem. 55*, 987–991.

McGhie, A. R., Chin, J., Fair, P. G., Blaine, R. L. (1983), *Thermochim. Acta 67*, 241–247.

Morgan, D. J. (1976), in: *First European Symposium on Thermal Analysis.* London: Heyden and Sons, pp. 355–360.

Morgan, D. J., Warrington, S. B., Warne, S. St. J. (1988), *Thermochim. Acta 135*, 207–225.

Morros, S. A., Stewart, D. (1976), *Thermochim. Acta 14*, 13–24.

Murat, M. (1983), in: *Treatise in Analytical Chemistry,* Vol. 12, Part 1. New York: John Wiley and Sons, Chap. 7.

Nassau, K., Gallagher, P. K., Miller, A. E., Graedel, T. E. (1987), *Corrosion Science 27*, 669–684.

Norem, S. D., O'Neill, M. J., Gray, A. P. (1970), *Thermochim. Acta 1*, 29–34.

O'Bryan, H. M., Gallagher, P. K. (1989), *Chem. Mater. 1*, 526–529.

O'Bryan, H. M., Gallagher, P. K., Brandle, C. D. (1985), *J. Am. Ceram. Soc. 68*, 493–486.

O'Bryan, H. M., Gallagher, P. K., Berkstresser, G. W., Brandle, C. D. (1990), *J. Mater. Res. 5*, 183–189.

O'Neill, M. J. (1964), *Anal. Chem. 36*, 1238–1244.

Palmour, H., Johnson, D. R. (1967), in: *Sintering and Related Phenomena.* New York: Gordon Breach Pub., pp. 779–791.

Paulik, F., Paulik, J. (1986), *Thermochim. Acta 100*, 23–59.

Pearce, E. M., Khanna, Y. P., Raucher, D. (1980), in: *Thermal Characterization of Polymeric Materials.* New York: Academic Press, Chap. 8 and references therein.

Plant, A. F. (1971), *Industrial Res. (July)*, 36–42.

Prime, R. B. (1980), in: *Thermal Characterization of Polymeric Materials.* New York: Academic Press, Chap. 5 and references therein.

Quinn, T. J (1990), *Temperature*, 2nd ed. New York: Academic Press.

Richardson, T. J., De Jonge, L. C. (1990), *J. Mater. Res. 5*, 2066–2074.

Roberts-Austen, W. C. (1899a), *Proc. Inst. Mech. Eng. 1*, 35–41.

Roberts-Austen, W. C. (1899b), *Metallographist 2*, 186–195.

Rouquerol, J. (1989), *Thermochim. Acta 144*, 209–224.

Rowe, M. W., Gallagher, P. K., Gyrogy, E. M. (1983), *J. Chem. Phys. 79*, 3534–3536.

Rowe, M. W., Edgerley, D. A., Hyman, M., Lake, S. M. (1979), *J. Mat. Sci. 14*, 999–1005.

Swarim, S. J., Wims, A. M. (1976), *Anal. Calorim. 4*, 155–171.

Vogel, E. M, Gallagher, P. K. (1985), *Mater. Lett. 4*, 5–9.

Warne, S. St. J., Gallagher, P. K. (1987), *Thermochim. Acta 110*, 269–280.

Warne, S. St. J., Morgan, D. J., Milodowski, A. E. (1981), *Thermochim. Acta 51*, 105–112.

Watson, E. S., O'Neill, M. J., Justin, J., Brenner, N. (1964), *Anal Chem. 36*, 1233–1237.

Wendlandt, W. W. (1970), *Thermochim. Acta 1*, 11–17.

Wendlandt, W. W. (1988), *Thermal Analysis*, 3rd ed. New York: John Wiley and Sons.

Wieboldt, R. C., Adams, G. E., Lowry, S. R., Rosenthal, R. J. (1988), *Am. Lab. 1988*, 70–77.

Wiedemann, H. G., Bayer, G. (1985), *Thermochim. Acta 121*, 479–485.

Wunderlich, B. (1990), *Thermal Analysis*. New York: Academic Press

Zhong, Z., Gallagher, P. K. (1991), *Thermochim. Acta 186*, 199–204.

General Reading

Books:

Brown, M. E. (1988), *Introduction to Thermal Analysis*. New York: Chapman and Hall.

Brown, M. E., Dollimore, D., Galway, A. K. (1980), *Reactions in the Solid State*. Amsterdam: Elsevier Scientific Publishers.

Comprehensive Analytical Chemistry, Vols. 7 and 12 (1975). Amsterdam: Elsevier Scientific Publishers.

Hemminger, W., Hohne, G. (1984), *Calorimetry – Fundamentals and Practice*. Basel: VCH Publishers.

Keattch, C. J., Dollimore, D. (1975), *An Introduction into Thermogravimetry*. London: Heyden and Sons.

Pope, M. I., Judd, M. D. (1977), *Differential Thermal Analysis*. London: Heyden and Sons.

Treatise on Analytical Chemistry, Vol. 12, Part I (1980). New York: J. Wiley and Sons.

Turi, E. A. (1981), *Thermal Characterization of Polymeric Materials*. New York: Academic Press.

Wendlandt, W. W. (1985), *Thermal Analysis*, 3rd ed. New York: John Wiley and Sons.

Wunderlich, B. (1990), *Thermal Analysis*. New York: Academic Press.

Journals:

Journal on Thermal Analysis. London: Heyden and Sons.

Thermochimica Acta. Amsterdam: Elsevier Scientific Publishers.

Series:

Proceedings of European Symposium on Thermal Analysis and Calorimetry (ESTAC).

Proceedings of North American Thermal Analysis Society (NATAS).

Progress in Vacuum Microbalance Techniques. London: Heyden and Sons.

Thermal Analysis (Proc. Int. Conf. Therm. Anal.), various publishers.

8 Application of Synchrotron X-Radiation to Problems in Materials Science

Andrea R. Gerson[1], Peter J. Halfpenny, Stefania Pizzini[2], Radoljub Ristić,
Kevin J. Roberts[3], David B. Sheen, and John N. Sherwood

Department of Pure and Applied Chemistry, University of Strathclyde, Glasgow, U.K.
[1] Department of Chemistry, King's College, University of London, U.K.
[2] LURE, Bâtiment 205 D, Centre Universitaire Paris-Sud, Orsay-Cedex, France
[3] and also at SERC Daresbury Laboratory, Warrington, U.K.

List of Symbols and Abbreviations

a, b, c	crystal unit cell parameters
c	velocity of light
B	bending magnet strength in Tesla
d	lattice plane spacing
E	energy of the beam
E_p	energy of the particle
F_i	amplitude of the backscattering factor
F_{nkl}	modulus of the structure factor
I	integrated diffracted intensity
I_0	incoming flux
I_t	transmitted flux
k	magnitude of the photoelectron wave vector
K_0	bulk modulus
L	sample to source distance
m	mass
m_0	rest mass of the electron
N_i	co-ordination number for atoms of type i
$P(\lambda)$	photon flux
q	wave vector (magnitude)
r	shell distance
R	refinement factor
$R(E)$	reflectivity coefficient
R_g	geometrical resolution factor in X-ray topography
r_i	radial distance from absorbing atom
R_s	radius of the synchrotron storage ring in meters
S	source size
S_0	damping term for multibody effects in EXAFS analysis
t	sample thickness
x	sample to film distance
z	number of molecules in the unit cell
α, β, γ	cell parameters
δ	deviation parameter for an incommensurate phase
Θ	vertical divergence of the beam
Θ_B	Bragg angle
λ	X-ray wavelength
λ_c	critical wavelength
λ_d	damping factor used in EXAFS analysis to allow for inelastic scattering effects
μ	linear absorption coefficient
μ_0	background absorption
σ_i	Debye-Waller type factor used in EXAFS analysis
ϕ	fixed incident glancing angle
ϕ_c	critical angle for total external reflection

Φ_i	phase shift function used in EXAFS analysis
$\chi(k)$	EXAFS function
ADP	ammonium dihydrogen phosphate
b.c.c.	body-centred cubic
EDXRD	energy dispersive X-ray diffraction
EXAFS	extended X-ray absorption fine structure
f.c.c.	face-centred cubic
ITO	indium/tin oxide
KZC	K_2ZnCl_4
MBA-NB	(-)-2-(α-methylbenzylamino)-5-nitropyridine
MBE	molecular beam epitaxy
ML	monolayers
NF	nickel formate dihydrate
PTS	2,4-hexadiynediol-bis-(p-toluene sulfonate)
QEXAFS	quick-scanning EXAFS
RDF	radial distribution function
ReflEXAFS	reflectivity EXAFS
SANS	small angle neutron scattering
TEM	transmission electron microscopy
XAS	X-ray absorption spectroscopy
XANES	X-ray absorption near-edge structure
XSW	X-ray standing waves

8.1 Introduction

8.1.1 Synchrotron Storage Rings

When charged particles travelling at relativistic energies are constrained to follow a curved trajectory, as for example when they pass through a magnetic field, they emit electromagnetic radiation. The total emitted intensity of the radiation is proportional to Γ^4, where

$$\Gamma = E_p/(mc^2) \tag{8-1}$$

E_p and m are the energy and mass of the particle and c is the velocity of light. From this expression it follows that, for a certain energy, the highest intensity will be emitted by the lightest moving particle, and hence that electrons and positrons will be the most efficient emitters. This type of radiation is produced in all circular particle accelerators (*synchrotrons*) and thus it is usually referred to as *synchrotron radiation*.

The emitted radiation shows not only a high intensity compared with conventional sources but also covers a broad spectral range from the infra-red to the hard X-ray region. The considerable potential of this type of source as demonstrated by early experiments using radiation emitted by synchrotrons built and operated primarily for nuclear physics research has led to the development of *synchrotron storage rings*

dedicated to the production of synchrotron radiation. These sources and their use are now so widespread that many countries in the world either have built or are building sources. Those which are not, often have ready access to such facilities.

The general form of a storage ring and the basic mode of its operation is shown schematically in Fig. 8-1. Electrons or positrons developed from a suitable source are first accelerated to energies around 60 MeV in a linear accelerator. From here they are injected into a booster synchrotron in which the energies are increased to 600 MeV levels by acceleration in the circular orbit. These high energy particles are then injected in bunches into the main storage ring until integrated particle currents of circa 200–300 mA are attained. The storage ring consists of straight sections linked by curved portions around which are situated the electromagnets (bending magnets) which cause the essential curvature of the beam.

In the storage ring the bunches of particles are still further accelerated to relativistic (2–8 GeV) energies. This stored current of the eletron or positron bunches continues to circulate under the accelerating field.

The current, and hence beam intensity, decays over a long period (10–20 hours) as a result of the loss of energy due to collisions of the particles with residual gas

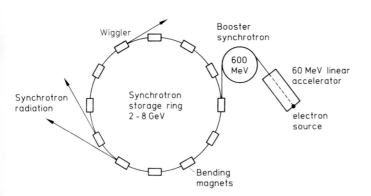

Figure 8-1. Schematic representation of the general form of an electron synchrotron storage ring and associated facilities.

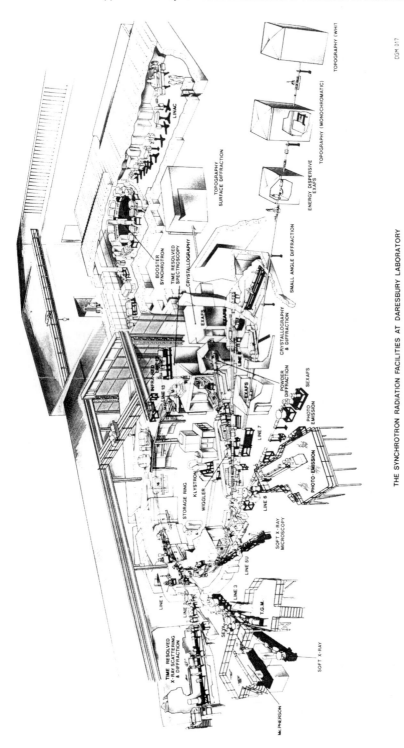

Figure 8-2. General view of the experimental hall of the SERC UK Synchrotron Radiation Source facility at Daresbury, U.K.

molecules and the storage ring walls. During this period, utilisable radiation is emitted as the electrons pass through the bending magnets. The radiation is emitted in the forward direction tangentially to the particle beam. It is extracted for use principally through some type of tangential beam port into a beam line, fully evacuated to the ring vacuum for the softer radiation or through a beryllium window into air for the harder radiation ($\lambda < 0.3$ nm).

The beam lines and their associated experimental hutches are accommodated in an experimental hall. A general view of such a facility is shown in Fig. 8-2.

8.1.2 Characteristics of Synchrotron Radiation

8.1.2.1 Radiation from Bending Magnets

The typical spectral distribution of synchrotron radiation is shown in Fig. 8-3. The horizontal scale is defined by a critical wavelength λ_c and the vertical scale of intensity by the electron current and the energy of the beam. λ_c is given by the formula:

$$\lambda_c = 5.6\, R_s/E^3 = 1.86/(B E^3)\,\text{(mm)} \quad (8\text{-}2)$$

where the energy of the beam, E, is expressed in GeV, R_s the radius of the ring in metres and B the bending magnet strength in Tesla. From this we see that in order to obtain a spectrum of energy extending from the infra-red to the hard X-ray regions then $\lambda_c \leq 0.1$ nm and hence E should be $2-5$ GeV and $R_s \approx 10-20$ m.

8.1.2.2 Radiation from Insertion Devices

In any particular system the energy range and beam intensity can be extended and modified by the use of insertion devices known as wigglers and undulators.

(a) *Wigglers*. A wiggler consists of a series of superconducting magnets $(3-5)$

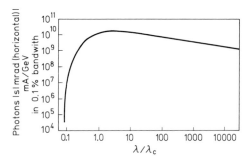

Figure 8-3. Synchrotron radiation spectrum emitted by an electron moving in a curved trajectory.

placed in a straight section of the ring. The purpose of this is to cause the electron to execute a path of shorter local radius of curvature than in the dipole bending magnets. The result [Eq. (8-2)] is to increase the critical energy and to shift the spectrum to higher energies by a parallel amount, thus providing even harder radiation than from the bending magnets. The radiation is again emitted tangentially to the smaller curve and can be extracted via a window and beam-pipe.

(b) *Undulators*. Undulators, which are again inserted in a linear section of the storage ring are multipole wigglers which develop the acceleration by causing either a transverse sinusoidal or helical magnetic field. The result, under some geometrical operating circumstances, is to produce a line spectrum of harmonics of increased power density and spectral brilliance.

Further details on the design, operation and properties of synchrotron storage rings and insertion devices will be found in Kunz (1979), Winick and Doniach (1980), Stuhrmann (1982) and Koch (1983).

8.1.2.3 Other Characteristics

In addition to this broad spectral range the beam shows the following general properties:

(a) *Low divergence*. For $\lambda = \lambda_c$ the vertical divergence of the beam $\Theta \approx 1/\Gamma$ where

$\Gamma = E_{\mathrm{p}}/(m_0 c^2)$ and $m_0 c^2$ is the rest mass energy of the electron. When E is of the order of 1 GeV, $\Theta \leq 0.1$ mrad (0.006°, 20″).

(b) *Defined polarisation.* The radiation is 100% polarised with the electric vector parallel to the orbital plane. Above and below the orbital plane the radiation is elliptically polarised. The degree of ellipticity depends on the angle of viewing.

(c) *A time structure.* The electron bunches can be injected with bunch lengths typically 0.05–1 ns and at intervals which can range from a few nanoseconds to a few microseconds.

(d) *High intensity and brightness.* The intensity of the emitted radiation is directly proportional to the current circulating in the synchrotron. It is typically $\approx 1.6 \times 10^{10} E$ (GeV) photons/[s · mrad (horizontal) · mA · 0.1% (bandwidth)]. Brightness depends on the dimensions of the radiating area as viewed along the beam. Typically this is $> 10^{10}$ photons/[mm^2 · s · mrad^2 · 0.1% (bandwidth)] at $\lambda = 1$ Å.

(e) *Large beam size.* The emitted beam has a relatively large cross-section, 1–2 cm diameter, depending on the synchrotron source.

8.1.3 Summary

In operational terms synchrotron X-ray sources offer the advantages of a broad spectrum or tunable monochromatic source of high and uniform spectral brilliance some 100–10000 times brighter than traditional sources depending on the particular use. The well-defined polarisation offers prospects for improving signal to noise ratio in some applications. The time structure coupled with the intensity allows dynamic experiments on a wide time range. The low beam divergence allows for improved resolution. In combination these characteristics offer a wide range of prospects for novel and extended experiments using both diffraction and spectroscopic techniques.

In the following pages we present a number of examples of the various ways in which synchrotron X-radiation can be used to examine problems in materials science. In each case we detail how the characteristics referred to above can be used to advantage. These examples are taken predominantly from our own studies of crystal growth, surface and interface science, and solid-state chemistry. They and the bibliography should allow extrapolation to a wider range of problems and experiments.

Background on X-ray diffraction generally will be found in this Volume, Chap. 4.

8.2 X-Ray Absorption Spectroscopy

In recent years X-ray absorption spectroscopy (XAS) has achieved an increasing importance in materials science due to its ability to obtain information on the local structure of selected atomic species in complicated systems which may not be suitable for conventional X-ray crystallographic analysis, e.g. amorphous and multi-component materials, alloys, surfaces, etc. The development of XAS as a tool for structural determination started effectively in the late 1970s when synchrotron radiation sources became widely available. XAS benefits from the high intensity, the collimation and the broad spectral range of synchrotron radiation. Since X-ray fluxes emitted by synchrotron sources are typically 3–4 orders of magnitude greater than those obtained with laboratory sources, the data acquisition times for absorption edge spectra are reduced from the order of weeks to minutes. The collimation of the

X-ray synchrotron beam is such that energy resolutions as good as 10^{-4} can be obtained. Finally the continuous synchrotron radiation spectrum (Fig. 8-3) allows L and K-edges of most elements to be accessed (Winick and Doniach 1980).

8.2.1 Basic Principles and Essential Theory

A typical X-ray absorption spectrum is given in Fig. 8-4 which shows the K-edge spectrum of an Fe foil and illustrates the modulation of the absorption cross section for the photoexcitation of a 1 s core shell electron. The strong oscillations which extend beyond the edge for about 30–40 eV, the so called X-ray absorption near-edge structure (XANES), involve the multiple scattering of the excited photoelectrons and are determined by the geometrical arrangement of atoms in a local cluster around the absorbing atom (Bianconi, 1988; Durham, 1988).

The structure observed in the absorption spectrum beyond the XANES region is known as EXAFS (extended X-ray absorption fine structure). The oscillations are a final state electron effect, arising from the interference between the outgoing photoelectron wave ejected in the absorption process and the wave backscattered from the neighbours of the absorbing atom (Sayers et al., 1971; Lee et al., 1981; Hayes and Boyce, 1982; Köningsberger, 1988). These techniques are also treated in this Volume, Chap. 4, Section 4.9.

The primary objective of EXAFS investigations is to determine the local atomic environment of the absorbing atom, by analysing the measured structure. The EXAFS function is defined as:

$$\chi(k) = (\mu - \mu_0)/\mu_0 \qquad (8\text{-}3)$$

where k is the photoelectron wave vector and μ_0 is the background absorption i.e. the absorption coefficient of an isolated atom. In the so called "plane wave approximation" the EXAFS function can be expressed (Stern, 1974) as:

$$\chi(k) = -\sum_i \{[N_i/(k\,r_i^2)]\,F_i(k)\,S_0^2(k)$$
$$\cdot \exp[-r_i/\lambda_d(k)]$$
$$\cdot \exp(-2k^2\sigma_i^2)\sin(2k\,r_i + \Phi_i)\} \qquad (8\text{-}4)$$

which applies to the photoexcitation of a 1 s electron (K-edge) in a polycrystalline sample. This equation expresses the EXAFS function as a sum of contributions, one for each "shell" of neighbouring atoms. Each atomic shell contains N_i atoms of type i at a radial distance r_i from the absorbing atom. A shell centred at r_i contributes with a sinusoidal function of wavelength $\approx \pi/r_i$ in k space. The phase of each sinusoidal function is related to the distance r_i of the neighbouring atoms from the central atom, and to the phase shift Φ_i introduced by the potentials of central and backscattering atoms. The envelope of each sinusoidal is determined by the number of atoms in the shell and by the amplitude of the backscattering factor F_i. The damping terms in Eq. (8-4) take into account multibody effects, S_0^2, inelastic scattering, $\exp(-r_i/\lambda_d)$,

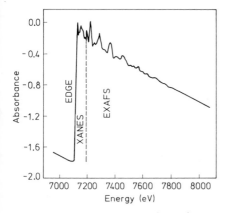

Figure 8-4. Fe K edge X-ray absorption spectrum of a Fe foil illustrating the main regions of interest in an absorption spectrum (after Pizzini, 1990 d).

and thermal and structural disorder, $\exp(-2\sigma^2 k^2)$.

A simple Fourier transform (see also Vol. 9, Chap. 4) of the EXAFS data essentially provides a radial distribution function centred on the absorbing central atom. For more detailed analysis we can model the EXAFS data [i.e., using the kind of relationship given in Eq. (8-4)] to provide information on the distance, the number and the type of the neighbours of the absorbing atom. Since only elastically scattered electrons can interfere, and the elastic mean free path of electrons is short, EXAFS only contains information on the *local* structure of the absorbing atom. Since crystallinity is not a prerequisite for this technique, amorphous materials can also be investigated. The atom-specific nature of the EXAFS process makes it suitable for the study of complex (e.g., multi-component) systems or for the determination of the environment of dilute species.

8.2.2 Methods of Detection

X-ray absorption spectra can be measured directly in a transmission experiment by monitoring the attenuation of the X-ray beam incident on the sample as a function of photon energy, or indirectly by measuring the energy dependence of physical processes which depend on absorption, such as fluorescence emission, photoelectron yield or X-ray reflectivity (e.g., Heald, 1988).

In a transmission experiment the absorption of the sample is measured by monitoring the incoming (I_0) and the transmitted (I_t) flux, which are related by the expression:

$$I_t = I_0 \exp(-\mu t) \tag{8-5}$$

where μ is the linear absorption coefficient of the sample and t is the sample thickness.

When the EXAFS signal is only a small fraction of the total absorption (typically less than 5%) it is more advantageous to measure processes which are specific to the absorbing atom of interest, such as fluorescence emission. In the case of dilute analytes, the fluorescence intensity is proportional to the absorption coefficient μ and to a good approximation is given by:

$$I_f \approx \mu I_0 \tag{8-6}$$

A schematic diagram of the experimental equipment for XAS measurements in transmission and in fluorescence modes is shown in Fig. 8-5.

Information obtained from the transmission and fluorescence measurements reflects the local structure of the *bulk* material. By detecting the photoelectrons (which have a short escape depth in a solid material) emitted as a consequence of the absorption process, surface structure can also be specifically investigated. Alternatively, surface sensitivity can be enhanced by working in glancing incidence geometry. Since the refractive index of X-rays is slightly less than unity (James, 1958), an incident beam at a glancing angle under-

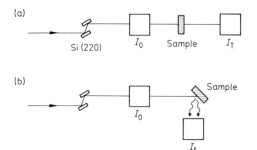

Figure 8-5. Sketch of the experimental set-up for an X-ray absorption spectroscopy measurement. The white synchrotron beam is monochromated typically by a Si (220) channel cut crystal and recorded with ion chambers (I_0 and I_T) and a fluorescence detector (I_f). (a) Transmission detection geometry; (b) fluorescence detection geometry.

goes total external reflection where the X-rays penetrate only 3–5 nm into the material. EXAFS spectra recorded under such conditions (ReflEXAFS) provides a highly effective structural probe for surface studies.

It can be shown (Parratt, 1954; Bosio et al., 1984) that the X-ray absorption coefficient μ of a condensed medium can be obtained from the energy dependence of the reflectivity coefficient $R(E)$, measured for a fixed incident glancing angle ϕ close to the critical angle. The absorption coefficient so obtained is specific to the surface layer probed by the X-ray beam. If the measurement is carried out for photon energies beyond an absorption edge, the reflectivity spectrum exhibits a structure, which can be related to the EXAFS spectrum (Martens and Rabe, 1980).

If the atomic species of interest is diluted at the surface, fluorescence detection, still at glancing incidence, gives a more favourable signal to noise ratio. A sketch of the equipment for glancing angle XAS measurements is shown in Fig. 8-6. Since the typical critical angles for X-rays are of the order of a few mrad, an appropriate precision goniometer system has to be used. The instrumentation used for glancing angle EXAFS at the Daresbury SRS has been described by Pizzini et al. (1989).

An overview of the application of EXAFS to the characterisation of materials systems has been given by Gurman (1982). In this review we consider some specific case studies of structural characterisation of dilute metallic alloys, amorphous metals, semiconductor surfaces, corrosion, phase transformations, solid-state reactivity and heterogeneous catalysis.

8.2.3 Structure of Cu-Rich Precipitates in α-Fe

Understanding the role of irradiation-induced second phase precipitation in the embrittlement of steel is of crucial importance in defining the long term stability of nuclear reactor pressure vessels. Through its ability to determine the local order around the precipitated impurities, EXAFS can provide important structural information which cannot be obtained using other techniques.

α-Fe based alloys containing 1.3 at.% Cu were investigated using fluorescence detection above the Cu K-edge (Pizzini et al., 1990a and 1990b). The precipitation of small Cu-rich clusters was studied in a series of thermally aged model alloys, well characterised with transmission electron microscopy (TEM) (Buswell and Brock 1987; Phythian and English, 1991) and small angle neutron scattering (SANS) measurements (Beaven et al., 1986; Lucas and Odette, 1986; Buswell et al., 1986). These studies show that at peak hardness condition (2 h of thermal ageing) small precipitates of mean diameter 2.5 nm are formed. Interpretation of SANS magnetic ratio and positron annihilation data also

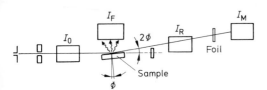

Figure 8-6. Sketch of the experimental set-up for a glancing angle XAS experiment (after Pizzini et al., 1989): the X-ray beam, collimated by entrance slits, hits the sample at an angle ϕ. The incident and reflected beams are monitored by ionisation chambers. For measurements on diluted species, a fluorescence detector is positioned above the sample. A third detector (ionisation chamber, scintillation counter as solid-state detector) can be used to measure the transmission spectrum of a model compound to achieve energy calibration.

indicate a possible association of copper in the precipitates with Fe and/or vacancies. Although the TEM measurements suggest that the Cu-rich precipitates might have a b.c.c. structure, the structural properties of the clusters and their evolution on ageing can be better investigated using X-ray absorption spectroscopy.

The local structure around Cu in Fe-Cu alloys was investigated in the "as quenched" condition and after 2 h, 10 h and 760 h of thermal ageing at 550 °C. The fluorescence signal was collected using a multi-element solid-state detector which provided an energy resolution of about 300 eV at 10 keV and allowed the Cu K emission lines to be discriminated from the intense Fe K background emission. The results of the EXAFS experiments are summarised in Fig. 8-7, where the EXAFS spectra for the thermally aged alloys (data acquisition time ≈ 5 h) are compared with the transmission EXAFS spectra of metallic Fe and Cu. An analysis of the EXAFS data indicates that in the "as quenched" material Cu is in a typically b.c.c. environment. This is consistent with Cu being in solid solution in the Fe matrix, i.e., Cu being substitutional in the host b.c.c. lattice. Analysis of the data, least-squares fitted using calculated phase shifts (Gurman, 1988) shows a reduction of the Cu-Fe distance (0.246 nm) with respect to the Fe-Fe distance in Fe (0.248 nm). After 2 h of thermal ageing the EXAFS spectra still retain the structure observed for the data of the "as quenched" samples. Previous SANS and TEM data (Buswell et al., 1988) showed that at this stage of the thermal ageing about 50% of the Cu is retained in the matrix, while the remaining copper precipitates in small clusters with an average diameter of about 2–3 nm. Therefore about a half of the Cu atoms probed by EXAFS are expected to be coordinated to Fe atoms in the α-Fe matrix and to exhibit a b.c.c.-like radial distribution function. The remaining Cu atoms are coordinated to Cu atoms in the small clusters. The maintenance of a b.c.c.-like EXAFS spectrum strongly suggests that at "peak hardness" condition the precipitates have b.c.c. structure. Least-squares fitting analysis of the EXAFS data confirms these qualitative indications. Although Cu and Fe neighbouring atoms cannot clearly be distinguished by EXAFS, because of the closeness of their atomic numbers, least-squares fitting of the experimental data seem to indicate that Cu is largely coordinated to Cu nearest neighbours, pointing to the formation of Cu rich clusters.

The results indicate that as the precipitates grow with increasing annealing time, their structure changes from b.c.c. to f.c.c. The EXAFS spectra recorded for the 760 h-aged samples more closely resemble those typical of an f.c.c. structure than a b.c.c. structure. This is consistent with previous observations by TEM of f.c.c. precipitates produced in over-aged materials (greater than 2 h at 550 °C).

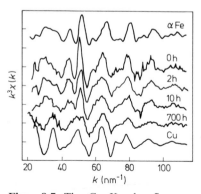

Figure 8-7. The Cu K edge fluorescence EXAFS spectra of the Fe-Cu alloy thermally aged for 0 h, 2 h, 10 h and 700 h are compared with the transmission EXAFS spectra of α-Fe and Cu foils (after Pizzini et al., 1990).

8.2.4 Amorphisation of $Ni_{50}Mo_{50}$ by Mechanical Alloying

The formation of novel amorphous metallic alloys (Samwer, 1988) obtained by the mechanical reaction of the pure constituent elements is currently of significant interest in materials science (Weeber and Bakker, 1988; see also Vol. 15, Chap. 5, Sec. 5.6.3). It has been shown (Schultz, 1987) that glassy binary metals with unexplored composition ranges can be created by mechanical alloying. $Ni_{50}Mo_{50}$ alloys, obtained by mixing together and milling equimolar quantities of pure crystalline elemental powders of Ni and Mo, were investigated using transmission EXAFS at both Ni K- and Mo K-edges (Cocco et al., 1989 and 1990).

The Ni and Mo K-edge EXAFS spectra of the samples as a function of milling time are shown in Fig. 8-8. These spectra show a significant reduction of EXAFS amplitude with milling time, due to the gradual loss of long range order associated with the onset of amorphisation.

The environment around the Ni atoms changes significantly during milling with the f.c.c. lattice becoming more disordered after 2 h milling and rapidly losing its crystallity between 2 and 10 h. The Fourier transforms of the EXAFS data (Cocco et al., 1989, 1990, 1992) show the loss of the 4th neighbour shell after 5 h milling time. At 10 h milling time the 2nd and 3rd shells are also smeared-out and the 1st shell shifted to a shorter bond length. The latter effect is commonly observed in glassy alloys (Teo et al., 1983) and in this case is indicative of Ni-Mo interactions.

The Mo K-EXAFS data reveal that, after milling, the local environment of Mo is more crystalline than that of Ni. An amplitude reduction of the EXAFS oscillations is still observed, but the overall b.c.c. struc-

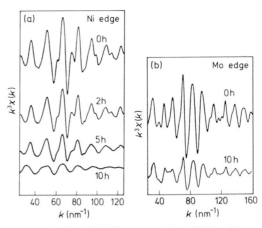

Figure 8-8. The EXAFS spectra of the Ni-Mo alloys, recorded as a function of milling time (after Cocco et al., 1990). (a) Ni K edge; (b) Mo K edge.

ture remains, even after 10 h milling. The Fourier transform shows (Cocco et al., 1990) that the 4 b.c.c. shells essentially remain. X-ray diffraction measurements (Cocco et al., 1989) reveal no change in the Mo b.c.c. lattice parameter with milling and confirm the low solubility of Ni in Mo (Hansen, 1958). However, as Mo is up to 20 wt.% soluble in Ni, one can presume that amorphisation follows a substantial insertion of Mo in the Ni lattice. The stress and defect generation associated with this is probably reflected in the loss of long range order which takes place after mechanical alloying (Johnson, 1986; Dubois, 1988).

8.2.5 Structure of the Surface Oxide on GaAs (100)

The characterisation of surface oxides of group III-V semiconductors is important in the definition of device fabrication techniques. In this study the surface oxide structure on a commercial GaAs (100) wafer was investigated using X-ray absorption spectroscopy in glancing angle geometry and reflectivity detection (Barrett et al.,

1990; Pizzini et al., 1990c). ReflEXAFS spectra were recorded above the As K and Ga K absorption edges, so that the local structure around both atomic sites could be obtained separately. The spectra were recorded for two incident angles (ϕ_1 and ϕ_2) below the critical angles, corresponding to X-ray penetration depths of about 2 nm and 3 nm. The structural parameters obtained from least-squares fits of the ReflEXAFS data (Table 8-1) indicate that the surface of the GaAs (100) wafer is partially oxidised. The ratio of the coordination numbers of bulk-like structures in the surface layer, to those typical of GaAs, indicates that the oxide thickness is 0.7–0.9 nm thick.

The results summarised in Table 8-1 show that Ga and As atoms have different oxygen environments in the surface oxide. For both Ga and As atoms there is an oxygen neighbour shell at ≈ 0.17 nm, close to that found for the tetrahedral coordination in $GaAsO_4$. However, Ga atoms have an additional oxygen neighbour shell at ≈ 0.195 nm, which is typical of the octahe-

dral coordination of Ga_2O_3. Thus while As atoms seem to be exclusively in a tetrahedral environment at the surface, Ga seems to be present in both tetrahedral and octahedral co-ordinations.

The two matched cation shells (Ga or As) at the same distances (≈ 0.28 nm and ≈ 0.31 nm for both Ga and As central atoms) are probably associated with the two different oxygen coordinations for Ga. The shell at ≈ 0.28 nm possibly arises from the coordination between cations when both are in a tetrahedral environment whilst the shell at ≈ 0.31 nm is related to the coordination between octahedral Ga and a tetrahedral cation (As or Ga). The matching of these second shell distances for both Ga and As central atoms shows the oxide to be a single component which can thus be modelled as a *microscopically random mixture* of tetrahedral and octahedral sites linked via oxygen bridges, with Ga occupying both and As only the former 4-fold site. A sketch of the structural model based on these results is shown in Fig. 8-9 a.

Comparison between the Ga and As environments as a function of incident angle (i.e., of depth) shows no significant variation for As concentration while it indicates a distinct increase in Ga coordination at the surface. This trend indicates a greater association of Ga atoms with O in the outermost oxide layer and a corresponding As depletion.

A structural model (Pizzini et al., 1990c) of this oxide on GaAs (100) based on a continuous random network (Polk and Boudreaux, 1973; Greaves and Davies, 1974) with oxygens attached to the dangling bonds at the surface, has been built up from a mixture of octahedral and tetrahedral Ga's, tetrahedral As's and bridging O's. The surface oxide so constructed results in an open structure with an atomic density much smaller than that of any of

Table 8-1. Coordination numbers and shell distances extracted from least-squares fits of the As K and Ga K-edge ReflEXAFS spectra of the GaAs(100) spectra (angle ϕ_1) (after Barrett et al., 1990).

r (nm)	N	r (nm)	N	r (nm)	N
Ga K-edge					
Ga-O		Ga-O-M		Ga-As	
0.172	0.6	0.288	3.1	0.246	3.0
0.195	2.2	0.315	3.4		
As K-edge					
As-O		As-O-M		As-Ga	
0.168	0.9	0.284	1.1	0.243	3.1
		0.309	1.0		
Bulk GaAs					
				Ga-As	
				0.2446	

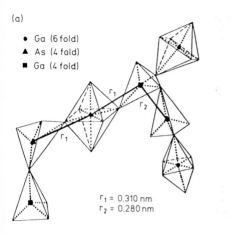

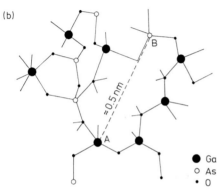

Figure 8-9. (a) Sketch of the local coordination of Ga and As atoms in the oxide coating on GaAs(100): 6-fold and 4-fold coordinated Ga atoms are linked to 4-fold coordinated As atoms via oxygen bridges (after Barrett et al., 1990); (b) sketch of the section of a typical fissure in the oxide on GaAs(100), obtained from a three-dimensional molecular model of the oxide (after Pizzini, 1990d).

the crystalling oxides of Ga and/or As. The presence of 2-fold coordinated oxygens and the variable cation coordination combine to encourage the formation of micro-voids (typical size $\approx 0.5-0.8$ nm) whose internal surfaces are oxygen rich (Fig. 8-9).

8.2.6 Surface Corrosion of Borosilicate Glasses

A knowledge of the leaching mechanism of metallic ions from glass surfaces is impor-

tant to the understanding of materials requirements for the vitrification processes associated with nuclear waste management. Detailed studies have been carried out of the leaching of Fe (Sacchi et al., 1989) and U (Greaves et al., 1989) through the glass matrix. In the study of the corrosion-induced transport of Fe^{3+} ions in borosilicate glasses loaded with Fe_2O_3 (Sacchi et al., 1989) fluorescence X-ray absorption spectroscopy was used to follow the changes in the ionic environment during the leaching process. The measurements were carried out using a glancing angle geometry commensurate with a penetration depth of about 1 μm.

The results of this study show that after leaching in H_2O at 373 K, with the sample surface still in the "wet" state, dramatic changes in the edge structures and in the EXAFS spectra are observed. These changes are seen both as a function of depth and of the extent of leaching (Fig. 8-10) and give some insight into the transport properties of Fe^{3+} ions through the glass network during corrosion.

In contrast to most minerals where ferric sites are octahedral, in glasses Fe^{3+} is commonly tetrahedrally coordinated (Binsted et al., 1986). This configurational change can be readily recognised in EXAFS by a shortening of the Fe-O distance. The two different geometries can also be distinguished in the edge structure from the magnitude of the "white line" (peak A in Fig. 8-10) which is considerable for octahedrally coordinated Fe^{3+}, but much reduced when the metal occupies tetrahedral sites. A converse behaviour is exhibited by the pre-edge feature (peak B in Fig. 8-10) which is larger for tetrahedrally coordinated Fe^{3+} in oxide glasses (Binsted et al., 1986). Both types of ferric coordination can be identified at the surface of borosilicate glass (Sacchi et al., 1989).

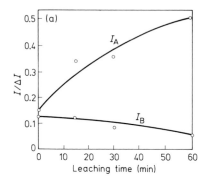

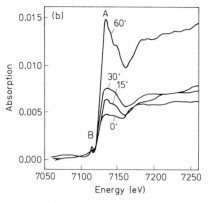

Figure 8-10. The changing near-edge structure of Fe K edge measured in glancing angle fluorescence mode for borosilicate glass as a function of leaching time (after Sacchi et al., 1989). The inset shows the increase in the white line (peak A) and the decrease in the pre-edge feature (peak B) with advancing corrosion. Both are normalised to the step height I.

The increase in the absorption step height as leaching progresses demonstrates that ferric ions migrate to the glass surface, quadrupling in concentration over a period of ≈ 1 h. Initially the white line (A) is weak and the pre-edge feature strong, indicative of extensive tetrahedral bonding. After 1 h of leaching the converse is true. pointing to the developement of a substantial fraction of octahedrally coordinated ferric ions. The same picture emerges from the EXAFS data which show the presence of a single Fe-O distance at first, followed by the appearance of a longer distance as corrosion continues. From what is known

about the bulk structure of Fe^{3+} in aegerine glass and in the mineral, the glancing angle XAS behaviour of the borosilicate glass surface strongly suggests that leaching promotes dissolution of Fe from the glass network and its transport towards the corrosion interface where it precipitates in a form similar to that of a ferric silicate (see also Vol. 10, Chap. 12, Sec. 12.3.3).

8.2.7 Oxidation of Polished Stainless Steels

The detection of the early stages of corrosion together with the structural characterisation of any surface phases formed can provide a useful insight into our understanding of fundamental and applied aspects of surface reactivity. X-ray absorption spectroscopy was used to characterise changes in the environment around Fe in the surfaces of polished polycrystalline stainless steel samples which had been oxidised at 1000 °C for up to 4 min (Barrett et al., 1989 a and 1989 b). Angle dependent reflectivity curves were recorded for a fixed photon energy beyond the Fe K edge. XAS spectra were recorded above the Fe-K edge for an incident angle below the critical angle, corresponding to an X-ray penetration depth of about 3.5 nm. The results, summarised in Fig. 8-11 demonstrate that self-consistent information can be achieved by combining the near edge, edge height, chemical shift and reflectivity data. Figure 8-11 a shows that with increasing oxidation time the slope of the angle-dependent reflectivity curves dramatically decreases. This behaviour reflects the increased absorption of the X-ray beam (Parratt, 1954) and is evidence for the migration of Fe ions to the surface layer.

The variation of the absorption edge height as a function of oxidation time (Fig. 8-11 b) indicates an increase in the

Fe^{3+} content near the surface, which probably relates to Fe diffusion through the protective Cr_2O_3 layer normally present on stainless steel. For steel measured in transmission the edge position coincides with that of metallic Fe. The increase in chemical shift with corrosion time (Fig. 8-11 c) indicates that the oxidation state is changing as corrosion proceeds. The edge position is shifted by 1 eV for 1 min corrosion time and after 4 min the shift is 5 eV, corresponding to that observed for Fe_2O_3.

In Fig. 8-11 d the near-edge spectra for the uncorroded, 1 min, 3 min and 4 min corroded steel surfaces are compared to bulk steel, Fe_3O_4 and Fe_2O_3 model spectra. The bulk steel has a near edge structure similar to that of metallic iron (Grunes, 1983). A white line, typical of oxide spectra, appears as corrosion proceeds, as does the pre-edge feature ≈ 10 eV below the edge. The uncorroded steel surface shows evidence for an oxide-like white line. After 1 min the pre-edge structure resembles that found for Fe_3O_4. As corrosion proceeds, Fe_2O_3 becomes the more dominant phase, as indicated by the splitting of the pre-edge and main peaks. After 3 min the near edge structure is virtually identical to that of Fe_2O_3.

The EXAFS results (Barrett et al., 1989 a and 1989 b) confirm these indications. In agreement with the near-edge data, the polished steel surface is metal-like. The dominant structure is that of the bulk steel although there is some reduction of the first shell (f.c.c.) amplitude. After 4 min corrosion time the f.c.c crystal structure of the metallic steel has disappeared and the Fourier transform is virtually identical to octahedral Fe_2O_3.

These results indicate that the polished surface, although retaining its metallic character, shows signs of initial oxidation. As corrosion proceeds the surface content

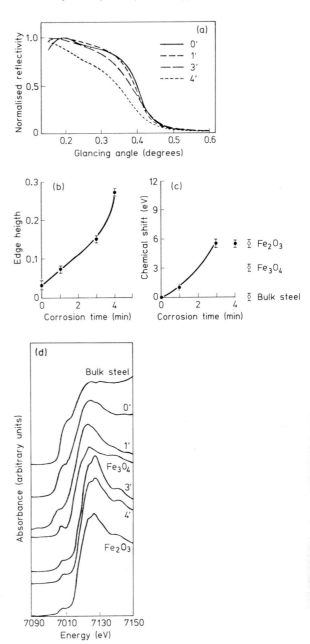

Figure 8-11. Summary of the measurements on corroded stainless steels (after Barrett et al., 1989 a). (a) Normalised reflectivity curves as a function of oxidation time. (b) The step height at the Fe K edge as a function of corrosion time. (c) The chemical shift as a function of corrosion time. (d) Near-edge spectra of the Fe K edge for steel surfaces corroded for: (B) 0 min, (C) 1 min, (E) 3 min, (F) 4 min; compared to the near-edge spectra for: (A) the bulk steel, (D) Fe_3O_4 and (G) Fe_2O_3.

of Fe is enriched and the oxidation state increases. The first oxide phase to be formed on oxidation is probably similar to Fe_3O_4 but further reaction yields preferential oxidation to Fe(III) in the form of Fe_2O_3.

8.2.8 Changes in the Atomic Environment Around Zn in K_2ZnCl_4 during Phase Transformations

The formation of modulated and incommensurate crystalline phases is often accompanied by very subtle changes in atomic structure which may not be detectable using conventional analysis. EXAFS can be used to probe such changes and has been applied in situ to analyse changes to the local environment around Zn atoms in potassium tetrachlorozincate K_2ZnCl_4 (KZC). KZC crystallises with a ferroelectric (Gesi, 1978) structure (orthorhombic $Pna2_1$, Mikhail and Peters, 1979) which is commensurate with a wave vector with the magnitude $q = a*/3$ and which undergoes phase transitions at temperatures close to 127 °C and 282 °C (Gesi, 1978; Gesi and Iizumi, 1979; Jacobi, 1972; Itoh et al., 1980; Milia et al., 1983; Quilichini et al., 1982). Above the lock-in transition at ≈ 127 °C the structure is incommensurately modulated along the pseudo-hexagonal a-axis with a wave vector of magnitude $q = 1/3 \cdot (1 - \delta) q*$, where the deviation parameter δ increases with temperature (Kucharczyk et al., 1982) until close to 284 °C at which the structure transforms to a paraelectric phase with the space group $Pnam$.

The environment around the Zn atoms in the room temperature structure forms a distorted tetrahedron (Mikhail and Peters, 1979) where the Zn-Cl distances range from 0.2241 to 0.2289 nm. The structural changes around the Zn atoms in the KZC structure associated with the 127 °C and 282 °C phase transitions have been studied using EXAFS. The samples were mounted in a compact furnace (Bhat et al., 1990a) and transmission EXAFS spectra were recorded as a function of temperature (Bhat et al., 1990b).

Figure 8-12 shows the Fourier transforms of the EXAFS data as a function of temperature. In Table 8-2 the structural parameters obtained from least-squares fits of the EXAFS data are reported. The expected increase in disorder associated with the phase transition can be seen in the

Table 8-2. EXAFS characterisation of the tetrahedral distortion in KZC using a two shell model (after Bhat et al., 1990b).

T ($^\circ$C)	N	r (nm)	A (nm^2)	N	r (nm)	A (nm^2)
25	2.0	0.221	0.00007	2.1	0.229	0.00007
127	2.0	0.220	0.00012	2.0	0.227	0.00008
212	1.4	0.216	0.00009	2.7	0.228	0.00008
286	1.4	0.218	0.00018	2.7	0.227	0.00013
(25	2.0	0.221	0.00007	2.0	0.229	0.00007)

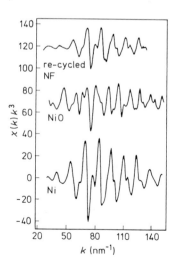

Figure 8-12. Fourier transforms of the EXAFS data recorded for KZC as a function of temperature (after Bhat et al., 1990b).

modelled Debye-Waller factors and is noticeable by a reduction in amplitude in the
Fourier transforms. The change in tetrahedral distortion as a function of temperature
is also evidenced in Table 8-2. Although
the Zn radial distribution function (RDF)
is the summation of the 12 closely distributed Zn-Cl bond lengths, a 2-shell model
provides a useful insight to the change in
the relative weighting of the neighbouring
shells with temperature. Below 212 °C the
data can be modelled by 2 shells equally
weighted (2:2) at Zn-Cl distances of about
0.220 nm and 0.229 nm whereas above
212 °C the weighting changes to about
1.5:2.5 respectively.

Upon recycling to room temperature the
structure returns close to that observed before transformation. Thus, although X-ray
topography (see Sec. 8.6) shows that these
phase transformations do induce changes
in macro-structure, i.e., the generation of
defects (El Korashy, 1988), this study indicates that these changes are not directly
correlated with changes in the local atomic
structure.

8.2.9 Dynamic Investigation of the Reduction of Nickel Formate

It has been demonstrated (Frahm,
1989a and 1989b) that high quality X-ray
absorption spectra can be obtained using
the quick-scanning or QEXAFS technique.
QEXAFS uses a continuous scan of the
monochromator axis and time integration
of the data rather than the conventional
step/count approach. The improvement in
data acquisition time is significant and
opens the possibility of the application of
the technique to the characterisation of
structural changes in chemical systems,
particularly in cases where the kinetics of
transformation may be comparatively slow
(≈ 1 min).

The applicability of the technique to
studies of thermally-induced solid-state decomposition is here demonstrated by examining the decomposition of nickel formate dihydrate [$Ni(COOH)_2 \cdot 2H_2O$, here-
inafter NF]. The sample was mounted in a
small transmission micro-furnace (Bhat
et al., 1990a) which can operate in the temperature range 25–320 °C. QEXAFS spectra were recorded as a function of temperature (Edwards et al., 1990).

Differential scanning calorimetry of NF
reveals three noticable features at temperatures of 170 °C, 210 °C and 270 °C. These
have been ascribed to transitions from the
hydrated to anhydrous formate (A), from
anhydrous formate to oxide (B) and from
oxide to metal (C) (Vecher et al., 1985; Iglesia and Boudart, 1984; Waiti et al., 1974).
The reaction to the metallic state progresses by a nucleation and growth mechanism (Waiti et al., 1974) and this is evidenced from the isothermal ($T = 257$ °C)
Fourier transformed QEXAFS data (Fig.
8-13) taken in the metastable region close
to transition C.

The nature of the transformed phase is
clearly shown in the room temperature
EXAFS data shown in Fig. 8-14. It can be
seen that the re-cycled material resembles
the metal rather than the oxide structure
from which it has transformed. The main
difference is the significant amplitude reduction observed in the recycled formate
which is indicative of the formation of
rather small metal clusters. The data were
modelled using curved wave theory (Gurman, 1988). Modelling the refined data
against hypothetical radial distributions
around Ni based on spherical clusters of
various radii (Table 8-3) yields a particle
size of around 0.5 nm. This is in good
agreement with measurements of particle
size by diffraction line broadening (Waiti
et al., 1974).

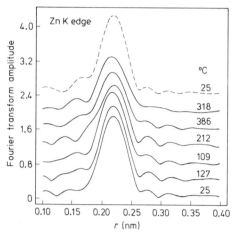

Figure 8-13. Ni K edge Fourier transformed QEXAFS data of nickel formate dihydrate (100 ms integration time and scan speed 20 mdeg s^{-1}) recorded in situ and isothermally in the metastable region close to transition C showing the slow thermal reduction of NiO to metallic Ni (after Edwards et al., 1990).

Table 8-3. Ni f.c.c. radial distributions calculated as a function of cluster size compared to the refined data for the cycled formate (after Edwards et al., 1990).

Cluster rad. (nm)	Atoms in cluster	1st shell	2nd shell	3rd shell	4th shell
0.3	13	5.54	1.85	3.69	0.92
0.5	55	7.85	3.27	9.60	4.15
0.8	201	9.43	3.94	14.57	6.86
∞	∞	12.00	6.00	24.00	12.00
This data		5.8	3.1	10.9	4.7

8.2.10 Role of the Sulfidation Process in the Operation of Pt/Re Catalysts

The preparation of reforming catalysts involves a number of classic but poorly understood basic steps such as impregnation, calcination and reduction followed by modifying steps such as sulfiding. The utility of EXAFS in the characterisation and optimising the sulfidation stage for a Pt/Re catalyst has been demonstrated by Oldman (1986). Figure 8-15 shows Pt L$_3$ absorption edges for differential oxidation states. At the edge, absorption occurs via a fully allowed pseudo-atomic transition, $2p \rightarrow 5d$, with the intensity depending on the number of unoccupied valence levels. An excellent correlation exists between the normalised white line amplitude and the oxidation state.

The edge data, the Fourier transforms (Fig. 8-16) and least-squares fitting of the EXAFS data show that a poor catalyst performance is associated with the com-

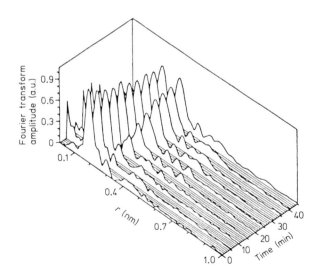

Figure 8-14. The Ni K edge k^3-weighted EXAFS data recorded at room temperature showing the comparison between thermally cycled nickel formate dihydrate, NiO and Ni metal (after Edwards et al., 1990).

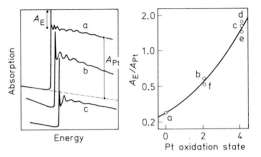

Figure 8-15. L_3 edge XAS for some model materials showing the correlation between normalised white line intensity (A_E/A_{Pt}) and oxidation state (after Oldman, 1986): (a) Pt foil, (b) $PtCl_2$, (c) $PtO_2 \cdot 2H_2O$, (d) $Na_2Pt(OH)_6$, (e) H_2PtCl_6, (f) PtS.

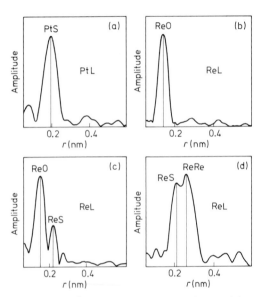

Figure 8-16. k^3 weighted Fourier transforms without phase shifts for 2.5% Pt/2.5% Re on alumina (a) Pt L_3 edge (b) Re L_3 edge both with original sulfiding procedure followed by air exposure. (c) Re L_3 edge revised procedure and air exposed, (d) Re L_3 edge reduced in H_2 at 450 °C (after Oldman, 1986).

plete conversion of Pt to PtS (4S atoms at 2.31 Å) while Re is not sulfided at all. A revised sulfiding procedure shows partial conversion of both Pt and Re. PtS forms a coating on Pt preventing air oxidation, whilst Re is clearly not protected in this way. Under in situ reduction conditions at 450 °C PtS is completely reduced to Pt f.c.c. crystallites while ReS_2 remains.

8.3 Structural Studies Using X-Ray Diffraction

8.3.1 Introduction

The combination of high flux, wide spectral range (tunability) and low natural divergence makes synchrotron radiation an optimum source for the investigation of long range order in crystalline solids. Over the past decade these properties have been utilised for studies using both white-beam and monochromatic diffraction techniques in the examination of crystalline powders, single crystals and the surfaces of ordered interfaces.

Conventional X-ray analysis is routinely used by crystallographers for the definition of the 3-D molecular structure of small and medium size molecules. For most crystal structures synchrotron radiation offers no significant advantages over laboratory based techniques. Thus the application of synchrotron radiation to single crystal studies has focussed on research areas where the specific and unique properties of this photon source can be utilised for the solution of specific problems. Such areas of application fall into a number of groups; studies of macromolecular biological systems (e.g., proteins and viruses), high resolution powder diffraction and energy dispersive diffraction, dynamical diffraction studies using nearly-perfect single crystals, X-ray standing wave spectroscopy and diffraction from surfaces. Applications to studies of biological systems are beyond the scope of this review and the reader is best referred to a number of reviews (e.g., Helliwell, 1984 and 1991) on this subject.

Studies of surfaces and interfaces (see Sec. 8.4), X-ray standing waves (see Sec. 8.5) and the topographic assessment of nearly perfect crystals (see Sec. 8.6) are described elsewhere in this review.

8.3.2 High Resolution Powder Diffraction

8.3.2.1 Introduction

Conventional powder diffraction analysis using Bragg/Brentano geometry is prone to peak displacement and deformation due to sample misalignment or beam focussing errors. The use of synchrotron radiation sources enable excellent angular resolution to be achieved (Cox et al., 1983; Hastings et al., 1984). Diffractometers designed specifically for use with synchrotron radiation have been developed by Hart, Parrish and co-workers (Hart and Parrish, 1989; Parrish et al., 1986; Hart and Parrish, 1986; Lim et al., 1987).

Figure 8-17 shows a schematic view of the powder diffractometer available on station 8.3 (Cernik et al., 1990 b) at the Daresbury SRS. The wavelength is selected by a Si(111) monochromator which operates in the wavelength range 0.07–0.25 nm. The

incident beam cross-section is defined by horizontal and vertical entrance slits. The sample is contained in a flat-plate sample stage which can rotate the specimen about an axis orthogonal to the diffractometer axis and the axis of the synchrotron orbit. The angular resolution of the diffractometer is defined by a set of parallel stainless steel foils which are housed in a vacuum vessel to reduce air scatter. These foils are attached to the 2Θ arm. The diffracted radiation is monitored by an Ar/Xe proportional detector mounted after the collimator at the end of the 2Θ arm.

8.3.2.2 High Resolution Refinement of the Lattice Parameters of Tungsten

The high resolution attainable with the synchrotron powder diffraction technique has been demonstrated through the work of Hart et al. (1990) who demonstrated the accuracy achievable when measuring the lattice parameters of tungsten. To do this four sets of data using two wavelengths (0.1 nm and 0.125 nm) were collected on station 8.3 at the Daresbury SRS. A Si internal (SRM640B obtained from the Na-

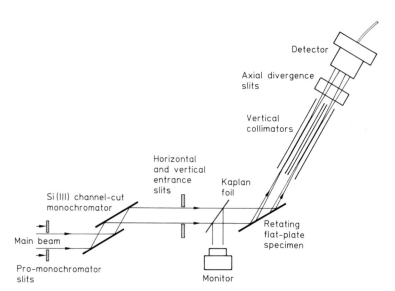

Figure 8-17. The parallel-beam powder diffractometer on station 8.3 at the SRS (after Cernik et al., 1990 b) which comprises Si (111) monochromator, slits, a rotating flat-plate sample holder, two-circle optically-encoded diffractometer, diffracted-beam collimator and an Ar/Xe proportional detector.

tional Institute of Standards and Technology) standard was mixed with the samples in the ratio 3:1 Si:W. Data was collected between 60° and 100° for 2Θ. Peak positions were determined to within 0.1 mdeg for 2Θ by fitting the experimental data to a pseudo-Voight function (Fig. 8-18). All fitted functions were perfectly symmetrical with halfwidths between 0.05° and 0.07° for 2Θ. The mean deviation between calculated and measured 2Θ values was at most 0.0005° to give a lattice parameter accuracy of parts per million. The lattice parameter for tungsten was determined to be 0.3165629(19) nm.

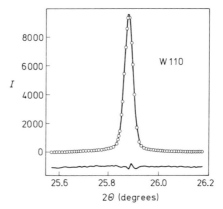

Figure 8-18. High resolution powder diffraction study of tungsten (after Hart et al., 1990). 110 Bragg reflection profile; circles: experimental data; full line: theoretical fit using a pseudo-Voight function; bottom line: difference between theory and experiment.

8.3.2.3 Determination of the Structure of Cimetidine Using Rietveld Refinement

The use of conventional single crystal methods for the determination of molecular structure is often limited by the difficulties associated with preparing crystals of adequate quality. Cimetidine $[Cd_3(OH)_5(NO_3)]$ is a powerful histamine antagonist of pharmaceutical interest. Its molecular structure has been determined using ab inito methods (Cernik et al., 1990a). Data were collected between 6° and 105° for 2Θ using $\Theta/(2\Theta)$ geometry. Atom positions for non-hydrogen atoms were obtained by direct methods followed by iterative cycles of least squares refinement and Fourier synthesis. These positions were then refined using Rietveld analysis (Rietveld, 1969; Murray et al., 1990, see this Volume, Chap. 4, Section 4.14.8). After convergence was achieved a difference Fourier synthesis showed positive electron density at the expected positions for the hydrogen atoms. These positions were then included in the model structure. Figure 8-19a shows the collected diffraction pattern together with the simulated pattern based on the Rietveld method. The resulting crystal structure (Fig. 8-19b) was found to be monoclinic with four molecules to the unit cell and a space group of $P2_1/n$. That the structure of a low symmetry organic system including unconstrained hydrogen positions can be solved by ab initio methods is an extreme test of the reliability of the synchrotron radiation powder method.

8.3.2.4 Influence of Growth Environment on the Structure of Long Chain Hydrocarbons

The crystal structure of homologous mixtures of n-alkanes is of considerable importance in view of the deleterious influence of wax crystallisation in diesel fuel. The potential influence of the crystallisation environment on the structure of the precipitated wax was investigated (Gerson et al., 1990; Cunningham et al., 1991a) for the model system of mixtures of n-eicosane $(C_{20}H_{42})$ and n-docosane $(C_{22}H_{46})$ recrystallised from the melt and from n-dodecane $(C_{12}H_{26})$. X-ray diffraction patterns obtained from samples containing multiple phases of low symmetry crystal systems

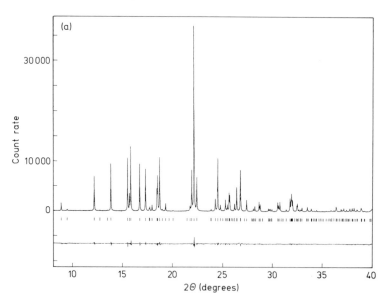

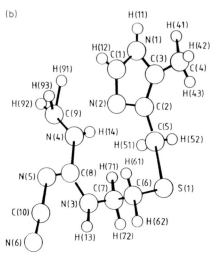

Figure 8-19. The crystal structure of Cimetidine; (a) showing experimental data overlaied with Rietveld refinement, the vertical lines indicate the positions of all expected diffraction peaks and the bottom curve shows the difference between experiment and theory; (b) model showing the refined molecular structure (after Cernik et al., 1990a).

are extremely complex and to index these patterns good quality high resolution data are needed. An indication of the complexity of the patterns produced is shown in Fig. 8-20a which presents the central sec-

tion of the diffraction pattern recorded from a sample comprising 18% $C_{20}H_{42}$, 82% $C_{22}H_{46}$ crystallising from $C_{12}H_{26}$. The number of phases present in such a mixture is clearly indicated by the 3 sets of (001) peaks present in Fig. 8-20b.

The phase system of the mixed homologues can be clearly seen from the refined unit cell parameters and lattice plane spacings summarised in Table 8-4. The structures of the samples crystallised from the melt confirm the observations of Lüth et al. (1974); the pure and mixed homologues crystallising with triclinic (one molecule in the unit cell, $z = 1$) and orthorhombic ($z = 4$) structures respectively. The orthorhombic phase observed in the mixed systems is thought to be of a related structure to that known for the single odd *n*-alkanes, the main difference being in the degree of disorder of the molecules (Lüth et al., 1974). A different behaviour was observed when the same mixtures were crystallised from *n*-dodecane where a rather complex phase behaviour occurs involving a number of crystalline structures including a new polymorphic form which phase-separates from the mixtures over the

(a)

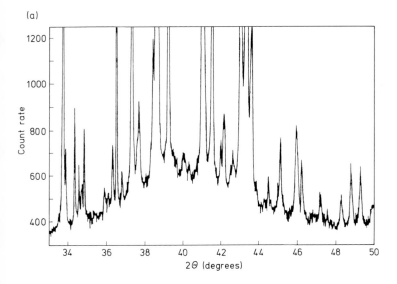

(b)

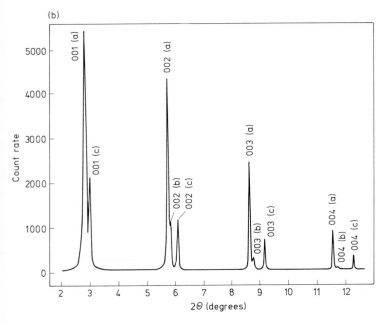

Figure 8-20. Synchrotron radiation diffraction patterns showing phase separation in a mixture comprising 18% $C_{20}H_{42}$ and 82% $C_{22}H_{46}$ crystallised from $C_{12}H_{26}$: (a) wide angle diffraction pattern; (b) the low angle diffraction pattern (after Cunningham et al., 1990b).

composition range from ≈ 45 to 90% n-docosane.

Crystallographically the orthorhombic and triclinic ($z = 2$) phases of mixed homologues may be able to be considered to exhibit positional disordered crystal structures with the n-eicosane and n-docosane molecules randomly distributed over the available crystal sites. Further disordering

may also be present due to partial molecular rotation. Molecular packing difficulties arising in this disordered orthorhombic structure result in unit cells less dense than those observed for single n-alkane systems.

The use of synchrotron X-rays to study phase transitions under high pressure is discussed in Volume 5, Chap. 8, Section 8.2.3.

Table 8-4. Unit cell parameters for homologous mixtures of $C_{20}H_{42}$ and $C_{22}H_{46}$ recrystallised from the melt and from dodecane solution, (*) is the average molecular volume based on the refined lattice parameters (after Cunningham et al., 1991 b).

C20:C22	a (nm)	b (nm)	c (nm)	α (°)	β (°)	γ (°)	* (nm³)	z	Growth method
100:0	0.428	0.482	2.552	91.00	93.54	107.39	0.501	1	melt/soln
99:1	0.429	0.483	2.555	91.42	93.49	107.41	0.503	1	soln
	0.499	0.769	5.797	90.00	90.00	90.00	0.536	4	
84:16	0.502	0.769	5.633	90.00	90.00	90.00	0.443	4	melt
64:36	0.504	0.767	5.747	90.00	90.00	90.00	0.550	4	soln
50:50	0.502	0.765	5.825	90.00	90.00	90.00	0.556	4	melt
40:60	0.504	0.762	5.852	90.00	90.00	90.00	0.561	4	soln
28:72	0.498	0.753	5.922	90.00	90.00	90.00	0.559	4	soln
	0.464	0.698	2.802	84.28	93.22	97.50	0.447	2	
	third phase, longest lattice plane spacing = 2.934 nm								
24:76	0.501	0.752	5.959	90.00	90.00	90.00	0.561	4	soln
	0.464	0.699	2.818	84.72	92.99	97.39	0.451	2	
18.82	0.495	0.753	5.925	90.00	90.00	90.00	0.558	4	soln
	0.464	0.696	2.802	84.72	93.16	97.46	0.446	2	
	third phase, longest lattice plane spacing = 2.921 nm								
14:86	0.498	0.750	5.923	90.00	90.00	90.00	0.553	4	soln
	0.464	0.703	2.802	84.51	93.11	97.29	0.449	2	
	third phase, longest lattice plane spacing = 2.920 nm								
8:92	0.463	0.707	2.803	84.66	93.18	97.03	0.453	2	soln
1:99	0.496	0.750	5.861	90.00	90.00	90.00	0.545	4	son
	0.429	0.481	2.791	91.77	91.96	107.09	0.550	1	
0:100	0.428	0.483	2.793	91.59	92.39	107.83	0.548	1	melt/soln

8.3.3 Energy Dispersive Powder Diffraction

8.3.3.1 Introduction

Monochromatic synchrotron radiation techniques suffer from the obvious disadvantage that only a fraction ($\approx 0.01\%$) of the available wavelength bandwidth provided by a synchrotron source is used. This disadvantage can be overcome by using energy dispersive techniques. Energy dispersive X-ray diffraction (EDXRD) is defined from the alternative formulation of Bragg's law in terms of energy (E);

$$E\,(\mathrm{keV}) \approx 6.2/(d \sin \Theta) \qquad (8\text{-}7)$$

in which each set (d) of lattice planes satisfies Bragg's law for a discrete energy and the synchrotron white radiation source is used with an energy sensitive detector to provide a diffraction spectrum as a function of energy at a fixed detector angle. The fixed detector scattering geometry means that diffraction patterns can be recorded in a few minutes. Also since it is a one beam in, one beam out technique small aperture environmental cells for in-situ studies can be used.

Figure 8-21 shows a schematic view of the dedicated instrument for EDXRD studies, station 9.7 at the Daresbury SRS

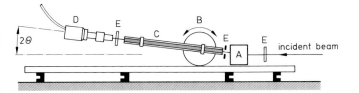

Figure 8-21. Energy dispersive X-ray diffraction facility on station 9.7 of the Daresbury SRS; sample area (A), detector axis (B), collimator (C), detector (D) and slits (E) (after Cunningham et al., 1990 b).

(Clark, 1989; Clark and Miller, 1990). The axial divergence and therefore the resolution is controlled by a set of Soller slits placed prior to the solid-state detector but is limited by the electronics of the detector system. These are composed of 25 metal foils 50 cm long each separated by 0.1 mm. Slits and detector are both mounted on a 2Θ rotation arm. Horizontal and vertical slits reduce the incident beam size to within the counting limits of the solid-state detector.

8.3.3.2 In Situ Characterisation of Solid-State Chemical Transformations

The fixed diffraction geometry of the EDXRD technique is particularly suitable for in situ studies of chemical transformations in the solid state (e.g., Anwar et al., 1990). Applications have been made to a number of systems including studies of the calcination of zirconium hydroxide (Mamott et al., 1991) and the hydration of cement (Barnes et al., 1991).

Zirconium hydroxide, an important chemical precursor used in the synthesis of zirconia, can be precipitated from a zirconium salt by means of base addition or the reaction of a slurry of an insoluble zirconium salt with a base (Houchin, 1987). One of the main structural units thought to be associated with zirconium hydroxide are tetramer units which join together during precipitation by a polymerisation reaction (Clearfield, 1964). Subsequent processing of the precipitated powder involves calci-

nation where the amorphous phase is converted to a crystalline form and conversion where this phase transforms to the desired monoclinic phase. Figure 8-22 shows energy dispersive data from a detailed study by Mamott et al. (1991) who carried out an in-situ investigation of how a change in pH of the liquid surrounding the zirconium hydroxide precipitate affected the dynamic parameters of the synthesis process and the kinetics of conversion from the tetragonal to monoclinic zirconia. Their study shows that the kinetics of the initial precipitation stage directly affects the kinetics of both the calcination and conversion reactions. In particular their data appear to confirm the mediating role played by OH^- ions which seem to control the balance between the hydration and dehydration stages of this reaction. They find that an increase in pH during precipitation generates excess OH^- ions which supress dehydration, enhancing the polymerisation rate and resulting in a more amorphous phase.

Despite its universal role in the construction industry there is surprising little information concerning structural changes associated with the hydration of cement. The action of hydraulic cement commences with a mixing of a finely ground (unhydrated) cement powder with water in a typical ratio of 0.5–0.4. The setting time of a cement depends on its nature and on the conditions but can be as short as a few minutes. The hydration of a fast setting cement, based on hydrates of calcium trisulpho-aluminate (often referred to as

(a)

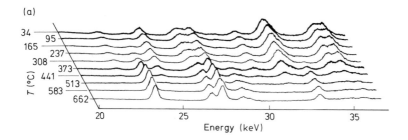

(b)

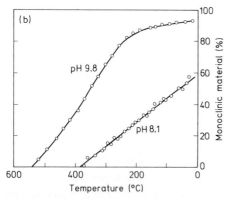

Figure 8-22. Dynamic energy dispersive diffraction data showing the conversion stage in the preparation of zirconia (after Mamott et al., 1991). (a) Part of the diffraction pattern taken during the cooling sequence, the transformation from the tetragonal to monoclinic phases at $I \approx 373$ K is clearly shown. (b) Plot of the growth of the monoclinic phase for samples prepared at pH values of 9.8 and 8.1 as determined from an assessment of the relative peak areas of the two phases.

Ettringite) has been investigated (Fig. 8-23) in-situ by Barnes et al. (1992). Figure 8-23 a shows a plot of the diffraction pattern obtained as a function of hydration time. From an analysis of the peak positions of the 110, 112 and 114 diffraction profiles it can be seen that there are significant changes in the a and c parameters in the hexagonal Ettringite unit cell as a function of hydration with, for example, the a parameter increasing by about 0.01 nm ($\approx 1\%$) during hydration. Barnes et al. (1992) ascribe the lattice dilation to the initially formed Ettringite being deficient in sulphate due to the relatively low solubility

of the anhydrous calcium sulfate. In the initial stage this is substituted by the more plentiful carbonate phases. The time dependent evolution of the hydration reaction is shown in Fig. 8-23 b which reveals that the development of the Ettringite phase begins immediately after mixing and continues to grow in a sigmoidal manner without any obvious induction time. Further experiments show that the maximum phase formation takes, typically, $1-2$ days to occur.

8.3.3.3 Structural Phase Transformations at High Pressure

Energy dispersive diffraction using synchrotron radiation is well suited to high pressure studies (see also Vol. 5, Chap. 8) since it permits the examination of minute volumes of samples in high pressure cells. Moderate pressure can be achieved in Drickamer cells and exceptionally high pressures in diamond anvil cells.

Hauserman and coworkers (Hauserman et al., 1989 and 1991) have used a Drickamer-type high pressure cell to measure the compressibility of TiB_2. Such measurements demand the use of high photon energy (i.e., > 50 KeV) part of the synchrotron radiation spectra as the X-rays need to penetrate a mixture of sample and pressure calibration material surrounded by several millimeters of pressure transmitting material. TiB_2 crystallises in a hexagonal crystal system with a space group $P6/mm$. Hauserman et al. (1989 and 1990) typically

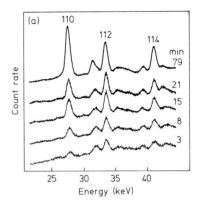

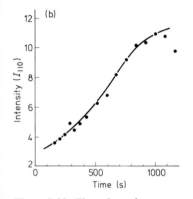

Figure 8-23. Time dependant energy dispersive diffraction data revealing the development of Ettringite phases during the in situ hydration of cement (after Barnes et al., 1991). (a) Diffraction patterns showing the 110, 112 and 114 Bragg reflections. (b) Plot of the evolution of the intensity of the 110 Bragg peak as a function of hydration time.

used 4 or 5 copper reflections to determine the pressure and 6 to 8 reflections from TiB_2 to calculate the changes in both the a and c lattice parameters. Figure 8-24 a shows some of the data recorded at pressures of 25.2 (±0.1) and 123.5 (±0.6) kbar. Measurements of the lattice constants as a function of pressure revealed (Fig. 8-24 b) the bulk modulus K_0, of TiB_2 to be 4.1 (±0.2) Mbar and showed TiB_2 to be one of the least compressible materials known (diamond has $K_{0'} \approx 4.4$ Mbar (Mc-Skimin and Bond, 1957) and thus a potentially useful industrial material.

In contrast to the moderate pressures in Drickamer cells very high pressures (up to ≈ 1 Mbar) can be obtained using diamond anvil cells. Diamond is an ideal window material due to its transparency not only to X-rays but also to optical frequencies. The optical transparency of the cell windom material can be applied by using the fluorescent lines of Ruby, which shift linearly with pressure, to calibrate the cell pressure.

Interest in the solid-state properties of mixed metal oxides has increased significantly since the discovery by Bednorz and Müller (1986) of superconductivity at temperatures up to 30 K in the La-Ba-Cu-O system. Chu et al. (1987) have shown that under pressures of up to 2 kbar the maximum temperature for superconductivity is increased to 52.5 K. This work has led to an increase in effort in both theoretical and experimental studies towards gaining an understanding of the mechanism responsible for high temperature superconductivity. Akhtar et al. (1988) studied the structure of La_2CuO_4, prepared by solid-state reaction (Longo and Raccah, 1973) at high pressure using a diamond avil and energy dispersive diffraction. The data, summarised in Fig. 8-25 show that up to ≈ 150 mbar the lattice constants decrease linearly with pressure whilst the near constancy of the c/a axial ratio demonstrates that the compressibility of this material is isotropic. Akhtar et al. (1988) correlate their data with band structure calculations (Stocks et al., 1988) which appear to suggest that the inter-ionic bonding for this material is dominated by the 3-d network of La-O interactions. Such behaviour contrasts with the known 2-d magnetic properties (Shirane et al., 1987) which originate from the $Cu-O_2$ layers.

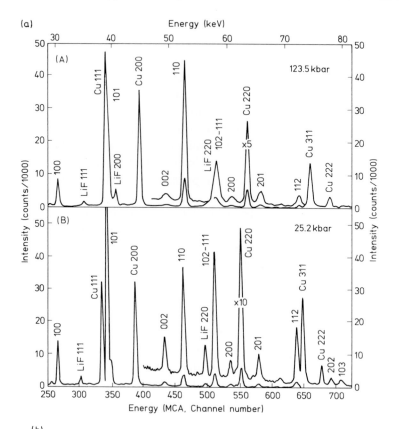

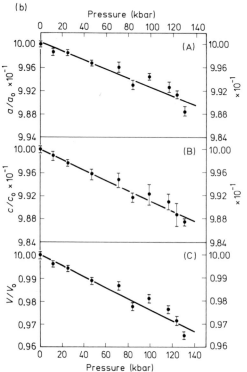

Figure 8-24. Studies of the compressibility of TiB_2 using energy dispersive diffraction and a Drickamer-type high pressure cell (after Hausermann et al., 1991). (a) Diffraction patterns taken at 123.5 and 25.2 kbar, the reflections labelled in black are of the added copper pressure calibration sample, some LiF contaminant peaks are also discernable; (b) reduced a and c lattice parameters (A and B) together with unit cell volumes (C) as a function of pressure.

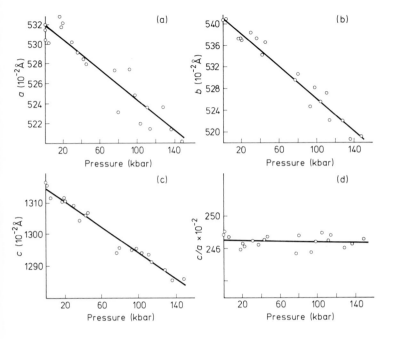

Figure 8-25. Energy dispersive data showing the isotropic compressibility of La_2CuO_4 using a diamond anvil cell (after Akhtar et al., 1988), $2\Theta = 9.81°$, open circles data points and full lines show the computed fit; (a)–(c) effect of pressure on the a, b and c lattice parameters respectively; (d) effect of pressure on the c/a axial ratio.

8.3.3.4 Particle Formation in a Liquid Environment

Very little information is available on the structural nature of crystalline particles as they form within solution and in particular the relationship between structure and growth kinetics. In principle X-rays should prove to be an effective probe for detecting crystal nucleation and growth. The much shorter wavelength of X-rays compared to visible radiation should allow the detection of crystal nuclei at much smaller sizes than when using conventional methods such as optical turbidity (e.g., Gerson et al., 1991 b). Using the high energy of synchrotron radiation the EDXRD technique has been applied (Gerson et al. 1990; Cunningham et al., 1990 and 1991) to collect powder diffraction patterns in situ on n-alkanes during crystallisation from solution. However, the general applicability of this technique has been limited to some extent by problems associated with solution heating.

Recent work by Doyle et al. (1991), summarised in Fig. 8-26, used a variation of the EDXRD technique in which a fast scanning monochromator is used to provide the wavelength range needed for the energy dispersion. The diffraction patterns Fig. 8-26a taken during the crystallisation of n-eicosane from a saturated solution of n-dodecane show the formation of well-resolved Bragg peaks demonstrating that in situ diffraction data can be obtained from n-eicosane crystals in the solution environment. Scans of diffraction intensity at constant energy Fig. 8-26b used to detect the onset of crystallisation revealed three distinct regions; firstly between 20 and 18 °C there is little change in the scattered intensity, indicating the absence of crystalline material. Between 18 and 14 °C the intensity increases approximately linearly as crystallisation occurs. Finally, below 12 °C the intensity becomes constant indicating effectively complete crystallisation of the material. The dotted line in Fig. 8-26 b represents the equivalent experiment carried out in the laboratory using an automated turbidometric apparatus (Gerson et al., 1991 a). Whilst the behaviour in the two experiments is similar the obvious discontinuity in the initial slope of the data

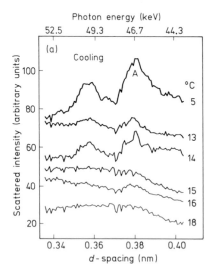

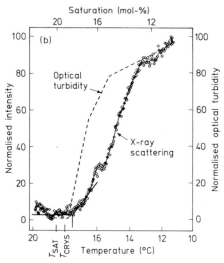

Figure 8-26. In situ data taken using the wavelength scanning EDXRD technique (after Doyle et al., 1991) with a detector angle 2Θ of $4°$ during the crystallisation of a saturated solution of 20% n-eicosane in n-dodecane; (a) a series of diffraction peaks at varying temperatures, (b) plot ($\diamond\diamond\diamond\diamond\diamond$) of the diffracted beam intensity for the diffraction peak labelled A in (a) as a function of cooling the solution at 0.5 °C/minute. Turbidometric data ($---$) are overlaied for comparison.

collected using X-rays provides potential direct evidence of significant ordering in the liquid phase prior to crystallisation.

8.3.4 Laue Diffraction

8.3.4.1 Introduction

Since the publication of the first Laue diffraction patterns in 1912 (Friedrich et al., 1912) this technique has been used routinely by metallurgists and mineralogists to define crystal orientation and symmetry. The general applications of the technique are well documented in the book of Amoros et al. (1975).

The Laue method uses a stationary crystal and a polychromatic X-ray beam. Synchrotron radiation is therefore an ideal source for this purpose and offers many advantages over traditional sources for routine use. The very broad spectral range provides diffraction patterns with a very high information content (many thousands of reflections) (Cruickshank et al., 1987) and the highly collimated beam offers a much superior resolution. Of greatest importance, however, is that the high intensity of the beam allows these patterns to be recorded from very small crystals ($\approx 100 \ \mu m^3$) (Andrews et al., 1988) in very short periods of time (0.1 – 10 s). This allows the technique to be used for the assessment of crystal structure and quality at the very earliest stages of materials preparation when only small crystals are available. It also offers an alternative prospect of dynamic and in situ examinations of structural changes during processing.

A further and great advantage of the synchrotron radiation Laue technique is that it can be used for molecular structure determination. In the early days, this prospect was discounted by Bragg (1949 and 1975) who argued that this was precluded because several orders of reflection could be superimposed on one Laue diffraction spot, thus preventing the evaluation of an electron density map. Cruickshank et al. (1987) have since shown that with synchrotron radiation the overlapping orders problem is not limiting. This has opened the way, in combination with the above properties, to molecular structure determinations using small crystals in short time scales (Andrews et al., 1988) and hence with labile materials. Dynamic changes in molecular structure can also be followed even in complicated systems such as catalysis in crystals of phosphorylase-b (Hajdu et al., 1987), and phase transitions between different molecular forms of insulin (Reynolds et al., 1988).

8.3.4.2 Instrumentation and Experimentation

The typical Laue diffraction system shown in Fig. 8-27 is that used on the Daresbury synchrotron radiation source. The front end consists of a water cooled tungsten carbide aperture followed by a set of foil attenuators to reduce radiation damage by the softer X-rays in the spectrum and an incident beam monitor.

This is followed by a collimation system comprising a fast shutter giving exposures down to a few milliseconds and a set of pinhole collimators. These provide a timed and defined beam to the crystal situated on a goniometer which provides two orthogonal axes of rotation. Data recording is carried out on high speed films such as Reflex 25. Exposure times are the orders of fractions of a second for normal size (≈ 1 mm) crystals and still only a few seconds or minutes for very small crystals.

8.3.4.3 Radiation Damage

With this method, as with all other synchrotron radiation methods which use the full spectral range of the synchrotron beam, the problems of radiation damage are ever present (Roberts et al., 1982; Bhat et al., 1982, 1984 and 1988). Radiation damage may be minimised or removed by correct use of the attenuator positioned in the front end of the system. With materials such as sodium chlorate, which we find to be ultrasensitive, the introduction of an aluminium absorber (200–1000 μm) removes sufficient of the softer radiation to allow reliable experiments for a reasonable experimental period; at least 10–100 times the experimental recording period. For less sensitive inorganic and organic materials, thinner absorbers yield satisfactory results. The most obvious consequence of radiation damage is the gradual development of a radial asterism around the diffraction spots. Assessment of the rate of development of this can give information on the rate of structural damage which can be correlated with other factors such as the formation of colour centres (Halfpenny et al., 1991). The assessment of this prob-

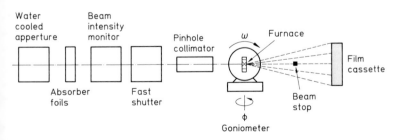

Figure 8-27. Schematic representation of the Laue diffraction camera at the Daresbury Synchrotron Radiation Source.

Table 8-5. The mosaic spread of some organic single crystals determined by synchrotron radiation Laue diffraction (after Roberts et al., 1983).

Material	Formula	Crystal system	Growth technique	Sample size (mm)			Accumulated dose (Mrad)	Mosaic spread	Remarks
				h	v	t			
benzoic acid	C_6H_5-COOH	monocl.	melt	23	21	6.0	0.5	$-24'$	uniformly strained
anthracene	$C_{14}H_{10}$	monocl.	melt	18	8	2.0	1.7	$-1.6°$	grossly strained, sample colurs
naphthalene	$C_{10}H_8$	monocl.	melt	12	6	2.0	5.2	$-2°$	grossly strained
p-terphenyl	$C_6H_5-C_6H_4-C_6H_4$	monocl.	melt	31	10	4.0	1.3	$-7°$	grossly strained
stilbene	$C_6H_5-(CH_2)_2-C_6H_5$	monocl.	melt	21	11	3.0	0.4	$-2.5'$	uniformly strained
terephthalic acid	$COOH-C_6H_4-COOH$	tricli.	solution	0.1	0.1	0.1	0.1	$-27'$	grossly strained
TTF−TCNQ[a]	$C_6S_4H_4 \cdot C_{12}N_4H_4$	monocl.	diffusion	10	3	0.1	9.9	$-70''$	no asterism, some resolution of micro-defects
benzophenone	$C_6H_5-CO-C_6H_5$	orthor.	solution	22	15	1.5	58.7	$-70''$	micro-defect resolution, some colouring of the sample
benzophenone	$C_6H_5-CO-C_6H_5$	orthor	melt	8	8	2.0	5.9	$-50'$	uniformly strained
benzil	$C_6H_5-(CO)_2-C_6H_5$	trigon.	melt	7	7	2.0	11.1	$-6'$	uniformly strained, sample colours
fluorenone	$C_6H_4-CO-C_6H_4$	orthor.	solution	15	12	1.5	47.8	$-70''$	no asterism, some resolution of micro-defects
carbazole	$C_6H_4-NH-C_6H_4$	orthor.	melt	13	5	1.0	12.9	$-70''$	no asterism, surface damage obscures bulk defects
adamantane	$(CH_2)_4(CH_2)_6$	cubic	solution	3	3	3.0	9.8	$-70''$	resolution of microdefects, region in sample misaligned by $-20'$
adamantane	$(CH_2)_4(CH_2)_6$	cubic	melt	8	8	2.0	7.0	$2.7°$	grossly strained
urea	$CONH_2$	tetrag.	solution	4	3	0.5	0.6	$-70''$	no asterism, some resolution of micro-defects
urea	$CONH_2$	tetrag.	prilled	2.5	2.5	2.5	0.3	$-1'$	several crystallites (<10), low asterism for individual grains
urea/NaCl	$CONH_2 \cdot NaCl$	orthor.	solution	13	6	1.0	1.0	$-10'$	some localised surface strain
pentaerythritol	$C-(CH_2OH)_4$	tetrag.	solution	12	8	2.0	2.4	$-8'$	low asterism, some resolution of macro-structure
oxalic acid dihydrate	$COOH-COOH \cdot 2H_2O$	orthor.	solution	21	5	3.0	0.3	$-11'$	thick sample, low asterism, some resolution of surface defects

n-eicosane	$CH_3-(CH_2)_{18}-CH_3$	melt	tricli.	5	5	2.0	5.2	$-2°$	grossly strained, significant radiation damage
PTS[b]	$R=CH_3-C_6H_5-SO_2-O-CH_2-$	solution	monocl.	5	3	2.0	3.5	$-10'$	low localised strain
TCDU[b]	$R=C_6H_5-NH-CO-O-(CH_2)_4-$	solution[c]	monocl.	3	3	1.0	5.9	$-70'$	uniformly strained
DDEU[b]	$R=C_2H_5-NH-CO-O-(CH_2)_3-$	solution	monocl.	6	2	0.1	1.7	$-1.9°$	grossly strained

[a] Tetrathiofulvalene-tetracyanoquinodimethane.
[b] Single-crystal diacetylene polymers; general formula of monomer $R-C\equiv C-C\equiv C-R$.
[c] The monomer crystals were grown from solution, before polymerisation.

lem is an essential preliminary to all such synchrotron experiments since many do involve the existence or development of strain in samples.

8.3.4.4 Assessment of the Quality of Organic Crystals

The Laue method is a useful and simple technique for the assessment of crystal quality. Beyond the limit of X-ray topography with synchrotron radiation ($< 500''$ mosaic spread) (see Sec. 8.6) there are few techniques available to the crystal grower which can be used to assess the quality of large crystals produced when developing a new growth technique. Such information is essential in order to assess the influence of changing growth parameters on increasing perfection. Small beam Laue diffraction can be used to search the volume of a crystal to produce information on the local perfection. The larger synchrotron beam (≈ 2 cm diameter at the Daresbury SRS) allows the complete assessment of crystals of this size in one exposure of around a few minutes duration. The overall perfection of the crystal is demonstrated by the quality of the image. A net and well-defined image is typical of a crystal of high quality which is potentially useful for more detailed X-ray topography. An increase in mosaic spread is evidenced by an asterism of the image streaking radially across the plate. Analysis of the degree of asterism (Amoros et al., 1975) can be used to calculate mosaic spread. Details of a typical study (Roberts et al., 1983) of a range of organic crystals are given in Table 8-5.

8.3.4.5 Solid-State Polymerisation of PTS

The monomer 2,4-hexadiynediol-bis-(p-toluene sulfonate) (PTS) undergoes polymerisation in the solid state by a single crystal to single crystal transformation. The

(i)

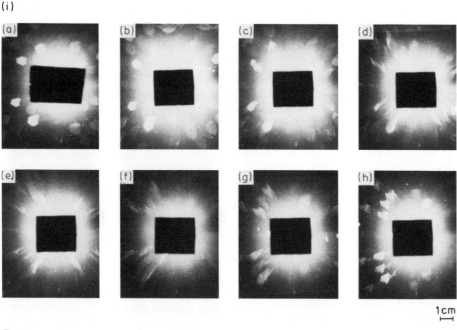

Figure 8-28. (i) Selected Laue patterns for PTS from a series taken on Polaroid film recording the progress of radiolytic polymerisation after cumulative radiation doses of; (a) 77 Mrad, (b) 185 Mrad, (c) 259 Mrad, (d) 458 Mrad, (e) 551 Mrad, (f) 619 Mrad, (g) 657 Mrad, (h) 716 Mrad (after Dudley et al., 1991). (ii) The kinetics of polymerisation of PTS expressed as a change in relative spot area (x) (324-reflection) versus accumulated time in the beam A. Unfiltered beam at 295 K: B. Unfiltered beam at 150 K: C. Filtered beam (715 mm Al) at 295 K.

reaction can be initiated radiolytically, photochemically and thermally. The progress of reaction can be followed by the use of either full beam or narrow beam "spot" Laue patterns. Figure 8-28a shows a range of Laue patterns from a typical experiment. The patterns were recorded on polaroid film and represent a selection from a total of 60 taken over a 30 minute period. The original and final patterns which represent those of the monomer and polymer respectively are distinctive and confirm the perfection of each. It will be noted that as the reaction initiates and proceeds strain develops due to the formation of polymer in the crystal. This causes an asterism. As the reaction proceeds to completion this strain relaxes and the pattern gradually develops to that of the polymer. This experiment is a radiolytic experiment carried out using the synchrotron beam as both the initiator and probe of the reaction.

Analysis of the rate of development of the asterism allows the definition of the kinetics of the process. Figure 8-28b shows that the influence of decreasing temperature on the induction time to nucleation of the reaction was less than that obained by filtering the softer X-ray components from the beam. By using the latter method, the induction time could be increased sufficiently to allow the study of thermal and UV polymerisation well before rapid radiolytic polymerisation was initiated. Since each crystal image is a topograph (see Sec. 8.6) this also allowed the study of the influence of lattice perfection on the polymerisation process (Dudley et al., 1985, 1986, 1990, and 1991).

8.3.4.6 Influence of Strain on the Growth Rate of Secondary Nucleated Particles

A major problem in the performance of industrial crystallisers is the growth rate dispersion of secondary nuclei (attrition fragments). Many of these fragments show a zero growth rate. Studies of the structural perfection of micro crystals (5–25 mm diameter) using Laue diffraction techniques have shown that the growth rate of the crystals at constant supersaturation is proportional to the mosaic spread of the crystal (Fig. 8-29). This influence has been found for sodium chlorate (Ristić et al., 1988), potash alum (Ristić et al., 1991), sodium nitrate and Rochelle Salt (Mitrovic et al., 1990).

These results parallel similar observations on other small crystals of more complicated materials, such as proteins and zeolites, for which a cessation of growth is found when they have attained a certain size (Andrews et al., 1987). Those which cease to grow were found to have the largest mosaic spread.

Figure 8-29. (a) Laue photographs of two different Rochelle salt crystals taken using the SRS "white" beam: wavelength range approximately 3–25 nm, collimator 0.2 mm, crystal film distance 50 mm. (i) Asterism obtained for a crystal with growth rate 114.2 nm s^{-1}; (ii) stronger asterism for a crystal with growth rate 52.3 nm s^{-1}. (b) The variation of growth rate for Rochelle salt crystals as a function of mosaic spread (after Mitrovic et al., 1990).

(a)

(i)

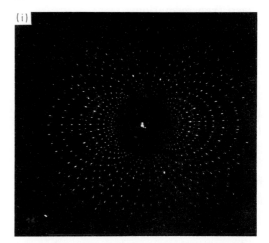

(ii)

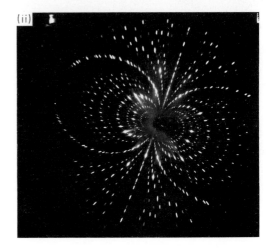

(b)

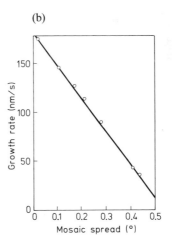

8.3.4.7 Probing Structural Changes During Phase Transformations

Although the Laue topographic technique (Tuomi et al., 1974) has been used for investigations of solid-state transformation (Bordas et al., 1975, Aleshko-Ozhevsij 1982 and 1983; Zarka, 1983; Gillespie et al., 1989), there have been surprisingly few applications of the use of synchrotron radiation Laue patterns to study phase transitions (El Korashy et al., 1987) or lattice pseudosymmetry (Docherty et al., 1988) in small molecular systems.

The applicability of this technique to solid-state phase transitions has recently been demonstrated (Bhat et al., 1990) for the three systems: potassium tetrachlorozincate (KZC), pentaerythritol and caesium periodate. These three systems were chosen to include a typical range, mechanistically, of the types of phase transition currently of interest and hence to demonstrate the broad applicability of synchrotron Laue techniques to the study of solid-state transitions.

As an example of this work we show (Fig. 8-30) four Laue photographs taken of

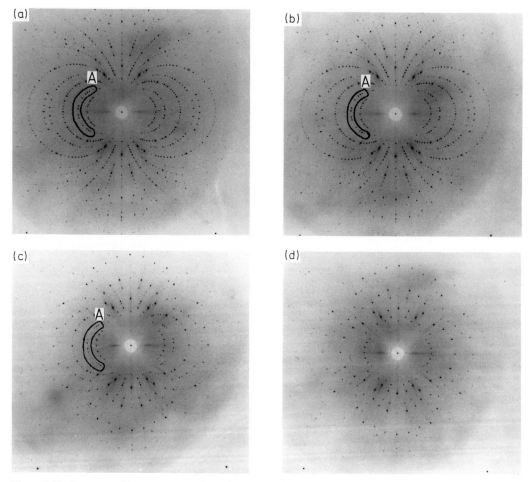

Figure 8-30. Sequence of Laue patterns taken of KZC as a function of temperature with the beam axis along the [010] lattice direction (a) 60.3 °C; (b) 139.8 °C; (c) 278.9 °C; (d) 328.0 °C (after Bhat et al., 1990a).

KZC as a function of temperature. The data were recorded for the room temperature ferroelectric phase (a), at the beginning of the incommensurate phase (b), towards the end of the incommensurate phase (c) and in the paraelectric phase (d). Zones of satellite spots (labelled A) are clearly visible for the ferroelectric phase (a). These super-lattice reflections change in relative intensities [(b) (c)] within the incommensurate phase and disappear completely in the paraelectric phase (d).

8.3.4.8 Application to Crystal Structure Determination

As Laue diffraction patterns do not suffer from the problems associated with overlapping reflections, and due to the excellent properties of synchrotron radiation, synchrotron Laue patterns can be used for crystallographic and molecular structure determination using microcrystals of considerably smaller size than are necessary for traditional methods of examination. In fact, specimens usually regarded as powder particles are potentially useful.

Combined with unit cell parameters, which still have to be obtained from a monochromatic pattern, Laue data have been used to calculate difference Fourier maps of proteins (Hajdu et al., 1987) and to solve small crystal structures by the Patterson method (Harding et al., 1988).

Film can be recorded in one second or less giving the intensities of 2000 or more reflections.

Typical of the complexity of the type of compounds which have been assessed are $C_{25}H_{20}N_2O_2$ ($R = 0.05$) (Helliwell et al., 1989), $Mo_5S_2O_{23}(NEt_4)_4PhCN$ ($R = 0.10$) (McGinn, 1990), $Rh_6(CO_{14}Ph_2PCH_2PPh_2$ ($R = 0.16$) (Clucas et al., 1988), $FeRhCl(CO)_5Ph_2PC(=CH_2)PPh_2$ ($R = 0.14$) (Harding et al., 1988). The accuracy of

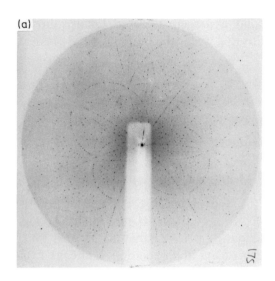

(a)

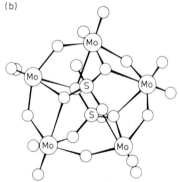

(b)

Figure 8-31. (a) Laue diffraction photograph of a small crystal of tetraethyl pentamolybdo-phosphate, taken on station 9.7 on the SRS Wiggler beam line at Daresbury Laboratory, U.K. Crystal size $100 \times 140 \times 15 \, mm^3$, crystal to film distance 48 mm, exposure time 0.2 s. (b) The molecular structure derived from analysis of these data. (Reproduction by permission of Drs. M. M. Harding and S. J. McGinn, Liverpool University.)

the determination in each case is defined by the refinement factor R. The crystal sizes were of the order of $100 \times 150 \times 15 \, \mu m^3$. Figure 8-31 shows a typical example of the Laue pattern and derived structure for one of these materials.

Some examples of synchrotron Laue studies of phase transitions under high pressure, for instance structural studies of solid hydrogen at room temperature at 25 GPa, are cited in Volume 5, Chap. 8, Section 8.2.3.

8.4 Diffraction and Scattering from Real Surfaces and Interfaces

8.4.1 Introduction

A comparison of the scattering lengths of X-rays (about 10 μm) and electrons (about 1 nm) shows that in a normal X-ray experiment the radiation scattered from the bulk material will dominate over that from a surface monolayer by about 6–7 orders of magnitude. This inherently low signal, specific to the surface regions, can be overcome by using synchrotron radiation. This has the additional advantage that the high energy and intensity of the radiation enables the structure of interfaces buried under dense overlayers to be examined non-destructively in a variety of environments in addition to high and ultra-high vacuum.

8.4.2 Information Content Available from Surface Diffraction Data

Surface diffraction can be used to assess and characterise the structural order in both "in-plane" and "normal" directions with respect to the interface under examination. In-plane structure can be deduced from an equatorial diffraction scan about an axis parallel to the surface normal. The depth sensitivity of such a measurement can be varied by changing the incident glancing angle and is dramatically enhanced under conditions of total external reflection where the X-rays are confined to the surface regions of the sample (≈ 0.2 nm) as a damped evanescent wave. Measurement of the width of Bragg surface diffrac-

tion features provides information on the coherence length or domain size of the interfacial structure. For solid-solid interfaces, scans in reciprocal space with respect to the underlying lattice of the substrate provide direct evidence of reduction in symmetry, as might be expected if the surface reconstructs. They also show whether the layer is incommensurate or not. Such equatorial scans together with scans of the reciprocal lattice normal to the interface, along the Bragg truncation rods, indicate the structural registry of substrate and overlayer. The intensity of Bragg diffracted surface reflections can be analysed using conventional crystallographic techniques. An inherent advantage of this technique is that the data can be readily analysed using kinematic scattering theory. Dynamic theory is not implicitly required.

Diffraction scans can be complemented by high resolution measurements of X-ray reflectivity, made by varying the angle of incidence of the X-rays on the sample surface. Measurements in the angular range $(0.5–10)\,\phi_c$, where ϕ_c is the critical angle, allow the atomic structure and structural variation normal to the surface to be determined. Structural information can be obtained directly from the interference of Kiessig fringes observed in reflectivity profiles (Kiessig, 1931; Koenig and Carron, 1967; Isherwood, 1977). This is shown schematically in Fig. 8-32 for a three layer system (air/thin-layer/substrate). A scan of reflectivity versus incidence angle firstly shows the total external reflection from the top surface followed by a series of Kiessig fringes due to interference between the specularly scattered X-rays from the "top" and "buried" interfaces. From the angular positions of the interference fringes the layer thickness (t) and the mean electron density of the film can be calculated (Koenig and Carron, 1967).

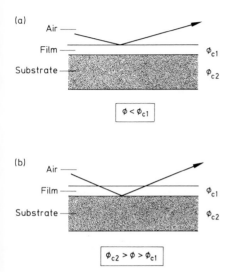

(a)

Air

Film

Substrate

ϕ_{c1}

ϕ_{c2}

$\phi < \phi_{c1}$

(b)

Air

Film

Substrate

ϕ_{c1}

ϕ_{c2}

$\phi_{c2} > \phi > \phi_{c1}$

Figure 8-32. Reflection geometry in a three layer (i.e., air/film/substrate) system in which the substrate is more dense than the film. ϕ – angle of incidence, ϕ_{c1} – critical angle for the film and ϕ_{c2} – critical angle for the substrate. (Note: $\phi_{c1} < \phi_{c2}$).

An analysis of fringe positions is only optimal where surface layers are chemically homogeneous as a function of depth. In systems exhibiting a graded interface the complete reflectivity profile needs to be fitted to a model which sums the reflectivity contributions from each sub-layer of the surface. Using such a model the reflectivity from the surface can be calculated from a classical Fresnel recursion relationship (e.g., Parratt, 1954; Wainfan et al., 1959).

8.4.3 Instrumentation

The experimental recording of surface diffraction and X-ray reflectivity data requires the use of a 5-circle diffraction geometry. Figure 8-33 shows a schematic view of the surface diffraction facility available on station 9.4 of the Daresbury SRS.

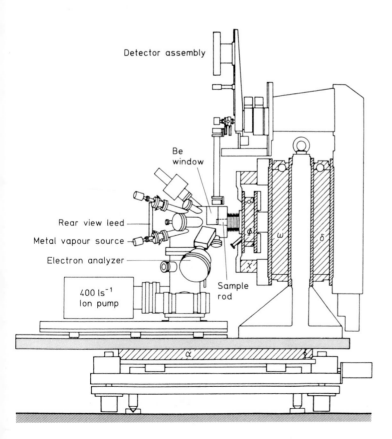

Detector assembly

Be window

Rear view leed

Metal vapour source

Electron analyzer

400 ls⁻¹ Ion pump

Sample rod

Figure 8-33. Schematic view of the 5-circle goniometer for surface X-ray diffraction used on station 9.4(b). This is shown with the UHV surface science analysis chamber (after Norris, Moore and Harris, unpublished).

This comprises a high resolution 5-circle diffractometer together with monochromator and a 2-circle analyser stage. This diffractometer has the capacity to hold extremely large and massive environmental stages. The experimental requirements for X-ray reflectivity measurement alone are less severe and 2-circle goniometry (see Sec. 8.3) can also be used.

8.4.4 Lattice Coherency in Group III–V Quantum-Well Structures

The definition of materials characterisation procedures for the preparation of semiconductor multilayers is of crucial importance for the development of novel low dimensional structures for applications in integrated opto-electronics. Surface diffraction is being increasingly applied to the characterisation of lattice coherency between the substrate and its heteroepitaxial overlayer (Ryan et al., 1987; Lucas et al., 1988; Jedrecy et al., 1990).

Figure 8-34 summarises some of the results of a study by Ryan et al. (1987) in which the structure of 23 nm layer of InGaAs (nominal composition $In_{0.53}Ga_{0.47}As$) deposited on InP was examined. The sample (a) was prepared by molecular beam epitaxy (MBE) and was studied using a glancing angle of incidence ($\approx 2.5°$) corresponding to a penetration depth of ≈ 240 nm. The triple axis (i.e. monochromator, sample and analyser) diffraction configuration enables a very high resolution in the two dimensions of the scattering plane so that the 2-d intensity distribution (b) of scattered X-rays around a Bragg reflection can be seen. It can be seen that the main Bragg peak itself is split due a slight lattice mismatch between the Fe doped InP substrate and its MBE grown InP buffer layer which induces a slight (0.03%) tetragonal distortion in the epitaxial overlayer. The slight skewing of the line of satellite peaks from the crystallographic axis indicates that the crystal face has been cut at an angle of $\approx 2°$ InP(100) in the [110] direction.

Examination of the environment around the InP 440 Bragg reflection of the sample reveals a series of weak satellite peaks arising from the capping layer of InP. The amplitude and periodicity of these satellites are modulated by a similar set of weaker and broader satellites (c) with roughly three times the period of the capping layer fringes originating from the InGaAs quantum-well layer. Fitting this experimental X-ray pattern produces a structural model for the interface which is in good agreement with that predicted from MBE growth experiments viz. InP cap thickness 68 (± 1) nm, InGaAs layer thickness 23 (± 1) nm, lattice parameter mismatch ($\Delta a/a$) $2.73(1) \times 10^{-3}$, rms surface roughness 2 (± 0.2) nm and rms interlayer interface roughness 1.2 (± 0.2) nm.

8.4.5 Ge/Si (001) Strained-Layer Semiconductor Interfaces

The properties of hetero-epitaxial thin films prepared by MBE for electronic device applications depend on a number of factors including layer perfection. Crystal strain provided by deliberately mismatching the lattices of the substrate and epitaxial film can be used to modify and optimise the electronic or optical properties of the material (e.g., O'Reilly, 1989). The degree of strain that can be accommodated depends inherently upon the layer thickness, with mismatch dislocations being formed for thicker overlayers. Studies have been made of systems such as lattice-matched $NiSi_2/Si(111)$ (Robinson et al., 1988) and Ge/Si (001) (Williams et al., 1989; Macdonald et al., 1990; Macdonald, 1990).

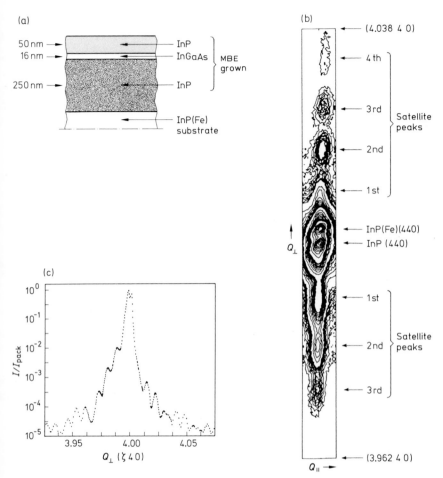

Figure 8-34. Surface diffraction investigation of a single quantum well structure after Ryan et al. (1987). (a) Sample details comprising InP capping layer, InGaAs quantum well and InP buffer layer, epitaxially grown by MBE on Fe doped InP. The layer thicknesses were estimated from the growth conditions. (b) An iso-intensity contour plot of the observed X-ray scattering in the region of the InP 440 Bragg reflection. The data are presented on a quasi-logarithmic scale over four decades of intensity to avoid over-contouring the main peak. $Q_\perp$, $Q_\parallel$ are the scattering vectors normal, parallel to the surface. (c) The intensity distribution along the [100] axis showing the modulation of the satellite Bragg intensities.

Figure 8-35 shows some of the results of a surface diffraction study of the effects of coverage of Ge films on Si (001) substrates (Williams et al., 1989; Macdonald et al., 1990; Macdonald, 1990). The data show a series of radial scans through the 200 Bragg peak as a function of the numbered monolayers (ML) of Ge deposited in situ from a Knudsen cell using a dedicated MBE system (Vlieg et al., 1987). Note that 1 ML ≈ 0.14 nm. It can be seen that the diffraction peak profile remains unchanged for a coverage less than 3 ML which indicates that the Ge epitaxy is coherent to the Si substrate. The lack of peak broadening illustrates the high crystalline quality of the overlayer. At layer thicknesses of ≈ 4 ML the development of a weak secondary feature at a lower scattering vector demonstrates the onset of strain relaxation. This

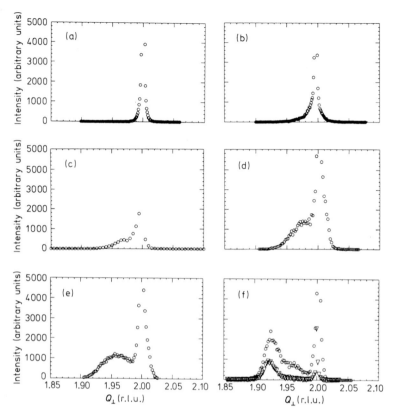

Figure 8-35. Surface X-ray diffraction scans as a function of coverage of MBE grown epitaxial Ge on Si (001) (after Macdonald, 1990). The intensity is plotted on an arbitrary scale with the Bragg peak intensity being $\approx 10^5$ for 11 ML, circles denote data for angles of incidence and exit of 0.09° and 0.13° respectively whilst triangles represent the same scan for 0.07° and 0.05° respectively leading to a reduced effective penetration depth. (a) Clean Si, (b) 3.9 ML of Ge, (c) 4.7 ML of Ge, (d) 5.5 ML of Ge, (e) 7.1 ML of Ge, (f) 11.0 ML of Ge. "r.l.u." stands for "reciprocal lattice unit".

increases in intensity and in separation from the first peak upon further deposition for coverages up to ≈ 11 ML, at which point the peak closely matches that expected for bulk Ge. Such a gradual shift indicates a gradual relaxation of strain after exceeding the 3–4 ML critical thickness. The lack of definition in the diffraction pattern between these two features is evidence for a strain profile within the epilayer. By reducing the effective penetration depth from 10 to 4 nm (Fig. 8-35f – the scan profile is denoted by triangles) the atomic layers close to the interface are still strained while the uppermost layers are almost fully relaxed.

8.4.6 In Situ Structure of Au (100) Under Electrochemical Potential Control

The structure of the electrode interface under conditions of potential control remains one of the fundamental goals for electrochemists. In an electrochemical environment, the electric field or charge at the surface can be controlled by changing the potential across the polarised "double layer" with fields as high as 10^7 V/cm being easily accessible. In electrochemical experiments the corresponding change in surface charge can easily exceed 0.1 electron per surface atom and can provide the driving force for the kind of surface reconstruction which is well known to exist on well-characterised metal surfaces under ultra-high vacuum conditions.

Surface diffraction through its capability to probe in situ through the electrolyte environment is the only technique which can examine surface structure at the electrochemical interface. Recent developments have seen the use of surface diffraction to probe a number of electrochemical systems (Melroy et al., 1988; Samant et al., 1988; Ocko et al., 1990).

Ocko et al. (1990) have investigated the surface structure of an Au (100) electrode surface in an aqueous solution of perchloric acid, as a function of the applied electrode potential. Cyclic voltammetry of the Au (100) surface in $HClO_4$ solution exhibits an anodic current peak at ≈ 0.7 V versus a Ag/AgCl electrode. From a number of other measurements (Hamelin, 1982; Kolb and Schneider, 1985, Friedrich et al., 1989; Zei et al., 1989) it has been inferred that a change in potential can induce a relaxation of the expected (e.g., Palmberg and Rhodin, 1967) hexagonal surface reconstruction. Figure 8-36 shows some of the data from the study of Ocko et al. (1990). Reflectivity data (Fig. 8-36a) confirms that the hexagonal reconstruction, as observed in ultra high vacuum, does indeed take place in solution at negative electrode potentials but, as implied from the cyclic voltametry, this reconstruction disappears at positive potentials to yield a lower density surface layer. Returning to a positive potential results in the recovery of the hexagonal reconstruction.

Additional and complementary information is obtained from surface diffraction measurements (Fig. 8-36b) which show formation of surface domains which have a coherence length of ≈ 30 nm. These are misoriented $\approx 0.8°$ with respect to the underlying Au (100) lattice.

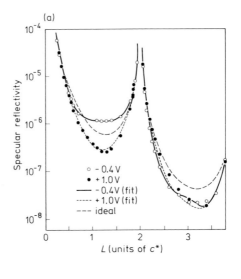

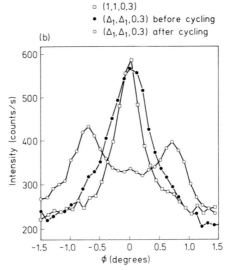

Figure 8-36. In situ surface examination of Au (100) under conditions of potential control (after Ocko et al., 1990). (a) Absolute reflectivity data at an electrode potential of -0.4 V (open circles) and 1.0 V (solid circles). The long-dashed line is for ideal termination with no rms displacement amplitude. The solid and short-dashed lines are fits as described in the text. (b) A comparison between glancing-angle X-ray diffraction rocking curves at an electrode potential of -0.4 V before (closed circles) and after cycling. The potential induces the formation of surface domains rotated with respect to the uncycled lattice of $\approx \pm 0.8°$.

8.4.7 Crystal/Solution Interface Structure of ADP (100) During Growth

Despite the seminal work of Burton, Cabrera and Frank (1951) on the structure of the crystal growth interface there is a critical absence of detailed experimental data to verify many of the currently proposed models. This, for the most part, has been due to the lack of a suitable in situ structural probe of the crystal structure of the growing crystalline interface under a layer of solution. Whilst microscopic observations can reveal screw dislocation sources of surface steps and surface roughening the influence of surface chemistry and interface structure on these processes remains totally unknown.

Experiments by Cunningham et al. (1990a and 1990b) have demonstrated that surface X-ray diffraction can be successfully used for the in situ assessment of growth and dissolution processes. This investigation involved an in situ examination of the (100) surfaces of ammonium dihydrogen phosphate ($NH_4H_2PO_4$ or ADP). X-ray rocking curves of the crystal were recorded under a variety of surface conditions; in air, under a covering of n-hexane and under a saturated aqueous solution of ADP. In air, the rocking curves measured for the ADP crystals were found to be extremely sharp with a half width of slightly less than 4 seconds of arc. Allowing for the slight angular dispersion in the triple-axis mode when the diffraction planes of the monochromator and sample crystals are not exactly matched this compares favourably with a theoretically predicted value of 2.5 seconds of arc. When covered with an insoluble solvent, in this case n-hexane, there was no noticeable variation in the diffraction profile, but on replacing the n-hexane with a saturated solution of ADP the in situ reflection curve

broadened asymmetrically to a width of about 14 seconds of arc.

On dissolution of the crystal induced by a step increase in the growth cell temperature, the reflection width decreased again showing the process to be reversible. Rapid recording of the rocking curve as dissolution proceeded showed this process to occur within several minutes and to reach equilibrium within one hour. Figure 8-37 shows the data taken during the subsequent regrowth of the crystal surface initiated through cooling of the cell. The results mirror observations of the dissolution experiment.

These experiments show evidence for an interfacial layer, of substantial thickness, when the crystal surface is in equilibrium with its saturated solution. An interfacial layer has also been observed by other groups using light scattering (Steininger and Bilgram, 1990). It might be expected

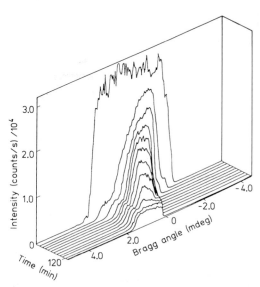

Figure 8-37. 200 rocking curves (wavelength ≈ 0.062 nm) recorded in situ from a nearly perfect single crystal of ADP under ≈ 6 mm of saturated aqueous solution showing the reduction of mosaic spread during crystal growth (after Cunningham et al., 1990a).

that such an interface would consist of weakly bonded layers. Additionally these layers would be highly labile and at the outset of crystallisation or dissolution thermal fluctuations would destroy the interfacial ordering and hence the diffraction conditions for this region. This supposition is supported by the study of the dynamic change in reflection curve shape and width. The transformation is comparatively rapid. During both growth and dissolution experiments the rocking curves reduce to half their initial width in under three minutes. After this initial reduction in width there follows a longer period as the interface returns to an equilibrium state – possibly involving an ordered interfacial structure.

8.4.8 Thin Polyphenylene Films Prepared from Different Precursors

Polyphenylene films formed on indium/tin oxide (ITO) electrode layers have been found to be effective substrates for super birefringent-effect liquid crystal displays (Scheffer and Nehring, 1984). These displays have been found to offer a significant advance in performance, such as improved contrast ratio and viewing angle, than that provided by the twisted nematic displays used in current technology. The polyphenylene inner substrate pre-orientates the liquid crystal molecules up to 25° (Nerhing et al., 1987; Holmes et al., 1989). This pre-tilting enables twist angles between the liquid crystal layers of up to 270° to be achieved without causing two dimensional instabilities in the liquid crystal matrix. An important benchmark in assessing the likely polymer film performance is the film density which reflects the packing density of the polymer molecules. A high density implies a tighter molecular packing, a smaller pre-tilt angle and hence less optimum device performance.

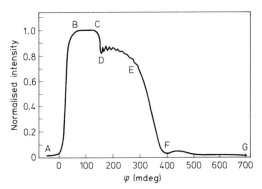

Figure 8-38. Reflectivity curve for polyphenylene film (carbonate precursor) at 0.1286 nm (after Roberts et al., 1990).

Figure 8-38 shows some of the results (Roberts et al., 1990) of the structural examination of polyphenylene thin films on ITO coated glass substrates deposited from either methylacetate and methylcarbonate precursors (Ballard et al., 1988). The reflectivity profile exhibits a number of distinct regions, i.e., total reflection (see also this Volume, Chap. 9) from the polyphenylene film surface (AB), Kiessig fringes from the polyphenylene film (BC) and Kiessig fringes from the ITO substrate. The fringe positions were fitted using the method of Koenig and Carron (1967) and the film thickness and densities calculated (Table 8-6). The film density for the films prepared from the methylacetate precursor is much lower than that from the methylcarbonate precursor which implies that the latter has a much lower molecular packing density. This analysis is also in good agreement with the optical characterisation of the pre-tilt angles and demonstrates the utility of this non-destructive technique.

8.4.9 Photodissolution of Silver on Chalcogenide Glasses

Cowley and Lucas (1990) used X-ray surface reflectivity to characterise the diffu-

Table 8-6. Critical angles, thicknesses ([a]), electron densities and film densities for polyphenylene thin films prepared from carbonate and acetate precursors (after Roberts et al., 1990). Thickness data produced by elipsometry measurements ([b]) are also presented.

| Precursor | Critical angles | | Film thickness | | Electron density (Ne) | Density | Pre-tilt angle |
	(mdeg) 0.1286 nm	(mdeg) 0.1349 nm	([a]) (nm)	([b]) (nm)	(m^{-3})	$(kg\ m^{-3})$	(°)
carbonate	147.5	150.7	166.9	170.0	3.99×10^{-30}	$1260\ (+/-2)$	16
acetate	140.1	143.1	153.0	157.0	3.60×10^{-30}	$1130\ (+/-2)$	20

sion of silver into chalcogenide glasses under optical stimulation which is of topical interest for its potential as a lithographic system. Figure 8-39 shows a series of measurements (Lucas, 1989) made as a function of the photo-stimulation of a sample comprising 200 Å of silver deposited onto a

320 nm thick layer of $As_{30}S_{70}$ on float glass. Analysis of the reflectivity spectra using Parratt's method (Parrat, 1954) indicate that before illumination (a) the surface layer comprises ≈ 2.3 nm of oxide and 18.4 nm of silver on the chalcogenide glass.

After 100 s of illumination (b) the reflectivity oscillations are reduced in amplitude but maintain essentially the same angular positions, which implies that whilst the silver layer thickness has not significantly changed during initial illumination the density of this layer has. This reflects the initial diffusion of silver into the chalcogenide glass through grain boundaries.

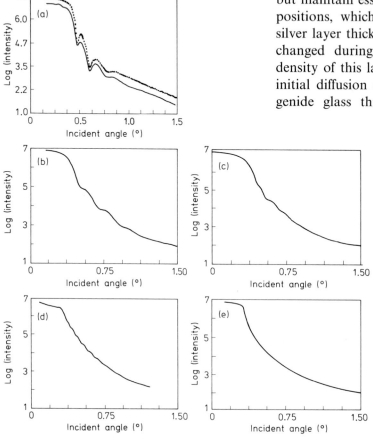

Figure 8-39. X-ray reflectivity studies of the photo-stimulated diffusion of silver into $As_{30}S_{70}$ glass (after Cowley and Lucas, 1990). (a) Experimental reflectivity curve (●●●●) before illumination overlaied with computer fit (———) using Parratt's equation; (b–e) reflectivity curves after illumination with a 200 W mercury lamp for 100, 160, 320 and 600 s respectively.

With increasing illumination times (c–e) the change in the fringe positions clearly shows the reduction in silver film thickness associated with the inter-diffusion process and the resultant formation of a silver/ glass reacted layer which is ≈ 4 times thicker than the initial silver film.

On the basis of this work Cowley and Lucas (1990) suggested that the photostimulated interdiffusion takes place initially at grain boundaries or voids. The average thickness of the silver layer remains constant but its average electron density is reduced to give a very inhomogeneous reaction product layer. As the reaction proceeds from 100 to 300 s, silver islands are formed, while the product layer develops into a reasonably homogeneous layer. Even at the end, silver or silver-oxide islands remain. Upon further illumination the reaction product layer moves into the chalcogenide region destroying any sharp interface between these layers.

8.4.10 Freely Supported Films of Octylcyanobiphenyl

The use of reflectivity techniques need not be restricted to investigations of sur-

faces. Gierlokta et al. (1990) have demonstrated an alternative way of characterising the structure of the liquid/air interfaces using freely suspended films. Using this approach films with thicknesses varying from a few to many thousands of molecular layers can be assessed. Figure 8-40 shows the kind of data that can be obtained from systems such as thin films of octylcyanobiphenyl.

The data shows two characteristic features in these reflectivity profiles; firstly the smectic interlayer separation gives rise to a Bragg-like central peak which is broadened due to the finite number of layers. Secondly, away from the Bragg peak well resolved Kiessig fringes due to interference of the X-rays reflected from the two liquid/ air interfaces can be seen and can be used to determine the total thickness of the layer. As this method allows independent determination of both intermolecular layer separation and the total film thickness the approach is very sensitive to the out-of-plane structure of the surface layers which in Gierlotka et al.'s (1990) study indicates an extended top layer at both of the interfaces with a thickness ≈ 1.36 times the spacing of the interior layers.

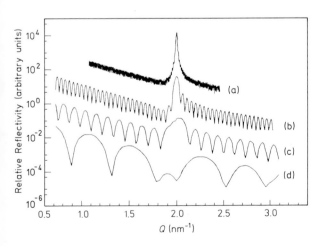

Figure 8-40. Reflectivity data for free standing films of octylcyanobiphenyl for 160 (a), 43 (b), 16 (c) and 4 (d) molecular layers in thickness; (a), (b) and (c) are shifted for reasons of clarity (after Gierlokta et al., 1990).

8.5 X-Ray Standing Wave Spectroscopy

8.5.1 Introduction

X-ray standing waves (XSW) are generated through dynamical interference between the incident and diffracted X-rays during Bragg diffraction from nearly perfect single crystals (Batterman, 1964). This is shown schematically in Fig. 8-41 which also illustrates that by adjusting the crystal over its range of diffraction causes the X-ray standing waves to lie between the atomic layers (inset A) of the structure or to be coincident with them (inset B). Experimentally (c), by recording the reflected and fluorescent radiation emitted from the crystal *at the same time*, the position of the incorporated atoms, surface films etc. with respect to the crystal lattice can be determined (Cowen et al., 1980; Bedzyk et al., 1984; Materlik and Zegenhagen, 1984; Materlik et al., 1984; Hertel et al., 1985; Funke and Materlik, 1985). In this respect XSW and EXAFS (see Sec. 8.2) are complementary techniques as they provide information concerning both the long and short range correlations for specific atoms within crystalline solids.

8.5.2 Surface Coordination of Adsorbed Br on Si (111)

Funke and Materlik (1985) have used XSW to investigate the atomic coordination of adsorbed Br atoms on Si (111). The result of the standing wave measurement is shown in Fig. 8-42 in which the Bragg reflectivity is shown superimposed with total bromine $K\alpha$ fluorescence yield. The best fit of the experimental data is to a model which assumes a single dominant adsorption site with an atomic coverage of 35 ($\pm$ 2)% with the Br atoms occupying a position ≈ 0.201 (± 0.003) nm from the Si (111) diffraction

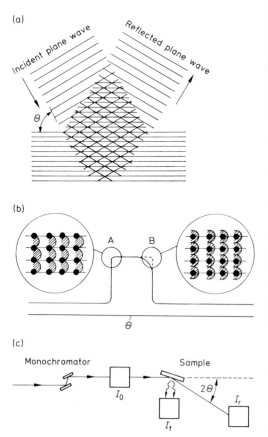

Figure 8-41. X-ray standing waves: (a) interference of incident and Bragg diffracted X-rays to give rise to a standing wave field; (b) changes to the angle of incidence move the standing waves from lying between the atomic layers (inset A) of the structure to be coincident with them (inset B); (c) experimental set up for standing wave measurements. I_0, I_r and I_f are the incident, reflected and fluorescent intensities respectively, Θ is the Bragg angle (after Cunningham et al., 1991 b).

planes. This result shows that bromine is not adsorbed at the position on-top of the outermost surface Si atoms since assuming a Si-Br covalent bond length of 0.218 nm would give a coherent position of 0.257 nm with respect to the Si (111) surface.

Figure 8-43 shows a (112) atomic projection of the Si lattice indicating the position of the incorporation site. The 0.201 nm position determined from the XSW data

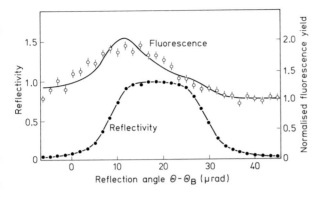

Figure 8-42. Angular dependence of the Si(111) reflectivity yield overlaied with fluorescence yield. In this case 45 L (2.10⁻⁷ mbar for 300 s) Br were deposited through the gas inlet valve. Measuring time was 30 minutes.

is indicated by point-dashed lines. The dashed circle is centred at this distance above an open threefold coordination site and illustrates the close agreement between the measured adsorption site and a model in which bromine is adsorbed ionically as Br⁻ with an ionic radius of 0.196 nm. Such a model is also in agreement with the measured coherent coverage of 0.35 ML, which due to steric hindrance or electronic repulsion factors, allows at most, one third of a monolayer to be adsorbed as long as adjacent sites are not occupied.

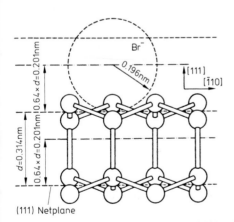

Figure 8-43. Side view of the unrelaxed Si(111)-(1×1) surface (after Funke and Materlik, 1985). The threefold open site of the adsorbed Br⁻ above the outermost Si(111) diffraction plane is indicated by a dashed circle.

8.5.3 Structural Environment Around Cu²⁺ Ionic Habit Modifiers in Ammonium Sulfate

In industrial crystallisers, control of particle size and shape is often accomplished with the aid of habit modifiers. The selection and definition of the additives, which in ionic systems take the form of trace metallic ions, is currently restricted by the lack of quantitative structural data concerning the bonding of the ionic additive to the host system. Whilst fluorescence EXAFS can be used to assess the concentration and the local atomic environment around the habit modifying ions (Barrett et al., 1989c; Armstrong et al., 1990; Cunningham et al., 1991; Cunningham, 1991a), X-ray standing waves need to be used to define the relationship between the ion and its underlying crystal lattice.

The addition of trace amounts of Cu²⁺ ions to crystallising solutions of ammonium sulfate results in the habit modification of the product from a prismatic morphology elongated along the c-axis to one elongated along the pseudo-hexagonal a-axis. X-ray fluorescence analysis of the habit faces confirmed that the adsorbed ions were not sectorially distributed but were uniformly segregated throughout the crystals whilst EXAFS and ab initio quantum chemistry calculations revealed that

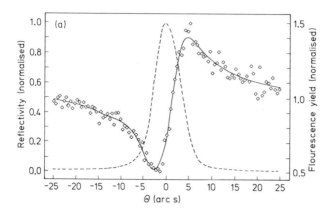

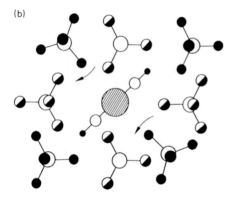

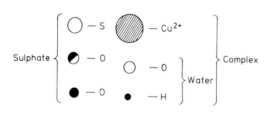

Figure 8-44. XSW data from examination of Cu^{2+} in ammonium sulfate (after Armstrong et al., 1991). (a) Normalised 002 X-ray standing wave spectra. The data show the overlay of the X-ray rocking curve (- - - - -) and the angular dependance of the Cu K fluorescence intensity (◇◇◇◇). The latter is displaced to the right with respect to the former which is indicative of a substitutional incorporation of Cu^{2+} for SO^{2-}. (b) Proposed structural model (the NH_4^+ ions have been removed for clarity) for the incorporation of Cu as a $Cu(H_2O)_4^{2+}$ complex which removes the SO_4^{2-} and four NH_4^+ ions to maintain a charge neutrality.

Cu^{2+} is closely correlated with six oxygen atoms in a distorted octahedral configuration. The radial distribution function similar to that around Cu^{2+} in copper sulfate pentahydrate in that it appears strongly coordinated to four water molecules in a square planar arrangement with weaker interactions with two sulfate ions.

XSW spectra (Fig. 8-44) recorded from the doped ammonium sulfate showed (a) that the copper ion occupies a lattice site close to the plane of highest electron density. This is consistent with an incorporation site close to the SO_4^{2-} ion and indicates that adsorption results in a substantial structural rearrangement and an associated redistribution of ionic charge. The exact mechanism of incorporation is associated with the similarity in charge distributions which allows the water ions to successfully mimic the ammonium ions during adsorption. This substitution of one SO_4^{2-} by Cu^{2+} requires the removal of four NH_4^+ ions to maintain strict charge neutrality. The resulting structure (b) forms a distorted octahedral environment around the copper ion with weak binding to the two sulphate ions above and below a copper tetrahydrate square planar complex which lies close to the (101) crystal plane.

8.6 Characterisation of Lattice Defects Using X-Ray Topography

8.6.1 Introduction

The group of techniques which go under the general title of X-ray diffraction topography (or more usually just X-ray topography) are used for imaging and characterising dislocations and other extended defects in relatively large crystals with a defect content typically less than 10^4 to 10^5 dislocations cm^{-2}. These techniques are applicable to a diverse range of crystalline materials satisfying the above criteria, but have proved particularly successful for analysis in electronic materials where a high level of perfection is an essential requirement for optimal performance. Unlike comparable defect imaging methods used in electron microscopy, these techniques do not require magnification and are therefore not restricted to a narrow field of view, do not require ultra thin samples and can in principle be used with the sample crystal in a variety of environments. The complementary nature of X-ray topography and electron microscopy when used to investigate processes occuring over a wide dimensional range is intimated below.

X-ray topographic techniques take advantage of the remarkable sensitivity of X-ray diffraction intensity to spatial and orientation distortion of the crystal lattice. Depending on the perfection of the crystal sample and its thickness, images can be produced as a consequence of kinematical or dynamical scattering processes leading to direct or dynamical images respectively. The theoretical basis for image formation is to be found in comprehensive review articles such as that of Batterman and Cole (1964) or in the book by Tanner (1976). The techniques, developed 30 to 40 years ago, employed the broad band monochromatic radiation characteristic of standard laboratory sources. These are still used extensively today. Although some of the methodology when using synchrotron radiation is essentially the same, there are significant differences in the characteristics of synchrotron radiation which allow a variety of incomparable materials science experiments to be carried out.

Synchrotron radiation techniques fall into two broad categories, white radiation topography and double crystal or narrow band monochromatic topography. Each is capable of providing its own unique brand of information.

8.6.2 White Radiation Synchrotron Topography

8.6.2.1 Introduction

The principal advantages of white radiation synchrotron topography are illustrated by the following features which emphasise the important differences between synchrotron and conventional characteristic X-ray sources, specifically: spectral range, high photon flux, beam divergence and time structure.

8.6.2.2 Instrumentation

In many respects the experimental arrangement employed for white radiation synchrotron topography bears a closer resemblence to that for the Laue method than that used for conventional characteristic radiation topography. Figure 8-45 shows a schematic diagram of the white radiation topographic camera (Bowen et al., 1982) on station 7.6 at the Daresbury synchrotron source. The four circle instrument is large enough to enable the mounting of large and heavy apparatus to achieve specific sample environments. Images are recorded on X-ray film or plates and at

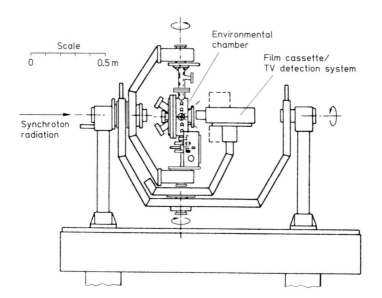

Figure 8-45. White radiation topographic camera on station 7.6 at Daresbury Laboratory (after Bowen et al., 1982).

lower resolution on a TV imaging system. Unlike a conventional Lang Camera, the large dimensions of the synchrotron beam eliminate the need for scanning mechanisms for most samples. Furthermore, the diffracted beam slit necessary to shield the film from the primary beam in topographs recorded on laboratory sources, is not generally required for synchrotron topography. The large specimen to film distances which can be used in the latter case adequately separate most images from the transmitted beam. To capture simultaneously the maximum number of images from different reflections, the film may be placed normal to the primary beam. As discussed below, however, optimum resolution is achieved by setting the film normal to the diffracted beam from one selected reflection.

8.6.2.3 Advantages of the Broad Spectral Range

As a consequence of the continuous nature of the synchrotron spectrum, white radiation topographs almost inevitably comprise superimposed images from several harmonic reflections. The integrated diffracted intensity (I) due to a given harmonic, and therefore its relative contribution to the topographic image, can be calculated from (Tuomi et al., 1974)

$$I = [|F_{hkl}| P(\lambda) \lambda^3 \exp(-\mu t)]/\sin^2 \Theta_B \quad (8\text{-}8)$$

where $|F_{hkl}|$ is the modulus of the structure factor, $P(\lambda)$ is the photon flux, λ is the wavelength, Θ_B is the Bragg angle, μ is the linear absorption coefficient and t is the sample thickness. The presence of harmonics within the topographic image has two major consequences. Firstly, since the intrinsic reflection range decreases with decreasing wavelength, the strain sensitivity is greater for high order reflections. The presence of high order harmonics in the image therefore results in a loss of resolution. The second major effect of harmonic contamination is observed in highly perfect crystals exhibiting Pendellösung fringes. In such cases complex fringe patterns can occur due to interference between wavefields of different harmonics and overlap of fringes associated with the various reflections (Tuomi et al., 1978). In this context it should be noted that the synchrotron radiation is highly polarised in the plane of the electron orbit. Consequently the periodic fading on Pendellösung fringes which is observed in conventional topographs (Hart and Lang, 1965), due to incoherent superposition of the σ and π polarisation components, does not occur in synchrotron radiation topographs when the plane of polarisation is either parallel or normal to the plane of incidence (Hart and Lang, 1965).

An important advantage over characteristic radiation topographs is the fact that the continuous spectrum employed in white radiation topography images warped and misoriented crystals in their entirety. In the conventional Lang technique, for example, only those regions of such samples which lie within the angular range defined by the beam divergence can be imaged. Increasing the beam divergence in characteristic source experiments allows more of the crystal to be imaged but at the cost of inferior resolution.

The contrast of defects in continuous radiation synchrotron topographs is substantially more complex than that observed in conventional Lang topography, comprising both diffraction (extinction) and orientation contrast components. As a consequence of orientation contrast effects the nature of the image is strongly dependent upon the sample to film distance since the diffracted beams arising from misorientated regions within the crystal are spatially separated. As demonstrated by Hart (1975) for lithium fluoride, this image separation can provide an extremely sensitive measure of subgrain misorientations, with tilts of less than 25 seconds of arc readily detectable.

The potential for wavelength tuning, possible because of the broad synchrotron

spectrum, greatly enhances the flexibility of topography. It enables, for example, optimisation of diffraction geometries and conditions (Petroff and Sauvage 1978; Halfpenny and Sherwood, 1990) and depth profiling in reflection topography (Dudley, 1990). The need for the flexibility which wavelength tuning provides is well demonstrated in the case of synchrotron section topography of large uncut crystals (Halfpenny and Sherwood, 1990). In such studies two competing factors determine the optimum diffraction conditions. Firstly, to minimise the geometrical distortion of the section topograph, the Bragg angle should be as close to 45° as possible, either through the use of high order reflections or long wavelengths. The former are generally weak with narrow intrinsic reflection ranges and so yield inferior spatial resolution. The second controlling factor is absorption which for a given sample thickness will limit the wavelength which can be employed. In general the wavelength achieving the optimum compromise between these two factors will not correspond to a characteristic line. For many samples it is only through the wavelength tunability of synchrotron topography that such conditions can be realised. Figure 8-46 shows a white radiation synchrotron section topograph of a crystal of the organic non-

Figure 8-46. Synchrotron white radiation section topograph of a 2 cm thick crystal of $(-)$-2-(α-methyl-benzylamino)-5-nitropyridine (MBA-NP) (after Halfpenny and Sherwood, 1990).

linear optical material (–)-2-(α-methylbenzylamino)-5-nitropyridine (MBA-NP). The low linear absorption coefficients of organic materials such as this enable relatively thick samples to be topographed (up to 2 cm in thickness in this case). The image in Fig. 8-46 is formed principally from the 040 reflection at 0.08 nm and the 060 at 0.055 nm at a Bragg angle of 15°, the relative contributions from these harmonics being approximately 79% and 21% respectively. At 0.08 nm the linear absorption coefficient of MBA-NP is about 1.7 cm^{-1} giving a product μt of around 3. The image is compressed parallel to the diffraction vector by a factor of $\sin 2\Theta$ (i.e., 50%). Longer wavelengths, though decreasing image distortion substantially degrade defect resolution through increased dynamical contrast contributions. The conditions employed represent an optimum compromise in the present case, yielding well defined images of dislocations, inclusions, growth sector boundaries and delineating clearly the seed crystal from which the sample was grown.

The observation of misfit dislocations in (GaAlAsP)/GaAs epitaxial layers (Petroff and Sauvage, 1978) provides another important illustration of the use of wavelength tuning in white radiation synchrotron topography. In this case the wavelength was tuned to just above the K absorption edge of gallium. A substantial reduction in absorption was thus achieved, similar to that for silver $K \alpha$ radiation. The longer wavelength employed, however, resulted in considerably narrower image widths and therefore enhanced resolution.

8.6.2.4 Influence of Beam Flux, Dimensions and Divergence

A crystal placed in the synchrotron beam yields a pattern of diffraction spots, as in the conventional Laue method. Because of the extended beam in the synchrotron case, however, each spot is in fact a topograph of the crystal. In this way the time required for defect characterisation by invisibility criteria in different reflections can be greatly reduced. The X-ray flux from a synchrotron source is some 3 to 4 orders of magnitude greater than a conventional sealed tube X-ray source, producing dramatic reductions in exposure times for topography. In contrast to Lang topography using a laboratory source, the large dimensions of the incident synchrotron beam allow large areas to be imaged without the need for sample scanning, thereby further enhancing the speed advantage of synchrotron topography.

It can be shown that the geometrical resolution R_g is given by

$$R_g = S x / L \tag{8-9}$$

where S is the source size, x the sample to film distance and L the sample to source distance. With a source size of $0.15 \text{ mm} \times 1.55 \text{ mm}$ and a sample to source distance of 80 metres (topography station 7.6 at the SRS, Daresbury U.K.) it is apparent that a geometrical resolution comparable with that of laboratory Lang topography can be achieved with relatively large sample to film distances. The geometrical resolution defined above relates to a diffracted beam normal to the X-ray film. For a film normal to the incident beam, the geometrical resolution then depends primarily upon the Bragg angle of the reflection in question. Those with large Bragg angles are incident on the film at shallow angles and therefore suffer substantial distortion and loss of resolution.

The high photon flux, together with the fact that no sample scanning mechanism is required and the possibility of high resolution despite large specimen to film distances, together make white radiation synchrotron topography ideal for dynamic and in situ studies of a wide range of physical and chemical phenomena, including recrystallisation (Gastaldi and Jourdan, 1978 and 1979; Jourdan and Gastaldi, 1979), plastic deformation (Miltat and Bowen, 1979; Ahamad et al., 1987; Ahamad and Whitworth, 1988), motion of magnetic domains (Tanner et al., 1976; Safa and Tanner, 1978; Sery et al., 1978; Clark et al., 1979; Chikaura and Tanner, 1979; Stephenson et al., 1979; Toumi et al., 1979; Clark et al., 1983) and solid-state reactivity (Dudley et al., 1983 and 1986).

Studies of the recrystallisation of aluminium (Gastaldi and Jourdan, 1978 and 1979; Jourdan and Gastaldi, 1979) provide a clear demonstration of the power of white radiation synchrotron topography. The principal alternatives to X-ray methods for such studies are TEM and optical microscopy. While the high resolution of TEM enables the observation of the nucleation stage of the recrystallisation, the relatively small area imaged (about 10 μm across) precludes observation of the growth of nuclei into large grains. Furthermore, the growth behaviour of grains in the thin samples required for TEM is likely to differ substantially from that in bulk samples. Optical microscopy suffers from the disadvantage of being a purely surface technique. Grain growth rates can be as high as a millimetre per minute so for X-ray studies the rapid exposure times achievable with white radiation synchrotron topography are essential for dynamic observations. The continuous spectrum of the synchrotron source is also vital since the orientation of the growing grains cannot be predicted. Utilising these advantages, white radiation topographic studies of grain growth in aluminium have provided information on the formation and evolution of

growth defects and have revealed a distinction between the structure of moving and stationary grain boundaries. Although the resolution of X-ray topography is insufficient to observe nucleation, the transmission from nucleation to grain growth has been examined.

Short exposure times and the ability to record several reflections simultaneously are also important advantages in studies of plastic deformation and dislocation motion. The work of Whitworth and co-workers (Ahamad et al., 1987; Ahamad and Whitworth, 1988) on dislocation motion in ice illustrates well the capabilities of synchrotron topography in such studies. Rapid exposures (around 20 seconds) enabled dynamic observations of dislocation motion. Dislocation loops gliding on the basal plane form hexagonal loops comprising screw and 60° segments, indicating the existence of a Peierls barrier in these orientations. Straight dislocations in both orientations were found to move at approximately equal velocities which varied linearly with applied stress. The velocity of glide of edge dislocations on non-basal planes, however, was an order of magnitude higher. Of particular note is the fact that substantial recovery was observed after periods as short as 18 minutes, thereby illustrating the need for rapid exposures and the dynamic, in situ capabilities of white radiation synchrotron topography.

The additional space available around the sample, without loss of resolution, in the case of white radiation synchrotron topography offers considerable flexibility in the control of sample environment, whether it be temperature (furnaces or cryostats), pressure, magnetic or electric fields or the introduction of corrosive or other reactive ambients. Figure 8-47 shows an environmental chamber used for in situ X-ray topography under well defined environmen-

Figure 8-47. Environmental chamber for synchrotron radiation topography (after Gillespie et al., 1989).

tal conditions on the topography station 7.6 at the Daresbury Synchrotron Radiation Source (Gillespie et al., 1989). The water cooled cylindrical chamber is constructed from stainless steel UHV components. Two ports fitted with beryllium windows form the X-ray entrance and exit windows of the chamber. In addition a further 12 radial and 4 angled ports provide access for the furnace, sample holder, inputs for gas supplies and the pumping system as well as electrical connections. The furnace, constructed from tubular stainless steel with thermocoax (nichrome in iconel) windings, fits directly over the sample holder and has two 2 cm apertures to allow passage of the X-ray beam. The chamber/furnace combination can operate up to

1270 K at a pressure of around 10^{-7} torr. At higher pressures the maximum operating temperature is limited to about 770 K to avoid damage to the beryllium windows. The best achievable vacuum is 10^{-7} torr.

This chamber has been employed in a range of X-ray topographic studies of solid state chemical reactivity. Figure 8-48 shows the changes in structural perfection of a crystal of nickel sulfate hexahydrate during dehydration at reduced pressure. Figure 8-48 a shows a white radiation synchrotron topograph of the crystal at room temperature and atmospheric pressure. After 30 minutes at a pressure of 0.1 Pa (also at room temperature) regions of surface strain have developed, associated with numerous dehydration sites, as shown in Fig. 8-48 b. It is apparent that the strain centres and therefore the decomposition sites are not associated with the grown-in dislocations visible in the topograph, in contradiction, in this case, to the earlier suggestions that dislocations play a key role in solid-state decomposition. Further studies performed using this chamber include observations of the initial stages of the thermal decomposition of calcite at temperatures up to 748 K; thermal dehydration of nickel sulfate hexahydrate at atmospheric pressure and an examination of the alpha/beta phase transformation in quartz at low pressure. Despite considerable potential, however, the application of white radiation synchrotron topography to studies of the role of defects and structural perfection in solid state chemical reactivity has been rather limited.

A significant problem which can seriously limit the application of synchrotron white radiation topography is that of radiation-induced damage. Exposure of many materials to an unfiltered white synchrotron beam can result in the formation of colour centres, sample curvature, genera-

Figure 8-48. White radiation topograph of nickel sulfate hexahydrate recorded using the environmental chamber (a) at atmospheric pressure and (b) after pumping at 0.1 Pa for 30 minutes (after Gillespie et al., 1989).

also accounts for the curvature observed in many radiation damaged samples. Since the X-ray flux reaching the exit surface of a crystal is diminished by absorption, the induced damage also decreases through the sample thickness. Differing levels of decomposition and point defect generation through the sample therefore result in a non-uniform stress which relaxes by generating sample curvature and in some instances dislocations. The use of, for example, an aluminium filter can substantially reduce the level of radiation damage by attenuating the strongly absorbed long wavelength components of the beam which are largely responsible for the radiation damage. A major drawback of this, however, is that the relative contribution of high order harmonics (resulting from the short wavelengths) to the topographic image is increased, thereby impairing resolution.

8.6.2.5 Experiments Involving the Time Structure

A unique aspect of synchrotron radiation which has been greatly underexploited is its time structure. The radiation generated by a synchrotron storage ring occurs in the form of a regular sequence of pulses as the electron bunches orbit around the ring. The frequency of these pulses is determined by the number of electron bunches and the dimensions of the ring. Stable operating frequencies in the range from a few MHz to GHz can be achieved. This well defined time structure has been successfully exploited in stroboscopic topography of periodic phenomena including the observation of surface acoustic waves in lithium niobate and quartz (Whatmore et al., 1982; Goddard et al., 1983). Since X-ray topography also permits the simultaneous observation of crystal defects, interactions

tion of slip dislocations and ultimately, complete radiolytic decomposition, leading to void formation. The extent of radiation damage depends upon the sensitivity of the material and its X-ray absorption coefficient. It is interesting to note that many organic materials, despite their relative chemical instability, exhibit surprisingly greater resistance to radiation induced damage than inorganic materials. This is simply a result of the lower absorption coefficients of the former. Absorption between these and surface acoustic waves can also be examined.

8.6.3 Double Crystal Topography

8.6.3.1 Introduction

In a white radiation experiment, the intrinsic angular width and wavelength dispersion of a reflection defined by the rocking curve is generally less than the divergence of the incident beam and its wavelength spread. The resulting overlap of diffracted beams from adjacent parts of the crystal results in an integrated intensity which reduces strain sensitivity and prevents quantitative strain mapping. Slight misorientation or dilation of the lattice can occur between growth sectors of a crystal as a consequence of the different growth mechanism, or within a single growth sector because of the fluctuations of temperature or supersaturation during growth. This is usually associated with impurity incorporation. Although X-ray topography is inherently capable of detecting strain to below 1 part in 10^5 these differences will not generally be imaged in a white beam experiment. This is unfortunate since impurities at low concentrations may be of significance to the performance of a device into which such crystalline material is incorporated. Therefore, although white beam topography is probably the preferred technique for defect analysis and for studying for example the temporal evolution of defects, it is not sensitive enough for certain applications.

In order to improve the strain sensitivity of X-ray topography, the double crystal arrangement is used. Bonse (1958) has provided the theoretical background for the technique and the criteria for optimal use has been reiterated in a review on the use of synchrotron radiation in topography by Miltat and Sauvage-Simkin (1984). The de-

tailed diffractometry associated with rocking curve analysis which can provide information about the mosaic spread, curvature, defect content and thickness (particularly with multilayer structures) of a specimen and the topographic image obtained with the double crystal configuration are intimately linked.

8.6.3.2 Instrumentation

In general use, the system is set up in the $(+\,-)$ parallel configuration as shown in Figs. 8-49 a and 8-49 b. When the two crystals are of the same material and the same reflections are used, the doubly diffracted beam is parallel to the incident beam and there is no wavelength dispersion. As the second (specimen) crystal is also illuminated by what is essentially a narrow band monochromatic plane wave coming from the first crystal, maximum strain sensitivity is achieved. The technique has been much

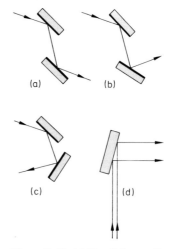

Figure 8-49. (a) Parallel non-dispersive double crystal geometry in $(+\,-)$ configuration when identical crystals and reflections are used; (b) non-parallel dispersive double crystal geometry in $(+\,-)$ configuration with non identical crystals; (c) double crystal geometry in the $(+\,+)$ configuration; (d) use of an assymmetric reflection for beam expansion.

used for studies of silicon because of the high degree of crystal perfection which can be attained with this material, a necessary requirement for obtaining rocking curve widths close to the intrinsic value. Although ideally the same material should be used for both crystals, this is not possible for general use, bearing in mind the necessity for a high degree of perfection in the monochromatic crystal. This means that there will be a mismatch between the lattice plane spacings for the two reflections and there will be some wavelength dispersion introduced, thereby reducing strain sensitivity. However, provided reflections are chosen in which the difference in lattice plane spacings is small, then the reduction in strain sensitivity is kept to a minimum and the technique still proves highly satisfactory. One bonus is that since higher order reflections may not coincide for the first and second crystal, harmonic content may be reduced or even eliminated. Because of the high levels of perfection available, germanium and sometimes quartz in addition to silicon are suitable as first crystal materials.

The use of synchrotron radiation should permit in situ double crystal experiments to be carried out using environmental stages. Up to the present time, few experiments have been reported, perhaps due to the general perception that imaging crystals which are being distorted due to tensile or compressive loading, thermal anisotropy and chemical processes is difficult with monochromatic radiation. However, George and Michot (1982) report using a high temperature tensile stage operating under a controlled atmosphere to study dislocation velocities in silicon and the movement of dislocations from a crack tip under load. High strain sensitivity was not a particular issue, so the wavelength dispersive $(++)$ crystal configuration appears to have been used. Also, Michot et al. (1984) have employed a hot stage to follow the evolution of fluid inclusions in synthetic quartz up to decrepitation.

Double crystal diffraction from two silicon crystals can be used as a tuneable non dispersive source of plane wave radiation for topographic studies of a specimen crystal positioned on a third axis. Kuriyama and his group at National Institute for Science and Technology, U.S.A., have developed this technique for their dynamical diffraction imaging (Kuriyama et al., 1989). In addition the use of assymetrically cut crystals can be used to expand (or decrease) the cross section of the incident beam. With synchrotron radiation, the advantages of video recording and processing techniques for real time experiments cannot be overstated. Unfortunately the resolution of even the best videocon detector is limited to about 10 μm. By using two orthogonal, asymetrically cut crystals to diffract the imaging beam from the sample, Kuriyama et al. have been able to produce images magnified up to 150× thus overcoming the resolution problem. As these authors state, the magnifier can be used as a zoom lens since a pair of asymetrically cut crystals provides a continuously changing magnification as the energy of the incident beam is changed.

8.6.3.3 Applications

Strain mapping using synchrotron double crystal techniques, with a major emphasis on differences between growth sectors, has been reported for quartz (Fig. 8-50, Zarka, 1984), strontium nitrate (Miltat and Sauvage-Simkin, 1984) and diamond (Lang et al., 1987). Such differences can be very small. By rotating the sample crystal about the Bragg reflection, topographs can be recorded at various positions on the rock-

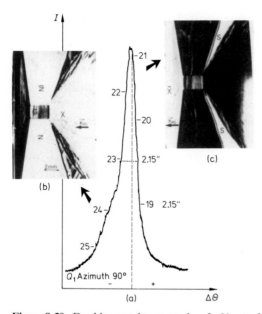

Figure 8-50. Double crystal topographs of a Y-cut of a quartz crystal (1000 reflection) showing contrast changes in images recorded at two different positions on the rocking curve (after Zarka, 1984). Such changes allow variations in the lattice parameter $\approx 10^{-5} - 10^{-6}$ and lattice tilts of $\approx 0.2''$ to be determined.

ing curve. Using this technique, the variations of contrast were used to estimate changes of lattice parameter to less than 1 part in 10^5. If both lattice dilation and tilting are present, the effects of each can be separated by rotating the sample crystal about the surface normal. At the other extreme, melt grown compound semiconductor crystals such as GaAs contain misorientations which are larger than the rocking curve width. Strain associated with these crystals has been assessed by mapping the contours of misorientation recorded at a series of fixed angular settings (Kitano et al., 1985).

Double crystal diffractometry and topography have also been applied successfully to epitaxial layers. An analysis of rocking curves gives the lattice mismatch, lattice tilt between epilayer and substrate,

layer thickness and compositional variation with depth, substrate and layer perfection, and curvature. As with the white radiation technique, topography can be used to study misfit dislocations. In addition moiré patterns observed in topographs of GaAlAs/GaAs layer structures have been used to determine lattice mismatch to 1 part in 10^6 (Chu and Tanner 1987).

8.7 Conclusions and Future Perspective

The foregoing examples demonstrate well the breadth of novel experiments which can be carried out using synchrotron X-radiation and the improved working which can be achieved compared with traditional radiation. There is obviously much scope for broadening these applications to a wider range of experiments.

For the future, however, current developments in synchrotron sources would lead to the possibility of an even greater range of opportunities. The principal potential advances are two-fold.

For the machines themselves the aim is to build sources of even greater intensity and energy and with tunable radiation available principally from insertion devices. This will allow considerably reduced exposure and recording times; sub seconds rather than minutes/hours, and the easier use of the time structure of the beam for stroboscopic type experiments. Quasi-forbidden reflections will become accessible. Weaker scattering processes can be studied more satisfactorily.

The increased energy range will give a greater penetrability for in situ dynamic studies of a variety of diffraction processes. It will also allow diffraction studies of higher atomic number materials and provide access to a wider range of experimentation using anomalous dispersion X-ray diffraction.

In parallel, developments in electronic recording will result in greater ease of image and pattern detection thus contributing to improved and more accurate operation.

There is no doubt that synchrotrons will continue to develop and occupy an essential place in the armoury of the materials scientists.

8.8 Acknowledgements

The authors gratefully acknowledge the support of the U.K. Science and Engineering Research Council for a number of the projects described in this article. More particularly, however, we thank the Director and staff of the SERC's Daresbury Laboratory for the provision of much of the wide range of facilities described and of which we have had use in carrying out these experiments.

8.9 References

Ahamad, S., Otomo, M., Whitworth, R. W. (1987), *J. Phys. (Paris) Coll. 48*, 175.

Ahamad, S., Whitworth, R. W. (1988), *Phil. Mag. A 57*, 749.

Akhtar, M. J., Catlow, C. R. A., Clark, S. M., Temmerman, W. M. (1988), *J. Phys. C: Solid State Phys. 21*, L917.

Aleshko-Ozhevskij, O. P., (1982), *Sov. Phys. Cryst. 27*, 673.

Aleshko-Ozhevskij, O. P. (1983), *Ferroelectrics 48*, 157.

Amoros, J. L., Buerger, M. J., Amoros, M. C. (1975), *The Laue Method*. New York: Academic Press.

Andrews, S. J., Hails, J. E., Harding, M. M., Cruickshank, D. W. J. (1987), *Acta Cryst. A 43*, 70.

Andrews, S. J., Papiz, M. Z., McMeeking, R., Blake, A. J., Lowe, B. M., Frankling, K. R., Helliwell, J. R., Harding, M. M. (1988), *Acta Cryst. B 44*, 73.

Anwar, J., Barnes, P., Clark, S. M., Dooryhee, E., Hausermann, D., Tarling, S. E. (1990), *J. Mat. Sci. Lett. 9*, 436.

Armstrong, D. R., Cunningham, D. A. H., Roberts, K. J., Sherwood, J. N. (1991), *Proc. 6th Int. Conf. on X-ray Absorption Fine Structure (XAFS VI)*. Ellis Horwood, Chichester, pp. 435–437.

Ballard, D. G. H., Courtis, A., Shirley, I. M., Taylor, S. C. (1988), *Macromolecules 21*, 294.

Barnes, P., Clark, S. M., Hauserman, D., Fentiman, C. H., Rashid, S., Muhamad, N. N., Henderson, E. (1992), *Phase Transitions*, in press.

Barrett, N. T., Gibson, P. N., Greaves, G. N., Mackle, P., Roberts, K. J., Sacchi, M. (1989a), *J. Phys. D: Appl. Phys. 22*, 542.

Barrett, N. T., Gibson, P. N., Greaves, G. N., Roberts, K. J., Sacchi, M. (1989b), *Physica B 158*, 690.

Barrett, N. T., Lamble, G. M., Roberts, K. J., Sherwood, J. N., Greaves, G. N., Davey, R. J., Oldman, R. J., Jones, D. (1989c), *J. Crystal Growth 94*, 689.

Barrett, N. T., Greaves, G. N., Pizzini, S., Roberts, K. J. (1990), *Surf. Sci. 227*, 337.

Batterman, B. W. (1964), *Phys. Rev. 133*, 759.

Batterman B. W., Cole, H. (1964), *Rev. Mod. Phys. 36*, 681.

Beaven, P. A., Frisius, F., Kampmann, R., Wagner, R. (1986), *Proc. of the Second International Symposium on Environmental Degradation of Materials in Nuclear Power Systems – Water Reactors;* Monterey, Sept. 1985, (ANS).

Bednorz, J. G., Mueller, K. A. (1986), *Z. Phys. B: Cond. Matter 64*, 189.

Bedzyk, M. J., Materlik, G., Kovalchuk, M. V. (1984), *Phys. Rev. B 30*, 2453.

Bhat, H. L., Sheen, D. B., Sherwood, J. N. (1982), *The Application of Synchrotron Radiation to Problems in Materials Science:* Proc. Daresbury Study Weekend DL/SCI/R19, pp. 90–95.

Bhat, H. L., Herley, P. J., Sheen, D. B., Sherwood, J. N. (1984), in: *Applications of X-ray Topographic Methods to Materials Science:* Weissmann, S., Balibar, F., Petroff, J. (Eds.). New York: Plenum Press, pp. 401–411.

Bhat, H. L., Littlejohn, A., McAllister, J. M. R., Shaw, J., Sheen, D. B., Sherwood, J. N. (1985), *Materials Science Monographs 286*, 707.

Bhat, H. L., Clark, S. M., El Korashy, A., Roberts, K. J. (1990a), *J. Appl. Cryst. 23*, 545.

Bhat, H. L., Roberts, K. J., Sacchi, M. (1990b), *J. Phys.: Condens. Matter 2*, 8557.

Bianconi, A. (1988), in: *X-ray Absorption, Principles, Applications, Techniques of EXAFS, SEXAFS and XANES:* Köningsberger, D. C., Prins, R. (Eds.). New York: Wiley, pp. 573–662.

Binsted, N., Greaves, G. N., Henderson, C. M. B. (1986), *J. de Physique C8*, 837.

Bonse, U. (1958), *Zeit. Phys. 153*, 278.

Bordas, J., Glazer, A. M., Hauser, H. (1975), *Phil. Mag., 32*, 471.

Bosio, L., Cortes, R., Defrain, A., Froment, M. J. (1984), *J. Electroan. Chem. 180*, 265.

Bowen, D. K., Clark, G. F., Davies, S. T., Nicholson, J. R. S., Roberts, K. J., Sherwood, J. N., Tanner, B. K. (1982), *Nucl. Instrum. Methods 192*, 277.

Bragg, W. L. (1949), *The Crystalline State*, Volume 1, *General Survey.* Bell and Sons, p. 27.

Bragg, W. L. (1975), in: *The Development of X-ray Analysis*, p. 137.

Burton, W. K., Cabrera, N., Frank, F. C. (1951), *Philos. Trans. Soc. London 243*, 299.

Buswell, J. T., Brock, J. M. (1987), *CEGB Report TPRD/B/0922/R87.*

Buswell, J. T., Little, E. A., Jones, R. B., Sinclair, R. N. (1986), *Proceedings of the Second International Symposium on Environmental Degradation of Materials in Nuclear Power Systems – Water Reactors;* Monterey, Sept. 1985, (ANS).

Buswell, J. T., English, C. A., Hetherington, M. H., Phythian, W. J., Smith, G. D. W., Worrall, G. M. (1988), *Proc. of the 14th International Symposium on Effects of Radiation on Materials, Andover, Massachusetts*, Vol. 2. American Society for Testing Metals.

Cernik, R. J., Clark, S. M., Pattison, P. (1989), to be published in: *Advances in X-ray Analysis.*

Cernik, R. J., Cheetham, A. K., Prout, C. K., Watkin, D. J., Wilkinson, A. P., Willis, B. T. M. (1990a), to be published in J. Appl. Cryst.

Cernik, R. J., Murray, P. K., Pattison, P. (1990b), *J. Appl. Cryst. 23*, 292.

Chikaura, Y., Tanner, B. K. (1979), *Jap. J. Appl. Phys. 18*, 1389.

Chu, X., Tanner, B. K. (1987), *Mater. Lett. 5*, 153.

Chu, C. W., Hor, P. H., Gao, L., Huang, Z. J. (1987), *Science 235*, 567.

Clark, S. M. (1989), *Nucl. Instrum. Meth. A 276*, 381.

Clark, S. M., Miller, M. C. (1990), to be published in: *Rev. Sci. Inst.*

Clark, G. F., Tanner, B. K., Sery, R. S., Savage, H. T. (1979), *J. de Physique 40 C5*, 183.

Clark, G. F., Goddard, P. A., Nicholson, J. R. S., Tanner, B. K., Wanklyn, B. M. (1983), *Phil. Mag. B47*, 307.

Clearfield, A. (1964), *Inorganic Chemistry 3*, 146.

Clucas, J. A., Harding, M. M., Maginn, S. J. (1988), *J. Chem. Soc. Chem. Comm.*, 185.

Cocco, G., Enzo, S., Barrett, N. T., Roberts, K. J. (1989), *J. Less Comm. Metals 154*, 177.

Cocco, G., Enzo, S., Barrett, N. T., Roberts, K. J. (1990), in: *Proc. of the 2nd European Conference on Progress in Synchrotron Radiation Research.* Società Italiana di Fisica, Conference Proceedings, Vol. 25, pp. 693–696.

Cocco, G., Enzo, S., Barrett, N. T., Roberts, K. J. (1992), *Phys. Rev. B*, in press.

Cowen, P. L., Golovenchko, J. A., Robbins, M. F. (1980), *Phys. Rev. Lett. 44*, 1680.

Cowley, R. A., Lucas, C. A. (1990), *Faraday Discuss. Chem. Soc. 89*, 181.

Cox, D. E., Hasting, J. B., Thomlinson, W., Prewitt, C. T. (1983), *Nucl. Instrum. Methods. 208*, 573.

Cruickshank, D. W. J., Helliwell, J. R., Moffat, K. (1987), *Acta Cryst. A43*, 656.

Cunningham, D. A. H. (1991), *Ph.D. Thesis, University of Strathclyde*, Glasgow.

Cunningham, D. A. H., Davey, R. J., Roberts, K. J., Sherwood, J. N., Shripathi, T. (1990a), *J. Cryst. Growth 99*, 1065.

Cunningham, D. A. H., Davey, R. J., Doyle, S. E., Gerson, A. R., Hausermann, D., Herron, M., Robinson, J., Roberts, K. J., Sherwood J. N., Shripathi, T., Walsh, F. C. (1990b), in: *Proc. 2nd Workshop on Synchrotron Light: Applications and Related Techniques:* Craievitch, A. (Ed.). Singapore: World Scientific Press, pp. 230–244.

Cunningham, D. A. H., Davey, R. J., Gerson, A. R., Ristić, R., Roberts, K. J., Sherwood, J. N., Shripathi, T. (1991a), in: *Particle Design via Crystallisation, Amer. Inst. Chem. Eng. Symposium Series, 84*, Vol. 87: Ramanarayanan, R., Kern, W., Larsen, M., Sikdar, S. (Eds.); pp. 104–113.

Cunningham, D. A. H., Gerson, A. R., Roberts, K. J., Sherwood, J. N., Wojciochowski, K. (1991b), in: *Advances in Industrial Crystallisation:* Garside, J., Davey, R. J., Jones, A. G. (Eds.). London: Butterworths, pp. 105–130.

Docherty, R., El Korashy, A., Jennissen, H.-D., Klapper, H. Roberts, K. J., Scheffen-Lauenroth, T. (1988), *J. Appl. Cryst. 21*, 406.

Doyle, S. E., Gerson, A. R., Roberts, K. J., Sherwood, J. N., Wroblewski, T. (1991), *J. Crystal Growth 112*, 302–307.

Dubois, J. M. (1988), *J. Less Comm. Metals 145*, 309.

Dudley, M. (1990), *J. X-ray Sci. and Technol. 2*, 195.

Dudley, M., Sherwood, J. N., Ando, D. J., Bloor, D. (1983), *Mol. Cryst. Liq. Cryst. 93*, 223.

Dudley, M., Sherwood, J. N., Ando, D. J., Bloor, D. (1985), in: *Polydiacetylenes:* Bloor, D., Chance, R. R. (Eds.). Dordrecht: Martin Nijhoff, pp. 87–92.

Dudley, M., Sherwood, J. N., Bloor, D. (1986), *Proc. Am. Chem. Soc. Div. Polym. Mater. Sci. Eng. 54*, 426.

Dudley, M., Baruchel, J., Sherwood, J. N. (1990), *J. Appl. Crystallog. 23*, 186.

Dudley, M., Sherwood, J. N., Bloor, D. (1991), *Proc. Roy. Soc. A 434*, 243.

Durham, P. J. (1988), in: *X-ray Absorption. Principles, Applications, Techniques of EXAFS, SEXAFS and XANES:* Königsberger, D. C., Prins. R. (Eds.). New York: J. Wiley and Sons, pp. 53–84.

Edwards, B., Garner, C. D., Roberts, K. J. (1990), *Proc. 2nd European Conference on Progress in X-ray Synchrotron Radiation Research: Società Italiana di Fisica, Conference Proceedings*, Vol. 24, pp. 415–418.

El Korashy, A. (1988), *Ph.D. Thesis, University of Assiut*, Egypt.

El Korashy, A., Roberts, K. J., Scheffen-Lauenroth, T., Dam, B. (1987), *J. Appl. Cryst. 20*, 512.

Frahm, R. (1989a), *Rev. Sci. Instrum. 60 (7)*, 2515.

Frahm, R. (1989b), *Physica B 158*, 342.

Friedrich, W., Knipping, P., von Laue, M. (1912), *Ber. Bayer. Acad. Wiss.*, 303.

Friedrich, A., Pettinger, B., Kolb, D. M., Lupke, G., Steinhoff, R., Marowsky, G. (1989), *Chem. Phys. Lett. 163*, 123.

Funke, P., Materlik, G. (1985), *Solid State Comm. 54*, 921.

Gastaldi, J., Jourdan, C. (1978), *Phys. Stat. Sol. (a) 49*, 529.

Gastaldi, J., Jourdan, C. (1979), *Phys. Stat. Sol. (a) 52*, 139.

George, A., Michot, G. (1982), *J. Appl. Cryst. 15*, 412.

Gerson, A. R. (1990), Ph.D. Thesis, University of Strathclyde.

Gerson, A. R., Roberts, K. J., Sherwood, J. N. (1991 a), *Acta Cryst B 47*, 280.

Gerson, A. R., Roberts, K. J., Sherwood, J. N. (1991 b), *Powder Technology 65*, 243.

Gesi, K. (1978), *J. Phys. Soc. Japan 45*, 1431.

Gesi, K., Iizumi, M. (1979), *J. Phys. Soc. Japan 46*, 697.

Gierlokta, S., Lambooy, P., de Jeu, W. H., (1990), *Europhys. Lett. 12*, 341.

Gillespie, K., Litlejohn, A., Roberts, K. J., Sheen, D. B., Sherwood, J. N. (1989), *Rev. Sci. Instrum. 60*, 2498.

Goddard, P. A., Clark, G. F., Tanner, B. K., Whatmore, R. W. (1983), *Nucl. Instrum., Methods 208*, 705.

Greaves, G. N., Davis, E. A. (1974), *Phil. Mag. 29*, 1201.

Greaves, G. N., Barrett, N. T., Antonini, M., Thornley, R., Willis, B. T., Steele, A. (1989), *J. Amer. Chem. Soc. 111*, 4313.

Grunes, L. A. (1983), *Phys. Rev. B 27*, 2111.

Gurman, S. J. (1982), *J. Mat Sci. 14*, 1541.

Gurman, S. J. (1988), *J. Phys. C 21*, 3699.

Hajdu, J., Machin, P. A., Campbell, J. W., Greenough, T. J., Clifton, J. J., Zurek, S., Glover, S., Johnson, L. N., Elder, M. (1987), *Nature 329*, 178.

Hajdu, J., Acharya, K. R., Stuart, D., Barford, D., Johnson, L. N. (1988), *Trends Biochem. Sci. 13*, 104.

Halfpenny, P. J., Sherwood, J. N. (1990), *Phil. Mag. Lett. 62*, 1.

Halfpenny, P. J., Ristić, R. I., Sherwood, J. N. (1991), in preparation.

Hamelin, A. (1982), *J. Electroanal. Chem. 142*, 299.

Hansen, M. (1958), *Constitution of Binary Alloys*. New York: McGraw-Hill.

Harding, M. M., Magin, S. J., Campbell, J. W., Clifton, I., Machin, P. A. (1988), *Acta Cryst. B 44*, 142.

Hart, M. (1975), *J. Appl. Cryst. 8*, 436.

Hart, M., Lang, A. R. (1965), *Acta Cryst. 19*, 73.

Hart, M., Parrish, W. (1986), *Materials Science Forum 9*, 39.

Hart, M., Parrish, W. (1989), *J. Mater. Res.*, in press.

Hart, M., Cernik, R. J., Parrish, W., Toraya, H. (1990), *J. Appl. Cryst. 23*, 286.

Hastings, J. B., Thomlinson, W., Cox, D. E. (1984), *J. Appl. Cryst. Growth 17*, 85.

Hausermann, D., Salvador, G., Sherman, W. F. (1989), in: *High Pressure Science and Technology: Proc. XIth AIRAFT Int. Conf., Kiev:* Novikov, N. V. (Ed.); pp. 186–194.

Hausermann, D., Daghoogghi, M. R., Sherman, W. F. (1990), *High Pressure Research 4*, 414.

Hayes, T. M., Boyce, J. B. (1982), *Solid State Phys. 37*, 173.

Heald, S. (1988), in: *X-ray Absorption: Principles, Applications, Techniques of EXAFS, SEXAFS and XANES:* Köningsberger, D. C., Prins, R. (Eds.). New York: Wiley, pp. 87–118.

Helliwell, J. R. (1984), *Reports on Prog. in Physics 47*, 1403.

Helliwell, J. R. (1991), *Macromolecular Crystallography with Synchrotron Radiation*. Cambridge: Harvard University Press.

Helliwell, J. R., Habash, J., Cruickshank, D. W. J., Harding, M. M., Greenough, T. J., Campbell, J. W., Clifton, I. J., Elder, M., Machin, P., Papiz, M., Zurek, S. (1989), *J. Appl. Cryst. 22*, 483.

Hertel, N., Materlik, G., Zegenhagen, Z. (1985), *J. Z. Phys. B 58 (3)*, 199.

Holmes, P. A., Nevin, A., Nehring, J., Amstutz, H. (1989), *European Patent Application No. 0299664*.

Houchin, M. R. (1987), *Patent wo87/07885*, U.S.A.

Iglesia, E., Boudart, M. (1984), *Journal of Catalysis 88*, 325.

Isherwood, B. J. (1977), *GEC Journal of Science and Technology 43*, 111.

Itoh, K., Kataoka, T., Matsunaga, H., Nakamura, E. (1980), *J. Phys. Soc. Japan. 48*, 1039.

Jacobi, H. (1972), *Z. Krist. 135*, 467.

James, R. J. (1958), *The Crystalline State*, Vol. 2, *The Optical Principles of the Diffraction of X-rays*. G. Bell and Sons.

Jedrecy, N., Sauvage-Simkin, M., Pinchaux, R., Geiser, N., Massies, J., Etgens, V. H. (1990), *Società Italiana di Fisica Conference Proceedings 25*, 481.

Johnson, W. L. (1986), *Prog. Mat. Sci. 30*, 81.

Jourdan, C., Gastaldi, J. (1979), *Scripta Met. 13*, 55.

Kiessig, H. (1931), *Ann. der Physik 10*, 769.

Kitano, T., Matsui, J., Ishikawa, T. (1985), *Jap. J. Appl. Phys. 24*, L948.

Koch, E. E. (Ed.) (1983), *Handbook on Synchrotron Radiation*, Vol. 1. Amsterdam: North Holland.

Koenig, J. H., Carron, J. G. (1967), *Mat. Res. Bull. 2*, 509.

Kolb, D. M., Schneider, J. (1985), *Surf. Sci. 162*, 764.

Köningsberger, D. C. (1988), *X-ray Absorption: Principles, Applications, Techniques of EXAFS, SEXAFS and XANES:* Köningsberger, D. C., Prins, R. (Eds.). New York: Wiley.

Kucharczyk, D., Paciorek, W., Kalincinska-Karut, J. (1982), *Phase Transitions 2*, 277.

Kunz, C. (Ed.) (1979), *Synchrotron Radiation: Techniques and Applications, Topics in Current Physics*, Vol. 10. Berlin: Springer.

Kuriyama, M., Steiner, B. W., Dobbyn, R. C. (1989), *Ann. Rev. Mater. Sci. 19*, 183.

Lang, A. R., Kowalski, G., Makepiece, A. P. W., Moore, M., Clackson, S., Yacoot, A. (1987), *Synchrotron Radiation, Appendix to the Daresbury Annual Report 1986/1987*, SERC Daresbury Laboratory.

Lee, P. A., Citrin, P. H., Eisenberger, P., Kincaid, B. M. (1981), *Rev. Mod. Phys. 53*, 769.

Lim, G., Parrish, W., Oritz, C., Bellotto, M., Hart, M. (1987), *J. Mater. Res. 2*, 471.

Longo, J. M., Raccah, P. M. (1973), *J. Solid State Chem. 6*, 526.

Lucas, C. A. (1989), *Ph.D. Thesis, Edinburgh University*.

Lucas, G. E., Odette, G. R. (1986), *Proceedings of the Second International Symposium on Environmental Degradation of Materials in Nuclear Power Systems – Water Reactors*, Monterey, Sept. 1985, (ANS).

Lucas, C. A., Hatton, P. D., Bates, S., Ryan, T. W., Miles, S., Tanner, B. K. (1988), *J. Appl. Phys. 63 (6)*, 1936.

Luth, H., Nyburg, S. C., Robinson, P. M., Scott, H. G. (1974), *Mol. Cryst. Liq. Cryst. 27*, 337.

Macdonald, J. E. (1990), *Faraday Discuss. Chem. Soc. 89*, 191.

Macdonald, J. E., Williams, A. A., van Filfhout, R., van der Veen, J. F., Finney, M. S., Johnson, A. D., Norris, C. (1990), in: *Proceedings of NATO ARW, Kinetics of Ordering at Surfaces*. New York: Plenum Press, in press.

Mamott, G. T., Barnes, P., Tarling, S. E., Jones, S. L., Norman, C. J. (1991), *J. Mater. Sci.*, in press.

Martens, G., Rabe, P. (1980), *Phys. Stat. Sol. (a) 58*, 415.

Materlik, G., Zegenhagen, Z. (1984), *J. Phys. Lett. 104 A*, 47.

Materlik, G., Frahm, A., Bedzyk, M. J. (1984), *Phys. Rev. Lett. 52*, 441.

McGinn, S. T. (1990), Ph.D. Thesis, Liverpool University.

McSkimin, H. J., Bond, W. L. (1957), *Phys. Rev. 105*, 116.

Melroy, O. R., Toney, M. F., Borges, G. L., Samant, M. G., Blum, L., Kortright, J. B., Ross, P. N. (1988), *Phys. Rev. B38*, 10962.

Michot, G., Weil, B., George, A. (1984), *J. Cryst. Growth 69*, 627.

Mikhail, I., Peters, K. (1979), *Acta Cryst. B35*, 1200.

Milia, F., Kind, R., Slak, J. (1983), *Phys. Rev. B27*, 6662–6668.

Miltat, J., Bowen, D. K. (1979), *J. de Physique 40*, 389.

Miltat, J., Sauvage-Simkin, M. (1984), in: *Applications of X-ray Topographic Methods to Materials Science*: Weissman, S., Balibar, F., Petroff, J. F. (Eds.). New York: Plenum Press, pp. 185–209.

Mitrovic, M. M., Ristić, R. I., Ciric, I. (1990), *Appl. Phys. A 51*, 374.

Murray, A. D., Crockroft, A. J., Fitch, A. N. (1990), *MPROF – a Program for multiplatten Rietveld refinement of crystal structures from X-ray powder data*. To be published.

Nehring, J., Amstutz, H., Holmes, P. A., Nevin, A. (1987), *Appl. Phys. Lett. 51*, 1283.

Nyburg, S. C., Potworowski, J. A. (1973), *Acta. Cryst. B29*, 347.

Ocko, B. M., Wong, J., Davenport, A., Isaacs, H. (1990), *Phys. Rev. Lett. 65*, 1466.

Oldman, R. J. (1986), *J. de Physique C8*, 321.

O'Reilly, E. P. (1989), *Semicond. Sci. and Tech. 4*, 121.

Palmberg, W. W., Rhodin, T. N. (1967), *Phys. Rev. 161*, 586.

Parratt, L. G. (1954), *Phys. Rev. 95*, 359.

Parrish, W., Hart, M., Huang, T. C. (1986), *J. Appl. Cryst. Growth 19*, 92.

Petroff, J. F., Sauvage, M. (1978), *J. Cryst. Growth 43*, 628.

Phythian, W. J., English, C. A. (1991), *UKAEA Harwell Report AERE-R-13632*, to be published.

Pizzini, S., Roberts, K. J., Greaves, G. N., Harris, N., Moore, P., Pantos, E., Oldman, R. J. (1989), *Rev. Sci. Instrum. 60*, 2525.

Pizzini, S., Roberts, K. J., Phythian, W. J., English, C. A. (1990a), *Phil. Mag. Lett. 61*, 223.

Pizzini, S., Roberts, K. J., Phythian, W. J., English, C. A. (1990b), *Proc. of the 6th International Conference on X-Ray Absorption Fine Structure (XAFS VI)*. Ellis Horwood, Chichester, pp. 530–532.

Pizzini, S., Roberts, K. J., Greaves, G. N., Barrett, N. T., Dring, I., Oldman, R. J. (1990c), *Faraday Discuss. Chem. Soc. 89*, 51.

Pizzini, S. (1990d), Ph.D. Thesis, University of Strathclyde.

Polk, D. E., Bordeaux, D. S. (1973), *Phys. Rev. Lett. 31*, 92.

Quilichini, M., Mathieu, J. P., Le Postollec, M., Toupry, N. (1982), *J. de Physique 43*, 787.

Reynolds, C. D., Stowell, B., Joshi, K. K., Harding, M. M., Maginn, S. J., Dodson, G. G. (1988), *Acta Cryst. B44*, 512.

Rietveld, H. M. (1969), *J. Appl. Cryst. 2*, 65.

Ristić, R. I., Sherwood, J. N., Wojciechowski, K. (1988), *J. Crystal Growth 91*, 163.

Ristić, R. I., Sherwood, J. N., Shripathi, T. (1991), in: *Advances in Industrial Crystallisation*: Garside, J., Davey, R. J. (Eds.). London: Butterworths, pp. 77–91.

Roberts, K. J., Sherwood, J. N., Bowen, D. K., Davies, S. T. (1982), *Materials Letters 2*, 300.

Roberts, K. J., Sherwood, J. N., Bowen, D. K., Davies, S. T. (1983), *Materials Letters 1*, 104.

Roberts, K. J., Sherwood, J. N., Shripathi, T., Oldman, R. J., Holmes, P. A., Nevin, A. (1990), *J. Physics D: Applied Physics 23*, 255.

Robinson, I. K., Tung, R. T., Feidenhans'l, R. (1988), *Phys. Rev. B38*, 3632.

Ryan, T. W., Hatton, P. D., Bates, S., Watt, M., Sotomayor-Torres, C., Claxton, P. A., Roberts, J. S. (1987), *Semicond. Sci. Technol.*, 241.

Sacchi, M., Antonini, G. M., Barrett, N. T., Greaves, G. N., Thornley, F. R. (1989), *Proc. 2nd Int. Work-*

shop on *Non-Crystalline Solids, San Sebastian (Spain)*, July 1989.

Safa, M., Tanner, B. K. (1978), *Phil. Mag. B 37*, 739.

Samant, M. G., Toney, M. F., Borges, G. L., Blum, L., Melroy, O. R. (1988), *J. Phys. Chem. 92*, 220.

Samwer, K. (1988), *Phys. Rep. 161*, 1.

Sayers, D. E., Stern, E. A., Lytle, F. M. (1971), *Phys. Rev. Lett. 27*, 1204.

Scheffer, T. J., Nehring, J. (1984), *Appl. Phys. Lett. 45*, 1021.

Schultz, L. (1987), *Mater. Sci. Engin. 93*, 213.

Sery, R. S., Savage, H. T., Tanner, B. K., Clark, G. F. (1978), *J. Appl. Phys. 49*, 2010.

Shirane, G., Endoh, Y., Birgeneau, R. J., Kastner, M. A., Hidaka, Y., Oda, M., Suzuki, M., Muarakami, T. (1987), *Phys. Rev. Lett. 59*, 1613.

Steininger, R., Bilgram, J. H. (1990), *J. Crystal Growth 99*, 98.

Stephenson, J. D., Kelha, V., Tilli, M., Tuomi, T. (1979), *Phys. Stat. Sol. (a) 51*, 93.

Stern, E. A. (1974), *Phys. Rev. B 10*, 3027.

Stocks, G. M., Temmerman, W. M., Szotek, Z., Sterne, P. A. (1988), *Supercond. Sci. Technol. 1*, 57.

Stuhrmann, H. B. (Ed.) (1982), *Uses of Synchrotron Radiation in Biology*. London: Academic Press.

Tanner, B. K. (1976), *X-ray Diffraction Topography*. Oxford: Pergamon Press.

Tanner, B. K., Safa, M., Midgley, D., Bordas, J. (1976), *J. Magn. Mat. 1*, 337.

Teo, B. K., Chen, H. S., Wang, R., D'Antonio, M. R. (1983), *J. Non-Crystalline Solids 58*, 249.

Tuomi, T., Naukkarnin, K., Rabe, P. (1974), *Phys. Status Solidi A 25*, 93.

Tuomi, T., Tilli, M., Kelha, V., Stephenson, J. D. (1978), *Phys. Stat. Sol. (a) 50*, 427.

Tuomi, T., Stephenson, J. D., Tilli, M., Kelha, V. (1979), *Phys. Stat. Sol. (a) 53*, 571.

Vecher, A. A., Dalidovich, S. V., Gusev, E. A. (1985), *Thermochimica Acta 89*, 383.

Vlieg, E., van't Ent, A., de Jong, A. P., Neerings, H., van der Veen, J. F. (1987), *Nucl. Instrum, and Methods A 262*, 522.

Wagner, C. J., Boldrick, M. S., Keller, L., (1988), *Advances in X-ray Analysis 31*, 129.

Wainfan, N., Parratt, L. G. (1960), *J. Appl. Phys. 31*, 1331.

Wainfan, N., Nancy, J., Scott, N. J., Parratt, L. G. (1959), *J. Appl. Phys. 30*, 1604.

Waiti, G. C., Lundu, M. L., Ghosh, S. K., Banerjee, B. K. (1974), *Thermal Analysis*, Vol. 2, *Proc. 4th Intern. Conf. Therm. Anal.*, Budapest 1974.

Weeber, A. W., Bakker, H. (1988), *Physica B 153*, 93.

Whatmore, R. W., Goddard, P. A., Tanner, B. K., Clark, G. F. (1982), *Nature 299*, 44.

Williams, A. A., Macdonald, J. E., van Silfhout, R. G., van der Veen, J. F., Johnson, A. D., Norris, C. (1989), *J. Phys. Condens. Matt. 1*, 273.

Winick, H., Doniach, S. (1980), *Synchrotron Radiation Research*. New York: Plenum Press.

Wood, I. G., Thompson, P., Matthewman, J. C. (1983), *Acta Cryst. B 39*, 543.

Zarka, A. (1983), *J. Appl. Cryst. 16*, 354.

Zarka, A. (1984), in: *Applications of X-ray Topographic Methods to Materials Science:* Weissmann, S., Balibar, F., Petroff, J. F. (Eds.). New York: Plenum Press, pp. 487–500.

Zei, M. S., Lehmpfuhl, G., Kolb, D. M. (1989), *Surf. Sci. 221*, 23.

General Reading

Amoros, J. L., Buerger, M. J., Amoros, M. C. (1975), *The Laue Method*. New York: Academic Press.

Bowen, D. K. (Ed.) (1983), *The Application of Synchrotron Radiation to Problems in Materials Science*, SERC Daresbury Laboratory Report DL/SCI/R19.)

Koch, E. E. (Ed.) (1983), *Handbook on Synchrotron Radiation*, Vol. 1. Amsterdam: North Holland.

Köningsberger, D. C., Prins, R. (Eds.) (1988), *X-ray Absorption, Principles, Applications, Techniques, of EXAFS, SEXAFS and XANES*. New York: Wiley.

Kunz, C. (Ed.) (1979), *Synchrotron Radiation: Techniques and Applications, Topics in Current Physics*, Vol. 10. Berlin: Springer.

Lee, P. A., Citrin, P. H., Eisenberger, P., Kincaid, B. M. (1981), *Rev. Modern Phys. 53*, 769.

Stuhrmann, H. B. (Ed.) (1982), Uses of Synchrotron Radiation in Biology. London: Academic Press.

Tanner, B. K. (1976), *X-ray Diffraction Topography*. Oxford: Pergamon Press.

Tanner, B. K. (1977), *Progress in Crystal Growth and Characterisation 1*, 23.

Tanner, B. K., Bowen, D. K. (Eds.) (1979), *Characterisation of Crystal Growth Defects by X-ray Methods*. New York: Plenum Press.

Weissmann, S., Balibar, F., Petroff, J.-F. (Eds.) (1984), *Application of X-ray Topographic Methods to Materials Science*. New York: Plenum Press.

Winick, H., Doniach, S. (Eds.) (1980), *Synchrotron Radiation Research*. New York, London: Plenum Press.

9 X-Ray Fluorescence Analysis

Ron Jenkins

JCPDS-International Centre for Diffraction Data, Swarthmore, PA, U.S.A.

List of Symbols and Abbreviations

B, b	background
C	concentration
E	energy
e_i	energy to produce one ion pair
F	Fano factor
$I(\lambda)$	photon intensity
J	total angular momentum
L	orbital angular momentum
LLD	lower limit of detection
l	angular quantum number
M	gas gain
m	magnetic quantum number
m	sensitivity of X-ray fluorescence method
N	number of electrons
N	number of measurements
n	principal quantum number
n_b, n_p	number of counts on peak (p) and background (b)
R	counting rate
R_b, R_p	background and peak counting rates
R_t	theoretical resolution
s	spin quantum number
t	time
t_b	background counting time
t_p	peak counting time
V	voltage
W	weight fraction
Z	atomic number
α	total absorption
λ	wavelength
μ	mass absorption coefficient
ϱ	density
σ	shielding constant
$\sigma_{(N)}$	random error
σ_{net}	net counting error
ϕ	binding energy
EDS	electron diffraction spectrometry
FET	field effect transistor
FWHM	full width half maximum
LOD	lead octadeconate
LSM	layered synthetic micro-structure
MDL	minimum detectable limits
NRLXRF	Naval Research Laboratory X-ray fluorescence

p-i-n diode	diode with p-doped, isolating and n-doped layers
PIXE	proton excited X-ray fluorescence
SRM	standard reference material
SSXRF	synchrotron source X-ray fluorescence
TAP	thallium acid phtalate
TRXRF	total reflection X-ray fluorescence

9.1 Introduction

X-rays are a short wavelength form of electromagnetic radiation discovered by Wilhelm Röntgen just before the turn of the century (Röntgen, 1898). X-ray photons are produced following the ejection of an inner orbital electron from an excited atom, and subsequent transition of atomic orbital electrons from states of high to low energy. A beam of X-rays passing through matter is subject to three processes, *absorption, scatter,* and *fluorescence*. The *absorption* of X-rays varies as the third power of the atomic number of the absorber. Thus, when a polychromatic beam of X-rays is passed through a heterogeneous material, areas of high average atomic number will attenuate the beam to a greater extent than areas of lower atomic number. Thus the beam of radiation emerging from the absorber has an intensity distribution across the irradiation area of the specimen, which is related to the average atomic number distribution across the same area. It is upon this principle that all methods of X-ray radiography are based. Study of materials by use of the X-ray absorption process is the oldest of all of the X-ray methods in use and Röntgen himself included a radiograph of his wife's hand in his first published X-ray paper. Today, there are many different forms of X-ray absorptiometry in use, including industrial radiography, diagnostic medical and dental radiography, and security screening. X-rays are *scattered* mainly by the loosely bound outer electrons of an atom. Scattering of X-rays may be coherent (same wavelength) or incoherent (longer wavelength). Coherently scattered photons may undergo subsequent mutual interference, leading in turn to the generation of diffraction maxima. The angles at which the diffraction maxima occur can be related to

the spacings between planes of atoms in the crystal lattice and hence, X-ray diffraction patterns can be used to study the structure of solid materials. Following the discovery of the diffraction of X-rays by Friedrich, Knipping and Von Laue (Friedrich et al., 1912) the use of this method for materials analysis has become very important both in industry and research, to the extent that, today, it is one of the most useful techniques available for the study of structure dependent properties of materials. *Fluorescence* occurs when the primary X-ray photons are energetic enough to create electron vacancies in the specimen, leading in turn to the generation of secondary (fluorescence) radiation produced from the specimen. This secondary radiation is characteristic of the elements making up the specimen. The technique used to isolate and measure individual characteristic wavelengths is following excitation by primary X-radiation, is called X-ray fluorescence spectrometry.

X-ray spectrometric techniques provided important information for the theoretical physicist in the first half of this century and since the early 1950's they have found an increasing use in the field of materials characterization. While most of the early work in X-ray spectrometry was carried out using electron excitation (e.g., von Hevesey, 1932), today use of electron-excited X-radiation is restricted mainly to X-ray spectrometric attachments to electron microscopes, and most stand-alone X-ray spectrometers use X-ray excitation sources rather than electron excitation. X-ray fluorescence spectrometry typically uses a polychromatic beam of short wavelength X-radiation to excite longer wavelength characteristic lines from the sample to be analyzed. Modern X-ray spectrometers use either the diffracting power of a single crystal to isolate narrow wavelength bands, or a proportional detector to isolate narrow energy bands, from the polychromatic radiation (including characteristic radiation) excited in the sample. The first of these methods is called *wavelength dispersive spectrometry* and the second, *energy dispersive spectrometry*. Because the relationship between emission wavelength and atomic number is known, isolation of individual characteristic lines allows the unique identification of an element to be made and elemental concentrations can be estimated from characteristic line intensities. Thus this technique is a means of materials characterization in terms of chemical composition.

9.2 Historical Development of X-Ray Spectrometry

X-ray fluorescence spectrometry provides the means of the identification of an element by measurement of its characteristic X-ray emission wavelength or energy. The method allows the quantization of a given element by first measuring the emitted characteristic line intensity and then relating this intensity to elemental concentration. While the roots of the method go back to the early part of this century, it is only during the last thirty years or so that the technique has gained major significance as a routine means of elemental analysis. The first use of the X-ray spectrometric method dates back to the classic work of Henry Moseley (Moseley, 1912). In Moseley's original X-ray spectrometer, the source of primary radiation was a cold cathode tube in which the source of electrons was residual air in the tube itself, the specimen for analysis forming the target of the tube. Radiation produced from the specimen then passed through a thin gold window, onto an analyzing crystal whence

it was diffracted to the detector. One of the major problems in the use of electrons for the excitation of characteristic X-radiation is that the process of conversion of electron energy into X-rays is relatively inefficient; about 99% of the electron energy is converted to heat energy. This means in turn that it may be difficult to analyze specimens which are volatile or tend to melt. Nevertheless, the technique seemed to hold some promise as an analytical tool and one of the first published papers on the use of X-ray spectroscopy for real chemical analysis appeared as long ago as 1922, when Hadding (Hadding, 1922) described the use of the technique for the analysis of minerals. In 1925, a practical solution to the problems associated with electron was suggested by Coster and Nishina (Coster and Nishina, 1925), who used primary X-ray photons for the excitation of secondary characteristic X-ray spectra.

The use of X-rays, rather than electrons, to excite characteristic X-radiation avoids the problem of the heating of the specimen. It is possible to produce the primary X-ray photons inside a sealed X-ray tube under high vacuum and efficient cooling conditions, which means the specimen itself need not be subject to heat dissipation problems or the high vacuum requirements of the electron beam system. Use of X-rays rather than electrons represented the beginnings of the technique of X-ray fluorescence as we know it today. The fluorescence method was first employed on a practical basis in 1928 by Glocker and Schreiber (Glocker and Schreiber, 1928). Unfortunately, data obtained at that time were rather poor because X-ray excitation is even less efficient than electron excitation. Also the detectors and crystals available at that time were rather primitive and thus the fluorescence technique did not seem to hold too much promise. In the event, widespread use of the technique had to wait until the mid 1940's when X-ray fluorescence was rediscovered by Friedman and Birks (Birks, 1976). The basis of their spectrometer was a diffractometer that had been originally designed for the orientation of quartz oscillator plates. A Geiger counter was used as a means of measuring the intensities of the diffracted characteristic lines and quite reasonable sensitivity was obtained for a very large part of the atomic number range.

The first commercial X-ray spectrometer became available in the early 1950's and although these earlier spectrometers operated only with an air path, they were able to provide qualitative and quantitative information on all elements above atomic number 22 (titanium). Later versions allowed use of helium or vacuum paths that extended the lower atomic number cut-off. Most modern commercially available X-ray spectrometers have a range from about 0.4 to 20 Å (40 to 0.6 keV) and this range will allow measurement of the K series from fluorine ($Z = 9$) to lutecium ($Z = 71$), and the L series from manganese ($Z = 25$) to uranium ($Z = 92$). Other line series arise from the M and N levels but these have little use in analytical X-ray spectrometry. In practice, the number of vacancies in electron levels resulting in the production of characteristic X-ray photons is less than the toal number of vacancies created in the excitation process, because the atom can also regain its initial state by re-organization of atomic electrons without the emission of X-ray photons (the Auger process). This is an important factor in determining the absolute number of counts that an element will give, under a certain set of experimental conditions. It is mainly for this reason that the sensitivity of the X-ray spectrometric technique is rather poor for

the very low atomic number elements, since fluorescent yields for these low atomic numbers are very small.

Today, nearly all commercially available X-ray spectrometers use the fluorescence excitation method and employ a sealed X-ray tube as the primary excitation source. Some of the simpler systems may use a radio-isotope source, because of considerations of cost and/or portability. While electron excitation is generally not used in stand-alone X-ray spectrometers, it is the basis of X-ray spectrometry carried out on electron column instruments. The ability to focus the primary electron beam allows analysis of extremely small areas down to a micron or so in diameter, or even less in specialized instruments. This, in combination with imaging and electron diffraction, offers an extremely powerful method for the examination of small specimens, inclusions, grain boundary phenomena, etc. The instruments used for this type of work may be in the form of a specially designed *electron microprobe analyzer* (e.g., Birks, 1963) or simply an energy or wavelength dispersive attachment to a scanning electron microscope.

Over the past thirty years or so, the X-ray fluorescence method has become one of the most valuable methods for the qualitative and quantitative analysis of materials. Many methods of instrumental elemental analysis are available today and among the factors that will generally be taken into consideration in the selection of one of these methods are accuracy, range of application, speed, cost, sensitivity and reliability. While it is certainly true that no one technique can ever be expected to offer all of the features that a given analyst might desire, the X-ray method has good overall performance characteristics. In particular, the speed, accuracy and versatility of X-ray fluorescence are the most important features among the many that have made it the method of choice in over 15 000 laboratories all over the world. Both the simultaneous wavelength dispersive spectrometer and the energy dispersive spectrometers lend themselves admirably to the qualitative and quantitative analysis of solid materials and solutions. Because the characteristic X-ray spectra are so simple, the actual process of allocating atomic numbers to the emission lines is also relatively simple, and the chance of making a gross error is rather small. The relationship between characteristic line intensity and elemental composition is also now well understood, and if intensities can be obtained that are free from instrumental artifacts, excellent quantitative data can be obtained. Today, conventional X-ray fluorescence spectrometers allow the rapid quantitation of all elements in the periodic table from fluorine (atomic number 9) and upwards. Recent advances in wavelength dispersive spectrometers have extended this element range down to carbon (atomic number 6). Over most of the measurable range, accuracies of a few tenths of one percent are possible, with detection limits down to the low ppm level.

9.3 Relationship Between Wavelength and Atomic Number

9.3.1 Continuous Radiation

When a high energy electron beam is incident upon a specimen, one of the products of the interaction is an emission of a broad wavelength band of radiation called *continuum*, also referred to as *white radiation* or *bremsstrahlung*. This white radiation is produced as the impinging high energy electrons are decelerated by the atomic electrons of the elements making up the specimen. The intensity/wavelength distribution of this radiation is typified by a minimum wavelength λ_{min}, which is proportional to the maximum accelerating potential V of the electrons, i.e., $12.4/V$ (keV). The intensity distribution of the continuum reaches a maximum at a wavelength 1.5 to 2 times greater than λ_{min}. Increasing the accelerating potential causes the intensity distribution of the continuum to shift towards shorter wavelengths. Most commercially available spectrometers utilize a sealed X-ray tube as an excitation source, and these tubes typically employ a heated tungsten filament as a source of electrons, and a layer of pure metal such as chromium, rhodium or tungsten, as the anode. The broad band of white radiation produced by this type of tube is ideal for the excitation of the characteristic lines from a wide range of atomic numbers. In general, the higher the atomic number of the anode material, the more intense the beam of radiation produced by the tube. Conversely, however, because the higher atomic number anode elements generally require thicker exit windows (to provide adequate dissipation of heat from the electrons scattered by the anode) in the tube, the longer wavelength output from such a tube is rather poor and so these high atomic number anode tubes are less satisfactory for excitation of longer wavelengths from low atomic number samples.

9.3.2 Characteristic Radiation

In addition to electron interactions leading to the production of white radiation, there are also electron interactions which produce characteristic radiation. If a high energy particle, such as an electron, strikes a bound atomic electron, and the energy of the particle is greater than the binding energy of the atomic electron, it is possible that the atomic electron will be ejected from its atomic position, departing from the atom with a kinetic energy $(E - \phi)$ equivalent to the difference between the energy E of the initial particle and the binding energy ϕ of the atomic electron. Where the exciting particles are X-ray photons, the ejected electron is called a photoelectron and the interaction between primary X-ray photons and atomic electrons is called the *photoelectric effect*. As long as the vacancy in the shell exists, the atom is in an unstable state and there are two processes by which the atom can revert back to its original state. The first of these involves a re-arrangement that does not result in the emission of X-ray photons, but in the emission of other photoelectrons from the atom. The effect is known as the Auger effect (Auger, 1925a, b), and the emitted photoelectrons are called Auger electrons.

The second process by which the excited atom can regain stability is by transference of an electron from one of the outer orbitals to fill the vacancy. The energy difference between the initial and final states of the transferred electron may be given off in the form of an X-ray photon. Since all emitted X-ray photons have energies proportional to the differences in the energy states of atomic electrons, the lines from a given element will be characteristic of that element. The relationship between the wavelength of a characteristic X-ray photon and the atomic number Z of the excited element was first established by Moseley. Moseley's law is written:

$$1/\lambda = K(Z - \sigma)^2 \tag{9-1}$$

in which K is a constant that takes on different values for each spectral series. σ is the shielding constant that has a value of just less than unity. The wavelength of the X-ray photon is inversely related to the energy E of the photon according to the relationship:

$$\lambda(\text{Å}) = 12.4/E \text{ (keV)} \tag{9-2}$$

Since there are two competing effects by which an atom may return to its initial state, and since only one of these processes will give rise to the production of a characteristic X-ray photon, the intensity of an emitted characteristic X-ray beam will be dependent upon the relative effectiveness of the two processes within a given atom. As an example, the number of quanta of K series radiations emitted per ionized atom, is a fixed ratio for a given atomic number, this ratio being called the fluorescent yield. The fluorescent yield varies as the fourth power of atomic number and approaches unity for higher atomic numbers. Fluorescent yield values are several orders of magnitude less for the very low atomic numbers. In practice this means that if, for example, one were to compare the intensities obtained from pure barium ($Z = 56$) and pure aluminium ($Z = 13$), all other things being equal, pure barium would give about 50 times more counts than would pure aluminium. The L fluorescent yield for a given atomic number is always less by about a factor of three than the corresponding K fluorescent yield.

9.3.3 Selection Rules

An excited atom can revert to its original ground state by transferring an electron from an outer atomic level to fill the vacancy in the inner shell. An X-ray photon is emitted from the atom as part of this de-excitation step, the emitted photon having an energy equal to the energy difference between the initial and final states of the transferred electron. Each unique atom has a number of available electrons which can take part in the transfer and since millions of atoms are typically involved in the excitation of a given specimen, all possible de-excitation routes are taken. These de-excitation routes can be defined by a simple set of selection rules that account for the majority of the observed wavelengths. Each electron in an atom can be defined by four quantum numbers. The first of these quantum numbers is the principal quantum number n, that can take on all integral values. When n is equal to 1, the level is referred to as the K level; when n is 2, the L level, and so on. ℓ is the angular quantum number and this can take all values from $(n-1)$ to zero. m is the magnetic quantum number that can take values of $+\ell$, zero and $-\ell$. s is the spin quantum number with a value of $\pm 1/2$. The total momentum "J" of an electron is given by the vector sum of $(\ell + s)$. Since no two electrons within a given atom can have the same set of quantum numbers, a series of levels or *shells* can be constructed. Table 9-1 gives the atomic structures of the first three principal shells. The first shell, the K shell, has

Table 9-1. Atomic structures of the first three principal shells.

Shell/Electrons	n	m	l	s	Orbitals	J
K (2)	1	0	0	$\pm 1/2$	1s	1/2
L (8)	2	0	0	$\pm 1/2$	2s	1/2
	2	0	0	$\pm 1/2$	2p	1/2
	2	-1	1	$\pm 1/2$	2p	1/2, 3/2
	2	$+1$	1	$\pm 1/2$	2p	1/2, 3/2
M (18)	3	0	0	$\pm 1/2$	3s	1/2
	3	1	0	$\pm 1/2$	3p	1/2
	3	-1	1	$\pm 1/2$	3p	1/2, 3/2
	3	$+1$	1	$\pm 1/2$	3p	1/2, 3/2
	3	0	0	$\pm 1/2$	3d	1/2
	3	-1	1	$\pm 1/2$	3d	1/2, 3/2
	3	$+1$	1	$\pm 1/2$	3d	1/2, 3/2
	3	-2	2	$\pm 1/2$	3d	3/2, 5/2
	3	$+2$	2	$\pm 1/2$	3d	3/2, 5/2

a maximum of two electrons and these are both in the 1s level (orbital). Note that the "s" refers, in this context, to the orbital shape (s = *sharp*) taken by the electrons and not the spin quantum number! Since the value of J must be positive in this instance the only allowed value is 1/2. In the second shell, the L shell, there are eight electrons, two in the 2s level and six in the 2p levels. In this instance J has a value of 1/2 for the 1 s level and 3/2 or 1/2 for the 2p level. Thus giving a total of three possible L transition levels. These levels are referred to as L_I, L_{II} and L_{III} respectively. In the M level, there are a maximum of eighteen electrons, two in the 3s level, eight in the 3p level and ten in the 3d level. Again with the values of 3/2 or 1/2 for J in the 3p level; and 5/2 and 3/2 in the 3d level, a total of five M transition levels are possible. Similar rules can be used to build up additional levels, N, O etc.

The selection rules for the production of *normal* (diagram) lines require that the principal quantum number must change by at least one, the angular quantum number must change by only one, and the J quantum number must change by zero or one. Transition groups may now be constructed, based on the appropriate number of transition levels. Application of the selection rules indicates that in, for example, the K series, only L_{II} to K and L_{III} to K transitions are allowed for a change in the principal quantum number of one. There are an equivalent pairs of transitions for $n = 2$, $n = 3$, $n = 4$, etc. Figure 9-1 shows the lines that are observed in the K series. Three groups of lines are indicated. The normal lines are shown on the left hand side of the figure and these consist of three pairs of lines from the L_{II}/L_{III}, M_{II}/M_{III} and N_{II}/N_{III} sub-shells respectively. While most of the observed fluorescent lines are normal, certain lines may also occur in X-ray

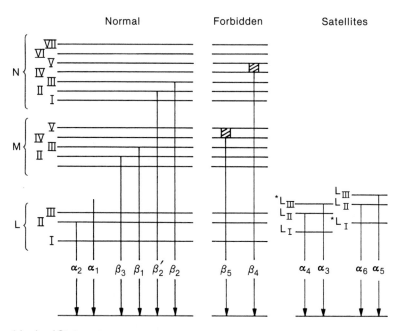

Figure 9-1. Observed lines in the K series.

* Ionized States

spectra that do not at first sight fit the basic selection rules. These lines are called *forbidden* lines and are shown in the center portion of the figure. Forbidden lines typically arise from outer orbital levels where there is no sharp energy distinction between orbitals. As an example, in the transition elements, where the 3d level is only partially filled and is energetically similar to the 3p levels, a weak forbidden transition (the β_5) is observed. A third type of line may also occur – called satellite lines, which arise from dual ionizations. Following the ejection of the initial electron in the photoelectric process, a short, but finite, period of time elapses before the vacancy is filled. This time period is called the lifetime of the excited state. For the lower atomic number elements, this lifetime increases to such an extent that there is an significant probability that a second electron can be ejected from the atom before the first vacancy is filled. The loss of the second electron modifies the energies of the electrons

in the surrounding sub-shells, and other pairs of X-ray emission lines are produced, corresponding to the α_1/α_2. In the K series the most common of these satellite lines are the α_3/α_4 and the α_5/α_6 doublets. These lines are shown at the right hand side of the figure. Although, because they are relatively weak, neither forbidden transitions nor satellite lines have great analytical significance, they may cause some confusion in qualitative interpretation of spectra and may even be misinterpreted as coming from trace elements.

9.3.4 Nomenclature for X-Ray Wavelengths

The classically accepted nomenclature system for the observed lines is that proposed by Siegbahn in the 1920's. Figure 9-2 shows plots of the reciprocal of the square root of the wavelength, as a function of atomic number, for the K, L and M series. As indicated by Moseley's law

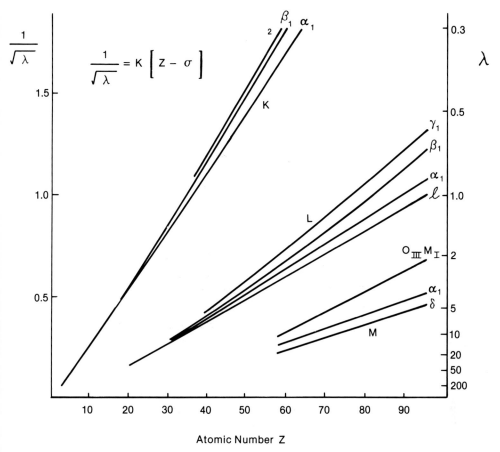

Figure 9-2. Moseley diagrams for the K, L and M series.

[Eq. (9-1)] such plots should be linear. A scale directly in wavelength is also shown, to indicate the range of wavelengths over which a given series occurs. In practice, the number of lines observed from a given element will depend upon the atomic number of the element, the excitation conditions and the wavelength range of the spectrometer employed. Generally, commercial spectrometers cover the range 0.3 to 20 Å (newer instruments may allow measurements in excess of 100 Å) and three X-ray series are covered by this range, the K series, the L series and the M series, corresponding to transitions to K, L and M levels respectively. Each series consists of a number of

groups of lines. The strongest group of lines in the series is denoted α, the next strongest β, and the third γ. There are a much larger number of lines in the higher series and for a detailed list of all of the reported wavelengths the reader is referred to the work of Bearden (Bearden, 1964). In X-ray spectrometry most of the analytical work is carried out using either the K or the L series wavelengths. The M series may, however, also be useful, especially in the measurement of higher atomic numbers.

Both the simultaneous (scanning) wavelength dispersive spectrometer and the energy dispersive spectrometers lend themselves admirably to the qualitative analysis of materials, since there is a simple relationship between the wavelength or energy of a characteristic X-ray photon, and the atomic number of the element from which the characteristic emission line occurs. Thus by measuring the wavelengths, or energies, of a given series of lines from an unknown material, the atomic numbers of the excited elements can be established. The inherent simplicity of characteristic X-ray spectra make the process of allocating atomic numbers to the emission lines relatively easy, and the chance of making a gross error is rather small. There are only 100 or so elements, and within the range of the conventional spectrometer each element gives, on an average, only half a dozen lines. If one compares the X-ray emission spectrum with the ultra-violet emission spectrum, since the X-ray spectrum arises from a limited number of inner orbital transitions, the number of X-ray lines is similarly rather few. Ultra-violet spectra, on the other hand, arise from transitions to empty levels, of which there may be many, leading to a significant number of lines in the UV emission spectrum. A further benefit of the X-ray emission spectrum for qualitative analysis is that because transitions do arise from inner orbitals, the effect of chemical combination, or valence state is almost negligible.

9.4 Properties of X-Radiation

9.4.1 Absorption of X-Rays

When a beam of X-ray photons of intensity $I_0(\lambda)$ falls onto a specimen a fraction of the beam will pass through the absorber this fraction being given by the expression:

$$I(\lambda) = I_0(\lambda) \exp(\mu \varrho x) \tag{9-3}$$

where μ is the mass absorption coefficient of absorber for the wavelength and ϱ the density of the specimen. x is the distance travelled by the photons through the specimen. It will be seen from Eq. (9-3) that a number $(I_0 - I)$ of photons has been lost in the absorption process. Although a significant fraction of this loss may be due to scatter, by far the greater loss is due to the photoelectric effect. Photoelectric absorption occurs at each of the energy levels of the atom, thus the total photoelectric absorption is determined by the sum of each individual absorption within a specific shell. Where the absorber is made up of a number of different elements, as is usually the case, the total absorption is made up of the sum of the products of the individual elemental mass absorption coefficients and the weight fractions of the respective elements. This product is referred to as the total matrix absorption. The value of the mass absorption referred to in Eq. (9-3) is a function of both the photoelectric absorption and the scatter. However, the photoelectric absorption influence is usually large in comparison with the scatter and to all intents and purposes the mass absorption coefficient is equivalent to the photoelectric absorption. A plot of the mass absorption coefficient as a function of wavelength contains a number of discontinuities, called absorption edges, at wavelengths corresponding to the binding energies of the electrons in the various subshells. Between absorption edges, as the wavelength of the incident X-ray photons become longer, the absorption increases. This particular effect is very important in quantitative X-ray spectrometry because the intensity of a beam of characteristic photons leaving a specimen is dependent upon the relative absorption effects of the different atoms making up the specimen. This effect is called a matrix effect and is one of the reasons why a curve of characteristic line intensity as a function of element concentration may not be a straight line.

9.4.2 Coherent and Incoherent Scattering

Scattering occurs when an X-ray photon interacts with the electrons of the target element. Where this interaction is elastic, i.e., no energy is lost in the collision process, the scattering is referred to as coherent (Rayleigh) scattering. Since no energy change is involved, the coherently scattered radiation will retain exactly the same wavelength as that of the incident beam. It can also happen that the scattered photon gives up a small part of its energy during the collision, especially where the electron with which the photon collides is only loosely bound. In this instance, the scatter is referred to as incoherent (Compton scattering). Compton scattering is best presented in terms of the corpuscular nature of the X-ray photon. In this instance, an X-ray photon collides with a loosely bound outer atomic electron. The electron recoils under the impact, removing a small portion of the energy of the primary photon, which is then deflected with the corresponding loss of energy, or increase of wavelength.

9.4.3 Interference and Diffraction

X-ray diffraction is a combination of two phenomena – coherent scatter and interference. At any point where two or more waves cross one another, they are said to interfere. Interference does not imply the impedance of one wave train by another, but rather describes the effect of superposition of one wave upon another. The principal of superposition is that the resulting displacement, at any point and at any instant, may be formed by adding the instantaneous displacements that would be produced at the same point by independent wave trains, if each were present alone. Under certain geometric conditions, wavelengths that are exactly in phase may add to one another, and those that are exactly out of phase, may cancel each other out. Under such conditions coherently scattered photons may constructively interfere with each other giving diffraction maxima.

As illustrated in Fig. 9-3a, a crystal lattice consists of a regular arrangement of atoms, with layers of high atomic density existing throughout the crystal structure. Planes of high atomic density means, in turn, planes of high electron density. Since scattering occurs between impinging X-ray photons and the loosely bound outer orbital atomic electrons, when a monochromatic beam of radiation falls onto the high atomic density layers scattering will occur. In order to satisfy the requirement for constructive interference, it is necessary that the scattered waves originating from the individual atoms i.e. the scattering points, be in phase with one another. The geometric conditions for this condition to occur are illustrated in Fig. 9-3b. Here, a series of parallel rays strike a set of crystal planes at an angle θ and are scattered as previously described. Reinforcement will occur when the difference in the path lengths of the two interfering waves is equal to a whole number of wavelengths. This path length difference is equal $2d \sin \theta$, where d is the interplanar spacing; hence the overall condition for reinforcement is that:

$$n\lambda = 2d \sin \theta \qquad (9\text{-}4)$$

where n is an integer. Equation (9-4) is a statement of Bragg's law. Bragg's law is important in wavelength dispersive spectrometry since by using a crystal of fixed $2d$, each unique wavelength will be diffracted at a unique diffraction angle (Bragg

(a)

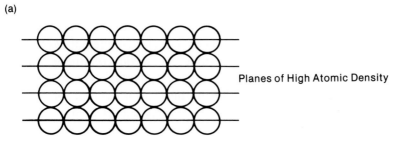

Planes of High Atomic Density

(b)

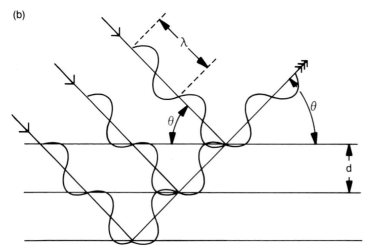

Figure 9-3. The crystal lattice and origin of X-ray diffraction. The structure of an ideal crystal is shown in (a). (b) Bragg condition: In order to ensure that the scattered waves remain in phase, the path length difference between successive waves ($2d \cdot \sin\theta$) must equal a whole number (n) of wavelengths (λ), i.e. $n\lambda = 2d \cdot \sin\theta$.

angle). Thus, by measuring the diffraction angle θ, knowledge of the d-spacing of the analyzing crystal allows the determination of the wavelength. Since there is a simple relationship between wavelength and atomic number, as given by Moseley's law [Eq. (9-1)], one can establish the atomic number(s) of the element(s) from which the wavelengths were emitted.

9.5 Instrumentation for X-Ray Fluorescence

9.5.1 Sources

Several different types of source have been employed for the excitation of characteristic X-radiation, including those based on electrons, X-rays, γ-rays, protons and synchrotron radiation. Sometimes a bremsstrahlung X-ray source is used to generate specific X-radiation from an intermediate pure element sample called a *secondary fluorescer*. All of the early work in X-ray spectrometry was done using electron excitation and this technique is still used very successfully today in electron column applications. Not too much use is made of electron excitation in classical X-ray fluorescence systems, owing mainly to the inconvenience of having to work under high vacuum and because of problems of heat dissipation. By far the most common source today is the X-ray photon source. This source is used in primary mode in the wavelength and primary energy dispersive systems, and in secondary fluorescer mode in secondary target energy dispersive spectrometers. A γ-source is typically a radio-isotope that is used either directly, or in a mode equivalent to the secondary fluorescer mode in energy dispersive spectrometry. Most conventional wavelength dispersive X-ray spectrometers use a high power (2 to 4 kW) X-ray bremsstrahlung source. Energy dispersive spectrometers use either a high power or low power (0.5 to 1.0 kW) primary source, depending on whether the spectrometer is used in the secondary or primary mode. In all cases, the primary source unit consists of a very stable high voltage generator, capable of providing a potential of typically 40 to 100 kV. The current from the generator is fed to the filament of the X-ray tube that is typically a coil of tungsten wire. The applied current causes the filament to glow, emitting electrons in all directions. A portion of this electron cloud is accelerated to the anode of the X-ray tube, which is typically a water cooled block of copper with the required anode material plated or cemented to its surface. The impinging electrons produce X-radiation, a significant portion of which passes through a thin beryllium window to the specimen.

In order to excite a given characteristic line the source must be run at a voltage V_0 well in excess of the critical excitation potential V_c of the element in question. The relationship between the measured intensity of the characteristic line I, the tube current i and the operating and critical excitation potentials is as follows:

$$I = K\, i\, (V_0 - V_c)^{1.6} \qquad (9\text{-}5)$$

The product of i and V_0 represents the maximum output of the source in kilowatts. The optimum value for V_0/V_c is 3 to 5. This optimum value occurs because at very high operating potentials, the electrons striking the target in the X-ray tube penetrate so deep into the target that self

absorption of target radiation becomes significant.

Since it is the intention to eventually equate the value of I for a given wavelength or energy to the concentration of the corresponding analyte element, it is vital that, over the short term (1 to 2 hours), both the tube current and voltage be stabilized to better than a tenth of percent. The current from the generator is fed to the tungsten filament of the X-ray tube. A sealed X-ray tube has an anode of Cr, Rh, W, Ag, Au or Mo and delivers an intense source of continuous and characteristic radiation, which then impinges onto the analyzed specimen, where characteristic radiation is generated. In general, most of the excitation of the longer wavelength characteristic lines comes from the longer wavelength characteristic lines from the tube, and most of the short wavelength excitation, from the continuous radiation from the tube. Since the relative proportions of characteristic to continuous radiation from a target increase with decrease of the atomic number of the anode material, optimum choice of a target for the excitation of a range of wavelengths can present some problems. One way around this problem is to use a dual target tube and many different varieties of such tubes have been employed over the years. A recent manifestation of this is a dual anode tube in which the second (low atomic number) material is plated on top of the first (high atomic number) material. At high tube voltages the electrons penetrate beyond the thin layer of low atomic number material and the output is mainly continuum from the high atomic number substrate anode. At low tube voltages the electrons dissipate their energy mainly in the low atomic number surface layer with a resulting output biased in favor of longer wavelength radiation. Combinations of Sc/Mo, Cr/Ag, and Sc/W have all proven useful in this regard (Kikkert and Hendry, 1983).

9.5.2 Detectors

An X-ray detector is a transducer for converting X-ray photon energy into voltage pulses. Detectors work through a process of photo-ionization in which interaction between the entering X-ray photon and the active detector material produces a number of electrons. The current produced by these electrons is converted to a voltage pulse by a capacitor and resistor, such that one digital voltage pulse is produced for each entering X-ray photon. In addition to being sensitive to the appropriate photon energies, i.e, being applicable to a given range of wavelengths or energies, there are two other important properties that an ideal detector should possess. These properties are *proportionality and linearity*. Each X-ray photon entering the detector produces a voltage pulse and where the size of the voltage pulse is proportional to the photon energy, the detector is said to be *proportional*. Proportionality is needed where the technique of pulse height selection is to be used. Pulse height selection is a means of electronically rejecting pulses of voltage levels other than those corresponding to the characteristic line being measured. This technique is a very powerful tool in reducing background levels and the influence of overlapping lines from elements other than the analyte (Jenkins and de Vries, 1970). X-ray photons enter the detector at a certain rate and where the output pulses are produced at this same rate the detector is said to be *linear*. Linearity is important where the various count rates produced by the detector are to be used as measures of the photon intensity for each measured line. The properties of proportionality and linearity are, to a cer-

tain extent, controllable by the electronics associated with the actual detector. To this extent, while it is common practice to refer to the characteristics of detectors, the properties of the associated pulse processing chain should always be included.

A gas-flow proportional counter consists of a cylindrical tube about 2 cm in diameter, carrying a thin (25 to 50 μm) wire along its radial axis. The tube is filled with a mixture of inert gas and quench gas – typically 90% argon/10% methane (P-10). The cylindrical tube is grounded and about + 1800 volts is applied to the central wire. The anode wire is connected to a resistor shunted by a capacitor forming the pulse producing circuit. An X-ray photon entering the detector produces a number of ion pairs, each comprising one electron and one argon positive ion. The average energy e_i required to produce one ion pair in argon is equal to 26.4 eV. Thus the number of ion pairs n produced by a photon of energy W will equal (E/e_i). Following ionization, the charges separate with the electrons moving towards the (anode) wire and the argon ions to the grounded cylinder. As the electrons approach the high field region close to the anode wire they are accelerated sufficiently to produce further ionization of argon atoms. Thus a much larger number N of electrons will actually reach the anode wire. This effect is called gas gain, or gas multiplication, and its magnitude is given by M, where M is equal to N/n. For gas flow proportional counters used in X-ray spectrometry, M typically has a value of around 10^5. Provided that the gas gain in constant, the size of the voltage pulse V produced is directly proportional to the energy E of the incident X-ray photon.

In practice not all photons arising from photon energy E will be exactly equal to V. There is a random process associated with the production of the voltage pulses and the resolution of a counter is related to the variance in the average number of ion pairs produced per incident X-ray photon. The resolution is generally expressed in terms of the full width at half maximum of the pulse amplitude distribution. The theoretical resolution R_t of a flow counter can be derived from:

$$R_t(\%) = \frac{38.3}{\sqrt{E}} \qquad (9\text{-}6)$$

While the gas flow proportional counter is ideal for the measurement of longer wavelengths it is rather insensitive to wavelengths shorter than about 1.5 Å. For this shorter wavelength region it is common to use the scintillation counter. The scintillation counter consists of two parts, the phosphor (scintillator) and the photo-multiplier. The phosphor is typically a large single crystal of sodium iodide that has been doped with thallium. When X-ray photons fall onto the phosphor, blue light photons are produced, where the number of blue light photons is related to the energy of the incident X-ray photon. These blue light photons produce electrons by interaction with a photo-surface in the photo-multiplier, and the number electrons of electrons is linearly increased by a series of secondary surface *dynodes*, electrically charged with respect to each other, in the photo-multiplier. The current produced by the photo-multiplier is then coverted to a voltage pulse, as in the case of the gas flow proportional counter. Since the number of electrons is proportional to the energy of the incident X-ray photon, the scintillation counter is a *proportional counter*. Because of inefficiencies in the X-ray/blue-light/electron conversion processes, the average energy to produce a single event in the scintillation counter is more than a magnitude greater than the equivalent process in the flow counter. For

this reason, the resolution of the scintillation counter is about four times worse than that of the flow counter.

The Si(Li) detector consists of a small cylinder (about 1 cm diameter and 3 mm thick) of p-type silicon that has been compensated by lithium to increase its electrical resistivity. A Schottky barrier contact on the front of the silicon disk produces a p-i-n type diode. In order to inhibit the mobility of the lithium ions and to reduce electronic noise, the diode and its preamplifier are cooled to the temperature of liquid nitrogen. While some progress has recently been reported on the use of miniature refrigerator compressors for this cooling process (Kevex Corp., 1987 a, b), the common practice remains to mount the detector assembly on a *cold finger* that is in turn kept cool with a reservoir of liquid nitrogen. By applying a reverse bias of around 1000 volts, most of the remaining charge carriers in the silicon are removed. Incident X-ray photons interact to produce a specific number of electron hole pairs. The average energy to produce one ion pair is equal to about 3.8 eV for cooled silicon. The charge produced is swept from the diode by the bias voltage to a charge sensitive preamplifier. A charge loop integrates the charge on a capacitor to produce an output pulse as in the case of the flow proportional counter, although in this case the gain is equal to unity since the Si(Li) detector does not have an equivalent property to gas gain.

The resolution R of the Si(Li) detector is given in eV by:

$$R = \{(\sigma_{\text{noise}})^2 + [2.35\,(e_i\,F\,E)]^2\}^{1/2} \qquad (9\text{-}7)$$

F is a variable called the Fano factor, which for the Si(Li) detector has a value of about 0.12 (Walter, 1971). The noise contribution is about 100 eV. Using a value of 3.8

Table 9-2. Characteristics of common X-ray detectors.

Detector type	Useful range (Å)	Average energy per ion pair (eV)	Data for Cu Kα				
			(Number of electrons)			(Resolution)	
			Initial	Gain	Final	(eV)	(%)
Si(Li)	0.5–8	3.8	2116	1	2×10^3	160	2.0
Gas flow proportional	1.5–50	26.4	305	6×10^4	2×10^7	1086	13.5
Scintillation	0.2–2	350	23	10^6	2×10^7	3638	45.3
Crystal spectrometer [a]							
LiF (200)						31	0.39
LiF (220) 1st order						22	0.27
LiF (220) 2nd order						12	0.15

[a] Based on a 10 mm long primary collimator of 150 μm spacing and a 5 cm long secondary collimator with 120 μm spacing.

for e_i, the calculated resolution for Mn Kα radiation ($E = 5.895$ eV) is about 160 eV.

Table 9-2 summarizes some of the characteristics of the three detectors commonly employed in X-ray spectrometry and also compares the resolution of these detectors with that of the crystal spectrometer. As was previously shown, the number of ion pairs produced is directly proportional to the energy of the X-ray photon and inversely proportional to the average energy to produce one ion pair. Going from the Si(Li) detector, to the flow counter to the scintillation counter, the average energy to produce an ion pair increases by roughly an order of magnitude each time. Since the resolution of the detector is related to the square root of the number of electrons per photon, the resolution of the three detectors varies by about the square root of 10 or about three times, as indicated in the data for Cu Kα shown in the table. Another important practical parameter is the actual number of electrons produced for one photon of Cu Kα that eventually arrive at the anode. The size of the voltage pulse produced by the cathode follower of the detector is proportional to the current

produced, i.e., to the number of electrons reaching the collector. In a detector that has internal gain, this number of electrons will be the product of the initial number of electrons and the gain of the detector. In the case of the gas flow proportional counter, this is the gas gain, and in the scintillation counter, the photo-multiplier gain. It will be remembered that the Si(Li) detector has no internal gain. The Si(Li) detector has a high number of initial electrons/photon (therefore, a high resolution), but, relative to the other two detectors, has a small number of final electrons. This means that normal external amplification of voltage pulses from the Si(Li) detector cannot be used because this would also amplify noise from the detector. For this reason a cooled charge-sensitive preamplifier is used in Si(Li) spectrometers rather than a simple linear electronic amplifier. Finally, the table shows the actual resolution of the detectors used on their own, in comparison with a crystal spectrometer. While the resolution of the Si(Li) detector for Cu Kα is worse than the crystal spectrometer, as the energy of the analyte photon increases, the resolution difference between detectors alone and crystal spectrometers decreases. Note too that the resolution of both gas flow and scintillation counters is almost never sufficient for use without a crystal spectrometer.

9.5.3 Types of Spectrometer Used for X-Ray Fluorescence

The basic function of the spectrometer is to separate the polychromatic beam of radiation coming from the specimen in order that the intensities of each individual characteristic line can be measured. A spectrometer should provide sufficient resolution of lines to allow such data to be taken, at the same time providing a sufficiently large response above background to make the measurements statistically significant, especially at low analyte concentration levels. It is also necessary that the spectrometer allow measurements over the wavelength range to be covered. Thus, in the selection of a set of spectrometer operating variables, four factors are important: *resolution, response, background level and range*. Owing to many factors, optimum selection of some of these characteristics may be mutually exclusive; as an example, attempts to improve resolution invariably cause lowering of absolute peak intensities. There is a wide variety of instrumentation available today for the application of X-ray fluorescence techniques and it is useful to break these instrument types down into three main categories: wavelength dispersive spectrometers (sequential and simultaneous); energy dispersive spectrometers (primary or secondary) and special spectrometers (including total reflection, synchrotron source and proton induced). The wavelength dispersive system was introduced commercially in the early 1950's and probably around 15 000 or so such instruments have been supplied commercially,

roughly half of these in the U.S.A. Energy dispersive spectrometers became commercially available in the early 1970's and today, there are several thousands of these units in use. There are far less of the specialized spectrometers in use and these generally incorporate an energy dispersive rather than a wavelength dispersive spectrometer.

Wavelength dispersive spectrometers employ diffraction by a single crystal to separate characteristic wavelengths emitted by the sample. Energy dispersive spectrometers use the proportional characteristics of a photon detector, typically lithium drifted silicon, to separate the characteristic photons in terms of their energies. Since there is a simple relationship between wavelength and energy, these techniques each provide the same basic type of information, and the characteristics of the two methods differ mainly in their relative sensitivities and the way in which data is collected and presented. Generally speaking the wavelength dispersive system is roughly one to two orders of magnitude more sensitive than the energy dispersive system. Against this, however, the energy dispersive spectrometer measures all elements within its range at essentially the same time, whereas the wavelength dispersive system identifies only those elements for which it is programmed. To this extent, the energy dispersive system is more useful in recognizing unexpected elements. Both wavelength and energy dispersive spectrometers typically employ a primary X-ray photon source operating at 0.5 to 3 kW. A disadvantage with this type of source is that the specimen scatters the white from the source leading to significant background levels that tend to become one of the major limitations in the determination of low concentration levels. Typical analysis times vary from about 10 seconds to three minutes per element. The minimum sample size required is of the order of a few milligrams, although typical sample sizes are probably around several grams. Good accuracy is obtainable and in favorable cases, standard deviations of the order of a few tenths of one percent are possible. This is because the matrix effects in X-ray spectrometry are well understood and relatively easy to overcome. The sensitivity is fair and determinations down to the low parts per million level are possible for most elements.

X-ray spectrometers also differ in the number of elements which they are able to measure at one time and the speed at which they collect data. All of the instruments are, in principle at least, capable of measuring all elements in the periodic classification from $Z = 9$ (F) and upwards, and most modern wavelength dispersive spectrometers can do some useful measurements down to $Z = 6$ (C). Most systems can be equipped with multi-sample handling facilities and can be automated by use of minicomputers. Typical spectrometer systems are capable of precisions of the order of a few tenths of one percent with sensitivities down to the low ppm level. Single channel wavelength dispersive spectrometers are typically employed for both routine and non-routine analysis of a wide range of products, including ferrous and non-ferrous alloys, oils, slags and sinters, ores and minerals, thin films, and so on. These systems are very flexible but, relative to multichannel spectrometers, are somewhat slow. The multi-channel wavelength dispersive instruments are used almost exclusively for routine, high throughput, analysis where the great need is for fast accurate analysis, but where flexibility is of no importance. Energy dispersive spectrometers have the great advantage of being able to display information on all elements at the same time. They lack somewhat in resolution compared with the wavelength dispersive spectrometer, but the ability to reveal elements absent as well as elements present make the energy dispersive spectrometer ideal for general trouble-shooting problems. They have been particularly effective in the fields of scrap alloy sorting, in forensic science and in the provision of elemental data to supplement X-ray powder diffraction data.

While the various types of specialized X-ray spectrometers are not generally available to the average user, they do have important roles to play in special areas of application. Included within this category are *total reflection spectrometers* (TRXRF), *synchrotron source spectrometers* (SSXRF), and *proton induced X-ray emission spectrometers* (PIXE). Two things that each of these three special systems have in common are a very high sensitivity and ability to work with extremely low concentrations and/or small specimens. The TRXRF system makes use of the fact that at very low glancing angles, primary X-ray photons are almost completely absorbed within thin specimens and the high background that woud generally occur due to scatter from the sample support is absent. The recent development of high intensity synchrotron radiation beams has led to interest in their application for X-ray fluorescence analysis. The high intensity available in synchrotron beams allows use of very narrow bandpath monochromators between source and specimen, giving, in turn, a high degree of selective excitation. This selectivity overcomes one of the major disadvantages of the classical EDS approach and allows excellent detection limits to be obtained. The proton induced X-ray emission system differs from conventional energy dispersive spectroscopy in that a proton source is used in place of the photon source. The proton source is typically a Van de Graf generator or a cyclotron, giving protons in the energy range of about 2 to 3 MeV. In addition to being an intense source relative to the conventional photon source, the proton excitation system generates relatively much lower backgrounds. In addition, the cross section for characteristic X-ray production is quite large and good excitation efficiency is possible.

One of the problems with any X-ray spectrometer system is that the absolute sensitivity (i.e., the measured c/s per % of analyte element) decreases significantly as the lower atomic number region is approached. The three main reasons for this are: decrease in the fluorescence yield with decrease in atomic number; the absolute number of useful long wavelength X-ray photons from a bremsstrahlung source decreases with increase of wavelength; and thirdly, absorption effects generally become more severe with increase of the wavelength of the analyte line. The first two of these problems are inherent to the X-ray excitation process and to constraints in the basic design of conventional X-ray tubes. The third, however, is a factor that depends very much on the instrument design, and, in particular, upon the characteristics of the detector. The detector that is used in long wavelength spectrometers is typically a gas flow proportional counter, in which an extremely thin, high-transmission window is employed. The detector typically employed in energy dispersive systems is the Si(Li) diode which has an electrical contact layer on the front surface – typically a 0.02 μm thick layer of gold, followed by a 0.1 μm thick dead layer of silicon. The absorption problems caused by these two layers become most significant for low energy X-ray photons, which have a high probability of being absorbed in the dead layer. Probably the biggest source of absorption loss in the Si(Li) detector is that due to the thin beryllium window that is part of the liquid nitrogen cryostat. A combination of these facts causes a loss in the sensitivity of a typical energy dispersive system of almost an order of magnitude, for the K lines of sulfur ($Z = 16$) to sodium ($Z = 11$). The equivalent number with a gas flow counter would be about a factor of two. Until very recently, the lowest atomic number usefully detectable with a typical energy dispersive spectrometer has been magnesium ($Z = 12$). New developments in ultra-thin windows for the Si(Li) detector (e.g., Kevex Corp., 1987 a, b) now allows measurements down to oxygen ($Z = 8$). In comparison, the lower atomic number limit for the conventional wavelength dispersive system is fluorine ($Z = 9$), and by use of special crystals this can be extended down to beryllium ($Z = 4$).

A compromise that must always be made in the design and setup of any spectrometer is that between intensity and resolution, resolution being defined as the ability of the spectrometer to separate lines. In flat-crystal wavelength dispersive system this resolution is dependent upon the angular dispersion of the analyzing crystal and the divergence allowed by the collimators (Jenkins, 1974). In the energy dispersive system the resolution is dependent only upon the detector and detector amplifier. In absolute terms, the resolution of the wavelength dispersive system typically lies in the range 10 to 100 eV, compared to a value of 150 to 200 eV for the energy dispersive system. An advantage of the wavelength dispersive spectrometer in this context is that the resolution/intensity measuring selection is much more controllable. In the case of secondary fluorescer systems some local modifications to the measured spectrum can be made by use of filters either in the primary beam or in the secondary fluoresced beam and this does offer some flexibility.

9.5.3.1 Wavelength Dispersive Systems

A wavelength dispersive spectrometer may be a single channel instrument in which a single crystal and a single detector are used for the measurement of a series of wavelengths sequentially; or a multi-channel spectrometer in which many crystal/detector sets are used to measure elements simultaneously. Of these two basic types, the sequential systems are the most common. A typical sequential spectrometer system consists of the X-ray tube, a specimen holder support, a primary collimator, an analyzing crystal and a tandem detector. The gas flow counter is ideal for the measurement of the longer wavelengths and the scintillation counter is best for the short wavelengths. The geometric arrangement of these components is shown in Fig. 9-4. A portion of the characteristic *fluorescence* radiation from the specimen is passed via a collimator or slit onto the surface of an analyzing crystal, where individual wavelengths are diffracted to the detec-

tor in accordance with the Bragg law. A goniometer is used to maintain the required θ-to-2θ relationship between crystal and detector. Typically, six or so different analyzing crystals and two different collimators are provided in this type of spectrometer, giving the operator a wide range of choice of dispersion conditions. In general, the smaller the $2d$-spacing of the crystal, the better the separation of the lines but the smaller the wavelength range that can be covered. Since the maximum achievable angle θ on a wavelength dispersive spectrometer is generally around $73°$, the maximum wavelength that can be diffracted by a crystal of spacing $2d$ is equal to about $1.9\,d$. The separating power of a crystal spectrometer is dependent upon the divergence allowed by the collimators (that to a first approximation determines the width of the diffracted lines), and the angular dispersion of the crystal (Jenkins, 1974). Since mechanical limitations prevent wide selectability of line shape just by selection of collimator divergence, in practice, the resolution of the spectrometer is typically determined by the angular dispersion of the analyzing crystal, albeit with some influ-

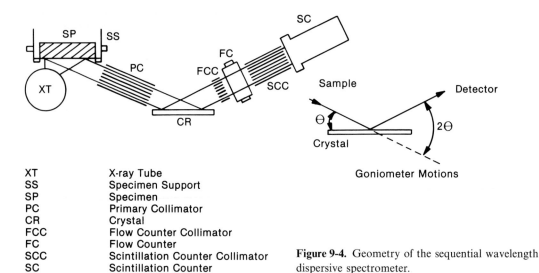

XT	X-ray Tube
SS	Specimen Support
SP	Specimen
PC	Primary Collimator
CR	Crystal
FCC	Flow Counter Collimator
FC	Flow Counter
SCC	Scintillation Counter Collimator
SC	Scintillation Counter

Figure 9-4. Geometry of the sequential wavelength dispersive spectrometer.

ence of the breadth of the diffracted line profile. The angular dispersion $d\theta/d\lambda$ of a crystal of spacing $2\,d$ is given by:

$$d\theta/d\lambda = n/(2\,d\cos\theta) \qquad (9\text{-}8)$$

It will be seen from Eq. (9-8) that the angular dispersion will be high when the $2d$-spacing is small. This is unfortunately as far as the range of the spectrometer is concerned, because a small value of $2\,d$ means in turn a small range of wavelengths coverable. Thus, as with the resolution and peak intensities, the obtaining of high dispersion can only be obtained at the expense of cutting down the wavelength range covered by a particular crystal. In order to circumvent this problem, it is likely that several analyzing crystals will be employed in the coverage of a number of analyte elements. Many different analyzing crystals are available, each having its own special characteristics, but three or four crystals will generally suffice for most applications. Table 9-3 gives a short list of the more commonly used crystals. While the maximum wavelength covered by traditional spectrometer designs is about 20 Å, recent developments now allow the extension of the wavelength range significantly beyond this value.

Classically, large single crystals have been used as diffracting structures in the wavelength dispersive spectrometer. The three dimensional lattice of atoms is oriented and fabricated such that Bragg planes form the interatomic 2d-spacing for the wavelength in question. The selection of crystals for the longer (> 8 Å) wavelength region is difficult, however, mainly because there are not too many crystals available for work in this region. The most commonly employed crystal is probably TAP, thallium acid phthalate ($2\,d = 26.3$ Å), and this allows measurement of the K lines of elements including magnesium, sodium, fluorine and oxygen. Several alternatives to single crystals as diffracting structures have been sought over the years including the use of complex organic materials with large 2d-spacings, gratings, and specular reflectors, metal disulphides, organic intercalation complexes such as graphite, molybdenum disulphide, mica and clays. Moderate success for long wavelength measurements was achieved by the use of soap films (Henke, 1965) having spacings in the range 80 to 120 Å. These layered (multi-layer) structures are composed of planes of heavy metal cations separated by chains of organic acids. The basis for their usefulness as diffracting structures is the periodic electron density contrast between the heavy metals sites and the lower density organic material. Lead octadeconate (LOD) with a 2d-spacing around 100 Å is one such film that has been used for carbon and oxygen

Table 9-3. Analyzing crystals used in wavelength dispersive X-ray spectrometry. For a comprehensive list of analyzing crystals see Bertin (1975).

Crystal	Planes	$2\,d\,(\text{Å})$	Atomic number range	
			K lines	L lines
Lithium fluoride	(220)	2.848	$>$ Ti (22)	$>$ La (57)
Lithium fluoride	(200)	4.028	$>$ K (19)	$>$ Cd (48)
Penta erythritol	(002)	8.742	Al (13) – K (19)	–
Thallium acid phthalate	(001)	26.4	F (9) – Na (11)	–
Layered synthetic microstructures		50 – 120	Be (4) – F (9)	–

analysis. Unfortunately, this diffracting medium lacks either adequate angular dispersion, or reflectivity, for the elements fluorine, sodium and higher atomic numbers. In addition to crysals and multilayer films, a third alternative has recently become available and this is the *layered synthetic micro-structure*, LSM. LSM's are constructed by applying successive layers of atoms or molecules on a suitably smooth substrate. In this manner, both the 2*d*-spacing and composition of each layer are selected for optimum diffraction characteristics, thus, to a certain extent, they can be designed and fabricated to give optimum performance for special applications (Barbee, 1985). Typical, experimental data shows factors of about four to six times improvement in peak intensities compared to TAP, for the range of elements measured (Nicolosi et al., 1985).

The output from a wavelength dispersive spectrometer may be either analog or digital. For qualitative work an analog output is traditionally used and the digital output from the detector amplifier is fed through a *d/a* converter, called a *rate meter*, to an *x/t* recorder which synchronously coupled with the goniometer scan speed. The recorder thus records an intensity/time diagram, in terms of an intensity/(2θ) diagram. It is generally more convenient to employ digital counting for quantitative work and a timer/scale combination is provided which will allow pulses to be integrated over a period of several tens of seconds and then displayed as count or count rate. In more modern spectrometers a scaler/timer may also take the place of the rate meter, using *step-scanning*. In this process the contents of the scaler are displayed on the *x*-axis of the *x/t* recorder as a voltage level. The scaler/timer is then reset and started to count for a selected time interval. At the end of this time the timer sends a stop pulse to the scaler that now holds a number of counts equal to the product of the counting rate and the count time. The contents of the scaler are then displayed as before, the goniometer stepped to its next position and the whole cycle repeated. Generally, the process is completely controlled by a microprocessor or a minicomputer.

Qualitative analysis has traditionally been performed by scanning the goniometer synchronously at a fixed angular speed. However, most scanning spectrometers are slow in this sequential angular/intensity data collection mode, not only because the data are taken sequentially, but also because in order to cover the full range of elements a series of scans must be made with different conditions. In addition to this, scanning at a fixed speed is somewhat inefficient because in the crystal dispersive system, atomic number varies as a function of one over the square root of the angle. In effect this means that in the low atomic number region much scanning time is wasted in scanning angular space that contains no characteristic line data. Although some unique designs have brought about some reduction in data acquisition times (e.g., Jenkins et al., 1985), this remains a major limitation of wavelength dispersive spectrometer. Data interpretation involves identifying each line in the measured spectrogram. However, since the relationship between atomic number and Bragg angle is rather complicated, it is common practice to use sets of tables for interpretation relating wavelength and atomic number, with diffraction angle for specific analyzing crystals (e.g., White and Johnson, 1970). Some automated wavelength dispersive spectrometers provide the user with software programs for the interpretation and labelling of peaks (e.g., Garbauskas and Goehner, 1983).

9.5.3.2 Energy Dispersive Systems

The energy dispersive spectrometer consists of the excitation source and the spectrometer/detection system. The spectrometer/detector is typically a Si(Li) detector which is a proportional detector of high intrinsic resolution. A multi-channel analyzer is used to collect, integrate and display the resolved pulses. While similar properties are sought from the energy dispersive system as with the wavelength dispersive system, the means of selecting these optimum conditions are very different. Since the resolution of the energy dispersive system is equated directly to the resolution of the detector, this feature is of paramount importance. The output from an energy dispersive spectrometer is generally displayed on a cathode ray tube and the operator is able to dynamically display the contents of the various channels as an energy spectrum and provision is generally made to allow zooming in on portions of the spectrum of special interest, to overlay spectra, to subtract background, and so on. Generally, some form of mini-computer is available for spectral stripping, peak identification, quantitative analysis, and other useful functions.

Even though the Si(Li) detector is the most common detector used in energy dispersive X-ray spectrometry, it is certainly not the only one. As an example, the higher absorbing power of germanium makes it an alternative for the measurement of high energy spectra. Both cadmium telluride, CdTe, and mercuric iodide, HgI_2, show some promise as detectors capable of operating satisfactorily at room temperature. As an example, a HgI_2 based energy dispersive spectrometer, in which both detector and FET were cooled using a Peltier cooler, has been used in a scanning electron microscope and a resolution of 225 eV

(FWHM) obtained for Mn Kα (5.9 keV) and 195 eV for Mg Kα (1.25 keV) (Iwanczyk et al., 1986).

All of the earlier energy dispersive spectrometers were operated in what is called the *primary* mode and consisted simply of the excitation source, typically a closely coupled low-power, end-window X-ray tube, and the detection system. In principle, this primary excitation system offered the possibility of a relatively inexpensive instrument, with two significant advantages over the wavelength dispersive system. Firstly the ability to collect and display the total emission spectrum from the sample at the same time, giving great speed in the acquisition and display of data. Secondly, offering mechanical simplicity since there is almost no need at all for moving parts. In practice, however, there is a limit as to the maximum count rate that the spectrometer can handle and this led, in the mid-1970's, to the development of the *secondary* mode of operation. In the secondary mode, a carefully selected pure element standard is interposed between primary source and specimen, along with absorption filters where appropriate, such that a selectable energy range of secondary photons is incident upon the sample. This allows selective excitation of certain portions of the energy range, thus increasing the ratio of useful to unwanted photons entering the detector. While this configuration does no completely eliminate the count rate and resolution limitations of the primary system, it certainly does reduce them.

9.5.3.3 Total Reflection Spectrometers (TRXRF)

One of the major problems that inhibits the obtaining of good detection limits in small samples is the high background due to scatter from the sample substrate support material. The suggestion to overcome this problem by using total reflection of the primary beam was made as long ago as 1971 (Yoneda and Horiuchi, 1971) but the absence of suitable instrumentation prevented real progress being made until the late 1970's (Knoth and Schwenke, 1978; Schwenke and Knoth, 1982). Mainly owing to the work of Schwenke and his co-workers, good sample preparation and presentation procedures are now available, making TRXRF a valuable technique for trace analysis (Michaelis et al., 1984). As illustrated in Fig. 9-5, the TRXRF method is essentially an energy dispersive technique in which the Si(Li) detector is placed close to (about 5 mm), and directly above, the sample. Primary radiation enters the sample at a glancing angle of a few seconds of arc. The sample itself is typically presented as a thin film on the optical flat surface of a quartz plate. In the instrument described by Michaelis a series of reflectors is employed to aid in the reduction of background. Here, a beam of radiation from a sealed X-ray tube passes through a fixed aperture onto a pair of reflectors that are placed very close to each other. Scattered radiation passes through then first aperture to impinge on the sample at a very low glancing angle. Because the primary radiation enters the sample at an angle barely less than the critical angle for total reflection, this radiation barely penetrates the substrate media; thus scatter and fluorescence from the substrate are minimal. Because the background is so low, picogram amounts can be measured or concentrations in the range of a few tenths of a ppb can be obtained without recourse to pre-concentration (Aiginger and Wobrauschek, 1985).

One area in which the TRXRF technique has found great application is in the analysis of natural waters (West and Nurrenberg, 1988). The concentration levels of, for example, transition metals in rain, river and sea waters, are normally too low to allow estimation by standard X-ray fluorescence techniques, unless pre-concentration is employed. Using TRXRF concentration levels down to less than $10 \ \mu g \ l^{-1}$ are achievable. While the TRXRF method is most applicable to homogeneous liquid samples in which the sample is evaporated onto the optical flat, success has also been achieved in the application of the method to solids including particulates, sediments, air dusts and minerals. In these instances the sample is first digested in concentrated nitric acid and then diluted to a calibrated

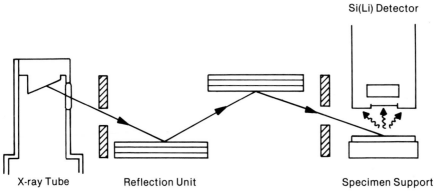

Si(Li) Detector

X-ray Tube   Reflection Unit   Specimen Support

Figure 9-5. The total reflection X-ray spectrometer – TRXRF.

volume with ultra-pure water, after the addition of an internal standard. Where undissolved material is still present, this may be dispersed using an ultrasonic bath before the specimen is taken. In addition to the advantage of ease of specimen preparation and the ability to handle milligram quantities of material, the TRXRF method is also relatively simple to apply quantitatively. Because the specimen is only a few microns thick, one generally does not observe the rather complicated matrix effects usually encountered with thick samples. Thus the only standard required is that to establish the sensitivity of the spectrometer, in terms of c/s per %, for the element(s) in question. This standard is generally added ot the sample to be analyzed during the specimen preparation procedure.

9.5.3.4 Synchrotron Source X-Ray Fluorescence (SSXRF)

The availability of intense, linearly polarized synchroton radiation beams (Sparks, 1980) has prompted workers in the fields of X-ray fluorescence (e.g., Gilfrich et al., 1983) and X-ray diffraction (e.g., Parrish et al., 1986), to explore what the source has to offer over more conventional excitation media. In the synchrotron, electrons with kinetic energies of the order of several billion electron volts (typically 3 GeV at this time), orbit in a high vacuum tube between the poles of a strong (about 10^4 Gauss) magnets. A vertical field accelerates the electrons horizontally, causing the emission of synchrotron radiation. Thus synchrotron source radiation can be considered as magnetic bremsstrahlung in contrast to normal electronic bremsstrahlung produced when electrons are decelerated by the electrons of an atom. In the case of both fluorescence and diffraction it has been found that because the primary source of radiation is so intense, it is possible to use a high degree of monochromatization between source and specimen, giving a source that is wavelength (and, therefore, energy) tunable, as well as being highly monochromatic. There are several different excitation modes that can be used using SSXRF including: direct excitation with continuum, excitation with absorber modified continuum, excitation with source crystal monochromatized continuum, excitation with absorber modified continuum, excitation with source crystal monochromatized continuum, excitation with source radiation scattered by a mirror, and reflection and transmission modes.

The intensity of the synchrotron beam is probably 4 to 5 orders of magnitude greater than the conventional bremsstrahlung source sealed X-ray tubes. This, in combination with its energy tunability and polarization in the plane of the synchroton ring, allows very rapid bulk analyses to be obtained on small areas. Because the synchrotron beam has such a high intensity and small divergence, it is possible to use it as a microprobe of high spacial resolution (about 10 μm). Absolute limits of detection around 10^{-14} μg have been reported using such an arrangement (Petersen et al., 1986). Synchrotron source X-ray fluorescence has also been used in combination with TRXRF. Very high signal/background ratios have been obtained employing this arrangement for the analysis of small quantities of aqueous solutions dried on the reflector, with detection limits of < 1 ppb or 1 pg. Additional advantages accrue because synchrotron radiation is higly polarized and background due to scatter can be greatly reduced by placing the detector at 90° to the path of the incident beam and in the plane of polarization. A disadvantage of the SSXRF technique is that the source intensity decreases with time, but this can be overcome by bracketing analytical measurements between source standards and/or by continuously monitoring the primary beam. In addition to this, problems can also arise from the occurrence of diffraction peaks from highly ordered specimens. Giauque et al. have described experiments (Giauque et al., 1986), using the Stanford Synchrotron Radiation Laboratory, to establish what minimum detectable limits (MDL's) could be obtained under optimum excitation conditions. Using thin film standards on stretched tetrafluoro-polyethylene mounts, it was found that for counting times of the order of a few hundred seconds, MDL's of the order of 20 ppb could be obtained on a range of elements including Ca, Ti, V, Mn, Fe, Cu, Zn, Rb, Ge, Sn and Pb. The optimum excitation energy was found to be about one and a quarter times the absorption edge energy of the element in question. By working with a fixed excitation energy of 18 keV, the average MDL was found to be about 100 ppb.

9.5.3.5 Proton Excited X-Ray Fluorescence (PIXE)

While the use of protons as a potential source for the excitation of characteristic X-rays has been recognized since the early 1960's, it is only over the past several years that the technique has come into its own (Garten, 1984). The PIXE method uses a beam of fast ions (protons) of primary energies in the range 1 to 4 MeV. In addition to the ion accelerator, the system contains an energy defining magnetic deflection field, a magnetic or electrostatic lens along the excitation beam pipe, a high vacuum target chamber for the specimen(s) and an energy dispersive detector/analyzer. The great advantage of the PIXE method over other sources is that it generates only a

small amount of background and is thus applicable to very low concentration levels and the analysis of very small samples. As an example, the use of conventional X-ray fluorescence and PIXE have been compared with special reference to applications in art and archeology. Together, they seem to offer the museum scientist and archaeologist excellent tools for non-destructive testing (Malmqvist, 1986). Comparison of the PIXE method has been made with many other spectroscopic techniques for the analysis of twenty-two elements in ancient pottery (Bird et al., 1986). The high sensitivity of PIXE has also been of great use in the field of forensic science and in medicine (Cesareo, 1982). Applications include in vitro analysis of trace elements in human body fluids and normal pathological tissues, in vivo analysis of iodine in the thyroid, lead in the skeleton and cadmium in the kidney. Trace elements in blood serum of patients with liver cancer has also been studied by PIXE. The copper/zinc ratio was found to be significantly higher as compared to normal patients (Lin et al., 1985). Serum copper/zinc ratio is potentially useful in the diagnosis and prognosis of liver cancer.

9.6 Accuracy of X-Ray Fluorescence

9.6.1 Counting Statistical Errors

The production of X-rays is a random process that can be described by a Gaussian distribution. Since the number of photons counted is nearly always large, typically thousands of hundreds of thousands, rather than a few hundred, the properties of the Gaussian distribution can be used to predict the probable error for a given count measurement. There will be a random error $\sigma(N)$ associated with a measured value of N, this being equal to $\sqrt{N}$. As an example, if 10^6 counts are taken, the 1σ standard deviation will be $\sqrt{10^6} = 10^3$, or 0.1%. The measured parameter in wavelength dispersive X-ray spectrometry is generally the counting rate R and, based on what has been already stated, the magnitude of the random counting error associated with a given datum can be expressed as:

$$\sigma(\%) = 100/\sqrt{N} = 100/\sqrt{R\,t} \qquad (9\text{-}9)$$

Care must be exercised in relating the counting error (or indeed any intensity related error) with an estimate of the error in terms of concentration. Provided that the sensitivity of the spectrometer in c/s per %, is linear, a count error can be directly related to a concentration error. However, where the sensitivity of the spectrometer changes over the range of measured response, a given fractional count error may be much greater when expressed in terms of concentration.

In many analytical situations the peak lies above a significant background and this adds a further complication to the counting statistical error. An additional factor that must also be considered is that whereas with the scanning wavelength dispersive spectrometer the peaks and background are measured sequentially, in the case of the energy dispersive and the multichannel wavelength dispersive spectrometers, a *single* counting time is selected for the complete experiment, thus all peaks and all backgrounds are counted for the same time. To estimate the net counting error in the case of a sequential wavelength dispersive spectrometer it is necessary to consider the counting error of the net response of peak counting rate R_p, and background counting rate R_b, since the analyte element is only responsible for $(R_p - R_b)$.

Equation (9-9) must then be expanded to include the background count rate term:

$$\sigma(R_p - R_b) = \frac{100}{\sqrt{t}} \cdot \frac{1}{\sqrt{R_p} - \sqrt{R_b}} \qquad (9\text{-}10)$$

One of the conditions for Eq. (9-10) is that the total counting time "t", must be correctly proportioned between time spent counting on the peak t_p, and time spent counting on the background t_b:

$$t_p/t_b = \sqrt{(R_p/R_b)} \qquad (9\text{-}11)$$

Several points are worth noting with reference to Eq. (9-10). Firstly, where the count time is limited – which is usually in case in most analyses, the net counting error is a minimum when $(\sqrt{R_p} - \sqrt{R_b})$ is maximum. This expression can, therefore, be used as a *figure of merit* for the setting up of instrumental variables (Jenkins and de Vries, 1970). Secondly, it will be noted that as R_b becomes small relative to R_p, Eq. (9-10) approximates to Eq. (9-9). In other words, as the background becomes less significant relative to the peak, its effect on the net counting error becomes smaller. The point at which the peak to background value exceeds 10:1 is generally taken as that where background can be ignored completely. A third point to be noted is that as R_b approaches R_p the counting error becomes infinite, and this will be a major factor in determining the lowest concentration limit that can be detected.

In the case of the energy dispersive spectrometer, the peak and background are recorded simultaneously and the question of division of time between peak and background does not arise. A choice does, however, have to be made as to what portion of the complete recorded spectrum should be used for the measurement of peak and background. The associated net counting error σ_{net} associated is given by:

$$\sigma_{net} = \sqrt{P + B(1 + n_p/n_b)} \qquad (9\text{-}12)$$

Where each number of background channels is chosen to be one half of the total number of peak channels, Eq. (9-12) reduces to $\sigma_{net} = \sqrt{(P + 2B)}$, or expressed as a percentage of the peak as:

$$\sigma_{net} = \frac{100\sqrt{P + 2B}}{P} \qquad (9\text{-}13)$$

9.6.2 Matrix Effects

In the conversion of net line intensity to analyte concentration it is necessary to correct for any absorption and/or enhancement effects which occur. Absorption effects include both primary and secondary absorption, and enhancement effects include direct enhancement, involving the analyte element and one enhancing element, plus third element effects, that involve additional element(s) beyond the analyte and enhancer. Primary absorption occurs because all atoms of the specimen matrix will absorb photons from the primary source. Since there is a competition for these primary photons by the atoms making up the specimen, the intensity/ wavelength distribution of these photons available for the excitation of a *given* analyte element may be modifed by other matrix elements. Secondary absorption refers to the effect of the absorption of characteristic analyte radiation by the specimen matrix. As characteristic radiation passes out from the specimen in which it was generated, it will be absorbed by all matrix elements, by amounts relative to the mass absorption coefficients of these elements. The mass absorption coefficient is a parameter which defines the magnitude of the absorption of a certain element for a specific X-ray wavelength. The total absorption α of

a specimen, is dependant on both primary and secondary absorption. The total absorption by element i for an analyte wavelength λ_j is given by the following relationship:

$$\alpha_i = \mu_i(\lambda) + A[\mu_i(\lambda_j)] \qquad (9\text{-}14)$$

The factor A is a geometric constant equal to the ratio of the sines of the incident and take-off angles of the spectrometer. This factor is needed to correct for the fact that the incident and emergent rays from the sample have different path lengths. The term λ in the equation refers to the primary radiation. Since most conventional X-ray spectrometers use a bremsstrahlung source, in practice λ is a range of wavelengths, although in simple calculations it may be acceptable to use a single *equivalent* wavelength value (Stephenson, 1971), where the equivalent wavelength is defined as a single wavelength having the same excitation characteristics as the full continuum.

There are a number of routes by which the analyte element can be excited or enhanced and these are illustrated in Fig. 9-6. Figure 9-6a shows the direct excitation of an analyte element i by the primary continuum P_1. There may also be excitation of the analyte by characteristic lines from the source, designated in Fig. 9-6b by P_2. Both the continuous and characteristic radiation from the source may be somewhat modified by Compton scatter and the excitation by this modified source radiation is indicated in Fig. 9-6c by P_3. Enhancement effects, Fig. 9-6d, occur when a non-analyte matrix element A emits a characteristic line that has an energy just in excess of the absorption edge of the analyte element. This means that the non-analyte element in question is able to excite the analyte, giving characteristic photons over and above those produced by the primary continuum. This gives an increased, or *enhanced*, signal

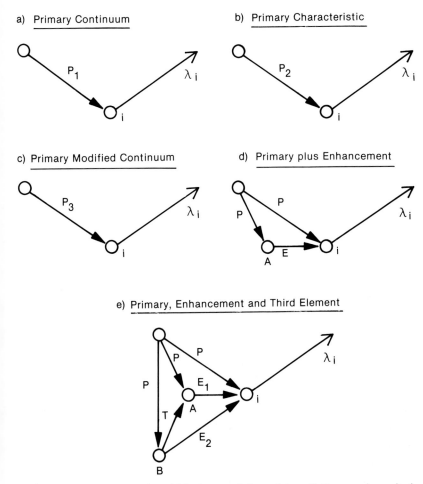

a) Primary Continuum

b) Primary Characteristic

c) Primary Modified Continuum

d) Primary plus Enhancement

e) Primary, Enhancement and Third Element

Figure 9-6. Various routes by which characteristic analyte radiation can be excited.

from the analyte. The *third element effect* is also shown in the figure. Here, a third element B, is also excited by the source. Not only can B directly enhance *i*, but it can also enhance A, thus increasing the enhancing effect of A on *i*. This last effect is called the third element effect. Table 9-4 shows some of the data published for the chromium/iron/nickel system (Shiraiwa and Fujino, 1967) and illustrates the relative importance of the various excitation routes. Relating these data to Fig. 9-6, *i* is the element chromium, iron is the enhancer A, and nickel is the third element.

Table 9-4. Relative importance of primary excitation, enhancement and third element effects in the Cr/Fe/Ni system (Shiraiwa and Fujino, 1967).

		Percentage of counts for Cr Kα (from the sources indicated) (element proportions given for Ni : Fe : Cr)		
	Route	25 : 25 : 50	10 : 40 : 50	40 : 10 : 50
Primary	P	87.5	87.2	87.6
Secondary	E1	6.7	10.6	2.6
	E2	5.3	2.0	9.4
Third element	T	0.5	0.2	0.4

ment B. The data given are for three different alloy compositions: 25/25/50; 10/40/50 and 40/10/50, nickel/iron/chromium in each case. Note that in each case, roughly 87% of the actual measured chromium $K\alpha$ radiation comes from direct excitation by the source. In the case of direct enhancement by iron, each one percent of iron added increases the chromium radiation by about 0.26%. Direct excitation of chromium by nickel gives about 0.21% increase in chromium radiation per percent nickel added. It will also be seen that the third element effect is much less important and here the chromium intensity is increased by only about 0.02% per percent of nickel added. From this one would predict that a *fourth* element effect would be negligible.

9.7 Quantitative Analysis

In the X-ray analytical laboratory the quantitative method of analysis employed will be typically predicated by a number of circumstances of which probably the four most common are: the complexity of the analytical problem; the time allowable; the computational facilities at the disposal of the analyst; and the number of standards available. It is convenient to break quantitative analytical methods down into two major categories: single element methods, and multiple element methods, as shown in Table 9-5. The simplest quantitative analysis situation to handle, is the determination of a single element in a known matrix. A slightly more difficult case might be the determination of a single element where the matrix is unknown. As shown in the table, three basic methods are commonly employed in this situation; use of internal standards, use of standard addition, or use of a scattered line from the X-ray source. The most complex case is the analysis of

Table 9-5. Quantitative procedures employed in X-ray fluorescence analysis.

Single element methods	internal standardization
	standard addition
	use of scattered source radiation
Multiple element methods	type standardization
	use of influence coefficients
	fundamental parameter techniques

all, or most, of the elements in a sample, about which little or nothing is known. In this case a full qualitative analysis would be required before any attempt is made to quantitate the matrix elements. Once the qualitative composition of the sample is known, again, one of three general techniques is typically applied; use of type standardization, use of an influence coefficient method, or use of a fundamental parameter technique. Both the influence coefficient and fundamental parameter technique require a computer for their application.

The great flexibility, sensitivity and range of the various types of X-ray fluorescence spectrometer make them ideal for quantitative analysis. In common with all analytical methods, quantitative X-ray fluorescence analysis is subject to a number of random and systematic errors that contribute to the final accuracy of the analytical result. Like all instrumental methods of analysis, the potentially high precision of X-ray spectrometry can only be translated into high accuracy if the various systematic errors in the analysis process are taken care of. The precision of a wavelength dispersive system, for the measurement of a single, well separated line, is typically of the order of 0.1%, and about 0.25% for the energy dispersive system. The major source of the random error is the X-ray source, with additional count time dependent errors arising from the statistics of the actual counting process. The random error can be significantly worse in the case of the energy dispersive system in those cases where full of partial line overlap occurs. Even though good peak and background stripping programs are available to reduce this problem, the statistical limitations of dealing with the difference of two large numbers remain. This is probably the biggest hindrance to obtaining precise count data from complex mixtures and this error can reach several percent in worst cases.

A good *rule-of-thumb* which can be used in X-ray fluorescence analysis to estimate the expected standard deviation σ at an analyte concentration level C, is given by:

$$\sigma = K \sqrt{C + 0.1} \qquad (9\text{-}15)$$

where K varies between 0.005 and 0.05 (Jenkins, 1988). For example, at a concentration level $C = 25\%$, the expected value of σ would be between about 0.025% and 0.25%. A K value of 0.005 would be considered very high quality analysis and a value of 0.05 rather poor quality. The value of K actually obtained under routine laboratory conditions depends upon many factors but with reasonably careful measurements a K value of around 0.02 to 0.03 can be obtained.

Table 9-6 lists the four main categories of random and systematic error encountered in X-ray fluorescence analysis. The first category includes the selection and preparation of the sample to be analyzed. Two stages are generally involved before the actual prepared specimen is presented to the spectrometer, these being sampling and specimen preparation. The actual sampling is rarely under the control of the spectroscopist and it generally has to be assumed that the container containing the material for analysis does, in fact, contain

Table 9-6. Sources of error in X-ray fluorescence analysis.

Source	Random (%)	Systematic (%)
1 Sample preparation	0–1	0–5
Sample inhomogeneity	–	0–50
2 Excitation source	0.05–0.2	0.05–0.5
Spectrometer	0.05–0.1	0.05–0.1
3 Counting statistics	Time dependent	–
Dead time	–	0–25
4 Primary absorption	–	0–50
Secondary absorption	–	0–25
Enhancement	–	0–15
Third element	–	0–2

a representative sample. It will be seen from the table that in addition to a relatively large random error, inadequate sample preparation and residual sample heterogeneity can lead to very large systematic errors. For accurate analysis these errors must be reduced by use of a suitable specimen preparation method. The second category includes errors arising from the X-ray source previously discussed. Source errors can be reduced to less than 0.1% by use of the ratio counting technique, provided that high frequency transients are absent. The third category involves the actual counting process and these errors can be both random and systematic. System errors due to detector dead time can be corrected either by use of electronic dead time correctors of by some mathematical approach. The fourth category includes all errors arising from inter-element effects. Each of the effects listed can give large systematic errors that must be controlled by the calibration and correction scheme.

The correlation between the characteristic line intensity of an analyte element and the concentration of that element, is typically non-linear over wide ranges of concentration, due to inter-element effects between the analyte element and other elements making up the specimen matrix. However, the situation can be greatly simplified in the case of homogeneous specimens, where severe enhancement effects are absent, and here, the slope of a calibration curve is inversely proportional to the total absorption α of the specimen for the analyte wavelength. In this instance the slope K of the calibration curve is taken as W/I where I is the line intensity and W the weight fraction of the analyte element.

Thus, the following relationship holds:

$$W = I/K \qquad (9\text{-}16)$$

Single element techniques *reduce* the influence of the absorption term in Eq. (9-16), generally by referring the intensity of the analyte wavelength to a similar wavelength, arising either from an added standard, or from a scattered line from the X-ray tube. In certain cases, limiting the concentration range of the analyte may allow the assumption to be made that the absorption value does not significantly change over the concentration range that the calibration curve is essentially linear. This assumption is applied in the traditional *type standardization technique.* Type standardization was widely employed in the 1960's and 1970's, but now that computers are generally available, it is usually considered more desirable to work with general purpose calibration schemes, which are applicable to a variety of matrix types, over wide concentration ranges. Sherman (Sherman, 1955) showed that it was possible to express the intensity/concentration relationship in terms of independently determined *fundamental* parameters. Unfortunately, fundamental type methods require a fair degree of computation and suitable computational facilities were not generally available until the early 1970's.

By the early 1960's, limited computational facilities were available at reasonable costs, and because of the clear need for some degree of fast mathetmatical correction of matrix effects, a number of so-called *empirical correction* techniques were developed which required far less computation and were, therefore, usable by the computers available at the time.

In principle, an empirical correction procedure can be described as the correction of an analyte element intensity for the influence of an interfering element(s) using the product of the intensity from the interfering element line and a constant factor, as the correction term (Beattie and Brissey, 1954). This constant factor is, today, generally referred to as an *influence coefficient,* since it is assumed to represent the influence of the interfering element on the analyte. Commonly employed influence coefficient methods may use either the intensity, or the concentration, of the interfering element as the correction term. These methods are referred to as *intensity correction* and *concentration correction* methods, respectively. Intensity correction models give a series of linear equations which do not require much computation, but they are generally not applicable to wide ranges of analyte concentration. Various versions of the intensity correction models found initial application in the analysis of non-ferrous metals where correction constants were applied as look-up tables. Later versions (Lucas-Tooth and Pyne, 1964) were supplied on commercially available computer controlled spectrometers and were used for a wider range of application. The Lachance-Traill model (Lachance and Traill, 1966) is a concentration model, which in effect requires the solving of a series of simultaneous equations, by regression analysis or matrix inversion techniques. This approach is more rigorous than the intensity models, and so they became popular in the early 1970's as suitable low-cost mini-computers became available.

9.7.1 Measurement of Pure Intensities

The intensity of an analyte line is subject not only to the influence of other matrix elements, but also to random and systematic errors due to the spectrometer and counting procedure employed. Provided that a sufficient number of counts is taken, and provided that the spectrometer source is adequately calibrated, *random* errors from these sources are generally insignificant, relative to other errors. *Systematic* errors from these sources are, however, by no means insignificant, and effects such as counting dead time, background, and line overlap, can all contribute to the total experimental error in the measured intensity. A problem may arise in that incorrect conclusions about potential matrix effects may be drawn from data, which is subject to instrument systematic errors. As an example, under a certain set experimental conditions, a series of binary alloys may give a calibration curve of decreasing slope. The conclusion may be drawn that radiation from the analyte element was enhanced by the other matrix element, where in point of fact, the problem could have also been due to dead time loss in the counting circuitry because the count rate was too high. If a correction were applied for an enhancement effect, the procedure would break down if the count rates were changed, for example, by varying the source conditions. Problems of this type can be particularly troublesome in the application of influence correction methods. Unless the instrument dependent errors are completely separated from the matrix dependent terms, the instrument effects will tend to become associated with the influence correction terms. In practice, this may not be completely disastrous for a specific spectrometer calibrated for a particular application, since the method will probably work, provided that the experimental conditions do not change. However, one major consequence is that it will probably be impossible to *transport* a set of correction constants from one spectrometer to another. Mainly for this reason, it is common practice today to attempt to obtain intensities as free from systematic instrumental errors as possible. Such intensities are referred to as *pure* intensities.

9.7.2 Use of Internal Standards

One of the most useful techniques for the determination of a single analyte element in a known or unknown matrix, is to use an internal standard. The technique is one of the oldest methods of quantitative analysis and is based on the addition of a known concentration of an element which gives a wavelength close to that of the analyte wavelength. The assumption is made that the effect of the matrix on the internal standard is essentially the same as the effect of the matrix on the analyte element. Internal standards are best suited to the measurements of analyte concentrations below about 10%. The reason for this limit arises because it is generally advisable to add the internal standard element at about the same concentration level as that of the analyte. When more than 10% of the internal standard is added, it may significantly change the specimen matrix and introduce errors into the determination. Care must also be taken to ensure that the particle sizes of specimen and internal standard are about the same, and that the two components are adequately mixed. Where an appropriate internal standard cannot be found it may be possible to use the analyte itself as an internal standard. This method is a special case of standard addition, and it is generally referred to as *spiking*.

9.7.3 Type Standardization

As has been previously stated, provided that the total specimen absorption does not vary significantly over a range of analyte concentrations, and provided that enhancement effects are absent and that the specimen is homogeneous, a linear relationship will be obtained between analyte concentration and measured characteristic line intensity. Where these provisos are met, type standardization techniques can be employed. It will also be clear from previous discussion, that by limiting the range of analyte concentration to be covered in a given calibration procedure, the range in absorption can also be reduced. Type standardization is probably the oldest of the quantitative analytical methods employed, and the method is usually evaluated by taking data from a well characterized set of standards, and, by inspection, establishing whether a linear relationship is indeed observed. Where this is not the case, the analyte concentration range may be further restricted. The analyst of today is fortunate in that many hundreds of good reference standards are commercially available. While the type standardization method is not without its pitfalls, it is nevertheless extremely useful and is especially useful for quality control type applications where a finished product is being compared with a desired product.

Special reference standards may be made up for particular purposes, and these may serve the dual purpose of instrument calibration as well as establishing working curves for analysis. As an example, two thin glass film standard reference materials specially designed for calibration of X-ray spectrometers are available from the National Bureau of Standards in Washington, as Standard Reference Materials (SRM 1832 and 1833; Pella et al., 1986). They consists of a silica-base film deposited by focussed ion-beam coating onto a polycarbonate substrate. SRM 1832 contains aluminium, silicon, calcium, vanadium, manganese, cobalt and copper; and SRM 1833 contains silicon, potassium, titanium, iron, zinc and rhodium. The standards are especially useful for the analysis of particulate matter.

9.7.4 Influence Correction Methods

It is useful to divide influence coefficient correction procedures into three basic types: Fundamental, Derived and Regression. Fundamental models are those which require starting with concentrations, then calculating the intensities. Derived models are those which are based on some simplification of a fundamental method but which still allow concentrations to be calculated from intensities. Regression models are those which are semi-empirical in nature, and which allow the determination of influence coefficients by regression analysis of data sets obtained from standards. All regression models have essentially the same form and consist of a weight fraction term, W (or concentration C); an intensity (or intensity ratio) term, I; an instrument dependent term which essentially defines the sensitivity of the spectrometer for the analyte in question, and a correction term, which corrects the instrument sensitivity for the effect of the matrix. The general form is as follows:

$$(9\text{-}17)$$

$$W/R = \text{constant} + \text{model dependent term}$$

Where W is the weight fraction of the analyte and R is the ratio of the analyte intensity measured from a pure elemental standard to that of the analyte line intensity measured from the specimen, with each intensity corrected for instrumental effects.

The different models vary only in the form of this correction term. Figure 9-7 shows several of the more important of the commonly employed influence coefficient methods. All of these models are concentration correction models in which the product of the influence coefficient and the *concentration* of the interfering element are used to correct the slope of the analyte calibration curve. The Lachance-Traill model was the first of the concentration correction models to be published. Some years after the Lachance-Traill paper appeared, Heinrich and his co-workers at the National Bureau of Standards, suggested an extension to the Lachance-Traill approach (Rasberry and Heinrich, 1974) in which absorbing and enhancing elements are separated as α and β terms. These authors suggested that the enhancing effect

cannot be adequately described by the same hyperbolic function as the absorbing effect. A thorough study (Tertian, 1986) of Lachance-Traill coefficients based on theoretically calculated fluorescence intensities, show that all binary coefficients vary systematically with composition. Both the Claisse-Quintin (Claisse-Quintin, 1967) and Lachance-Claisse models use higher order terms to correct for so-called *crossed effects*, which includes enhancement and third element effects. These models are generally more suited for very wide concentration range analysis.

In all of these methods one of three basic approaches is used to determine the values of the influence coefficients, following the initial measurement of intensities using a series of well characterized standards. The first approach is to use multiple regression analysis techniques to give the best fit for slope, background and influence coefficient terms. Alternatively, the same data set can be used to graphically determine individ-

---

Linear Model:

$$W_i/R_i = K_i$$

Lachance-Traill (1966):

$$W_i/R_i = K_i + \sum_j a_{ij}W_j$$

Claisse-Quintin (1967):

$$W_i/R_i = K_i + \sum_j a_{ij}W_j + \sum_j \gamma_{ij}W_j^2$$

Rasberry-Heinrich (1974):

$$W_i/R_i = K_i + \sum_j a_{ij}W_j + \sum_{k \neq j} \beta_{ik}(W_k/1 + W_i)$$

Lachance-Claisse (1980):

$$W_i/R_i = 1 + \sum_j a_{ij}W_j + \sum_j \sum_{k > j} a_{ijk}W_j W_k$$

Figure 9-7. Influence coefficient models.

ual influence coefficients. As an example, in the case of the Lachance-Traill equation, for a binary mixture a/b the expression for the determination of a would be:

$$W_a/R_a = 1 + (\alpha_{ab} W_b) \qquad (9\text{-}18)$$

By plotting data from a range of analyzed standards in terms of W_a/R_a as a function of W_b a straight line should be observed with a slope of α_{ab} and an intercept of unity. This approach is especially useful for visualizing the form of the influence coefficient correction (Lachance, 1985). Thirdly, the influence coefficient can be calculated using a fundamental type equation based on physical constants.

The major advantage to be gained by use of influence coefficient methods is that a wide range of concentration ranges can be covered using a relatively inexpensive computer for the calculations. A major disadvantage is that a large number of well analyzed standards may be required for the initial determination of the coefficients. However, where adequate precautions have been taken to ensure correct separation of instrument and matrix dependent terms, the correction constants are transportable from one spectrometer to another and, in principle, need only be determined once.

9.7.5 Fundamental Methods

Since the early work of Sherman there has been a growing interest in the provision of an intensity/concentration algorithm which would allow the calculation of the concentration values without recourse to the use of standards. Sherman's work was improved upon first by the Japanese team of Shiraiwa and Fujino (Shiraiwa and Fujino, 1967) and later, by the Americans, Criss and Birks (Criss and Birks, 1968; Criss, 1980) with their program NRLXRF. The same group also solved the problem of

describing the intensity distribution from the X-ray tube (Gilfrich and Birks, 1968). The problem for the average analyst in the late '60's and early '70's, however, remained that of finding sufficient computational power to apply these methods. In the early 1970's, de Jongh suggested an elegant solution (de Jongh, 1973) in which he proposed the use of a large main-frame computer for the calculation of the influence coefficients, then use of a small minicomputer for their actual application using a concentration correction influence model. One of the problem areas remains that of adequately describing the intensity distribution form the X-ray tube. Gilfrich and Birks demonstrated an experimental approach to this problem by measuring the spectral distribution from the tube in an independent experiment. More recently, this work has been extended to calculate spectral distributions using data obtained from the electron microprobe (Pella et al., 1985). While software packages are available for fundamental type calculations using data obtained with the energy dispersive system, one major drawback remains in their application system which use a modified primary excitation spectrum. Most fundamental quantitative approaches in use today employ measured or calculated continuous radiation functions in the calculation of the primary absorption effect. Where sharp discontinuities or "breaks" in this primary spectrum occur, as in the case of the energy dispersive system, the calculation becomes very complicated.

9.8 Trace Analysis

9.8.1 Analysis of Low Concentrations

The X-ray fluorescence method is particularly applicable to the qualitative and quantitative analysis of low concentrations of elements in a wide range of samples, as well as allowing the analysis of elements at higher concentrations in limited quantities of materials. The measured signal in X-ray analysis is a distribution of counting rate R as a function of either 2θ angle (wave-lenght dispersive spectrometers), or as counts per channel as a function of energy (energy dispersive spectrometers). A measurement of a line at peak position gives a counting rate which, in those cases where the background is insignificant, can be used as a measure of the analyte concentration. However, where the background is *significant* the measured value of the analyte line at the peak position now includes a count rate contribution from the background. The analyte concentration in this case is related to the net counting rate. Since both peak and background count rates are subject to statistical counting errors, the question now arises as to the point at which the net peak signal is statistically significant. The generally accepted definition for the lower limit of detection is *that concentration equivalent to two standard deviations of the background counting rate.* A formula for the lower limit of detection *LLD* can now be derived (Jenkins and de Vries, 1970):

$$LLD = \frac{3}{m}\sqrt{\frac{R_b}{t_b}} \qquad (9\text{-}20)$$

Note that in Eq. (9-20) t_b represents one half of the total counting time. The detection limit expression for the energy dispersive spectrometer is similar to that for the wavelenght dispersive system except that t_b now becomes the live-time of the energy dispersive spectrometer.

The sensitivity "m" of the X-ray fluorescence method is expressed in terms of the intensity of the measured wavelength per unit concentration, expressed in c/s per percent. Figure 9-8 shows the sensitivity *(excitation factor)* of a wavelength dispersive spectrometer and indicates that the sensitivity varies by about four orders of magnitude over the measurable element range, when expressed in terms of rate of change in response per rate of change in concentration. For a fixed analysis time the detection limit is proportional to $(m/\sqrt{R_b})$ and this is taken as a *figure of merit* for trace analysis. The value of m is determined mainly by the power loading of the source, the efficiency of the spectrometer for the appropriate wavelength and the fluorescent yield of the excited wavelength. The value of R_b is determined mainly by the scattering characteristics of the sample matrix and the intensity/wavelength distribution of the excitation source.

It is important to note that not only does the sensitivity of the spectrometer vary significantly over the wavelength range of the spectrometer, but so too does the background counting rate. In general, the background varies by about two orders of magnitude over the range of the spectrometer. By inspection of Eq. (9-20) it will be seen that the detection limit will be best when the sensitivity is high and the background is low. Both the spectrometer sensitivity and the measured background vary with the average atomic number of the sample. While detection limits over most of the atomic number range lie in the low part per million range, the sensitivity of the X-ray spectrometer falls off quite dramatically towards the long wavelength limit of the spectrometer due mainly to low fluorescence yields and the increased influence of absorption. As a result, poorer detection limits are found at the long wavelength extreme of the spectrometer, that corresponds to the lower atomic numbers. Thus the detection limits for elements such as

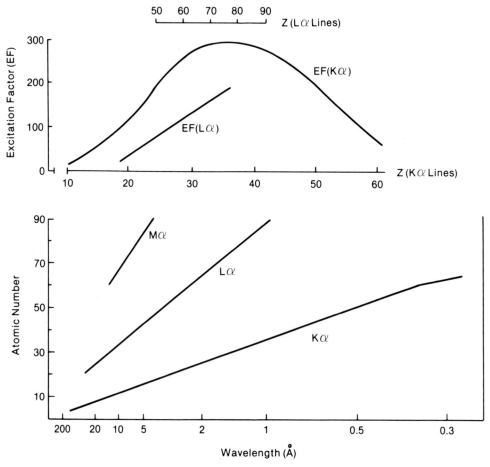

Figure 9-8. Sensitivity of the X-ray spectrometer as a function of atomic number.

fluorine and sodium are at the levels of hundredths of one percent rather than parts per million. The detection limits for the very low atomic number elements such as carbon ($Z = 6$) and oxygen ($Z = 8$) are, however, very poor and are typically of the order of 3 to 5%.

The major factors effect the detection limit for a given element: first, the sensitivity of the spectrometer for that element in terms of the counting rate per unit concentration of the analyte element; second, the background (blank) counting rate; and third, the available time for counting peak and background photons. In comparing

the energy and wavelength dispersive type systems, the absolute sensitivity of the wavelength dispersive system is almost always higher than the equivalent value for an energy dispersive system – perhaps by one to three orders of magnitude. This is because modern wavelength dispersive systems are able to handle count rates up to 1 000 000 c/s compared to about 40 000 c/s (for the total output of the selected excitation range) in the case of most energy dispersive systems. This difference is due to the fact that in the case of the energy dispersive spectrometer the electronic pulses created by each X-ray must be sufficiently long for precise energy measurement, thus the maximum count rate because of pulse pile up effects. In the case of the wavelength dispersive system the energy selection is done by the analyzing crystal and much faster detector electrons can be used. The ability of the wavelength dispersive system to work at high counting rates allows use of a high loading at the primary source.

Since it is not possible to measure the background directly, it is common practice to make a background measurement at a selected position close to the peak, the assumption being that the background at this position is the same as the background under the peak. This assumption will, of course, break down where the peak is superimposed on top of a variable background. In this case it is usual to measure the background either side of the peak. In energy dispersive X-ray fluorescence it is common practice to select a number of channels representing a net number of counts on the peak, superimposed on counts from the background. Two ranges of channels and are chosen on either side of the peak giving the total background. A further complication occurs where the analyte line is partially overlapped by another line. In this instance the measured value of the peak includes a contribution from the background as before, but in addition, a contribution from the interfering peak and a line overlap correction must be included.

The background that occurs at a selected characteristic wavelength or energy, arises mainly from scattered source radiation. Since scatter increases with decrease in the average atomic number of the scatterer, it is found that backgrounds are much higher from low average atomic number specimens, than from specimens of high average atomic number. To a first approximation, the background in X-ray fluorescence varies as $1/Z^2$. Since the spectral intensity from the X-ray source increases quite sharply as one approaches a wavelength equal to one half the minimum wavelength of the continuum, backgrounds from samples excited with bremsstrahlung sources are generally very high at short wavelengths (high energies), again, especially in the case of low averge atomic number samples.

The influence of the background is rather complicated because so many variables come into play. One of the major advantages of the secondary target energy dispersive system over the conventional system is that backgrounds are dramatically lower in the secondary excitation mode, because much of the background in a fluoresced X-ray spectrum comes from scattered primary continuum. The measured backgrounds for the wavelength and energy dispersive systems are very similar in the low atomic number regions of the X-ray spectrum, and for the secondary target energy dispersive system, are lower by up to an order of magnitude in the higher energy (mid-range atomic number) region. All of these factors taken together lead to detection limits for the secondary target energy dispersive system that are typically a factor of 3 to 8 times worse than the wavelength dispersive spectrometer. The actual detection limit measurable also depends upon the characteristics of the specimen itself, including such things as absorption, scattering power etc.

9.8.2 Analysis of Small Amounts of Sample

Conventional X-ray fluorescence spectrometers are generally designed to handle rather large specimens with surface areas of the order of several square centimeters and problems occur where the sample to be analyzed is limited in size. The modern wavelength dispersive system is especially inflexible in the area of sample handling, mainly because of geometric constraints arising from the need for close coupling of the sample to X-ray tube distance, and the need to use an airlock of some kind to bring the sample into the working vacuum. The sample to be analyzed is typically placed inside a cup of fixed external dimensions that is, in turn, placed in the carousel. This presentation system places constraints not only on the maximum dimensions of the sample cup, but also, on the size and shape of samples that can be placed into the cup itself. Primary source energy dispersive systems do not require the same high degree of focussing and to this extent, are more easily applicable to any sample shape or size, provided that the specimen will fit into the radiation protected chamber. In some instances the spectrometer can even be brought to the object to be analyzed. Because of this flexibility, the analysis of odd shaped specimens have been almost exclusively the purview of the energy dispersive systems.

In the case of secondary target energy dispersive systems, while the geometric constraints are still less severe than the wavelength system, they are much more critical than in the case of primary systems. This is due not just to the additional mechanical movements in the secondary target system, but also to the limitations imposed by the extremely close coupling of X-ray tube to secondary target. This is probably the reason that most energy dispersive spectrometer manufacturers generally offer a primary system for bulk sample analysis, and retain the secondary target system, generally equipped with a multiple specimen loader, for the analysis of samples that have been constrained to the internal dimensions of a standard sample cup during the specimen preparation procedure.

Where practicable, the best solution for the handling of limited amounts of material is invariable found in one of the specialized spectrometer systems. This is also generally true for the analysis of low concentrations. The TRXRF system is ideally suited for the analysis of small samples in those cases where the specimen can be dispersed as a thin film onto the surface of the reflector substrate. The high specific intensity synchrotron source offers sensitivities many times greater than the sealed X-ray tube source and because of the small beam divergence it is possible to obtain good signals from very small specimens.

9.9 References

Aiginger, H., Wobrauschek, P. (1985), *Adv. X-ray Anal. 28*, 1

Auger, P. (1925a), *Compt. rend. 180*, 65

Auger, P. (1925b), *Journal de Physique 6*, 205.

Barbee, T. W. (1985), *Supperlattices Microstruct. 1*, 311.

Bearden, J. A. (1964), *X-ray Wavelengths, U.S. Atomic Energy Commision Report NYO-10586*, p. 533.

Beattie, M. J., Brissey, R. M. (1954), *Anal. Chem. 26*, 980.

Bertin, E. P. (1975), *Principles and Practice of X-ray Spectrometric Analysis*, 2nd. ed. New York: Plenum, Appendix 10.

Bird, J. R., Duerden, P., Clayton, E., Wilson, D. J., Fink, D. (1986), *Nucl. Instrum. and Phys. Res. Sect. B 15*, 86.

Birks, L. S. (1963), *Electron Probe Microanalysis, Chemical Analysis Series*, Volume XVII. New York: Interscience.

Birks, L. S. (1976), *History of X-ray Spectrochemical Analysis*. Washington DC: American Chemical Society Centenial Volume, ACS.

Bloch, J. M. (1985), *Brookhaven Natl. Lab. Rep., BNL-51847*, p. 36.

Cesareo, M. (1982), *X-ray Fluorescence in Medicine*. Rome: Field Educational Italia.

Claisse Inc., Sainte-Foy, Quebec.

Claisse, F., Quintin, M. (1967), *Can. Spectrosc. 12*, 159.

Coster, D., Nishina, J. (1925), *Chem. News 130*, 149.

Criss, J. W. (1980), *Adv. X-ray Anal. 23*, 93

Criss, J. W., Birks, L. S. (1968), *Anal. Chem. 40*, 1080.

de Jongh, W. K. (1973), *X-ray Spectrom. 2*, 151.

Friedrich, W., Knipping, P., Von Laue, M. (1912), *Ann. Physik. 41*, 971.

Garbauskas, M. F., Goehner, R. P. (1983), *Adv. X-ray Anal. 26*, 345.

Garten, R. P. H. (1984), *Trends in Anal. Chem. 3 no. 6*, 152.

Giauque, R. D., Jaklevic, J. M., Thompson, A. C. (1986), *Anal. Chem. 58*, 940–944.

Gilfich, J. V., Birks, L. S. (1968), *Anal. Chem. 40*, 1077.

Gilfrich, J. V. (1983), *Anal. Chem. 55*, 187.

Glocker, R., Scheiber, H., (1928), *Ann. Physik. 85*, 1085.

Hadding, A., (1922), *Z. Anorg. Allgem. Chem. 122*, 195.

Henke, B. L. (1965), *Adv. X-ray Anal. 8*, 269.

Iwanczyk, J. S., Dabrowski, A. J., Huth, G. C., Bradley, J. G., Conley, J. M., Albee, A. L. (1986), *IEEE Trans. Nucl. Sci., NS-33*, 355.

Jenkins, R. (1974), *An Introduction to X-ray Spectrometry*. London: Wiley/Heyden, Chap. 4.

Jenkins, R. (1988), *X-ray Fluorescence Spectrometry*. New York: Wiley-Interscience.

Jenkins, R., de Vries, J. L. (1970), *Practical X-ray Spectrometry*, 2nd. ed. London: MacMillan, Chap. 4.

Jenkins, R., Gould, R. W., Gedcke, D. A. (1981), *Quantitative X-ray Spectrometry*. New York: Dekker, Sec. 4.3.

Jenkins, R., Hammell, B., Cruz, A., Nicolosi, J. A. (1985), *Norelco Reporter 32*, 1.

Kevex Corporation (1987a), *Cryoelectric Detector*. Foster City (CA): Kevex Corporation.

Kevex Corporation (1987b), *New Product Bulletin, Spring*. Foster City (CA): Kevex Corporation.

Kikkert, J. N., Hendry, G. (1983), *Adv. X-ray Anal. 27*, 423.

Knoth, J., Schwenke, H. (1978), *Fresenius Z. Anal. Chem. 301*, 200.

Lachance, G. R., Traill, R. J. (1966), *Can Spectrosc. 11*, 43.

Lachance, G. R. (1985), *Introduction to Alpha Coefficients*, Paris: Corporation Scientifique.

Lin, X. (1985), *Zhonhna Zhongliu Zazhi 7*, 411.

Lucas-Tooth, H. J., Pyne, C. (1964), *Adv. X-ray Anal. 7*, 523.

Malmqvist, K. G. (1986), *Nucl. Instrum. Meth. Phys. Res. Sect. B14*, 86.

Michaelis, W., Knoth, J., Prange, A., Schwenke, H. (1984), *Adv. X-ray Anal. 28*, 75.

Moseley, H. G. J. (1912), *Phil. Mag. 26*, 1024.

Moseley, H. G. J. (1913), *Phil. Mag. 27*, 703.

Nicolosi, J. A., Jenkins, R., Groven, J. P., Merlo, D. (1985), *Proc. SPIE 563*, p. 378.

Parrish, W., Hart, M., Huang, T. C. (1986), *J. Appl. Cryst. 19*, 92.

Pella, P. A., Feng, L. Y., Small, J. A. (1985), *X-ray Spectrom. 14*, 125.

Pella, P. A., Newbury, D. E., Steel, E. B., Blackburn, D. H. (1986), *Anal. Chem. 56*, 1133.

Petersen, W., Ketelsen, P., Knoechel, A., Pausch, R. (1986), *Nucl. Instrum. Methods Phys. Sect. A, A246* (1-3) 731.

Rasberry, S. D., Heinrich, K. F. J. (1974), *Anal. Chem. 46*, 81.

Röntgen, W. C. (1898), *Ann. Phys. Chem. 64*, 1.

Schwenke, H., Knoth, J. (1982), *Nucl. Instrum. and Methods 193*, 239.

Sherman, J. (1955), *Spectrochim. Acta 7*, 283.

Shiraiwa, T., Fujino, N. (1967), *Bull. Chem. Soc. Japan. 40*, 2289.

Sparks, C. J. Jr. (1980), *Synchrotron Radiation Research:* Winnick, H., Doniach, S. (Eds.). New York: Plenum Press, p. 459.

Stephenson, D. A. (1971), *Anal. Chem. 43*, 310.

Tertian, R. (1986), *X-ray Spectrom. 15*, 177.

Von Hevesey, G. (1932), *Chemical Analysis by X-rays and its Application*. New York: McGraw-Hill.

Walter, F. J. (1971), *Energy dispersion X-ray analysis, ASTM Special Technical Publication STP 485*. Philadelphia: ASTM, p. 125.

West, T. S., Nurrenberg, H. W. (Eds.) (1988), *The Determination of Trace Metals in Natural Waters*. Oxford: Blackwell.

White, E. W., Johnson, G. G. Jr. (1970), *X-ray Emission and Absorption Wavelengths and Two-Theta Tables, ASTM Data Series DS-37A*. Philadelphia: ASTM.

Yoneda, Y., Horiuchi, T. (1971), *Rev. Sci. Instrum. 42*, 1069.

General Reading

A Selection of Books Dealing with X-Ray Spectrometry

Anderson, C. A. (Ed.), (1972), *Microprobe Analysis*. New York: Wiley-Interscience.

Azaroff, L. V. (Ed.), (1974), *X-ray Spectroscopy*. New York: McGraw Hill.

Bertin, E. P. (1975), *Principles and Practice of X-ray Spectrometry Analysis*, 2nd. ed. New York: Plenum.

Bertin, E. P. (1985), *Introduction to X-ray Spectrometry Analysis*. New York: Plenum.

Birks, L. S. (1969), *X-ray Spectrochemical Analysis, 2nd ed*. New York: Wiley-Interscience.

Cauchois, Y., Bonnelle, C., Mande, C. (1982), *Advances in X-ray Spectroscopy*. Oxford: Pergamon.

Jenkins, R. (1974). *An Introduction to X-ray Spectromery*. London: Heyden.

Jenkins, R., de Vries, J. L. (1970), *Practical X-ray Spectrometry*, 2nd. ed. London: MacMillan.

Jenkins, R., Gould, R. W., Gedcke, D. (1981), *Quantitative X-ray Spectrometry*. New York: Dekker.

Jenkins, R. (1988), *X-ray Fluorescence Spectrometry*. New York: Wiley-Interscience.

Liebhafsky, H. A., Pfeiffer, H. G., Winslow, E. H., Zemany, P. D. (1972), *X-rays, Electrons, and Analytical Chemistry – Spectrochemical Analysis with X-rays*. New York: Wiley-Interscience.

Pattee, H. H., Cosslett, V. E., Engstrom, (1963), *X-ray Optics and X-ray Microanalysis*. New York: Academic Press.

Russ, J. C., Shen, R. B., Jenkins, R. (1978), EXAM: *Principles and Practice*. Prairie View: Edax International.

Woldseth, R. (1973), *X-ray Energy Spectrometry*. Burlingame: Kevex Corporation.

10 Polymer Molecular Structure Determination

Elizabeth A. Williams

GE Corporate Research and Development, Schenectady, NY, U.S.A.

List of Symbols and Abbreviations

a_1	thermodynamic activity of the solvent in a two-component system
A	absorbance
b	path length of a cell
c	weight concentration of a polymer in solution
c	velocity of light
C	concentration
D	diffusion coefficient
D_T	thermal diffusion coefficient
D_0	diffusion coefficient at infinite dilution
$\overline{DP}$	average degree of polymerization
E	energy
g	gravitational constant
h	Planck constant
H_0	static magnetic field
H_1	rf magnetic field
I	intensity of light passing through the sample
I	spin quantum number
I_0	intensity of the incident beam (light scattering)
$i_{ex(\theta)}$	excess scattered intensity of light
K_m	constant, in MHS equation
k_a	specific absorptivity of a molecule
M	molecular weight
M_i	molecular weight of species i
$\overline{M}_n$	number average molecular weight
$\overline{M}_v$	viscosity average molecular weight
$\overline{M}_w$	weight average molecular weight
M_z, M_{z+1}	higher molecular weight averages
n	molar concentration
n_i	number of moles of i present
R	gas constant
R	retention ratio
ΔR	electric resistance difference (or bridge imbalance) in a Wheatstone bridge
r_g	radius of gyration
R_θ	differential scattering cross-section (Rayleigh ratio)
ΔR_θ	difference between R_θ of a polymer solution and the solvent
S_0	sedimentation coefficient at infinite dilution
T	absolute temperature
t	time
T_1	spin-lattice relaxation time
V, V_r	retention volume
V^0	void volume (of the channel)
w_i	weight fraction of species i
X_2	mole fraction of the solute in a two-component system

γ	gyromagnetic ratio
$[\eta]$	intrinsic viscosity
$\eta_{1,2}$	viscosity of a two-phase polymer solution (1), the solvent (2)
η_r	relative viscosity
η_{sp}	specific viscosity
$[\eta]M$	hydrodynamic volume
θ	scattering angle
λ	wavelength
λ_T	retention parameter
μ	magnetic moment
v	partial specific volume of the polymer solute
v	frequency of radiation
π	osmotic pressure
ϱ	density
ω	angular velocity
v_0	Larmor frequency

ABS	acrylonitrile/butadiene/styrene
ATR	attenuated total (or internal) reflection (spectroscopy)
BPA	bisphenol A
CW	continuous wave
D	dimethylsiloxy (group)
D^H	methyl, hydrogen siloxy (group)
DP	degree of polymerization
DRI	differential refractive index
FFF	field flow fractionation
FID	free induction decay
FT-IR	Fourier transform IR
GPC	gel permeation chromatography
HP	high performance
IR	infrared spectroscopy
IV	intrinsic viscosity
LALLS	low-angle laser light scattering
m	meso
MHS	Mark-Houwink-Sakurada (equation)
NMR	nuclear magnetic resonance (spectroscopy)
NOE	nuclear Overhauser effect
PAS	photoacoustic spectroscopy
PBD	polybutadiene
PBT	poly(1,4-butylene terephthalate)
PXE	polyxylylene ether
r	racemic
rf	radiofrequency
THF	tetrahydrofuran
TMS	trimethylsilyl
UV-Vis	ultraviolet-visible
VPO	vapor phase osmometry

10.1 Introduction

The complete characterization of a synthetic organic polymer typically involves making a variety of measurements which provide information about the composition, structure and physical and chemical properties of the material. The object is usually to explore the structure-property relationships and to apply this information to the development of new or improved materials. Among the areas of study normally included are rheology, physical testing, spectroscopy, chemical analysis, microscopy, thermal analysis and chromatography. Other techniques may also be employed, such as X-ray diffraction, for example, for studies of polymer crystallinity. In order to limit the scope of this chapter, only those methods which provide information about the molecular structure of the polymer chain will be discussed. These include the various methods for molecular weight determination, and nuclear magnetic resonance spectroscopy (NMR), infrared (IR) and Raman spectroscopies. The reader is referred to several available texts on polymer characterization (Craver, 1983; Koenig, 1980; Siesler and Holland-Moritz, 1980; Baijal, 1980; Bovey, 1972; Randall, 1977; Tonelli, 1989; Cooper, 1989) for discussions of the areas not included in this chapter, or for more detailed discussions of each technique. In addition, the discussion will be limited to information specific for polymer structure determination. It will not include, for example, those aspects of spectroscopy used to characterize small organic molecules. The reader is again referred to the many texts available on each analytical technique for introductory material or more detailed discussions.

10.2 Polymer Structure

Several structural features differentiate polymers from lower molecular weight compounds. The unreacted starting material from which a polymer is made is referred to as the monomer. A polymer is a chemical compound consisting essentially of repeating structural units which are bonded together to form a chain:

$$--\{-A-A-A-A-A-A-A-A-A-A-A-A-\}--$$

The individual repeating units -(A-)- which consist of the reacted monomer and comprise the polymer chain are referred to as *monomer units*. Polymers containing repeating units which are chemically identical are called *homopolymers*. Since the polymerization process is statistical in nature, most polymers are comprised of mixtures of molecules having a range of degrees of polymerization (DP), and therefore different molecular weights:

$$A-A-A-A-A-A-A-A-A-A-A-A \quad + \quad A-A-A-A$$
$$+ \quad A-A-A-A-A-A-A \quad + \quad A-A-A$$
$$+ \quad A-A-A-A-A-A-A-A-A-A-A-A-A-A-A-A$$

Thus a key structural parameter for characterizing polymers but not applicable to other compounds is the molecular weight distribution.

The situation becomes more complicated when two different monomers are coreacted to form a *copolymer*. For example, monomers A and B can combine in several fashions to produce materials which may have dramatically different properties. If there is a preference for either A or B to react with itself rather than the other component, a *block* copolymer (1) may be formed:

$$-\{-A-A-A-A-A-B-B-B-B-B-B-A-A-A-A-A-B-B-B-B-B-\}-$$

1

A block copolymer comprises of long sequences of A followed by long sequences of B. The amount of "blocking" obtained depends upon a variety of experimental and chemical parameters which include reaction conditions, the amounts of A and B present and the strength of the preference for A (or B) to react with itself. Note that in block copolymers synthesized in this fashion there is a distribution of block lengths in the polymer molecules and it is the *average* block length which is usually measured. The average block length can affect the properties of the polymer almost as much as the gross chemical composition (amount of A and B present) and is an important structural parameter to be able to measure.

The opposite situation to blocking occurs when the monomers react preferentially with each other; in this case an *alternating* copolymer (2) is formed:

$$-\{-A\text{-}B\text{-}A\text{-}B\text{-}A\text{-}B\text{-}A\text{-}B\text{-}A\text{-}B\text{-}\}-$$

2

Finally, if there is little or no preferential reactivity, a *random* copolymer (3) is formed. The structure of this type of copolymer is just as the name implies, a statistical distribution of each monomer unit in the polymer chain. The molecular structure is obviously dependent upon the relative amounts of A and B present. For example, in a 50:50 copolymer of A and B, there would be equal probability that A would react with A as with B. This polymer would contain chains with a distribution of A and B units similar to the following:

$$-\{-B\text{-}A\text{-}B\text{-}A\text{-}A\text{-}B\text{-}A\text{-}B\text{-}B\text{-}A\text{-}B\text{-}A\text{-}A\text{-}B\text{-}A\text{-}B\text{-}B\text{-}B\text{-}A\text{-}B\text{-}A\text{-}A\text{-}\}-$$

3

Increasing the amount of A present in the polymer would result in an increase in the

number of longer sequences of A units in the resulting random copolymer.

In addition to linear polymers, two- and three-dimensional polymer structures are possible. These include *branched* polymers (4):

$$-\{-A\text{-}A\text{-}A\text{-}A\text{-}A\text{-}A\text{-}A\text{-}A\text{-}A\text{-}A\text{-}A\text{-}A\text{-}A\text{-}A\text{-}A\text{-}A\text{-}A\text{-}A\text{-}\}-$$

$$(A)_x \qquad (A)_y \qquad (A)_z$$
$$A' \qquad\quad A' \qquad\quad A'$$

4

in which there are random branches off the main polymer chain, *graft* copolymers (5) in which a different homopolymer is attached at random points off the main polymer chain:

$$-\{-A\text{-}\}-$$

$$(B)_x \qquad (B)_y \qquad (B)_z$$
$$B' \qquad\quad B' \qquad\quad B'$$

5

and *ladder* polymers (6) in which two parallel polymer chains are joined at frequent intervals by short connecting chains:

6

A three-dimensional copolymer has a network of randomly placed branched points connecting the chains.

The molecular structure becomes substantially more complicated when three or

more monomer units are involved in the polymer. In addition, other structures, such as rearrangement products of the monomer, stereoisomers or geometric isomers of the monomer, can also be present in the polymer chain. For example, monosubstituted vinyl polymers such as polypropylene may exist in a variety of configurational stereoregular sequences (*tacticity*). For two adjacent monomer units, two possibilities exist:

$$\begin{array}{ccc}
& CH_3 & CH_3 \\
& | & | \\
-CH_2-C-CH_2-C- \\
& H \diagdown \quad \diagup H \\
& m
\end{array}$$

7

$$\begin{array}{ccc}
& CH_3 & H \\
& | & | \\
-CH_2-C-CH_2-C- \\
& H \diagdown \quad \diagup CH_3 \\
& r
\end{array}$$

8

The dyad structures in which adjacent monomer units have the same relative configuration are *meso* (denoted "m") (7), whereas a *racemic* dyad ("r") (8) consists of a pair of monomer units with opposite configuration. If the polymer chain consists exclusively of carbons with the methyl groups on the same side (all m dyads), the polymer is *isotactic* (9). If the chain consists of monomer units with the methyl group on alternating sides the polymer is *syndiotactic* (10), and if there is a random distribution of methyl groups on either side of the chain polymer is *atactic* (11):

$$\begin{array}{cccc}
CH_3 & CH_3 & CH_3 & CH_3 \\
| & | & | & | \\
-CH_2-C-CH_2-C-CH_2-C-CH_2-C- \\
| & | & | & | \\
H \; m & H \; m & H \; m & H
\end{array}$$

9

$$\begin{array}{cccc}
CH_3 & H & CH_3 & H \\
| & | & | & | \\
-CH_2-C-CH_2-C-CH_2-C-CH_2-C- \\
| & | & | & | \\
H & CH_3 \; r & H \; r & CH_3 \\
& r
\end{array}$$

10

$$\begin{array}{cccc}
CH_3 & H & CH_3 & CH_3 \\
| & | & | & | \\
-CH_2-C-CH_2-C-CH_2-C-CH_2-C- \\
| & | & | & | \\
H \; r & CH_3 \; r & H \; m & H
\end{array}$$

11

Similarly, the carbon-carbon double bond in polybutadiene may be either the *cis* or *trans* isomer. The features discussed above are referred to as the chain microstructure. Both tacticity and geometric isomerism affect the physical properties of the polymer. It is often advantageous to be able to identify not only these variations on the monomer structure, but branching structures, graft points and chain end groups.

Fortunately, several analytical techniques, including IR, Raman and particularly NMR spectroscopy can furnish information about the composition of the monomer units and how they are distributed in the chain. Using these techniques it is often possible to identify other structural features even though end groups and graft points, for example, may be present in quite low concentration.

In summary, the parameters that define molecular structure in polymers include molecular weight distribution, monomer composition and structure, chain microstructure, end groups, branching structures and graft points. The analytical techniques used to elucidate these structures will be discussed individually, but it should be understood that they are most often used in concert, each method providing a differnt piece of information in the overall structure determination.

10.3 Molecular Weight Determination

Molecular weight is a major factor (second only to chemical composition) in determining the properties of polymers and it is imperative to have some knowledge of this parameter in order to evaluate a material. There are a number of methods for obtaining molecular weight information. These methods differ depending upon (1) whether an average molecular weight or distribution of molecular weights is measured, (2) wether a number average or weight average molecular weight is determined, and the technique provides *absolute* (independent of the nature of the polymer) or *relative* molecular weight information. It is the intent of this section to briefly describe the more common methods used and the scope and limitations of each.

10.3.1 Molecular Weight Averages

A complete description of the molecular weight distribution of a polymer would require a method which would permit the number of molecules at each DP to be measured. This is impractical, if not impossible, to accomplish with currently available technology. In practice the molecular weight distribution is considered as a continuous function rather than integer-valued in order to facilitate the mathematical treatment. Some techniques, such as gel permeation chromatography (GPC), do provide a complete molecular weight distribution. The molecular weight is most often measured, however, as either the number average molecular weight ($\bar{M}_n$) or the weight average molecular weight ($\bar{M}_w$).

The number average molecular weight is the total weight of the polymer divided by the total number of molecules (moles) present:

$$\bar{M}_n = \sum n_i M_i / \sum n_i = \sum w_i / \sum (w_i/M_i) \quad (10\text{-}1)$$

where M_i is the molecular weight of species i and n_i is the number of moles of i present; w_i is the weight of component i. $\bar{M}_n$ may be obtained through a technique such as osmometry which is sensitive to the number of molecules of polymer present (colligative properties). Typical number average molecular weights for commercial polymers range from 10 000 to 100 000. In general, the physical properties associated with high polymers are not present below $\bar{M}_n$ of 10 000.

The weight average molecular weight is defined as:

$$\bar{M}_w = \sum w_i M_i / \sum w_i =$$
$$= \sum n_i M_i^2 / \sum (n_i M_i) \quad (10\text{-}2)$$

Note that it is necessary to have a measurable quantity proportional to the square of the mass of the molecules in order to measure $\bar{M}_w$. Since each molecule contributes to $\bar{M}_w$ in proportion to the square of its mass, heavier molecules contribute more to $\bar{M}_w$ than light ones. Conversely, $\bar{M}_n$ is more heavily influenced by low molecular weight species. The ratio $\bar{M}_w/\bar{M}_n$ is a measure of the width of the molecular weight distribution and is called the *polydispersity* of the polymer. For a hypothetical monodisperse polymer, $\bar{M}_w/\bar{M}_n = 1$, but actual values of polydispersity range from between 1 and 2 for linear addition and condensation polymers, to as high as 50 for highly branched polymers.

In addition to $\bar{M}_n$ and $\bar{M}_w$, higher order molecular weight averages, such as $\bar{M}_z$ and $\bar{M}_{z+1}$ may be defined:

$$\bar{M}_z = \sum n_i M_i^3 / \sum (n_i M_i^2) =$$
$$= \sum w_i M_i^2 / \sum (w_i M_i) \quad (10\text{-}3)$$
$$\bar{M}_{z+1} = \sum n_i M_i^4 / \sum (n_i M_i^3) =$$
$$= \sum w_i M_i^3 / \sum (w_i M_i^2) \quad (10\text{-}4)$$

These higher molecular weight averages are of significance in determining certain rheological properties, such as melt elasticity, which are drastically affected by the presence of small amounts of higher molecular weight species. Finally, a linear relationship is obtained between the logarithms of the intrinsic viscosities of a series of fractionated polymers and the logarithms of their molecular weights. The simplest expression of this relationship is the Mark-Houwink-Sakurada (MHS) equation, Eq. (10-5), which relates intrinsic viscosity $[\eta]$ (the solution viscosity extrapolated to infinite dilution) to molecular weight M:

$$[\eta] = K_m M^a \qquad (10\text{-}5)$$

If the viscosity average molecular weight is defined:

$$\bar{M}_v = [\sum n_i M_i^{1+a} / \sum (n_i M_i)]^{1/a} =$$
$$= [\sum w_i M_i^a / \sum w_i]^{1/a} \qquad (10\text{-}6)$$

the expression relating intrinsic viscosity measurements to the viscosity average molecular weight of a polydisperse linear polymer is:

$$[\eta] = K_m \bar{M}_v^a \qquad (10\text{-}7)$$

In order to calculate the viscosity average molecular weight, the constant K_m and exponent a of the MHS relationship must be known. Both K_m and a are functions of the solvent and temperature as well as the specific polymer involved. Note that $\bar{M}_v$ is always larger than $\bar{M}_n$, but as can be seen in Eq. (10-6), when $a = 1$, $\bar{M}_v = \bar{M}_w$. In general, however, a varies between 0.5 and 1.0 for randomly coiled polymers and $\bar{M}_v$ is less than $\bar{M}_w$.

10.3.2 Absolute vs. Secondary Methods

The molecular weight determination methods which depend upon properties of a polymer solution that vary monotonically with molecular weight and are independent of the chemical structure of the polymer are called *absolute* methods. These methods include end-group analysis, membrane and vapor phase osmometry (VPO), cryoscopy and ebulliometry, ultracentrifugation and light scattering. Unlike these techniques, determinations of intrinsic viscosity (IV) and gel permeation chromatography (GPC) retention volume are dependent on molecular volume and therefore produce data which are influenced by the molecular structure and chain configuration. As a result, the relationship between the data obtained by these *indirect* techniques and the molecular weight must be determined independently for each polymer.

10.3.3 Number Average Molecular Weight ($\bar{M}_n$)

Number average molecular weight ($\bar{M}_n$) may be determined by any technique that permits calculation of the number of molecules present. This may be carried out by either using a technique that is dependent upon the number of molecules present (colligative properties), or which measures the number of end groups present in a known amount of polymer.

10.3.3.1 End Group Analysis

If the concentration of end groups present in the polymer chain can be measured, $\bar{M}_n$ can be calculated from the known structure of the polymer. There are several methods available to determine end groups. Condensation polymers typically contain reactive end groups such as acids, amines and hydroxyl groups which may be titrated by suitable reagents. Spectroscopic methods such as ultraviolet-visible (UV-Vis), IR and NMR can often be used to measure end groups. For example, Fig. 10-1 shows the ^{29}Si NMR spectrum of a trimethylsilyl endcapped polydimethylsiloxane fluid (for a further account of this technique see Sec. 10.5 below). In this case the technique provides a measure of both end groups (peak A) and mid-chain repeat units (peak B). The relative amounts of B and A were measured using the integrated intensities of the peaks from which the av-

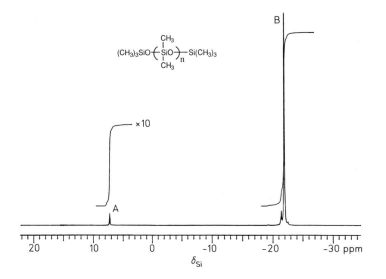

Figure 10-1. Silicon-29 NMR spectrum of a trimethylsilyl endcapped polydimethylsiloxane polymer. The average degree of polymerization $(\overline{DP})$ is calculated from the integrations of the areas under peaks A and B: $\overline{DP} = 2\,[(B/A) + 1] = 45$.

erage degree of polymerization, $\overline{DP}$, was determined:

$$\overline{DP} = 2\,[(B/A) + 1] \tag{10-8}$$

M_n can then be calculated from $\overline{DP}$ since the molecular weight of the polymer repeat unit can be calculated from the structure. In some cases, reacting the end group with a suitable reagent can enhance the ability to detect chain ends by introducing a more sensitive probe. For example, a UV-absorbing functional group with a high extinction coefficient can be reacted with an end group to enhance the detectability of the end group.

10.3.3.2 Colligative Properties

The techniques of ebulliometry, cryoscopy, vapor phase osmometry (VPO) and membrane osmometry all rely on the same thermodynamic principle which relates the activity of a solvent to the mole fraction of solute present. These "colligative properties" are those properties of a solution that depend only on the number of species present in solution, not on their chemical nature. In a two-component system at constant temperature and pressure, the ther-

modynamic activity of the solvent (a_1) in a dilute ideal solution depends on the mole fraction of the solute (X_2) according to the expression:

$$\ln a_1 = -X_2 \tag{10-9}$$

If the weight fraction of the solute, w_2, is known, its molecular weight (M_2) may be calculated:

$$X_2 = n_2/n_1 = w_2\,M_1/(w_1\,M_2) \tag{10-10}$$

Thus the activity of the solvent, which may be measured by one of the methods outlined below, allows the number average molecular weight to be calculated.

10.3.3.3 Experimental Methods

Vapor phase osmometry and membrane osmometry are complementary in the range of molecular weights which can be measured. VPO is generally used for $\overline{M}_n$ in the $100-10\,000$ range. The technique relies on isothermal distillation from a saturated vapor phase to a drop of solution containing the polymer under examination. The presence of the polymer in the drop decreases the solvent activity (and thus the mole fraction of solvent in the vapor phase) causing the transfer of solvent to take place. The amount of solvent transfer is measured indirectly from the heat of condensation of the solvent in the polymer solution. In order to do this the steady-state temperature difference between a drop of pure solvent and a drop of polymer solution is measured using two thermistors connected to a Wheatstone bridge. Drops of solvent and solution are placed on the thermistors and the temperature difference is detected as a resistance difference (ΔR), or bridge imbalance, that is linearly related to ΔT. The molecular weight, $\bar{M}_n$, can be obtained from a plot of $\Delta R/c$ versus c, the weight concentration of the polymer solution, extrapolated to infinite dilution:

$$(\Delta R/c)_{c=0} = K\bar{M}_n^{-a} \tag{10-11}$$

where K and a are constants for a given solvent and temperature. Deviation of a from unity signifies nonideal behavior.

Membrane osmometry is an absolute method for determining $\bar{M}_n$ in the range $10\,000-200\,000$. The technique has been used extensively for molecular weight determinations of synthetic polymers. A simple capillary osmometer consists of two cells, one containing solvent and the other containing the polymer solution. The cells are each connected to vertical capillaries and separated by a semipermeable membrane through which solvent may flow. The solvent flows from the solvent cell into the cell containing the polymer solution owing to the reduced activity of the solvent in the solution. The osmotic pressure (π) may then be determined from the excess height (h) of the solution in the capillary on the polymer side:

$$\pi = h\varrho g \tag{10-12}$$

where g is the gravitational constant (980 cm/s^2) and ϱ is the density of the solvent. To determine $\bar{M}_n$, the osmotic

pressure is determined for a series of polymer concentrations, c, (grams per cubic centimeter) and a plot of π/c versus c is prepared and extrapolated to infinite dilution. The molecular weight is obtained from Eq. (10-14) which is derived from the van't Hoff Eq. (10-13) showing that π is proportional to the molar concentration, n, of the solute:

$$\pi = nRT = cRT/M \tag{10-13}$$
$$\bar{M}_n = RT/(\pi/c)_{c=0} \tag{10-14}$$

where R is the gas constant, T is the absolute temperature and M is the molecular weight of the solute.

An advantage of osmometry over other molecular weight determination methods is that the effect of low molecular weight impurities may be eliminated by allowing these materials to diffuse through the membrane prior to making the measurements. As a practical technique, osmometry has largely been superseded by size exclusion chromatography (gel permeation chromatography).

Cryoscopy and ebulliometry are additional techniques that have limited practical application for synthetic polymers. Cryoscopy, or freezing point depression, and ebulliometry (boiling point elevation) are generally successful only for low molecular weight species, usually below 5000. All the methods based on colligative properties produce $\bar{M}_n$ by measuring the decrease in solvent activity due to the number of solute molecules present; the sensitivity, therefore, *decreases* with increasing molecular weight.

10.3.4 Light Scattering

Light scattering is an experimentally difficult but powerful tool providing information about the molecular weight and radius of gyration (molecular size) of a polymer in solution. The scattering of light by polymer molecules may be measured as a difference ΔR_θ between the differential scattering cross-section (Rayleigh ratio) of the polymer solution versus the solvent:

$$\Delta R_\theta = R_\theta(\text{solution}) - R_\theta(\text{solvent}) =$$
$$= i_{\text{ex}(\theta)}\, r^2/I_0 \qquad (10\text{-}15)$$

where $i_{\text{ex}(\theta)}$ is the excess scattered intensity of the light, r is the distance from the particle, θ is the angle (scattering angle) between the scattered beam and the incident light beam and I_0 is the intensity of the incident beam. Assuming the absence of interparticle interactions (dilute solution), the excess scattered intensity may be written:

$$\Delta R_\theta = K(1 + \cos^2\theta)\sum c_i M_i \qquad (10\text{-}16)$$

where K is a constant which depends on readily measured parameters such as the wavelength of the light, refractive index of the solvent and the change in refractive index with concentration of dissolved polymer. By definition $\bar{M}_w = \sum c_i M_i / \sum c_i$, therefore

$$\Delta R_\theta = K(1 + \cos^2\theta)\, c\, \bar{M}_w \qquad (10\text{-}17)$$

where c is the concentration of the solution in gms/cm^3. At $\theta = 0$ Eq. (10-17) reduces to:

$$\Delta R_\theta = 2\,K\,c\,\bar{M}_w \qquad (10\text{-}18)$$

Simultaneous extrapolation to both zero angle to eliminate effects of intraparticle interference and zero concentration to obtain $\bar{M}_w$ directly may be carried out using a graphical method described by Zimm (1948).

An example of a Zimm plot is shown in Fig. 10-2. The quantity $K\,c/\Delta R_\theta$ is plotted against $[\sin^2(\theta/2) + k\,c]$ where k is a scaling factor to make $\sin^2(\theta/2)$ and c of the same order of magnitude. For each concentration, extrapolation is made to zero angle, and for each angle θ, extrapolation is made to zero concentration. The point at which they both intercept is the quantity $1/\bar{M}_w$ [see Eq. (10-18)]. Further information is available from the slope of the zero concentration line which gives the second virial coefficient, and the slope of the zero angle line which gives the radius of gyration.

The use of laser light sources has facilitated measurements by eliminating the

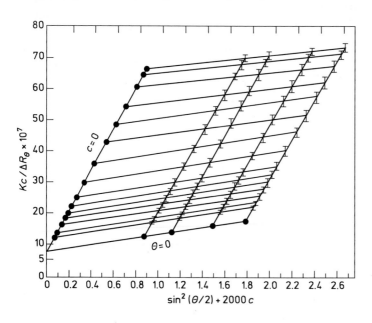

Figure 10-2. Zimm plot showing extrapolation of light scattering data to zero concentration and zero scattering angle.

need for extrapolation to $\theta = 0°$ since current low-angle laser light scattering photometers permit observation of the light intensity at angles as low as $2°$. In this case, a plot of $c/\Delta R_\theta$ is made for several concentrations of the polymer and extrapolated to zero concentration to obtain $1/\bar{M}_w$. For a more detailed explanation of the theory of light scattering the reader is referred to Baijal (1980), Cooper (1989) and Flory (1953).

10.3.5 Ultracentrifugation

The ultracentrifuge is capable of spinning a polymer solution at speeds up to 70 000 rpm. The equipment consists of a thermostatted rotor made of aluminum or titanium which is spun in vacuum at high, controlled speeds. The cells used to contain the polymer solution are available in a variety of designs depending on the type of experiment involved, the solvent and the detection system employed. Concentration gradients in the cell containing the solution are usually detected by measuring refractive index changes compared with solvent; UV absorption can also provide a direct concentration measurement for UV absorbing polymers. The molecular weight and molecular weight distribution of a polymer can be measured using two different approaches.

10.3.5.1 Sedimentation Velocity

The sedimentation velocity approach (Elmgren, 1982) involves spinning the sample at high speeds and measuring the rate of sedimentation which depends upon the mass and shape of the particles. The relationship between molecular weight and sedimentation coefficient at infinite dilution, S_0, may be written:

$$M = S_0 R T/[D_0 (1 - v \varrho)] \qquad (10\text{-}19)$$

where D_0 is the diffusion coefficient at infinite dilution, v is the partial specific volume of the polymer solute, and ϱ is the density of the solvent. Both D_0 and v must be obtained experimentally. A semiempirical equation is also available for the sedimentation coefficient at infinite dilution S_0:

$$S_0 = k M^a \qquad (10\text{-}20)$$

where k and a are constants dependent upon the polymer, solvent and temperature of the system. Relatively few values of k and a are available, however, and the technique is subject to a number of complications arising from the polydispersity of most synthetic polymers and the effect of the increased pressure on parameters such as viscosity and density of the solvent.

10.3.5.2 Equilibrium Sedimentation

The equilibrium sedimentation approach involves spinning the sample at much lower speeds, and allowing a steady state to be reached in which the sedimentation force is balanced by molecular diffusion. The technique is time consuming, often requiring one or two weeks for equilibrium to be reached, but the advantages are that it is much less susceptible to experimental variables and provides absolute molecular weight determinations. As with the sedimentation velocity technique, the concentration is measured as a function of the position in the cell. At equilibrium, for an ideal solution, the weight average molecular weight can be written:

$$(10\text{-}21)$$

$$\bar{M}_w = 2 R T \ln (c_2/c_1)/[(1 - v \varrho) \omega^2 (r_2^2 - r_1^2)]$$

where c_1 and c_2 are the polymer concentrations at distances r_1 and r_2 from the axis of rotation, ω is the angular velocity, v is the partial specific volume of the polymer and ϱ is the density of the solution. By measuring the concentration as a function of r

throughout the cell, higher molecular weight averages, such as M_z and M_{z+1}, may also be obtained.

In a variation of this technique, the polymer is dissolved in a mixture of two miscible solvents which differ in density. At equilibrium, a density gradient of solvent is established and the polymer remains at the position where its effective buoyant density matches that of the solvent mixture. Among the applications of this technique are separating homopolymer components of a blend, establishing the composition distribution in copolymers, and detecting residual homopolymers in graft copolymers. For example, density gradient centrifugation has been shown to be an effective method for detecting polystyrene homopolymer in graft copolymers of polystyrene and cellulose acetate, and for determining the composition of the graft copolymer (Ende and Stannett, 1964).

Another variation of the equilibrium sedimentation technique is the *approach to equilibrium* method in which the concentration of the polymer is determined at the meniscus and at the bottom of the cell. Mathematical methods for treating the data (Archibald, 1947) allow $\bar{M}_w$ and the second virial coefficient, which is a measure of the polymer-solute interaction, to be determined from measurements made in less than an hour.

10.3.6 Viscometry

Compared with other techniques, viscosity measurements are easy, rapid and require inexpensive equipment and simple sample preparation. The disadvantage is that the theoretical relationship between viscosity and molecular weight has not been firmly established. Nonetheless, the extensive use of this technique in all areas of polymer science has led to the availabil-

ity of a large volume of experimental data which has been used to develop empirical relationships between viscosity of dilute solutions and polymer molecular weights.

Polymer molecules can greatly increase the viscosity of a solvent even at low concentration. Furthermore, for a given polymer and solvent, the higher the molecular weight of the polymer, the greater the increase in viscosity. Thus the intrinsic viscosity, or capability of the polymer to increase the viscosity of the solvent, is a measure of the molecular weight of the material. The measurement does not provide an absolute molecular weight determination, however, and for each polymer the viscosity-molecular weight relationship must be established by examining a carefully fractionated polymer using molecular weights determined by an absolute method. Once this relationship has been established, however, the viscosity average molecular weight may be determined by measuring the intrinsic viscosity of any polymer with the same composition regardless of heterogeneity or molecular weight, as long as it is in the range of molecular weights included in developing the empirical equation.

The MHS equation:

$$[\eta] = K_m \bar{M}_v^a \qquad (10\text{-}22)$$

allows the viscosity average molecular weight to be determined from measurement of the intrinsic viscosity if the values of K_m and a are available for the given polymer-solvent combination. Fortunately, extensive compilations of these values are available in the literature (Brandrup and Immergut, 1975) to facilitate this process.

Viscosity measurements are typically made using a capillary viscometer such as the modified Ostwald type depicted in Fig. 10-3. If the time (t) it takes a dilute polymer solution to fall from point A to

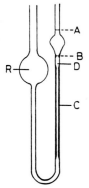

Figure 10-3. Modified Ostwald type capillary viscometer in which R is the reservoir and C is the capillary and D is the diffuser channel which helps prevent turbulent flow (Cooper, 1989, p. 178).

point B in the viscometer is compared with the time required for the solvent alone, the relationship between the viscosities of the two solutions is:

$$\eta_1/\eta_2 = t_1/t_2 \qquad (10\text{-}23)$$

where the subscript 1 refers to the polymer solution and 2 refers to the solvent. The ratio η_1/η_2 is the relative viscosity, η_r. The specific viscosity, $\eta_{sp} = \eta_r - 1$, is the incremental viscosity produced by the presence of the polymer solute. The ratio η_{sp}/c provides a measure of the specific capacity of the polymer to increase the relative viscosity. At infinite dilution, the value of this ratio is the intrinsic viscosity, $[\eta]$:

$$[\eta] = (\eta_{sp}/c)_{c=0} = [(\eta_r - 1)/c]_{c=0} =$$
$$= [(\ln \eta_r/c)]_{c=0} \qquad (10\text{-}24)$$

In practice the relative viscosity is measured at several concentrations and one of the quantities in Eq. (10-24) is plotted against concentration. Extrapolation to infinite dilution gives the value of η.

10.3.7 Gel Permeation Chromatography

The development of gel permeation chromatography (GPC) (Moore, 1964) has revolutionized the determination of molecular weight distributions in polymers (Yau et al., 1979; Janca, 1984). The technique is a liquid chromatographic separation in which the column is filled with a rigid porous gel and the separation occurs according to size of the polymer molecules by differential pore permeation (hence the more general term "size exclusion chromatography" is also applicable). As the solvent passes through the column, larger molecules are eluted from the column first, followed by smaller molecules. The concentration of the polymer in the eluting solvent is determined by measurement with a suitable detector, such as refractive index, UV or IR. A typical GPC chromatogram is shown in Fig. 10-4.

10.3.7.1 Instrumentation

The instrumentation consists essentially of a liquid chromatograph using a gel stationary phase to carry out the separation. A schematic of a typical high performance (HP) GPC is shown in Fig. 10-5. Recent

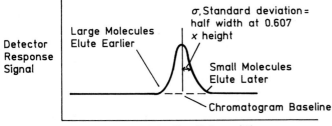

Figure 10-4. GPC chromatogram.

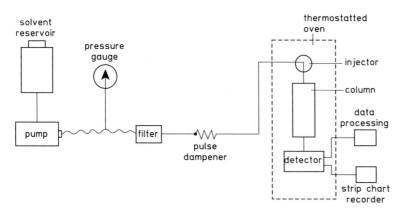

Figure 10-5. A typical high performance (HP) GPC apparatus.

developments include the availability of a wide range of gels for a variety of polymers and solvents, improvements in solvent delivery systems and high pressure capabilities of the equipment, and more varied and sensitive detectors. It is particularly important that the solvent flow rate be constant and reproducible in order to avoid significant errors in molecular weight measurements. The result of these improvements is that complete polymer molecular weight and molecular weight determinations can often be accomplished in well under an hour.

10.3.7.2 Stationary Phases and Solvents

Selection of a stationary phase is critical since it is responsible for the separation process. In order to effect separations over a range of molecular weights, the manufacturer must be able to control pore size distribution for each packing material. It is important that the solvent swell the gel slightly, yet if it is too aggressive a solvent the integrity of the substrate may be compromised. The most commonly used stationary phases for synthetic organic polymers are the Styragels® and μ-Styragels® (Waters Associates) which are crosslinked polystyrene resin beads. The latter are micron-sized beads which greatly increase

resolution and efficiency in conducting the analysis. Typical solvents for these resins include tetrahydrofuran (THF), chloroform and toluene; orthodichlorobenzene is often used for materials requiring warmer solvents.

10.3.7.3 Detectors

A number of different detectors are available for GPC. By far the most commonly used are detectors that measure the differential refractive index (DRI) between the solvent and column effluent. DRI detectors have the advantage of very general applicability, maximum sensitivity being obtained when the solute and solvent differ as much as possible in refractive index. A disadvantage, however, is their relatively low sensitivity and resultant inability to measure low concentrations of solute. UV and IR detectors which are fixed wavelength or tunable are also available for GPC of polymers containing suitable chromophores. The UV detectors are very sensitive and used more frequently than the IR detectors. Both suffer from the problems associated with finding a solvent that is transparent to the incident light. All of these detectors measure the mass concentration of the polymer. This approach assumes that the refractive index is independent of molecular weight (DRI detector), and that the number of chromophores per gram of polymer is constant over the molecular weight range (UV and IR detectors). It is important, therefore, in multi-component polymers that the composition of the polymer be constant over the molecular weight range.

Recently, the use of both concentration and molecular weight detectors in concert has become possible in GPC. In order to do this effectively, the molecular weight must be measured quickly and on small amounts of solute. This requirement has led to the development of viscometric and light scattering detectors for this application. The low-angle laser light scattering (LALLS) photometer developed by Ouano and Kaye (1974) and commercially available from Chromatix (KMX-6) consists of a flowthrough cell and a laser light source. The light scattering measured in the cell is a function of both concentration and molecular weight of the solute. If another detector such as a DRI is connected in line with the LALLS, the absolute molecular weight of the solute can be determined across the whole chromatogram. Similar information is available with viscosity detection. Noise problems in the earliest viscometric detectors were overcome with the development of a capillary bridge viscometer now available from Viscotek. When this is used in tandem with a concentration detector the molecular weight may be evaluated for the entire chromatogram.

10.3.7.4 Calibration

GPC is usually carried out without the benefit of a molecular weight detector. In this case, a calibration procedure must be used to correlate the peak eluting at a specific retention volume with the molecular weight obtained by an absolute method.

The easiest method for carrying this out is to use molecular weight standards that have been fractionated to produce very narrow distributions of molecular weights to generate an unequivocal calibration curve of the elution time or "retention volume", V, as a function of molecular weight: $V = f(M)$. Polystyrene and poly(ethylene oxide) standards covering a wide range of molecular weights are commercially available, for example. This procedure is only valid for the same polymer and solvent combination, however, since polymers of the same molecular weight are likely to have different chain dimensions (and the separation is based on molecular size). For most polymers suitable standards are not available, however, and molecular weights are usually reported as apparent molecular weights relative to a standard (generally polystyrene). It has also been shown that the retention volume is proportional to the hydrodynamic volume, $[\eta] M$ (Grubisic et al., 1967). Thus for a given solvent and temperature, polymer solutes of varying composition and size produce a common plot of $\log [\eta] M_w$ against V for a given set of columns. This is the most commonly used of several "universal" calibrations and has received considerable support in the literature. Its effectiveness has been rationalized on the basis of the Flory-Fox equation (Flory and Fox, 1951) which shows that the product $[\eta] M$ is proportional to the radius of gyration (r_g), the size parameter which determines the extent of pore penetration in GPC separations.

10.3.8 Field Flow Fractionation

Field flow fractionation (FFF) is a separation method similar to GPC but differing in the mechanism of separation [see Chap. 12 in Cooper (1989) for an excellent summary]. In FFF the principal component involved in the separation is a solution flow channel in which the molecules are segregated by the action of an external applied field or gradient. The flow profile in the channel is such that the low flow region is near one of the walls of the channel. The sample is introduced as a plug into the flow stream at the head of the channel. Application of the field drives the molecules toward the accumulation wall. After the field is applied and a short time period (usually on the order of seconds) has elapsed to allow a steady state to be achieved, diffusion away from the wall balances the field-driven motion toward the wall. At this point flow is initiated and the separation process begins. The separation is dependent upon the molecules being driven to different depths in the slow flow region near the wall. The flow drops to a very low velocity near the wall; as a result the components that are driven close to the wall move very slowly downstream and elute last. In general, the force exerted by the field increases with molecular weight driving the higher molecular weight molecules closer to the wall. Thus FFF differs from GPC in that the higher molecular weight molecules elute later than the lower molecular weight molecules. Figure 10-6 illustrates the structure of a FFF channel and the flow profile within it (Cooper, 1989).

The fields and gradients available to induce separations in macromolecules include a sedimentation field, electric field, thermal gradient and a cross-flow or hydraulic gradient. The most useful technique for synthetic organic polymers is

thermal FFF in which copper bars serve as a heater on one side of the channel, and a cold plate on the other side.

The retention characteristics of the solute are generally measured as the retention ratio:

$$R = V^0/V_r \tag{10-25}$$

in which V^0 is the void volume of the channel, and V_r is the retention volume. The compression of the solute against the wall in thermal FFF is described by the retention parameter λ_T:

$$\lambda_T = D/D_T \Delta T \tag{10-26}$$

where D is the diffusion coefficient, D_T is the thermal diffusion coefficient and ΔT is the temperature drop between the two walls of the channel. The relationship (Giddings, 1973) between R and λ_T is:

$$R = 6\lambda \quad (\lambda \ll 1) \tag{10-27}$$
$$R = 6\lambda - 12\lambda^2 \quad (\lambda > 0.05) \tag{10-28}$$

Experimentally it has been observed that D_T is independent of molecular weight for a given polymer type (Schimpf and Giddings, 1987) and D/D_T can be written:

$$D/D_T = A' M^b \tag{10-29}$$

where A' and b are constants for each polymer type and solvent pair. Equation (10-29) provides the relationship between the retention parameter and the molecular weight of the solute.

Unlike GPC, retention in FFF is not dependent exclusively on molecular size, but is also affected by the specific solvent-

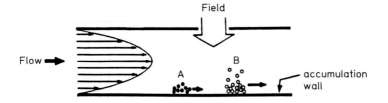

Figure 10-6. The structure of a FFF channel and flow profile.

solute interactions which influence D_T. Thus it was found that although polystyrene and poly(methyl methacrylate) polymer samples with similar diffusivities (D) (hence hydrodynamic size) gave no separation by GPC, close to baseline resolution was achieved by thermal FFF (Schimpf and Giddings, 1987).

FFF appears to be an extremely versatile method for polymer molecular weight characterization. One limitation is that it requires relatively small samples (< 1 mg). In addition, the technique has only recently begun to be extensively applied and the literature base is currently limited to results from a few laboratories.

10.4 Vibrational Spectroscopy

Vibrational spectroscopy, particularly infrared (IR), has been used extensively to study polymers. IR and Raman spectra provide information about the chemical composition of the polymer, and in many cases can give information about the chain structure, degree of branching, stereoregularity, geometric isomerism, conformation, crystallinity and type of end groups present.

10.4.1 Basic Principles

The total energy of a molecule is composed of the translational, rotational, vibrational and electronic energies. Transitions may be induced between different energy levels providing that the frequency of the incident radiation, v, corresponds to an energy transition according to the relationship $\Delta E = h v$, where h is Planck's constant. If this is expressed in terms of the wavelength λ:

$$\Delta E = h(c/\lambda) \tag{10-30}$$

where c is the velocity of light. The most commonly used measure is the inverse of the wavelength, $1/\lambda$, or wavenumber, v (cm^{-1}). The infrared spectrum is that portion of the electromagnetic spectrum between wavelengths 2 and 25 microns which corresponds to $5000-40$ cm^{-1}.

When IR radiation is passed through a sample, certain frequencies are absorbed by the molecule that correspond to vibrational changes in the molecule. In order for this to occur, there must be a mode of interaction between the incident radiation and the vibrational energy levels. This mode of interaction is an oscillating electric dipole induced by the vibration which interacts with the oscillating electric field of the electromagnetic radiation. IR absorption occurs for each vibrational degree of freedom of the molecule providing that a change in the dipole moment of the molecule takes place during the vibration. Obviously, for a large molecule with many vibrational degrees of freedom there may be many IR bands observed. Since each molecule has individual sets of energy levels, the absorption spectrum is characteristic of the functional groups that are in the molecule. For example, a carbon-carbon triple bond ($-C\equiv C-$) absorbs in the region $2300-2000$ cm^{-1}, whereas a double bond ($>C=C<$) is weaker and absorbs at lower frequency, $1900-1500$ cm^{-1}. Furthermore, vibrational motions exist in which the whole molecule is involved. These "fingerprint" bands serve to make the entire spectrum unique for the specific molecule under examination. Although both IR and Raman spectroscopy arise from the same physical phenomenon, that is, the vibrations of the atoms corresponding to quantum mechanically allowed transitions between vibrational energy levels, the interaction between the incident radiation and the sample is different. In

IR spectroscopy the sample is exposed to polychromatic radiation from which specific frequencies are absorbed and detected. In Raman spectroscopy, the sample is exposed to monochromatic (usually visible) radiation which is then scattered either elastically with the same frequency (Rayleigh scattering) or inelastically at higher and lower frequencies to produce Raman scattering.

The distinction between Raman, IR and Rayleigh scattering is shown schematically in Fig. 10-7. In IR spectroscopy the energy is absorbed by promoting a molecule into a higher vibrational energy state v_1 from the ground state v_0. Rayleigh scattering, however, involves the absorption of light in the visible region by a molecule followed by the immediate emission of radiation of the same wavelength, shown as a transition to and relaxation from a hypothetical energy state. If, however, excitation is followed by relaxation to the vibrational state v_1, then the emitted radiation will be less energetic than the exciting line. Alternatively, if excitation from v_1 to an excited state is followed by relaxation to the ground state v_0, the emitted radiation will be more energetic than the incident radia-

tion. The latter case, known as anti-Stokes behavior, is much less likely than the former, or Stokes behavior, owing to the larger number of molecules in the ground state, and it is the Stokes lines that are usually measured experimentally. Raman spectroscopy complements IR since the latter requires that a change must occur in the overall molecular dipole moment, whereas the former is associated with a change in the polarizability of the molecule. Therefore, comparison of the IR and Raman spectra shows that although many of the same frequencies occur in both spectra, there are also many that are found in IR spectra that are not seen in Raman spectra and vice versa. For example, the symmetrical stretching mode of the carbon disulfide molecule is not active in the infrared since no net displacement of the charge center occurs, but it is active in Raman spectroscopy since there is a resultant polarization of the electron charge in the chemical bond. In general, therefore, IR produces strong absorption bands for polar functional groups such as carbonyls and hydroxyl groups, whereas Raman spectra show strong lines for nonpolar functional groups such as carbon-carbon double bonds and disulfide linkages $(-S-S-)$.

10.4.2 Instrumentation

The infrared spectrometer consists of three essential features: a source of IR radiation, a monochromator and a detector (see Fig. 10-8). The source of IR radiation consists of a wire or rod heated to temperatures of 1200–1500 °C. The heating element may be Nichrome wire, a silicon carbide rod (Globar) or a Nernst glower, which is a rod containing a mixture of yttrium, zirconium and erbium oxides. The dispersing element of the spectrometer is

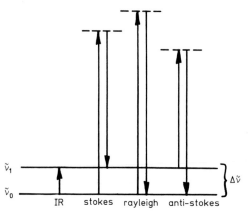

Figure 10-7. Schematic distinction between IR, Raman and Rayleigh scattering.

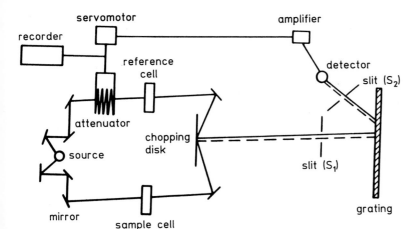

Figure 10-8. Schematic of a grating infrared spectrometer.

designed to separate the incident black-body radiation into its component frequencies. In earlier instruments a prism was used, but current instruments incorporate diffraction gratings. Most detectors operate on the thermocouple principle. Incident radiation causes a temperature rise at the thermocouple junction and generates a potential difference that increases with temperature. The light beam is interrupted by a chopper to produce an ac, rather than dc, signal.

Figure 10-8 shows a simplified schematic of a typical IR spectrometer. The light from the source is split into two identical beams, one of which passes through the sample cell and the other which passes through the reference cell. The beams are then passed through the chopper which allows them to pass through the slit S_1 in alternating time periods and into the monochromator (grating). The grating rotates slowly and sends individual frequencies to the detector. If radiation has been absorbed, the alternately weaker or stronger beams give rise to a pulsating or ac current from the detector to the amplifier. The ac portion of the signal is amplified and the strength of the reference beam is matched to that of the sample beam. The

resultant spectrum is usually recorded as percent transmittance as a function of frequency (in wavenumbers) or wavelength.

A tremendous improvement in sensitivity is afforded through the use of Fourier transform (FT) IR spectroscopy. This is particularly useful in polymer analysis where the bands may be broad and weak, and in which small components, such as end groups, are often the target of analysis. FT-IR spectroscopy makes use of all the frequencies from the source simultaneously, rather than individually as in the scanning instruments, through the use of a Michelson interferometer (shown schematically in Fig. 10-9). The Michelson interfer-

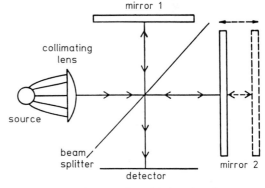

Figure 10-9. Diagram of a Michelson interferometer.

ometer consists of two perpendicular mirrors, one of which is stationary (mirror 1) and one which moves at constant velocity along the path between it and the source. Light from the source passes through the collimating lens and is split into two equal beams. The beams are then reflected by mirrors 1 and 2, and recombined. If the two mirrors are equidistant from the beam splitter, the two light beams will be in phase when they return and will constructively interfere. If, however, mirror 2 is moved further away, the recombining beams reaching the detector will differ in phase and destructive interference results. Monochromatic radiation produces a cosine wave as the optical paths are varied since, as the mirror is continuously moved, the signal oscillates from strong to weak for each quarter wavelength movement of the mirror. When broad spectrum light is used, the output seen at the detector is the summation of all the cosine oscillations produced by the individual frequency components of the light. The output time domain signal is an interferogram which can be converted into the infrared spectrum by Fourier transformation.

The experimental setup for Raman spectroscopy is relatively straightforward as shown in block diagram form in Fig. 10-10. The components required include a light source, a chamber for illuminating the sample, a light dispersion system to resolve the intense signal from the elastically scattered light from the weaker inelastically scattered Raman signal, a detector, amplifier and recorder. Since the Raman effect is very inefficient and produces only a very small amount of inelastic scattering, powerful light sources such as lasers are required to generate enough signal to detect. In order to detect the low level of scattered photons generated, high performance photomultiplier tubes are needed. The electri-

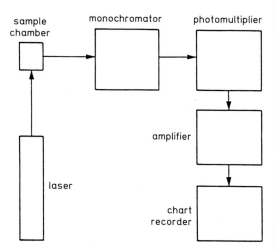

Figure 10-10. Block diagram of a Raman spectrometer.

cal pulses generated in the photomultiplier tube by the incident photons are amplified by either a DC amplifier or pulse counting system. The intensity of the scattered light is recorded as a function of wavelength or frequency.

10.4.3 Sampling

Sample requirements for IR and Raman differ considerably owing to the nature of each technique. Since IR spectroscopy involves a light transmission process, the transparency of the sample and sample container is critical. Raman spectroscopy, however, is a light scattering technique, and the sample form is relatively unimportant. Infrared samples may be prepared as solutions, hot pressed or solvent cast films, pressed into a transparent disk in a matrix such as KBr (IR transparent), or dispersed as a mull in a suitable material such as paraffin oil (Nujol). Since solution IR spectra must be obtained in cells which transmit in the IR region of interest, expensive, hygroscopic and fragile materials such as NaCl, KBr or CsI must be used for the cells. The solvent, too, must be relatively transparent in the region of interest. This problem is somewhat alleviated by the subtraction techniques available with FT spectrometers.

In contrast to IR spectroscopy, the Raman spectra of polymers can often be obtained without any prior sample preparation. This is important in cases where pretreatment of the material before spectroscopy examination is undesirable, such as in studies of the effects of thermal history on a material. Raman samples can be opaque, and cells can be made of glass since the excitation source is visible light. Powders may be packed in small glass capillary tubes or pressed into a pellet.

10.4.4 Other Techniques

10.4.4.1 Attenuated Total Reflection

Another technique which is available for materials that fail to produce good IR transmission data is attenuated total reflection (ATR) or internal reflection spectroscopy. This method is based on the phenomenon of total internal reflection. The sample is brought into good optical contact with the reflecting surface of a high refractive index prism. Depending on the difference in refractive index between the sample and the prism, a critical angle of incidence can be found beyond which total internal reflection takes place in the sample. The ATR spectrum differs only slightly from transmission spectra. Since the penetration depth increases with increasing wavelength, the absorption bands at longer wavelength have greater intensities. The technique allows good spectra to be obtained of fibers, paints and fabrics, materials which are traditionally very difficult by transmission IR spectroscopy. An advantage of this technique is that only a few microns of the surface are sampled by the beam, so that spectra from surface coatings are enhanced relative to the bulk phase. ATR is the most commonly used IR technique for sampling surfaces, but several other methods such as normal specular reflectance, grazing angle reflectance and diffuse reflectance are also used for specialized applications (Vidrine and Lowry, 1983). ATR attachments are available for commercial IR spectrometers.

10.4.4.2 Photoacoustic Spectroscopy

Photoacoustic spectroscopy (PAS) is used for obtaining IR spectra of samples which defy normal sample preparation methods, or for which surface information or depth profiles are desired. The basis of the technique is the photoacoustic effect which causes some materials to give off audible sound when exposed to modulated light. If a sample is exposed intermittently to modulated light, the sample surface and the air adjacent to the sample are heated during these intervals. The intermittent expansion of air causes a pressure wave, or sound, to be emitted. If a microphone is used to detect the sound emitted by the sample, it can be directly related to the absorption of energy by the sample surface. The use of FT-IR spectrometers and their higher measurement efficiency overcomes the inherent low sensitivity of the photoacoustic microphone cell, thus allowing spectra to be obtained in reasonable times. The advantage of PAS is the relative insensitivity to surface morphology.

10.4.5 Applications

10.4.5.1 Compositional Analysis

Infrared spectroscopy has been used most extensively to characterize polymer composition. The appearance of absorption bands at frequencies characteristic of

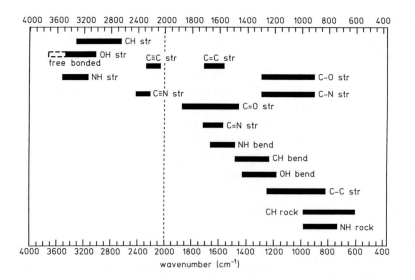

Figure 10-11. Some infrared group frequencies.

specific functional groups permits identification of the components of the polymer chain from the group frequencies observed in the spectrum. A collection of some typical group frequencies is shown in Fig. 10-11. Within the general groups shown, specific types of each functional group resolve into smaller regions which can further identify the structure. For example, carbonyl-containing compounds exhibit a strong stretching absorption band in the 1870–1540 cm^{-1} region of the IR spectrum. For unconjugated aliphatic ketones, however, this carbonyl stretching band lies within the range 1705–1750 cm^{-1}. In addition, because of the unique arrangement of the groups in a molecule, and the presence of additional bands arising from coupling between adjacent groups, the spectrum may be viewed as a fingerprint of the specific polymer structure. Collections of spectra are published by both Sadtler Research Laboratories and Aldrich Chemical Company, and most instrument manufacturers offer li-

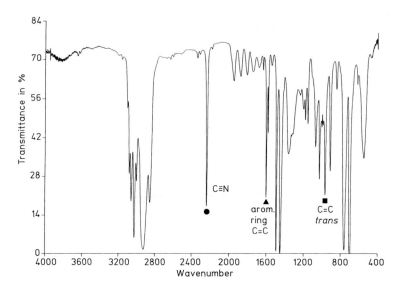

Figure 10-12. Infrared spectrum of an acrylonitrile/butadiene/styrene (ABS) terpolymer. The symbols are: ● acrylonitrile; ■ butadiene; ▲ styrene.

braries compatible with their data systems. An example of a typical infrared spectrum used for compositional analysis is shown in Fig. 10-12. The characteristic frequencies used to identify the individual components of the acrylonitrile/butadiene/styrene (ABS) terpolymer are labelled on the spectrum. The peak at 2237 cm^{-1} corresponds to the carbon-nitrogen triple bond stretching mode and is representative of the acrylonitrile component. The absorption at 966 cm^{-1} arises from the carbon-carbon double bond stretch (*trans*) of the butadiene component, and the peak at 1602 cm^{-1} is due to the aromatic ring of the styrene component.

In general, the first step in identifying a material from its IR spectrum is to look for specific functional groups. It is easiest to start at the high frequency end of the spectrum where there are fewer overlapping bands. In the case of the spectrum in Fig. 10-12, for example, the nitrile band stands out and is easily identified. Once enough information has been obtained to guess at the structure, comparison with library spectra can be made. If IR spectral libraries are included in the software of the instrument, searches for a match to the unknown spectrum can be carried out by the computer.

IR can also be used to determine if chemical changes have taken place in a polymer. Bromination of poly(2,6-dimethyl-1,4-phenylene oxide), or polyxylylene ether (PXE) (12), occurs to produce up to 100% monobrominated polymer (White and Orlando, 1975) (13):

Figure 10-13 shows a comparison between the IR spectra of the PXE homopolymer (a) and a copolymer (b) prepared to contain 50% monobrominated rings. Significant changes occur in the spectrum. Bands at 1400, 1171 and 962 cm^{-1} are new in the brominated material. These correspond to a new aromatic ring C=C stretching vibration (1400 cm^{-1}), a new C—O stretching vibration at 1171 cm^{-1} and an additional out of plane CH deformation of the aromatic ring at 962 cm^{-1}. The C—Br absorbance is probably obscured at 400 cm^{-1}. Although qualitative evaluations of the degree of bromination are feasible by IR, quantitative determination is more easily carried out by wet chemical analysis or ^{13}C NMR (Williams et al., 1990) (see section on NMR).

The simplicity of preparing Raman samples belies the difficulties frequently encountered in attempting to obtain a Raman spectrum of a polymer. Problems arise if a sample strongly absorbs the laser beam, resulting in the rapid destruction of the material by photolysis or pyrolysis. Even weak absorption can cause problems, however, since the resulting fluorescence emission often presents a serious background interference to observation of the Raman spectrum. The fluoresence most often arises from an impurity or additive in the polymer. One method for eliminating this problem is to soak the sample in the laser until the background decays to a satisfactory level. If this does not work, or if the polymer is adversely affected by the

12   13

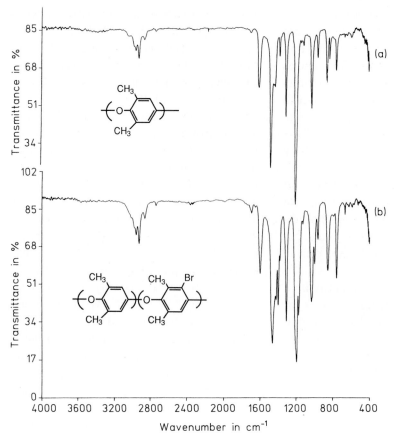

Figure 10-13. Infrared spectra of (a) poly(2,6-dimethyl-1,4-phenylene oxide) (PXE) and (b) a sample of PXE in which 50% of the aromatic rings are monobrominated.

process, other purification steps may be necessary.

As mentioned above, Raman spectroscopy is particularly useful for characterizing nonpolar groups such as hydrocarbons. For example, the polymerization of 1,3-butadiene proceeds to give 1,2 and 1,4 addition products. In the case of 1,4 addition, either *cis* or *trans* isomers are possible in the resulting olefin:

The physical properties of the polymer depend to a large degree on the distribution of the monomer units among 14, 15 and 16. The Raman C=C vibrational modes in *cis*-(14) and *trans*-1,4-polybutadiene (15) and the vinyl group of 1,2-polybutadiene (16) are sufficiently resolved to allow the composition of butadiene rubbers to be characterized using Raman spectroscopy

$$
\begin{array}{cc}
\underset{\text{H}}{\overset{-\text{CH}_2}{\diagdown}}\text{C}=\text{C}\underset{\text{H}}{\overset{\text{CH}_2-}{\diagup}} \quad cis\text{-PBD} &
\underset{\text{H}}{\overset{-\text{CH}_2}{\diagdown}}\text{C}=\text{C}\underset{\text{CH}_2-}{\overset{\text{H}}{\diagup}} \quad trans\text{-PBD} \\
\mathbf{14} & \mathbf{15}
\end{array}
$$

$$
\begin{array}{c}
-\text{CH}-\text{CH}_2- \\
| \\
\text{CH} \quad\quad \text{1,2-PBD} \\
\| \\
\text{CH}_2 \\
\mathbf{16}
\end{array}
$$

Cornell and Koenig, 1969). Expansions of the C=C stretching region in the Raman spectra of four polymers prepared using different polymerization methods and which differ substantially in composition of these structural units are shown in Fig. 10-14. The band at 1639 cm^{-1} arises from the vinyl group of 16 whereas the 1650 and 1664 cm^{-1} bands are representative of the *cis*- and *trans*-PBD, structures 14 and 15, respectively. The spectra show clearly that the anionic polymerization using sodium (Fig. 10-14d) produces a large amount of 1, whereas the emulsion polymer (Fig. 10-14a) contains mostly *trans* units (10).

Note that the symmetrical carbon-carbon double bond stretching vibration in 15 is *infrared inactive*, so other modes must be used for IR analysis of these polymers. The CH out-of-plane vibrations at 910, 967 and 740 cm^{-1} in the infrared arise from 16, 15 and 14, respectively, and have been used by Silas et al. (1959) to determine unsaturation distribution in these polymers.

0.4.5.2 End Groups and Branching

The end group of a polymer chain often differs substantially in structure from the main repeat unit of the chain. Spectroscopic techniques such as IR can be used to identify and/or quantify the end groups present. For example, the IR spectrum of polyethylene (Snyder, 1967) shows peaks corresponding to the vinyl end groups present (Fig. 10-15). In order to see end groups, thicker films may be used to enhance the lower intensity peaks in the analysis. Similarly, the hydroxyl end groups in PXE (12) (White and Loucks, 1984), poly(2-methyl-6-phenyl-1,4-phenylene oxide) (17) (White and Klopfer, 1972) and poly(2,6-diphenyl-1,4-phenylene oxide) (18) (White and Klopfer, 1970) have also been determined by FT-IR by using the hy-

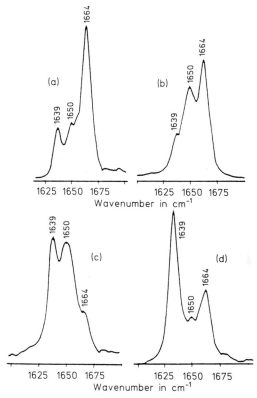

Figure 10-14. Raman spectra (C=C region) of butadiene rubbers prepared under different polymerization conditions: (a) emulsion polybutadiene (21% *cis*, 59% *trans*, 20% vinyl), (b) lithium polybutadiene (36% *cis*, 47% *trans*, 17% vinyl), (c) liquid polybutadiene (41% *cis*, 16% *trans*, 43% vinyl), (d) sodium polybutadiene (17% *cis*, 26% *trans*, 57% vinyl). (Reprinted, in part, with permission from Cornell and Koenig, 1969. Copyright 1969 American Chemical Society.)

droxyl peaks at 3610, 3554 and 3550 cm^{-1}, respectively, of the polymers dissolved in carbon disulfide and comparing the spectra with those of polymers at the same concentration in which all the hydroxyl groups had been capped with acetic anhydride.

17

18

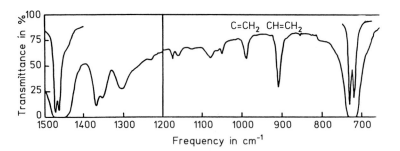

Figure 10-15. Infrared spec
trum of polyethylene show
ing vinyl end groups
(Reprinted with permission
from Snyder, 1967. Copy-
right by the American In-
stitute of Physics.)

If a high molecular weight polymer is available in which there is little or no branching, the spectrum may be used to subtract the base polymer peaks from a spectrum of the same material in which it is desired that end groups or branched structures be enhanced. For example, Fig. 10-16 shows the effect of subtracting the spectrum of high-density polyethylene film from low-density polyethylene film; the resulting spectrum provides a determination of the methyl groups in polyethylene (Siesler and Holland-Moritz, 1980, p. 180).

10.4.5.3 Quantitative Analysis

The absorbance, A, is the quantity mea
sured in infrared spectroscopy:

$$A = \log(I_0/I) \tag{10-31}$$

$$\log(I_0/I) = k_a b C \tag{10-32}$$

where I_0 is the intensity of incident light and I is the intensity of light passing through the sample. The relationship be
tween this measured value and the concen
tration of the molecule is given by the Beer
Lambert law [Eq. (10-32)] where k_a is the

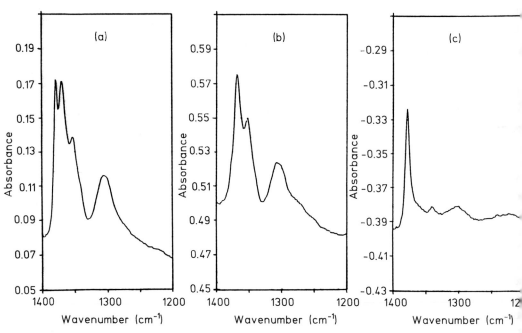

Figure 10-16. IR spectra of (a) low-density polyethylene film, (b) high-density polyethylene film and (c) the difference spectrum, (a)–(b). (Reprinted with permission from Siesler and Holland-Moritz, 1980, p. 180. Copy right by Marcel Dekker, Inc.)

specific absorptivity of the molecule, C is the concentration and b is the path length (cm) of the cell. In practice, the most commonly applied technique is the comparison of a material containing an unknown amount of a component with standards of known composition of the component:

$$\frac{A_1}{A_2} = k\frac{c_1}{c_2} \qquad (10\text{-}33)$$

For polymers, specific care must be taken that the analytical and reference materials differ *only* in composition since differences in crystallinity or sequence distribution can affect the intensities of the bands in the spectrum.

The intensities of Raman signals depend on a number of properties of the sample, such as refractive index and fluorescence, as well as several instrumental factors. In spite of the difficulties of quantitative analysis by Raman spectroscopy, copolymer composition can be carried out by the same relative band ratio method used in IR spectroscopy.

10.5 Nuclear Magnetic Resonance Spectroscopy

Nuclear magnetic resonance (NMR) spectroscopy is an extremely powerful method for observing individual structural features in a polymer chain. Since Bovey and Tiers first reported high resolution NMR spectra of polymers (Bovey et al., 1959), literally thousands of papers concerning applications of NMR spectroscopy to polymers have appeared in the literature. Early studies were concerned with the use of ^{1}H NMR. Subsequent technological developments led to the availability of high resolution ^{13}C NMR in the late 1960's. The

initial report by Schaefer (1969) was followed by another explosion of papers reporting applications of this technique to the analysis of polymer microstructure.

NMR studies often yield exceptionally detailed information about chain structure and it is not unusual to be able to resolve pentad (five monomer units) stereosequences, for example. In addition to stereoregularity in the polymer chain, NMR can be used to identify and measure composition, end groups, comonomer sequences, branching and chain defects. NMR offers the further advantage that no response factors are necessary for quantitative analysis. The data can be obtained, therefore, under conditions such that the area under each resonance is directly proportional to the number of nuclei contributing to the signal.

10.5.1 Basic Principles

The NMR phenomenon is dependent upon the magnetic properties of the nucleus. Certain atomic nuclei possess spin angular momentum which gives rise to different spin states in the presence of a magnetic field. The nucleus can have $2I + 1$ distinct energy states, where I is the spin quantum number and can have values of zero, half integers and whole integers. Nuclei with $I = 0$, including all nuclei with both an even atomic number and an even mass number such as ^{12}C and ^{16}O, do not have a magnetic moment and therefore do not exhibit the NMR phenomenon. All other nuclei can, in theory, be observed by NMR. Fortunately for polymer chemists, both ^{1}H and ^{13}C are nuclei with $I = 1/2$.

If nuclei with $I = 1/2$ are placed in a magnetic field, H_0, they may align either with or against the field giving rise to two spin states. The separation between the two energy levels is directly proportional to the magnitude of H_0 according to the equation:

$$\Delta E = \mu H_0/I = \gamma \hbar H_0 \qquad (10\text{-}34)$$

where μ and γ are the magnetic moment and gyromagnetic ratio, respectively, of the nucleus. The populations of the two states are given by the Boltzmann distribution law. Since they differ by only a few millijoules in energy, there is a very slight excess of nuclei in the lower spin state. The nuclei behave as spinning tops and precess about the axis of the applied field, H_0, with a frequency known as the Larmor frequency, $\bar{\omega}_0 = 2\pi\nu_0$. The NMR experiment consists of supplying energy of the proper frequency to induce transitions between the energy levels. The two spin states for a nucleus with $I = 1/2$ are depicted in Fig. 10-17. A resonance condition, and thus transitions between the two energy levels, occurs when the nucleus is irradiated with electromagnetic energy of radiofrequency ν_0 corresponding to a magnetic component, H_1, perpendicular to the applied magnetic field, and fitting the condition $\Delta E = h\nu_0$. The Larmor frequency, ν_0, is quite different for each type of nucleus observed. Thus in a 7.1 T static magnetic field, $\nu_0 = 75$ MHz for ^{13}C and 300 MHz for ^{1}H.

If all nuclei of a given type (e.g., all ^{13}C nuclei in a sample) absorbed exactly the same frequency radiation, very little information would be obtained from a high-resolution NMR spectrum. The precession frequencies are usually not the same, however, and the precise value depends upon the local chemical environment of each nucleus in the molecule. For example, the ^{1}H and ^{13}C NMR spectra of poly(1,4-butylene terphthalate) PBT) (Figs. 10-18 and 10-19) show different peaks arising from each different type of carbon or proton in the polymer repeat unit. Fortunately, the peak positions or "chemical shifts" of the nuclei are generally predictable using substituent additivity calculations and model systems. The chemical shift is measured in frequency units (Hz) from a reference peak (usually tetramethylsilane), but generally reported in the dimensionless unit of parts per million (ppm) in which the dependence of the peak position on magnetic field strength has been factored out (Becker 1980).

Both ^{1}H and ^{13}C NMR are important in polymer studies but there are significant differences between the two nuclei. The gyromagnetic ratio (γ) of ^{13}C is only one fourth that of ^{1}H. Since the sensitivity is proportional to γ^3, ^{13}C has 1/64 the inherent sensitivity of ^{1}H. Furthermore, ^{13}C has a natural abundance of only 1.1% as opposed to almost 100% for ^{1}H. This further lowers the relative sensitivity of the ^{13}C experiment to 1/6000 that of ^{1}H. Conversely, the resolution of structural features is often much greater in ^{13}C NMR when compared with ^{1}H NMR and the total chemical shift range for ^{13}C is approximately 20 times that for ^{1}H. In spite of the greater amount of information available from ^{13}C NMR, its use was restricted to a few research groups until the development of Fourier transform (FT) NMR (Ernst

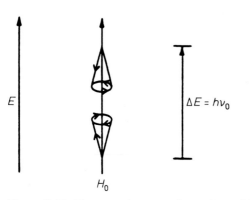

Figure 10-17. The two spin states of a nucleus with $I = 1/2$ in a magnetic field H_0.

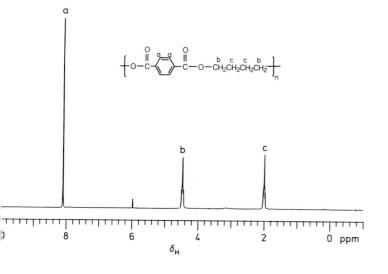

Figure 10-18. ¹H NMR (200 MHz) spectrum of poly(1,4-butylene terephthalate).

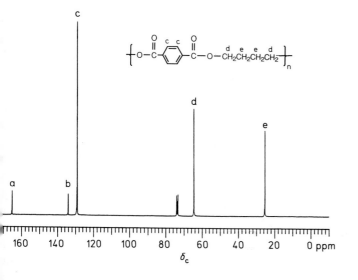

Figure 10-18. ¹H NMR (200 MHz) spectrum of poly(1,4-butylene terephthalate).

nd Anderson, 1966) in the late 1960's. FT-NMR facilitated observation of lower sensitivity nuclei, such as ^{13}C and ^{29}Si, by greatly decreasing spectral accumulation times. Today both ¹H and ^{13}C NMR are routine tools in polymer characterization.

0.5.2 Instrumentation

Figure 10-20 shows a greatly simplified schematic of a NMR spectrometer. The ba-

sic components include a magnet of very uniform field, a radiofrequency (rf) source and a method of detecting absorption of rf energy by the sample, and a recorder for displaying the spectrum. Early NMR systems employed permanent magnets or electromagnets. Modern NMR spectrometers use superconducting magnets which allow much higher fields to be obtained. This has the advantage of producing both higher sensitivity and higher resolution of the peaks in the spectrum.

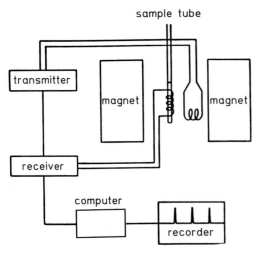

Figure 10-20. Basic components of a NMR spectrometer.

The source of rf power may be either a crystal controlled at a single frequency or a frequency synthesizer for use over a range of frequencies. In the continuous wave (CW) method of achieving the resonance condition in NMR spectroscopy, the range of frequencies of the nuclei in the sample are swept by varying either the static field, H_0, or the rf field, H_1. As each nucleus is brought into resonance, an induced volt-age in the rf pickup coil is amplified and recorded directly as a plot of voltage (peak intensity) versus frequency.

The Fourier transform (FT) method involves simultaneous excitation of all frequencies by the application of a short rf pulse near v_0. Immediately following the rf pulse, the signal (voltage) is detected as a function of time as the spins reestablish equilibrium population distributions. This pattern, shown in Fig. 10-21, is called a free induction decay (FID), and consists of one exponentially decaying sine wave for each frequency component in the spectrum. Fourier transformation (Fig. 10-21) produces the normal frequency spectrum identical to that obtained in the CW experiment. The advantage is that the FT method saves a considerable amount of time by exciting all the nuclei simultaneously, rather than sweeping slowly through the frequency range of the nucleus. Thus a single trace may be obtained in seconds rather than the several minutes required for a CW scan. FT-NMR also is well suited to signal averaging and multiple FIDs may be collected before Fourier transformation of the data. As mentioned

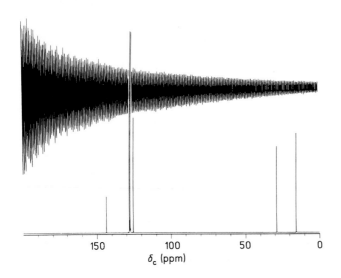

Figure 10-21. Free induction decay (FID) and resulting frequency spectrum after Fourier transformation.

previously, it was the tremendous signal-to-noise improvement afforded by FT-NMR that led to the dramatic upsurge in studies involving nuclei of low natural abundance and allowed low level structures, such as end groups in polymers, to be detected.

More recently, the application of several specialized techniques has extended high-resolution NMR studies to include materials in the solid state (Fyfe, 1983). Thus highly crosslinked and insoluble polymers may now be examined by methods similar to those that have proved invaluable for characterizing soluble polymers for more than 30 years (Komoroski, 1986).

10.5.3 Applications

10.5.3.1 Compositional Analysis

NMR, particularly ^{13}C NMR, is an indispensable tool for determining the composition of polymers. A great advantage of ^{13}C NMR is that each magnetically unique carbon atom in the structure of the molecule is usually resolved into a single line in the spectrum. Compare, for example, the ^{13}C NMR spectrum of PBT (Fig. 10-19) with that of the polyetherimide (19) shown in Fig. 10-22. It can be deduced immediately from

information about the local environment of the carbon atoms in the molecule. For example, the peak at 64 ppm in Fig. 10-19 is characteristic of an aliphatic carbon atom with a single bond to oxygen (i.e., an alcohol or an ether). A general grouping of carbon chemical shifts is shown in Fig. 10-23. Extensive literature compilations of chemical shift data are available to facilitate structure elucidation. In addition, certain ^{13}C NMR experiments can differentiate methine (CH), methylene (CH$_2$) and methyl (CH$_3$) carbon signals to further characterize the material.

Once the spectra have been obtained and the structures of the monomer units comprising the polymer have been established, NMR can be used to determine the composition of the material. This technique may be applied to any number of components as long as peaks representing each component can be identified. The different monomer units may arise, either from the addition of two comonomers in the initial synthesis of the polymer, or by chemical modification of the polymer. An example of the latter case is the bromination of PXE discussed in Sec. 10.4.5. The ^{13}C NMR spectrum of a 57 mol% brominated polymer is shown in Fig. 10-24 along with assignments of the pertinent peaks. The degree of bromination can be calcu-

19

examining the two spectra that the structure of 19, which produces 18 different lines in the ^{13}C spectrum, contains many more unique carbon atoms than the structure of PBT, which produces only 5 lines in the ^{13}C spectrum. Furthermore, knowledge of the chemical shifts of the peaks provides

lated from a direct comparison of integrated areas of the peaks labelled g and h which represent the unbrominated and brominated aromatic rings, respectively, or by using peak d to represent the unbrominated ring and peaks (e + f) to represent the brominated ring (Williams et al., 1990).

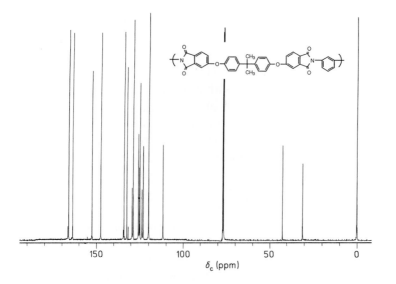

Figure 10-22. ^{13}C NMR (125 MHz) spectrum of polyetherimide 19.

^{1}H NMR spectra are more complicated in that the peaks usually appear as multiplets for each proton rather than as single lines. This multiplet structure arises from interactions between proximate proton nuclei in the molecule which cause splitting of energy levels and thus several transitions where one might be expected. Since these splittings are characteristic of the types of protons involved in the coupling, and the bond lengths and angles between the cou-

pled protons, knowledge of the magnitude of the splitting may assist in assigning the structure. Detailed discussions of spin-spin coupling are available in basic NMR texts (Becker, 1980; Emsley et al., 1965). These couplings are not observed between ^{13}C nuclei owing to the low natural abundance of this isotope of carbon (only 1.1 percent of the neighbors of a ^{13}C nucleus are other ^{13}C nuclei – the rest are ^{12}C isotope which does not have a spin).

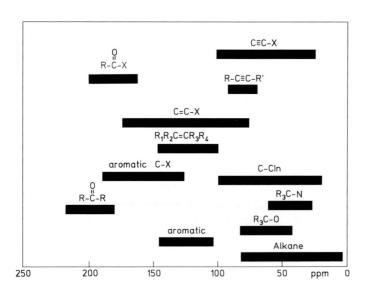

Figure 10-23. General carbon-13 chemical shifts.

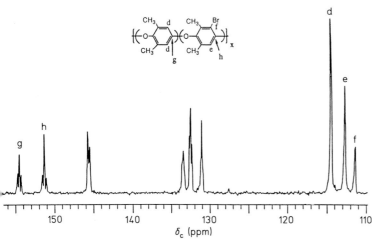

Figure 10-24. Carbon-13 NMR spectrum (75.4 MHz) of the aromatic region of a 57 mol% brominated poly(2,6-dimethyl-1,4-phenylene oxide). Assignments are shown for peaks used in determining degree of bromination.

Coupling does occur, however, between the ^{13}C nuclei and nearby protons. This is eliminated in normal ^{13}C spectra by irradiating the protons (*proton decoupling*) at their Larmor frequency to induce rapid transitions between proton spin states. Thus the carbon nuclei see only the average spin state of the protons, and single lines are produced. The use of proton decoupling in obtaining ^{13}C spectra does, however, have ramifications in quantitative analysis.

The ^1H and ^{13}C NMR spectra of poly(1,4-phenylene ether sulfone) are shown in Fig. 10-25. In the ^1H spectrum, the coupling between H_1 and H_2 is evident in the doublet seen for each peak. The magnitude of the splitting in frequency units (Hz) is the coupling constant and is independent of the magnetic field strength of the spectrometer. In the ^{13}C spectrum, the four unique carbon atoms in the polymer are resolved into single peaks owing to proton decoupling. Note that the intensities of the peaks corresponding to the protonated aromatic carbons (1, 4) are more than twice that of the peaks for the non-protonated aromatic carbons (2, 3). This arises from

perturbation of the carbon populations due to irradiation of the nearby protons which are coupled to the carbons through dipole-dipole interactions. This is called the *nuclear Overhauser effect* (NOE) and results in an enhancement of the signal arising from the protonated carbons. Furthermore, not all protonated carbons are enhanced to the same degree. The NOE is only operative under conditions of proton decoupling, and thus is not seen in normal ^1H spectra. Although it provides another tool for characterization, the NOE must be eliminated or taken into consideration when performing quantitative analysis by NMR.

10.5.3.2 Quantitative Analysis

NMR can be readily made quantitative since in modern spectrometers the absorption of rf by all the nuclei in the sample is identical over normal spectral widths. In principle this means that the area under each peak in a ^{13}C spectrum, for example, is proportional to the number of carbons of that type that are present in the molecular structure. In practice, however, two fac-

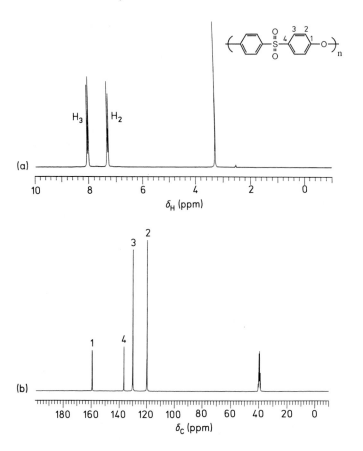

Figure 10-25. (a) ^{1}H and (b) ^{13}C NMR spectra of poly(1,4-phenylene ether sulfone) in dimethylsulfoxide (DMSO)-d$_6$. The peak at 3.3 ppm in the proton spectrum is due to water in the DMSO-d$_6$ solvent. The peak at 40 ppm in the ^{13}C spectrum is due to solvent.

tors must be taken into account to obtain quantitative data. The NOE, mentioned above, must be eliminated in cases where nuclei other than protons are observed and proton decoupling is used. This is achieved by turning the decoupler *off* during the delay between RF pulses (*gated decoupling*) and allowing the normal equilibrium populations to be established.

The second factor that must be considered for quantitative analysis is the time required for the nuclei to reestablish a Boltzmann population distribution after the rf pulse has disturbed the equilibrium. The process that allows the nuclei to return to the lower spin state is called *spin-lattice relaxation* and involves the transfer of energy from the nuclei to their surroundings. For quantitative analysis, it is necessary to

use a sufficient delay between rf pulses to ensure that all nuclei are fully relaxed before the next rf pulse is applied. This may be many seconds, or even minutes, and could hinder rapid acquisition of spectra. Frequently, a small amount of a *paramagnetic relaxation reagent*, such as tris(acetylacetonato) chromium (III), is added to shorten the relaxation times of nuclei other than protons. Relaxation reagents provide an extremely efficient pathway for nuclear spins to return to the ground state through electron-nuclear dipole-dipole relaxation effected by the unpaired electron of the reagent. This has the added benefit of eliminating most or all of the NOE since dipolar relaxation of nuclei, from which the NOE is derived, is overwhelmed by the presence of the unpaired electron.

Thus for quantitative analysis by NMR of nuclei other than proton, a small amount (50–75 mg) of a relaxation reagent should be added to the solution if possible to shorten the T_1's, gated decoupling should be used to eliminate any residual NOE, and a sufficiently long relaxation delay (typically a few seconds) must be employed to allow for complete relaxation of the nuclei. For protons, it is sufficient to simply use a relaxation delay of appropriate length. Frequently the T_1's of the material are measured and a relaxation delay of several times the longest T_1 is used.

10.5.3.3 End Groups and Branching

An example of an end group analysis by ^{29}Si NMR was shown in Fig. 10-1 for a trimethylsilyl endcapped polydimethylsiloxane fluid (Sec. 10.3.3.1). A similar analysis may be performed on a hydroxy-endcapped polydimethylsiloxane fluid (Fig. 10-26) in which the chemical shift of the silicon atom at the end of the polymer chain is now −10.5 ppm rather than 7 ppm as in Fig. 10-1 (−SiMe$_2$OH end group vs. −SiMe$_3$). Note that the silanol end group is present in less than 1 mole% in Fig. 10-26. Levels of less than 0.01 mole% may be detected under ideal conditions (concen-

trated solution, end group that is well resolved from the other polymer peaks) by NMR.

In spite of the lower sensitivity of ^{13}C and ^{29}Si NMR, these nuclei are often the only choice for detecting end groups by NMR. The ^{1}H chemical shift range is a factor of 10–20 times smaller with the result that very often the end groups are buried under the main polymer peaks. For example, the ^{13}C NMR spectrum of a bisphenol A (BPA) polycarbonate [poly-bis(4-hydroxyphenyl-2,2-propane) carbonate] chain terminated with an aromatic acetylenic end group is shown in Fig. 10-27. The peaks for the acetylenic carbons are well out of the region of those for the polymer peaks and provide a clear identification of the end group. In the ^{1}H NMR spectrum, however, the aromatic protons of the end group are buried under the polymer peaks.

Branching in a polymer chain can also be identified if it is present at a detectable level. Both ^{29}Si and ^{13}C NMR are very effective methods for detecting chain branching in polymers. In Fig. 10-28, the ^{29}Si NMR spectrum of a polydimethylsiloxane fluid is shown with the trimethylsilyl endcaps at 7–9 ppm, the main chain dimethylsiloxy repeat units at −19 to

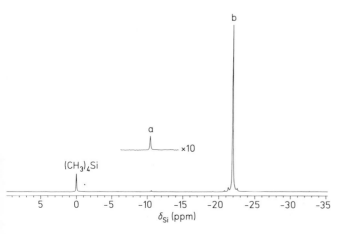

(CH$_3$)$_4$Si

a ×10

b

5   0   −5   −10   −15   −20   −25   −30   −35
δ_{Si} (ppm)

Figure 10-26. ^{29}Si (39.6 MHz) NMR spectrum of a hydroxy endcapped polydimethylsiloxane fluid with average degree of polymerization $\overline{DP} \sim 300$. Peak a corresponds to the silicon atoms at the end of the polymer chain. Peak b corresponds to the main chain dimethylsiloxy repeat units.

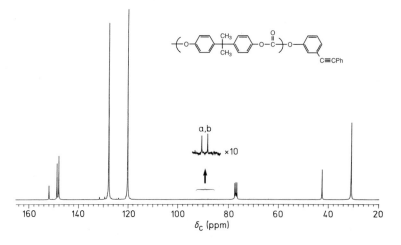

Figure 10-27. ^{13}C (75.4 MHz) NMR spectrum of bisphenol A polycarbonate showing peaks arising from the aromatic acetylenic end groups shown.

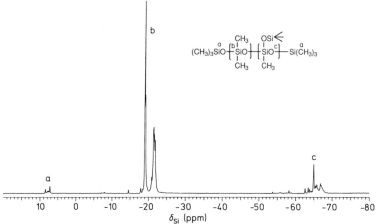

Figure 10-28. ^{29}Si (59.6 MHz) NMR spectrum of a trimethylsilyl end-capped polydimethylsiloxane fluid showing resonances arising from branch points in the polymer chain. Peaks a arise from the trimethylsilyl endcaps, peaks b from the main chain repeat units, and peaks c from the branch points in the polymer chain.

− 22 ppm, and a group of peaks in the − 63 to − 68 ppm region of the spectrum. The last group of peaks arises from the presence of branch points in the polymer chain (20):

$$\mathrm{\sim\sim SiO-\underset{\underset{OSi\sim\sim}{|}}{\overset{\overset{CH_3}{|}}{Si}}-OSi\sim\sim}$$

20

The multiple peaks for each type of silicon species in the polymer arises from the sen-

sitivity of the chemical shift to neighboring groups as well as those that are several bonds removed. Thus the chemical shift of the trimethylsilyl groups at the end of the polymer chain vary depending upon whether the neighboring group is a normal dimethylsiloxy chain repeat unit or a branch point.

Detailed analyses of branching in polyethylene have also been conducted. For example, Fig. 10-29 shows the ^{13}C NMR spectrum of low-density polyethylene (Bovey et al., 1976) with ethyl, *n*-butyl and *n*-amyl branches identified.

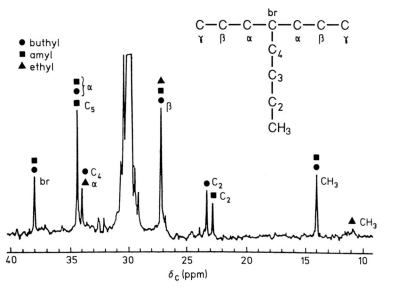

Figure 10-29. ^{13}C (25.2 MHz) NMR spectrum of low-density polyethylene showing ethyl, n-butyl and n-amyl branches (Bovey et al., 1976).

10.5.3.4 Tacticity

The advantage of ^{13}C over ^{1}H NMR is perhaps most clearly demonstrated in tacticity studies of polypropylene (Tonelli, 1989). The ^{1}H and ^{13}C spectra of isotactic, atactic and syndiotactic polypropylene are shown in Fig. 10-30. Both stereoregular polypropylenes (syndiotactic and isotactic) give simple spectra since only one type of

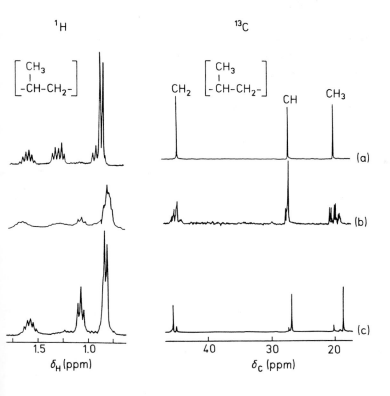

Figure 10-30. ^{1}H (220 MHz) and ^{13}C (25 MHz) NMR spectra of (a) isotactic (all m dyads), (b) atactic (random stereochemistry) and (c) syndiotactic (all r dyads) polypropylene (Tonelli, 1989).

dyad is present in each polymer. In the case of the isotactic polypropylene, *meso* dyads are present whereas the syndiotactic polypropylene contains *racemic* dyads (Sec. 10.2). Note that the spectrum of the syndiotactic polymer shows additional small peaks which indicate that the stereochemistry is not as clean as in the case of the isotactic polymer.

In the proton spectrum of the atactic polypropylene the presence of a large number of overlapping peaks due to the stereo-irregularity of the polymer chain gives rise to a poorly resolved spectrum. In the methyl region of the ^{13}C spectrum, however, as shown in Fig. 10-31, *9 out of 10*

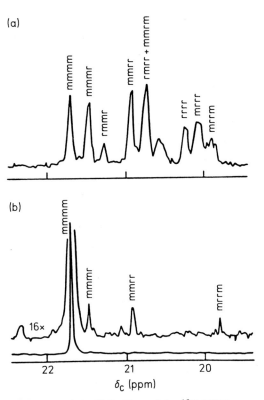

Figure 10-31. Methyl region of the ^{13}C NMR spectrum of atactic polypropylene showing chemical shift sensitivity to pentad sequences. (Reprinted with permission from Tonelli and Schilling, 1981. Copyright 1981 American Chemical Society.)

individual pentad sequences are resolved. Clearly ^{13}C NMR is a very powerful technique for studying stereosequences in polymers.

10.5.3.5 Comonomer Sequences

As mentioned in Sec. 10.2, when two (or more) monomers are copolymerized, the individual monomer units may be arranged in a number of ways in the polymer chain to produce an alternating, block or random copolymer. The sensitivity of chemical shifts to the nature of neighboring groups provides a means of assessing which type of polymer structure has been obtained. This is analogous to the resolution of different stereosequences in the polymer chain, except that it arises from compositional, rather than stereochemical, differences. For example, a portion of the ^{29}Si NMR spectrum of a 1:1 siloxane copolymer containing dimethylsiloxy [(Me)$_2$SiO] and methyl, hydrogen siloxy groups [(Me)(H)SiO], ("D" and "D^H" groups, respectively) is shown in Fig. 10-32. The region of the spectrum shown arises *only* from the silicon atom of the dimethylsiloxy groups (D). The triplet structure with the largest separation between the three groups of peaks is due to the triad structure in the chain. For example, the silicon nucleus of a D group will have a different chemical shift if it is surrounded on either side by two D^H groups, one each of D and D^H, and two D groups. Similarly, each of those peaks will differ slightly in chemical shift depending upon the structure of the next two siloxy groups further out in the chain (pentad structure) giving rise to the triplet structure within the main triplet. Since these units are further away, the effect on the chemical shift is smaller. Finally, these triplets are further split into triplets due to the resolution of heptad

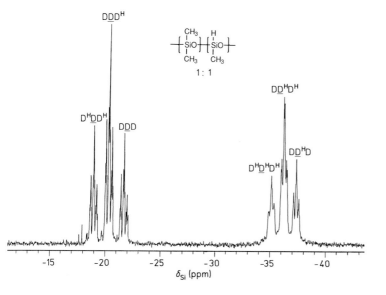

Figure 10-32. ^{29}Si (99.4 MHz) NMR spectrum (di-methylsiloxy region) of a 1:1 copolymer containing dimethylsiloxy groups (D) and methyl, hydrogen siloxy groups (D^H) in random sequence. Assignment of peaks corresponding to the three different triad structures are noted (Gray et al., 1989). Additional multiplicity arises from pentad and heptad sequences.

structure in this system. Again, the spacing between the peaks is now just barely resolveable since the chemical shift effects are occurring over a long distance in the polymer chain. Similar patterns are observed for the D^H groups with peaks centered around -36 ppm.

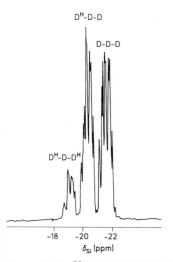

Figure 10-33. ^{29}Si (59.6 MHz) NMR spectrum (D region) of a 1:1 copolymer of dimethylsiloxane and methyl, hydrogen siloxane (DD^H) which contains longer homopolymer sequences than a statistical distribution (partial blocking).

Analysis of the spectrum in Fig. 10-32 shows that the polymer is completely random since a statistical distribution of the different sequences is observed. If blocking occurs, longer homopolymer sequences occur preferentially which is reflected in the spectrum. Figure 10-33 shows the same region of the spectrum for a polymer which is also 1:1 in D and D^H groups, but which has longer D and D^H homopolymer sequences. Thus the peaks arising from DDD triads are increased in intensity at the expense of the D^{H}DD^H and D^{H}DD triads. This polymer is also a lower molecular weight material and the additional fine structure in the spectrum arises from chemical shifts due to the proximity of end groups.

Copolymers prepared by the oxidative coupling of 2,6-xylenol and 2-alkyl-6-methylphenol to produce 21 may be similarly characterized by ^{13}C NMR (Williams et al., 1982). The ^{13}C NMR spectrum of a copolymer prepared from 70% 2,6-xylenol and 30% 2-alkyl-6-methylphenol in which the alkyl group is a C_{16} hydrocarbon moiety (2-hexadecyl) is shown in Fig. 10-34. The peaks at 154–156 ppm arise from the aromatic ring carbon atom (C-4) which is directly attached to an oxygen atom and is *meta* to the alkyl groups. The chemical shift of this carbon is sensitive to the substituents on the adjacent ring bonded to the same oxygen atom. Thus C-4 of a dimethyl substituted aromatic ring has a chemical shift of 154.8 ppm if the adjacent ring is another dimethyl substituted ring (AA), and 155.3 ppm if the adjacent ring is substituted with a methyl group and

trum. The juncture units (BA and AB) are present only in very small amounts. This is shown in Fig. 10-35a which is an expansion of the 154–156 ppm region of the spectrum shown in Fig. 10-34 (a 70:30 block copolymer). In a completely random copolymer a statistical distribution of these dyads is present. Figure 10-35b shows the identical region of the spectrum for a copolymer which comprises the same ratio of the two monomer units (70:30), but is a random copolymer. The intensities of the AB and BA dyad peaks are consistent with a statistical distribution of the B monomer units throughout the polymer chain.

The applications presented above often rely on small chemical shift differences aris-

154.8

CH_3   CH_3
 —O—⟨ring⟩—O—⟨ring⟩—
CH_3   CH_3
   A   A
         22

155.3

CH_3   C_16H_33
 —O—⟨ring⟩—O—⟨ring⟩—
CH_3   CH_3
   A   B
         23

156.0

C_16H_33   C_16H_33
 —O—⟨ring⟩—O—⟨ring⟩—
CH_3   CH_3
   B   B
         24

155.4

C_16H_33   CH_3
 —O—⟨ring⟩—O—⟨ring⟩—
CH_3   CH_3
   B   A
         25

a 2-hexadecyl group (AB). Similarly, C-4 of a methyl, 2-hexadecyl substituted ring has a chemical shift of 156.0 ppm if it is adjacent to another methyl, 2-hexadecyl substituted ring (BB) versus 155.4 ppm if it sees a dimethyl substituted ring on the other side of the oxygen atom (BA).

In a blocked copolymer with long blocks, homopolymer-type sequences $(A)_n$ and $(B)_n$ predominate and may be measured by the amount of AA and BB dyads in the spec-

ing from minor or remote structural changes in the polymer molecule. The data can become quite complicated and may cause confusion unless care is taken to obtain the appropriate model systems or control samples. For example, significant spectral differences may arise from such diverse causes as molecular weight degradation, changes in the structure of the polymer chain, and the presence of different additives in the polymer. If used properly, how-

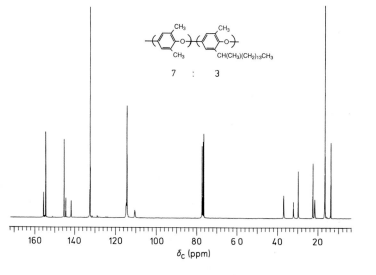

Figure 10-34. ^{13}C (75.4 MHz) NMR spectrum of a copolymer prepared from 2,6-xylenol (70%) and 2-(2-hexadecyl)-6-methylphenol (30%).

ever, NMR spectroscopy is unsurpassed in providing detailed information on polymer molecular structure.

10.6 Acknowledgements

The author wishes to thank Mr. P. Gundlach, Ms. E. Parks, Mr. P. Donahue and Ms. J. Smith for their assistance in obtaining data presented in this chapter.

10.7 References

Archibald, W. J. (1947), *J. Phys. Coll. Chem. 51*, 1204.

Baijal, M. D. (Ed.) (1980), *Plastic Polymer Science and Technology*. New York: John Wiley and Sons.

Becker, E. D. (1980), *High Resolution NMR*. New York: Academic Press.

Bovey, F. A. (1972), *High Resolution NMR of Macromolecules*. New York: Academic Press.

Bovey, F. A., Tiers, G. V. C., Filipovich, G. (1959), *J. Polym. Sci. 37*, 73.

Bovey, F. A., Schilling, F. C., McCrackin, F. L., Wagner, H. L. (1976), *Macromolecules 9*, 76.

Brandrup, J., Immergut G. H. (Eds.) (1975), *Polymer Handbook*, 2nd. ed. New York: John Wiley and Sons, Chap. 4.

Cooper, A. R. (1989), *Determination of Molecular Weight*. New York: John Wiley and Sons.

Cornell, S. W., Koenig, J. L. (1969), *Macromolecules 2*, 540.

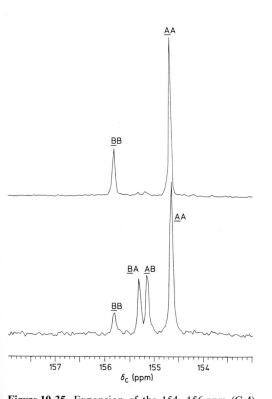

Figure 10-35. Expansion of the 154–156 ppm (C-4) region of the ^{13}C NMR spectrum for (a) the block copolymer shown in Fig. 10-34, and (b) a random copolymer comprised of the same monomer units. Dyad sequences are as shown in the text.

Craver, C. D. (Ed.) (1983), *Polymer Characterization, Advances in Chemistry Series 203*. Washington: American Chemical Society.

Elmgren, H. (1982), *J. Polym. Sci., Polym. Lett. Ed. 20*, 57.

Emsley, J. W., Feeney, J., Sutcliffe, L. H. (1965), *High Resolution Nuclear Magnetic Resonance Spectroscopy*, Vol. 1. London: Pergamon Press.

Ende, H. A., Stannett, V. (1964), *J. Polym. Sci. A 2*, 4047.

Ernst, R. R., Anderson, W. A. (1966), *Rev. Sci. Inst. 37*, 93.

Flory, P. J. (1953), *Principles of Polymer Chemistry*. Ithaca, New York: Cornell University Press.

Flory, P. J., Fox, T. G. (1951), *J. Amer. Chem. Soc. 73*, 1904.

Fyfe, C. A. (1983), *Solid State NMR for Chemists*. Guelph: C.F.C. Press.

Giddings, J. C. (1973), *J. Chem. Ed. 50*, 667.

Gray, G. W., Hawthorne, W. D., Lacey, D., White, M. S. (1989), *Liquid Crystals 6*, 503–513.

Grubisic, Z., Rempp, P., Benoit, M. (1967), *J. Polym. Sci. B 5*, 753.

Janca, J. (Ed.) (1984), *Steric Exclusion Liquid Chromatography of Polymers*. New York: Marcel Dekker.

Koenig, J. L. (1980), *Chemical Microstructure of Polymer Chains*. New York: John Wiley and Sons.

Komoroski, R. A. (Ed.) (1986), *High Resolution NMR Spectroscopy of Synthetic Polymers in Bulk*. Deerfield Beach: VCH Publishers.

Moore, J. C. (1964), *J. Polym. Sci. A 2*, 835.

Ouano, A. C., Kaye, W. (1974), *J. Polym. Sci. A-1 12*, 1151.

Randall, J. C. (1977), *Polymer Sequence Determination: Carbon-13 NMR Method*. New York: Academic Press.

Schaefer, J. (1969), *Macromolecules 2*, 210.

Schimpf, M. E., Giddings, J. C. (1987), *Macromolecules 20*, 1561.

Siesler, H. W., Holland-Moritz, K. (1980), *Infrared and Raman Spectroscopy of Polymers*. New York: Marcel Dekker.

Silas, R. S., Yates, J., Thornton, V. (1959), *Analytical Chemistry 31*, 529, and references therein.

Snyder, R. G. (1967), *J. Chem. Phys. 47*, 1316.

Tonelli, A. E. (1989), *NMR Spectroscopy and Polymer Microstructure: The Conformational Connection*. New York: VCH Publishers, pp. 28–30.

Tonelli, A. E., Schilling, F. C. (1981), *Accts. Chem. Res. 14*, 233.

Vidrine, D. W., Lowry, S. R. (1983) in: *Polymer Characterization*, Ch. 34, *Adv. Chem. Ser. No. 203:* Craver, C. D. (Ed.). Washington: American Chemical Society, pp. 595–613.

White, C. M., Orlando, C. M. (1975), in: *Polyethers, ACS Symp. Ser. No. 6*. Washington: American Chemical Society, p. 169.

White, D. M., Klopfer, H. J. (1970), *J. Polym. Sci. Part A-1 8*, 1427.

White, D. M., Klopfer, H. J. (1972), *J. Polym. Sci. Part A-1 10*, 1565.

White, D. M., Loucks, G. R. (1984), *ACS Polym. Preprints, 25 (1)*, 129.

Williams, E. A., Donahue, P. E., Loucks, G. R., Davis, G. C. (1982), unpublished results.

Williams, E. A., Skelly Frame, E. M., Donahue, P. E., Marotta, N. A., Kambour, R. P. (1990), *Applied Spectroscopy 44*, 1107.

Yau, W. W., Kirkland, J. J., Bly, D. D. (1979), *Modern Size-Exclusion Liquid Chromatograph: Practice of Gel Permeation and Gel Filtration Chromatography*. New York: John Wiley and Sons.

Zimm, B. H. (1948), *J. Chem. Phys. 16*, 1099.

General Reading

Allcock, H. R., Lampe, F. W. (1990), *Contemporary Polymer Chemistry*. Englewood Cliffs (NJ): Prentice-Hall.

Baijal, M. D. (Ed.) (1980), *Plastic Polymer Science and Technology*. New York: John Wiley and Sons.

Becker, E. D. (1980), *High Resolution NMR*. New York: Academic Press.

Bovey, F. A. (1972), *High Resolution NMR of Macromolecules*. New York: Academic Press.

Cooper, A. R. (1989), *Determination of Molecular Weight*. New York: John Wiley and Sons.

Flory, P. J. (1953), *Principles of Polymer Chemistry*. Ithaca, New York: Cornell University Press.

Koenig, J. L. (1980), *Chemical Microstructure of Polymer Chains*. New York: John Wiley and Sons.

Randall, J. C. (1977), *Polymer Sequence Determination: Carbon-13 NMR Method*. New York: Academic Press.

Siesler, H. W., Holland-Moritz, K. (1980), *Infrared and Raman Spectroscopy of Polymers*. New York: Marcel Dekker.

Index

© VCH Verlagsgesellschaft mbH, D-6940 Weinheim (Federal Republic of Germany), 1992

Distribution:

VCH, P.O. Box 101161, D-6940 Weinheim (Federal Republic of Germany)

Switzerland: VCH, P.O. Box, CH-4020 Basel (Switzerland)

United Kingdom and Ireland: VCH (UK) Ltd., 8 Wellington Court, Cambridge CB1 1HZ (England)

USA and Canada: VCH, Suite 909, 220 East 23rd Street, New York, NY 10010-4606 (USA)

ISBN 3-527-26815-4 (VCH, Weinheim)   ISBN 0-89573-690-X (VCH, New York)
Set ISBN 3-527-26813-8 (VCH, Weinheim)   Set ISBN 1-56081-190-0 (VCH, New York)